Symmetry Measures on Complex Networks

Special Issue Editor

Angel Garrido

MDPI • Basel • Beijing • Wuhan • Barcelona • Belgrade

MDPI

Special Issue Editor
Angel Garrido
Department of Fundamental Mathematics
Spain

Editorial Office
MDPI AG
St. Alban-Anlage 66
Basel, Switzerland

This edition is a reprint of the Special Issue published online in the open access journal *Symmetry* (ISSN 2073-8994) from 2011–2017 (available at: http://www.mdpi.com/journal/symmetry/special_issues/measures).

For citation purposes, cite each article independently as indicated on the article page online and as indicated below:

Author 1; Author 2. Article title. *Journal Name* **Year**, *Article number*, page range.

First Edition 2017

ISBN 978-3-03842-498-7 (Pbk)
ISBN 978-3-03842-499-4 (PDF)

Table of Contents

Chapter 1: Fundamentals

Chapter 2: Applications

About the Special Issue Editor

Angel Garrido Current situation: Chairman, Permanent Professor, Doctor, Faculty of Sciences UNED. Professional Experience: Polytechnic University of Madrid; University of Manchester (UK); UNED. Degree: Licenciado (Degree) in Exact Sciences (Pure Mathematics), Faculty of Sciences, Complutense University of Madrid. Programmer and Analyst at IBM. Master in Artificial Intelligence at UNED. Full PhD studies in Mathematics, Computer Science, and Philosophy. PhD (UNED), with Summa Cum Laude by unanimity, and First Extraordinary Prize.

Author of 22 books, published in some prestigious editorials:

- *LOGICS OF OUR TIME*. Editorial Dykinson, 2014.
- *APPLIED LOGIC. VAGUENESS AND UNCERTAINTY*. Editorial Dykinson, 2014.
- *MATHEMATICAL LOGIC AND ARTIFICIAL INTELLIGENCE*. Editorial Dykinson, 2015.
- *DISPUTES OF THOUGHT*. Editorial Dykinson, 2016.
- *PHILOSOPHY AND COMPUTATION*. Editorial Dykinson, 2017
- *LEIBNIZ AND NEW LOGICS*. Editorial Comares, Granada, 2018.
- *RETHINKING HEIDEGGER. SCIENCE AND TECHNOLOGY IN MARTIN HEIDEGGER THROUGH JACQUES DERRIDA*. Editorial Dykinson, 2018.
- *MODERN OPTIMIZATION*. With Vasile Postoliça (Bacau University). Matrix-Rom Editura, Bucuresti, 2007.

Editor-in Chief of *AXIOMS* journal. Editor of many other journals.
A total of 287 published papers, to date.
Gold Medal of the University 'Vasile Alecsandri', Bacau.

Preface to "Symmetry Measures on Complex Networks"

Here, we aim to analyze some very interrelated concepts regarding graphs, in relation to Symmetry/Asymmetry degrees, their Entropies, Clustering Coefficients, and so on. These interrelated concepts may be applied when we study different types of systems, in particular complex networks. A system can be easily defined as a set of components functioning together as a whole. A systemic point of view allows us to isolate one part of the world, and in doing so, we can focus on those aspects that interact more closely than others. Network Science is a new scientific field that analyzes the interconnection among diverse networks, for instance, among Physics, Semantics, and so on. Among its developers, we may recall Duncan Watts, who developed the Small-World Network, or Albert-László Barabasi and Réka Albert, who developed Scale-Free Networks. In this latter work, both authors found that websites constitute the network of the World Wide Web (WWW) and they have very interesting mathematical properties. Network Theory is a rapidly expanding area of Computer Sciences, and may be considered as part of Graph Theory.

Complex networks are everywhere. Many different phenomena in nature can be modeled as a network, such as brain structures, the brain as a network of neurons (their nodes), connected by synapses (their edges); and social interactions, or the WWW.

All such systems can be represented in terms of nodes and edges. On the Internet, nodes represent routers and edges are represented by wires, or physical connections between them. In transport networks, nodes can represent cities, and edges show the roads that connect them. These edges can have weights. Such networks are not random. The topology of very different networks may be very close. They are rooted in the Power Law, with a scale-free structure. How can very different, complex systems have the same underlying topological features? Searching the hidden laws of these networks, as well as modeling and characterizing them, constitute the current lines of research.

Here, it seems appropriate to study the most theoretical and applied aspects—currently of great relevance—of all these developments, in thirty-two interesting articles.

Angel Garrido
Special Issue Editor

Chapter 1:
Fundamentals

symmetry

MDPI

Article

Symmetry in Complex Networks

Angel Garrido

Department of Fundamental Mathematics, Faculty of Sciences UNED, Senda del Rey 9, 28040 Madrid, Spain;
agarrido@mat.uned.es (algbmv@telefonica.net)

Received: 16 November 2010; in revised form: 4 January 2011; Accepted: 7 January 2011;
Published: 10 January 2011

Abstract: In this paper, we analyze a few interrelated concepts about graphs, such as their degree, entropy, or their symmetry/asymmetry levels. These concepts prove useful in the study of different types of Systems, and particularly, in the analysis of Complex Networks. A System can be defined as any set of components functioning together as a whole. A systemic point of view allows us to isolate a part of the world, and so, we can focus on those aspects that interact more closely than others. Network Science analyzes the interconnections among diverse networks from different domains: physics, engineering, biology, semantics, and so on. Current developments in the quantitative analysis of Complex Networks, based on graph theory, have been rapidly translated to studies of brain network organization. The brain's systems have complex network features—such as the small-world topology, highly connected hubs and modularity. These networks are not random. The topology of many different networks shows striking similarities, such as the scale-free structure, with the degree distribution following a Power Law. How can very different systems have the same underlying topological features? Modeling and characterizing these networks, looking for their governing laws, are the current lines of research. So, we will dedicate this Special Issue paper to show measures of symmetry in Complex Networks, and highlight their close relation with measures of information and entropy.

Keywords: graph theory; applications of graph theory; group theory fuzzy sets; fuzzy logic; logic of vagueness; fuzzy topology; Fuzzy Measure theory; fuzzy real analysis; Small World graphs; Complex Networks; artificial intelligence

MSC Classification: 05C10; 97K30; 94C15; 54A99; 94D05; 05C82; 05B52; 28E10; 26E50; 05C82; 97R40.

1. Some Previous Concepts

A *graph* [1] may be defined as a pair, $G = (V, E)$, where $V = V(G)$ is the node set, and $E = E(G)$ is the edge set, *i.e.*, the set of 2-element subsets of V.

Given an edge $\{i, j\} \in E$, we say that the nodes i and j are *adjacent*; and we denote $i \sim j$.

The neighborhood of i will be:

$$N(i) = \{j \in V: j \sim i\}$$

And the *degree of i* can be expressed as:

$$deg(i) = d(i) = card\ \{N(i)\}$$

A graph, G, is *finite*, if the set of its nodes, $V(G)$, is finite.

And it is *locally finite*, if all of its nodes have finite degrees.

Two very important results may be considered now:

<u>*Handshaking Lemma (or Theorem).*</u>

In any graph, the sum of the degrees of all nodes (or "total degree") is equal to twice the number of edges.

<u>*Degree Theorem.*</u>

In any graph there is an even number of nodes with an odd degree.

The adjacency matrix is a convenient representation of the interaction between nodes. Several Complex Networks measures can be defined over adjacency matrices; for instance: clustering coefficient (local or global), diameter, average degree of the network, and so on. All of them play a key role in network theory.

The *distance* between two nodes is defined as the length of the shortest path connecting them.

The *diameter* of a network is the maximal distance between any pair of their nodes.

The *average path length* is the average of the distance over all pairs of nodes. Thus, it determines the "size" of the network.

An *automorphism* of a graph, G, is any bijection:

$$a: V(G) \rightarrow V(G)$$

that applies edges onto edges, and non-edges onto non-edges.

The set of all automorphisms of a graph, G, is denoted by *Aut (G)*. It is the *automorphism group* of G. We will come back later to this concept.

Succinctly, the more symmetry a graph has the larger its automorphism group will be, and *vice versa*.

2. Symmetry and Networks

Pierre Curie stated [2]:

It is asymmetry that creates a phenomenon.

Paul Renaud generalized Curie's idea and stated [3]:

If an ensemble of causes is invariant with respect to any transformation,
the ensemble of their effects is invariant with respect to the same transformation.

Joe Rosen has stated the <u>Symmetry Principle</u> as [4]:

The symmetry group of the cause is a subgroup of the symmetry group of the effect.

Or less precisely:

The effect is at least as symmetric as the cause (and might be greater).

Also from Joe Rosen is the quote:

Recognized causal relations in nature are expressed as laws.
Laws impose equivalence relations in the state sets of causes and of effects.

So,

Equivalent states of a cause are mapped to (i.e., correlated with)
equivalent states of its effect.

This is the Equivalence Principle.

Somewhat less precisely, this principle may be expressed as:

Equivalent causes are associated with equivalent effects.

Concerning the *Equivalence Principle for Processes* on isolated physical systems, we can say that:

Symmetry **2011**, *3*, 1–15

Equivalent initial states <u>must</u> evolve into equivalent states
(while inequivalent states <u>may</u> evolve into equivalent states).

And the General Symmetry Evolution Principle:

The "initial" symmetry group is a subgroup of the "final" symmetry group.

This assertion can also be stated as:

For an isolated physical system the degree of symmetry
cannot decrease as the system evolves; instead,
it either remains constant or increases.

Finally, we have the Special Symmetry Evolution Principle:

As an isolated system evolves, the populations of the equivalence
classes of the sequence of states through which it passes cannot
decrease, but either remain constant or increase.

Equivalently,

The degree of symmetry of the state of an isolated system
cannot decrease during evolution; instead, it either
remains constant or increases.

As a further implication, Joe Rosen proposes this general theorem:

The degree of symmetry of a macrostate of stable equilibrium must be relatively high.

In the following, we will draw on the concepts and intuitions by Prof. Rosen, summarized in his paper "The Symmetry Principle" [4].

According to the traditional viewpoint, higher symmetry is related to higher order, less entropy and less stability.

In Prigogine's theory, symmetry has been regarded as order, or reduction of entropy. But this idea is *incorrect*. Rosen's Principle of Symmetry is the opposite of such theory.

Shu-Kun Lin [5] has proved both: the Symmetry Principle, around a continuous higher similarity-higher entropy relation; and the Rosen's Symmetry Principle, around a higher symmetry-higher stability relation. He proposed that entropy is the degree of symmetry and information is the degree of asymmetry of a structure.

According to Shu-Kun Lin [5], "symmetry is in principle ugly, because it is related to entropy and information loss"

With the motto:

Ugly Symmetry-Beautiful Diversity

This contradicts the more usual and commonplace vision of symmetry as a concept equivalent to desirable beauty, proportion and harmony. This can be surprising, but Shu-Kun Lin's arguments are really strong and convincing: Symmetric structure is stable but not necessarily beautiful. All spontaneous processes lead to the highest symmetry, which is the equilibrium or a state of "death".

"Life is beautiful but full of asymmetry"

It concludes [6] that

Beauty = Stability + Information

Intuitively, symmetry, like perfection or beauty, up to a certain level, is precious, but above that—apart from inexistent in the real world—would mean an end to the human thing, which is by nature- and fortunately- imperfect.

3. Symmetry as Invariance

Symmetry [1,6,7] in a system means invariance of its elements under a group of transformations, *i.e.*, the mathematical definition of symmetry of a graph is the set of transformations that leave the properties of the graph unchanged. When we focus on Network Structures, it means invariance of adjacency of nodes under the permutations on the node set [8,11].

A graph isomorphism is an equivalence, or equality, as relation on the set of graphs. Therefore, it partitions the class of all graphs into equivalence classes. The underlying idea of isomorphism is that some objects have the same structure, if we omit some individual characteristics of their components. A set of graphs isomorphic to each other is called an *isomorphism class of graphs* [1,8,10].

An *automorphism* of a graph, $G = (V, E)$, is an isomorphism from G onto itself. The family of all automorphisms of a graph, G, is a permutation group on $V(G)$. The inner operation of such a group is the composition of permutations. Its name is very well-known, the *Automorphism Group of G*, denoted by *Aut (G)*. And conversely, all groups may be represented as the automorphism group of some connected graph.

The automorphism group is an algebraic invariant of a graph. So, we can say that an automorphism of a graph is a form of symmetry in which the graph is mapped onto itself while preserving the edge-node connectivity. Such an automorphic tool may be applied both on Directed Graphs (DGs), and on Undirected Graphs (UGs), or Mixed Graphs.

Graphs are discrete mathematical constructs. Also, they are topological objects, not geometrical entities. And they may exhibit symmetries under transformations that are not node permutations: e.g., by scale invariance on fractals [31].

Another interesting concept in mathematics, the word "genus", has different, but strongly related, meanings. So, in Topology it depends on whether we consider orientable or non-orientable. In the case of connected and orientable surfaces, it is an integer that represents the maximum number of cuttings, along closed simple curves, without rendering the resultant manifold disconnected. Visually, we can imagine that it is the number of "handles" on the manifold. Usually, it is denoted by the letter *g*.

It is also definable through the Euler number, or *Euler Characteristic*, denoted χ. Such a relationship will be expressed, for closed surfaces, by $\chi = 2 - 2g$. When the surface has *b* boundary components, this equation transforms to $\chi = 2 - 2g - b$, which obviously generalizes the above equation. For example, a sphere, an annulus, or a disc have genus $g = 0$. Instead of this, a torus has $g = 1$.

In the case of non-orientable surfaces, the genus of a closed and connected surface will be a positive integer, representing the number of cross-caps attached to a sphere.

Recall that a cross-cap is a two-dimensional surface that is topologically equivalent to a Mobius string.

As in the precedent analysis, it can be expressed in terms of the Euler characteristic, by $\chi = 2 - 2k$, where *k* is the non-orientable genus. For example, a projective plane has a non-orientable genus $k = 1$. And a Klein bottle has a non-orientable genus $k = 2$.

Turning to graphs [1], the corresponding genus will be the minimal integer, *n*, such that the graph can be drawn without crossing itself on a sphere with *n* handles. So, a planar graph has genus $n = 0$, because it can be drawn on a sphere without self-crossing.

In the non-orientable case, the genus will be the minimal integer, *n*, such that the graph can be drawn without crossing itself on a sphere with n cross-caps.

Moving on to topological graph theory, we will define as genus of a group, G, the minimum genus of any of the undirected and connected Cayley graphs for G.

From the viewpoint of Computational Complexity, the problem of "graph genus" is NP-complete. Recall that a problem, L, is NP-complete if it has two properties: It is in the set of NP (nondeterministic polynomial time) problems, *i.e.*, any given solution to L can be verified quickly (in polynomial time); and it is also in the set of NP-hard problems, *i.e.*, any NP problem can be converted into L by a transformation of the inputs in polynomial time.

A *graph invariant*, or graph property, is a property that depends only on the abstract structure of the graph, not on its representations, such as a particular labeling or drawing of the graph. So, we may define a graph property as any property that is preserved under all possible isomorphisms of the graph. Therefore, it is a property of the graph itself, independent of the representation of the graph.

The semantic difference between invariant and property also consists in its quantitative or qualitative character. For instance, when we say that "the graph has no directed edges", this is a property, because it is a qualitative statement. While when we say "the number of nodes of degree two in such a graph", this is an invariant, because it is a quantitative statement.

From a strictly mathematical viewpoint, a graph property can be interpreted as a class of graphs, composed by the graphs that have in common the accomplishment of some conditions. Hence, a graph property can also be defined as a function whose domain would be the set of graphs, and which range would be the bi-valued set, {true, false}; the value of the property depending on whether it is verified or violated for the graph.

A graph property is called *hereditary*, if it is inherited by its induced subgraphs.

And a graph property will be *additive*, if it is closed under disjoint union.

For example, the property of a graph being planar is both additive and hereditary. And the property of being connected is neither.

The computation of certain graph invariants is very useful to discriminate whether two graphs are isomorphic or non-isomorphic. For any particular invariant, two graphs with different values cannot be isomorphic. However, two graphs with the same invariant value may or may not be isomorphic.

It is possible to prove that every group is the automorphism group of a graph.

If the group is finite, the graph may be taken to be finite.

G. Polya observed that not every group is the automorphism group of a tree.

Many reasons are behind the current popularity of Complex Networks [13]. To cite but a few, their generality and flexibility for representing any natural structure, including those structures that reveal dynamical changes of topology [11,14,20,22].

Before turning to Complex Networks, it is very convenient to introduce some concepts which are useful in understanding Networks, as measures of their principal characteristics [1,11].

The *characteristic path length* measures the distance from every node to every other node. It is calculated by the median of the shortest paths from each node to every other node. So, as a derived measure, the diameter gives us the maximum possible distance between all pairs of reachable nodes.

Another commonly used value is the *Clustering Coefficient*. It is the mean of the clustering indices of all the nodes in the graph. It is usually denoted C. It tells us how well connected the neighborhood of the node is. So, it is the answer to this question: How close is the neighborhood of a node to be a clique (*i.e.*, a complete subgraph). Finding C, we look for the neighbors of the corresponding node, and then find the number of existing edges between them. The ratio of the number of existing edges to the number of all possible edges is the clustering index of the node.

If the neighborhood is fully connected, then the clustering coefficient must be equal to one, $C = 1$. In the opposite situation, a value of $C = 0$ signifies that the neighborhood is fully disconnected. And any intermediate value is a measure of the graph's degree of connectedness. Values close to zero mean that there are hardly any connections in the neighborhood.

This measure has been used to summarize features of undirected and unweighted networks in Complexity Science.

An interesting type of graph is Regular Networks, where each node is connected to all other nodes; *i.e.*, they are fully connected. Because of such a type of structure, they have the lowest path length (L), and the lowest diameter (D), being $L = D = 1$. Also, they have the highest clustering coefficient (C). So, it holds that $C = 1$.

Furthermore, they have highest possible number of edges, given by

$$Card\ (E) = n\ (n-1)/2 \sim n^2;$$

4. Random Graphs

In *Random Graphs (RGs)*, each pair of nodes is connected with probability p. They have a low average path length [8,17], following that:

$$L \sim (ln\ n)\ /\ n<k> \sim ln\ n,\ for\ n \gg 1$$

Therefore, the total network may be covered in $<k>$ steps, from which

$$n \sim <k>^L$$

Moreover, Random Graphs possess a low clustering coefficient, when the graph is sparse. Thus,

$$C = p = <k> /n \ll 1$$

The reason is that the probability of each pair of neighboring nodes to be connected is precisely equal to p.

The Small-World effect is observed on a network when it has a low average path length:

$$L << n,\ for\ n >> 1$$

Recall [15,21,24,25] the now very famous *"six degrees of separation"*, which also may be called *"small-world phenomenon"*. The subjacent idea is that two arbitrarily selected people may be connected by only six degrees of separation, or six handshakes (in average, and it is not much larger than this value). Therefore, the diameter of the corresponding graph is not much larger than six.

The usual example is social connections. So, the Small-World property [11,15] can be interpreted as that despite its large size (of the corresponding graph), the shortest path between two nodes is small, as for example on the WWW, or on the Internet.

5. Self-Similarity

Self-similarity on a network [11] indicates that it is approximately similar to any part of itself, and therefore, it is fractal. In many cases, real networks possess all these properties, *i.e.*, they are Fractal, Small-World, and Scale-Free.

Fractal dimensions describe self-similarity of diverse phenomena: Images, temporal signals, *etc.* Such fractal dimension gives us an indication of how completely a fractal appears to fill the space, as one zooms down to finer and finer scales. It is so a statistical measure.

The most important of such measures are Renyi dimension, Hausdorff dimension, and Packing dimension.

A Fuzzy set approach also may produce some very consistent models [26–28].

6. Small-World Model

The *Watts-Strogatz Small-World Model*, proposed in 1998, is a hybrid case between a Random Graph and a Regular Lattice [14,20,22]. So, Small-World models share with Random Graphs some common features, such as: The Poisson or Binomial degree distribution, near to Uniform degree distribution; network size: It does not grow; each node has approximately the same number of edges.

Therefore, it shows a *homogeneous nature*. Because of their ease of implementation, the more usual procedures to compute such measures are correlation dimension and box counting.

Watts-Strogatzmodels show the low average path length typical of Random Graphs,

$$L \sim ln\ n,\ for\ n >> 1$$

And also such models give us the usual high clustering coefficient of Regular Lattices, being

$$C \approx 0.75, \text{ for } k >> 1$$

In consequence, WS-models have a small-world structure, being well clustered. The Random Graphs coincide on the small-world structure, but they are poorly clustered. This model (WS) has a peak degree distribution, of Poisson type.

7. Scale-Free Networks

With reference to the last analyzed model [16,17,23,24], called *Scale-Free Network*, this appears when the degree distribution follows a Power-Law:

$$P(k) \sim k^{-\gamma}$$

In such a case, there exist a small number of highly connected nodes, called *Hubs*, which are the tail of the distribution.

On the other hand, the great majority of the sets of their nodes have few connections, representing the head of such distribution.

Such a model was introduced [8,14,16] by Albert-Laszlo; Barabasi and Reka Albert, in 1999.

Some of their essential features are these: non-homogeneous nature, in the sense that some (few) nodes have many edges from them, and the remaining nodes only have very few edges, or links; as related to the network size, it continuously grows; and regarding to the connectivity, it obeys a Power-Law distribution.

Many massive graphs, such as the WWW graph, share certain characteristics, described as such aforementioned Power-Law.

Bela Bollobas and Oliver Riordan [9,11] consider a Random Graph process in which nodes are added to the graph one at a time, and joined to a fixed number of earlier nodes, chosen with probability proportional to their degree. After n steps, the resulting graph has diameter approximately equal to $\log n$. This affirmation is true for $n = 1$. But for $n \geqslant 2$, the diameter value would show asymptotical convergence to $(\log n)/\log(\log n)$.

Another very interesting mechanism is the so-called *Preferential Attachment process (PA)*. This would be any class of processes in which some quantity is distributed among a number of sets (for instance, objects or individuals), according to how much they already have, so that intuitively "the rich get richer" (the more interrelated get more new connections than those who are not).

The principal scientific interest in PA is that they may produce interesting power law distributions.

A very notable example of a Scale-Free Network is the World Wide Web (WWW). As we know [23,25], it is a collection of many, possibly very different, sub-networks. Related to the Web graph characteristics, we notice the Scale Invariance as being very important [10].

8. Diameter of the Web

Another interesting feature is the possibility of obtaining a measurement of the World Wide Web, its *diameter*, i.e., the shortest distance between any pair of nodes into the system, or at least some adequate bound, either a mean value [18,19], etc.

The WWW representation is made by a very large digraph, whose nodes are documents, and whose edges are links (URLs), pointing from one document to another [14,20,22].

Reka Albert *et al.* [18] found that the average of the shortest path between two nodes will be

$$<d> = 0.35 + 2.06 \log N$$

where N is the number of nodes in the Random Graph considered. This shows that the WWW is a Small-World network.

In particular, if we take

$$N = 8 \times 10^8$$

we will obtain

$$<d_{Web}> = 18.59$$

This important result signifies that two randomly chosen nodes (documents), on the graph which represent the WWW, are only on average 19 clicks (or steps into the WWW graph) from each other.

For a given value of the number of nodes, N, the distribution associated to d is of Gaussian (Normal) type. It is also very remarkable the logarithmic dependence of such diameter on the value of N. In this sense, R. Albert *et al.* indicate that the future evaluation of $<d>$, with the increasing of the WWW, would change from 19 to only 21.

9. Community Structure

The Community Structure can also be called *Modularity*. It is a very frequent characteristic in many real networks. Therefore, it has become a key problem in the study of networked systems [11,12,21].

Giving out its deterministic definition is nontrivial because of the complexity of networks. The concept of *modularity* (Q) can be used as a valid measure for community structure.

Some current models have proposed to capture the basic topological evolution of Complex Networks by the hypothesis that highly connected nodes increase their connectivity faster than their less connected peers, a phenomenon denoted as *PA* (preferential attachment). So, we can find a class of models that view networks as evolving dynamical systems, rather than static graphs.

Most evolving network models are based on two essential hypotheses, *growth and preferential attachment*.

Growth suggests that networks continuously expand through the addition of new nodes and links between the nodes. And *preferential attachment* states that the rate at which a node with k links acquires new links will be a monotonically increasing function of k.

We can consider an undirected n-graph, or network, G, with adjacency matrix denoted as $A = (a_{ij})$, where $a_{ij} = 1$, if nodes i and j are connected; otherwise, $a_{ij} = 0$. Then, the *modularity function*, denoted by Q, will be defined as:

$$Q(P_k) = \Sigma \left[\left\{ L(V_j, V_j) / L(V, V) \right\} - \left\{ L(V_j, V) / L(V, V) \right\}^2 \right]$$

where P_k is a partition of the nodes into k groups, and:

$$L(V', V'') = \Sigma_{i \in V', i \in V''} a_{ij}$$

The modularity function, Q, provides a way to determine whether a partition will be valid to decipher the community structure in a network. Maximization of such modularity function, over all the possible partitions of a network, is indeed a highly effective method.

An important case in community detection is that some nodes may not belong to a single community, and then placing them into more than one group may be much more reasonable. Such nodes can provide a *"fuzzy" categorization* [25], and hence, they may take a special role, such as signal transduction in biological networks.

10. Fuzzy Symmetry

Recall that according to Klaus Mainzer, "Symmetry and Complexity determine the spirit of nonlinear science". And "the universal evolution is caused by symmetry break, generating diversity, increasing complexity and energy" [26].

Graph theory has emerged as a primary tool for detecting numerous hidden structures in various information networks, including Internet graphs, social networks, biological networks, or more generally, any graph representing relations in massive data sets. Analyzing these structures is very useful to introduce concepts such as Graph Entropy and Graph Symmetry.

We consider a function on a graph, $G = (V, E)$, with P a probability distribution on its node set, V. The mathematical construct called Graph Entropy will be denoted by G, E. It will be defined as

$$H\,(G,\ P) = min\ \Sigma\ p_i\ log\ p_i$$

Observe that such a function will be convex. It tends to $+\infty$ on the boundary of the non-negative orthant of R^n. And monotonically to $-\infty$ along rays from the origin. So, such a minimum is always achieved and it will be finite.

The entropy of a system represents the amount of uncertainty one observer has about the state of the system. The simplest example of a system will be a random variable, which can be shown by a node into the graph, where their edges represent the mutual relationship between them. Information measures the amount of correlation between two systems, and it reduces to a mere difference in entropies. So, the entropy of a graph is a measure of graph structure, or lack of it. Therefore, it may be interpreted as the amount of Information, or the degree of "surprise", communicated by a message. And as the basic unit of Information is the bit, entropy also may be viewed as the number of bits of "randomness" in the graph, verifying that the higher the entropy, the more random is the graph.

It is possible to introduce some new asymmetry and symmetry level measures as by [27,28]. Note that our results may also be applied to some different classes of spaces.

Recall some very useful definitions from Fuzzy Measure Theory.

Definition 1: Let U be the universe of discourse, with $\wp$ a σ-algebra on U. Then, given a function

$$m : \wp \rightarrow [0,1]$$

we describe m as a Fuzzy Measure, if it verifies:

(I) $m\,(\theta) = 0$;
(II) $m\,(U) = 1$;
(III) If $A, B \in \wp$, with $A \subseteq B$, then $m\,(A) \leqslant m\,(B)$ *[monotonicity]*.

When we take the Entropy concept, we attempt to measure the fuzziness, *i.e.*, the degree of being fuzzy for each element in $\wp$.

Definition 2: The *Entropy measure* can be designed as the function.

$$H{:}\wp \rightarrow [0,1]$$

verifying:

(I)   If A is a crisp set, then $H\,(A) = 0$;
(II)   If $H\,(x) = 1/2$, for each $x \in A$, then $H\,(A)$ is maximal (total uncertainty);
(III) If A is less fuzzified than B, it holds that $H\,(A) \leqslant H\,(B)$;
(IV) $H\,(A) = H\,(U/A)$.

Definition 3: The *Specificity Measure* will be introduced as a measure of the tranquility when we take decisions. Such Specificity Measure (denoted by Sp) will be a function:

$$Sp{:}\ [0,1]^U \rightarrow [0,1] \qquad\qquad ()$$

where

(I) *Sp* (θ) = 0;
(II) *Sp* (*k*) = 1 if and only if *k* is a unitary set (singleton);
(III) If *V* and *W* are normal fuzzy sets in *U*, with $V \subset W$, then *Sp* (*V*) ⩾ *Sp* (*W*);

Note. $[0,1]^U$ denotes the class of fuzzy sets in U; Let *(E, d)* be a fuzzy metric space.
We proceed to define our new fuzzy measures. Such functions might be defined as some of the type

$$\{L_i\}_{i \in \{s,a\}}$$

where *s* denotes symmetry, and *a* denotes asymmetry.

Suppose that from here we denote by *c* (*A*) the cardinal of a fuzzy set, *A*. We denote by *H* (*A*) its *entropy measure*, and by *Sp* (*A*) its corresponding *specificity measure*.

Theorem 1. Let *(E, d)* be a fuzzy metric space, with *A* as a subset of *E*, and let *H* and *Sp* be both above fuzzy measures defined on *(E, d)*. Then, the first function, operating on *A*, may be defined as

$$L_s(A) = Sp(A)((1 - c(A))/(1 + c(A))) + (1/(1 + H(A))$$

and will be also a fuzzy measure. This measure is called *Symmetry Level Function.*

Theorem 2. Let *(E, d)* be a fuzzy metric space, with *A* as any subset of *E*, and let *H* and *Sp* be both above fuzzy measures defined on *(E, d)*. Then, the function

$$L_a(A) = 1 - \left\{ Sp(A)\Big((1 - c(A))/(1 + c(A)) + (1/(1 + H(A))\Big) \right\}$$

This measure is called Asymmetry Level Function.

Corollary 1. In the same precedent hypotheses, the Symmetry Level Function is a Normal Fuzzy Measure.

Corollary 2. Also, in such conditions the Asymmetry Level Function will be a Normal Fuzzy Measure.

Recall that the values of a fuzzy measure, *Sp*, are decreasing when the size of the considered set is increasing. And also that the Range of the Specificity Measure, *Sp*, will be [0,1].

11. New Lines of Research

An important fact, but commonly forgotten, is that an element can belong to more than a fuzzy set at the same time. This admits new generalizations on the theoretical basis of important topics [30], as may be Clustering and Community structures.

And recall the line which was open by the *Three Laws of Similarity* of Shu-Kun Lin [30], according to which, in parallel to the first and the second laws of thermodynamics, we have:

(i) *The first law of information theory.* The logarithmic function *L* = *ln w*, or the sum of entropy and information, *L* = *S* + *I*, of an isolated system remains unchanged, where *S* denotes the entropy and *I* the information content of the system.

(ii) *The second law of information theory.* Information of an isolated system decreases to a minimum at equilibrium.

(iii) *The third law of information theory.* For a perfect crystal (at zero absolute thermodynamic temperature), the information is zero and the static entropy is at the maximum. Or in a more general form, "for a perfect symmetric *static* structure, the information is zero and the *static entropy* is the maximum".

Analyzing the Gibbs' paradox, Dr. Lin arrives to its well-known:

(iv) *Similarity principle.* The higher the similarity among the components is, the higher the value of entropy will be and the higher the stability will be.

By these three laws and such principle, Dr. Lin has clarified the relation of symmetry to several other concepts, as higher symmetry, higher similarity, higher entropy, less information and less diversity, related to higher stability. Upon these deep foundations, the tracks of mutual relationships between such fuzzy measures can be traced: as it is the case with Symmetry, Entropy, Similarity, and so on, which can lead in the future to advances for innovative fields connected to them.

The paper of Prof. Joel Ratsaby is also very inspiring [31]. He introduces an algorithmic complexity framework for representing Lin's concepts of static entropy, stability and their connection to the second law of thermodynamic. Instead of static entropy, according to Ratsaby, the *Kolmogorov complexity* of a static structure may be the proper measure of disorder. Consider one static structure in a surrounding perfectly-random universe in which it acts as an interfering entity which introduces a local disruption of randomness. This is modeled by a selection rule, *R*. So, we may clearly explain why more complex static structures are less stable. To continue in this line of promising investigation can be very interesting in the future.

According to Garlaschelli *et al.* [32], "while special types of symmetries (e.g., automorphisms) are studied in detail within discrete mathematics for particular classes of deterministic graphs, the analysis of more general symmetries in real Complex Networks is far less developed".

They argued that real networks, as any entity characterized by imperfections or errors, necessarily require a stochastic notion of invariance. So, they propose a definition of stochastic symmetry based on graph ensembles.

But we suggest that in addition, they can and must try theoretical approximations from the field of fuzzy measures, since it is those of symmetry and entropy, really interrelated between them. Thus, to regulate mathematically, by modulating, the diverse degrees with which one will find these types of characteristics in reality, when we consider networks and systems.

12. Conclusions

Our initial purpose was to provide a comprehensive vision on principal aspects, and essential properties, of Complex Networks, from a new Mathematical Analysis point of view, and in particular to show the promise of the new functions of Symmetry/Asymmetry Levels.

The essential idea was to obtain an as wide as possible perspective of certain aspects of Complex Networks, as well as of the fuzzy measures when they are acting on them. With the new results and the pointed lines of advance, we think that it will be possible to penetrate into the aforementioned problems, to come to a deeper comprehension of the symmetry, of the entropy and of other similar fuzzy measures, interesting not only from a theoretical viewpoint, but promising for many scientific applications.

Acknowledgments: I wish to express my gratitude to Joel Ratsaby, from Ariel University Center (Israel), who proposed me to take charge of this Special Issue of the journal *Symmetry*, to which this paper belongs. Likewise, to Shu-Kun Lin, who asked for it later, as well as to the support received from the Editorial Board for this publication, and very especially to Cathy Wang, for her assistance. Also I want to be grateful for their wise advices to the anonymous referees of my paper.

References

1. Bornholdt, S.; Schuster, H.G. *Handbook of Graphs and Networks: From the Genome to the Internet*; Wiley: Weinheim, Germany, 2003.
2. Curie, P. *Symmetry in Physics*; Rosen, J., Ed.; American Society of Physics Teachers Series 3; American Institute of Physics: Melville, NY, USA, 1982; pp. 17–25.
3. Renaud, P. *Symmetry in Physics*; Rosen, J., Ed.; American Society of Physics Teachers; American Institute of Physics: Melville, NY, USA, 1982; p. 26.
4. Rosen, J. The Symmetry Principle. In *Entropy Journal, Symmetry in Science: An Introduction to the General Theory*; Springer-Verlag: New York, NY, USA, 1995; pp. 308–314.
5. Lin, S.K. Correlation of Entropy with Similarity and Symmetry. *J. Chem. Inf. Comput. Sci.* **1996**, *36*, 367–376.
6. Lin, S.K. Ugly Symmetry. In Division of Organic Chemistry, Proceedings of Tetrahedral Carbon's 125th Anniversary Symposium—The 218th ACS National Meeting, New Orleans, LA, USA, 1999.

7. Weyl, H. *Symmetry*; Princeton University Press: Princeton, NJ, USA, 1983.
8. Barabasi, A.-L. *Linked: How Everything is Connected to Everything Else*; Plume Publisher: New York, NY, USA, 2004. http://www.nd.edu/~alb/ (updated 06/04/02, Department of Physics, University of Notre Dame, USA).
9. Bollobas, B. Cambridge Studies in Advanced Mathematics 73. In *Random Graphs*; Cambridge University Press: Cambridge, UK, 2001.
10. Bollobas, B. *Modern Graph Theory*; Springer Verlag: Berlin, Germany, 1998.
11. Bollobas, B. *Handbook of Large-Scale Random Networks*; Springer Verlag: Berlin, Germany, 2009.
12. Newman, M. The structure and function of Complex Networks. *SIAM Rev.* **2003**, *45*, 167–256.
13. Newman, M. *The structure and Dynamics of Complex Networks*; Princeton University Press: Princeton, NJ, USA, 2006.
14. Albert, R.; Barabasi, A.L. Statistical Mechanics of Complex Networks. *Rev. Mod. Phys.* **2002**, *74*, 47–72.
15. Watts, D.J. *Six Degrees: The Science of a Connected Age*; W. W. Norton and Company: New York, NY, USA, 2003.
16. Barabasi, A.L. , Bonabeau, R. Scale-Free Networks. *Scient. Am.* **2003**, *288*, 50–59.
17. Calderelli, G. *Scale-Free Networks*; Oxford University Press: Oxford, UK, 2007.
18. Bollobas, B. The diameter of a Scale-Free Random Graph. In *Combinatorica*; Springer Verlag: Berlin, Germany, 2004.
19. Albert, R.; Jeong, H.; Barabasi, A.L. Diameter of the World-Wide Web. *Nature* **1999**, *401*, 129–130.
20. Barrat, A. *Dynamical processes in Complex Networks*; Cambridge University Press: Cambridge, UK, 2008.
21. Strogatz, S.H. Exploring Complex Networks. *Nature* **2001**, *410*, 268–276.
22. Bocaletti, S. Complex Networks: Structure and Dynamics. *Phys. Rep.* **2006**, *424*, 175–308.
23. Dorogotsev, S.N.; Mendes, J.F.F. Evolution of Networks. *Adv. Phys.* **2002**, *51*, 1079–1094.
24. Dorogotsev, S.N.; Mendes, J.F.F. *Evolution of Networks: From Biological Networks to the Internet and WWW*; Oxford University Press: Oxford, UK, 2003.
25. Dorogotsev, S.N.; Goltsev, A.V.; Mendes, J.F.F. Critical phenomena in Complex Networks. *Rev. Mod. Phys.* **2008**, *80*, 1275–1284.
26. Mainzer, K. *Symmetry and Complexity—the Spirit and Beauty of Nonlinear Science*; World Scientific Publ. Company: California, NJ, USA, 2005.
27. Garrido, A. Asymmetry and Symmetry Level Measures. *Symmetry* **2010**, *2*, 707–721.
28. Garrido, A. Asymmetry level as a fuzzy measure. *Acta Univ. Apulensis Math. Inf.* **2009**, *18*, 11–18.
29. Garrido, A. Entropy, Genus and Symmetry on Networks. *ROMAI J.* **2010**, *6*, 23–38.
30. Lin, S.K. The Nature of the Chemical Process. *1. Symmetry Evolution—Revised Information Theory, Similarity Principle and Ugly Symmetry. Int. J. Mol. Sci.* **2001**, *2*, 10–39.
31. Ratsaby, J. An algorithmic complexity interpretation of Lin's third law of information theory. *Entropy J.* **2008**, *10*, 6–14.
32. Garlaschelli, D.; Ruzzenenti, F.; Barossi, R. Complex Networks and Symmetry I: A Review. *Symmetry* **2010**, *2*, 1683–1709.

symmetry

MDPI

Article

Long Time Behaviour on a Path Group of the Heat Semi-group Associated to a Bilaplacian

Remi Leandre

Institut de Mathématiques de Bourgogne, Université de Bourgogne, 21078, Dijon Cedex, France;
E-Mail: Remi.leandre@u-bourgogne.fr

Received: 16 February 2011 / Accepted: 7 March 2011 / Published: 21 March 2011

Abstract: We show that in long-time the heat semi-group on a path group associated to a Bilaplacian on the group tends to the Haar distribution on a path group.

Keywords: heat semigroup; Haar distribution; path group

1. Introduction

Let us consider a compact connected Lie group G of dimension d endowed with its normalized biinvariant Haar measure dg. Let us consider the Laplacian Δ on it. It is equal to $\sum(\partial_{e_i})^2$ where e_i is an orthonormal basis of the Lie algebra of G. It generates a Markov semi-group P_t:

$$\frac{\partial}{\partial t}P_t f = \Delta P_t f \tag{1}$$

if f is smooth. Moreover there is a **strictly positive** heat kernel

$$P_t f(g) = \int_G p_t(g, g') f(g') dg' = \int_G p_t(e, g^{-1} g') f(g') dg' \tag{2}$$

when $t \to \infty$

$$P_t f(g) \to \int_G f(g) dg \tag{3}$$

Let us consider a Bilaplacian on G, this means a power Δ^k $k > 1$. It generates still a semi-group P_t^k. P_t^k is not a Markovian semi-group. This means that the heat kernel $p_t^k(g, g')$ associated to P_t^k can **change sign.** We have still when $t \to \infty$

$$P_t^k f(g) \to \int_G f(g) dg \tag{4}$$

In the first case, the heat semi-group is represented by the Brownian motion on G. In the second case, there is until now no stochastic process associated to it. In the case of $\mathbb{R}^d$, the path integral involved with the semi-group P_t^k is defined as a distribution in [1].

We are motivated in this work by an extension in infinite dimension of these results, by considering the case of the path-group $C([0,1], G)$ from continuous path from $[0,1]$ into G starting from e.

Let us recall that the Haar measure $d\tilde{g}$ on a topological group $\tilde{G}$ exists as a full measure **if and only if** the group is locally compact. Haar measure means that for all bounded measurable function $\tilde{F}$

$$\int_{\tilde{G}} \tilde{F}(\tilde{g}_1 \tilde{g}) d\tilde{g} = \int_{\tilde{G}} \tilde{F}(\tilde{g}) d\tilde{g} \tag{5}$$

Symmetry **2011**, 3, 72–83

The difficult requirement to satisfy is the Lebesgue dominated convergence: Let $\tilde{F}_n$ be a bounded increasing sequence of measurable functions tending almost surely to $\tilde{F}$. Then

$$\int_{\tilde{G}} \tilde{F}_n d\tilde{g} \to \int_{\tilde{G}} \tilde{F} d\tilde{g} \tag{6}$$

Haar measures in infinite dimension were studied by Pickrell [2] and Asada [3]. We have defined the Haar distribution on a path group by using the Hida-Streit approach of path integrals as distribution [4–7]. We refer to the review of Albeverio for various rigorous approaches to path integrals [8] and our review on geometrical path integrals [4,9].

In the case of a path group, we can consider the Wiener process on a path-group $t \to \{s \to g_{s,t}\}$ starting from the unit path (See the work of Airault-Malliavin ([10]), the work of Baxendale [11] and the review paper of Léandre on that topic [12]). We have shown that

$$E[F(g_{.,t})] \to \int_{C([0,1],G)} F(g(.))dD \tag{7}$$

when $t \to \infty$ where dD is the Haar distribution on the path group and F is a test functional of Hida type [7].

Recently we are motivated by extending stochastic analysis tools in the non-Markovian context ([13–17]). Especially in [1], we are interested in constructing the sheet and martingales problem in distributional sense for a big-order differential operator on $\mathbb{R}^d$. We consider for that the Connes test algebra.

Let us recall what is the main difference between the Hida test algebra and the Connes test algebra.

(1) Hida considers Fock spaces and tensor product of Hilbert spaces.

(2) Connes, motivated by his work on entire cyclic cohomology, considers Banach spaces. Tensor product of Banach spaces whose theory (mainly due to Grothendieck) is much more complicated than the theory of tensor product of Hilbert spaces.

In [1], we are motivated by the generalization of martingale problems in the non-Markovian context. We consider Connes spaces in [1]. In the present context, we are not motivated by that and we return in the original framework of [7].

We consider the heat semi-group on a path group associated to a bilaplacian on the group in the manner of [1]. In [1], we look at the case of $\mathbb{R}^d$. Here we consider the case of the compact Lie group G. The analysis is similar because we have analog estimates of the heat-kernel [18–20].

In order to resume, we consider an element σ of an Hida Fock space, we associate a functional $\Psi(\sigma)$ on the path group. The heat-semi group (in the **distributional** sense) Q_t^k satisfies the three next properties:

(1) $Q_t^k\Psi(\sigma)$ is still in the considered space
(2) $Q_t^k \circ Q_{t'}^k = Q_{t+t'}^k$
(3) When $t \to \infty$

$$Q_t^k\Psi(\sigma)(g.) \to \int_{C([0,1],G)} \Psi(\sigma)dD \tag{8}$$

Q_t^k is not a Markovian semi-group on $C([0,1], G)$. Especially, Q_t^k is not represented by a stochastic process. However we expect to extend in this context (7).

2. A Brief Review on the Haar Distribution on a Path Group

Let us recall what is the Brownian motion $t \to B_t$ on $\mathbb{R}$. We consider the set of continuous path $t \to B_t$ issued from 0 from $\mathbb{R}^+$ into $\mathbb{R}$. We consider the sigma-algebra $\mathbb{F}_t$ spanned by $B_s, s \leq t$. The Brownian motion probability measure dP is characterized as the solution of the following martingale problem: if f is any bounded smooth function on $\mathbb{R}$,

$$t \to f(B_t) - \int_0^t \Delta f(B_s) ds \qquad (9)$$

is a martingale associated to the filtration $\mathbb{F}_t$. This means that

$$E[(f(B_t) - \int_0^t \Delta f(B_s) ds)G] = E[(f(B_{t'}) - \int_0^{t'} \Delta f(B_{s'}) ds')G] \qquad (10)$$

where G is a bounded functional $\mathbb{F}_{t'}$ measurable ($t' < t$).

The Brownian motion is only continuous. However we can define stochastic integrals (as it was done by Itô). Let $s \to h_s$ be a bounded continuous process. We suppose that h_s is $\mathbb{F}_s$ measurable. Then the Itô integral is defined as follows:

$$\int_0^1 h_s \delta B_s = \lim_{k \to \infty} \sum_{l \leq k} h(\frac{l}{k})(B_{\frac{(l+1)}{k}} - B_{\frac{l}{k}}) \qquad (11)$$

Moreover we have the Itô isometry

$$E[(\int_0^1 h_s \delta B_s)^2] = E[\int_0^1 h_s^2 ds] \qquad (12)$$

Associated to the Brownian motion is classically associated the Bosonic Fock space.

Let H_2 be the Hilbert space of L^2 functions $h(.)$ from $\mathbb{R}^+$ into $\mathbb{R}$. We consider the symmetric tensor product $H_2^{\hat{\otimes}^n}$ of H_2. It can be realized as the set of symmetric maps h^n from $(\mathbb{R}^+)^n$ into $\mathbb{R}$ such that

$$\int_{(\mathbb{R}^+)^n} |h^n(s_1, .., s_n)|^2 ds_1 .. ds_n = \|h^n\|_2^2 < \infty \qquad (13)$$

The symmetric Fock space WN_0 coincides with the set of formal series $\sigma = \sum h^n$ such that $\sum n! \|h^n\|^2 < \infty$. To each h^n we associate the nth Wiener chaos

$$\Psi(h^n) = \int_{(\mathbb{R}^+)^n} h^n(s_1, .., s_n) \delta B_{s_1} ... \delta B_{s_n} \qquad (14)$$

if B_s is the standard $\mathbb{R}$-valued Brownian motion. The definition of the Wiener chaos $\Psi(h^n)$ is a small improvement of the stochastic integral $\int_0^1 h(s) \delta B_s$. By using the symmetry of h^n, we have:

$$\Psi(h^n) = n! \int_{0 < s_1 < s_2 < ... < s_n} h^n(s_1, .., s_n) \delta B_{s_1} ... \delta B_{s_n} \qquad (15)$$

Moreover $E_P[|\Psi(h^n)|^2] = n! \|h^n\|_2^2$ and $\Psi(h^n)$ and $\Psi(h^m)$ are orthogonal in $L^2(dP)$. The L^2 of the Brownian motion can be realized as the symmetric Fock space through the isometry Ψ.

We introduce the Laplacian Δ^+ on $(\mathbb{R})^+$ and we consider the Sobolev space $H_{2,k}$ associated to $(\Delta^+ + I)^k$. On the set of formal series $\sigma = \sum h^n$, we choose a slightly different Hilbert structure:

$$\|\sigma\|_{k,C}^2 = \sum_{n=0}^{\infty} n! C^n \|h^n\|_{2,k}^2 < \infty \qquad (16)$$

We get another symmetric Fock space denoted $WN_{k,C}$. We remark that if $k' \geq k, C' \geq C$

$$\|\sigma\|_{k',C'} \geq \|\sigma\|_{k,C} \qquad (17)$$

The Hida test function space $W.N._{\infty-}$ is the intersection of $W.N_{k,C}$ $k \geq 1$, $C \geq 1$ endowed with the projective topology. A sequence σ_n of the Hida Fock space converges to σ for the topology of the Hida Fock space if σ_n converges to σ in all $W.N_{k,C}$. The map Wiener chaos Ψ realized a map from $W.N_{\infty-}$ into the set of continuous Brownian functional dense in $L^2(dP)$. We refer to the books [21] and [22] for an extensive study between the Fock space and the L^2 of the Wiener measure.

In infinite dimensional analysis, there are basically 3 objects:

(i) An algebraic model.
(ii) A mapping space and a map Ψ from the algebraic model into the space of functionals on this mapping space.
(iii) A path integral μ which is an element of the topological dual of the algebraic model.

In the standard case of the Brownian motion, μ is the vacuum expectation:

$$\mu[\Psi(\sigma)] = h^0 \tag{18}$$

A distribution on the Hida Fock space is a linear map μ from $W.N_{\infty-}$ into $\mathbb{R}$ which satisfies the following requirement: there exists k, C, K such that for all $\sigma \in W.N_{\infty-}$

$$|\mu(\sigma)| \leq C\|\sigma\|_{k,C} \tag{19}$$

Getzler in his seminal paper [23] is the first author who considered another map than the map Wiener chaos. Getzler is motivated by the heuristic considerations of Atiyah-Bismut-Witten relating the structure of the free loop space of a manifold and the Index theorem on a compact spin manifold. Getzler used as algebraic space a Connes space and as map Ψ the map Chen iterated integrals.

Getzler's idea was developed by Léandre ([9]) to study various path integrals in the Hida-Streit approach with a geometrical meaning. Especially Léandre ([5–6]) succeeded to define the Haar measure dD as a distribution on a current group. Let us recall quickly the definition on it. We consider a compact Riemannian manifold M ($S \in M$) and a compact Lie group G ($g \in G$). We consider the current group $C(M, G)$ of continuous maps $S \rightarrow g(S)$ from M into G. We consider the cylindrical functional $h(g(S_1), .., g(S_r))$ on the current group. We have

$$\int_{C(M,G)} h(g(S_1), .., g(S_r))dD = \int_{G^r} h(g_1, .., g_r)dg_1..dg_r \tag{20}$$

We would like to close this operation consistently. It is the object of [5-6].

(1) *Construction of the algebraic model.* We consider the positive self-adjoint Laplacian on $M \times G$ $\Delta^{M \times G}$. We consider the Sobolev space H_k of maps from h $M \times G$ into $\mathbb{R}$ such that

$$\int_{M \times G} ((\Delta^{M \times G} + 2)^k h)^2 dS dg = \|h\|_k^2 \tag{21}$$

We consider the tensor product $H_k^{\otimes n}$ associated to it and we consider the natural Hilbert norm on it (dS and dg are normalized Riemannian measures on M and G respectively). $W.N_{k,C}$ is the set of formal series $\sigma = \sum h^n$ such that

$$\sum C^n \|h^n\|_k^2 = \|\sigma\|_{k,C}^2 < \infty \tag{22}$$

The Hida test functional space is the space $W.N_{\infty-} = \cap W.N_{k,C}$ endowed with the projective topology.

(2) *Construction of the map* Ψ. To h^n we associate

$$\Psi(h^n)(g(.)) = \int_{[0,1]^n} h^n(g(S_1), .., g(S_n), S_1, .., S_n)dS_1...dS_n \tag{23}$$

We put if $\sigma = \sum h^n$

$$\Psi(\sigma) = \sum_{n=0}^{\infty} \Psi(h^n) \tag{24}$$

The map Ψ realizes a continuous map from $W.N_{\infty-}$ into the set of continuous functional on $C(G, M)$.

(3) *Construction of the path integral.* We put if h^n belongs to all the Sobolev Hilbert spaces H_k

$$\int_{G,M} \Psi(h^n)dD = \int_{M^n \times G^n} h^n(g_1, .., g_n, S_1, .., S_n)dg_1..dg_n dS_1..dS_n \tag{25}$$

This map can be extended into a linear continuous application from $W.N_{\infty-}$ (We say it is a Hida distribution) into $\mathbb{R}$. This realizes our definition ([5–7]) of the Haar distribution on the current group $C(M, G)$.

Let $I \in [0, 1]^n$. We consider the normalized Lebesgue measure dv^n on $[0, 1]^n$. Let L_i be the ith partial Laplacian on G^n. We consider the total operator

$$L_t^n = \prod_{i=1}^{n}(L_i + 2) \prod_{i=1}^{n}(-\frac{\partial^2}{\partial s_i^2} + 2) \tag{26}$$

which operates on function h^n on $G^n \times [0, 1]^n$ and we consider its power $(L_t^n)^k$. Let $h^n(g^n, I)$ be a function on $G^n \times [0, 1]^n$. We put

$$\|h^n\|_{C,k}^2 = C^n \int_{G^n \times [0,1]^n} |(L_t^n)^k h^n(g^n, I)|^2 dg^n dv^n(I) \tag{27}$$

(dg^n is the normalized Haar measure on G^n and dv^n the normalized Lebesgue measure on $[0, 1]^n$).

We put

$$\sigma = \sum h^n \tag{28}$$

and we consider the Hilbert norm

$$\|\sigma\|_{k,C}^2 = \sum \|h^n\|_{k,C}^2 \tag{29}$$

Definition 1. *The Hida Fock space $W.N_{\infty-}$ is the space constituted of the σ defined above such that for all $k \in \mathbb{N}, C > 0 \|\sigma\|_{k,C}^2 < \infty$*

If σ belongs to $W.N_{\infty-}$, we associate

$$\Psi(\sigma)(g(.)) = \sum_{n=0}^{\infty} \int_{[0,1]^n} h^n(g(s_1), .., g(s_n), I)dv^n(I) \tag{30}$$

where $s \to g(s)$ belongs to $C([0, 1], G)$.

Theorem 2. *If $\sigma \in W.N_{\infty-}$, $\Psi(\sigma)$ is a continuous bounded function on $C([0, 1], G)$.*

We put

$$\int_{C([0,1],G)} \Psi(h^n)dD = \int_{[0,1]^n \times G^n} h^n(g_1, .., g_n, s_1, .., s_n)dg_1..dg_n ds_1..ds_n \tag{31}$$

Let us recall three of the main theorems of [7]:

Theorem 3. *dD can be extended as a distribution on the Hida Fock space. This means that there exists k, C, K such that for all $\sigma \in W.N_{\infty-}$*

$$|\int_{C([0,1],G)} \Psi(\sigma)dD| \leq K\|\sigma\|_{k,C} \tag{32}$$

Theorem 4. *If* $\Psi(\sigma) \geq 0$, $\int_{C([0,1];G)} \Psi(\sigma)dD \geq 0$.

Theorem 5. *If* $\Psi(\sigma) = 0$, $\int_{C([0,1];G)} \Psi(\sigma)dD = 0$.

3. A Non-Markovian Semi-group on a Path Group

In the sequel, we will suppose that $4k \geq d$. In such a case ([20]), we have

$$|p_t^k(g,g')| \leq \frac{C}{t^{d/4k}} G_{2k,a}\left(\frac{d(g,g')}{t^{1/4k}}\right) \tag{33}$$

where $G_{m,a}(u) = \exp[-au^{2m/2m-1}]$. $p_t^k(g,g')$ is the heat-kernel associated to the heat semi-group P_t^k and d is the biinvariant Riemannian distance on G.

$$P_t^k f(g) = \int_G p_t^k(g,g')f(g')dg' \tag{34}$$

Moreover, since Δ^k is biinvariant

$$p_t^k(gg^1, g'g^1) = p_t^k(g^1 g, g^1 g') = p_t^k(g,g') \tag{35}$$

Since it is an heat kernel associated to a semi-group, it satisfies the Kolmogorov equation:

$$p_{t+s}^k(g,g') = \int_G p_t^k(g,g^1)p_s^k(g^1,g')dg^1 \tag{36}$$

This shows that if $t \in [0,1]$ that

$$\|P_t^k f\|_\infty \leq C\|f\|_\infty \tag{37}$$

and that

$$|P_t^k|[|d(e,.)|^p](e) \leq Ct^{\alpha(k,p)} \tag{38}$$

Remark: *We could get in the sequel more general convolution semi-groups [20] with generators of degree* $2k$ *whose associated heat-kernels satisfied still (33).*

Let us divide the interval time $[0,1]$ into in time intervals $[t_l, t_{l+1}]$ of length $1/m$. Let F be a cylindrical functional $h(g_{t_1}, g_{t_2}, g_{t_m})$. Let us introduce

$$P_t^{k,m} h(g_{t_1}, ..., g_{t_m}) = \int_{G^m} h(g_{t_1}g_1, ..., g_{t_m}g_m) \prod_{i=1}^{m} p_{t/m}^k(g_{i-1}, g_i)dg_i \tag{39}$$

$(g_0 = e)$. This defines a semi-group on G^m. Let us show this statement. We remark

$$P_s^{k,m} P_t^{k,m} F^m(g_{t_1}, ..., t_m) =$$

$$\int_{G^m \times G^m} h(g_{t_1}\bar{g}_1 g_1, ..., g_{t_m}\bar{g}_m g_m) \prod_{i=0}^{m-1} p_{t/m}^k(g_i, g_{i+1}) \prod_{i=0}^{m-1} p_{s/m}^k(\bar{g}_i, \bar{g}_{i+1})dg_i d\bar{g}_i \tag{40}$$

We do the change of variable $\tilde{g}_i = \bar{g}_i g_i$; $g_i = g_i$. We recognize in the last expression

$$\int_{G^m \times G^m} h(g_{t_1}\tilde{g}_1, ..., g_{t_m}\tilde{g}_m) \prod_{i=0}^{m-1} p_{t/m}^k(g_i, g_{i+1}) \prod_{i=0}^{m-1} p_{s/m}^k(\tilde{g}_i g_i^{-1}, \tilde{g}_{i+1} g_{i+1}^{-1})dg_i d\tilde{g}_i \tag{41}$$

But

$$\int_{G^m} \prod_{i=0}^{m-1} p_{t/m}^k(\check{g}_i, \check{g}_{i+1}) \prod_{i=0}^{m-1} p_{s/m}^k(\check{g}_i \check{g}_i^{-1}, \check{g}_{i+1} \check{g}_{i+1}^{-1}) dg_i =$$

$$\int_{G^m} \prod_{i=0}^{m-1} p_{t/m}^k(g_i, g_{i+1}) p_{s/m}^k(\check{g}_i, \check{g}_{i+1} \check{g}_{i+1}^{-1} \check{g}_i) dg_i =$$

$$\int_{G^m} \prod_{i=0}^{m-1} p_{s/m}^k(\check{g}_i, \overline{g}_i) p_{t/m}^k(\overline{g}_i, \check{g}_{i+1}) d\overline{g}_i = \prod_{i=0}^{m-1} p_{\frac{s+t}{m}}^k(\check{g}_i, \check{g}_{i+1}) \quad (42)$$

We have used the semi-group property (36) of P_t^k and the fact that P_t^k is biinvariant (35).

We would like to extend by continuity this formula for functionals which depend on an infinite number of variables $\Psi(\sigma)$ of the previous type. We put for h^n:

$$\mu[\Psi(h^n)] =$$

$$\int_{G^n \times [0,1]^n} h^n(g_1, \ldots, g_n, s_1, \ldots, s_n) \prod_{i=0}^{n-1} p_{s_{i+1}-s_i}^k(g_i, g_{i+1}) dg_i d\nu^n(s_1, \ldots, s_n) \quad (43)$$

($s_0 = 0$). We order $s_1 < s_2 < \ldots < s_n$ without to loose generality.

We extend μ by linearity.

Theorem 6. *μ is a Hida distribution . Moreover if $\Psi(\sigma) = 0$, $\mu[\Psi(\sigma)] = 0$.*

Proof: By the property of the cylindrical semi-group listed in the beginning of this part, we have

$$\int_{G^n} |h^n(g_1, \ldots, g_n, s_1, \ldots, s_n)| \prod_{i=0}^{n-1} |p_{s_{i+1}-s_i}^k(g_i, g_{i+1})| dg_i \leq C^n \|h^n\|_\infty \quad (44)$$

where $\|\ \|_\infty$ is the uniform norm of h^n. This uniform norm can be estimated by Sobolev imbedding theorem by $\|h^n\|_{k,C}$ for some big k and C independent of n. It follows clearly from that μ is an Hida distribution.

Let us give some details in order to estimate $\|h^n\|_\infty$. We introduce the ordered set of eigenvalues λ_i of Δ. Let $(\alpha) = (i_1, \ldots, i_n)$. Let ϕ_i be the normalized eigenvectors associated to λ_i. We consider $\mathbb{C}$-valued functions to do that. We introduce $\phi_{(\alpha)}(g_1, \ldots, g_n) = \prod \phi_{i_j}(g_j)$. We get

$$h^n = \sum_{(\alpha)} \lambda_{(\alpha)} \phi_{(\alpha)} \quad (45)$$

Therefore

$$\|h^n\|_\infty \leq \sum_{(\alpha)} \|\lambda_{(\alpha)}\|_\infty \|\phi_{(\alpha)}\|_\infty \quad (46)$$

By Garding and Sobolev inequality, the right-hand side of the previous inequality is smaller than

$$C^n \sum_{(\alpha)} K_{(\alpha)}^{-l} \|\lambda_{(\alpha)}\|_{k,C} \|\phi_{(\alpha)}\|_{k,C} \quad (47)$$

for some big k, some big C and some big l.

$$K_{(\alpha)} = \prod_{i \in (\alpha)} (2 + \lambda_i) \quad (48)$$

Let us recall that $\lambda_i \geq 0$ and that $\lambda_i \geq Ci^m$ for some m ([24]). We apply Cauchy-Schwartz inequality in (47). We deduce that

$$\|h^n\|_\infty \leq C^n \{\sum_{(\alpha)} K_{(\alpha)}^{-l}\}^{1/2} \{\sum_{(\alpha)} \|\lambda_{(\alpha)}\|_{k,C}^2 \|\phi_{(\alpha)}\|_{k,C}^2\}^{1/2} \tag{49}$$

But

$$\sum_{(\alpha)} \|\lambda_{(\alpha)}\|_{k,C}^2 \|\phi_{(\alpha)}\|_{k,C}^2 = \|h^n\|_{k,C}^2 \tag{50}$$

Moreover, by [8], $\lambda_i \geq Ci^m$ for some i. Therefore if l is big enough, $\sum_{(\alpha)} K_{(\alpha)}^{-l}$ is finite bounded independently of n.

Let us consider the polygonal approximation of mesh $1/l$ g^l of g.. If $\Psi(\sigma) = 0$, we get $\Psi(\sigma)(g^l_.) = 0$. But $\Psi(\sigma)(g^l_.)$ is a cylindrical functional which depends only of $g_{t_1}, .., g_{t_l} = x_1$. We use the properties listed in the beginning of this part. We get

$$P_1^{k,l}[\Psi(\sigma)(g^l_.)](e) = 0 \tag{51}$$

by the property listed of the beginning of the cylindrical semi-group $P_t^{k,l}$. It remains to show that when $l \to \infty$ that $P_1^{k,l}[\Psi(\sigma)(g^l_.)]$ is very close from $\mu[\Psi(\sigma)(g_.)]$. This follows from the next consideration. Let h^n be an elementary tensor product. We get clearly

$$|\mu[\Psi(h^n)(g^l_.)] - \mu[\Psi(h^n)(g_.)]| \leq C^n \|h^n\|_{1,\infty}$$
$$\sum_{i=0}^{n} \int_{[0,1]^n \times G} |d(e,g_i)|(|p_{s_i - [s_i]_-}^k(e,g_i)| + |p_{[s_i]_+ - s_i}^k(e,g_i)|)dg_i dv^n(s_1,..,s_n) \tag{52}$$

where $[s]_-$ denotes the supremum of the time of the subdivision smaller to s and $[s]_+$ denotes the infimum of the time of the subdivision larger to s. $\|h^n\|_{1,\infty}$ is the uniform C^1 norm of h^n. This norm can be estimated by the Sobolev imbedding theorem by $\|h^n\|_{k^1,C_1}$ for k^1 and C^1 independent of n as in (50).

It remains to use the inequality (35) to conclude.$\Diamond$

Definition 7: μ is called the Wiener distribution issued from the unit path associated to Δ^k.

Let h^n be a smooth function from $G^n \times [0,1]^n$ into $\mathbb{R}$. We suppose that $0 < s_1 < s_2.. < s_n$ in order to simplify the exposition. We put

$$P_t^{k,n} F^n(g_1, .., g_n, s_1, .., s_n) =$$
$$\int_{G^m} h^n(g_1 y_1, .., g_m y_n, s_1, .., s_n) \prod_{i=0}^{n-1} p_{t(s_{i+1} - s_i)}^k (y_i, y_{i+1})dy_i \tag{53}$$

$P_t^{k,n}$ is the cylindrical semi-group on cylindrical functional associated to $g_{s_1}, .., g_{s_n}$.

lemma 8: *There exist a C' bounded when t is bounded and which depend not of n, a k' which depend only of k and not on n such that*

$$\|P_t^{k,n} h^n\|_{C,k} \leq \|h^n\|_{C',k'} \tag{54}$$

Proof: If we take derivative in g_i, the result comes by taking derivative under the sign integral in (43). The result arises then from (37). Let us take first of all derivative in s_i. Either we take derivative of h^n and the result goes by the same way. Or we take derivative in s_{i+1} or s_i of the heat kernel p_s^k. We represent in the way (43) the integral, we remark that the heat kernel satisfies the heat-equation and we integrate by parts in order to conclude.$\Diamond$

Let us suppose that the time subdivision is fixed. Clearly

$$P_{t'}^{k,n} \circ P_t^{k,n} = P_{t+t'}^{k,n} \tag{55}$$

Let h^n be a function from $G^n \times [0,1]^n$ into $\mathbb{R}$. We put

$$Q_t^k[\Psi(h^n)](g.) = \int_{[0,1]^n} P_t^{k,n} h^n(g_{s_1},..,g_{s_n},s_1,..,s_n) dv^n(s_1,..,s_n) \tag{56}$$

Theorem 9: Q_t^k *can be extended by linearity as a continuous linear operator on the Hida Fock space. If* $\Psi(\sigma)(g.) = 0$, $Q_t^k[\Psi(\sigma)](g.) = 0$ *and we get the semi-group property*

$$Q_t^k[Q_{t'}^k[\Psi(\sigma)]](g.) = Q_{t+t'}^k[\Psi(\sigma)](g.) \tag{57}$$

if σ belong to $W.N_{\infty-}$.

Proof: The fact that Q_t^k can be extended by linearity follows from the previous lemma. $Q_t^k[\Psi(\sigma)](x.) = 0$ if $\Psi(\sigma) = 0$ holds exactly as in the proof of Theorem 6. For a simple element h^n of the Hida Fock space, we have clearly:

$$Q_t^k[Q_{t'}^k[\Psi(h^n)]](g.) = Q_{t+t'}^k[\Psi(h^n)](g.) \tag{58}$$

This result can be extended by continuity.◇

4. Long Time Behaviour

The main theorem of this paper is the following:

Theorem 10: *If σ belong to $W.N_{\infty-}$, then when $t \to \infty$*

$$Q_t^k[\Psi(\sigma)](e.) \to \int_{C([0,1];G)} \Psi(\sigma) dD \tag{59}$$

where $e.$ is the unit path.

Proof: Let us decompose $L^2(G)$ in an orthonormal basis of eigenvectors ϕ_i of Δ associated to the eigenvalues λ_i. Classically [24], $\sup_g |\phi_i(g)| \leq Ci^{m_0}$ and $\lambda_i \geq Ci^{m_1}$ for some positive m_0 and m_1. Classically the heat kernel is given by

$$p_t^k(g,g') = 1 + \sum_{i>0} \exp[-\lambda_i^k] \phi_i(g) \phi_i(g') \tag{60}$$

>From the previous bound, we deduce if $t \geq 1$

$$\sup_{g,g'} |p_t^k(g,g')| \leq C < \infty \tag{61}$$

$P_t^{k,n}$ is associated if $s_1 < s_2 < .. < s_n < 1$ to a invariant elliptic operator on G^n. It has therefore the unique invariant measure $\otimes dg_i$. This shows that if h^n is an element of the Hida Fock space that

$$P_t^{k,n} h^n(e,..,e,s_1,s_n) \to \int_{G^n} h^n(g_1,...,g_n,s_1,...,s_n) \prod_{i=1}^n dg_i \tag{62}$$

provided all s_i are different.

By the previous estimates, if $t \geq 1$

$$\sup |P_t^{k,n} h^n| \leq C^n \|h^n\|_\infty \tag{63}$$

where $\|h^n\|_\infty$ is the supremum norm of h^n which can be estimated by Sobolev imbedding theorem by $\|h^n\|_{C',k'}$ for some C', some k' independent of n. Therefore

$$Q_t^k[\Psi(\sigma)](e_.) = \sum_{n=0}^{\infty} \int_{[0,1]^n} P_t^k[h^n](e,..,e,s_1,..s_n) ds_1..ds_n \tag{64}$$

By the dominated Lebesgue convergence, this tends when $t \to \infty$ to

$$\sum_{n=0}^{\infty} \int_{G^n \times [0,1]^n} h^n(g_1,..,g_n,s_1,..,s_n) \prod_{i=1}^{n} dg_i \prod_{i=1}^{n} ds_i = \int_{C([0,1];G)} \Psi(\sigma) dD \tag{65}$$

$\Diamond$

5. Conclusions

We define a non-Markovian semi-group on a path group which acts on a Hida type test algebra on the path group and we study its long time behaviour related to the Haar distribution on the path group.

References

1. Léandre, R. Stochastic analysis without probability: Study of some basical tools. *J. Pseudo. Differ. Oper. Appl.* **2010**, *1*, 389-400.
2. Pickrell, D. *Invariant Measures for Unitary Groups Associated to Kac-Moody Algebras*; A.M.S.: Providence, RI, USA, 2000; Volume 693.
3. Asada, A. Regularized calculus: An application of zeta regularization to infinite dimensional geometry and analysis. *Int. J. Geometry. Mod. Phys.* **2004**, *1*, 107-157.
4. Léandre, R. Path integrals in noncommutative geometry. In *Encyclopedia of Mathematical Physics*; Naber, G., Ed.; Elsevier: Oxford, UK, 2006; pp. 8-12.
5. Léandre, R. Lebesgue measure in infinite dimension an an infinite distribution. *J. Math. Sci.* **2009**, *159*, 833-836.
6. Léandre, R. Infinite Lebesgue distribution on a current group as an invariant distribution. In *Foundations of Probability and Physics IV*; Vaxjoe, Sweden, 2006; Khrennikov, A, Ed.; A.I.P.: Melville, NY, USA, 2007; Volume 889, pp. 332-337.
7. Léandre, R. Long time behaviour of the Wiener process on a path group. In *Group Theory: Classes, Representations and Connections and Applications*; Danellis, C.H., Ed.; Nova Publisher: New York, NY, USA, 2010; pp. 313-323.
8. Albeverio, S. Wiener and Feynman path integrals and their applications. *Proc. Symp. Appl. Math.*, **1997**, *52*, 163-194.
9. Léandre, R. Connes-Hida Calculus in index theory. In *14th International Congress on Mathematical Physics*; Lisboa, Portugal, 2003; Zambrini, J.C., Ed.; World Scientific: Singapore, 2003; pp. 493-498.
10. Airault, H.; Malliavin, P. *Analysis Over Loop Groups*; University Paris VI: Paris, France, 1991.
11. Baxendale, P. Wiener processes on manifolds of maps. *Proc. Roy. Soc. Edimburg. A* **1980**, *87*, 127-52.
12. Léandre, R. The geometry of Brownian surfaces. *Probab. Surv.* **2006**, *3*, 37-88.
13. Léandre, R. Itô-Stratonovitch formula for a four order operator on a torus. *Acta Phys. Debrecina.* **2008**, *42*, 133-138.
14. Léandre, R. Itô-Stratonovitch formula for the Schroedinger equation associated to a big order operator on a torus. *Phys. Scr.* **2009**, *136*, 014028.
15. Léandre, R. Itô-Stratonovitch formula for the wave equation on a torus. In *Computations of Stochastic Systems*; El Tawil, M.A., Ed.; Springer: Heidelberg, Germany, 2010; pp. 68-75.
16. Léandre, R. Itô formula for an integro differential operator without a stochastic process. In *ISAAC 2009*; London, UK, 2010; Wirth, J., Ed.; World Scientific: Singapore, 2011; pp. 226-232.
17. Léandre, R. The Itô transform for a general class of pseudo-differential operators. *Stochastic Models and Data Analysis*; Skiadas, C., Ed.; Chania, Greece, 2010.

Symmetry **2011**, 3, 72–83

18. Auscher, P.; Tchamitchian P. *Square Root Problem for Divergence Operators and Related Topics.*; S.M.S.: Paris, France, 1998; Volume 249.
19. Davies, E.B. Uniformly elliptic operators with measurable coefficients. *J. Funct. Ana.* **1995**, *132*, 141-169.
20. Ter Elst, A.; Robinson, D.W. Subcoercive and Subelliptic Operators on Lie Groups: Variable Coefficients. *Publ. Inst. Math. Sci.* **1993**, *29*, 745-801.
21. Hida, T.; Kuo, H.H.; Potthoff J.; Streit L. *White Noise: An Infinite Dimensional Calculus*; Kluwer: Dordrecht, The Netherlands, 1993.
22. Meyer, P.A. *Quantum Probability for Probabilists*; Springer: Heidelberg, Germany, 1993.
23. Getzler, E. Cyclic homology and the path integral of the Dirac operator, 1988, Unpublished Preprint.
24. Gilkey, P. *Invariance Theory, the Heat Equation and the Atiyah-Singer Index Theorem*, 2nd ed.; C.R.C. Press: Boca Raton, FL, USA, 1995.

symmetry

Article

On Symmetry of Independence Polynomials

Vadim E. Levit [1],* and Eugen Mandrescu [2]

[1] Department of Computer Science and Mathematics, Ariel University Center of Samaria, Kiryat HaMada, Ariel 40700, Israel

[2] Department of Computer Science, Holon Institute of Technology, 52 Golomb Street, Holon 58102, Israel; E-Mail: eugen_m@hit.ac.il

* Author to whom correspondence should be addressed; E-Mail: levitv@ariel.ac.il; Tel.: +972-3-9066163; Fax: +972-3-9066692.

Received: 27 April 2011; in revised form: 20 June 2011 / Accepted: 22 June 2011 /Published: 15 July 2011

Abstract: An *independent* set in a graph is a set of pairwise non-adjacent vertices, and $\alpha(G)$ is the size of a maximum independent set in the graph G. A matching is a set of non-incident edges, while $\mu(G)$ is the cardinality of a maximum matching. If s_k is the number of independent sets of size k in G, then $I(G;x) = s_0 + s_1 x + s_2 x^2 + ... + s_\alpha x^\alpha$, $\alpha = \alpha(G)$, is called the *independence polynomial* of G (Gutman and Harary, 1986). If $s_j = s_{\alpha-j}$ for all $0 \leq j \leq \lfloor \alpha/2 \rfloor$, then $I(G;x)$ is called *symmetric* (or *palindromic*). It is known that the graph $G \circ 2K_1$, obtained by joining each vertex of G to two new vertices, has a symmetric independence polynomial (Stevanović, 1998). In this paper we develop a new algebraic technique in order to take care of symmetric independence polynomials. On the one hand, it provides us with alternative proofs for some previously known results. On the other hand, this technique allows to show that for every graph G and for each non-negative integer $k \leq \mu(G)$, one can build a graph H, such that: G is a subgraph of H, $I(H;x)$ is symmetric, and $I(G \circ 2K_1;x) = (1+x)^k \cdot I(H;x)$.

Keywords: independent set; independence polynomial; symmetric polynomial; palindromic polynomial

MSC: 05C31; 05C69

---

1. Introduction

Throughout this paper $G = (V, E)$ is a simple (*i.e.*, a finite, undirected, loopless and without multiple edges) graph with vertex set $V = V(G)$ and edge set $E = E(G)$. If $X \subset V$, then $G[X]$ is the subgraph of G spanned by X. By $G - W$ we mean the subgraph $G[V - W]$, if $W \subset V(G)$. We also denote by $G - F$ the partial subgraph of G obtained by deleting the edges of F, for $F \subset E(G)$, and we write shortly $G - e$, whenever $F = \{e\}$.

The *neighborhood* of a vertex $v \in V$ is the set $N_G(v) = \{w : w \in V \text{ and } vw \in E\}$, while $N_G[v] = N_G(v) \cup \{v\}$; if there is no ambiguity on G, we write $N(v)$ and $N[v]$.

K_n, P_n, C_n denote, respectively, the complete graph on $n \geq 1$ vertices, the chordless path on $n \geq 1$ vertices, and the chordless cycle on $n \geq 3$ vertices.

The *disjoint union* of the graphs G_1, G_2 is the graph $G = G_1 \cup G_2$ having as vertex set the disjoint union of $V(G_1), V(G_2)$, and as edge set the disjoint union of $E(G_1), E(G_2)$. In particular, nG denotes the disjoint union of $n > 1$ copies of the graph G.

If G_1, G_2 are disjoint graphs, $A_1 \subseteq V(G_1)$, $A_2 \subseteq V(G_2)$, then the *Zykov sum* of G_1, G_2 with respect to A_1, A_2, is the graph $(G_1, A_1) + (G_2, A_2)$ with $V(G_1) \cup V(G_2)$ as vertex set and

$$E(G_1) \cup E(G_2) \cup \{v_1 v_2 : v_1 \in A_1, v_2 \in A_2\}$$

as edge set [1]. If $A_1 = V(G_1)$ and $A_2 = V(G_2)$, we simply write $G_1 + G_2$.

The *corona* of the graphs G and H with respect to $A \subseteq V(G)$ is the graph $(G, A) \circ H$ obtained from G and $|A|$ copies of H, such that every vertex belonging to A is joined to all vertices of a copy of H [2]. If $A = V(G)$ we use $G \circ H$ instead of $(G, V(G)) \circ H$ (see Figure 1 for an example).

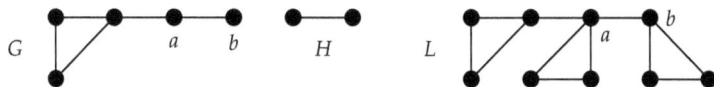

Figure 1. G, H and $L = (G, A) \circ H$, where $A = \{a, b\}$.

Let G, H be two graphs and C be a cycle on q vertices of G. By $(G, C) \triangle H$ we mean the graph obtained from G and q copies of H, such that each two consecutive vertices on C are joined to all vertices of a copy of H (see Figure 2 for an example).

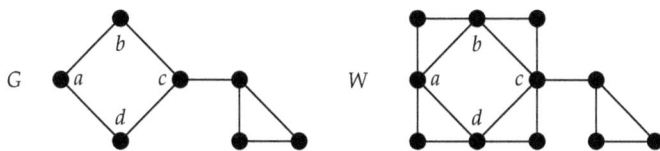

Figure 2. G and $W = (G, C) \triangle H$, where $V(C) = \{a, b, c, d\}$ and $H = K_1$.

An *independent* (or a *stable*) set in G is a set of pairwise non-adjacent vertices. By $Ind(G)$ we mean the family of all independent sets of G. An independent set of maximum size will be referred to as a *maximum independent set* of G, and the *independence number* of G, denoted by $\alpha(G)$, is the cardinality of a maximum independent set in G.

Let s_k be the number of independent sets of size k in a graph G. The polynomial

$$I(G; x) = s_0 + s_1 x + s_2 x^2 + \ldots + s_\alpha x^\alpha, \quad \alpha = \alpha(G)$$

is called the *independence polynomial* of G [3,4], the *independent set polynomial* of G [5]. In [6], the *dependence polynomial* $D(G; x)$ of a graph G is defined as $D(G; x) = I(\overline{G}; -x)$.

A matching is a set of non-incident edges of a graph G, while $\mu(G)$ is the cardinality of a maximum matching. Let m_k be the number of matchings of size k in G.

The polynomial

$$M(G; x) = m_0 + m_1 x + m_2 x^2 + \ldots + m_\mu x^\mu, \quad \mu = \mu(G)$$

is called the *matching polynomial* of G [7].

The independence polynomial has been defined as a generalization of the matching polynomial, because the matching polynomial of a graph G and the independence polynomial of its line graph are identical. Recall that given a graph G, its *line graph* $L(G)$ is the graph whose vertex set is the edge set of G, and two vertices are adjacent if they share an end in G. For instance, the graphs G_1 and G_2 depicted in Figure 3 satisfy $G_2 = L(G_1)$ and, hence, $I(G_2; x) = 1 + 6x + 7x^2 + x^3 = M(G_1; x)$.

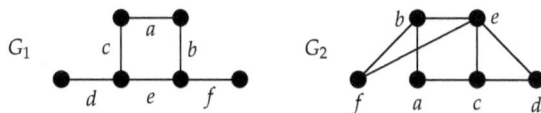

Figure 3. G_2 is the line-graph of and G_1.

In [3] a number of general properties of the independence polynomial of a graph are presented. As examples, we mention that:

$$I(G_1 \cup G_2; x) = I(G_1; x) \cdot I(G_2; x), \quad I(G_1 + G_2; x) = I(G_1; x) + I(G_2; x) - 1.$$

The following equalities are very useful in calculating of the independence polynomial for various families of graphs.

Theorem 1. *Let $G = (V, E)$ be a graph of order n. Then the following identities are true:*

(i) $I(G; x) = I(G - v; x) + x \cdot I(G - N[v]; x)$ *holds for each $v \in V$ [3].*

(ii) $I(G \circ H; x) = (I(H; x))^n \cdot I\left(G; \frac{x}{I(H;x)}\right)$ *for every graph H [8].*

A finite sequence of real numbers $(a_0, a_1, a_2, ..., a_n)$ is said to be:

- *unimodal* if there is some $k \in \{0, 1, ..., n\}$, such that $a_0 \leq ... \leq a_{k-1} \leq a_k \geq a_{k+1} \geq ... \geq a_n$;
- *log-concave* if $a_i^2 \geq a_{i-1} \cdot a_{i+1}, i \in \{1, 2, ..., n-1\}$;
- *symmetric* (or *palindromic*) if $a_i = a_{n-i}, i = 0, 1, ..., \lfloor n/2 \rfloor$.

It is known that every log-concave sequence of positive numbers is also unimodal.

A polynomial is called *unimodal (log-concave, symmetric)* if the sequence of its coefficients is unimodal (log-concave, symmetric, respectively).

For instance, the independence polynomial:

- $I(K_{42} + 3K_7; x) = 1 + 63x + 147x^2 + 343x^3$ is log-concave;
- $I(K_{43} + 3K_7; x) = 1 + 64x + 147x^2 + 343x^3$ is unimodal, but it is not log-concave, because $147 \cdot 147 - 64 \cdot 343 = -343 < 0$;
- $I(K_{127} + 3K_7; x) = 1 + 148x + \mathbf{147}x^2 + 343x^3$ is non-unimodal;
- $I(K_{18} + 3K_3 + 4K_1; x) = 1 + 31x + 33x^2 + 31x^3 + x^4$ is symmetric and log-concave;
- $I(K_{52} + 3K_4 + 4K_1; x) = 1 + 68x + 54x^2 + 68x^3 + x^4$ is symmetric and non-unimodal.

It is easy to see that if $\alpha(G) \leq 3$ and $I(G; x)$ is symmetric, then it is also log-concave.

For other examples, see [9–14]. Alavi *et al.* proved that for every permutation π of $\{1, 2, ..., \alpha\}$ there is a graph G with $\alpha(G) = \alpha$ such that $s_{\pi(1)} < s_{\pi(2)} < ... < s_{\pi(\alpha)}$ [9].

The following conjecture is still open.

Conjecture 1. *The independence polynomial of every tree is unimodal [9].*

Hence to prove the unimodality of independence polynomials is sometimes a difficult task. Moreover, even if the independence polynomials of all the connected components of a graph G are unimodal, then $I(G; x)$ is not for sure unimodal [15]. The following result shows that symmetry gives a hand to unimodality.

Theorem 2. *If P and Q are both unimodal and symmetric, then $P \cdot Q$ is unimodal and symmetric [16].*

A *clique cover* of a graph G is a spanning graph of G, each connected component of which is a clique. A *cycle cover* of a graph G is a spanning graph of G, each connected component of which is a vertex, an edge, or a proper cycle. In this paper we give an alternative proof for the fact that the polynomials $I(G \circ 2K_1; x)$, $I(\Phi(G); x)$, and $I(\Gamma(G); x)$ are symmetric for every clique cover Φ, and every cycle cover Γ of a graph G, where $\Phi(G)$ and $\Gamma(G)$ are graphs built by Stevanović's rules [17]. Our main finding claims that the polynomial $I(G \circ 2K_1; x)$ is divisible both by $I(\Phi(G); x)$ and $I(\Gamma(G); x)$.

The paper is organized as follows. Section 2 looks at previous results on symmetric independence polynomials, Section 3 presents our results connecting symmetric independence polynomials derived by Stevanović's rules [17], while Section 4 is devoted to conclusions, future directions of research, and some open problems.

2. Related Work

The symmetry of the matching polynomial and the characteristic polynomial of a graph were examined in [18], while for the independence polynomial we quote [17,19,20]. Recall from [18] that G is called an *equible graph* if $G = H \circ K_1$ for some graph H. Both matching polynomials and characteristic polynomials of equible graphs are symmetric [18]. Nevertheless, there are non-equible graphs whose matching polynomials and characteristic polynomials are symmetric.

It is worth mentioning that one can produce graphs with symmetric independence polynomials in different ways. For instance, the independence polynomial of the disjoint union of two graphs having symmetric independence polynomial is symmetric as well. Another basic graph operation preserving symmetry of the independence polynomial is the Zykov sum of two graphs with the same independence number. We summarize other constructions respecting symmetry of the independence polynomial in what follows.

2.1. Gutman's Construction [21]

For integers $p > 1, q > 1$, let $J_{p,q}$ be the graph built in the following manner [21]. Start with three complete graphs K_1, K_p and K_q whose vertex sets are disjoint. Connect the vertex of K_1 with $p - 1$ vertices of K_p and with $q - 1$ vertices of K_q (see Figure 4 as an example).

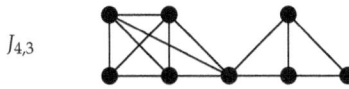

$J_{4,3}$

Figure 4. $I(J_{4,3}; x) = 1 + 8x + 14x^2 + x^3$ and $I(J_{4,3} + K_6; x) = 1 + 14x + 14x^2 + x^3$.

The graph thus obtained has a unique maximum independent set of size three, and its independence polynomial is equal to

$$I(J_{p,q}; x) = 1 + (p + q + 1)x + (pq + 2)x^2 + x^3.$$

Hence the independence polynomial of $G = J_{p,q} + K_{pq-p-q+1}$ is

$$I(G; x) = I(J_{p,q}; x) + I(K_{pq-p-q+1}; x) - 1 = 1 + (2 + pq)x + (2 + pq)x^2 + x^3,$$

which is clearly symmetric and log-concave.

2.2. Bahls and Salazar's Construction [20]

The K_t-path of length $k \geq 1$ is the graph $P(t, k) = (V, E)$ with $V = \{v_1, v_2, ..., v_{t+k-1}\}$ and $E = \{v_i v_{i+j} : 1 \leq i \leq t + k - 2, 1 \leq j \leq \min\{t - 1, t + k - i - 1\}\}$. Such a graph consists of k copies of K_t, each glued to the previous one by identifying certain prescribed subgraphs isomorphic to K_{t-1}. Let $d \geq 0$ be an integer. The d-augmented K_t path $P(t, k, d)$ is defined by introducing new vertices $\{u_{i,1}, u_{i,2}, ..., u_{i,d}\}_{i=0}^{t+k-2}$ and edges $\{v_i u_{i,j}, v_{i+1} u_{i,j} : j = 1, ..., d\}_{i=1}^{t+k-2} \cup \{v_1, u_{0,j} : j = 1, ..., d\}$. Let $G = (V, E)$ and $U \subseteq V$ be a subset of its vertices. Let $v \notin V$ and define the *cone* of G on U with vertex v, denoted $G^*(U, v) = (G, U) + K_1$, where $K_1 = (\{v\}, \emptyset)$. Given G and U and a graph H, we write $H + (G, U)$ instead of $(H, V(H)) + (G, U)$.

Theorem 3. *Let $t \geq 2, k \geq 1$, and $d \geq 0$ be integers, and let $G = (V, E)$ be a graph with $U \subseteq V$ a distinguished subset of vertices. Suppose that each of the graphs $G, G - U$, and $(G, U) + K_1$ has a symmetric and unimodal independence polynomial, and $\deg(I(G; x)) = \deg(I((G, U) + K_1; x)) = \deg(I(G - U; x)) + 2$. Then the independence polynomial of the graph $P(t, k, d) + (G, U)$ is symmetric and unimodal [20].*

2.3. Stevanović's Constructions [17]

Taking into account that $s_0 = 1$ and $s_1 = |V(G)| = n$, it follows that if $I(G; x)$ is symmetric, then $s_0 = s_\alpha$ and $s_1 = s_{\alpha-1}$, i.e., G has only one maximum independent set, say S, and $n - \alpha(G)$ independent sets, of size $\alpha(G) - 1$, that are not subsets of S.

Theorem 4. *If there is an independent set S in G such that $|N(A) \cap S| = 2|A|$ holds for every independent set $A \subseteq V(G) - S$, then $I(G; x)$ is symmetric [17].*

The following result is a consequence of Theorem 4.

Corollary 1. (i) *If $\alpha(G) = \alpha, s_\alpha = 1, s_{\alpha-1} = |V(G)|$, and for the unique stability system S of G it is true that $|N(v) \cap S| = 2$ for each $v \in V(G) - S$, then $I(G; x)$ is symmetric [17]; (ii) If G is a claw-free graph with $\alpha(G) = \alpha, s_\alpha = 1, s_{\alpha-1} = |V(G)|$, then $I(G; x)$ is symmetric.*

Corollary 1 gives three different ways to construct graphs having symmetric independence polynomials [17].

- **Rule 1.** For a given graph G, define a new graph H as: $H = G \circ 2K_1$.

  For an example, see the graphs in Figure 5: $I(G; x) = 1 + 6x + 9x^2 + 3x^3$, while

  $$I(H_1; x) = (1+x)^6 \left(1 + 12x + 48x^2 + 77x^3 + 48x^4 + 12x^5 + x^6\right) = 1 + 18x + 135x^2 + 565x^3$$

  $$+ 1485x^4 + 2601x^5 + 3126x^6 + 2601x^7 + 1485x^8 + 565x^9 + 135x^{10} + 18x^{11} + x^{12}.$$

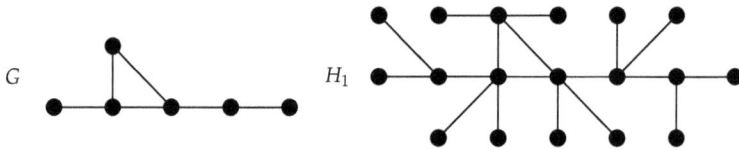

Figure 5. G and $H_1 = G \circ 2K_1$.

- A *cycle cover* of a graph G is a spanning graph of G, each connected component of which is a vertex (which we call a *vertex-cycle*), an edge (which we call an *edge-cycle*), or a proper cycle. Let Γ be a cycle cover of G.

  Rule 2. Construct a new graph H from G, denoted by $H = \Gamma(G)$, as follows: if $C \in \Gamma$ is
  (i) a vertex-cycle, say v, then add two vertices and join them to v;
  (ii) an edge-cycle, say uv, then add two vertices and join them to both u and v;
  (iii) a proper cycle, with

  $$V(C) = \{v_i : 1 \le i \le s\}, E(C) = \{v_i v_{i+1} : 1 \le i \le s-1\} \cup \{v_1 v_s\},$$

  then add s vertices, say $\{w_i : 1 \le i \le s\}$ and each of them is joined to two consecutive vertices on C, as follows: w_1 is joined to v_s, v_1, then w_2 is joined to v_1, v_2, further w_3 is joined to v_2, v_3, etc.
  Figure 6 contains an example, namely, $I(G; x) = 1 + 6x + 9x^2 + 3x^3$, while

  $$I(H_2; x) = 1 + 13x + 60x^2 + 125x^3 + 125x^4 + 60x^5 + 13x^6 + x^7 =$$

  $$= (1+x)\left(1 + 12x + 48x^2 + 77x^3 + 48x^4 + 12x^5 + x^6\right).$$

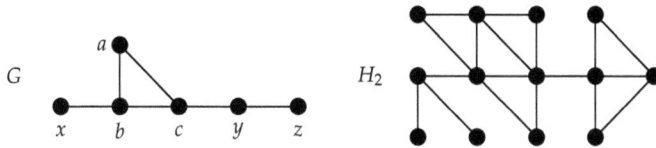

Figure 6. G and $H_2 = \Gamma(G)$, where $\Gamma = \{\{x\}, \{a, b, c\}, \{y, z\}\}$.

- A *clique cover* of a graph G is a spanning graph of G, each connected component of which is a clique. Let Φ be a clique cover of G.

Rule 3. Construct a new graph H from G, denoted by $H = \Phi(G)$, as follows: for each $Q \in \Phi$, add two non-adjacent vertices and join them to all the vertices of Q.

Figure 7 contains an example, namely, $I(G; x) = 1 + 6x + 9x^2 + 3x^3$, while

$$I(H_3; x) = 1 + 12x + 48x^2 + 77x^3 + 48x^4 + 12x^5 + x^6.$$

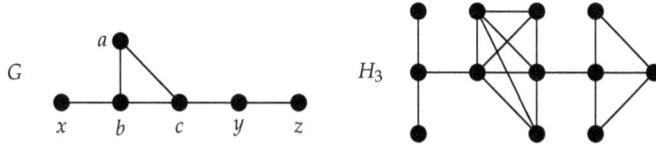

Figure 7. G and $H_3 = \Phi(G)$, where $\Phi = \{\{x\}, \{a, b, c\}, \{y, z\}\}$.

Theorem 5. *Let H be the graph obtained from a graph G according to one of the **Rules 1, 2** or **3**. Then H has a symmetric independence polynomial [17].*

Let us remark that $I(H_1; x) = (1 + x)^6 \cdot I(H_3; x)$ and $I(H_2; x) = (1 + x) \cdot I(H_3; x)$, where H_1, H_2 and H_3 are depicted in Figures 5, 6, and 7, respectively.

2.4. Inequalities and Equalities Following from Theorem 5

When inequalities connecting coefficients of the independence polynomial is under consideration, the symmetry mirrors the area, where they are already established. The following results illustrate this idea.

Proposition 1. *Let $G = H \circ 2K_1$ be with $\alpha(G) = \alpha$, and (s_k) be the coefficients of $I(G; x)$. Then $I(G; x)$ is symmetric, and [22]*

$$s_0 \leq s_1 \leq \ldots \leq s_p, \text{ for } p = \lfloor (2\alpha + 2)/5 \rfloor, \text{ while}$$

$$s_t \geq \ldots \geq s_{\alpha-1} \geq s_\alpha, \text{ for } t = \lceil (3\alpha - 2)/5 \rceil.$$

Theorem 6. *Let H be a graph of order $n \geq 2$, Γ be a cycle cover of H that contains no vertex-cycles, G be obtained by **Rule 2**, and $\alpha(G) = \alpha$. Then $I(G; x)$ is symmetric and its coefficients (s_k) satisfy the subsequent inequalities [22]*

$$s_0 \leq s_1 \leq \ldots \leq s_p, \text{ for } p = \lfloor (\alpha + 1)/3 \rfloor, \text{ and}$$

$$s_q \geq \ldots \geq s_{\alpha-1} \geq s_\alpha, \text{ for } q = \lceil (2\alpha - 1)/3 \rceil.$$

Let $H_n, n \geq 1$, be the graphs obtained according to **Rule 3** from P_n, as one can see in Figure 8.

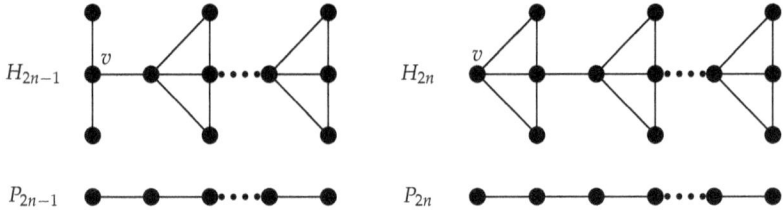

Figure 8. P_n and $H_n = \Omega\{P_n\}$.

Theorem 7. *If* $J_n(x) = I(H_n; x), n \geq 0$, *then* [23]

(i) $J_0(x) = 1, J_1(x) = 1 + 3x + x^2$ *and* $J_n, n \geq 2$, *satisfies the following recursive relations:*

$$J_{2n}(x) = J_{2n-1}(x) + x \cdot J_{2n-2}(x), \quad n \geq 1,$$

$$J_{2n-1}(x) = (1+x)^2 \cdot J_{2n-2}(x) + x \cdot J_{2n-3}(x), \quad n \geq 2;$$

(ii) J_n *is both symmetric and unimodal.*

It was conjectured in [23] that $I(H_n; x)$ is log-concave and has only real roots. This conjecture has been resolved as follows.

Theorem 8. *Let* $n \geq 1$. *Then* [24]

(i) *the independence polynomial of* H_n *is*

$$I(H_n; x) = \prod_{s=1}^{\lfloor (n+1)/2 \rfloor} \left(1 + 4x + x^2 + 2x \cdot \cos \frac{2s\pi}{n+2} \right);$$

(ii) $I(H_n; x)$ *has only real zeros, and, therefore, it is log-concave and unimodal.*

3. Results

The following lemma goes from the well-known fact that the polynomial $P(x)$ is symmetric if and only if it equals its reciprocal, *i.e.,*

$$P(x) = x^{\deg(P)} \cdot P\left(\frac{1}{x}\right). \tag{1}$$

Lemma 1. *Let* $f(x)$, $g(x)$ *and* $h(x)$ *be polynomials satisfying* $f(x) = g(x) \cdot h(x)$. *If any two of them are symmetric, then the third is symmetric as well.*

For $H = 2K_1$, Theorem 1 gives

$$I(G \circ 2K_1; x) = (1+x)^{2n} \cdot I\left(G; \frac{x}{(1+x)^2} \right).$$

Since

$$\frac{x}{(1+x)^2} = \frac{\frac{1}{x}}{\left(1+\frac{1}{x}\right)^2} \quad and \quad \deg(I(G \circ 2K_1; x)) = 2n,$$

one can easily see that the polynomial $I(G \circ 2K_1; x)$ satisfies the identity (1). Thus we conclude with the following.

Symmetry **2011**, 3, 472–486

Theorem 9. *For every graph G, the polynomial $I(G \circ 2K_1; x)$ is symmetric [17].*

3.1. Clique Covers Revisited

Lemma 2. *If A is a clique in a graph G, then for every graph H*

$$I((G, A) \circ H; x) = I(H; x)^{|A|-1} \cdot I((G, A) + H; x).$$

Proof. Let $G_1 = (G, A) \circ H$ and $G_2 = ((G, A) + H) \cup ((|A| - 1)H)$.
   For $S \in Ind(G)$, let us define the following families of independent sets:

$$\Omega_S^{G_1} = \{S \cup W : W \subseteq V(G_1 - G), S \cup W \in Ind(G_1)\},$$

$$\Omega_S^{G_2} = \{S \cup W : W \subseteq V(G_2 - G), S \cup W \in Ind(G_2)\}.$$

Since A is a clique, it follows that $|S \cap A| \le 1$.
   Case 1. $S \cap A = \emptyset$.
   In this case $S \cup W \in \Omega_S^{G_1}$ if and only if $S \cup W \in \Omega_S^{G_2}$. Hence, for each size $m \ge |S|$, we get that

$$\left| \{S \cup W \in \Omega_S^{G_1} : |S \cup W| = m\} \right| = \left| \{S \cup W \in \Omega_S^{G_2} : |S \cup W| = m\} \right|.$$

Case 2. $S \cap A = \{a\}$.
   Now, every $S \cup W \in \Omega_S^{G_1}$ has $W \cap V(H) = \emptyset$ for exactly one H, namely, the graph H whose vertices are joined to a. Hence, W may contain vertices only from $(|A| - 1)H$.
   On the other hand, each $S \cup W \in \Omega_S^{G_2}$ has $W \cap V(H) = \emptyset$ for the unique H appearing in $(G, A) + H$. Therefore, W may contain vertices only from $(|A| - 1)H$.
   Hence for each positive integer $m \ge |S|$, we obtain that

$$\left| \{S \cup W \in \Omega_S^{G_1} : |S \cup W| = m\} \right| = \left| \{S \cup W \in \Omega_S^{G_2} : |S \cup W| = m\} \right|.$$

Consequently, one may infer that for each size, the two graphs, G_1 and G_2, have the same number of independent sets, in other words, $I(G_1; x) = I(G_2; x)$.
   Since $G_2 = ((G, A) + H) \cup ((|A| - 1)H)$ has $|A| - 1$ disjoint components identical to H, it follows that $I(G_2; x) = I(H; x)^{|A|-1} \cdot I((G, A) + H; x)$. $\square$

Corollary 2. *If A is a clique in a graph G, then*

$$I((G, A) \circ 2K_1; x) = (1 + x)^{2|A|-2} \cdot I((G, A) + 2K_1; x).$$

Theorem 10. *If G is a graph of order n and Φ is a clique cover, then*

$$I(G \circ 2K_1; x) = (1 + x)^{2n-2|\Phi|} \cdot I(\Phi(G); x).$$

Proof. Let $\Phi = \{A_1, A_2, ..., A_q\}$. According to Corollary 2, each
   (a) vertex-clique of Φ yields $(1 + x)^{2-2} = 1$ as a factor of $I(G \circ 2K_1; x)$, since a vertex defines a clique of size 1;
   (b) edge-clique of Φ yields $(1 + x)^2$ as a factor of $I(G \circ 2K_1; x)$, since an edge defines a clique of size 2 (see Figure 9 as an example);
   (c) clique $A_j \in \Phi, |A_j| \ge 3$, produces $(1 + x)^{2|A_j|-2}$ as a factor of $I(G \circ 2K_1; x)$ (see Figure 10 as an example).
   Since the cliques of Φ are pairwise vertex disjoint, one can apply Corollary 2 to all the q cliques one by one.

Figure 9. $G_1 = K_2 \circ 2K_1$, $I(G_1; x) = (1+x)^2 \cdot I(\Phi(K_2); x) = (1+x)^2 \cdot (1+4x+x^2)$.

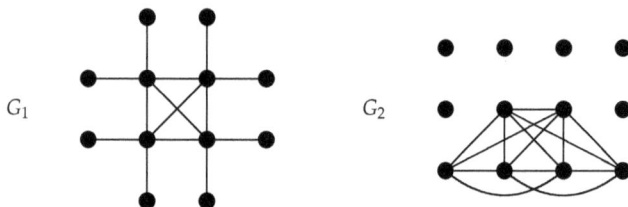

Figure 10. $G_1 = K_4 \circ 2K_1$, $G_2 = 6K_1 \cup \Phi(K_4)$, and $I(G_1; x) = (1+x)^6 \cdot I(\Phi(K_4); x)$.

Using Corollary 2 and the fact that $A_1 \cap A_2 = \varnothing$, we have

$$I((G, A_1 \cup A_2) \circ 2K_1; x) = I((((G, A_1) \circ 2K_1), A_2) \circ 2K_1; x) =$$

$$= (1+x)^{2|A_2|-2} \cdot I((((G, A_1) \circ 2K_1), A_2) + 2K_1; x) =$$

$$= (1+x)^{2|A_2|-2} \cdot I((((G, A_2) + 2K_1), A_1) \circ 2K_1; x) =$$

$$= (1+x)^{2(|A_1|+|A_2|)-2} \cdot I((((G, A_2) + 2K_1), A_1) + 2K_1; x).$$

Repeating this process with $\{A_3, A_4, ..., A_q\}$, and taking into account that all the cliques of Φ are pairwise disjoint, we obtain

$$I((G \circ 2K_1; x) = I((G, A_1 \cup A_2 \cup ... \cup A_q) \circ 2K_1; x) =$$

$$= (1+x)^{2(|A_1|+|A_2|+...+|A_q|)-2q} \cdot I(((((G, A_1) + 2K_1), A_2...), A_q) + 2K_1; x) =$$

$$= (1+x)^{2n-2|\Phi|} \cdot I(\Phi(G); x),$$

as required. □

Lemma 1 and Theorem 10 imply the following.

Corollary 3. *For every clique cover* Φ *of a graph G, the polynomial* $I(\Phi(G); x)$ *is symmetric* [17].

Clearly, for every $k \leq \mu(G)$ there exists a clique cover containing k non-trivial cliques, namely, edges. Consequently, we obtain the following.

Theorem 11. *For every graph G and for each non-negative integer* $k \leq \mu(G)$, *one can build a graph H, such that: G is a subgraph of H, $I(H; x)$ is symmetric, and* $I(G \circ 2K_1; x) = (1+x)^k \cdot I(H; x)$.

3.2. Cycle Covers Revisited

Lemma 3. *If C is a proper cycle in a graph G, then for every graph H*

$$I((G, C) \circ 2H; x) = I(H; x)^{|C|} \cdot I((G, C) \triangle H; x).$$

Proof. Let $C = (V(C), E(C))$, $q = |V(C)|$, $G_1 = (G, C) \circ 2H$, and $G_2 = ((G, C) \triangle H) \cup (qH)$.

For an independent set $S \subset V(G)$, let us denote:

$$\Omega_S^{G_1} = \{S \cup W : W \subseteq V(G_1) - V(G), S \cup W \in Ind(G_1)\},$$

$$\Omega_S^{G_2} = \{S \cup W : W \subseteq V(G_2) - V(G), S \cup W \in Ind(G_2)\}.$$

Case 1. $S \cap V(C) = \emptyset$.

In this case $S \cup W \in \Omega_S^{G_1}$ if an only if $S \cup W \in \Omega_S^{G_2}$, since W is an arbitrary independent set of $2qH$. Hence, for each size $m \geq |S|$, we get that

$$\left| \{S \cup W \in \Omega_S^{G_1} : |S \cup W| = m\} \right| = \left| \{S \cup W \in \Omega_S^{G_2} : |S \cup W| = m\} \right|.$$

Case 2. $S \cap V(C) \neq \emptyset$.

Then, we may assert that

$$\left| \Omega_S^{G_1} \right| = |\{S \cup W : W \text{ is an independent set in } 2(q - |S \cap V(C)|)H\}| = \left| \Omega_S^{G_2} \right|,$$

since W has to avoid all the "H-neighbors" of the vertices in $S \cap V(C)$, both in G_1 and G_2.

Hence, for each positive integer $m \geq |S|$, we get that

$$\left| \{S \cup W \in \Omega_S^{G_1} : |S \cup W| = m\} \right| = \left| \{S \cup W \in \Omega_S^{G_2} : |S \cup W| = m\} \right|.$$

Consequently, one may infer that for each size, the two graphs, G_1 and G_2, have the same number of independent sets. In other words, $I(G_1; x) = I(G_2; x)$.

Since G_2 has $|C|$ disjoint components identical to H, it follows that

$$I(G_2; x) = (1 + x)^{|C|} \cdot I((G, C) \triangle H; x),$$

as required. ☐

Corollary 4. *If C is a proper cycle in a graph G, then*

$$I((G, C) \circ 2K_1; x) = (1 + x)^{|C|} \cdot I((G, C) \triangle K_1; x).$$

Theorem 12. *If G is a graph of order n and Γ is a cycle cover containing k vertex-cycles, then*

$$I(G \circ 2K_1; x) = (1 + x)^{n-k} \cdot I(\Gamma(G); x).$$

Proof. According to Corollaries 2 and 4, each

(a) vertex-cycle of Γ yields $(1 + x)^{2-2} = 1$ as a factor of $I(G \circ 2K_1; x)$, since each vertex defines a clique of size 1;

(b) edge-cycle of Γ yields $(1 + x)^2$ as a factor of $I(G \circ 2K_1; x)$, since every edge defines a clique of size 2;

(c) proper cycle $C \in \Gamma$ produces $(1 + x)^{|C|}$ as a factor (see Figure 11 as an example).

Let $\Gamma = \{C_j : 1 \leq j \leq q\} \cup \{v_i : 1 \leq i \leq k\}$ be a cycle cover containing k vertex-cycles, namely, $\{v_i : 1 \leq i \leq k\}$.

Using Corollary 4 and the fact that $C_1 \cap C_2 = \emptyset$, we have

$$I((G, C_1 \cup C_2) \circ 2K_1; x) = I((((G, C_1) \circ 2K_1), C_2) \circ 2K_1; x) =$$

$$= (1 + x)^{|C_2|} \cdot I((((G, C_1) \circ 2K_1), C_2) \triangle K_1; x) =$$

$$= (1 + x)^{|C_2|} \cdot I((((G, C_2) \triangle K_1), C_1) \circ 2K_1; x) =$$

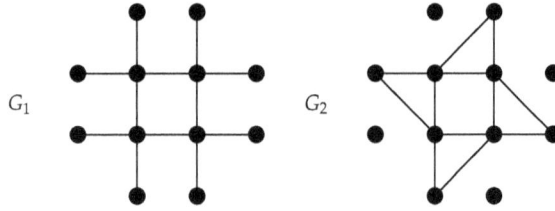

Figure 11. $G_1 = C_4 \circ 2K_1$, $G_2 = 4K_1 \cup \Gamma(C_4)$ and $I(G_1; x) = (1+x)^4 \cdot I(\Gamma(C_4); x)$

$$= (1+x)^{|C_1|+|C_2|} \cdot I((((G, C_2) \triangle K_1), C_1) \triangle K_1; x).$$

Repeating this process with $\{C_3, C_4, ..., C_q\}$, and taking into account that all the cycles of Γ are pairwise vertex disjoint, we obtain

$$I((G \circ 2K_1; x) = I((G, C_1 \cup C_2 \cup ... \cup C_q) \circ 2K_1; x) =$$

$$= (1+x)^{|C_1|+|C_2|+...+|C_q|} \cdot I(((((G, C_1) \triangle K_1), C_2...), C_q) \triangle K_1; x) =$$

$$= (1+x)^{n-k} \cdot I(\Gamma(G); x),$$

as claimed. $\square$

Lemma 1 and Theorem 12 imply the following.

Corollary 5. *For every cycle cover Γ of a graph G, the polynomial $I(\Gamma(G); x)$ is symmetric* [17].

4. Conclusions

In this paper we have given algebraic proofs for the assertions in Theorem 5, due to Stevanović [17]. In addition, we have shown that for every clique cover Φ, and every cycle cover Γ of a graph G, the polynomial $I(G \circ 2K_1; x)$ is divisible both by $I(\Phi(G); x)$ and $I(\Gamma(G); x)$.

For instance, the graphs from Figure 12 have: $I(G; x) = 1 + 6x + 9x^2 + 2x^3$, while

$$I(G \circ 2K_1; x) = (1+x)^6 \left(1 + 12x + 48x^2 + 76x^3 + 48x^4 + 12x^5 + x^6\right) =$$

$$= (1+x)^5 \cdot I(\Gamma(G); x) = (1+x)^6 \cdot I(\Phi(G); x),$$

$$I(\Gamma(G); x) = 1 + 13x + 60x^2 + 124x^3 + 124x^4 + 60x^5 + 13x^6 + x^7,$$

$$I(\Phi(G); x) = 1 + 12x + 48x^2 + 76x^3 + 48x^4 + 12x^5 + x^6.$$

The characterization of graphs whose independence polynomials are symmetric is still an open problem [17].

Let us mention that there are non-isomorphic graphs with the same independence polynomial, symmetric or not. For instance, the graphs G_1, G_2, G_3, G_4 presented in Figure 13 are non-isomorphic, while

$$I(G_1; x) = I(G_2; x) = 1 + 5x + 5x^2 , \text{ and}$$

$$I(G_3; x) = I(G_4; x) = 1 + 6x + 10x^2 + 6x^3 + x^4.$$

Recall that a graph having at most two vertices with the same degree is called *antiregular* [25]. It is known that for every positive integer $n \geq 2$ there is a unique connected antiregular graph of order n,

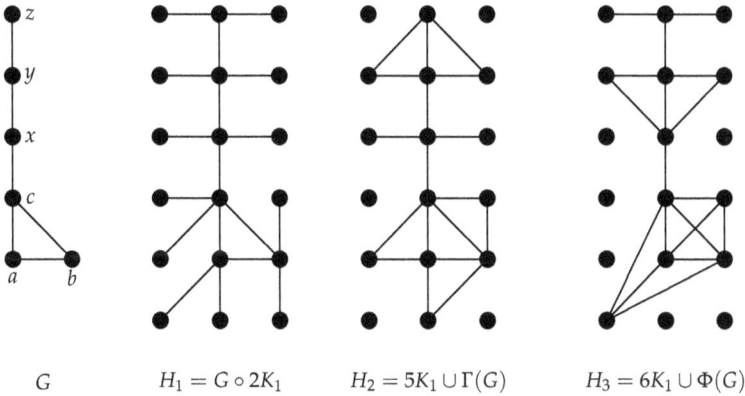

Figure 12. G with $\Gamma(G) = \{\{y,z\}, \{x\}, \{a,b,c\}\}$ and $\Phi(G) = \{\{z\}, \{x,y\}, \{a,b,c\}\}$.

Figure 13. Non-isomorphic graphs.

denoted by A_n, and a unique non-connected antiregular graph of order n, namely $\overline{A_n}$ [26]. In [27] we showed that the independence polynomial of the antiregular graph A_n is:

$$I(A_{2k-1}; x) = (1+x)^k + (1+x)^{k-1} - 1, \quad and$$

$$I(A_{2k}; x) = 2 \cdot (1+x)^k - 1, \quad k \geq 1.$$

Let us mention that $I(A_{2k}; x) = I(K_{k,k}; x)$ and $I(A_{2k-1}; x) = I(K_{k,k-1}; x)$, where $K_{m,n}$ denotes the complete bipartite graph on $m + n$ vertices. Notice that the coefficients of the polynomial

$$I(A_{2k}; x) = 2 \cdot (1+x)^k - 1 = \sum_{j=0}^{k} s_j x^j$$

satisfy $s_j = s_{k-j}$ for $1 \leq j \leq \lfloor k/2 \rfloor$, while $s_0 \neq s_k$, i.e., $I(A_{2k}; x)$ is "almost symmetric".

Proposition 2. *Characterize graphs whose independence polynomials are almost symmetric.*

It is known that the product of a polynomial $P(x) = \sum_{k=0}^{n} a_k x^k$ and its reciprocal $Q(x) = \sum_{k=0}^{n} a_{n-k} x^k$ is a symmetric polynomial. Consequently, if $I(G_1; x)$ and $I(G_2; x)$ are reciprocal polynomials, then the independence polynomial of $G_1 \cup G_2$ is symmetric, because $I(G_1 \cup G_2; x) = I(G_1; x) \cdot I(G_2; x)$.

Proposition 3. *Describe families of graphs whose independence polynomials are reciprocal.*

Acknowledgments: We would like to thank one of the anonymous referees for helpful comments, which improved the presentation of our paper.

References

1. Zykov, A.A. *Fundamentals of Graph Theory*; BCS Associates: Scottsdale, AZ, USA, 1990.
2. Frucht, R.; Harary, F. On the corona of two graphs. *Aequ. Math.* **1970**, *4*, 322–325.

3. Gutman, I.; Harary, F. Generalizations of the matching polynomial. *Utilitas Math.* **1983**, *24*, 97–106.
4. Arocha, J.L. Propriedades del polinomio independiente de un grafo. *Rev. Cienc. Mat.* **1984**, *V*, 103–110.
5. Hoede, C.; Li, X. Clique polynomials and independent set polynomials of graphs. *Discrete Math.* **1994**, *125*, 219–228.
6. Fisher, D.C.; Solow, A.E. Dependence polynomials. *Discrete Math.* **1990**, *82*, 251–258.
7. Farrell, E.J. Introduction to matching polynomials. *J. Combin. Theory* **1979**, *27*, 75–86.
8. Gutman, I. Independence vertex sets in some compound graphs. *Publ. Inst. Math.* **1992**, *52*, 5–9.
9. Alavi, Y.; Malde, P.J.; Schwenk, A.J.; Erdős, P. The vertex independence sequence of a graph is not constrained. *Congr. Numerantium* **1987**, *58*, 15–23.
10. Levit, V.E.; Mandrescu, E. On unimodality of independence polynomials of some well-covered trees. In *LNCS* 2731; Calude, C.S., Dinneen, M.J., Vajnovszki, V., Eds.; Springer: Berlin, Germany, 2003; pp. 237–256.
11. Levit, V.E.; Mandrescu, E. A family of well-covered graphs with unimodal independence polynomials. *Congr. Numerantium* **2003**, *165*, 195–207.
12. Levit, V.E.; Mandrescu, E. Very well-covered graphs with log-concave independence polynomials. *Carpathian J. Math.* **2004**, *20*, 73–80.
13. Levit, V.E.; Mandrescu, E. Independence polynomials of well-covered graphs: Generic counterexamples for the unimodality conjecture. *Eur. J. Comb.* **2006**, *27*, 931–939.
14. Levit, V.E.; Mandrescu, E. The independence polynomial of a graph-a survey. In *Proceedings of the 1st International Conference on Algebraic Informatics, Aristotle University of Thessaloniki, Greece*, Thessaloniki, Greece, 2005; pp. 233–254. Available online: cai05.web.auth.gr/papers/20.pdf (accessed on 15 July 2011)
15. Keilson, J.; Gerber, H. Some results for discrete unimodality. *J. Am. Stat. Assoc.* **1971**, *334*, 386–389.
16. Andrews, G.E. *The Theory of Partitions*; Addison-Wesley: Reading, Boston, MA, USA, 1976.
17. Stevanović, D. Graphs with palindromic independence polynomial. *Graph Theory Notes N. Y. Acad. Sci.* **1998**, *XXXIV*, 31–36.
18. Kennedy, J.W. Palindromic graphs. *Graph Theory Notes N. Y. Acad. Sci.* **1992**, *XXII*, 27–32.
19. Gutman, I. A contribution to the study of palindromic graphs. *Graph Theory Notes N. Y. Acad. Sci.* **1993**, *XXIV*, 51–56.
20. Bahls, P.; Salazar, N. Symmetry and unimodality of independence polynomials of path-like graphs. *Australas. J. Combin.* **2010**, *47*, 165–176.
21. Gutman, I. Independence vertex palindromic graphs. *Graph Theory Notes N. Y. Acad. Sci.* **1992**, *XXIII*, 21–24.
22. Levit, V.E.; Mandrescu, E. Graph operations and partial unimodality of independence polynomials. *Congr. Numerantium* **2008**, *190*, 21–31.
23. Levit, V.E.; Mandrescu, E. A family of graphs whose independence polynomials are both palindromic and unimodal. *Carpathian J. Math.* **2007**, *23*, 108–116.
24. Wang, Y.; Zhu, B.X. On the unimodality of independence polynomials of some graphs. *Eur. J. Comb.* **2011**, *32*, 10–20.
25. Merris, R. *Graph Theory*; Wiley-Interscience: New York, NY, USA, 2001.
26. Behzad, M.; Chartrand, D.M. No graph is perfect. *Am. Math. Mon.* **1967**, *74*, 962–963.
27. Levit, V.E.; Mandrescu, E. On the independence polynomial of an antiregular graph. 2010, arXiv:1007.0880v1 [cs.DM]. Available online: arxiv.org/PS_cache/arxiv/pdf/1007/1007.0880v1.pdf (accessed on 15 July 2011)

symmetry

MDPI

Article

Classifying Entropy Measures

Angel Garrido

Fundamental Mathematics Department, Faculty of Sciences UNED, Paseo Senda del Rey 9. 28040-Madrid, Spain; agarrido@mat.uned.es

Received: 27 April 2011; in revised form: 6 July 2011; Accepted: 6 July 2011; Published: 20 July 2011

Abstract: Our paper analyzes some aspects of Uncertainty Measures. We need to obtain new ways to model adequate conditions or restrictions, constructed from vague pieces of information. The classical entropy measure originates from scientific fields; more specifically, from Statistical Physics and Thermodynamics. With time it was adapted by Claude Shannon, creating the current expanding Information Theory. However, the Hungarian mathematician, Alfred Renyi, proves that different and valid entropy measures exist in accordance with the purpose and/or need of application. Accordingly, it is essential to clarify the different types of measures and their mutual relationships. For these reasons, we attempt here to obtain an adequate revision of such fuzzy entropy measures from a mathematical point of view.

Keywords: mathematical analysis; measure theory; fuzzy systems; information theory

---

1. Introduction

The *Shannon Entropy* is a measure of the average information content one is missing when one does not know the value of the random variable. This concept proceeds from the famous Shannon paper [1]. It represents an absolute limit on the best possible lossless compression of any communication under certain constraints, treating messages to be encoded as a sequence of independent and identically distributed random variables.

Usually, we define the Shannon Entropy by the following expression:

$$H(P) = -\Sigma_i \, p_i log p_i = \Sigma_i \, p_i log(1/p_i)$$

H_n will be a function of n non-negative random variables that add up to 1, and represent probabilities. H_n acts on the n-tuple of values on the sample, $(p_i)_{i=1,2,...,n}$.

The information that we receive from an observation is equal to the degree to which uncertainty is reduced.

Among its main *properties*, we have:

Continuity. The measure H should be *continuous*, in the sense that changing the values of the probabilities by a very small amount, should only change the H value by a small amount.

Maximality. The measure H will be maximal, if all the outcomes are equally likely, *i.e.*, the uncertainty is highest when all the possible events are equiprobable; thus,

$$H_n(p_1, p_2, ..., p_n) \leq H_n(1/n, 1/n, ..., 1/n)$$

And the entropy will increase with the number of outcomes,

$$H_n(1/n, 1/n,...,1/n) < H_{n+1} (1/(n + 1), 1/(n + 1),...,1/(n + 1))$$

Additivity. The amount of entropy should be independent of how the process is considered, as being divided into parts. Such a functional relationship characterizes the entropy of a system with

Symmetry **2011**, 3, 487–502

respect to the sub-systems. It demands that the entropy of every system can be identified and, then, computed from the entropies of their sub-systems.

i.e., if $S = \smile_{i=1,2,...,n} S_i$, then $H(S) = \Sigma_{i=1,2,...,n} H(S_i)$.

This is because *statistical entropy* is a probabilistic measure of uncertainty, or ignorance about data, whereas *Information* is a measure of a reduction in that uncertainty.

Entropy and related information measures provide descriptions of the long term behavior of random processes [2], and that this behavior is a key factor in developing the Coding Theorems of IT (Information Theory).

The contributions of Andrei Nikolaievich Kolmogorov (1903–1987) to this mathematical theory provide great advances to the Shannon formulations, proposing a new complexity theory, now translated to Computer Sciences. According to such theory, the complexity of a message is given by the size of the program necessary to enable the reception of such a message. From these ideas, Kolmogorov analyzes the entropy of literary texts and the subject Pushkin poetry. Such entropy appears as a function of the semantic capacity of the texts, depending on factors such as their extension and also the flexibility of the corresponding language.

It should also be mentioned that Norbert Wiener (1894–1964), considered the founder of Cybernetics, who in 1948 proposed a similar vision for such a problem. However, the approach used by Shannon differs from that of Wiener in the nature of the transmitted signal and in the type of decision made by the receiver.

In the Shannon model, messages are firstly encoded, and then transmitted, whereas in the Wiener model the signal is communicated directly through the channel without need of being encoded.

Another measure conceptualized by R. A. Fischer (1890–1962), the so called *Fisher Information* (*FI*), applies statistics to estimation, representing the amount of information that a message carries concerning an unobservable parameter.

Certainly the initial studies on IT were undertaken by Harry Nyquist (1889–1976) in 1924, and later by Ralph Hartley (1888–1970), who in 1928 recognized the logarithmic nature of the measure of information. This was later essential the key in Shannon and Wiener's papers.

The contribution of the Romanian mathematician and economist Nicholas Georgescu-Roegen (1906–1994), who studied in London with Karl Pearson, is also very interesting, whose great work was *The Entropy Law and the Economical Process*. In this memorable book, he proposed that the second law of thermodynamics also governs economic processes. Such ideas permitted the development of some new fields, such as Bioeconomics or Ecological Economics.

Also some others should be noted, studying a different kind of measure, the so called *inaccuracy measure*, involving two probability distributions.

R. Yager [3], and M. Higashi and G. J. Klir [4] showed the entropy measure as the difference between two fuzzy sets. More specifically, this is the difference between a fuzzy set and its complement, which is also a fuzzy set.

The *Shannon Entropy* is a measure of the average information content one is missing when one does not know the value of the random variable. These ideas proceed from their famous seminal paper [1]. It represents an absolute limit on the best possible lossless compression of any communication, under constraints, treating messages to be encoded as a sequence of independent and identically distributed random variables. The information that we receive from an observation is equal to the degree to which uncertainty is reduced. So,

$$I = H(before) - H(after)$$

Finally, we may define *I*(*Information*) in terms of the probability, *p*, by the following *Properties of Information Measure, I.*

(1) $I(P) \geq 0$, *i.e.*, information is a non-negative quantity;
(2) $I(1) = 0$, *i.e.*, if an event has probability 1, we get no information from the occurrence of the event;

(3)   If two independent events occur, the information we get from observing the events is the sum of both informations;

(4)   Information measure must be continuous, and also a monotonic function of the probability. So, slight changes in probability should result in slight changes in the information.

2. Graph Entropy

Graph theory has emerged as a primary tool for detecting numerous hidden structures in various information networks, including Internet graphs, social networks, biological networks, or more generally, any graph representing relations in massive data sets. Analyzing these structures is very useful to introduce concepts such as Graph Entropy and Graph Symmetry.

We consider a functional on a graph, $G = (V, E)$, with P a probability distribution on its node set, V, and we suppose varying random samples, $P = (p_i)_{i=1,2,...,n}$, on the probabilistic space.

The mathematical construct called a Graph Entropy will be denoted by GE. It will be defined as

$$H(G, P) = \min_P \left[\Sigma_{i=1,2,...,n} \, p_i log p_i \right]$$

Observe that such a function is convex. It tends to $+\infty$ on the boundary of the non-negative orthant of R^n. Also, monotonically tends to $-\infty$ along rays departing from the origin. So, such a minimum is always achieved and will be finite.

The entropy of a system represents the amount of uncertainty one observer has about the state of the system. The simplest example of a system will be a random variable, which can be shown by a node in the graph, being their edges representative of the mutual relationship between them. Information measures the amount of correlation between two systems, and reduces in entropies to a mere difference. So, *the Entropy of a Graph* (will be denoted by GE) *is a measure of graph structure, or lack of it.*

Therefore, it may be interpreted as the amount of Information, or the degree of "surprise", communicated by a message. As the basic unit of Information is the bit, Entropy also may be viewed as the number of bits of "randomness" in the graph, verifying that the higher the entropy, the more random the graph.

Let G now be an arbitrary finite rooted Directed Acyclic Graph (or DAG, in acronym). For each node, v, we denote $i(v)$ the number of edges that terminates at v. Then, the Entropy of the graph is definable as

$$H(G) = \Sigma \left[i(v) - 1 \right] log_2 \left[\left((Card(E) - Card(V) + 1) / (i(v) - 1) \right) \right]$$

$H(X)$ may be interpreted in different ways. For instance, given a random variable, X, it informs us about how random X is, how uncertain we should be about X, or how much variability X has.

In a variant of the Graph Coloring Problem, we take the objective function to minimize the Entropy of such coloring. So, it is called the Minimum Entropy Coloring. In Chromatic Entropy, we understand the minimum Entropy of a coloring. Its role is essential in the problem of coding. If we consider this problem from a computational viewpoint, it will be of NP-hard type; for instance, on Interval Graphs.

The study of different concepts of Entropy will be very interesting, and not only on Physics, but also on Information Theory, and other Mathematical Sciences, considered in its more general vision. Also it may be a very useful tool for Biocomputing, for instance, or in many others, such as studying Environmental Sciences. This is because, among other interpretations with important practical consequences, the law of Entropy means that energy cannot be fully recycled.

Many quotations have been made until now referring to the content and significance of this fuzzy measure, for example:

"Gain in Entropy always means loss of Information, and nothing more" [5].

"Information is just known Entropy. Entropy is just unknown Information" [6].

Mutual Information and Relative Entropy, also called Kullback-Leibler divergence, among other related concepts, have been very useful in Learning Systems, both in supervised and unsupervised cases.

We attempt to analyze the mutual relationship between the distinct types of entropies, such as:

- The Quantum Entropy, also called Von Neumann Entropy;
- The KS-Entropy (from Kolmogorov and Sinai), which is also called Metric Entropy [7,8];
- The Topological Entropy; or
- The Graph Entropy, among others.

3. Quantum Entropy

This entropy was first defined by the Hungarian mathematician Janos Neumann (a.k.a. John von Neumann) in 1927, with the purpose of showing the irreversible behavior of quantum measurement processes. In fact, *Quantum Entropy* (will be denoted as *QE*) is an extension of the precedent Gibbs Entropy to the quantum realm [8]. It will be interpreted as the average information the experimenter obtains when he makes many copies of a series of observations, on an identically prepared mixed state. It plays a very important role in studying correlated systems, and also for defining entanglement measures. Recall that "Entanglement" is one of the properties of Quantum Mechanics that caused Einstein to dislike this theory. However, from then on, Quantum Mechanics has reached high success predicting experimental results and has also been proven on the correlation predicted by the theory of such entanglement.

We can apply the notion of QE to Networks. As QE is defined for quantum states, we need a method to map graphs into states. Such states for a quantum mechanical system are described by a density matrix. Usually, it is denoted as ρ. It is a positive semi-definite matrix with unitary *trace* $(\rho) = 1$. There are many different ways, however, to associate graphs to density matrices. Until now, we have eliminated several problems through certain interesting results, but many open questions still remain.

Between the known results, we can see that the entropy for a d-regular graph tends to be in the limit when $n \rightarrow \infty$ to the entropy of K_n, i.e., the *n-complete n-graph*.

Another result may be that the entropy of graphs increases as a function of the cardinality of their edges.

Between the open problems, we can list some of them as relative to an interesting tool, a related matrix, called the *Normalized Laplacian.* This is defined by

$$\mathcal{L}(G) = \triangle^{-1/2} L(G) \triangle^{-1/2}$$

The Combinatorial Laplacian Matrix of G (abridged, Laplacian of G) is given as

$$L(G) = \triangle(G) - A(G)$$

Hence being computable as the difference between matrix degree, $\triangle(G)$, and adjacency matrix, $A(G)$.

The *degree* of a node, v, is the number of edges adjacent to v. Usually it is denoted by $d\ (v)$. The *degree sum* of the graph G is d_G, and it will be given by $d_G = \Sigma\ d(v)$. The *average degree* of G is expressed as $d_G{}^* = m\ \Sigma\ d(v)$, where m is the number of non-isolated nodes.

A graph, G, is *d-regular*, if $d(v) = d$, for all $v \in V(G)$.

The *degree matrix* of G is a $(n \times n)$-matrix with entries given as

$$[\triangle(G)](u,v) = d(v), if\ u = v; and otherwise\ 0$$

So, the *Laplacian* of a graph, G, scaled by its degree-sum is a density matrix,

$$\rho_G = ((L(G))/(d_G)) = ((L\,(G))/(tr(\triangle(G)))) = ((L(G))/(m\,d_G{}^*))$$

With the well-known expression for the entropy of a density matrix, ρ, by $S(\rho) = -\,tr(\rho \log_2 \rho)$. Hence, departing from the concept of Laplacian of a Graph, we can say that $S(\rho_G)$ is the QE of G. If we suppose two decreasing sequences of eigenvalues of $L(G)$ and ρ_G, respectively given by

$$\lambda_1 \geq \lambda_2 \geq ... \geq \lambda_n = 0, \text{and } \mu_1 \geq \mu_2 \geq ... \geq \mu_n = 0$$

mutually related by a scaling factor, *i.e.*,

$$\mu_i = ((\lambda_i)/(d_G)) = ((\lambda_i\})/(md_G{}^*))$$

Therefore, the *Entropy of a density matrix* ρ_G can also be written as

$$S(G) = -\Sigma\mu_i log_2\mu_i$$

with the notational convention $0 \log 0 = 0$. Since its rows sum up to 0, we can conclude that the smallest eigenvalue of the density matrix must also be equal to zero, and the number of connected components of the graph is given by the multiplicity of 0 as an eigenvalue.

The QE is a very useful tool for problems such as when it is applied to the Enumeration of Spanning Trees.

4. Algorithmic Entropy

Algorithmic Entropy is the size of the smallest program that generates a string. It is denoted by $K(x)$, or AE. It receives many different names, for instance, Kolmogorov-Chaitin Complexity, or only Kolmogorov Complexity, also Stochastic Complexity, or Program-size complexity [9,10].

AE is a *measure of the amount of information in an object*, x. Therefore, it also measures its randomness degree. The AE of an object is a measure of the computational resources needed to specify such an object. *i.e.*, the AE of a string is the length of the shortest program that can produce this string as its output. So, the Quantum Algorithmic Entropy (QAE), also called *Quantum Kolmogorov Complexity* (QKC) is the length of the shortest quantum input to a Universal Quantum Turing Machine (UQTM) that produces the initial "qubit" string with high fidelity. Hence, the concept is very different of the Shannon Entropy, because, whereas this will be based on probability distributions, the AE is based on the size of programs.

All strings used may be elements of $\Sigma^* = \{0,1\}^*$, being ordered lexicographically. The length of a string x is denoted by $|x|$.

Let U be a fixed prefix-free Universal Turing Machine. For any string x of $\Sigma^* = \{0,1\}^*$, the *Algorithmic Entropy of x* will be defined by

$$K(x) = min_p\{|p| : U(p) = x\}$$

From this concept, we can introduce the *t-time-Kolmogorov Complexity*, or t-time-bounded algorithmic entropy [11].

For any time constructible t, we introduce a refinement by

$$K^t(x) = min_p\{|p| : U(p) = x, in\,at\,most\,t(|x|)steps\}$$

From these, we may obtain that for all x and y,

$$(i)K(x) \leq K^t(x) \leq |x| + O(1), \tag{1}$$

and also

$$(ii) K^t(x/y) \leq K^t(x) + O(1)$$

The Kolmogorov-Chaitin (KC, by acronym) as new tool possesses many applications, in fields as diverse as Combinatorics, Graph Theory, Analysis of Algorithms, or Learning Theory, among others [10,11].

5. Metric Entropy

We consider now the Metric Entropy, also called *Kolmogorov Entropy*, or Kolmogorov-Sinai Entropy, in acronym *K-S Entropy*. Its name is associated with Andrei N. Kolmogorov, and his disciple, Yakov Sinai [4].

Let (X, Ω, μ) be a probability space, or in a more general way, a fuzzy measurable space [12]. Recall that a measurable partition of X is such that each one of their elements is a measurable set, therefore, an element of the fuzzy σ-algebra, Ω. And let I^X be the set of mappings from X to the closed unit interval, $I = [0,1]$.

A *fuzzy σ-algebra*, Σ, on a nonempty set, X, is a subfamily of I^X satisfying that

(1) $1 \in \Sigma$;
(2) If $\alpha \in \Sigma$, then $1 - \alpha \in \Sigma$;
(3) If $\{\alpha i\}$ is a sequence in Σ, then $\vee \alpha i = \sup i \in \Sigma$;

A fuzzy probability measure, on a fuzzy σ-algebra, Σ, is a function

$$m : \Sigma \to [0,1]$$

which holds

[1] m (1) = 1
[2] for all $\alpha \in \Sigma$, m(1 - α) = 1 - m(α)
[3] for all α, $\beta \in \Sigma$, m($\alpha \vee \beta$) + m($\alpha \wedge \beta$) = m (α) + m (β)
[4] If $\{\alpha i\}$ is a sequence in Σ, such that $\alpha i \uparrow \alpha$, being $\alpha \in \Sigma$, then m(α) = sup $\{m(\alpha i)\}$

We call (X, Ω, μ) a fuzzy-probability measure space, and the elements of Ω are called measurable fuzzy sets.

The notion of "fuzzy partition" was introduced by E. Ruspini. Given a finite measurable partition, $\wp$, we can define its Entropy by

$$H_\mu(\wp) = \sum_{p \in \wp} - \mu(p) \log \mu(p) \tag{1}$$

As usual in these cases, we take as convention that $0 \log 0 = 0$.

Let $T: X \to X$ be a measure-preserving transformation. Then, the *Entropy of T w.r.t. a finite measurable partition*, $\wp$, is expressed as

$h_\mu(T, \wp) = \lim_{n \to \infty} H_\mu(\wp^{-k} \wp)$

with $H\mu$ the entropy of a partition, and where $\vee$ denotes the join of partitions. Such a limit always exists.

Therefore, we may define the Entropy of T as

$$h_\mu(T) = \sup_{\wp} h_\mu(T, \wp) \tag{}$$

by taking the supremum over all finite measurable partitions.

Many times $h\mu(T)$ is named the Metric Entropy of T. So, we may differentiate this mathematical object from the well-known as Topological Entropy.

We may investigate the mutual relationship of the Metric Entropy and the Covering Numbers.

Let (X, d) be a metric space, and let $Y \subseteq X$ be a subset of X. We say that $Y^* \subseteq X$ is an ε-cover of Y, if for each $y \in Y$, there exists a $y^* \in Y^*$ such that d $(y, y^*) \leq \varepsilon$. It is clear that there are many different covers of Y. But we are especially interested here in one which contains the lesser number of elements.

We call the cardinal, or size, of such a cover its Covering Number. Mathematically expressed, the ε-covering number of Y is

$$N(\varepsilon, Y, d) = \min\{\text{card } (Y^*) : Y^* \text{ is an } \varepsilon\text{-cover}\} \qquad ()$$

A *proper cover* is one where $Y^* \subseteq Y$. And a *proper covering number* is defined in terms of the cardinality of the minimum proper cover. Both, covering numbers and proper covering numbers, are related by

$$N(\varepsilon, Y) \leq N_{\text{proper}}(\varepsilon, Y) \leq N((\varepsilon/2), Y)$$

Furthermore, we recall that the Metric Entropy, H(ε, Y), is a natural representation of the cardinal of the set of bits needed to send, in order to identify an element of the set up to precision ε. It will be expressed by

$$H(\varepsilon, Y) = \log N(\varepsilon, Y) \qquad ()$$

In a dynamical system, the metric entropy is equal to zero for non-chaotic motion. And it is strictly greater than zero for chaotic motion. So, it will be interpreted as a simple indicator of the complexity of a dynamical system.

6. Topological Entropy

Let (X, d) be a compact metric space, and let f: $X \rightarrow X$ be a continuous map. For each $n > 0$, we define a new metric, d_n, by

$$d_n(x, y) = max\{d(f^i(x), f^i(y)) : 0 \leq i < n\}$$

Two points, x and y, are close with respect to this metric, if their first n iterates (given by f^i, $i = 1, 2,...$) are close.

For $\varepsilon > 0$, and $n \in N^*$, we say that $S \subset X$ is an (n, ε)-*separated set*, if for each pair, x, y, of points of S, we have $d_n(x, y) > \varepsilon$. Denote by $N(n, \varepsilon)$ the maximum cardinality of a (n, ε)-separated set. It must be finite, because X is compact. In general, this limit may exist, but it could be infinite. A possible interpretation of this number is as a measure of the average exponential growth of the number of distinguishable orbit segments. So, we could say that the higher the topological entropy is, the more essentially different orbits we have [2,7].

From an analytical viewpoint, the topological entropy is a continuous and monotonically increasing function.

N(n,ε) shows the number of "distinguishable" orbit segments of length n, assuming we cannot distinguish points that are less than ε apart.

The topological entropy of f is then defined by

$$H_{\text{top}} = \lim_{\varepsilon \rightarrow 0} \lim \sup_{n \rightarrow \infty} [(1/n) \log N(n, \varepsilon)] \qquad ()$$

Therefore, TE is a non-negative number measuring the complexity degree of the system. So, it gives the exponential growth of the cardinality for the set of distinguished orbits, according to time advances [13–16].

7. Chromatic Entropy

A system can be defined as a set of components functioning together as a whole. A systemic point of view allows us to isolate a part of the world, and so, we can focus on those aspects that interact more closely than others. The entropy of a system represents the amount of uncertainty one observer has about the state of the system [10,12]. The simplest example of a system will be a random variable, which can be shown by a node in the graph being their edges representative of the mutual relationship between them. Information measures the amount of correlation between two systems, and reduces to a mere difference between entropies. So, the Entropy of a Graph (will be denoted by GE) is a measure of graph structure, or lack of it. Therefore, it may be interpreted as the amount of Information, or the degree of "surprise", communicated by a message. Further, as the basic unit of Information is the bit, Entropy also may be viewed as the number of bits of "randomness" in the graph, verifying that the higher the entropy, the more random the graph.

We consider a functional on a graph, $G = (V,E)$, with P a probability distribution on its node (or vertex) set, V. This mathematical construct will be denoted by GE and defined as

$$H(G, P) = min\Sigma p_i log p_i$$

Let G now be an arbitrary finite rooted Directed Acyclic Graph (DAG, in acronym). For each node, v, we denote $i(v)$ the number of their edges that terminates at v. Then, the Entropy of the graph is

$$H(G) = \Sigma[i(v) - 1]log_2[((Card(E) - Card(V) + 1)/(i(v) - 1))]$$

H(X) may be interpreted in some different ways. For instance, given a random variable, X, it informs us about how random X is, how uncertain we should be about X, or how much variability X possesses.

In a variant of the "Graph Coloring Problem", we take the objective function to minimize the Entropy of such coloring. So, it is called the Minimum Entropy Coloring.

In Chromatic Entropy, we understand the minimum Entropy of a coloring. Its role is essential in the problem of coding. If we consider this problem from a computational viewpoint, it is NP-hard; for instance, on Interval Graphs.

8. Mutual Relationship between Entropies

In the mid 1950s, the Russian mathematician Andrei N. Kolmogorov imported Shannon's probabilistic notion of entropy into the theory of dynamical systems, and showed how entropy can be used to tell whether two dynamical systems are non-conjugate, i.e., non-isomorphic. His work inspired a whole new approach in which entropy appears as a numerical invariant of a class of dynamical systems. Because the Kolmogorov's metric entropy is an invariant of measure theoretical dynamical systems, it is therefore closely related to Shannon's source entropy [14].

Ornstein showed that metric entropy suffices to completely classify two-sided Bernoulli processes, a basic problem which for many decades appeared completely intractable. Recently, has been shown how to classify one-sided Bernoulli processes; this turns out to be quite a bit harder. In 1961, Adler et al. introduced [17,18] the aforementioned topological entropy, which is the analogous invariant for topological dynamical systems. There exists a very simple relationship between these quantities, because maximizing the metric entropy over a suitable class of measures defined on a dynamical system, gives its topological entropy. The relationship between TE and the Entropy in the sense of Measure Theory (K-S) is given by the so-called Variational Principle, which established that

$$h(T) = sup\{h_\mu(T)\}_{\mu \in P(X)}$$

This may be interpreted as TE is equal to the supremum of Kolmogorov-Sinai (or K-S) entropies, $h\mu(T)$, with μ belonging to the set of all T-invariant Borel probability measures on X.

The mutual relationship between Algorithmic Entropy and Shannon Entropy is that the expectation of the former gives us the latter, up to a constant depending on the distribution.

Also we may express, departing of P(x) as a recursive probability distribution, that

$$0 \leq \Sigma P(x)K(x) - H(P) \leq K(P)$$

Finally, we recall that given a random variable, X, its Shannon Entropy is given by

$$H(X) = -\Sigma P(x)log_2 P(x)$$

whereas the Renyi Entropy of order $\alpha \neq 1$ of such random variable will be

$$H_\alpha(X) = (1/(1 - \alpha)) \log_2 (\textstyle\sum P(x)^\alpha)$$

The Renyi Entropy of order α converges to the Shannon Entropy, when α tends to one, *i.e.,*

$$lim_{\alpha \to 1}\{(1/(1 - \alpha)) \log_2 (\textstyle\sum P(x)^\alpha)\} = -\textstyle\sum P(x) \log_2 P(x)$$

Hence,

$$lim_{\alpha \to 1} H_\alpha(X) = H(X) \qquad\qquad ()$$

Therefore, the Renyi Entropy may be considered as a generalization of the Shannon Entropy, or dually, the Shannon Entropy will be a particular case of Renyi Entropy [13,14].

9. Graph Symmetry

As we know, Symmetry in a system means invariance of its elements under a group of transformations. When we take Network Structures, it means invariance of adjacency of nodes under the permutations on node set.

Let G and H be two graphs. An isomorphism from G to H will be a bijection between the node sets of both graphs, i.e., a f: G → H, such that any two nodes, u and v, of G are adjacent in G if and only if f(u) and f(v) are also adjacent in H. Usually, it is called "edge-preserving bijection". If an isomorphism exists between two graphs, G and H, then such graphs are called Isomomorphic Graphs.

The graph isomorphism is an equivalence, or equality, as relation on the set of graphs. Therefore, it partitions the class of all graphs into equivalence classes. The underlying idea of isomorphism is that some objects have the same structure, if we omit the individual character of their components. A set of graphs isomorphic to each other is denominated an isomorphism class of graphs.

An automorphism of a graph, G = (V, E), will be an isomorphism from G onto itself. So, a graph-automorphism of a simple graph, G, is simply a permutation on the set of its nodes, V (G), f: G → G, such that the image of any edge of G is always an edge in G. That is, if e = {u, v} ∈ V(G), then f(e) = {f(u), f(v)} ∈ V(G). Either expressed in group theoretical way, we have

$$u \sim v \text{ if and only if } ug \sim vg \text{ if and only if } u^g \sim v^g \qquad\qquad ()$$

The family of all automorphisms of a graph G is a permutation group on $V(G)$. The inner operation of such group is the composition of permutations. Its name is very well-known, the *Automorphism Group of G*, and abridgedly, it is denoted by Aut(G). Conversely, all groups may be represented as the automorphism group of a connected graph. The automorphism group is an algebraic invariant of a graph. So, we can say that an automorphism of a graph is a form of symmetry in which the graph is mapped onto itself while preserving the edge-node connectivity. Such automorphic tool may be applied both on Directed Graphs (DGs) and on Undirected Graphs (UGs).

We will say either graph invariant or graph property, when it depends only of the abstract structure, not on graph representations, such as particular labeling or drawing of the graph. So, we may define a graph property as every property that is preserved under all their possible isomorphisms of the graph. Therefore, it will be a property of the graph itself, not depending on the representation of the graph.

The semantic difference also consists in its quantitative or quantitative character. For instance, when we said that "the graph does not possess directed edges", this will be a property, because it is a

Symmetry **2011**, *3*, 487–502

qualitative statement. While when we say "the number of nodes of degree two in such graph", this would be an invariant, because it is a quantitative statement.

From strictly a mathematically viewpoint, a graph property can be interpreted as a class of graphs, composed by graphs that have the accomplishment of some conditions in common. Hence, a graph property can also be defined as a function of whose domain would be the set of graphs, and its range will be the bi-valued set composed of two options, true and false, {T, F}, according to which a determinate condition is either verified or violated for the graph. A graph property is called hereditary, if it is inherited by its induced subgraphs. And it is additive, if it is closed under disjoint union. For example, the property of a graph to be planar is both additive and hereditary. Instead of this, the property of being connected is neither.

The computation of certain graph invariants may be very useful for the purpose of discriminating when two graphs are isomorphic, or rather non-isomorphic. The support of these criteria will be that for any invariant at all, two graphs with different values cannot be isomorphic between them. However, two graphs with the same invariants may or may not be isomorphic between them. So, we will arrive to the notion of completeness.

Let I(G) and I(H) be invariants of two graphs, G and H. It will be considered complete if the identity of the invariants ever implies the isomorphism of the corresponding graphs, i.e., if I(G) = I(H), then G will be isomorphic to H.

A directed graph, or digraph, is the usual pair G = (V,E), but now with an additional condition: it has at most one directed edge from node i to node j, being $1 \leq i, j \leq n$. We add the term "acyclic" when there are no cycles of any length. Usually, we use the acronym DAG to denote an acyclic directed graph. A very important result may be this: For each n, the cardinality of the n-DAGs, or DAGs with n labeled nodes, is equal to the number of (n × n)-matrices of 0's and 1's whose eigenvalues are positive real numbers.

It is possible to prove that every group is the automorphism group of a graph. If the group is finite, the graph may be taken to be finite. Further, George Polya observed that not every group must be the automorphism group of a tree.

10. Symmetry as Invariance

One of the more fundamental results in Physics and in any Science [12,14–16,19] is that obtained by the great mathematician *Emmy Noether* (1882–1935). This was proved in 1915, and published in 1918. It states that any differentiable symmetry of the action of a physical system has a corresponding conservation law. Hence, for each continuous symmetry of a physical theory there is a corresponding conserved quantity, *i.e.*, a physical quantity that does not change with time. So, Symmetry under translation corresponds to conservation of momentum; Symmetry under rotation to conservation of angular momentum; Symmetry in time to conservation of energy. Also it is present in Relativity Theory, Quantum Mechanics and so on. It is a very important result, because it allows us to derive conserved quantities from the mathematical form of our theories. Recall that the action of a physical system is an integral of a Lagrangian function, from which the behavior of the system can be determined by the Principle of Least Action. Note that this theorem does not apply to systems that cannot be modeled with a Lagrangian, for instance to dissipative systems.

The *Noether Theorem* has become essential not only in modern Theoretical Physics, but in the Calculus of Variations, and therefore, in fields such as Modeling and Optimization. In fact, all modern Physics is based on a bunch of Symmetry Principles, from which the rest follows. So, we can say that *the Laws of Nature are constrained by Symmetry*. Such theorem admits distinct but essentially equivalent, statements, as may be *"to every differentiable symmetry generated by local actions, there corresponds a conserved current"*. This connects today with many evolving subjects of modern Physics, such as Gauge Symmetry, in Quantum Mechanics, the results of Witten (String Theory) and many others. Noether is remembered not only by this theorem (actually, they are two results, with many consequences), but by many contributions to Abstract Algebra. There is also a quantum version of this Noether's theorem, known as the Ward-Takahasi Identity.

The conservation law of a physical quantity is expressed by a continuity equation, where the conserved quantity is named Noether's Charge, and the flow carrying that "charge" is the Noether's Current. In Quantum Mechanics, invariance under a change of phase of the wave function leads to the Conservation of Particle Number.

11. Fuzzy Entropies

In recent decades, the expansion of fuzzy mathematics and its applications are very formidable [17, 20]. The parallel version of different mathematical fields, but adapted to degrees of truth, is in advance. The basic idea according to which an element not necessarily belongs totally, or does not belong in absolute, to a set, but it can belong more or less, that is, in some degree it signifies a modern revolution in scientific thinking, adapting the sometimes hieratic mathematics to the features of the real world. So, it produces new fields, such as Fuzzy Measure Theory, which generalizes the classical Measure Theory of Lebesgue and other authors. It must be very useful as a tool in our own papers and occurs in every mathematical field. In Fuzzy Modeling we attempt to construct Fuzzy Systems. Many times, it will be a very difficult task, because it is necessary to identify many parameters. It offers a great potential for analyzing structures with non-stochastic imprecise input information.

In Fuzzy Optimization [17,21], our objective is to maximize or minimize a fuzzy set submitted to some fuzzy constraints, but we cannot make this directly with the "value" of a fuzzy set. For this reason, in areas such as Finance, we wish to maximize/minimize the value of a discrete/continue random variable, being restricted by a probability mass/density function. So, we change the multi-objective problem into a single crisp objective subject to the fuzzy constraints and it is possible to generate good approximate solutions by Genetic Algorithms. Also there are different fuzzy optimization problems, which include learning a Fuzzy Neural Network, useful to solve fuzzy linear programming problems (FLP), and fuzzy inventory control, using such Genetic Algorithms.

12. About Negentropy

Negentropy is essential for the axiomatized concept of entropy (denoted by H). Many of its seminal ideas were derived from Claude E. Shannon [1], and Alfred Renyi [17,22]. It is also related to the coding length of the random variable. In fact, with some simple assumptions, H is the coding length of the random variable. Entropy is the basic concept of Information Theory. It can be interpreted, for a random variable, as the degree of information that the observation of the variable produces. The more "randomness" presented in the variable, the larger the entropy. It is defined, for a discrete random variable, Y, as

$$H(Y) = -\Sigma P(Y = y_i) log P(Y = y_i)$$

where the y_i are the possible values of Y.

This may be generalized for the continuous case, being then called *Differential Entropy* (also named *continuous entropy*). It will be defined by

$$H(y) = -\int f(y) \, log \, f(y) \, dy$$

with $f(y)$ density function, associated with the continuous random variable Y.

There exists a very important result of Information Theory, according to *a Gaussian random variable has the largest entropy, among all random variables of the same variance*. So, the Normal, or Gaussian distribution is the "least structured", or equivalently, the "most random" among all distributions. But we have a second and very important measure of non-gaussianity (departure from the Normal). It is called with distinct names, such as *Negentropy*, either Negative Entropy or *Syntropy*, denoted by *J*. Actually, it is a slightly modified version of differential entropy, defined by

Symmetry **2011**, 3, 487–502

$$J(y) = H(y_{gauss}) - H(y)$$
being y_{gauss} a Gaussian random variable of the same covariance matrix as y.
Some of its properties are interesting, as

$$J(y) \geq 0, \text{ for each } y \qquad\qquad ()$$

That is, Negentropy is always non-negative. And it is null in the case of the Normal distribution:

$$J = 0 \text{ if and only if it is Gaussian} \qquad\qquad ()$$

According to Schrodinger's classical book *What is Life?*
"Negentropy of a living system is the entropy that it exports, to maintain its own entropy low"
And Brillouin [23] says that
"A living system imports negentropy, and stores it"
The Curie Principle of Symmetry, due to Pierre Curie, postulates that *the symmetry group of the cause is a subgroup of the symmetry group of the effect*. This idea may produce deep ramifications on Causality Theory, and also analyzing relationships among the foundations of physical theories.

13. Conclusions

Statistical entropy is a probabilistic measure of uncertainty, or ignorance about data. However, Information should be the measure of the reduction in that uncertainty. The Entropy of a probability distribution is just the expected value of the information of such a distribution. All these improved tools must allow us to advance not only in fields such as Optimization Theory, but also on Generalized Fuzzy Measures, Economics, modeling in Biology, and so on [17,24,25]. Here, we have shown some different entropy measures, more or less useful depending on its context and their need of applications, according to ideas suggested by the Hungarian mathematician Alfred Renyi many years ago [22,26–28].

Acknowledgments: I wish to express my gratefulness to Joe Rosen, Shu-Kun Lin, and Joel Ratsaby, which have proposed to collaborate newly with this paper in *SYMMETRY*. And also to my anonymous referees for their very wise commentaries.

References

1. Shannon, C.E. A mathematical theory of communication. *Bell Syst. Tech. J.* **1948**, 27, 379–423.
2. Wehrl, A. General properties of entropy. *Rev. Mod. Phys.* **1978**, 50, 221–260.
3. Yager, R. On the Measure of Fuzziness and Negation. *Int. J. General Syst.* **1979**, 5, 221–229.
4. Higashi, M. , and Klir, G.J. Measures of Uncertainty and Information based on possibility distributions. *Int. J. General Syst.* **1982**, 9, 43–58.
5. Lewis, G. N. The Entropy of Radiation. *Proc. Natl. Acad. Sci. USA* **1927**, 13, 307–314.
6. Frank, M. P. Approaching Physical Limits of Computing. *Multiple-Valued Logic* **2005**, doi:10.1109/ISMVL.2005.9. [CrossRef]
7. Simonyi, G. Graph Entropy: A Survey. *DIMACS* **1995**, 20, 399–441.
8. Sinai, G. On the concept of Entropy of a Dynamical System. *Dokl. Akad. Nauk SSSR* **1959**, 124, 768–771.
9. Passarini, F.; Severini, S. *The von Neumann Entropy of Networks*; University of Munich: Munich, Germany, 2009.
10. Volkenstein, M.V. *Entropy and Information (Progress in Mathematical Physics)*; Birkhauser Verlag: Berlin, Germany, 2009.
11. Devine, S. The insights of algorithmic entropy. *Entropy* **2009**, 11, 85–110.
12. Dehmer, M. Information processing in Complex Networks: Graph entropy and Information functionals. *Appl. Math. Comput.* **2008**, 201, 82–94.
13. Jozsa, R. Quantum Information and Its Properties. In *Introduction to Quantum Computation and Information*; Lo, H.K., Popescu, S., Spiller, T., Eds.; World Scientific: Singapore, 1998.

14. Titchener, M.R.; Nicolescu, R.; Staiger, L.; Gulliver, A.; Speidel, U. Deterministic Complexity and Entropy. *J. Fundam. Inf.* **2004**, Volume 64.

15. Titchener, M.R. A Measure of Information. In Proceedings of Data Compression Conference 2000, Snowbird, UT, USA; 2000; pp. 353–362.

16. Titchener, M.R. A Deterministic Theory of Complexity, Information and Entropy. In Proceedings of IEEE Information Theory Workshop, San Diego, CA, USA, February 1998.

17. Preda, V. Balcau, C. *Entropy Optimization with Applications*; Editura Academiei Romana: Bucuresti, Romania, 2010.

18. Li, M.; Vitanyi, P. *An Introduction to Kolmogorov Complexity and Its Applications*, 3rd ed.; Springer Verlag: Berlin, Germany, 2008.

19. Dumitrescu, D. Entropy of a fuzzy process. *Fuzzy Sets Syst.* **1993**, *55*, 169–177.

20. Wang, Z.; Klir, G.J. *Generalized Measure Theory*; Springer Verlag: Berlin, Germany and New York, NY, USA, 2008.

21. Garrido, A.; Postolica, V. *Modern Optimization*; Editura Matrix-Rom: Bucuresti, Romania, 2011.

22. Renyi, A. On measures of information and entropy. In Proceedings of the 4th Berkeley Symposium on Mathematics, Statistics and Probability, Berkeley, CA, USA, CA, USA, 20 June–30 July 1960; University of California Press: Berkeley; pp. 547–561.

23. Jaynes, E.T. Information theory and statistical mechanics. *Phys. Rev.* **1956**, *108*, 171–190.

24. Georgescu-Roegen, N. *The Entropy Law and the Economic Process*; Harvard University Press: Cambridge, MA, USA, 1971.

25. Liu, X. Entropy, distance and similarity measures of fuzzy sets and their relations. *Fuzzy Sets Syst.* **1992**, *52*, 305–318.

26. De Luca, A.; Termini, S. A definition of non-probabilistic entropy, in the setting of fuzzy theory. *Inf. Control* **1972**, *20*, 301–312.

27. You, C.; Gao, X. Maximum entropy membership functions for discrete fuzzy variables. *Inf. Sci.* **2009**, *179*, 2353–2361.

28. Dumitrescu, D. Fuzzy measures and the entropy of fuzzy partitions. *J. Math. Anal. Appl.* **1993**, *176*, 359–373.

symmetry

MDPI

Article

Information Theory of Networks

Matthias Dehmer

UMIT, Institute for Bioinformatics and Translational Research, Eduard Wallnöfer Zentrum 1, 6060, Hall in Tyrol, Austria; E-Mail: Matthias.Dehmer@umit.at; Tel./Fax: +43-050-8648-3851

Received: 26 October 2011; in revised form: 11 November 2011 / Accepted: 16 November 2011 / Published: 29 November 2011

Abstract: The paper puts the emphasis on surveying information-theoretic network measures for analyzing the structure of networks. In order to apply the quantities interdisciplinarily, we also discuss some of their properties such as their structural interpretation and uniqueness..

Keywords: information theory; networks; quantitative graph analysis

1. Introduction

Information theory has been proven useful to solve interdisciplinary problems. For example, problems in biology, chemistry, computer science, ecology, electrical engineering, and neuroscience have been tackled by using information-theoretic methods such as entropy and mutual information, see [1–5]. In particular, advanced information-measures such as the Jensen–Shannon divergence have also been used for performing biological sequence analysis [6].

In terms of investigating networks, information-theoretic techniques have been applied in an interdisciplinary manner [7–11]. In this paper, we put the emphasis on reviewing information-theoretic measures to explore the network structure and shed light on some of their strong and weak points. But note that the problem of exploring the dynamics of networks by using information theory has also been tackled, see [12].

Interestingly, the problem of exploring graphs quantitatively emerged in the fifties when investigating structural aspects of biological and chemical systems [13,14]. In this context, an important problem is to quantify the structural information content of graphs by using Shannon's information measure [14–19]. This groundbreaking work led to numerous measurements of network complexity by using Shannon's information measure [15,20–22]. Particularly this task firstly appeared when studying the complexity of chemical and biological systems [13,23–25]. Besides studying chemical and biological questions, the structural complexity of networks have been also explored in computer science [1,26], ecology [10,27–29], information theory [30], linguistics [31,32], sociology [33,34], and mathematical psychology [33,35]. Also, information-theoretic approaches for investigating networks have been employed in network physics, see [7,36,37].

As the measures have been explored interdisciplinarily, it is particularly important to understand their strong and weak points. Otherwise, the results of applications involving the measures can not be understood properly. Besides surveying the most important measures, the main contribution is to highlight some of their strong and weak points. In this paper, this relates to better understand their *structural interpretation* and to gain insights about their *uniqueness*. The uniqueness, often called the discrimination power or degeneracy of an information-theoretic graph measure (and of course of any graph measure) relates to the property how well it can discriminate non-isomorphic graphs by its values, see [38–40]. An important problem is to evaluate the degree of degeneracy of a measure by several quantities such as the sensitivity measure due to Konstantinova [40]. Note that the discrimination power of a measure clearly depends on the graph class in question, see [38,41].

2. Graph Entropies

2.1. Measures Based on Equivalence Criteria and Graph Invariants

To find such measures, seminal work was done by Bonchev [8,42], Mowshowitz [15–18], Rashevsky [19] and Trucco [14]. Chronologically, Rashevsky [19], MacArthur [27] and Trucco [14] were the first who applied Shannon's information measure to derive an entropy of a graph characterizing its topology. Then, Mowshowitz [15–18] called it *structural information content* of a graph and developed a theory to study the properties of such graph entropies under certain graph operations such as product, join *etc.* So far, numerous related quantities have been defined by applying the general approach of deriving partitions based on a graph invariant which is due to Mowshowitz [15]. As a result of these developments, Bertz [43], Basak *et al.* [44,45] and Bonchev [8,42,46] contributed various related measures which are all based on the idea of deriving partitions by using a graph invariant, e.g., vertices, edges, degrees, and distances.

Let $G = (V, E)$ be a graph, X be a graph invariant, and τ be an equivalence criterion. Then, G can be partitioned with respect to the elements of the graph invariant under consideration. From this procedure, one also obtains probability values for each partition [8,15] given by $p_i := \frac{|X_i|}{|X|}$. By applying Shannon's information measure [5], we yield the graph entropies as follows [8]:

$$I_t(G, \tau) := |X| \log(|X|) - \sum_{i=1}^{k} |X_i| \log(|X_i|) \tag{1}$$

$$I_m(G, \tau) := - \sum_{i=1}^{k} \frac{|X_i|}{|X|} \log\left(\frac{|X_i|}{|X|}\right) \tag{2}$$

where k equals the number of different partitions. I_t is called total information content and I_m is called the mean information content of G, respectively.

In the following, we survey graph entropy measures exemplarily by applying this principle. Besides well-known quantities, we also mention more recently developed indices.

1. Topological information content due to Rashevsky [19]:

$$I_a(G) := - \sum_{i=1}^{k} \frac{|N_i|}{|V|} \log\left(\frac{|N_i|}{|V|}\right) \tag{3}$$

$|N_i|$ denotes the number of topologically equivalent vertices in the i-th vertex orbit of G. k is the number of different orbits. This measure is based on symmetry in a graph as it relies on its automorphism group and vertex orbits. It can be easily shown that I_a vanishes for vertex transitive graphs. Also, it attains maximum entropy for asymmetric graphs. However, it has been shown [41] that these symmetry-based measures possess little discrimination power. The reason for this is that many non-isomorphic graphs have the same orbit structure and, hence, they can not be distinguished by this index. Historically seen, the term *topological information content* was proposed by Rashevski [19]. Then, Trucco [14] redefined the measure in terms of graph orbits. Finally, Mowshowitz [15] studied extensively mathematical properties of this information measure for graphs (e.g., the behavior of I_a under graph operations) and generalized it by considering infinite graphs [18].

2. Symmetry index for graphs due to Mowshowitz *et al.* [47]:

$$S(G) := (\log(|V|) - I_a(G)) + \log(|\text{Aut}(G)|) \tag{4}$$

In [47], extremal values of this index and formulas for special graph classes such as wheels, stars and path graphs have been studied. As conjectured, the discrimination power of S turned out to be higher than by using I_a as a discriminating term $\log\left(|\mathrm{Aut}(G)|\right)$ has been added, see Equation (4). In particular, we obtained this result by calculating S on a set of 2265 chemical graphs whose order range from four to nineteen. A detailed explanation of the dataset can be found in [48].

3. Chromatic information content due to Mowshowitz [15,16]:

$$I_c(G) := \min_{\hat{V}}\left\{ -\sum_{i=1}^{h} \frac{n_i(\hat{V})}{|V|} \log\left(\frac{n_i(\hat{V})}{|V|}\right) \right\} \tag{5}$$

where $\hat{V} := \{V_i | 1 \le i \le h\}$, $|V_i| := n_i(\hat{V})$ denotes an arbitrary chromatic decomposition of a graph G. $h = \chi(G)$ is the chromatic number of G. Graph-theoretic properties of I_c and its behavior on several graph classes have been explored by Mowshowitz [15,16]. To our knowledge, the structural interpretation of this measure as well as the uniqueness has not yet been explored extensively.

4. Magnitude-based information indices due to Bonchev *et al.* [49]:

$$I_D(G) := -\frac{1}{|V|} \log\left(\frac{1}{|V|}\right) - \sum_{i=1}^{\rho(G)} \frac{2k_i}{|V|^2} \log\left(\frac{2k_i}{|V|^2}\right) \tag{6}$$

$$I_D^W(G) := W(G)\log(W(G)) - \sum_{i=1}^{\rho(G)} ik_i \log(i) \tag{7}$$

where k_i is the occurrence of a distance possessing value i in the distance matrix of G. The motivation to introduce these measures was to find quantities which detect branching well, see [49]. In this context, branching of a graph correlates with the number of terminal vertices. By using this model, Bonchev *et al.* [49] showed numerically and by means of inequalities that these indices detect branching meaningfully. Also, it turned out that magnitude-based information indices possess high discrimination power for trees. But recent studies [50] have shown that the uniqueness of the magnitude-based information indices deteriorate tremendously when being applied to large sets of graphs containing cycles. More precisely, Dehmer *et al.* [50] evaluated the uniqueness of several graph entropy measures and other topological indices by using almost 12 million non-isomorphic, connected and unweighted graphs possessing ten vertices.

5. Vertex degree equality-based information index found by Bonchev [8]:

$$I_{\deg}(G) := \sum_{i=1}^{\bar{k}} \frac{|N_i^{k_v}|}{|V|} \log\left(\frac{|N_i^{k_v}|}{|V|}\right) \tag{8}$$

where $|N_i^{k_v}|$ is the number of vertices with degree equal to i and $\bar{k} := \max_{v \in V} k_v$. Note that this quantity is easy to determine as the time complexity of the calculation of the degrees is clearly polynomial. But it is intuitive that a simple comparison of the degree distribution of graphs is not meaningful to discriminate their structure. In [50], it has been shown that this measure possesses little discrimination power when applying the quantity to several sets of graphs.

6. Overall information indices found by Bonchev [46,51]:

$$OX(G) := \sum_{k=0}^{|E|} {}^kX; \quad \{X\} := \{{}^0X, {}^1X, \dots, {}^{|E|}X\} \tag{9}$$

$$I(G, OX) := OX \log(OX) - \sum_{k=0}^{|E|} {}^kX \log\left({}^kX\right) \tag{10}$$

The index calculates the overall value OX of a certain graph invariant X by summing up its values in all subgraphs, and partitioning them into terms of increasing orders (increasing number of subgraph edges k). In the simplest case, we have $OX = SC$, *i.e.*, it is equal to the subgraph count [51]. Several more overall indices and their informational functionals have been calculated, such as overall connectivity (the sum of total adjacency of all subgraphs), overall Wiener index (the sum of total distances of all subgraphs), the overall Zagreb indices, and the overall Hosoya index [51]. They all share (with some inessential variations) the property to increase in value with the increase in graph complexity. The properties of most of these information functionals will not be studied here in detail.

Clearly, we only surveyed a subset of existing graph entropy measures. Further measures which are based on the same criterion can be found in [51–53]. Also, we would like to mention that information measures for graphs based on other entropy measures have been studied [54]. For instance, Passerini and Severini [54] explored the von Neumann entropy of networks in the context of network physics. Altogether, the variety of existing network measures bears great potential for analyzing complex networks quantitatively. But in the future, the usefulness and ability of these measures must be investigated more extensively to gain further theoretical insights in terms of their properties.

2.2. Körner Entropy

The definition of the Körner entropy is rooted in information theory and has been introduced to solve a particular coding problem, see [30,55]. Simony [55] discussed several definitions of this quantity which have been proven to be equivalent. One definition thereof is

$$H(G, P) := \lim_{t \longrightarrow \infty} \min_{U \subseteq V^t, P^t(U) > 1 - \epsilon} \frac{1}{t} \log(\chi(G^t(U))) \tag{11}$$

For $V' \subseteq V(G)$, the induced subgraph on V' is denoted by $G(V')$ and $\chi(G)$ is the chromatic number [56] of G, G^t the t-th co-normal power [30] of G and

$$P^t(U) := \sum_{x \in U} P^t(x) \tag{12}$$

Note that $P^t(x)$ is the probability of the string x, see [55]. Examples and the interpretation of this graph entropy measure can be found in [30,55]. Due to the fact that its calculation relies on the stable set problem, its computational complexity may be insufficient. To our knowledge, the Körner entropy has not been used as a graph complexity measure in the sense of the quantities described in the previous section. That means, it does not express the structural information content of a graph (as the previously mentioned graph entropies) as it has been used in a different context, see [30,55]. Also, its computational complexity makes it impossible to apply this quantity on a large scale and to investigate properties such as correlation and uniqueness.

2.3. Entropy Measures Using Information Functionals

Information-theoretic complexity measures for graphs can also be inferred by assigning a probability value to each vertex of a graph in question [9,21]. Such probability values have been defined by using information functionals [9,21,48]. In order to define these information functionals, some key questions must be answered:

- What kind of structural features (e.g., vertices, edges, degrees, distances *etc.*) should be used to derive meaningful information functionals?
- In this context, what does "meaningful" mean?
- In case the functional is parametric, how can the parameters be optimized?
- What kind of structural information does the functional as well as the resulting entropy detect?

To discuss the first item, see [9,21,48] and note that metrical properties have been used to derive such information functionals. In order to prove whether a functional as well as the resulting entropy measures captures structural information meaningfully, an optimality criterion is needed. For example, suppose there exists a data set where the class labels of its entities (graphs) are known. By employing supervised machine learning techniques, the classification error can be optimized. Note that the last item relates to investigate the structural interpretation of the graph entropy measure. Indeed, this question could be raised for any topological index.

In order to reproduce some of these measures, we start with a graph $G = (V, E)$ and let f be an information functional representing a positive function that maps vertices to the positive reals. Note that f captures structural information of G. If we define the vertex probabilities as [9,21]

$$p(v_i) := \frac{f(v_i)}{\sum_{j=1}^{|V|} f(v_j)} \tag{13}$$

we yield the families of information-theoretic graph complexity measures [9,48]:

$$I_f(G) := -\sum_{i=1}^{|V|} \frac{f(v_i)}{\sum_{j=1}^{|V|} f(v_j)} \log\left(\frac{f(v_i)}{\sum_{j=1}^{|V|} f(v_j)}\right) \tag{14}$$

$$I_f^\lambda(G) := \lambda \left(\log(|V|) + \sum_{i=1}^{|V|} \frac{f(v_i)}{\sum_{j=1}^{|V|} f(v_j)} \log\left(\frac{f(v_i)}{\sum_{j=1}^{|V|} f(v_j)}\right) \right) \tag{15}$$

$\lambda > 0$ is a scaling constant. Typical information functionals are [9,21,48]

$$f_1(v_i) := \alpha^{c_1 |S_1(v_i, G)| + c_2 |S_2(v_i, G)| + \cdots + c_{\rho(G)} |S_{\rho(G)}(v_i, G)|}, \quad c_k > 0, 1 \leq k \leq \rho(G), \alpha > 0 \tag{16}$$

and

$$f_2(v_i) := c_1 |S_1(v_i, G)| + c_2 |S_2(v_i, G)| + \cdots + c_{\rho(G)} |S_{\rho(G)}(v_i, G)|, \quad c_k > 0, 1 \leq k \leq \rho(G) \tag{17}$$

The parameters $c_k > 0$ to weight structural characteristics or differences of G in each sphere have to be chosen such that at least $c_i \neq c_j$ holds. Otherwise the probabilities become $\frac{1}{|V|}$ leading to maximum entropy $\log(|V|)$. For instance, the setting $c_1 > c_2 > \cdots > c_{\rho(G)}$ have often been used, see [9,21,48]. Also, other schemes for the coefficients can be chosen but need to be interpreted in terms of the structural interpretation of the resulting entropy measure. As the measures are parametric (when using a parametric information functional), they can be interpreted as generalizations of the aforementioned partition-based measures.

By applying Equation (15), concrete information measures to characterize the structural complexity chemical structures have been derived in [48]. For example, if we choose the coefficients linearly decreasing, e.g.,

$$c_1 := \rho(G), c_2 := \rho(G) - 1, \ldots, c_{\rho(G)} := 1 \tag{18}$$

or exponentially decreasing, e.g.,

$$c_1 := \rho(G), c_2 := \rho(G)e^{-1}, \ldots, c_{\rho(G)} := \rho(G)e^{-\rho(G)+1} \tag{19}$$

the resulting measures are called $I_{f_{lin}^{fV}}^\lambda$ and $I_{f_{exp}^{fV}}^\lambda$, respectively. Importantly, it turned out that $I_{f_{lin}^{fV}}^\lambda$ and $I_{f_{exp}^{fV}}^\lambda$ possess high discrimination power when applying them to real and synthetic chemical graphs, see [48].

To obtain more advanced information functionals, the concept outlined above has been extended in [57]. The main idea for deriving these information functionals is based on the assumption that starting from an arbitrary vertex $v_i \in V$, information spreads out via shortest paths in the graph which can be determined by using Dijkstra's algorithm [58]. Then, more sophisticated information functionals as well as complexity measures have been defined [57] by using local property measures, e.g., vertex centrality measures [59]. In particular, some of them turned out to be highly unique when applying the measures to almost 12 million non-isomorphic, connected and unweighted graphs possessing ten vertices [50]. Interestingly, the just mentioned information-theoretic complexity measures showed a constantly high uniqueness that does not depend much on the cardinality of the underlying graph set. This property is desirable as we found that the uniqueness of the most existing measures deteriorates dramatically if the cardinality of the underlying graph set increases.

2.4. Information-Theoretic Measures for Trees

In this section, we sketch a few entropic measures which have been developed to characterize trees structurally. For example, Emmert-Streib *et al.* [60] developed an approach to determine the structural information content of rooted trees by using the natural partitioning of the vertices in such a tree. That means the number of vertices can be counted on each tree level which leads to a probability distribution and, thus, to an entropy characterizing the topology of a rooted tree. Dehmer [57] used this idea to calculate the entropy of arbitrary undirected graphs by applying a decomposition approach. Mehler [31] also employed entropic measures as balance and imbalance measures of tree-like graphs in the context of social network analysis. Other aspects of tree entropy have been tackled by Lions [61].

2.5. Other Information-Theoretic Network Measures

Apart from information-theoretic measures mostly used in mathematical and structural chemistry, several other entropic networks measures for measuring disorder relations in complex networks have been explored in the context of network physics, see [62]. If $P(k_v)$ denotes the probability of a vertex v possessing degree k, the distribution of the so-called remaining degree was defined by [62]

$$q(k_v) := \frac{(k+1)P_{k_v+1}}{<k>} \tag{20}$$

$<k> := \sum_k kP(k_v)$. By applying Shannon's information measure, the following graph entropy measure has been obtained [62]:

$$I(G) := \sum_{i=1}^{|V|} q(i) \log(q(i)) \tag{21}$$

It can be interpreted as a measure for determining the heterogeneity of a complex network [62]. In order to develop information indices for weighted directed networks, Wilhelm *et al.* [63] defined the measure called Medium Articulation that obtains its maximum for networks with a medium number of edges. It has been defined by [63]

$$MA(G) := R(G) \cdot I(G) \tag{22}$$

where

$$R(G) := -\sum_{i,j} T_{v_i v_j} \log\left(\frac{T^2_{v_i v_j}}{\sum_k T_{v_k v_j} \sum_l T_{v_i v_l}}\right) \tag{23}$$

represents the redundancy and [63]

$$I(G) := \sum_{i,j} T_{v_i v_j} \log\left(\frac{T_{v_i v_j}}{\sum_k T_{v_k v_j} \sum_l T_{v_i v_l}}\right) \tag{24}$$

the mutual information.

Finally, the normalized flux from v_i to v_j is

$$T_{v_i v_j} := \frac{t_{v_i v_j}}{\sum_{k,l} t_{v_k v_l}} \qquad (25)$$

$t_{v_i v_j}$ is the flux (edge weight) between v_i and v_j. It can be easily shown that R vanishes for a directed ring but attains its maximum for the complete graph [63]. The behavior of I is just converse. This implies that MA vanishes for extremal graphs and attains its maximum in between [63]. We remark that a critical discussion of MA and modified measures have been recently contributed by Ulanowicz *et al.* [64].

For finalizing this section, we also reproduce the so-called offdiagonal complexity (*OdC*) [65] that is based on determining the entropy of the offdiagonal elements of the vertex-vertex link correlation matrix [65,66]. Let $G = (V, E)$ be a graph and let $(c_{ij})_{ij}$ be the vertex-vertex link correlation matrix, see [65]. Here c_{ij} denotes the number of all neighbors possessing degree $j > i$ of all vertices with degree i [66]. $\bar{k} := \max_{v \in V} k_v$ stands for the maximum degree of G. If one defines [66]

$$a_{|V|} := \sum_{i=1}^{\bar{k}-|V|} c_{i,i+|V|} \qquad (26)$$

and

$$b_{|V|} := \frac{a_{|V|}}{\sum_{|V|=0}^{\bar{k}-1} a_{|V|}} \qquad (27)$$

OdC can be defined by [66]

$$OdC := \frac{-\left(\sum_{|V|=0}^{\bar{k}-1} b_{|V|} \log(b_{|V|}) \right)}{\log(|V|-1)} \in [0,1] \qquad (28)$$

As the measure depends on correlations between degrees of pairs of vertices [65], it is not surprising that its discrimination power is low, see [41].

3. Structural Interpretation of Graph Measures

We already mentioned the problem of exploring the structural interpretation of topological graph measures exemplarily in the preceding sections. In general, this relates to explore what kind of structural complexity a particular measure does detect. The following listing shows a few such *types of structural complexity* of measures which have already been explored:

- Branching in trees [49,67,68]. Examples for branching measures are the Wiener index [69], the magnitude-based measures also known as Bonchev–Trinajstić indices [49] and others outlined by Janežić *et al.* [68].
- Linear tree complexity depending on their size and symmetry [68]. Examples for such measures are the MI and MB indices, TC and TC1 Indices *etc.*, see [68].
- Balance and imbalance of tree-like graphs [31]. For examples, see [31].
- Cyclicity in graphs [23,38,68,70,71]. Note that in the context of mathematical chemistry, this graph property has been introduced and studied by Bonchev *et al.* [38]. Examples for branching measures are the BT and BI Indices, and the F index, see [70].
- Inner symmetry and symmetry in graphs [15,47,48,72]. Examples for such measures are I_a, S (see Section (2.1)) and I_{f_2} (see Section (2.3)).

In view of the vast amount of topological measures developed so far, determining their structural interpretation is a daunting problem. Evidently, it is important to contribute to this problem as measures could be then classified by this property. This might be useful when designing new measures or finding topological indices for solving a particular problem.

4. Summary and Conclusion

In this paper, we surveyed information-theoretic measures for analyzing networks quantitatively. Also, we discussed some of their properties, namely the structural interpretation and uniqueness. Because a vast number of measures have been developed, the former problem has been somewhat overlooked when analyzing topological network measures. Also, the uniqueness of information-theoretic and non-information-theoretic measures is a crucial property. Applications thereof might be interesting for applications such as problems in combinatorial chemistry [73]. In fact, many papers exist to tackle this problem [40,74–76] but not on a large scale. Interestingly, a statistical analysis has been recently shown [50] that the uniqueness of many topological indices strongly depends on the cardinality of a graph set in question. Also, it is clear that the uniqueness property depends on a particular graph class. This implies that results may not be generalized when the measure gives feasible results for a special class only, *e.g.*, trees, isomers *etc.*

Acknowledgements

Fruitful discussions with Danail Bonchev, Boris Furtula, Abbe Mowshowitz and Kurt Varuza are gratefully acknowledged. Matthias Dehmer thanks the Austrian Science Funds for supporting this work (project P22029-N13).

References

1. Allen, E.B. Measuring Graph Abstractions of Software: An Information-Theory Approach. In *Proceedings of the 8-th International Symposium on Software Metrics*, IEEE Computer Society: Ottawa, ON, Canada, 4–7 June 2002; p. 182.
2. Cover, T.M.; Thomas, J.A. *Elements of Information Theory*; Wiley & Sons: Hoboken, NJ, USA, 2006.
3. McDonnell, M.D.; Ikeda, S.; Manton, J.H. An introductory review of information theory in the context of computational neuroscience. *Biol. Cybern.* **2011**, *105*, 55–70.
4. Mathar, R.; Schmeink, A. A Bio-Inspired Approach to Condensing Information. In *Proceedings of the IEEE International Symposium on Information Theory (ISIT)*, IEEE Xplore: Saint-Petersburg, Russia, 31 July–5 August 2011; pp. 2524–2528.
5. Shannon, C.E.; Weaver, W. *The Mathematical Theory of Communication*; University of Illinois Press: Champaign, IL, USA, 1949.
6. Grosse, I.; Galván, P.B.; Carpena, P.; Roldán, R.R.; Oliver, J.; Stanley, H.E. Analysis of symbolic sequences using the Jensen-Shannon divergence. *Phys. Rev. E* **2002**, *65*, 041905:1–041905:16.
7. Anand, K.; Bianconi, G. Entropy measures for networks: Toward an information theory of complex topologies. *Phys. Rev. E* **2009**, *80*, 045102(R):1–045102(R):4.
8. Bonchev, D. *Information Theoretic Indices for Characterization of Chemical Structures*; Research Studies Press: Chichester, UK, 1983.
9. Dehmer, M. Information processing in complex networks: Graph entropy and information functionals. *Appl. Math. Comput.* **2008**, *201*, 82–94.
10. Hirata, H.; Ulanowicz, R.E. Information theoretical analysis of ecological networks. *Int. J. Syst. Sci.* **1984**, *15*, 261–270.
11. Kim, D.C.; Wang, X.; Yang, C.R.; Gao, J. Learning biological network using mutual information and conditional independence. *BMC Bioinf.* **2010**, *11*, S9:1–S9:8.
12. Barnett, L.; Buckley, L.; Bullock, C.L. Neural complexity and structural connectivity. *Phys. Rev. E* **2009**, *79*, 051914:1–051914:12.
13. Morowitz, H. Some order-disorder considerations in living systems. *Bull. Math. Biophys.* **1953**, *17*, 81–86.
14. Trucco, E. A note on the information content of graphs. *Bull. Math. Biol.* **1956**, *18*, 129–135.
15. Mowshowitz, A. Entropy and the complexity of the graphs I: An index of the relative complexity of a graph. *Bull. Math. Biophys.* **1968**, *30*, 175–204.
16. Mowshowitz, A. Entropy and the complexity of graphs II: The information content of digraphs and infinite graphs. *Bull. Math. Biophys.* **1968**, *30*, 225–240.

17. Mowshowitz, A. Entropy and the complexity of graphs III: Graphs with prescribed information content. *Bull. Math. Biophys.* **1968**, *30*, 387–414.
18. Mowshowitz, A. Entropy and the complexity of graphs IV: Entropy measures and graphical structure. *Bull. Math. Biophys.* **1968**, *30*, 533–546.
19. Rashevsky, N. Life, information theory, and topology. *Bull. Math. Biophys.* **1955**, *17*, 229–235.
20. Bonchev, D. Information Theoretic Measures of Complexity. In *Encyclopedia of Complexity and System Science*; Meyers, R., Ed.; Springer: Berlin, Heidelberg, Germany, 2009; Volume 5, pp. 4820–4838.
21. Dehmer, M. A novel method for measuring the structural information content of networks. *Cybern. Syst.* **2008**, *39*, 825–842.
22. Mowshowitz, A.; Mitsou, V. Entropy, Orbits and Spectra of Graphs. In *Analysis of Complex Networks: From Biology to Linguistics*; Dehmer, M., Emmert-Streib, F., Eds.; Wiley-VCH: Hoboken, NJ, USA, 2009; pp. 1–22.
23. Bonchev, D.; Rouvray, D.H. *Complexity in Chemistry, Biology, and Ecology*; Springer: New York, NY, USA, 2005.
24. Dancoff, S.M.; Quastler, H. Information Content and Error Rate of Living Things. In *Essays on the Use of Information Theory in Biology*; Quastler, H., Ed.; University of Illinois Press: Champaign, IL, USA, 1953; pp. 263–274.
25. Linshitz, H. The Information Content of a Battery Cell. In *Essays on the Use of Information Theory in Biology*; Quastler, H., Ed.; University of Illinois Press: Urbana, IL, USA, 1953.
26. Latva-Koivisto, A.M. *Finding a Complexity Measure for Business Process Models*. Helsinki University of Technology: Espoo, Finland, 2001; doi:10.1.1.25.2991.
27. MacArthur, R.H. Fluctuations of animal populations and a measure of community stability. *Ecology* **1955**, *36*, 533–536.
28. Solé, R.V.; Montoya, J.M. Complexity and Fragility in Ecological Networks. *Proc. R. Soc. Lond. B* **2001**, *268*, 2039–2045.
29. Ulanowicz, R.E. Information theory in ecology. *Comput. Chem.* **2001**, *25*, 393–399.
30. Körner, J. Coding of an Information Source Having Ambiguous Alphabet and the Entropy of Graphs. In *Proceedings of the Transactions of the 6th Prague Conference on Information Theory*; Academia: Prague, Czechoslovakia, 19–25 September 1973; pp. 411–425.
31. Mehler, A. Social Ontologies as Generalized Nearly Acyclic Directed Graphs: A Quantitative Graph Model of Social Tagging. In *Towards an Information Theory of Complex Networks: Statistical Methods and Applications*; Dehmer, M., Emmert-Streib, F., Mehler, A., Eds.; Birkhäuser Boston: MA, USA, 2011; pp. 259–319.
32. Mehler, A.; Weiß, P.; Lücking, A. A network model of interpersonal alignment. *Entropy* **2010**, *12*, 1440–1483.
33. Balch, T. Hierarchic social entropy: An information theoretic measure of robot group diversity. *Auton. Robot.* **2001**, *8*, 209–237.
34. Butts, C.T. The complexity of social networks: Theoretical and empirical findings. *Soc. Netw.* **2001**, *23*, 31–71.
35. Sommerfeld, E.; Sobik, F. Operations on Cognitive Structures—Their Modeling on the Basis of Graph Theory. In *Knowledge Structures*; Albert, D., Ed.; Springer: Berlin, Heidelberg, Germany, 1994; pp. 146–190.
36. Krawitz, P.; Shmulevich, I. Entropy of complex relevant components of Boolean networks. *Phys. Rev. E* **2007**, *76*, 036115:1–036115:7.
37. Sanchirico, A.; Fiorentino, M. Scale-free networks as entropy competition. *Phys. Rev. E* **2008**, *78*, 046114:1–046114:10.
38. Bonchev, D.; Mekenyan, O.; Trinajstić, N. Topological characterization of cyclic structures. *Int. J. Quantum Chem.* **1980**, *17*, 845–893.
39. Dehmer, M.; Sivakumar, L.; Varmuza, K. Uniquely discriminating molecular structures using novel eigenvalue-based descriptors. *MATCH Commun. Math. Comput. Chem.* **2012**, *67*, 147–172.
40. Konstantinova, E.V. The discrimination ability of some topological and information distance indices for graphs of unbranched hexagonal systems. *J. Chem. Inf. Comput. Sci.* **1996**, *36*, 54–57.
41. Dehmer, M.; Barbarini, N.; Varmuza, K.; Graber, A. A large scale analysis of information-theoretic network complexity measures using chemical structures. *PLoS One* **2009**, *4*, doi:10.1371/journal.pone.0008057.
42. Bonchev, D. Information indices for atoms and molecules. *MATCH Commun. Math. Comp. Chem.* **1979**, *7*, 65–113.

43. Bertz, S.H. The first general index of molecular complexity. *J. Am. Chem. Soc.* **1981**, *103*, 3241–3243.

44. Basak, S.C.; Magnuson, V.R. Molecular topology and narcosis. *Drug Res.* **1983**, *33*, 501–503.

45. Basak, S.C. Information-Theoretic Indices of Neighborhood Complexity and their Applications. In *Topological Indices and Related Descriptors in QSAR and QSPAR*; Devillers, J., Balaban, A.T., Eds.; Gordon and Breach Science Publishers: Amsterdam, The Netherlands, 1999; pp. 563–595.

46. Bonchev, D. Overall connectivities and topological complexities: A new powerful tool for QSPR/QSAR. *J. Chem. Inf. Comput. Sci.* **2000**, *40*, 934–941.

47. Mowshowitz, A.; Dehmer, M. A symmetry index for graphs. *Symmetry Cult. Sci.* **2010**, *21*, 321–327.

48. Dehmer, M.; Varmuza, K.; Borgert, S.; Emmert-Streib, F. On entropy-based molecular descriptors: Statistical analysis of real and synthetic chemical structures. *J. Chem. Inf. Model.* **2009**, *49*, 1655–1663.

49. Bonchev, D.; Trinajstić, N. Information theory, distance matrix and molecular branching. *J. Chem. Phys.* **1977**, *67*, 4517–4533.

50. Dehmer, M.; Grabner, M.; Varmuza, K. Information indices with high discrimination power for arbitrary graphs. *PLoS One* submitted for publication, **2011**.

51. Bonchev, D. The Overall Topological Complexity Indices. In *Advances in Computational Methods in Science and Engineering*; Simos, T., Maroulis, G., Eds.; VSP Publications: Boulder, USA, 2005; Volume 4B, pp. 1554–1557.

52. Bonchev, D. My life-long journey in mathematical chemistry. *Internet Electron. J . Mol. Des.* **2005**, *4*, 434–490.

53. Todeschini, R.; Consonni, V.; Mannhold, R. *Handbook of Molecular Descriptors*; Wiley-VCH: Weinheim, Germany, 2002.

54. Passerini, F.; Severini, S. The von Neumann entropy of networks. *Int. J. Agent Technol. Syst.* **2009**, *1*, 58–67.

55. Simonyi, G. Graph Entropy: A Survey. In *Combinatorial Optimization*; Cook, W., Lovász, L., Seymour, P., Eds.; ACM: New York, NY, USA, 1995; Volume 20, pp. 399–441.

56. Bang-Jensen, J.; Gutin, G. *Digraphs. Theory, Algorithms and Applications*; Springer: Berlin, Heidelberg, Germany, 2002.

57. Dehmer, M. Information-theoretic concepts for the analysis of complex networks. *Appl. Artif. Intell.* **2008**, *22*, 684–706.

58. Dijkstra, E.W. A note on two problems in connection with graphs. *Numer. Math.* **1959**, *1*, 269–271.

59. Brandes, U.; Erlebach, T. *Network Analysis*; Springer: Berlin, Heidelberg, Germany, 2005.

60. Emmert-Streib, F.; Dehmer, M. Information theoretic measures of UHG graphs with low computational complexity. *Appl. Math. Comput.* **2007**, *190*, 1783–1794.

61. Lyons, R. Identities and inequalities for tree entropy. *Comb. Probab. Comput.* **2010**, *19*, 303–313.

62. Solé, R.V.; Valverde, S. *Information Theory of Complex Networks: On Evolution and Architectural Constraints*; Springer: Berlin, Heidelberg, Germany, 2004; Volume 650, pp. 189–207.

63. Wilhelm, T.; Hollunder, J. Information theoretic description of networks. *Physica A* **2007**, *388*, 385–396.

64. Ulanowicz, R.E.; Goerner, S.J.; Lietaer, B.; Gomez, R. Quantifying sustainability: Resilience, efficiency and the return of information theory. *Ecol. Complex.* **2009**, *6*, 27–36.

65. Claussen, J.C. Characterization of networks by the offdiagonal complexity. *Physica A* **2007**, *365-373*, 321–354.

66. Kim, J.; Wilhelm, T. What is a complex graph? *Physica A* **2008**, *387*, 2637–2652.

67. Bonchev, D. Topological order in molecules 1. Molecular branching revisited. *J. Mol. Strut. THEOCHEM* **1995**, *336*, 137 – 156.

68. Janežić, D.; Milećević, A.; Nikolić, S.; Trinajstić, N. Topological Complexity of Molecules. In *Encyclopedia of Complexity and System Science*; Meyers, R., Ed.; Springer: Berlin, Heidelberg, Germany, 2009; Volume 5, pp. 9210–9224.

69. Wiener, H. Structural determination of paraffin boiling points. *J. Am. Chem. Soc.* **1947**, *69*, 17–20.

70. Balaban, A.T.; Mills, D.; Kodali, V.; Basak, S.C. Complexity of chemical graphs in terms of size, branching and cyclicity. *SAR QSAR Environ. Res.* **2006**, *17*, 429–450.

71. Finn, J.T. Measures of ecosystem structure and function derived from analysis of flows. *J. Theor. Biol.* **1976**, *56*, 363–380.

72. Garrido, A. Symmetry of complex networks. *Adv. Model. Optim.* **2009**, *11*, 615–624.

73. Li, X.; Li, Z.; Wang, L. The inverse problems for some topological indices in combinatorial chemistry. *J. Comput. Biol.* **2003**, *10*, 47–55.

74. Bonchev, D.; Mekenyan, O.; Trinajstić, N. Isomer discrimination by topological information approach. *J. Comput. Chem.* **1981**, *2*, 127–148.

75. Diudea, M.V.; Ilić, A.; Varmuza, K.; Dehmer, M. Network analysis using a novel highly discriminating topological index. *Complexity* **2011**, *16*, 32–39.

76. Konstantinova, E.V.; Paleev, A.A. Sensitivity of topological indices of polycyclic graphs. *Vychisl. Sist.* **1990**, *136*, 38–48.

symmetry

MDPI

Article

Defining the Symmetry of the Universal Semi-Regular Autonomous Asynchronous Systems

Serban E. Vlad

Street Zimbrului, Nr. 3, Bl. PB68, Ap. 11, 410430, Oradea, Romania;
E-Mail: serban_e_vlad@yahoo.com; Tel./Fax: 0040-259-479113

Academic Editor: name
Received: 1 November 2011; in revised form: 8 February 2012 / Accepted: 9 February 2012 / Published: 15 February 2012

Abstract: The regular autonomous asynchronous systems are the non-deterministic Boolean dynamical systems and universality means the greatest in the sense of the inclusion. The paper gives four definitions of symmetry of these systems in a slightly more general framework, called semi-regularity, and also many examples.

Keywords: asynchronous system; symmetry; semi-regularity

MSC: 94C10

1. Introduction

Switching theory has developed in the 1950s and the 1960s as a common effort of the mathematicians and the engineers of studying the switching circuits (a.k.a. asynchronous circuits) from digital electrical engineering. We are unaware of any existent mathematical work published after 1970 on what we call switching theory. The published works are written by engineers and their approach is always descriptive and unacceptable for the mathematicians. The label of *switching theory* has changed to *asynchronous systems* (or *circuits*) *theory*. One of the possible motivations of the situation consists in the fact that the important producers of digital equipments have stopped the dissemination of such researches.

Our interest in asynchronous systems had bibliography coming from the 1950s and the 1960s, as well as engineering works giving intuition, as well as mathematical works giving analogies. An interesting *rendez-vous* has happened when the asynchronous systems theory has met the dynamical systems theory, resulting in the so-called *regular* autonomous systems (a.k.a Boolean dynamical systems) where the vector field is $\Phi : \{0,1\}^n \to \{0,1\}^n$ and time is discrete or real, and we obtain the *unbounded delay model* of computation of Φ suggested by the engineers. The *synchronous* iterations of $\Phi : \Phi \circ \Phi, \Phi \circ \Phi \circ \Phi, \ldots$ of the dynamical systems are replaced by *asynchronous* iterations in which each coordinate $\Phi_1, \ldots, \Phi_n$ is iterated independently on the others, in arbitrary finite time.

We denote with $\mathbf{B} = \{0,1\}$ the binary Boolean algebra, together with the discrete topology and with the usual algebraic laws:

$$
\begin{array}{cc}
\begin{array}{c|cc}
- \\
\hline
0 & 1 \\
1 & 0
\end{array}
&
\begin{array}{c|cc}
\cdot & 0 & 1 \\
\hline
0 & 0 & 0 \\
1 & 0 & 1
\end{array}
\quad
\begin{array}{c|cc}
\cup & 0 & 1 \\
\hline
0 & 0 & 1 \\
1 & 1 & 1
\end{array}
\quad
\begin{array}{c|cc}
\oplus & 0 & 1 \\
\hline
0 & 0 & 1 \\
1 & 1 & 0
\end{array}
\end{array}
\tag{1}
$$

We use the same notations for the laws that are induced from **B** on other sets, for example $\forall x \in \mathbf{B}^n$, $\forall y \in \mathbf{B}^n$,

$$\overline{x} = (\overline{x_1}, \ldots, \overline{x_n})$$
$$x \cup y = (x_1 \cup y_1, \ldots, x_n \cup y_n)$$

etc. In Figure 1, we have drawn at (a) the logical gate NOT, *i.e.*, the circuit that computes the logical complement and at (b) a circuit that makes use of logical gates NOT. The asynchronous system that models the circuit from (b) has the state portrait drawn at (c). In the state portraits, the arrows show the increase of (the discrete or continuous) time. The underlined coordinates μ_i are these coordinates for which $\Phi_i(\mu_i) \neq \mu_i$ and they are called *excited*, or *enabled*, or *unstable*. The coordinates μ_i that are not underlined fulfill by definition $\Phi_i(\mu_i) = \mu_i$ and they are called *not excited*, or *not enabled*, or *stable*. The existence of two underlined coordinates in $(0,0)$ shows that $\Phi_1(0,0) = 1$ may be computed first, $\Phi_2(0,0) = 1$ may be computed first, or $\Phi_1(0,0)$, $\Phi_2(0,0)$ may be computed simultaneously, thus when the system is in $(0,0)$, it may run in three different directions, which results in non-determinism.

Our present purpose is to define the symmetry of these systems.

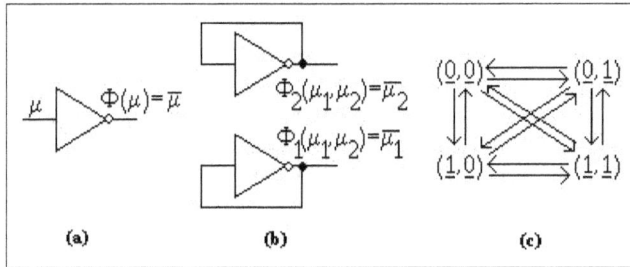

Figure 1. (a) the logical gate NOT; (b) circuit with logical gates NOT; (c) state portrait.

2. Semi-Regular Systems

Notation 1. *We denote* $\mathbf{N}_- = \{-1, 0, 1, 2, \ldots\}$.

Notation 2. $\chi_A : \mathbf{R} \to \mathbf{B}$ *is the notation of the characteristic function of the set* $A \subset \mathbf{R}$: $\forall t \in \mathbf{R}$,

$$\chi_A(t) = \begin{cases} 0, \text{if } t \notin A \\ 1, \text{if } t \in A \end{cases}.$$

Notation 3. *We denote with* $\overline{\Pi}_n$ *the set of the sequences* $\alpha = \alpha^0, \alpha^1, \ldots, \alpha^k, \ldots \in \mathbf{B}^n$.

Notation 4. *The set of the real sequences* $t_0 < t_1 < \ldots < t_k < \ldots$ *that are unbounded from above is denoted with* Seq.

Notation 5. *We use the notation* $\overline{P}_n$ *for the set of the functions* $\rho : \mathbf{R} \to \mathbf{B}^n$ *having the property that* $\alpha \in \overline{\Pi}_n$ *and* $(t_k) \in$ Seq *exist with* $\forall t \in \mathbf{R}$,

$$\rho(t) = \alpha^0 \chi_{\{t_0\}}(t) \oplus \alpha^1 \chi_{\{t_1\}}(t) \oplus \ldots \oplus \alpha^k \chi_{\{t_k\}}(t) \oplus \ldots \quad (2)$$

Definition 1. *Let* $\Phi : \mathbf{B}^n \to \mathbf{B}^n$ *be a function. For* $v \in \mathbf{B}^n, v = (v_1, \ldots, v_n)$ *we define the function* $\Phi^v : \mathbf{B}^n \to \mathbf{B}^n$ *by* $\forall \mu \in \mathbf{B}^n$,

$$\Phi^v(\mu) = (\overline{v_1}\mu_1 \oplus v_1 \Phi_1(\mu), \ldots, \overline{v_n}\mu_n \oplus v_n \Phi_n(\mu))$$

Remark 1. *For any $\mu \in \mathbf{B}^n, v \in \mathbf{B}^n$ and $i \in \{1,\dots,n\}$, if $v_i = 0$, then $\Phi_i^v(\mu) = \mu_i$ i.e., $\Phi_i(\mu)$ is not computed and if $v_i = 1$, then $\Phi_i^v(\mu) = \Phi_i(\mu)$ i.e., $\Phi_i(\mu)$ is computed. This is the meaning of asynchronicity.*

Definition 2. *Let $\alpha \in \overline{\Pi}_n$. The function $\widehat{\Phi}^\alpha : \mathbf{B}^n \times \mathbf{N}_- \to \mathbf{B}^n$ defined by $\forall \mu \in \mathbf{B}^n, \forall k \in \mathbf{N}_-$,*

$$
\begin{cases}
\widehat{\Phi}^\alpha(\mu,-1) = \mu, \\
\widehat{\Phi}^\alpha(\mu,k+1) = \Phi^{\alpha^{k+1}}(\widehat{\Phi}^\alpha(\mu,k))
\end{cases}
\tag{3}
$$

*is called **discrete time α−semi-orbit of** μ. We consider also the sequence $(t_k) \in Seq$ and the function $\rho \in \overline{P}_n$ from Equation (2), for which the function $\Phi^\rho : \mathbf{B}^n \times \mathbf{R} \to \mathbf{B}^n$ is defined by: $\forall \mu \in \mathbf{B}^n, \forall t \in \mathbf{R}$,*

$$
\Phi^\rho(\mu,t) = \widehat{\Phi}^\alpha(\mu,-1)\chi_{(-\infty,t_0)}(t) \oplus \widehat{\Phi}^\alpha(\mu,0)\chi_{[t_0,t_1)}(t) \oplus
$$
$$
\oplus \widehat{\Phi}^\alpha(\mu,1)\chi_{[t_1,t_2)}(t) \oplus \dots \oplus \widehat{\Phi}^\alpha(\mu,k)\chi_{[t_k,t_{k+1})}(t) \oplus \dots
\tag{4}
$$

*Φ^ρ is called **continuous time ρ−semi-orbit of** μ.*

Definition 3. *The **discrete time** and the **continuous time universal semi-regular autonomous asynchronous systems** associated to Φ are defined by*

$$
\widehat{\Xi}_\Phi = \{\widehat{\Phi}^\alpha(\mu,\cdot) | \mu \in \mathbf{B}^n, \alpha \in \overline{\Pi}_n\}
$$
$$
\overline{\Xi}_\Phi = \{\Phi^\rho(\mu,\cdot) | \mu \in \mathbf{B}^n, \rho \in \overline{P}_n\}
$$

Remark 2. *$\widehat{\Xi}_\Phi, \overline{\Xi}_\Phi$ and Φ are usually identified.*

Example 1. *In Figure 2 we have drawn at (a) the AND gate that computes the logical intersection, at (b) a circuit with two gates and at (c) the state portrait of $\Phi : \mathbf{B}^2 \to \mathbf{B}^2, \forall(\mu_1,\mu_2) \in \mathbf{B}^2, \Phi(\mu_1,\mu_2) = (0,1)$. We conclude that*

$$
\overline{\Xi}_\Phi = \{(\mu_1,\mu_2)\chi_{(-\infty,t_0)} \oplus (\mu_1\lambda_1,\mu_2 \cup \lambda_2)\chi_{[t_0,t_1)} \oplus
$$
$$
\oplus (\mu_1\lambda_1 v_1, \mu_2 \cup \lambda_2 \cup v_2)\chi_{[t_1,\infty)} | \mu,\lambda,v \in \mathbf{B}^2, t_0,t_1 \in \mathbf{R}, t_0 < t_1\}
$$

since the first coordinate might finally decrease its value and the second coordinate might finally increase its value, but the order and the time instant when these things happen are arbitrary.

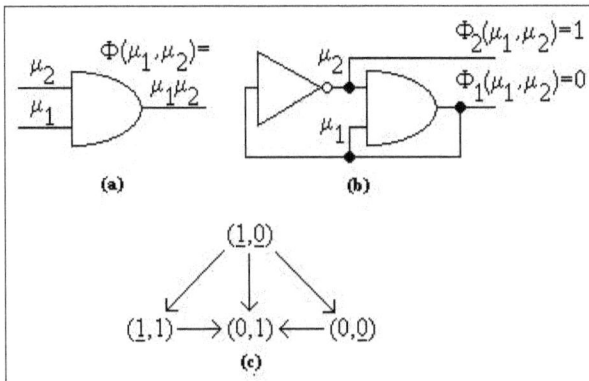

Figure 2. The semi-regular system $\overline{\Xi}_\Phi$ from Example 1.

3. Anti-Semi-Regular Systems

Definition 4. *Let* $\Phi : \mathbf{B}^n \to \mathbf{B}^n, \alpha \in \overline{\Pi}_n, (t_k) \in Seq$ *and* $\rho \in \overline{P}_n$ *from Equation (2). The function* $^*\widehat{\Phi}^\alpha : \mathbf{B}^n \times \mathbf{N}_- \to \mathbf{B}^n$ *that satisfies* $\forall \mu \in \mathbf{B}^n, \forall k \in \mathbf{N}_-$,

$$\begin{cases} ^*\widehat{\Phi}^\alpha(\mu, -1) = \mu \\ \Phi^{\alpha^{k+1}}(^*\widehat{\Phi}^\alpha(\mu, k+1)) = \ ^*\widehat{\Phi}^\alpha(\mu, k) \end{cases} \tag{5}$$

is called **discrete time** α−**anti-semi-orbit of** μ *and the function* $^*\Phi^\rho : \mathbf{B}^n \times \mathbf{R} \to \mathbf{B}^n$ *that satisfies* $\forall \mu \in \mathbf{B}^n, \forall t \in \mathbf{R}$,

$$^*\Phi^\rho(\mu, t) = \ ^*\widehat{\Phi}^\alpha(\mu, -1)\chi_{(-\infty, t_0)}(t) \oplus \ ^*\widehat{\Phi}^\alpha(\mu, 0)\chi_{[t_0, t_1)}(t) \oplus$$
$$\oplus \ ^*\widehat{\Phi}^\alpha(\mu, 1)\chi_{[t_1, t_2)}(t) \oplus \ldots \oplus \ ^*\widehat{\Phi}^\alpha(\mu, k)\chi_{[t_k, t_{k+1})}(t) \oplus \ldots \tag{6}$$

is called **continuous time** ρ−**anti-semi-orbit of** μ.

Remark 3. *We compare the semi-orbits and the anti-semi-orbits now and see that they run both from the past to the future, but the cause-effect relation is different: in* $\widehat{\Phi}^\alpha, \Phi^\rho$ *the cause is in the past and the effect is in the future, while in* $^*\widehat{\Phi}^\alpha, ^*\Phi^\rho$ *the cause is in the future and the effect is in the past.*

Definition 5. *The* **discrete time** *and the* **continuous time** *universal anti-semi-regular autonomous asynchronous systems associated to* Φ *are defined by*

$$^*\widehat{\Xi}_\Phi = \{^*\widehat{\Phi}^\alpha(\mu, \cdot) | \mu \in \mathbf{B}^n, \alpha \in \overline{\Pi}_n\}$$
$$^*\Xi_\Phi = \{^*\Phi^\rho(\mu, \cdot) | \mu \in \mathbf{B}^n, \rho \in \overline{P}_n\}$$

Example 2. *In Figure 3 we have drawn at (a) the circuit and at (b) the state portrait of* $\Psi : \mathbf{B}^2 \to \mathbf{B}^2$, $\forall(\mu_1, \mu_2) \in \mathbf{B}^2, \Psi(\mu_1, \mu_2) = (1, 0)$ *for which*

$$\overline{\Xi}_\Psi = \{(\mu_1, \mu_2)\chi_{(-\infty, t_0)} \oplus (\mu_1 \cup \lambda_1, \mu_2\lambda_2)\chi_{[t_0, t_1)} \oplus$$
$$\oplus (\mu_1 \cup \lambda_1 \cup \nu_1, \mu_2\lambda_2\nu_2)\chi_{[t_1, \infty)} | \mu, \lambda, \nu \in \mathbf{B}^2, t_0, t_1 \in \mathbf{R}, t_0 < t_1\}$$

The arrows in Figures 2(c) and 3(b) are the same, but with a different sense and we note that $\overline{\Xi}_\Psi = \ ^*\Xi_\Phi$, *where* Φ *is the one from Example 1.*

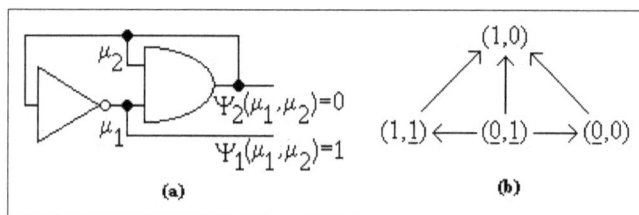

Figure 3. The semi-regular system $\overline{\Xi}_\Psi$ from Example 2.

4. Isomorphisms and Anti-Isomorphisms

Definition 6. *Let* $g : \mathbf{B}^n \to \mathbf{B}^n$. *It defines the functions* $\widehat{g} : \overline{\Pi}_n \to \overline{\Pi}_n, \forall \alpha \in \overline{\Pi}_n, \forall k \in \mathbf{N}$,

$$\widehat{g}(\alpha)(k) = g(\alpha^k)$$

$$\widetilde{g} : \overline{P}_n \to \overline{P}_n, \forall \rho \in \overline{P}_n, \forall t \in \mathbf{R},$$

$$\widetilde{g}(\rho)(t) = \begin{cases} (0,\ldots,0), & \text{if } \rho(t) = (0,\ldots,0) \\ g(\rho(t)), & \text{otherwise} \end{cases}$$

and $g : (\mathbf{B}^n)^{\mathbf{R}} \to (\mathbf{B}^n)^{\mathbf{R}}, \forall x \in (\mathbf{B}^n)^{\mathbf{R}}, \forall t \in \mathbf{R},$

$$g(x)(t) = g(x(t))$$

Theorem 1. *Let* $\Phi, \Psi, g, g' : \mathbf{B}^n \to \mathbf{B}^n$. *The following statements are equivalent:*

(a) $\forall v \in \mathbf{B}^n$, *the diagram*

$$
\begin{array}{ccc}
\mathbf{B}^n & \overset{\Phi^v}{\to} & \mathbf{B}^n \\
g \downarrow & & \downarrow g \\
\mathbf{B}^n & \overset{\Psi^{g'(v)}}{\to} & \mathbf{B}^n
\end{array}
$$

is commutative;

(b) $\forall \mu \in \mathbf{B}^n, \forall \alpha \in \overline{\Pi}_n, \forall k \in \mathbf{N}_-,$

$$g(\widehat{\Phi}^\alpha(\mu, k)) = \widehat{\Psi}^{\widetilde{g'}(\alpha)}(g(\mu), k)$$

(c) $\forall \mu \in \mathbf{B}^n,$

$$g(\mu) = \Psi^{g'(0,\ldots,0)}(g(\mu))$$

and $\forall \mu \in \mathbf{B}^n, \forall \rho \in \overline{P}_n, \forall t \in \mathbf{R},$

$$g(\Phi^\rho(\mu, t)) = \Psi^{\widetilde{g'}(\rho)}(g(\mu), t)$$

Proof. (a)$\Longrightarrow$(b): We fix arbitrarily $\mu \in \mathbf{B}^n$, $\alpha \in \overline{\Pi}_n$ and we use the induction on $k \geq -1$. For $k = -1$, (b) becomes $g(\mu) = g(\mu)$, thus we suppose that it is true for k and we prove it for $k + 1$:

$$g(\widehat{\Phi}^\alpha(\mu, k+1)) = g(\Phi^{\alpha^{k+1}}(\widehat{\Phi}^\alpha(\mu, k))) = \Psi^{g'(\alpha^{k+1})}(g(\widehat{\Phi}^\alpha(\mu, k)))$$
$$= \Psi^{g'(\alpha^{k+1})}(\widehat{\Psi}^{\widetilde{g'}(\alpha)}(g(\mu), k)) = \widehat{\Psi}^{\widetilde{g'}(\alpha)}(g(\mu), k+1)$$

(b)$\Longrightarrow$(c): The first statement results from (b) if we take $\alpha^0 = (0,\ldots,0)$ and $k = 0$. In order to prove the second statement, let $\mu \in \mathbf{B}^n$ and $\rho \in \overline{P}_n$ be arbitrary, thus Equation (2) holds with $(t_k) \in Seq, \rho(t_0),\ldots,\rho(t_k),\ldots \in \overline{\Pi}_n$. If $\forall t \in \mathbf{R}, \rho(t) = (0,\ldots,0)$ the statement to prove takes the form $g(\mu) = g(\mu)$ so that we can suppose now that a finite or an infinite number of $\rho(t_k)$ are $\neq (0,\ldots,0)$. In the case $\forall k \in \mathbf{N}, \rho(t_k) \neq (0,\ldots,0)$ that does not restrict the generality of the proof, we have that

$$\widetilde{g'}(\rho)(t) = g'(\rho(t_0))\chi_{\{t_0\}}(t) \oplus \ldots \oplus g'(\rho(t_k))\chi_{\{t_k\}}(t) \oplus \ldots \tag{7}$$

is an element of $\overline{P}_n$ and

$$g(\Phi^\rho(\mu, t)) = g(\mu\chi_{(-\infty,t_0)}(t) \oplus \widehat{\Phi}^\alpha(\mu, 0)\chi_{[t_0,t_1)}(t) \oplus \ldots \oplus \widehat{\Phi}^\alpha(\mu, k)\chi_{[t_k,t_{k+1})}(t) \oplus \ldots)$$
$$= g(\mu)\chi_{(-\infty,t_0)}(t) \oplus g(\widehat{\Phi}^\alpha(\mu, 0))\chi_{[t_0,t_1)}(t) \oplus \ldots \oplus g(\widehat{\Phi}^\alpha(\mu, k))\chi_{[t_k,t_{k+1})}(t) \oplus \ldots$$
$$= g(\mu)\chi_{(-\infty,t_0)}(t) \oplus \widehat{\Psi}^{\widetilde{g'}(\alpha)}(g(\mu), 0)\chi_{[t_0,t_1)}(t) \oplus \ldots \oplus \widehat{\Psi}^{\widetilde{g'}(\alpha)}(g(\mu), k)\chi_{[t_k,t_{k+1})}(t) \oplus \ldots$$
$$= \Psi^{\widetilde{g'}(\rho)}(g(\mu), t)$$

(c)$\Longrightarrow$(a): Let $v, \mu \in \mathbf{B}^n$ be arbitrary and fixed and we consider $\rho \in \overline{P}_n$ given by Equation (2), with $(t_k) \in Seq$ fixed, $\rho(t_0) = v$ and $\forall k \geq 1, \rho(t_k) \neq (0, \ldots, 0)$. We have

$$g(\Phi^\rho(\mu, t)) = g(\mu \chi_{(-\infty, t_0)}(t) \oplus \Phi^v(\mu) \chi_{[t_0, t_1)}(t) \oplus \widehat{\Phi}^\alpha(\mu, 1) \chi_{[t_1, t_2)}(t) \oplus \ldots)$$

$$= g(\mu) \chi_{(-\infty, t_0)}(t) \oplus g(\Phi^v(\mu)) \chi_{[t_0, t_1)}(t) \oplus g(\widehat{\Phi}^\alpha(\mu, 1)) \chi_{[t_1, t_2)}(t) \oplus \ldots \tag{8}$$

Case (i) $v = (0, \ldots, 0)$, the commutativity of the diagram is equivalent with the first statement of (c).

Case(ii) $v \neq (0, \ldots, 0)$,

$$\widetilde{g}'(\rho)(t) = g'(\rho(t))$$

$$= g'(v) \chi_{\{t_0\}}(t) \oplus g'(\rho(t_1)) \chi_{\{t_1\}}(t) \oplus \ldots$$

$$\Psi^{\widetilde{g}'(\rho)}(g(\mu), t) = g(\mu) \chi_{(-\infty, t_0)}(t) \oplus \Psi^{g'(v)}(g(\mu)) \chi_{[t_0, t_1)}(t) \oplus \widehat{\Psi}^{\widehat{g}'(\alpha)}(g(\mu), 1) \chi_{[t_1, t_2)}(t) \oplus \ldots$$

and from Equation (8), for $t \in [t_0, t_1)$, we obtain

$$g(\Phi^v(\mu)) = \Psi^{g'(v)}(g(\mu))$$

$\square$

Definition 7. *We consider the functions* $\Phi, \Psi : \mathbf{B}^n \to \mathbf{B}^n$. *If* $g, g' : \mathbf{B}^n \to \mathbf{B}^n$ *bijective exist such that one of the equivalent properties (a), (b) or (c) from Theorem 1 is satisfied, then we say that the couple* (g, g') *defines an* **isomorphism** *from* $\widehat{\Xi}_\Phi$ *to* $\widehat{\Xi}_\Psi$, *or from* Ξ_Φ *to* Ξ_Ψ, *or from* Φ *to* Ψ. *We use the notation* $\overline{Iso}(\Phi, \Psi)$ *for the set of these couples and we also denote with* $\overline{Aut}(\Phi) = \overline{Iso}(\Phi, \Phi)$ *the set of the* **automorphisms** *of* $\widehat{\Xi}_\Phi$, Ξ_Φ, *or* Φ.

Theorem 2. *For* $\Phi, \Psi, g, g' : \mathbf{B}^n \to \mathbf{B}^n$, *the following statements are equivalent:*

(a) $\forall v \in \mathbf{B}^n$, *the diagram is commutative;*

$$
\begin{array}{ccc}
\mathbf{B}^n & \overset{\Phi^v}{\to} & \mathbf{B}^n \\
g \downarrow & & \downarrow g \\
\mathbf{B}^n & \underset{\Psi^{g'(v)}}{\leftarrow} & \mathbf{B}^n
\end{array}
$$

(b) $\forall \mu \in \mathbf{B}^n, \forall \alpha \in \overline{\Pi}_n, \forall k \in \mathbf{N}_-$,

$$g(\mu) = {}^*\widehat{\Psi}^{\widehat{g}'(\alpha)}(g(\widehat{\Phi}^\alpha(\mu, k)), k)$$

(c) $\forall \mu \in \mathbf{B}^n$,

$$g(\mu) = \Psi^{g'(0, \ldots, 0)}(g(\mu))$$

and $\forall \mu \in \mathbf{B}^n, \forall \rho \in \overline{P}_n, \forall t \in \mathbf{R}$,

$$g(\mu) = {}^*\Psi^{\widetilde{g}'(\rho)}(g(\Phi^\rho(\mu, t)), t)$$

Proof. (a)$\Longrightarrow$(b): We fix arbitrarily $\mu \in \mathbf{B}^n$, $\alpha \in \overline{\Pi}_n$ and we use the induction on $k \geq -1$. In the case $k = -1$ the equality to be proved is satisfied

$$g(\mu) = g(\widehat{\Phi}^\alpha(\mu, -1)) = \widehat{\Psi}^{\widehat{g}'(\alpha)}(g(\widehat{\Phi}^\alpha(\mu, -1)), -1)$$

thus we presume that the statement is true for k and we prove it for $k + 1$. We have:

$$g(\mu) = {}^*\widehat{\Psi}^{\widehat{g}'(\alpha)}(g(\widehat{\Phi}^\alpha(\mu,k)),k)$$

$$= {}^*\widehat{\Psi}^{\widehat{g}'(\alpha)}(\Psi^{g'(\alpha^{k+1})}(g(\Phi^{\alpha^{k+1}}(\widehat{\Phi}^\alpha(\mu,k)))),k)$$

$$= {}^*\widehat{\Psi}^{\widehat{g}'(\alpha)}(g(\widehat{\Phi}^\alpha(\mu,k+1)),k+1)$$

The proof is similar with the proof of Theorem 1. $\square$

Definition 8. *Let* $\Phi, \Psi : \mathbf{B}^n \to \mathbf{B}^n$. *If* $g, g' : \mathbf{B}^n \to \mathbf{B}^n$ *bijective exist such that one of the equivalent properties (a), (b) or (c) from Theorem 2 is fulfilled, we say that the couple* (g, g') *defines an* **anti-isomorphism** *from* $\widehat{\Xi}_\Phi$ *to* ${}^*\widehat{\Xi}_\Psi$, *or from* $\overline{\Xi}_\Phi$ *to* ${}^*\overline{\Xi}_\Psi$, *or from* Φ *to* Ψ. *We use the notation* ${}^*\overline{Iso}(\Phi, \Psi)$ *for these couples and we also denote with* ${}^*\overline{Aut}(\Phi) = {}^*\overline{Iso}(\Phi, \Phi)$ *the set of the* **anti-automorphisms** *of* $\widehat{\Xi}_\Phi$, $\overline{\Xi}_\Phi$ *or* Φ.

5. Symmetry and Anti-Symmetry

Remark 4. *The fact that* $(1_{\mathbf{B}^n}, 1_{\mathbf{B}^n}) \in \overline{Aut}(\Phi)$ *implies* $\overline{Aut}(\Phi) \neq \emptyset$, *but all of* $\overline{Iso}(\Phi, \Psi)$, ${}^*\overline{Iso}(\Phi, \Psi)$ *and* ${}^*\overline{Aut}(\Phi)$ *may be empty.*

Definition 9. *Let* $\Phi, \Psi : \mathbf{B}^n \to \mathbf{B}^n$, $\Phi \neq \Psi$. *If* $\overline{Iso}(\Phi, \Psi) \neq \emptyset$, *then* $\widehat{\Xi}_\Phi$, $\widehat{\Xi}_\Psi$; $\overline{\Xi}_\Phi, \overline{\Xi}_\Psi$; Φ, Ψ *are called* **symmetrical**, *or* **conjugated**; *if* ${}^*\overline{Iso}(\Phi, \Psi) \neq \emptyset$, *then* $\widehat{\Xi}_\Phi$, ${}^*\widehat{\Xi}_\Psi$; $\overline{\Xi}_\Phi, {}^*\overline{\Xi}_\Psi$; Φ, Ψ *are called* **anti-symmetrical**, *or* **anti-conjugated**.

If $card(\overline{Aut}(\Phi)) > 1$, *then* $\widehat{\Xi}_\Phi$, $\overline{\Xi}_\Phi$ *and* Φ *are called* **symmetrical** *and if* ${}^*\overline{Aut}(\Phi) \neq \emptyset$, *then* $\widehat{\Xi}_\Phi$, $\overline{\Xi}_\Phi$ *and* Φ *are called* **anti-symmetrical**.

Remark 5. *The symmetry of* Φ, Ψ *means that* $(g, g') \in \overline{Iso}(\Phi, \Psi)$ *maps the transfers* $\mu \to \Phi^\nu(\mu)$ *in transfers* $g(\mu) \to g(\Phi^\nu(\mu)) = \Psi^{g'(\nu)}(g(\mu))$; *the situation when* Φ *is symmetrical and* $(g, g') \in \overline{Aut}(\Phi)$ *is similar. Anti-symmetry may be understood as mirroring:* $(g, g') \in {}^*\overline{Iso}(\Phi, \Psi)$ *maps the transfers (or arrows)* $\mu \to \Phi^\nu(\mu)$ *in transfers* $g(\mu) \longleftarrow g(\Phi^\nu(\mu)) = \Psi^{g'(\nu)}(g(\mu))$ *and similarly for* $(g, g') \in {}^*\overline{Aut}(\Phi)$.

Theorem 3. *Let* $\Phi, \Psi : \mathbf{B}^n \to \mathbf{B}^n$.
    (a) If $(g, g') \in \overline{Iso}(\Phi, \Psi)$, *then* $(g^{-1}, g'^{-1}) \in \overline{Iso}(\Psi, \Phi)$.
    (b) If $(g, g') \in {}^*\overline{Iso}(\Phi, \Psi)$, *then* $(g^{-1}, g'^{-1}) \in {}^*\overline{Iso}(\Psi, \Phi)$.

Proof. (a): The hypothesis states that $\forall \nu \in \mathbf{B}^n$, the diagram

$$
\begin{array}{ccc}
\mathbf{B}^n & \xrightarrow{\Phi^\nu} & \mathbf{B}^n \\
g \downarrow & & \downarrow g \\
\mathbf{B}^n & \xrightarrow{\Psi^{g'(\nu)}} & \mathbf{B}^n
\end{array}
$$

commutes, with g, g' bijective. We fix arbitrarily $\nu \in \mathbf{B}^n, \mu \in \mathbf{B}^n$. We denote $\mu' = g(\mu), \nu' = g'(\nu)$ and we note that

$$g^{-1}(\Psi^{\nu'}(\mu')) = \Phi^{g'^{-1}(\nu')}(g^{-1}(\mu')) \tag{9}$$

As ν, μ were chosen arbitrarily and on the other hand, when ν runs in $\mathbf{B}^n$, ν' runs in $\mathbf{B}^n$ and when μ runs in $\mathbf{B}^n$, μ' runs in $\mathbf{B}^n$, we infer that Equation (9) is equivalent with the commutativity of the diagram

$$
\begin{array}{ccc}
\mathbf{B}^n & \xrightarrow{\Psi^{\nu'}} & \mathbf{B}^n \\
g^{-1} \downarrow & & \downarrow g^{-1} \\
\mathbf{B}^n & \xrightarrow{\Phi^{g'^{-1}(\nu')}} & \mathbf{B}^n
\end{array}
$$

for any $v' \in \mathbf{B}^n$. We have proved that $(g^{-1}, g'^{-1}) \in \overline{Iso}(\Psi, \Phi)$.

(b): By hypothesis $\forall v \in \mathbf{B}^n$, the diagram

$$
\begin{array}{ccc}
\mathbf{B}^n & \overset{\Phi^v}{\to} & \mathbf{B}^n \\
g \downarrow & & \downarrow g \\
\mathbf{B}^n & \underset{\Psi^{g'(v)}}{\leftarrow} & \mathbf{B}^n
\end{array}
$$

is commutative, g, g' bijective and we prove that $\forall v' \in \mathbf{B}^n$, the diagram

$$
\begin{array}{ccc}
\mathbf{B}^n & \overset{\Psi^{v'}}{\to} & \mathbf{B}^n \\
g^{-1} \downarrow & & \downarrow g^{-1} \\
\mathbf{B}^n & \underset{\Phi^{g'^{-1}(v')}}{\leftarrow} & \mathbf{B}^n
\end{array}
$$

is commutative. $\quad\square$

Theorem 4. $\overline{Aut}(\Phi)$ *is a group relative to the law:* $\forall (g, g') \in \overline{Aut}(\Phi)$, $\forall (h, h') \in \overline{Aut}(\Phi)$,

$$(h, h') \circ (g, g') = (h \circ g, h' \circ g')$$

Proof. The fact that $\forall (g, g') \in \overline{Aut}(\Phi)$, $\forall (h, h') \in \overline{Aut}(\Phi)$, $(h \circ g, h' \circ g') \in \overline{Aut}(\Phi)$ is proved like this:
$\forall v \in \mathbf{B}^n$,

$$
\begin{aligned}
(h \circ g) \circ \Phi^v &= h \circ (g \circ \Phi^v) = h \circ (\Phi^{g'(v)} \circ g) = (h \circ \Phi^{g'(v)}) \circ g \\
&= (\Phi^{h'(g'(v))} \circ h) \circ g = \Phi^{(h' \circ g')(v)} \circ (h \circ g)
\end{aligned}
$$

the fact that $(1_{\mathbf{B}^n}, 1_{\mathbf{B}^n}) \in \overline{Aut}(\Phi)$ was mentioned before; and the fact that $\forall (g, g') \in \overline{Aut}(\Phi)$, $(g^{-1}, g'^{-1}) \in \overline{Aut}(\Phi)$ was shown at Theorem 3(a). $\quad\square$

Definition 10. *Any subgroup* $G \subset \overline{Aut}(\Phi)$ *with* $card(G) > 1$ *is called a **group of symmetry** of* $\widehat{\Xi}_\Phi$, *of* $\overline{\Xi}_\Phi$ *or of* Φ.

6. Examples

Example 3. $\Phi, \Psi : \mathbf{B}^2 \to \mathbf{B}^2$ *are given by, see Figure 4*

$$
\begin{aligned}
\forall (\mu_1, \mu_2) \in \mathbf{B}^2, \Phi(\mu_1, \mu_2) &= (\mu_1 \oplus \mu_2, \overline{\mu_2}) \\
\forall (\mu_1, \mu_2) \in \mathbf{B}^2, \Psi(\mu_1, \mu_2) &= (\overline{\mu_1}, \overline{\mu_1}\,\overline{\mu_2} \cup \mu_1 \mu_2)
\end{aligned}
$$

and the bijections $g, g' : \mathbf{B}^2 \to \mathbf{B}^2$ *are* $\forall (\mu_1, \mu_2) \in \mathbf{B}^2$,

$$
\begin{aligned}
g(\mu_1, \mu_2) &= (\overline{\mu_2}, \overline{\mu_1}) \\
g'(\mu_1, \mu_2) &= (\mu_2, \mu_1)
\end{aligned}
$$

(in order to understand the choice of g'*, to be remarked in Figure 4 the positions of the underlined coordinates for* Φ *and* Ψ*).* Φ *and* Ψ *are conjugated.*

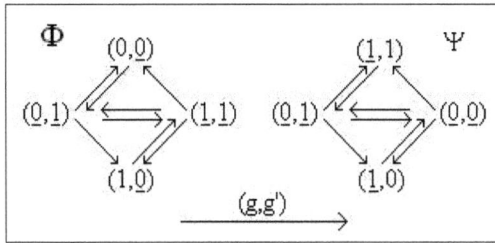

Figure 4. Symmetrical systems, Example 3.

Example 4. *The system from Figure 5 is symmetrical and a group of symmetry is generated by the couples* $(g, 1_{\mathbf{B}^3}), (u, 1_{\mathbf{B}^3}), (v, 1_{\mathbf{B}^3})$, *see Equation (10); g, u, v are transpositions that permute the isolated fixed points* $(1,0,0), (1,0,1), (1,1,1)$.

(μ_1, μ_2, μ_3)	$1_{\mathbf{B}^3}$	g	u	v	
$(0,0,0)$	$(0,0,0)$	$(0,0,0)$	$(0,0,0)$	$(0,0,0)$	
$(0,0,1)$	$(0,0,1)$	$(0,0,1)$	$(0,0,1)$	$(0,0,1)$	
$(0,1,0)$	$(0,1,0)$	$(0,1,0)$	$(0,1,0)$	$(0,1,0)$	
$(0,1,1)$	$(0,1,1)$	$(0,1,1)$	$(0,1,1)$	$(0,1,1)$	(10)
$(1,0,0)$	$(1,0,0)$	$(1,0,0)$	$(1,0,1)$	$(1,1,1)$	
$(1,0,1)$	$(1,0,1)$	$(1,1,1)$	$(1,0,0)$	$(1,0,1)$	
$(1,1,0)$	$(1,1,0)$	$(1,1,0)$	$(1,1,0)$	$(1,1,0)$	
$(1,1,1)$	$(1,1,1)$	$(1,0,1)$	$(1,1,1)$	$(1,0,0)$	

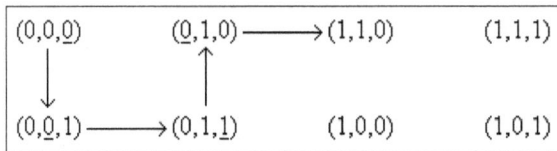

Figure 5. Symmetrical system, Example 4.

Example 5. *The function* $\Phi : \mathbf{B}^2 \to \mathbf{B}^2$ *defined by* $\forall \mu \in \mathbf{B}^2, \Phi(\mu_1, \mu_2) = (\overline{\mu_1}, \overline{\mu_2})$ *fulfills for* $v \in \mathbf{B}^2$:

$$\Phi^v(\mu_1, \mu_2) = (\overline{v_1}\mu_1 \oplus v_1\overline{\mu_1}, \overline{v_2}\mu_2 \oplus v_2\overline{\mu_2})$$

$$(\Phi^v \circ \Phi^v)(\mu_1, \mu_2) = (\overline{v_1}\Phi_1^{v_1}(\mu_1, \mu_2) \oplus v_1\overline{\Phi_1^{v_1}(\mu_1, \mu_2)}, \overline{v_2}\Phi_2^{v_2}(\mu_1, \mu_2) \oplus v_2\overline{\Phi_2^{v_2}(\mu_1, \mu_2)})$$

$$= (\overline{v_1}(\overline{v_1}\mu_1 \oplus v_1\overline{\mu_1}) \oplus v_1(\overline{v_1}\mu_1 \oplus v_1\overline{\mu_1} \oplus 1), \overline{v_2}(\overline{v_2}\mu_2 \oplus v_2\overline{\mu_2}) \oplus v_2(\overline{v_2}\mu_2 \oplus v_2\overline{\mu_2} \oplus 1))$$

$$= ((v_1 \oplus 1)\mu_1 \oplus v_1(\mu_1 \oplus 1) \oplus v_1, (v_2 \oplus 1)\mu_2 \oplus v_2(\mu_2 \oplus 1) \oplus v_2)$$

$$= (v_1\mu_1 \oplus \mu_1 \oplus v_1\mu_1 \oplus v_1 \oplus v_1, v_2\mu_2 \oplus \mu_2 \oplus v_2\mu_2 \oplus v_2 \oplus v_2)$$

$$= (\mu_1, \mu_2)$$

thus $(1_{\mathbf{B}^2}, 1_{\mathbf{B}^2}) \in {}^*\overline{Aut}(\Phi)$ *and* Φ *is anti-symmetrical. The state portrait of* Φ *was drawn in Figure 1(c).*

Notation 6. *Let* $\sigma : \{1, \ldots, n\} \to \{1, \ldots, n\}$ *be a bijection. We use the notation* $\pi_\sigma : \mathbf{B}^n \to \mathbf{B}^n$ *for the bijection given by* $\forall \mu \in \mathbf{B}^n$,

$$\pi_\sigma(\mu_1, \ldots, \mu_n) = (\mu_{\sigma(1)}, \ldots, \mu_{\sigma(n)})$$

Definition 11. *Any of* $\widehat{\widehat{\Xi}}_\Phi, \overline{\Xi}_\Phi$ *and* $\Phi : \mathbf{B}^n \to \mathbf{B}^n$ *is called* **symmetrical relative to the coordinates** *if the bijection* σ *exists,* $\sigma \neq 1_{\{1, \ldots, n\}}$ *such that* $(\pi_\sigma, \pi_\sigma) \in \overline{Aut}(\Phi)$.

71

Example 6. *We consider the function* $\Phi : \mathbf{B}^3 \to \mathbf{B}^3$ *defined by* $\forall \mu \in \mathbf{B}^3$, $\Phi(\mu_1, \mu_2, \mu_3) = (\mu_2 \mu_3 \oplus \mu_1 \oplus \mu_2, \mu_1 \mu_3 \oplus \mu_2 \oplus \mu_3, \mu_1 \mu_2 \oplus \mu_1 \oplus \mu_3)$ *and the permutation* $\sigma : \{1,2,3\} \to \{1,2,3\}$, $\sigma = \begin{pmatrix} 1 & 2 & 3 \\ \sigma(1) & \sigma(2) & \sigma(3) \end{pmatrix} = \begin{pmatrix} 1 & 2 & 3 \\ 3 & 1 & 2 \end{pmatrix}$. *A group of symmetry of* $\overline{\Xi}_\Phi$ *is represented by* $G = \{(1_{\mathbf{B}^3}, 1_{\mathbf{B}^3}),$ $(\pi_\sigma, \pi_\sigma), (\pi_{\sigma \circ \sigma}, \pi_{\sigma \circ \sigma})\}$. *We have given in Figure 6 the state portrait of* Φ.

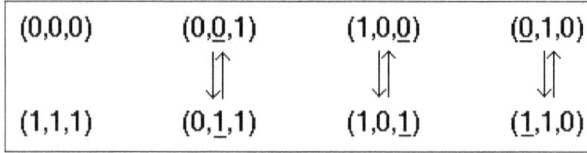

(0,0,0)	(0,0,1)	(1,0,0)	(0,1,0)
⇕	⇕	⇕	⇕
(1,1,1)	(0,1,1)	(1,0,1)	(1,1,0)

Figure 6. System that is symmetrical relative to the coordinates, Example 6.

Notation 7. *For* $\lambda \in \mathbf{B}^n$, *we denote by* $\theta^\lambda : \mathbf{B}^n \to \mathbf{B}^n$ *the translation of vector* $\lambda : \forall \mu \in \mathbf{B}^n$,

$$\theta^\lambda(\mu) = \mu \oplus \lambda$$

Definition 12. *If* $(\theta^\lambda, g') \in \overline{Aut}(\Phi)$ *holds for some* $(\theta^\lambda, g') \neq (1_{\mathbf{B}^n}, 1_{\mathbf{B}^n})$, *we say that any of* $\widehat{\Xi}_\Phi$, $\overline{\Xi}_\Phi$ *and* Φ *is* **symmetrical relative to translations**.

Example 7. *In Figure 7*

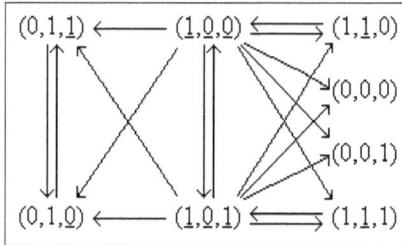

Figure 7. Φ has the automorphism $(\theta^{(0,0,1)}, 1_{\mathbf{B}^3})$, Example 7.

we have the system with Φ *given by Equation (11)*

$$
\begin{array}{cc}
(\mu_1, \mu_2, \mu_3) & \Phi \\
(0,0,0) & (0,0,0) \\
(0,0,1) & (0,0,1) \\
(0,1,0) & (0,1,1) \\
(0,1,1) & (0,1,0) \\
(1,0,0) & (0,1,1) \\
(1,0,1) & (0,1,0) \\
(1,1,0) & (1,0,0) \\
(1,1,1) & (1,0,1)
\end{array}
\tag{11}
$$

and $(\theta^{(0,0,1)}, 1_{\mathbf{B}^3}) \in \overline{Aut}(\Phi)$, *as resulting from the state portrait.*

Example 8. *In Equation (12) we have a function* $\Phi : \mathbf{B}^2 \to \mathbf{B}^2$ *for which four functions* $g'_1, g'_2, g'_3, g'_4 :$ $\mathbf{B}^2 \to \mathbf{B}^2$ *exist:*

$$
\begin{array}{cccccc}
(\mu_1, \mu_2) & \Phi & g_1' & g_2' & g_3' & g_4' \\
(0,0) & (0,0) & (0,0) & (1,0) & (0,0) & (1,0) \\
(0,1) & (0,1) & (0,1) & (0,1) & (1,1) & (1,1) \\
(1,0) & (1,1) & (1,0) & (0,0) & (1,0) & (0,0) \\
(1,1) & (1,0) & (1,1) & (1,1) & (0,1) & (0,1)
\end{array}
\tag{12}
$$

such that $(1_{\mathbf{B}^2}, g_1'), (1_{\mathbf{B}^2}, g_2'), (1_{\mathbf{B}^2}, g_3'), (1_{\mathbf{B}^2}, g_4') \in \overline{Aut}(\Phi)$. *The state portrait of* Φ *is drawn Figure 8.*

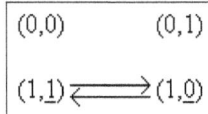

Figure 8. Φ is symmetrical relative to translations with $(0,0)$, Example 8.

Example 9. *The system from Figure 9 is symmetrical relative to translations, since it has the group of symmetry* $G = \{(1_{\mathbf{B}^2}, 1_{\mathbf{B}^2}), (\theta^{(1,1)}, 1_{\mathbf{B}^2})\}$. Φ *is self-dual* $\Phi = \Phi^*$, *where the dual* Φ^* *of* Φ *is defined by* $\Phi^*(\mu) = \overline{\Phi(\overline{\mu})}$.

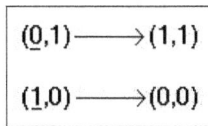

Figure 9. Function Φ that is self dual, $(\theta^{(1,1)}, 1_{\mathbf{B}^2}) \in \overline{Aut}(\Phi)$, Example 9.

Example 10. *Functions* $\Phi : \mathbf{B}^2 \to \mathbf{B}^2$ *exist, see Figure 10, that are symmetrical relative to the translations with any* $\lambda \in \mathbf{B}^2$, *thus their group of symmetry is* $G = \{(1_{\mathbf{B}^2}, 1_{\mathbf{B}^2}), (\theta^{(0,1)}, 1_{\mathbf{B}^2}), (\theta^{(1,0)}, 1_{\mathbf{B}^2}), (\theta^{(1,1)}, 1_{\mathbf{B}^2})\}$. *The fact that* $(\theta^{(1,1)}, 1_{\mathbf{B}^2}) \in G$ *shows that all these functions:* $\Phi(\mu) = (\mu_1, \mu_2)$, $\Phi(\mu) = (\mu_1, \overline{\mu_2})$, $\Phi(\mu) = (\overline{\mu_1}, \mu_2)$, $\Phi(\mu) = (\overline{\mu_1}, \overline{\mu_2})$ *are self-dual,* $\Phi = \Phi^*$.

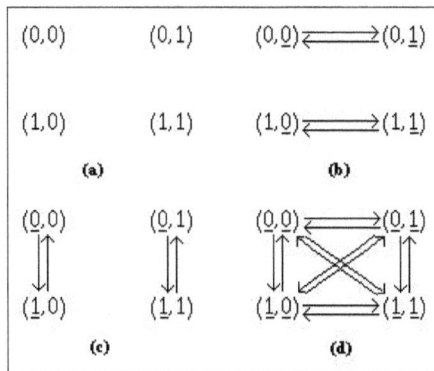

Figure 10. Functions Φ that are self dual, $(\theta^{(1,1)}, 1_{\mathbf{B}^2}) \in \overline{Aut}(\Phi)$, Example 10.

Example 11. *The group of symmetry G of the system from Figure 11 has four elements given by Equation (13)*

$$
\begin{array}{ccccc}
(\mu_1,\mu_2) & 1_{\mathbf{B}^2} & g & h & \theta^{(1,1)} \\
(0,0) & (0,0) & (0,1) & (1,0) & (1,1) \\
(0,1) & (0,1) & (1,1) & (0,0) & (1,0) \\
(1,0) & (1,0) & (0,0) & (1,1) & (0,1) \\
(1,1) & (1,1) & (1,0) & (0,1) & (0,0)
\end{array}
\tag{13}
$$

and we remark that $h = g^{-1}, \theta^{(1,1)} = (\theta^{(1,1)})^{-1}$ *hold, see also Equation (14).*

$$
\begin{array}{ccccc}
(\nu_1,\nu_2) & (1_{\mathbf{B}^2})' & g' & h' & (\theta^{(1,1)})' \\
(0,0) & (0,0) & (0,0) & (0,0) & (0,0) \\
(0,1) & (0,1) & (1,0) & (1,0) & (0,1) \\
(1,0) & (1,0) & (0,1) & (0,1) & (1,0) \\
(1,1) & (1,1) & (1,1) & (1,1) & (1,1)
\end{array}
\tag{14}
$$

We have $\Phi = \Phi^*$ *like previously.*

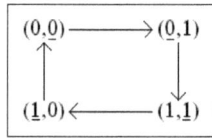

Figure 11. Symmetry including symmetry relative to translations, Example 11.

7. Conclusions

The paper defines the universal semi-regular autonomous asynchronous systems and the universal anti-semi-regular autonomous asynchronous systems. It also defines and characterizes the isomorphisms (automorphisms) and the anti-isomorphisms (anti-automorphisms) of these systems. Symmetry is defined as the existence of such isomorphisms (automorphisms), while anti-symmetry is defined as the existence of such anti-isomorphisms (anti-automorphisms). Many examples are given. A by-pass product in this study is anti-symmetry, which is related with systems having the cause in the future and the effect in the past. Another by-pass product consists in semi-regularity, since important examples of isomorphisms (automorphisms) are of semi-regular systems only and do not keep progressiveness and regularity [2,3].

References

1. Kuznetsov,Y.A. *Elements of Applied Bifurcation Theory, 2nd ed.*; Springer: Berlin, Germany, 1997.
2. Vlad, S.E. Boolean dynamical systems. *ROMAI J.* **2007**, *3*, 277–324.
3. Vlad, S.E.Universal regular autonomous asynchronous systems: Fixed points, equivalencies and dynamical bifurcations. *ROMAI J.* **2009**, *5*, 131–154.

Chapter 2:
Applications

symmetry

MDPI

Article

The Smallest Valid Extension-Based Efficient, Rare Graph Pattern Mining, Considering Length-Decreasing Support Constraints and Symmetry Characteristics of Graphs

Unil Yun [1,*], Gangin Lee [1] and Chul-Hong Kim [2]

[1] Department of Computer Engineering, Sejong University, 209, Neungdong-ro, Gwangjin-gu, Seoul 05006, Korea; ganginlee@sju.ac.kr
[2] Automotive IT Convergence Research Lab, Electronics and Telecommunications Research Institute, 218, Gajeong-ro, Yuseong-gu, Daejeon 34129, Korea; kch@etri.re.kr
* Correspondence: yunei@sejong.ac.kr; Tel.: +82-2-3408-2902; Fax: +82-2-3408-4321

Academic Editor: Angel Garrido
Received: 15 December 2015; Accepted: 26 April 2016; Published: 6 May 2016

Abstract: Frequent graph mining has been proposed to find interesting patterns (*i.e.,* frequent sub-graphs) from databases composed of graph transaction data, which can effectively express complex and large data in the real world. In addition, various applications for graph mining have been suggested. Traditional graph pattern mining methods use a single minimum support threshold factor in order to check whether or not mined patterns are interesting. However, it is not a sufficient factor that can consider valuable characteristics of graphs such as graph sizes and features of graph elements. That is, previous methods cannot consider such important characteristics in their mining operations since they only use a fixed minimum support threshold in the mining process. For this reason, in this paper, we propose a novel graph mining algorithm that can consider various multiple, minimum support constraints according to the types of graph elements and changeable minimum support conditions, depending on lengths of graph patterns. In addition, the proposed algorithm performs in mining operations more efficiently because it can minimize duplicated operations and computational overheads by considering symmetry features of graphs. Experimental results provided in this paper demonstrate that the proposed algorithm outperforms previous mining approaches in terms of pattern generation, runtime and memory usage.

Keywords: data mining; graph mining; graph symmetry; length-decreasing support; rare item problem

1. Introduction

Since the concept of data mining was proposed to find useful knowledge or information hidden in complicated large-scale data (also called big data), various approaches and applications for data mining have been researched [1–6]. After that, frequent pattern mining was proposed to find useful, hidden pattern information from such data and various mining techniques and applications have been developed [7–12]. Frequent graph mining approaches [13–17] have been proposed to satisfy the needs of users wanting to obtain mining results from large and complex graph data in the real world. It is hard to express recent data as simple structures, such as itemsets because of their complicated and multidimensional features. However, this data can easily be expressed in graph form since almost all data can be described as such. Previous traditional frequent pattern mining methods faced limitations that did not deal with such complicated databases because they were algorithms, focusing on processing item-based simple databases. For this reason, the concept of frequent graph pattern mining was suggested and studies for frequent graph mining have been increased dramatically. Since recent real world databases have become larger and more complicated, it is essential to deal with the modern data, rather than old. Due to the usefulness of graph data, a variety of relevant graph theories and applications have been studied [18–20]. Moreover, since graph pattern mining can draw useful data analysis results for various complicated graph databases, a variety of graph mining applications have been developed such as discovering objects based on graph mining [21], finding combinatorial splicing regulatory elements using graph mining [22], exploiting document information contents on graph mining [23], detecting intelligent malware based on graph mining [24] and analyzing market data using graph pattern mining [25].

However, previous frequent graph mining researches applied only support information for generated graph patterns (or sub-graphs) but did not consider the other valuable factors that could utilize various characteristics of graphs such as graph sizes and features of graph elements. In frequent graph mining, extracted sub-graphs have the following characteristics. Small sub-graphs with a few elements (vertices and edges) tend to be interesting if their supports are relatively high, while large sub-graphs with a large number of the elements can be interesting, although they have relatively low supports. However, previous graph mining methods cannot apply the above characteristics to mining processes since they use only one minimum support threshold, regardless of the graphs' sizes. Moreover, if we find large sub-graphs having many elements and low supports through existing methods, we have no choice but to lower a minimum support threshold more than required, causing generations of meaningless sub-graph patterns. In addition, each element composing a graph pattern can have its own support feature. However, traditional methods always use a single threshold regardless of the element characteristics. Hence, they cannot effectively consider the rare item problem [26–28], which means that not only do items or patterns have large supports but also ones with small values can contain useful knowledge or information. Accordingly, traditional approaches may fail to find rare but valuable patterns depending on settings of the minimum support threshold. If we lower the threshold more than needed in order to extract such pattern results, an enormous number of useless patterns may also be mined.

In contrast to traditional pattern mining methods that deal with item-based simple databases, graph pattern mining need more complicated mining operations to discover graph patterns. Especially, in order to prevent duplicated graph patterns from being mined, graph pattern mining has to perform works for deciding graph isomorphism, which is also known as a *NP-hard problem* that can cause enormous computational overheads. However, we can effectively solve such problems by applying a pattern growth technique based on graph symmetry features into our mining process. The symmetry features have been used to improve mining efficiency of various approaches [29–31]

Motivated by the aforementioned issues, we propose an efficient algorithm for Smallest Valid Extension-based Rare Graph pattern Mining considering length-decreasing support constraints and symmetry characteristics of graphs (called *SVE-RGM*), where we also propose and apply techniques for improving graph mining efficiency: symmetry feature-based graph pattern growth, a smallest valid extension (*SVE*) method for graphs and a *SVE*-based pruning strategy. Through the proposed algorithm, we can obtain a set of *SVE*-based Rare Graph patterns, called *SRGs*. By using the graph symmetry features, we can prevent duplicated graph patterns from being generated and reduce computational overheads for useless operations. We can also improve mining efficiency of the proposed method by employing the *SVE*-based pre-pruning technique, which does not cause any pattern loss. Experimental results in this paper show that *SVE-RGM* outperforms state-of-the-art algorithms.

The remainder of this paper is organized as follows. In Section 2, we provide related work regarding graph mining and in Sections 3 and 4 details of the proposed algorithm, *SVE-RGM* and performance analysis results are described, respectively. In Section 5, discussion for the proposed method is introduced. Finally in Section 6, we conclude this paper.

2. Related Work

2.1. Frequent Pattern Mining and Frequent Graph Pattern Mining

Since the *Apriori* algorithm [7] was developed, various works regarding frequent pattern mining have been suggested. The main goal of frequent pattern mining is to find all of frequent patterns from databases. If a frequency (or support) of a given pattern is higher than or equal to a minimum support threshold set by a user, it is considered as a frequent pattern. In the frequent pattern mining area, there is an important factor, called the *anti-monotone* property or the *downward closure* property. It contributes to improving mining efficiency by preventing invalid patterns from being generated. This property guarantees the following relation: if a pattern is an infrequent one, all the super patterns created from the pattern are also infrequent ones. Meanwhile, in order to overcome the drawbacks of the *Apriori* algorithm such as generating an enormous number of useless candidate patterns and database scanning works, a tree-based algorithm, *FP-growth*, was devised [32]. This algorithm mines frequent patterns without any candidate pattern generation, employing its own tree structure, called *FP-tree*. In addition, its mining process does not require excessive database scans, it needs only two database scans.

As in frequent pattern mining methods, frequent graph mining has also advanced through a similar process. Early studies have been researched on the basis of BFS (Breadth First Search) and subsequent researches have been conducted on the basis of DFS (Depth First Search). In addition, graph mining is also applied to extract a variety of valid patterns such as weighted frequent sub-graphs [33], closed and maximal frequent ones [13,15,33], approximate frequent ones [16,17], and so on. Similarly to frequent pattern mining, the main purpose of frequent graph pattern mining is to search for all the graph patterns satisfying a minimum support threshold from complicated databases composed of graph data. One of the major differences between them is that graph pattern mining has to consider extra conditions such as vertices, edges and graph isomorphism, in comparison to traditional pattern mining which deals only with simple items. There are several well-known fundamental graph mining algorithms such as *Gaston*, *gSpan*, *FFSM*, etc., where the *Gaston* algorithm [34,35] is most suitable for comparing the proposed algorithm, *SVE-RGM*, since as a state-of-the-art algorithm, *Gaston*, has the fastest runtime performance among these algorithms. The algorithm extracts frequent sub-graphs more efficiently by dividing mining process into three parts: path, free tree and cyclic graph steps, as well as by performing appropriate operations according to each step. In addition, an additional data structure used in the algorithm, named *embedding list*, makes it faster to conduct mining operations. However, such fundamental graph mining algorithms have limitations that only consider a single minimum support condition, regardless of various graph characteristics such as element types and lengths of graph patterns.

2.2. Pattern Mining On Multiple Minimum Support Constraints

In order to solve the *rare item problem* in the frequent pattern mining area, researchers have proposed various pattern mining algorithms based on multiple minimum support constraints [26–28]. Since *MSApriori* [28], an initial algorithm based on the framework of *Apriori*, was proposed, various methods have been developed. *CFP-growth* [26] is a tree-based algorithm that follows the basic process of *FP-growth* and *CFP-growth++* [27] is an enhanced version of *CFP-growth*. Although the above approaches have found solutions of the *rare item problem* by applying multiple minimum support constraints, they are item-based traditional algorithms that cannot deal with various characteristics of complicated graph data.

To solve the above problem, *FGM-MMS* [36] and *WRG-Miner* [37] were proposed. They are methods that consider multiple minimum support constraints in graph pattern mining processes. In contrast to traditional graph pattern mining that uses a single minimum support threshold regardless of characteristics of elements composing graphs, they employ different minimum support threshold values for the elements in a given graph database in order to overcome the *rare item problem* [26–28] in their graph mining processes. Recall that meaningful patterns with low support values cannot be mined if a given minimum support threshold is high. Meanwhile an enormous number of invalid patterns have to be extracted, if the threshold value becomes lower to find such pattern results. However, such approaches cannot consider important length or size characteristics of mined graph patterns. On the other hand, since the proposed algorithm can set various minimum support thresholds according to the lengths or sizes of graph patterns, it can mine graph pattern results with more practically useful information.

2.3. Pattern Mining on Length-Decreasing Support Constraints

In the frequent pattern mining area, *LPMiner/SLPMiner* [38] is the first algorithm applying different multiple support constraints for each length of patterns. After that, advanced algorithms applying weight conditions, *WLPMiner* [39] and *WSLPMiner* [40] were suggested. *LPMiner* and *WLPMiner* find frequent and weighted frequent patterns composed of itemsets, respectively and *SLPMiner* and *WSLPMiner* discover sequential frequent and weighted sequential frequent ones. However, the above algorithms are only limited to the general frequent pattern mining area dealing with simple itemset-based databases.

FGM-LDSC [41] is an algorithm applying length-decreasing support constraints on graph mining environments. Recall that Small graph patterns having a few vertices and edges tend to be interesting if their supports are relatively high, while large ones having many vertices and edges can be interesting even though their supports are relatively low. *FGM-LDSC* solves the above problem by applying a different minimum support for each length factor of found graph patterns (length-decreasing minimum supports). However, such an approach cannot consider characteristics of extracted graph elements. On the other hand, the proposed algorithm can set different minimum support threshold values according to the types of elements composing graph patterns. Such an advantage also leads to mining graph patterns with more meaningful information or knowledge.

3. Smallest Valid Extension-Based Rare Graph Pattern Mining, Considering Length-Decreasing Support Constraints and Symmetry Characteristics of Graphs

In this section, we introduce the basic concept and preliminaries of graph pattern mining that can help understanding of the proposed algorithm, *SVE-RGM*. Thereafter, we describe details of our method including an overall architecture, a graph pattern growth technique and various pattern pruning techniques. We also propose techniques for effectively applying multiple minimum support and length-decreasing support constraints into graph mining environments without any unintended errors such as pattern losses. In addition, we show how the proposed method, *SVE-RGM*, operates through an overall mining procedure of the algorithm.

3.1. Preliminaries

Graph data are a structural format that can effectively express various data such as network data, chemical data and genome data. There are various definitions and theories for explaining such graph data in a mathematical manner [34,35,42,43], where we introduce essential preliminaries related to the proposed algorithm, including the definitions of graph patterns and the concept of frequent graph patterns (further information on graph theories refer to the literature cited in this paper [18,20,42,43]). We first describe a fundamental concept and several important definitions of graph pattern mining for better understanding of the proposed method. A graph pattern consists of multiple vertices and edges. In addition, graph types are classified as directed or undirected graphs depending on whether or not there are directions of edges in graphs. They can also be classified as simple or multi graphs on the basis of the number of edges between any two vertices in graphs. Moreover, other graph types can be created through numerous factors such as labels and self-edges (or loops). In this paper, we explain the proposed contents on the basis of undirected and labeled simple graph forms. However, it is trivial to consider other graph forms into our graph mining operations since we only have to consider a few additional characteristics. Figure 1 shows an example of various graph types. Figure 1a is a simple, labeled and undirected graph without any self-edges, where each vertex and edge has its own name or label. Figure 1b is a multiple graph that has two or more edges between vertices. As shown in Figure 1b, edges labels may not be expressed if they do not need to be distinguished from one another or have the same label. Figure 1c is a directed graph having a self-edge.

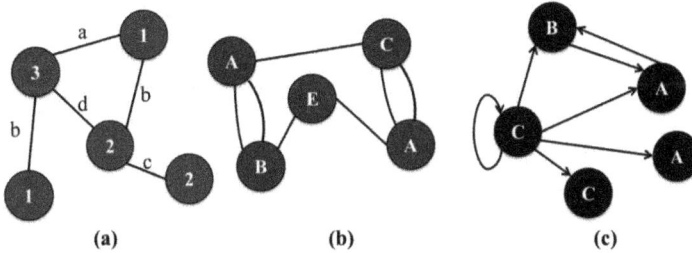

Figure 1. Example of various graph pattern forms. (**a**) A simple, labeled and undirected graph without any self-edges; (**b**) A multiple graph with multiple edges between vertices; (**c**) A directed graph with a self-edge.

Definition 1. *(Sub-graph) Let P be a sub-graph (or a graph pattern) composed of one or more elements (vertices and edges). Then, P can be denoted as two element groups. The first one is a set of vertices, $V(P) = \{v_1, v_2, \ldots, v_i\}$, and a set of edges, $E(P) = \{e_1, e_2, \ldots, e_j\}$.*

Definition 2. *(Graph isomorphism) Given a simple, labeled, and undirected graph pattern, P, its vertex and edge sets, $V(P)$ and $E(P)$ can also be denoted as follows:*

$$V(P) = \{v | v \in V(P)\},$$
$$E(P) = \{(v_1, v_2) | v_1, v_2 \in V(P) \text{ and } v_1 \neq v_2\} \tag{1}$$

Given two graph patterns, X and Y, we can say that X and Y are isomorphic, if their own $V(P)$ and $E(P)$ results are the same as each other on the basis of Equation (1) although the shapes of X and Y seem to be different from each other. Note that since all the edges in P have no directions, (v_1, v_2) and (v_2, v_1) are equal to each other.

All of the possible graph patterns have one of the following graph types: path, free tree, and cyclic graph. In addition, paths and free trees can be included in cyclic graphs and paths can be contained in free trees. In other words, the coverage of graph pattern types is denoted as path $\subseteq$ free-tree $\subseteq$ cyclic graph.

Definition 3. *(Degree of graph forms) all vertices except for both ends in a path have degree 2; meanwhile the end vertices have degree 1. Let X be a graph pattern. If X is a path with k vertices, X satisfies the following formula:*

$$D(v_l) = 2 \ (2 \leqslant l \leqslant k-1),$$
$$D(v_1) = 1, \ D(v_k) = 1, \tag{2}$$
$$|V(X)| = |E(X)| + 1$$

In Equation (2), D signifies a function that returns a degree number for an inputted vertex. v_1, v_n, $|V(P)|$ and $|E(P)|$ are the first and last vertices and the number of vertices and edges comprising P, respectively. A free tree should have at least one vertex of which the degree is 3 or more. In addition, there is no cyclic relation in all of its edges. If P is a free-tree with k vertices, the following conditions are satisfied:

$$D(v_l) \geqslant 3 \ (\exists v_l \in V(X)),$$
$$|V(X)| = |E(X)| + 1 \tag{3}$$

If X has one or more cyclic edges, X becomes a cyclic graph. Then, X has the following relation between the numbers of vertices and edges.

$$|V(X)| \leqslant |E(X)| \tag{4}$$

By using Equations (2)–(4), we can easily distinguish what type every graph pattern is.

Definition 4. *(Frequent graph pattern) Let $DB_G = \{Tr_1, Tr_2, \ldots, Tr_n\}$ be a given database storing n graph data records (also called graph transactions), where each graph transaction, Tr, is composed of multiple vertices and edges. Given a graph pattern, P, we can calculate the support of P, $S(P)$, as follows:*

$$Exist(P, Tr_k) = \begin{cases} 1, & \text{if } P \in Tr_k \\ 0, & \text{otherwise} \end{cases},$$

$$S(P) = \sum_{Tr_{k(1 \leqslant k \leqslant n)} \in DB_G} Exist(P, Tr_k) \tag{5}$$

In Equation (5), function *Exist* returns 1 if P is included in the corresponding Tr; otherwise, 0. Therefore, $S(P)$ is to add all the results of *Exist* with respect to every Tr in DB_G. In other words, the result of $S(P)$ signifies how many times P appears in DB_G. If $S(P)$ is not smaller than a user-given minimum support threshold, we regard P as a frequent sub-graph or a frequent graph pattern. Thus, the final goal of traditional frequent graph pattern mining is to extract all the possible graph patterns of which the support values are higher than or equal to this single minimum support threshold.

3.2. Overall Architecture of the Proposed Method

Figure 2 shows an overall architecture and flows of the proposed algorithm, *SVE-RGM*. It first scans a give graph database and then performs a series of works for mining *SRGs*. *SVE-RGM* conducts preprocessing works by reading the information of length-decreasing support and multiple minimum support constraints. After that, it computes a *Least Minimum Support (LMS)* factor for pre-pruning operations. Thereafter, the algorithm performs *SVE-RGM* growth for finding *SRGs* in a recursive manner. In this process, candidate patterns are generated and the algorithm checks whether or not they are valid by using the results of the proposed inverse function and the real rarity information corresponding to the candidates. These processes are conducted until we obtain all of the possible *SRGs* from the given graph database. When such recursive works are finished, we can have a complete set of *SRGs*.

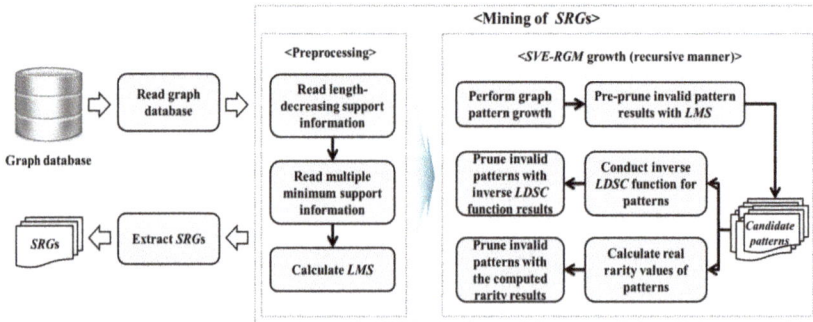

Figure 2. Overall architecture of *SVE-RGM*.

3.3. Mining SRGs from Graph Databases

Figure 3 is an example of a simple graph database. Graph pattern mining approaches including the proposed method find interesting graph pattern information from such types of graph data.

As shown in the procedure of Figure 2, we first scan a given graph database to calculate support values of the elements within the graph transactions, composing the database. Note that we assume that edge elements in the example database have the same edge label, for better understanding of the proposed method as shown in Figure 3. Therefore, support values for edges are not counted. Vertices that occur multiple times in a graph transaction are counted once [34,35,44]. After the database scanning work is finished, the proposed algorithm scans information of length-decreasing support constraints corresponding to the given graph database and multiple minimum support constraints for the elements composing the database.

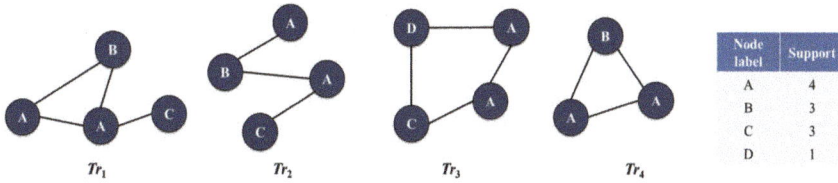

Figure 3. Example of a simple graph database.

3.3.1. Length-Decreasing Support Constraints and Smallest Valid Extension on Graph Mining

Recall that small sub-graphs having a few elements tend to be interesting if they have relatively high support values and large sub-graphs with many of elements can be interesting even though their support values are relatively low. It becomes important features supporting the reason why length-decreasing support constraints need to be applied into the graph mining operations. The easiest method for mining sub-graphs according to length-decreasing support constraints [41], tried to perform all of the possible pattern expansions, in order to confirm whether sub-graphs generated through the expansions satisfy each minimum support threshold, corresponding to their lengths. Therefore, this method causes fatal problems in terms of mining efficiency although correct results can be generated. To solve the problem, we define a length-decreasing support constraint function and its inverse function, and propose an *SVE* technique using these functions.

Definition 5. (*Length of graph*) *Let l be a length of a graph pattern, S. When S is a path or a free tree, l is the number of vertices in S. Meanwhile, if S is a cyclic graph, we consider l as follows. Let S_{prev} be a sub-graph pattern just before S becomes a cyclic graph, l_{prev} be a length of S_{prev}, and k be the number of cyclic edges inserted into S. Then, l becomes an addition of l_{prev} and k, where l can be denoted as $l = L(S)$.*

Figure 4 shows an example of length-decreasing support constraints. As shown in Figure 4a, there are various minimum support threshold values in the proposed algorithm, and one threshold value is set for each length factor. Especially, threshold settings become gradually lower according to the increase of graph pattern lengths.

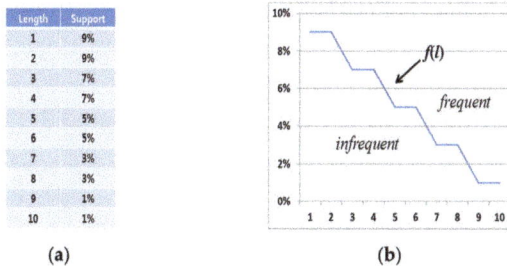

(a)   (b)

Figure 4. Example of length-decreasing support constraints. (**a**) A table with length and support information; (**b**) A graph corresponding to the length-decreasing support constraints in Figure 4a.

Definition 6. *(Length-decreasing support constraint (LDSC) function) For the length of graph pattern S, l, length-decreasing support constraint function, denoted as f(l) returns a minimum support threshold corresponding to l's current value. Since f(l) is constant or becomes lower as l comes to be larger, the inequality, $0 \leqslant f(l+1) \leqslant f(l) \leqslant 1$ is satisfied.*

Definition 7. *(Inverse function of LDSC) Given a support of graph pattern S, S(S), an inverse function of Definition 6 is denoted as $f^{-1}(S(S))$ and returns the minimum length that S must have in order to become a potentially frequent sub-graph pattern. Such a condition is also denoted as $f^{-1}(S(S)) = min(l \mid f(l) \leqslant S(S))$.*

Example 1. *Given length-decreasing support constraint information in Figure 4a, the corresponding LDSC function, f(l) is denoted as shown in Figure 4b. Since f(1) = 9% (=0.09), f(2) = 9% (=0.09), f(3) = 7% (=0.07) . . . and f(10) = 1% (=0.01), it is certain that the function satisfies the inequality, $0 \leqslant f(l+1) \leqslant f(l) \leqslant 1$. Let us assume that a sub-graph S has a support of 4% and a length of 5 respectively. Then, $f^{-1}(S(S))$ returns 7 since the minimum value is 7 among the lengths corresponding to the supports lower than or equal to 4%. Therefore, S must have more than length of 7 to be frequent. However, it is eventually infrequent since its length is 5.*

We can determine that certain sub-graphs included in the "infrequent" area as shown in Figure 4b are invalid while ones contained in the "frequent" area become valid, where $f(l)$ plays a role in distinguishing whether or not sub-graphs are frequent.

Through Definitions 5–7, we can draw the following *SVE* property for graph pattern mining based on length-decreasing support constraints, which helps perform the graph mining processes more efficiently by reducing the number of needless graph pattern expansions.

Definition 8. *(Smallest Valid Extension (SVE) property for graph mining) Given an infrequent graph pattern S, any super pattern of S, S' must have a length larger than the result of $f^{-1}(S(S))$ before it becomes a potentially frequent sub-graph pattern.*

Unlike traditional graph pattern mining, we need to consider the following additional characteristics in length-decreasing support constraint-based graph pattern mining. If a graph pattern is not valid in traditional graph pattern mining, we can omit the pattern and all of the corresponding operations related to the pattern because it and all of its possible super patterns become useless by the anti-monotone property. This property means that, if a certain pattern is infrequent, all the super patterns generated from the pattern are also infrequent. However, because the proposed algorithm applies different minimum support thresholds according to the length characteristics of generated graph patterns, the anti-monotone property cannot be maintained. In other words, although a certain sub-graph is infrequent in the current state, any of its super patterns may become frequent again as we conduct the graph pattern growth process. The previous approach [41] solved such a problem by applying an overestimation technique into its pattern pre-pruning factor. This technique can perform *LDSC*-based frequent graph pattern mining operations without any pattern loss, but it is a naïve technique that wastes computing resources in generating useless candidate patterns. However, based on the *SVE* property, we can find permanently invalid patterns. The following lemma supports such an advantage.

Lemma 1. *Let S and S' be a certain sub-graph pattern and a super pattern of S and L(S) and L(S') be the lengths of S and S' respectively. If $L(S') < f^{-1}(S(S))$ such that $S(S) < f(L(S))$, then S' is always an infrequent pattern.*

Proof. Depending on the characteristics of frequent graph mining, it is always true that $S(S) \geqslant S(S')$, and $S(S)$ is in inverse proportion to $f^{-1}(S(S))$. Therefore, we can induce the inequality, $f^{-1}(S(S)) \leqslant f^{-1}(S(S'))$. In order that S' expanded from the infrequent sub-graph S becomes frequent, these two conditions, $S(S) < f(L(S))$ and $S(S') \geqslant f(L(S'))$ must be satisfied. After we multiply the inverse function by the conditions, the result can be denoted as follows: $L(S) \leqslant f^{-1}(S(S)) \leqslant f^{-1}(S(S')) \leqslant L(S')$.

Therefore, S' becomes infrequent if it does not satisfy these conditions. Because the current mining step performed up to S, S' has not yet been expanded. Therefore, we can determine the values of $L(S)$, $L(S')$, and $f^{-1}(S(S))$ but cannot know the value of $f^{-1}(S(S'))$ ($L(S')$ can be inferred from $L(S)$). Therefore, if $L(S') \geqslant f^{-1}(S(S))$ is false, *i.e.*, $L(S') < f^{-1}(S(S))$ is true, S' becomes an infrequent graph pattern. For this reason, we can know whether or not S' is valid in advance even though any actual expansion process for S' is not performed. □

Note that the proposed overestimation technique is not an approximation method. Therefore, unlike the statistical approximation approach [45], our method does not mine any false positives. Our overestimation technique is employed to check and discard permanently meaningless graph patterns without any pattern loss during the mining process. However, since every pattern satisfying the overestimated condition is not the finally valid result (called a candidate pattern), we check the actual support of each candidate in order to mine actually meaningful graph patterns selectively. By doing this, we can obtain a complete set of frequent graph patterns considering the length-deceasing support constraints and rarity of graphs.

3.3.2. Pre-Pruning Infrequent Sub-Graphs by the SVE Property without Any Pattern Loss

Using the defined *SVE* property, we can determine information regarding what sub-graphs cause needless pattern expansions in advance. However, if they are directly pruned, fatal problems such as pattern losses can occur since applying the length-decreasing support constraints into graph mining generally breaks the anti-monotone property. That is, any infrequent sub-graphs can become frequent ones as their pattern expansion works are conducted. To solve the problems and maintain the anti-monotone property, we additionally consider the length information for graph transactions in graph databases as well as the *SVE* property.

Lemma 2. *Let S and S' be invalid graph patterns ($S' \supset \bigcap S$) and $SET_{S'} = \{Tr_1, Tr_2, \ldots, Tr_n\}$ be a set of graph transactions including S'. Then, if there is any element satisfying $L(Tr_i) < f^{-1}(S(S))$ among the elements of $SET_{S'}$ ($1 \leqslant I \leqslant n$), S' can permanently be pruned.*

Proof. In $SET_{S'} = \{Tr_1, Tr_2, \ldots, Tr_n\}$, each Tr is a graph transaction with S' in DB_G, and n becomes the support of S'. If there is any Tr_i such that $L(Tr_i) < f^{-1}(S(S))$ ($1 \leqslant i \leqslant n$), it means that lengths of all super patterns generated from S' are also smaller than $f^{-1}(S(S))$ because the super patterns cannot have more lengths than $L(Tr_i)$. Furthermore, since S' and the super patterns of S' do not satisfy the minimum length by the inverse function, neither of them naturally satisfies minimum support constraints. As a result, pruning S' does not have any negative effect on maintenance of the anti-monotone property. That is, we can obtain intended mining results without any problem. □

Example 2. *Let us consider the example in Figure 4 and assume that a certain sub-graph, S, has a length of 2 and a support of 5%, a super pattern of S, S', has a length of 3 and a support of 4% and a set of graph transactions for S', $SET_{S'}$ includes 4 graph transactions (denoted as $SET_{S'} = \{Tr_1, Tr_2, Tr_3, Tr_4\}$), where the length for each Tr is set to 7, 4, 10, and 5 respectively. Then, S' becomes an invalid pattern according to the SVE property and Lemma 1. Furthermore, since $L(Tr_2)$ is smaller than $f^{-1}(S(S))$, any super patterns of S' also become useless ones and therefore, S' can directly be pruned.*

Based on Lemma 2, we can prune all of the permanently useless patterns and omit the corresponding mining operations in advance without any pattern loss.

3.3.3. Multiple Minimum Supports of Vertex and Edge Elements on Graph Mining

In addition to the length-decreasing support constraints, we additionally consider multiple minimum supports of graph elements (vertices and edges) in this paper. Recall that meaningful graph patterns with low supports may not be extracted if a given minimum support threshold is high in traditional graph pattern mining, otherwise an enormous number of useless patterns should be mined if we lower the threshold to find such useful ones. By considering multiple, minimum support constraints of vertex and edge elements, as well as the length-decreasing support constraints, we can obtain a smaller number of more meaningful pattern results. In contrast to traditional graph pattern mining that has a single minimum support threshold, the proposed method has a different threshold for each element to consider the multiple minimum support constraints on graph pattern mining.

Definition 9. *(Minimum support constraints of vertices and edges) Given a graph database with multiple graph transactions Tr, $DB_G = \{Tr_1, Tr_2, \dots, Tr_n\}$, a set of x vertices and y edges comprising DB_G can be denoted as $V(DB_G) = \{v_1, v_2, \dots, v_x\}$ and $E(DB_G) = \{e_1, e_2, \dots, e_y\}$, respectively. Then, each of minimum support threshold, δ, is set for each element as shown in Table 1, where they are assigned by a user, respectively.*

Table 1. Multiple minimum supports of graph elements (vertices and edges).

Element	v_1	v_2	...	v_x	e_1	e_2	...	e_y
Threshold	δ_1	δ_2	...	δ_x	δ_{x+1}	δ_{x+2}	...	δ_{x+y}

In traditional graph pattern mining, there is only one factor for deciding whether or not a found graph pattern is frequent without the characteristics of its elements. Meanwhile, we need to consider a different way to apply the multiple minimum support constraints into our method.

Definition 10. *(Minimum support constraints of graph patterns) Let P be a graph pattern extracted from DB_G. Then, a set of vertices and edges can be denoted as $V(P) = \{v_1, v_2, \dots, v_i\}$ and $E(P) = \{e_1, e_2, \dots, e_j\}$, respectively. According to Definition 9, we know that each element has its own minimum support threshold set by a user, and P is composed of multiple elements. Hence, the minimum support threshold for P, T(P), is computed as the minimum value among the threshold values of P's elements.*

If $S(P)$ is not lower than $T(P)$, we can say that P is a valid graph pattern satisfying the rarity of graph elements based on the multiple minimum support constraints. The reason why we compute and use the minimum support threshold for each mined graph pattern is that we can consider the different rarity of each pattern in this way.

Definition 11. *(SVE-based Rare Graph pattern (SRG)) Given a graph pattern, X, we call X an SRG if $S(X) \geqslant f(L(X))$ and $S(X) \geqslant T(X)$. In other words, SRGs mean sub-graph patterns that satisfy both the length-decreasing support and multiple minimum support constraints.*

Consequently, the main goal of the proposed algorithm, *SVE-RGM*, is to mine all of the possible *SRGs* from a given graph database without any pattern loss.

3.3.4. Pre-Pruning Invalid Graph Patterns Based on Multiple Minimum Support Constraints

Recall that fatal pattern losses can be caused if we do not apply the additional considerations mentioned in Section 3.3.2. Similarly, we can also suffer from such a pattern loss problem if we directly prune graph patterns that do not satisfy their own multiple minimum support constraints. As mentioned above, elements of a graph pattern have their own threshold values set by a user. Therefore, the anti-monotone property is not satisfied with this situation. In other words, although a certain graph pattern has a support that does not satisfy the corresponding multiple minimum support constraint in the current state, any super pattern of it may become a valid result again in the process of graph pattern expansion. Hence, if we pre-prune such patterns without any additional consideration, fatal pattern losses can occur. Moreover, an enormous number of interesting patterns can be lost by unintended pruning of a few elements or graph patterns. Satisfying the anti-monotone property during the mining process is one of the most important rules to improve mining efficiency without any negative effect such as pattern losses. For this reason, we employ an overestimation method for maintaining the anti-monotone property without any pattern loss on the proposed algorithm.

Definition 12. *(Overestimated minimum support constraint) DB_G has multiple graph transactions and the corresponding elements as mentioned in Definition 9. Then, the overestimated minimum support constraint for DB_G, $O(DB_G)$, is computed as the smallest value among the valid minimum support constraints of all the elements comprising DB_G (it is also called Least Minimum Support (LMS)). In other words, let $SET_{T(DBG)} = \{\delta_1, \delta_2, \ldots, \delta_{x+y}\}$ ($\delta_1 \geqslant \delta_2 \geqslant \ldots \geqslant \delta_{x+y}$) be a sorted set of minimum support constraints for all the elements in DB_G (x and y are the numbers of vertices and edges, respectively). Then, we start comparing the smallest threshold δ_{x+y} with the real support of the element corresponding to δ_{x+y}. After that, $\delta_{(x+y)-1}$ is compared to the corresponding element support. Such a comparison is performed until we find the first element of which the support is higher than or equal to the corresponding minimum support constraint, δ_k ($1 \leqslant k \leqslant x + y$). Then, we consider δ_k as $O(DB_G)$.*

Consequently, *SRGs* extracted from the proposed algorithm are graph patterns that satisfy Lemmas 1 and 2 and the condition of Definition 12.

3.4. Improving Efficiency of Graph Mining Performance Based on Symmetry Features of Graphs

From the suggested definitions and constraints, we allowed the proposed method to mine a smaller number of meaningful graph pattern results, called *SRGs*. As mentioned above, the length-decreasing minimum support and multiple minimum support constraints also increase the mining efficiency of the proposed algorithm, *SVE-RGM*, by reducing the search space effectively. In addition, we can also raise the mining efficiency with the correctness of the algorithm maintained. Recall that the proposed method performs its own mining operations in a depth-first search manner. This also means that a few useless graph patterns may cause the proposed algorithm to generate an enormous number of invalid or duplicated pattern results. In contrast to the case of traditional frequent pattern mining that considers only an item-based simple format, a numerous number of duplicated graph patterns can be generated in graph pattern mining because of the complicated structures of graph data. In particular, we have to conduct graph isomorphism tests for the mined patterns in order to prevent duplicated ones from being extracted.

In order to perform the mining operations more efficiently, our algorithm applies the following order types of graph pattern growth: (1) path → cyclic graph and (2) path → free tree → cyclic graph. In other words, a certain vertex is selected as a prefix at first and a path is generated by adding another vertex and edge that can be attached to the prefix. Then, we can obtain a graph pattern in a path form. After that, there are three options for the next step. That is, it can be extracted as a longer path, a free tree, or a cyclic graph according to the attached vertex and edge types. Recall that a few useless graph patterns can cause an enormous number of invalid or duplicated pattern results. From the above features, we can determine that removing duplicated path creations has a large effect on reducing the number of useless pattern creations. In this regard, symmetry features of paths can be used as effective factors that can lead to correct choices not to cause any duplicated path result. Let $P = \{v_1, e_1, v_2, e_2, \ldots, e_{k-1}, v_k\}$ be a given path and $N = \{v, e'\}$ be a pair of one vertex and edge that are supposed to be attached to P. Then, when expanding P with N, we have two choices; the first one is to add N to the front of P and the second one is to add N to the rear of P because of the characteristics of paths. If we add N to P without any consideration, an enormous number of duplicated graph patterns can be generated as the graph pattern growth works are conducted during the mining process. Meanwhile, if we set a specific constraint for limiting expansion directions of paths, we can effectively prevent such a problem.

A path has at least two vertices and one edge. Then, we can determine whether or not the path is symmetric. In other words, given a path, $P = \{v_1, e_1, v_2, e_2, \ldots, e_{k-1}, v_k\}$, we can extract two strings from P as follows: v_1-e_1-v_2-e_2- $\ldots$ -e_{k-1}-v_k (original string) and v_k-e_{k-1}-v_k-e_{k-2}- $\ldots$ -e_1-v_1 (inverse string).

Then, if they are equal to each other, we consider P as a symmetric path. In this case, we do not need to consider what direction we have to choose because any selection leads to the same result. If the first string is lower than the second one in terms of a lexicographical order, we expand P by attaching new elements to the front of P. Meanwhile, if the first string is higher than the second one, we add the new ones to the rear of P. From the above path expansion technique based on the symmetry features of paths, we can omit any path expansion causing duplicated path creation. In addition, once the symmetry result of P is calculated, we can easily determine the symmetry result of its expanded path in a few additional computations. Let $Sym_{total}(P)$, $Sym_{front}(P)$, and $Sym_{rear}(P)$ be symmetry functions for the entire part of P ($\{v_1, e_1, v_2, e_2, \ldots, e_{k-1}, v_k\}$), the front part of P ($\{v_1, e_1, v_2, e_2, \ldots, e_{k-2}, v_{k-1}\}$) and the rear part of P ($\{v_2, e_2, v_3, e_3, \ldots, e_{k-1}, v_k\}$), where each function returns 0 when the corresponding string is symmetric, 1 when the corresponding original string is lower than the inverse one and -1 when the original one is higher than the inverse one. Using this method, we can easily know the symmetry result of super patterns of P. Let P' be a longer path that adds a new vertex and edge to P. Then, if the new elements have been attached to the front of P', we can determine that $Sym_{total}(P) = Sym_{rear}(P')$. Meanwhile, if the new ones have been added to the rear of P', it is true that $Sym_{total}(P) = Sym_{front}(P')$. Therefore, based on these characteristics, we can efficiently determine the symmetry results of mined patterns. By restricting directions of graph expansion based on the symmetry features of paths, we can improve the mining efficiency of the proposed method.

One of the most important considerations in frequent graph pattern mining is to enumerate all of the possible graph patterns without any redundancy. In contrast to the itemset format traditional frequent pattern mining focuses on, a graph pattern is composed of multiple vertices and edges, where the vertices can be ordered in many ways. Therefore, one graph pattern can also be denoted as a large number of topologically equivalent copies. Hence, it is essential to check graph isomorphism whenever a graph pattern is mined. Especially, checking graph isomorphism is a well-known NP-hard problem that can cause enormous computational overheads. However, as mentioned above, we do not have to check graph isomorphism for the path format because we established the symmetry-based constraint for paths in advance and allow paths to be enumerated on the constraint. When any path is expanded as a free-tree, we employ the backbone strategy of *Gaston*, which is different from the canonical representation used in *gSpan*. By using the technique, we can prevent any duplication of free-trees from being caused in the mining process without performing any works for graph isomorphism (the correctness of the backbone strategy was proved by showing that the *Gaston* algorithm extracted the same results as those of other approaches like *gSpan* [34,35]). When a path or a free-tree is expanded as a cyclic graph, we have no choice but to conduct graph isomorphism operations. However, we can reduce computational overheads by using the minimum spanning tree format when comparing cyclic graphs. A cyclic graph can be expressed as a minimum spanning tree, which is simpler than its original one. Therefore, we can compare graphs more quickly than doing in a naïve manner.

3.5. Algorithm Description: SVE-RGM

Figure 5 represents overall mining steps of the proposed algorithm, *SVE-RGM*. In the main procedure, *SVE-RGM*, the lowest value in *LDSC* is set as a minimum support threshold, δ and the algorithm computes *LMS* from the *MMS* data (lines 1–3). After that, it finds valid vertices and edges from DB_G through the calculated minimum support and *LMS* value (lines 3–6). Then, for each frequent vertex, the algorithm extracts valid sub-graph patterns according to length-decreasing support constraints and multiple minimum support thresholds as it performs a series of graph pattern expansion works (lines 7–11). When function *Expand_subgraphs* is called, *SVE-RGM* determines whether G is frequent or not and then assigns a flag, *true* or *false*, into the *isFrequent* variable (lines 1–4), where G is entered to P if G is frequent (line 3). Thereafter, for each edge in E, appropriate pattern expansion works are selectively conducted according to the state of G such as a path, a free tree, and a cyclic graph (lines 6–8). After that, if the support of the expanded pattern, G', is not smaller than *LMS*, the algorithm conducts the subsequent works (line 9). If *isFrequent* is *false*, then the algorithm decides whether to prune G' (lines 10–13). If G' is not pruned, *SVE-RGM* calls *Expand_subgraphs* recursively to perform the next pattern expanding operations (lines 14–16). After all of mining operations terminate, we can gain a complete set of *SRGs* considering the length-decreasing support constraints and the multiple minimum support constraints for rarity of graph patterns.

Input—DB_G: **a given graph database,** *LDSC*: **length-decreasing support constraint information,** *MMS*: **multiple minimum support information for elements in** DB_G **Output**—*P*: **a set of** *SRGs*
SVE-RGM(GDB, LDSC, MMS)   01. *minimum support*, $\delta \leftarrow$ **the lowest support constraint value in** *LDSC*   02. **calculate** *LMS* **from** *MMS*   03. **a set of vertices,** $V \leftarrow$ **all of the frequent vertices such that their supports** $\geq \delta$ 04. **delete vertices in** V **such that their supports** < *LMS*   05. **a set of edges,** $E \leftarrow$ **all of the frequent edges such that their supports** $\geq \delta$ 06. **delete edges in** E **such that their supports** < *LMS*   07. **for each vertex,** v_i, **in** V, **do**   08. **a sub-graph,** $G \leftarrow v_i$ 09. **a set of edges,** $E' \leftarrow$ **edges that can be attached to** v_i **from** E 10. $P = P \cup$ *Expand_subgraphs*(G, E') 11. **end for**
Expand_subgraphs(G, E)   01. **if** $S(G) \geq f(L(G))$ **and** $S(G) \geq T(G)$, **do**   02. *isFrequent* $\leftarrow$ *true*   03. $P = P \cup G$ 04. **else** *isFrequent* $\leftarrow$ *false*   05. **for each edge,** e_i, **in** E, **do**   06. **if** G **is** *path* **or** *free tree*, **do**   07. **an expanded graph,** $G' \leftarrow$ **adding** e_i **and the vertex,** v **included in** e_i **to** G 08. **else** $G' \leftarrow$ **adding only** e_i **such that it is a cyclic edge** //**a cyclic graph**   09. **if** $S(G') \geq LMS$, **do**   10. **if** *isFrequent* = *false*, **do**   11. $SET_{G'} \leftarrow$ **graph transactions including** G' **in** *GDB*   12. **if** $L(G') < f^{-1}(S(G))$ **and** $\forall L(g)$ $(g \in SET_{G'}) < f^{-1}(S(G))$, **do**   13. **go to line 5 with the next edge of** e_i // **the current** G' **is pruned**   14. $E' \leftarrow$ **a set of edges that can be attached to** G' 15. $P = P \cup$ *Expand_subgraphs*(G', E') 16. **end for**

Figure 5. *SVE-RGM* algorithm.

4. Performance Evaluation

4.1. Experimental Environment

To evaluate performance of the proposed algorithm, *SVE-RGM*, precisely and reasonably, the algorithm is compared with the following state-of-the-art algorithms, *Gaston* [34,35], *FGM-LDSC* [41], and *WRG-Miner* [37]. *Gaston* is a famous fundamental frequent graph pattern mining algorithm that features fast mining speed. *FGM-LDSC* is a *Gaston*-based approach that can consider length-decreasing support constraints into graph mining processes. *WRG-Miner* is an algorithm that extracts frequent graph patterns considering multiple minimum support constraints and different element weight factors. Note that we optimized *WRG-Miner* in order to compare it with ours fairly. They were written in C++ and performed in 4.0 GHz CPU, 16GB RAM, and WINDOWS 7 OS. We used real and synthetic datasets, PTE composed of chemical graph data, DTP composed of compound graph data and SYN100K composed of 100K synthetic graph data. Detailed characteristics of these two real graph datasets are available at the literature [34,35]. Figure 6 shows distributions regarding length-decreasing support constraints for the PTE, DTP, and SYN100K datasets.

(a)   (b)

(c)

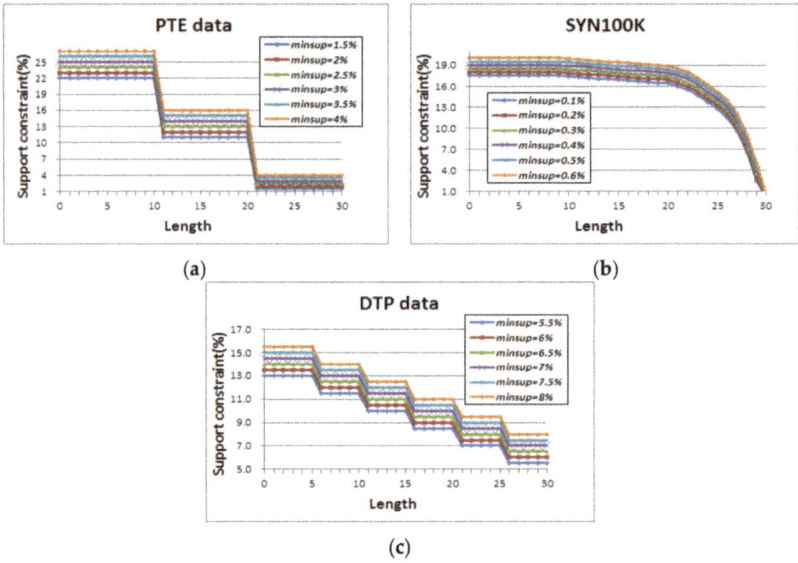

Figure 6. (a) Length-decreasing support constraints of PTE; (b) Length-decreasing support constraints of SYN100K; (c) Length-decreasing support constraints of DTP.

Each implementation of the algorithms receives a graph dataset and returns its own mining result as a file. The input format of a graph dataset is as follows. Let us express the example database shown in Figure 3 as the input format for the implementations. This graph dataset is composed of 4 graph transaction data, where they consist of 4 different vertex types and 1 edge type. Then, we denote each graph transaction as a set of relations among the links of vertices and edges. For example, the first graph transaction has 4 vertices, A, B, A and C and 1 edge (let us denote it as "a"). We first assign an index for each vertex. That is, the index of the first vertex A becomes 0 and that of the last one C becomes 3. A relation among vertices and an edge is denoted as follows. The relation corresponding to vertices A and B and edge a is denoted as 0 1 a. Table 2 is the result of changing the example dataset of Figure 3 to the input format. Note that labels of vertices and edges are expressed as integer values in real datasets. Based on the format shown in Table 2, each algorithm performs its own mining operations, where additional data for length-decreasing support constraints or multiple minimum support constraints are inputted according to the characteristics of the algorithms. Mining results are also stored like the format shown in Table 2 and the support information corresponding to each pattern is additionally stored.

Table 2. Input format of the graph dataset transformed from Table 1.

Tr_1	Tr_2	Tr_3	Tr_4
v 0 A	v 0 A	v 0 D	v 0 B
v 1 B	v 1 B	v 1 A	v 1 A
v 2 A	v 2 A	v 2 C	v 2 A
v 3 C	v 3 C	v 3 A	e 0 1 a
e 0 1 a	e 0 1 a	e 0 1 a	e 0 2 a
e 0 2 a	e 1 2 a	e 0 2 a	e 1 2 a
e 1 2 a	e 2 3 a	e 1 3 a	-
e 2 3 a	-	e 2 3 a	-

In order to assign δ values for the elements of given graph datasets, the following methodology used in the literature [26–28,36,37] was employed. Let e_i be an element within a given graph dataset.

Then, the corresponding δ value for each e_i is calculated as follows: $\delta_i = MAXIMUM(\beta \times S(e_i), LS)$ (*LS*: the smallest δ value in all the δ values). Note that the threshold of *Gaston* and *FGM-LDSC* is set to the same value as *LS* for fair comparisons. In the equation, β ($=1/\alpha$ ($0 < \beta \leqslant 1, 1 \leqslant \alpha$)) is a variable showing how closely the actual support value of each element is related to the corresponding δ value. In other words, each δ is more likely to have a value more similar to the real support of the corresponding element rather than *LS*, if β becomes closer to 1.

4.2. Analysis of Runtime Performance

Figure 7 shows runtime results of the PTE, DTP and SYN100K datasets on changing minimum support threshold settings, where the α values of *WRG-Miner* and *SVE-RGM* are fixed as 1. In Figure 7a, all the algorithms require more runtime resources as the given minimum support threshold becomes lower. However, the proposed algorithm guarantees the fastest runtime performance in all cases, while *Gaston* shows the slowest performance. Especially when the threshold is 1.5%, we can observe that there are large gaps among the compared algorithms. Since the proposed algorithm employs both the length-decreasing support constraints and the multiple minimum support constraints, it extracts a smaller number of interesting patterns compared to those of the others. From the result of Figure 7a, we can determine that the length-decreasing support constraints have a better effect on improving the runtime performance compared to the multiple minimum support constraints because *FGM-LDSC* is faster than *WRG-Miner* in this case. On the other hand, the SYN100K dataset shows a different tendency. In the case of Figure 7a, the proposed method shows the best result in every case, and *Gaston* is the worst among the compared ones. However, in this case, we can see that *WRG-Miner* is better than *FGM-LDSC*. From these different results, we can know that the multiple minimum support constraints are more effective on this synthetic dataset. In the case of the DTP dataset, we can see that the proposed method and *Gaston* have similar tendencies with those of the PTE dataset. However, the other algorithms have different results. In the DTP dataset, *WRG-Miner* has better runtime performance than that of *FGM-LDSC*, which means that the multiple minimum support constraints are more effective on reducing the execution time with respect to this dataset. The proposed method also has the best runtime results in this case. Meanwhile, the *Gaston* algorithm is slowest in almost all cases.

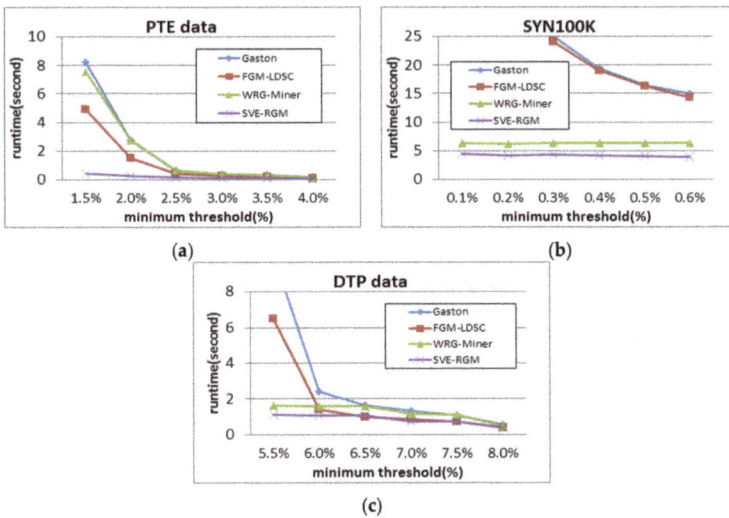

(a) (b)

(c)

Figure 7. (a) Runtime results of PTE on changing minimum support threshold; (b) Runtime results of SYN100K on changing minimum support threshold; (c) Runtime results of DTP on changing minimum support threshold

Figure 8 is the results of runtime performance on settings of the changing α value while the minimum support threshold is fixed. Recall that β is inversely proportional to α ($\beta = 1/\alpha$ ($0 < \beta \leqslant 1$, $1 \leqslant \alpha$)) and if β becomes closer to 1, δ is more likely to have a value more similar to the real support of the corresponding element rather than *LS*. Therefore, as α becomes lower, multiple minimum support constraint-based approaches extract a less number of pattern results. In Figure 8a,b, we can observe that all the algorithms spend stable runtime in mining their own graph patterns. However, the *Gaston* algorithm has the worst performance regardless of minimum support threshold or α settings. Meanwhile, the proposed method guarantees the best result in every case. The reason why the compared algorithms show such stable performance results is that PTE is not a dataset affected by the changes of the α value. For this reason, *WRG-Miner* does not show good performance in this case because it is a rare graph pattern mining algorithm based on multiple minimum support constraints. Meanwhile, the results of performance evaluation in SYN100K have a different tendency. In this case, *FGM-LDSC* outperforms *WRG-Miner* in almost all cases. As in the previous case, our algorithm, *SVE-RGM*, shows the best performance. Especially, the proposed method guarantees stable runtime performance regardless of α, while *WRG-Miner* becomes gradually slower as the α value becomes higher.

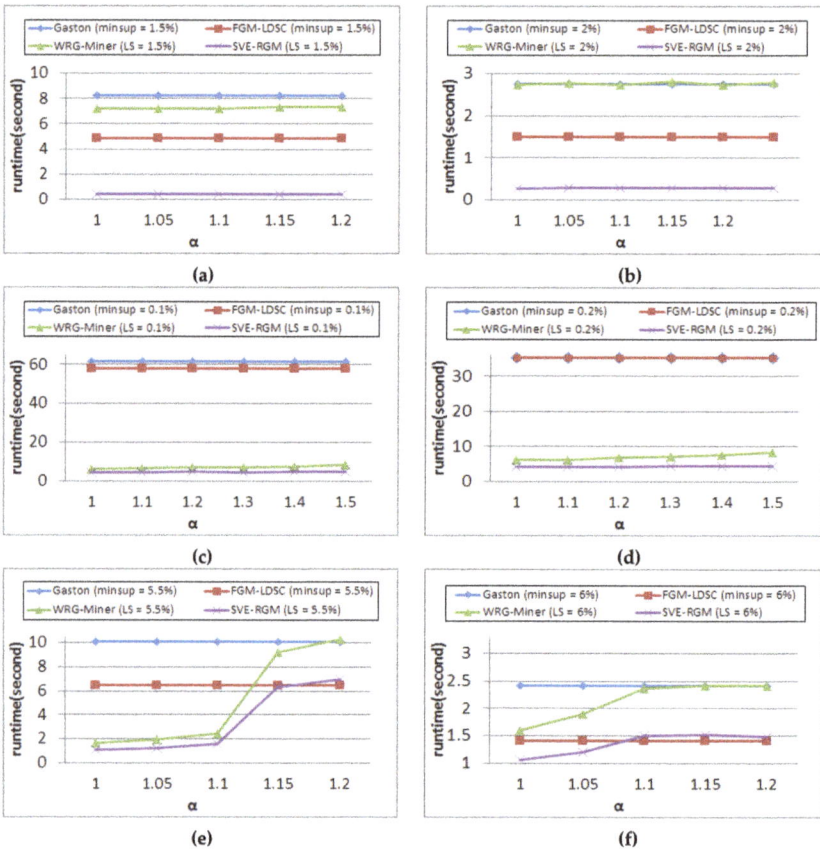

Figure 8. (**a**) Runtime results of PTE on changing α (*minsup* = 1.5%); (**b**) Runtime results of PTE on changing α (*minsup* = 2%); (**c**) Runtime results of SYN100K on changing α (*minsup* = 1.5%); (**d**) Runtime results of SYN100K on changing α (*minsup* = 2%); (**e**) Runtime results of DTP on changing α (*minsup* = 5.5%); (**f**) Runtime results of DTP on changing α (*minsup* = 6%).

In contrast to the results of the above two datasets, the proposed algorithm, *SVE-RGM* and the compared rare graph pattern mining algorithm, *WRG-Miner*, show noticeable changes depending on the settings of α. They are faster than the others at lower settings of α since they can better reflect the rarity features of graph elements into their own mining processes and reduce the number of mined patterns. In addition, the proposed algorithm also guarantees better runtime efficiency than that of *WRG-Miner* because our approach can selectively extract a smaller number of graph patterns than those of the competitor by considering the length-decreasing support constraints additionally. As α becomes larger, the performance of *WRG-Miner* and ours becomes similar to the other non-rare pattern mining algorithms.

4.3. Analysis of Memory Usage Performance

The next test is memory usage performance shown in Figure 9, where the basic parameter settings are equal to the previous test. When the threshold is relatively low in Figure 9a, the algorithms have similar memory consumption. However, as the threshold becomes lower, the gaps of their memory usage results become larger. In this case, all the algorithms except for ours have similar memory usage, regardless of threshold settings. Meanwhile, our algorithm has the most efficient memory usage in all the cases. In the case of Figure 9b, the proposed method guarantees the best performance. *Gaston* and *FGM-LDSC* have similar results, while *WRG-Miner* has relatively good performance compared to the tendency of Figure 9a because the multiple minimum support constraints of the algorithm have a positive effect on reducing memory usage in the SYN100K dataset. Nevertheless, *WRG-Miner* falls behind our method in every case. That is, the proposed algorithm *SVE-RGM*, shows the most efficient memory usage regardless of any environmental settings. As shown in Figure 9c, all the algorithms except for *WRG-Miner* have similar tendencies with those of the PTE dataset. Recall that the multiple minimum support constraints are more effective on the runtime performance with respect to this dataset. This advantage also leads the algorithm to save more memory resources as shown in the figure.

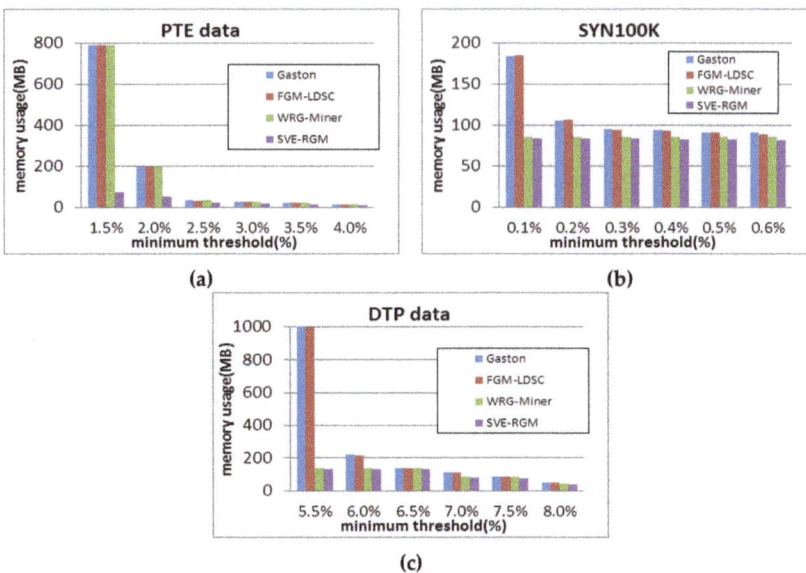

(a)

(b)

(c)

Figure 9. (a) Memory usage results of PTE on changing minimum support threshold; (b) Memory usage results of SYN100K on changing minimum support threshold; (c) Memory usage results of DTP on changing minimum support threshold.

Figure 10 shows memory usage results of the PTE and SYN100K datasets on changing α values while the threshold is fixed. As in the cases of Figure 9, the compared algorithms have similar tendencies. In the PTE dataset shown in Figure 10a,b, the first three algorithms, *Gaston*, *FGM-LDSC*, and *WRG-Miner*, consume similar memory in all cases, while the proposed one spends the smallest memory because of its own mining techniques and constraints. On the other hand, *WRG-Miner* has better performance than those of *Gaston* and *FGM-LDSC* in the SYN100K dataset because of its own constraints, multiple minimum supports for graph elements. However, its performance is still worse than our method as shown in the figure. In addition, all the algorithms show stable memory usage regardless of the α settings. This signifies that the SYN100K dataset has few effect on the α settings. Similarly to the runtime results shown in Figure 8e,f, memory usage of the rare graph pattern mining algorithms is changed according to the settings of α. However, the memory efficiency of the proposed algorithm is best among the compared ones in general. Meanwhile, since *Gaston* and *FGM-LDSC* do not consider the rarity of graph elements, they show the same results regardless of changes of α.

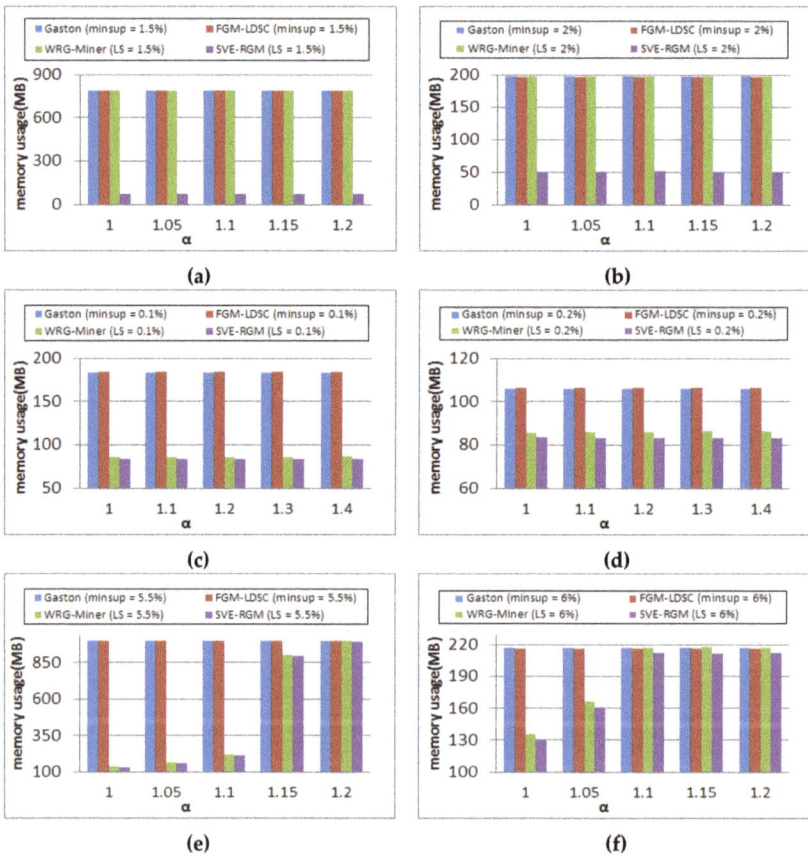

Figure 10. (**a**) Memory usage results of PTE on changing α (*minsup* = 1.5%); (**b**) Memory usage results of PTE on changing α (*minsup* = 2%); (**c**) Memory usage results of SYN100K on changing α (*minsup* = 1.5%); (**d**) Memory usage results of SYN100K on changing α (*minsup* = 2%); (**e**) Memory usage results of DTP on changing α (*minsup* = 5.5%); (**f**) Memory usage results of DTP on changing α (*minsup* = 6%).

4.4. Analysis of Pattern Generation Performance

Figure 11 shows results of graph pattern generation for the compared algorithms in a log-scale, where the basic settings of parameters are also the same as those of the previous tests. These figures support the results of the above runtime and memory usage performance evaluation. As mentioned above, the proposed algorithm can reduce an enormous number of less meaningful graph patterns by employing both the length-decreasing support constraints and the multiple minimum support constraints. As a result, we can obtain a smaller number of interesting patterns compared to the others. In the case of Figure 11b, *WRG-Miner* and our *SVE-RGM* generate the same pattern results since the multiple minimum support constraints have a less effect on the resulting patterns. However, the mining performance of the proposed method is better than that of *WRG-Miner* in spite of this pattern generation result. In the case of the DTP dataset, the proposed algorithm also mines the smallest number of frequent graph patterns.

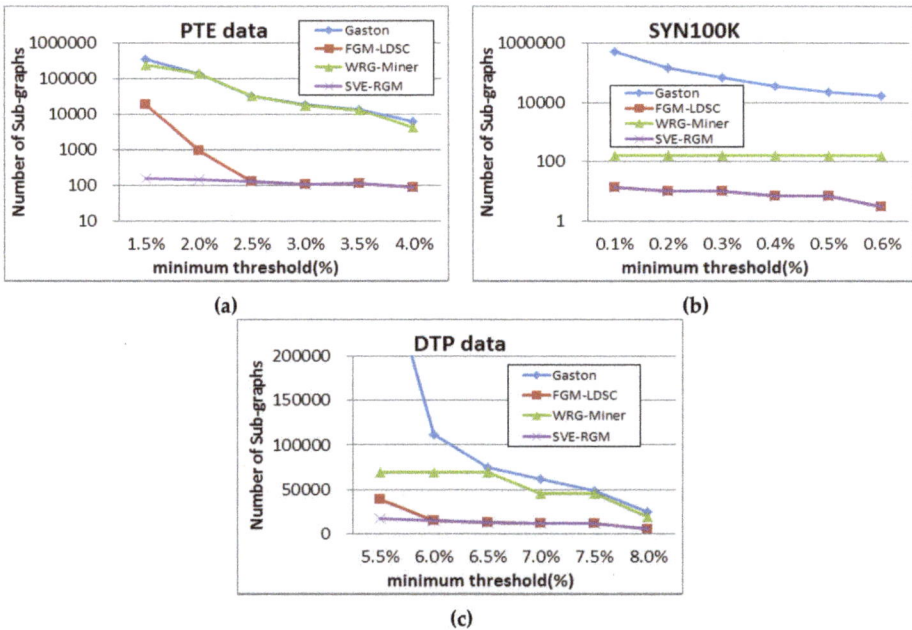

(a)   (b)

(c)

Figure 11. (**a**) Pattern generation results of PTE on changing minimum support threshold; (**b**) Pattern generation results of SYN100K on changing minimum support threshold; (**c**) Pattern generation results of DTP on changing minimum support threshold.

Figure 12 shows results of graph pattern generation on changing value α settings. In Figure 12a,b, the results of *Gaston* and *FGM-LDSC* are not changeable since they do not apply the multiple minimum support constraints for considering the *rare item problem*. Meanwhile, the other two algorithms, *WRG-Miner* and *SVE-RGM*, also show stable pattern generation results because this dataset has a less effect on the changes of α. On the other hand, we can see that the number of graph patterns mined from *WRG-Miner* is increased as the α value becomes higher. Meanwhile, *SVE-RGM* shows stable results regardless of the α value settings. In contrast to the cases of PTE and SYN100K, the algorithms considering the rarity of graph elements, *WRG-Miner* and our *SVE-RGM*, mine a different number of mining results as shown in the figure. As α becomes lower, they extract a smaller number of patterns because they can selectively mine patterns considering the rarity of graph elements.

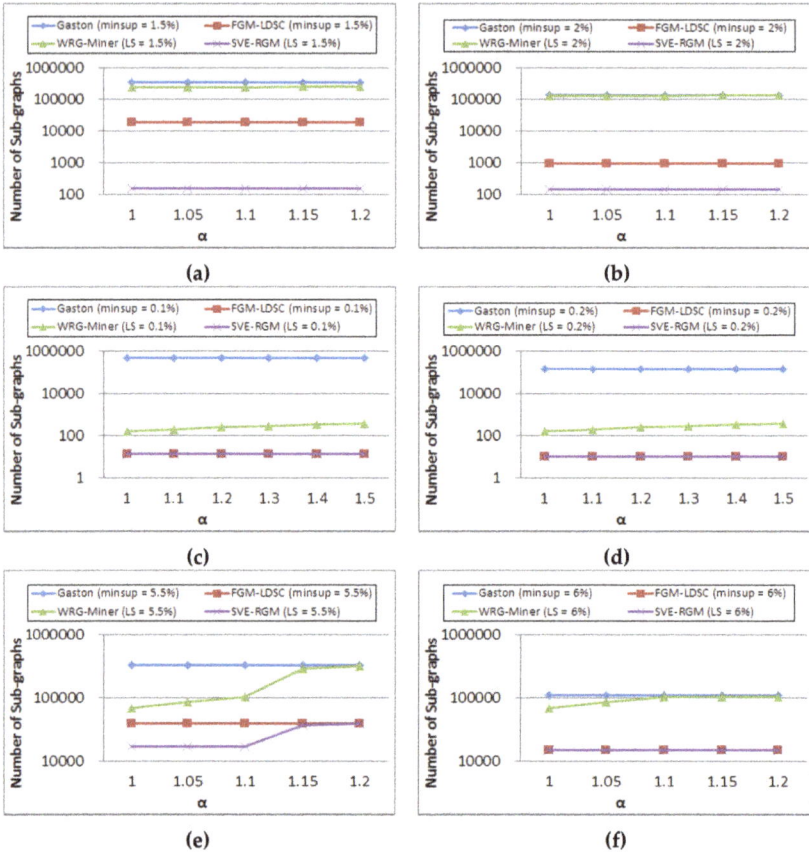

Figure 12. (a) Pattern generation results of PTE on changing α (*minsup* = 1.5%); (b) Pattern generation results of PTE on changing α (*minsup* = 2%); (c) Pattern generation results of SYN100K on changing α (*minsup* = 0.1%); (d) Pattern generation results of SYN100K on changing α (*minsup* = 0.2%); (e) Pattern generation results of DTP on changing α (*minsup* = 5.5%); (f) Pattern generation results of DTP on changing α (*minsup* = 6%).

4.5. Analysis of Algorithm Performance on Changing Length-Decreasing Support Constraints

Next, we test the mining performance of each algorithm on changing length-decreasing support constraints. Figures 13–15 show the results of the tests for each dataset, where α is fixed to 1 in order to reflect the rarity features into the mining processes of the rare graph pattern mining algorithms. Figure 13a shows the different settings of length-decreasing support constraints for the PTE dataset in this test, where the minimum support threshold is fixed to 1.5%. We show how the results of performance evaluation for the algorithms change according to these constraint settings. As shown in Figure 13b, *Gaston* and *WRG-Miner* have the same result regardless of the changing settings since they are algorithms not considering any length-decreasing support constraint. Meanwhile, the other algorithms have different runtime results depending on the constraint settings. As the condition is changed from cond. 1 to cond. 5, *FGM-LDSC* and *SVE-RGM* operate faster because they can mine a smaller number of pattern results as shown in Figure 13d. Meanwhile, memory consumption of the algorithms is different from the tendency of the runtime test. *Gaston* and *WRG-Miner* also spend the same memory regardless of the constraint settings.

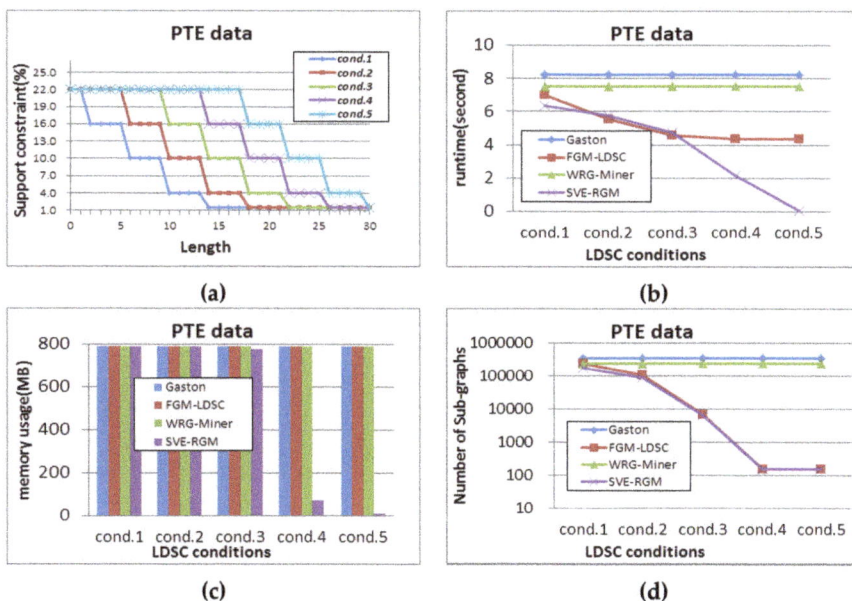

Figure 13. (**a**) Different settings of length-decreasing support constraints for PTE ($\alpha = 1$); (**b**) Runtime result for PTE ($\alpha = 1$); (**c**) Memory usage result for PTE ($\alpha = 1$); (**d**) Pattern generation result for PTE ($\alpha = 1$).

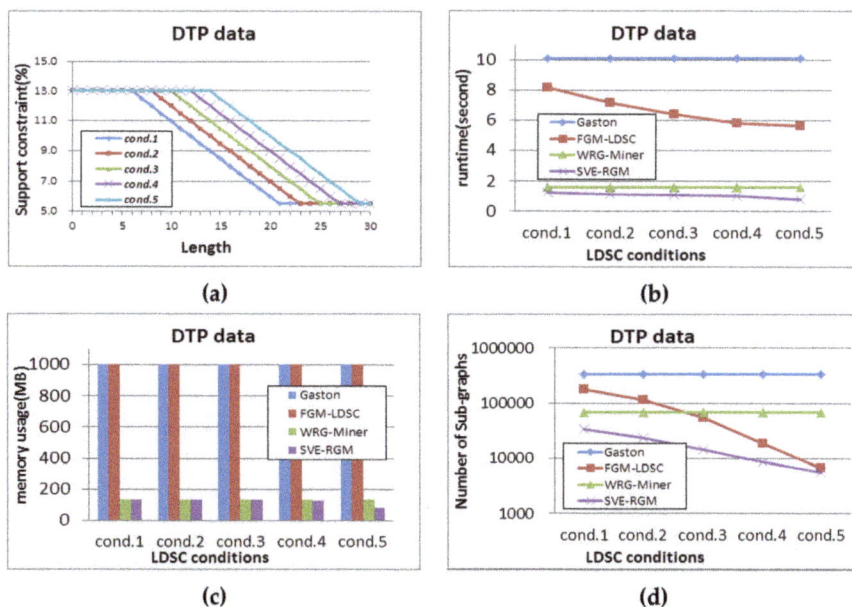

Figure 14. (**a**) Different settings of length-decreasing support constraints for DTP ($\alpha = 1$); (**b**) Runtime result for DTP ($\alpha = 1$); (**c**) Memory usage result for DTP ($\alpha = 1$); (**d**) Pattern generation result for DTP ($\alpha = 1$).

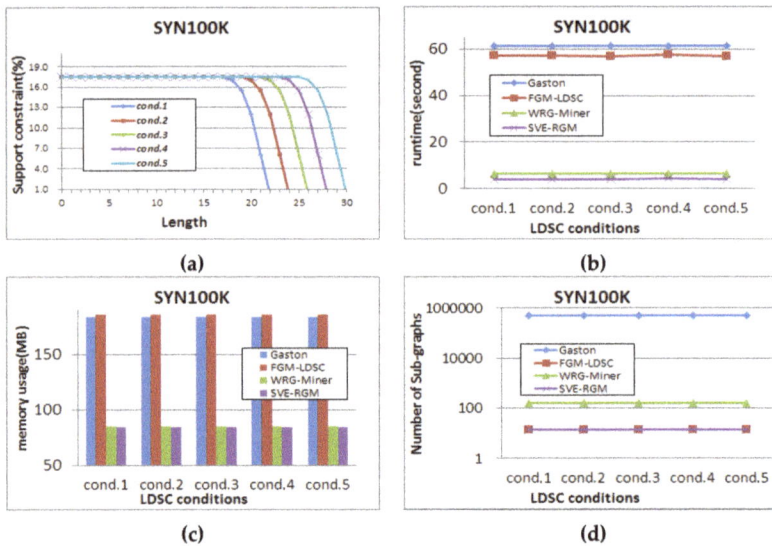

Figure 15. (**a**) Different settings of length-decreasing support constraints for SYN100K ($\alpha = 1$); (**b**) Runtime result for SYN100K ($\alpha = 1$); (**c**) Memory usage result for SYN100K ($\alpha = 1$); (**d**) Pattern generation result for SYN100K ($\alpha = 1$).

While the runtime results of *FGM-LDSC* are differently shown in Figure 13b, its memory usage results are not changeable as shown in Figure 13c. Meanwhile, the memory consumption of the proposed algorithm becomes lower as the condition is changed from cond. 1 to cond. 5.

Figure 14 is the results of the DTP dataset, where the minimum support threshold is fixed to 5.5%. As in the case of Figure 13, the performance evaluation results of *Gaston* and *WRG-Miner* are not changeable. The runtime results of *FGM-LDSC* and *SVE-RGM* are also changeable according to the different settings of the length-decreasing support constraints. In the case of memory usage performance, they also have better results compared to *Gaston* and *WRG-Miner*. Moreover, as α becomes lower, the corresponding memory performance of the proposed algorithm also becomes better.

Figure 15 shows the experimental results for the SYN100K dataset, where the minimum support threshold is fixed to 0.1%. In this test, all the compared algorithms have stable runtime, memory usage, and pattern generation results as shown in the figure. However, the proposed algorithm guarantees the best results in all cases. As shown in Figures 13–15 we can freely set the length-decreasing support constraints according to the use of given data and obtain the corresponding various results.

From the results of the above performance evaluation tests, we can determine that the proposed algorithm outperforms the state-of-the-art approaches in terms of runtime, memory usage, and pattern generation aspects.

5. Discussion

Since the concept of frequent pattern mining was considered, enormous approaches have been proposed in order to improve algorithm performance or discover more meaningful information and knowledge compared to previous methods. Since types of data are less and they have simple characteristics in the past, traditional frequent pattern mining methods were sufficiently able to analyze such data. However, as accumulated data become more complicated, the traditional ones faced technical limitations. Frequent graph pattern mining is a concept for dealing with various, complicated data that can be expressed as graph forms, and a variety of relevant works have been studied actively. We can consider a simple application example for frequent graph pattern mining. Let us assume that we are the marketing manager in a certain web-site for selling books. Then, we have to consider how to establish strategies for raising the sales more effectively. We can obtain information of web-site visits by various customers. In other words, if user A visits the main page, the book sales page and the e-book device sales page in sequence, we can regard this information as a graph pattern (each page becomes a vertex and each movement between the pages becomes an edge). Once such graph data are collected by various customers, we can construct a graph database and determine the characteristics of customer visits frequently occurring in the database by employing frequent graph pattern mining algorithms. By analyzing such mining results, we can also suggest and apply various, effective sales strategies such as exposing certain advertisements suitable for the relations of frequently occurring visits (*i.e.*, interval between the pages) and providing customers with special offers suitable for the analyzed results.

Meanwhile, the proposed method focuses on static graph database formats. However, in recent years, it has been important to deal with dynamic data on data streams as well as static data. Moreover, there are various environments of data accumulations and different purposes of them in the pattern mining area. Traditional pattern mining approaches were designed to process normal data (also called certain or precise data). However, there is another type of data called uncertain data, which mean each element composing data cannot be expressed clearly as presence or absence; instead, they have their own existential probabilities. By considering the above issues, we can expand the proposed method in order to operate on data streams and process uncertain graph data on the length-decreasing support constraints and multiple minimum support constraints. In addition, we can also consider how to find proper threshold settings in the proposed algorithm. In practice, it is hard to find and set appropriate threshold settings because of different characteristics of given data and purposes of users. However, if we employ the concept of top-k pattern mining, we do not need to consider how to find and set reasonable thresholds. Moreover, this technique can also be integrated effectively with the length-decreasing support constraints. By considering these issues, we can mine top-k graph patterns for each graph length.

6. Conclusions

In this paper, we proposed a new graph pattern mining algorithm for mining frequent sub-graph patterns on the basis of length-decreasing support constraints for considering different characteristics of graph pattern sizes and multiple minimum support constraints for considering rarity of graph elements. In addition, the SVE property for graph mining was suggested and applied into the proposed algorithm in order to improve the mining efficiency of the proposed algorithm. The suggested graph pattern pruning strategy based on the SVE property contributed to removing meaningless sub-graph patterns in advance with the anti-monotone property maintained. Moreover, an efficient overestimation method was devised to prevent unintended pattern losses from being caused. One of the fatal problems in traditional frequent graph pattern mining was that this approach only used a single minimum support threshold without any consideration of additional meaningful factors, such as pattern lengths and characteristics of graph elements. Thus, traditional methods were unable to find meaningful frequent graph patterns, or they had to generate a huge amount of unessential graph patterns according to threshold settings. In contrast, the proposed algorithm could prevent meaningless graph patterns from being generated and it also guaranteed efficient mining performance by using the length-decreasing support constraints and the multiple minimum supports constraints for considering the characteristics of graph pattern lengths and the rarity of graph patterns, respectively. The experimental results in this paper also supported that the proposed algorithm outperformed the state-of-the-art methods in various aspects such as runtime, memory usage and pattern generation. Moreover, as future work, it is also possible to expand the proposed techniques and algorithm by considering other interesting pattern mining areas that can be effectively integrated with graph pattern mining such as real time pattern mining on data streams, uncertain pattern mining, and top-k pattern mining.

Acknowledgments: This research was supported by the National Research Foundation of Korea (NRF) funded by the Ministry of Education, Science and Technology (NRF No. 20152062051, and NRF No. 20155054624), the Business for Academic-industrial Cooperative establishments funded Korea Small and Medium Business Administration in 2015 (Grants No. C0261068), and the Railway Technology Research Project of Korea Agency for Infrastructure Technology Advancement (No. 15RTRP-B082515-02).

Author Contributions: Unil Yun provided the main idea of this paper, designed the overall architecture of the proposed algorithm, and wrote the core contents of this paper. Gangin Lee implemented the proposed algorithm and wrote the main contents of this paper. Chul-Hong Kim investigated and reviewed references for graph theories and graph pattern mining applications to contribute to enhance the introduction and related work parts.

Conflicts of Interest: The authors declare no conflict of interest.

References

1.   Kim, J.; Yun, U.; Pyun, G.; Ryang, H.; Lee, G.; Yoon, E.; Ryu, K. A blog ranking algorithm using analysis of both blog influence and characteristics of blog posts. *Cluster Comput.* **2015**, *18*, 157–164. [CrossRef]

2.   Lee, G.; Yun, U.; Ryang, H. Mining weighted erasable patterns by using underestimated constraint-based pruning technique. *J. Intell. Fuzzy Syst.* **2015**, *28*, 1145–1157.

3.   Ryang, H.; Yun, U.; Pyun, G.; Lee, G.; Kim, J. Ranking algorithm for book reviews with user tendency and collective intelligence. *Multimedia Tools Appl.* **2015**, *74*, 6209–6227. [CrossRef]

4.   Ryang, H.; Yun, U.; Ryu, K. Discovering high utility itemsets with multiple minimum supports. *Intell. Data Anal.* **2014**, *18*, 1027–1047.

5.   Yun, U.; Lee, G. Sliding window based weighted erasable stream pattern mining for stream data applications. *Future Gener. Comp. Syst.* **2016**, *59*, 1–20. [CrossRef]

6.   Yun, U.; Pyun, G.; Yoon, E. Efficient Mining of Robust Closed Weighted Sequential Patterns Without Information Loss. *Int. J. Artif. Intell. Tools* **2015**, *24*, 1550007. [CrossRef]

7.   Agrawal, R.; Srikant, R. Fast Algorithms for Mining Association Rules. In Proceedings of the 20th International Conference on Very Large Data Bases, Santiago de Chile, Chile, 12–15 September 1994.

8.   Ryang, H.; Yun, U.; Ryu, K. Fast algorithm for high utility pattern mining with the sum of item quantities. *Intell. Data Anal.* **2016**, *20*, 395–415. [CrossRef]

9.   Ryang, H.; Yun, U. Top-k high utility pattern mining with effective threshold raising strategies. *Knowl. Based Syst.* **2015**, *76*, 109–126. [CrossRef]

10. Yun, U.; Lee, G. Incremental mining of weighted maximal frequent itemsets from dynamic databases. *Expert Syst. Appl.* **2016**, *54*, 304–327. [CrossRef]

11. Yun, U.; Ryang, H. Incremental high utility pattern mining with static and dynamic databases. *Appl. Intell.* **2015**, *42*, 323–352. [CrossRef]

12. Yun, U.; Kim, J. A fast perturbation algorithm using tree structure for privacy preserving utility mining. *Expert Syst. Appl.* **2015**, *42*, 1149–1165. [CrossRef]

13. Bifet, A.; Holmes, G.; Pfahringer, B.; Gavaldà, R. Mining frequent closed graphs on evolving data streams. In Proceedings of the 17th ACM SIGKDD International Conference on Knowledge Discovery and Data Mining, San Diego, CA, USA, 21–24 August 2011; pp. 591–599.

14. Hintsanen, P.; Toivonen, H. Finding reliable subgraphs from large probabilistic graphs. *Data Min. Knowl. Discov.* **2008**, *17*, 3–23. [CrossRef]

15. Thomas, L.T.; Valluri, S.R.; Karlapalem, K. MARGIN: Maximal Frequent Subgraph Mining. In Proceedings of the 6th IEEE International Conference on Data Mining, Hong Kong, China, 18–22 December 2006.

16. Zhang, S.; Yang, J.; Cheedella, V. Monkey: Approximate Graph Mining Based on Spanning Trees. In Proceedings of the 23rd International Conference on Data Engineering, Istanbul, Turkey, 11–15 April 2007.

17. Zou, Z.; Li, J. Mining Frequent Subgraph Patterns from Uncertain Graph Data. *IEEE Trans. Knowl. Data Eng.* **2010**, *22*, 1203–1218.

18. Dehmer, M.; Emmert-Streib, F. *Quantitative Graph Theory: Mathematical Foundations and Applications*; CRC Press: Boca Raton, FL, USA, 2014; pp. 1–34.

19. Dehmer, M.; Sivakumar, L; Varmuza, K. Uniquely Discriminating Molecular Structures Using Novel Eigenvalue—Based Descriptors. *Match-Commun. Math. Comput. Chem.* **2012**, *67*, 147–172.

20. Emmert-Streib, F.; Dehmer, M. *Information Theory and Statistical Learning*; Springer: New York, NY, USA, 2009; pp. 1–24.

21. Zhang, Q.; Song, X.; Shao, X.; Zhao, H.; Srikant, R. Object Discovery: Soft Attributed Graph Mining. *IEEE Trans. Pattern Anal. Mach. Intell.* **2016**, *38*, 532–545. [CrossRef] [PubMed]

22. Badr, E.; Heath, L.S. CoSREM: A graph mining algorithm for the discovery of combinatorial splicing regulatory elements. *BMC Bioinform.* **2015**, *16*, 1–15. [CrossRef] [PubMed]

23. Santosh, K.C. g-DICE: Graph mining-based document information content exploitation. *Int. J. Doc. Anal. Recognit.* **2015**, *18*, 337–355. [CrossRef]

24. Eskandari, M.; Raesi, H. Frequent sub-graph mining for intelligent malware detection. *Secur. Commun. Netw.* **2014**, *7*, 1872–1886. [CrossRef]

25. Videla-Cavieres, I.F.; Rios, S.A. Extending market basket analysis with graph mining techniques: A real case. *Expert Syst. Appl.* **2014**, *41*, 1928–1936. [CrossRef]

26. Hu, Y.H.; Chen, Y.L. Mining association rules with multiple minimum supports: A new mining algorithm and a support tuning mechanism. *Decis. Support Syst.* **2006**, *42*, 1–24. [CrossRef]

27. Kiran, R.U.; Reddy, P.K. Novel techniques to reduce search space in multiple minimum supports-based frequent pattern mining algorithms. In Proceedings of the 14th International Conference on Extending Database Technology, Uppsala, Sweden, 21–25 March 2011.

28. Liu, B.; Hsu, W.; Ma, Y. Mining association rules with multiple minimum supports. In Proceedings of the 5th ACM SIGKDD International Conference on Knowledge Discovery and Data Mining, San Diego, CA, USA, 15–18 August 1999.

29. Benhamou, B.; Jabbour, S.; Sais, L.; Salhi, Y. Symmetry Breaking in Itemset Mining. In Proceedings of the International Conference on Knowledge Discovery and Information Retrieval, Rome, Italy, 21–24 October 2014.

30. Desrosiers, C.; Galinier, P.; Hansen, P.; Hertz, A. Improving Frequent Subgraph Mining in the Presence of Symmetry. In Proceedings of the MLG Workshops, Firenze, Italy, 1–3 August 2007.

31. Vanetik, N. Mining Graphs with Constraints on Symmetry and Diameter. In Proceedings of the WAIM Workshops, Jiuzhaigou Valley, China, 15–17 July 2010.

32. Han, J.; Pei, J.; Yin, Y.; Mao, R. Mining frequent patterns without candidate generation: A frequent-pattern tree approach. *Data Min. Knowl. Discov.* **2004**, *8*, 53–87. [CrossRef]

33. Ozaki, T.; Etoh, M. Closed and Maximal Subgraph Mining in Internally and Externally Weighted Graph Databases. In Proceedings of the 25th IEEE International Conference on Advanced Information Networking and Applications Workshops, Singapore, Singapore, 22–25 March 2011.

34. Nijssen, S.; Kok, J.N. The Gaston Tool for Frequent Subgraph Mining. *Electr. Notes Theor. Comput. Sci.* **2005**, *127*, 77–87. [CrossRef]

35. Nijssen, S.; Kok, J.N. A quickstart in frequent structure mining can make a difference. In Proceedings of the Tenth ACM SIGKDD International Conference on Knowledge Discovery and Data Mining, Seattle, WA, USA, 22–25 August 2004.

36. Lee, G.; Yun, U. Frequent Graph Mining Based on Multiple Minimum Support Constraints. In Proceedings of the 4th International Conference on Mobile, Ubiquitous, and Intelligent Computing, Gwangju, Korea, 4–6 September 2013.

37. Lee, G.; Yun, U.; Ryang, H.; Kim, D. Multiple Minimum Support-Based Rare Graph Pattern Mining Considering Symmetry Feature-Based Growth Technique and the Differing Importance of Graph Elements. *Symmetry* **2015**, *7*, 1151–1163. [CrossRef]

38. Seno, M.; Karypis, G. Finding frequent patterns using length-decreasing support constraints. *Data Min. Knowl. Discov.* **2005**, *10*, 197–228. [CrossRef]

39. Yun, U. An efficient mining of weighted frequent patterns with length decreasing support constraints. *Knowl. Based Syst.* **2008**, *21*, 741–752. [CrossRef]

40. Yun, U.; Ryu, K.H. Discovering Important Sequential Patterns with Length-Decreasing Weighted Support Constraints. *Int. J. Inf. Technol. Decis. Mak.* **2010**, *9*, 575–599. [CrossRef]

41. Lee, G.; Yun, U. Frequent Graph Pattern Mining with Length-Decreasing Support Constraints. In Proceedings of the Multimedia and Ubiquitous Engineering, Seoul, Korea, 9–11 May 2013.

42. Dehmer, M.; Sivakumar, L. Recent Developments in Quantitative Graph Theory: Information Inequalities for Networks. *PLoS ONE* **2012**, *7*, e31395. [CrossRef] [PubMed]
43. Kraus, V.; Dehmer, M.; Emmert-Sreib, F. Probabilistic Inequalities for Evaluating Structural Network Measures. *Inf. Sci.* **2014**, *288*, 220–245. [CrossRef]
44. Samiullah, M.; Ahmed, C.F.; Fariha, A.; Islam, M.R.; Lachiche, N. Mining frequent correlated graphs with a new measure. *Expert Syst. Appl.* **2014**, *41*, 1847–1863. [CrossRef]
45. Sugiyama, M.; Llinares-López, F.; Kasenburg, N.; Borgwardt, K.M. Significant Subgraph Mining with Multiple Testing Correction. In Proceedings of the 2015 SIAM International Conference on Data Mining, Vancouver, BC, Canada, 30 April–2 May 2015.

symmetry

MDPI

Article

Optimal Face-Iris Multimodal Fusion Scheme

Omid Sharifi and Maryam Eskandari *

Department of Computer and Software Engineering, Toros University, Mersin 33140, Turkey;
omid.sharifi@toros.edu.tr
* Correspondence: maryam.eskandari@toros.edu.tr; Tel.: +90-324-325-3300

Academic Editor: Angel Garrido
Received: 9 May 2016; Accepted: 7 June 2016; Published: 15 June 2016

Abstract: Multimodal biometric systems are considered a way to minimize the limitations raised by single traits. This paper proposes new schemes based on score level, feature level and decision level fusion to efficiently fuse face and iris modalities. Log-Gabor transformation is applied as the feature extraction method on face and iris modalities. At each level of fusion, different schemes are proposed to improve the recognition performance and, finally, a combination of schemes at different fusion levels constructs an optimized and robust scheme. In this study, *CASIA Iris Distance* database is used to examine the robustness of all unimodal and multimodal schemes. In addition, Backtracking Search Algorithm (BSA), a novel population-based iterative evolutionary algorithm, is applied to improve the recognition accuracy of schemes by reducing the number of features and selecting the optimized weights for feature level and score level fusion, respectively. Experimental results on verification rates demonstrate a significant improvement of proposed fusion schemes over unimodal and multimodal fusion methods.

Keywords: multimodal biometrics; Backtracking Search Algorithm; match score level fusion; feature level fusion; decision level fusion; optimization

---

1. Introduction

The recognition of human beings based on physical and/or behavioral characteristics is a trend in places with high security needs. Unimodal biometric systems, which use single-source biometric traits, usually suffer due to several factors such as a lack of uniqueness, non-universality and noisy data [1]. In this respect, multimodality can be employed as a remedy in order to solve the limitations of unimodal systems and improve the system performance by extracting the information from multiple biometric traits.

The present work involves the consideration of face and iris biometric traits due to many similar characteristics of face and iris modalities. Face recognition performance may be affected by variations in terms of illumination, pose and expression [1]; on the other hand, non-cooperative situations lead to degradation of iris recognition performance [2]. In this study, we investigate the effect of information fusion on face- iris modalities at different levels of fusion in order to improve the recognition performance and solve the problems raised by unimodal face and iris traits.

The common biometric systems modules can be categorized as signal acquisition, feature extraction, and matching scores production. Generally, multimodal biometric systems fuse the information at four different fusion levels such as: sensor level; match score level; feature level; and; decision level fusion [1]. Match score level fusion is the most popular among all fusion levels due to the ease in accessing and fusing the scores. It general, three different categories are considered for match score level fusion, namely Transformation-based score fusion, Classifier-based score fusion, and Density-based score fusion. In Transformation-based score fusion, prior to fusion, the normalization of matching scores into a common domain and range is needed because of incompatibility of several biometric traits. In Classifier-based score fusion, the concatenated scores from different classifiers are treated as a feature vector and each matching score is seen as an element of feature vector. Density-based score fusion considers an explicit estimation of genuine and impostor score densities that causes increasing implementation complexity [3]. In feature level fusion, the original feature sets of different modalities are considered to involve richer information about the raw biometric data compared to matching score level fusion.

On the other hand, concatenating the feature sets leads to high dimensionality and redundant data, thus affecting the recognition performance [4]. Feature transformation and feature selection are usually applied as a remedy to reduce the effect of dimensionality and redundancy of feature level fusion [4]. Feature selection attempts to solve the problem by choosing an optimal subset of original feature sets based on a certain objective function. Several feature selection methods such as Particle Swarm Optimization (PSO), Genetic Algorithm (GA), and Sequential Forward Floating Selection (SFFS), have been implemented in different biometric systems for the purpose of optimization [1,5–8]. On the other hand, feature transformation attempts to represent the feature vector in another vector space by preserving the significant information and, subsequently, the dimension reduction. Principal Component Analysis (PCA), Linear Discriminant Analysis (LDA), Kernel Principal Component Analysis (KPCA), and Independent Component Analysis (ICA) have been applied in different biometric recognition systems as feature transformation methods [9].

In decision level fusion, each biometric matcher individually decides on the best match based on the provided input. In fact, the final decision is achieved by fusing the outputs of multiple matchers [10]. In general, a decision is represented by a logical number $d \in \{1,0\}$, where 1 means "accept" and 0 means "reject". From the classifiers' perspective, making any decision d_i is performed by comparing the matching scores s_i with a certain threshold T_i. Majority voting, behavior knowledge space, weighted voting based on the Dempster-Shafer theory of evidence, AND rule and OR rule, *etc.*, can be considered as decision level fusion techniques [11–14]. Mostly, this level of fusion is less studied in literature due to providing less information content compared to matching scores and features.

This study aims to investigate different fusion schemes at score level, feature level, decision level fusion and at a combination of aforementioned fusion levels using face and iris modalities. The face and iris can be considered as complementary biometric traits in which iris patterns are easily extracted from face images. They both can be acquired using a same camera simultaneously and can be considered as independent biometric traits. Face and iris fusion techniques involve specific feature extractors such as Discrete Wavelet Transform (DWT), Discrete Cosine Transform (DCT) and Gabor filters, the investigation of local and global feature extractors on the face and iris with the concentration on score and feature level fusion or combination of these two modalities, [1,6,15–17]. In [8,15,18], authors improve the recognition accuracy of face and iris modalities with different local and global feature extractors at score level and feature level fusion. Well-known techniques such as Weighted Sum Rule, Product Rule, Min Rule and Support Vector Machine (SVM) have been applied for face-iris combination at score level fusion and Particle Swarm Optimization (PSO) is used to explore the effect of feature selection on face and iris multimodal system performance. The authors of [17] proposed an intelligent 2v-support vector machine-based match score fusion algorithm. The proposed method integrates the quality of images in order to improve the recognition performance of face and iris modalities. A face-iris multimodal biometric system based on matching score level fusion using support vector machine (SVM) is applied in [1]. The authors implemented Discrete Cosine Transformation (DCT) to extract facial features and log-Gabor filter for extracting the iris pattern. The article improves the performance of multimodal face and iris biometrics through the selection of optimal features using the Particle Swarm Optimization (PSO) algorithm and the use of SVM for classification. A SVM-based fusion rule is also proposed in [16] to fuse two matching scores of face and iris modalities. Authors of [6] proposed an appropriate pattern representation strategy to extract the information using an over-complete global feature combination and, subsequently, the selection of the most useful features has been performed by Sequential Forward Floating Selection (SFFS).

In [19], an identification scheme has been proposed for improving the performance of face and iris multimodal biometric systems. The scheme is based on RBF (radial basis function) neural network fusion rules and applies both transformation-based score and classifier-based score fusion strategies. A new method has been proposed in [20] to fuse face and iris biometric traits with the weighted score level fusion technique to flexibly fuse the matching scores from these two modalities based on their weights availability. A more recent scheme has been proposed by [21], which uses matching score level and feature level fusion combination to improve the face and iris multimodal biometric systems. Optimized Weighted Sum Rule fusion has been applied in their work for score level fusion along with feature selection techniques such as Particle Swarm Optimization (PSO) and the Backtracking Search Algorithm (BSA) at feature level fusion.

The state-of-the-art literature review on face-iris multimodal biometric systems involves score level, feature level and/or a combination of these two levels of fusion. Therefore, this study investigates the effect of decision level fusion on a face-iris multimodal biometric system; in particular, the performance of the system is explored when considering threshold-optimized decision level fusion. We also aim to design a scheme to involve the consideration of matching score level, feature level, along with decision level fusion in order to investigate the effect of combining different fusion levels in designing robust fusion schemes for face and iris multimodal system. In this study, the facial and iris features are extracted using Log Gabor transform [22–24]. The Backtracking Search Algorithm (BSA) [25] as a feature selection method is applied to select the optimal set of facial and iris features at feature level fusion. At match score level fusion, the Weighted Sum Rule (WS) [26] is employed to combine the face and iris scores; additionally, BSA is applied to select the set of optimized weights for scores. Finally, at decision level fusion a threshold-optimized decision level fusion [27] is applied for improving the recognition performance of the multimodal system. The state-of-the-art performance of unimodal and multimodal schemes is reported on the *CASIA Iris Distance* [28] database in the verification context using Receiver Operator Characteristics (ROC) curves, Total Error Rate (TER) and Genuine Acceptance Rate (GAR) at a False Acceptance Rate (FAR) = 0.01%.

The contribution of the present work is to design a robust multimodal face-iris biometric system by combining the advantages of score level, feature level and decision level fusion. Human faces and irises can be considered as significant biometric traits in several surveillance, access control and forensic investigations applications such as airport control boards, criminal investigations, sexual dimorphism, and identity obfuscation applications. In addition, since the face and iris modalities are acquired simultaneously using the same camera, the proposed scheme is motivated to construct a robust multimodal biometric system. Therefore the proposed scheme can be applied practically in individual and multimodal face-iris recognition systems by extracting left and right iris patterns and then fusing them with facial features. The use of BSA as a robust feature and weight selection method in the proposed scheme is of interest for the performance enhancement of the system and overcoming the high computational time. On the other hand, the idea of using threshold-optimized points in the multimodal system is useful in the presence of outliers. Additionally, this work proposes to use the advantages of employing both irises with the face that provides higher verification performance while combining with facial information. In fact, the main difference between this work and prior work done on face and iris fusion is that it applies a hybrid scheme using score, feature and decision levels on the faces and irises of the same subjects, which can be applied practically in any surveillance, access control and forensic investigation applications.

The paper is organized as follows: Section 2 describes unimodal biometric systems, and the detailed implementation of different fusion levels. This section presents the architecture of the proposed scheme and the structure of implemented feature selection algorithm, as well. The detail of experimental results, including the database description and assessment protocols, is presented in Section 3, while Section 4 concludes this study.

2. Materials and Methods

2.1. Unimodal Biometric Systems

Face and iris as complementary biometric traits form the general structure of unimodal system in this study. They are considered as the most attractive areas for biometric schemes [29–35]. The unimodal face and iris system processing steps include preprocessing, feature extraction and producing matching scores. For face recognition systems, in the preprocessing step, Active Appearance Modeling (*AAM* toolbox) [36,37] is applied to detect face images based on the center position of the left and right irises. In fact, the precise center position of both irises is obtained by the toolbox to measure the angle of head roll that may happen during acquisition of face images. By using the center positions and the measured angle, both eyes are aligned in the face image. In addition, each image is resized to 60 × 60, and following this step the resized image undergoes histogram equalization (HE) and mean-variance normalization (MVN) [38] to reduce the effect of illumination. The facial features are then extracted using Log-Gabor transform. Generally, on the linear frequency scale, the structure of transfer function of the Log-Gabor transform is presented as [39]:

$$G\left(\omega\right) = exp\left\{\frac{-log\left(\frac{\omega}{\omega_0}\right)^2}{2 \times log\left(\frac{k}{\omega_0}\right)^2}\right\} \tag{1}$$

where ω_0 is the filter center frequency, ω is the normalized radius from the center and k is the standard deviation of the angular component. In order to achieve the constant shape filter, the ratio $\left(\frac{k}{\omega_0}\right)^2$ should be held constant for varying values of ω_0. In this work, the Log-Gabor transform includes four different scales and eight orientations. The values are fixed based on the different trial results. The produced Log-Gabor transformed image is then down-sampled by a fixed ratio on the trials as six. Therefore, the final size of Log-Gabor transformed image is reduced to 40 × 80. Finally, match scores are produced using the Manhattan distance measurement.

On the other hand, common processing steps of an iris recognition system are segmentation, normalization, feature extraction, and feature matching [39–41]. In this work, the Hough transform is applied in the segmentation stage of the iris recognition system for localizing the circular iris and pupil region, occluding eyelids and eyelashes, and reflections. The extracted iris region is then normalized into a fixed rectangular block. In feature extraction step, the unique pattern of irises is extracted using Log-Gabor transform with the same strategy as in face recognition. Therefore, the final size of the Log-Gabor transformed iris image is set to 40 × 80. Manhattan distance measurement is employed in feature matching step to produce the match scores.

2.2. Fusion Techniques on Face and Iris Biometrics

Multimodal face-iris biometric system development is one of the most significant steps in the present work. In this section, our aim is to describe the details of different fusion techniques for face and iris modalities. Since the proposed scheme involves the consideration of score level, feature level and decision level fusion, we describe each fusion technique separately at different subsections.

2.2.1. Feature Level Fusion

Feature level fusion concatenates the original feature sets of different modalities and, therefore, this level of fusion involves richer information about the raw biometric data. In this study, Log-Gabor transform is applied to face and iris biometric in order to extract rich and complex information on these two modalities. Indeed, the complementary details of face and iris biometrics, especially when both are acquired simultaneously with a same device, encourage us to fuse them using feature level fusion. On the other hand, the concatenation of face and iris Log-Gabor feature sets leads to high dimension vectors, resulting in the decrease of multimodal biometric system performance. Therefore, designing a scheme to retain the complementary information of the fused features of modalities with the capability to solve the dimensionality and redundancy problems is motivated. Designing a robust scheme needs the consideration of an effective feature selection method to select the optimized set of features by removing the redundant and irrelevant data. Several feature selection methods have been applied in the field of biometrics on fusion of face and iris modalities such as PSO. Recently, Backtracking Search Algorithm (BSA), a novel population-based iterative evolutionary algorithm, has been applied successfully on many numerical optimization benchmark problems [25]. BSA is compared with six widely used optimization methods, including PSO. The result of this comparison shows that BSA is more successful than the others [25]. Figure 1 depicts the block diagram of feature selection and fusion of face and iris modalities. The proposed scheme includes BSA optimization algorithm in order to select the optimized feature sets. The extracted texture features of face using Log-Gabor can be concatenated with extracted Log-Gabor features of the left or right iris as in Figure 1a, and then the best set of features is selected using BSA. In addition, we investigate the effect of considering both irises (left and right) feature sets while they are combined with face features and optimized using the BSA feature selection method, as in Figure 1b on recognition performance of the system. The final size of the face and iris Log-Gabor vector for each image after concatenating the corresponding filtered images is 32000 × 1. Thus, in this paper we project the Log-Gabor vector of face and iris modalities separately onto a linear discriminant space using Linear Discriminant Analysis (LDA) in order to reduce the dimensionality and computational cost prior to feature concatenation, as shown in Figure 1. In LDA, the eigenvectors used for projection is constrained by L-1, where L is number of subjects. We then perform BSA on the concatenated features to further reduce the dimension of each fused sample. Finally, the matching step is performed as depicted in the figure.

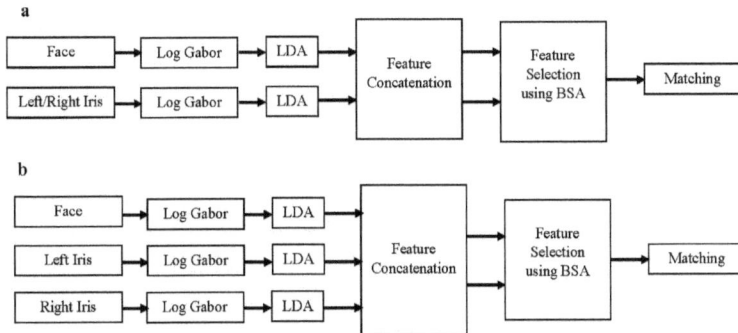

Figure 1. The block diagram of face-iris feature level fusion. LDA: Linear Discriminant Analysis; BSA: Backtracking Search Algorithm.

2.2.2. Match Score Level Fusion

Matching score level fusion techniques include different rules that combine the produced scores between the pattern vectors of different modalities. Generally, different matchers may produce different scores such as distances or similarity measures with different probability distributions or accuracies [3]. This kind of fusion technique covers several simple or complicated algorithms in order to fuse the scores such as Sum Rule, Weighted Sum Rule, Product Rule, classification using SVM and the estimation of scores density. Recent studies have shown similar and equivalent performance from the aforementioned fusion techniques [3,4,8,15,18]. Match score level fusion for this study involves the combination of left and/or right irises of a certain person with the same individual face scores. Figure 2 depicts the structure of match score level fusion scheme when face scores are fused with only one of the irises, as shown in Figure 2a, and when face scores are fused with both irises, as in Figure 2b.

a

b

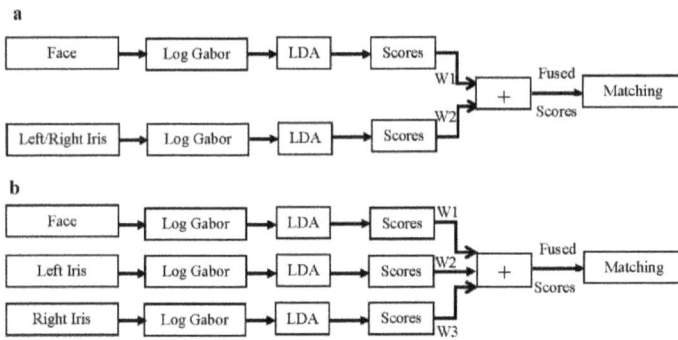

Figure 2. The block diagram of face-iris score level fusion. W1, W2 and W3: assigned weights for different modalities.

In this study, the Weighted Sum Rule technique is used in order to combine face and iris scores. Finding appropriate weights for different modalities is considered as an important issue to perform efficient fusion and, subsequently, for performance enhancement. In this respect, in [26], a user-specific weight strategy is used to compute the weighted sum of scores from different modalities. In general, the computation of weights is done based on the Equal Error Rate (EER), the distribution of scores, the quality of the individual biometrics or empirical schemes [5]. The Weighted Sum Rule (ws) of different score matchers can be calculated as:

$$ws = w_1 \times s_1 + w_2 \times s_2 + \cdots + w_n \times s_n \tag{2}$$

where $w_1, w_2, \ldots, w_n$ are the assigned weights for different modalities, and $s_1, s_2, \ldots, s_n$ are the computed scores using individual biometric systems. The present work assigns optimized weights to individual biometric systems using BSA feature selection algorithm.

2.2.3. Decision Level Fusion

In decision level fusion, each biometric matcher individually decides on the best match based on the provided input. In fact, the final decision is achieved by fusing the outputs of multiple matchers [10]. In general, a decision is represented by a logical number $d \in \{1,0\}$, where 1 means "accept" and 0 means "reject". From the classifiers' perspective, making any decision d_i is performed by comparing the matching scores s_i with a certain threshold T_i. Generally, this level of fusion is less studied in the literature and is not popular practically due to providing less information content compared to matching scores of different classifiers and the risk of performance degradation compared to score level fusion. Majority voting, weighted majority voting, Bayesian decision fusion, Dempster-Shafer theory of evidence, as well as the AND rule and OR rule can be considered as common decision level fusion techniques.

In this study, we apply the idea of threshold-optimized decision level fusion proposed in [27] to implement an optimized face-iris multimodal decision level fusion scheme. The threshold-optimized decision level fusion combines the decisions by AND and OR rules in an optimal way in which it guaranties to improve the fused classifiers in terms of error rates. The scheme is specifically useful in the presence of outliers when the proposed OR rule is applied [27]. In face and iris recognition systems, outliers can be caused by extraordinary expressions, poses, mis-registration, occlusions, reflections, contrast, luminosity, off angles, rotation, blurring and focus problems. Therefore in this work, we applied the threshold-optimized scheme using OR rule decision level fusion to combine face and iris modalities as depicted in Figure 3. In fact, the optimal operation points of face ROC can be fused with the optimal operation points of only one of the irises, as in Figure 3a, and also with the optimal operation points of both irises, as in Figure 3b.

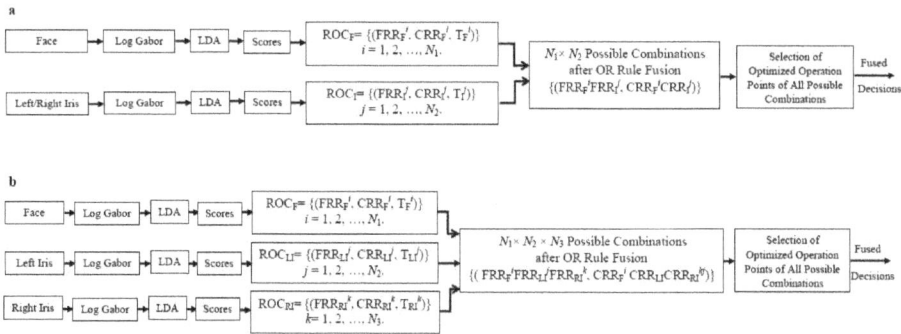

Figure 3. The block diagram of face-iris decision level fusion. ROC: Receiver Operator Characteristics; FRR: False Accept Rate; CRR: Correct Reject Rate

Generally, each biometric system is described by a ROC (Receiver Operator Characteristics), *i.e.*, the Genuine Accept Rate (GAR = 1 – False Reject Rate (FRR)) as a function of False Accept Rate (FAR), represented by GAR (FAR). The ROC is achieved by varying the threshold that discriminates the genuine and impostor matching scores, thus generating different GAR and FAR. Each point on the ROC, a certain pair (FAR, GAR) is called an operation point, corresponding to a specific threshold T of the matching scores. The threshold-optimized scheme fuses multiple ROCs together simply using the OR rule for performance enhancement. Therefore, the thresholds of matching scores are achieved when the optimal operation points on ROC are calculated.

Given N independent biometric systems, each characterized by its ROC, (FAR_i, GAR_i), $i = 1, ..., N$. In fact, the independency assumption is realistic in practice for fusion of different biometric modalities such as the face and iris. The optimized OR rule decision fusion under the independent assumption when the Correct Reject Rate for the impostors is defined as $CRR = 1 - FAR$ can be described by:

$$\max_{FRR_i \mid \prod FRR_i = FRR} \prod_{i=1}^{N} CRR_i \left(FRR_i \right) \tag{3}$$

That is the maximal value of the product of the correct rejection rates at a certain optimal combination of FRR_i, $i = 1, ..., N$, which satisfies $\prod_{i=1}^{N} FRR_i = FRR$. In other words, at a fixed FRR the optimal operation points of the component ROCs are achieved by optimizing Equation (3). In fact, the optimization problem defined in Equation (3) is solved in a recursive manner by fusing two arbitrary ROCs in order to generate a new optimal ROC. Then the computed threshold-optimized ROC is fused with the next arbitrary component ROC, and so on. Therefore, each operation point on the final fused ROC corresponds to N-optimized thresholds from N classifiers.

2.2.4. Architecture of the Proposed Scheme

This section describes the general structure of optimal proposed scheme for the fusion of face and iris biometrics. The scheme combines score level, feature level and decision level fusion to investigate the effect of combining different fusion levels in designing robust fusion schemes for a face and iris multimodal system. The block diagram of proposed scheme is depicted in Figure 4. In fact, our aim here is to design an optimal scheme by taking advantage of three aforementioned fusion modes, and eventually obtain a more reliable and robust biometric system. Therefore, the proposed scheme considers the combination of the face and left and right irises due to their complementary information. Our investigation on feature level fusion clarifies that combining facial features with both irises and then selecting an optimal set of features by using an appropriate feature selection method such as BSA leads to the involvement of rich and complex information of biometric data, and thus improves the recognition performance. Therefore, as shown in the block diagram of the proposed scheme in Figure 4, we first extract the optimal subset of face and both iris features at feature level fusion.

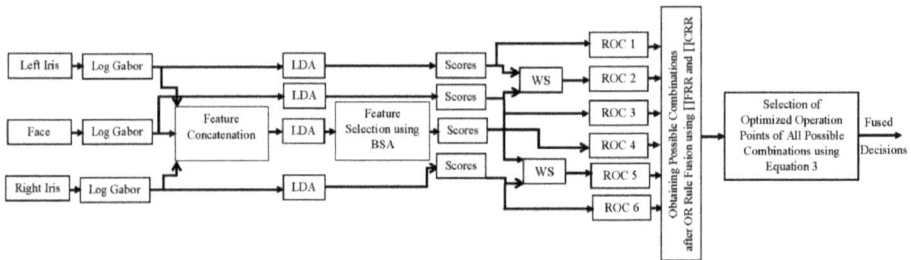

Figure 4. The block diagram of face-iris Proposed Scheme.

On the other hand, score level fusion contains rich information about the biometric input and is easy to process. In many applications, score-level fusion is able to achieve optimal performance. Therefore, the scheme attempts to fuse the complementary details of both irises with the face as shown in Figure 4. The Weighted Sum Rule fusion technique (WS) is applied to fuse the left and right iris scores separately with the face scores to achieve two optimal set of fused scores.

Decision level fusion schemes are simple and clear from a mathematical perspective. The proposed scheme in this study combines the decisions using the OR rule in an optimal way, and guaranties an improvement in the fused classifiers in terms of error rates. The produced scores from each modality separately, the produced scores after combining and selecting the optimized features at feature level fusion, along with the produced scores at match score level fusion using WS are considered as six different sets of scores to fuse threshold-optimized ROCs. Therefore, in a recursive manner, two arbitrary ROCs are fused to generate a new optimal ROC. Then the computed threshold-optimized ROC is fused with the next arbitrary component ROC, and so on.

2.2.5. BSA Feature Selection Algorithm

BSA has been introduced by Civicioglu [25] to solve numerical optimization problems. BSA tries to reduce the effect of problems faced in Evolutionary Algorithms such as excessive sensitivity to control parameters, premature convergence and slow computation. This algorithm aims to search local and global optimum in an optimization problem. It contains a single parameter, a simple, effective and fast structure that is capable of solving multimodal problems with the ability to adapt itself to different numerical optimization problems. BSA memory uses previous-generation experiences to generate trial populations. The algorithm includes five processes that include initialization, selection-I, mutation, crossover and selection-II. Algorithm 1 shows the general structure of BSA algorithm.

Algorithm 1 General Structure of Backtracking Search Algorithm [25].

---

1. Initialization
Repeat
2. Selection-I
Generation of Trial Population
3. Mutation
4. Crossover
End
5. Selection-II
Until stopping conditions are met

The *Initialization* step of the BSA algorithm initializes the population P of size n and dimension d randomly. The initial fitness value for each individual in P is calculated according to the fitness function. The direction of search is calculated in the *Selection-I* step and is called *historical population*. In fact, *historical population* is swarm-memory of BSA in which, initially, it is determined randomly and at the beginning of each iteration is updated through advantages of P based on two random numbers and randomly changing the order of individuals in *historical population*. BSA generates a trial population T using crossover and mutation strategies in order to recombine the crossover and mutation steps. The initial form of T is generated in the *Mutation* step that derives partial benefit from its experiences from former generations. The final form of T is generated in the Crossover step using a binary matrix (*map*) of size $n \times d$ and a recombination of the crossover and mutation steps. BSA considers a boundary control mechanism to regenerate the individuals beyond the search-space limit. In the *Selection-II* step, the fitness value for each individual in T is calculated according to the fitness function and, if the fitness values of T are better than fitness values in P, then P is updated by T to form new individuals. Besides exploring local optimum, BSA finds the global optimum by selecting the best individual and fitness value among all individuals in the current iteration. Therefore, the global optimum is updated to be P_{best} and the global optimum value is updated to be *fitness* P_{best}.

Indeed, above-mentioned explanations show the relatively simple structure of BSA and, according to its simple principle, it can be used in the implementation of different optimization problems. We use BSA at feature level and score level fusion in this study to select the optimized subset of features and weights. In score level fusion, we consider the idea of using BSA to select the optimized weights for Weighted Sum Rule fusion technique in order to have a better evaluation on the face-iris multimodal system. Basically, assigning appropriate weights in an efficient way to the scores produced using different individuals biometric systems may guarantee the performance improvement of multimodal biometric system.

BSA initialization step initializes population (P) and historical population (*oldp*) randomly between 0 and 1. The size of P and *oldp* is considered as the number of weights needed for fusing the scores of different modalities. The initialized weights are then normalized using the constraint $\sum_{i=1}^{k} w_i = 1$, where k is the number of weights and w is the weights. The fitness function is defined as follows for minimization:

$$F\left(\sum_{i=1}^{k} w_i EER_i\right) \tag{4}$$

where w_i is the set of optimized weights for different modalities and EER_i is a set of Equal Error Rates computed from the corresponding modalities scores.

The trial population T is considered as the original equation in [25] based on the following formula:

$$T = P + (map \times F) \times (oldp - P) \tag{5}$$

where F controls the amplitude of search direction matrix ($oldp - P$) and it is set experimentally.

In feature level fusion, the selection of features is based on a binary bit string of length M consisting of "0" and "1". The value of M indicates the number of features, "0" means the feature is not selected and "1" means the feature is selected. Therefore the dimension of initial population (P) and historical population (*oldp*) is equal to M, and both are randomly initialized using binary numbers. In this study, we compute the distance between reference and testing samples to find the match scores using the Manhattan distance measurement and then evaluate the lowest distance values. Therefore, the fitness function is defined to maximize GAR at FAR = 0.01%.

The original trial population T of BSA is modified in this study in order to generate binary numbers based on the following formula:

$$T = P \vee (map \wedge F) \wedge |(oldp - P)| \tag{6}$$

where F controls the amplitude of search direction matrix $|(oldp - P)|$ and it is set experimentally, $\vee$ and $\wedge$ are logical OR and AND operators.

The stopping condition for both binary and weight selection BSA is set to maximum number of iteration, or obtaining the optimal fitness value or failing to update the last best solution after 300 evaluations. If one of these three conditions is satisfied, the algorithm stops.

3. Results and Discussion

This section presents the detailed description of experimental setup, including database and assessment protocol applied in the current work for evaluating the proposed combined level fusion.

The experiments are carried out on a publicly available database called *CASIA-Iris-Distance*. The images in this database have been captured by a high-resolution camera, so both dual-eye iris and face patterns are available in the image region with detailed facial features that is appropriate for multimodal biometric information fusion [28]. Some samples of this database images are available in Figure 5.

Figure 5. Sample of CASIA-Iris-Distance images.

The full database contains the total number of 2567 images of 142 subjects and the images have been acquired at a distance of ~3 m from the camera [28]. The average size of extracted iris in this work is 170 × 150, and the average number of pixels between irises is 760. The availability of different variations on *CASIA-Iris-Distance* database is summarized in Table 1.

Table 1. Availability of Different Variations for Face and Irises in CASIA-Iris-Distance Database.

Face		Iris	
Pose variations	√	Occlusion-eyelash	√
Facial expressions	√	Occlusion-eyelid	√
Occlusion-glasses	√	Occlusion-glasses	√
Occlusion-mustache	√	Different noise factors (reflections, contrast, luminosity, off angle, rotation, blurring and focus problems)	√
Distance images	√	Distance images	√

In this work, we extract both irises of each subject from the corresponding face image to fuse the face and iris modalities. In order to validate the performance of unimodal and multimodal schemes in this study, the whole database is divided into two independent sets called Set-I and Set-II. The first set is used as the validation set to fix the parameters of feature level, score level and decision level fusion. BSA parameters (population size, iteration, F and *mix-rate*) to find optimized features and weights, and also estimation of the optimized thresholds, have been set using the validation set. This set (Set-I) consists of 52 subjects, and each subject possesses 10 samples. In this study, $F = 1$, *mix-rate-rate* = 1, population size and iteration are both set to 30 for binary BSA. On the other hand, for the weight selection BSA, $F = 1$, *mix-rate-rate* = 1, population size and iteration are set to 20 and 100, respectively. The dimension of search space for the binary BSA is the number of extracted features and for the weight selection BSA, it is number of weights needed for performing weighted sum in the range of [0.00, 1.00] with two-digit precision.

In addition, we consider 90 subjects for Set-II, and each subject possesses 10 samples. This set is divided into two equal partitions presenting five reference and five testing data for all the subjects in the database. The partitioning of these two subsets (reference and testing) is performed 10 times without any overlapping. Accordingly, in each trial 450 reference samples (90×5) and 450 testing samples (90×5) are considered. Therefore, 450 genuine scores and 40,050 ($90 \times 89 \times 5$) imposter matching scores are used to validate the verification performance analysis in this study. The results are averaged over 10 different runs and reported as mean and standard deviations runs in the verification context using ROC curves, Total Error Rate (TER) and Genuine Acceptance Rate (GAR) at False Acceptance Rate (FAR) = 0.01%. Generally, TER is the sum of FAR and FRR, which is equal to twice the value of EER. The implementation of all unimodal and multimodal biometric systems is done using Matlab.

The first set of experiments analyzes the results of the implementation of different unimodal recognition systems such as the left iris, right iris, and face. The experimental results are demonstrated in Table 2 using Log-Gabor.

Table 2. Verification Performance and Minimum Total Error Rates of Unimodal Systems. TER: Total Error Rate; GAR: Genuine Acceptance Rate; FAR: False Acceptance Rate.

Left Iris		Right Iris		Face	
Minimum TER (%)	GAR (at 0.01% FAR)	Minimum TER (%)	GAR (at 0.01% FAR)	Minimum TER (%)	GAR (at 0.01% FAR)
6.93 ± 1.24	69.67 ± 4.10	6.24 ± 1.01	71.55 ± 3.35	2.88 ± 0.75	83.22 ± 1.63

In Table 2, the best verification performance belongs to the face unimodal system at 83.22%. On the other hand, for the iris unimodal system, as shown in the table, the right iris achieves a better verification and TER compared to the left iris. We consider the fusion of face and iris modalities using different levels of fusion in order to observe the effect of fusion on the recognition performance. Thus, as shown in Table 3, we continue the experiments at feature level fusion of face and iris biometrics. Firstly, we perform the feature fusion of face and iris biometrics implemented in Figure 1 without applying BSA, and then BSA as a feature selection strategy is used to investigate the effect of an optimal feature selection algorithm on the verification performance.

Table 3. Verification Performance and Minimum Total Error Rates of Multimodal Biometric Systems at Feature Level Fusion.

Scheme	Minimum TER (%)	GAR (at 0.01% FAR)
Feature fusion scheme implemented in Figure 1a using left iris without BSA	2.15 ± 0.64	87.56 ± 2.94
Feature fusion scheme implemented in Figure 1a using left iris with BSA	2.06 ± 0.71	88.37 ± 2.68
Feature fusion scheme implemented in Figure 1a using right iris without BSA	2.06 ± 0.98	88.44 ± 3.05
Feature fusion scheme implemented in Figure 1a using right iris with BSA	1.84 ± 0.73	90.16 ± 2.88
Feature fusion scheme implemented in Figure 1b using both irises without BSA	1.03 ± 0.46	92.87 ± 1.65
Feature fusion scheme implemented in Figure 1b using both irises with BSA	0.86 ± 0.34	94.91 ± 1.83

It can be observed from Table 3 that the performance of BSA-based schemes is superior to the schemes without the BSA feature selection. The best performance in terms of TER and GAR is, respectively, 0.86% and 94.91%, and it is achieved using the feature level fusion scheme presented in

Figure 1b with BSA. In order to examine the effectiveness of score level fusion on combining face and iris modalities, the experiments are carried out for this level of fusion and the results are reported in Table 4. Comparing Tables 3 and 4 demonstrates that feature level fusion including optimal set of feature sets achieves a slightly better performance in terms of TER and verification when one of the irises is fused with facial features. However, as Table 4 indicates, match score level fusion outperforms feature level fusion when both irises are combined with the face. The best performance of match score level fusion in terms of TER and GAR is, respectively, 0.81% and 95.00%.

Table 4. Verification Performance and Minimum Total Error Rates of Multimodal Biometric Systems at Score Level Fusion.

Scheme	Minimum TER (%)	GAR (at 0.01% FAR)
Score fusion scheme implemented in Figure 2a using left iris	2.10 ± 0.63	88.17 ± 1.75
Score fusion scheme implemented in Figure 2a using right iris	2.01 ± 0.77	88.76 ± 2.18
Score fusion scheme implemented in Figure 2b using both irises	0.81 ± 0.48	95.00 ± 2.03

Table 5 shows the set of experiments at decision level fusion using the OR rule threshold-optimized scheme implemented in Figure 3. The best TER and GAR is obtained using Figure 3b when the face and both irises involved are at 0.58% and 96.87%. The optimized scheme achieves 1.87% and 1.96% improvement compared to the best verification performance of score level and feature level fusion schemes. Finally, the last set of experiment in Table 6 evaluates the proposed combined level fusion scheme and compares the corresponding result with achieved GAR and TER of each level of fusion separately.

Table 5. Verification Performance and Minimum Total Error Rates of Multimodal Biometric Systems at Decision Level Fusion.

Scheme	Minimum TER (%)	GAR (at 0.01% FAR)
Decision fusion scheme implemented in Figure 3a using left iris	0.96 ± 0.28	91.15 ± 2.38
Decision fusion scheme implemented in Figure 3a using right iris	0.94 ± 0.71	92.92 ± 3.01
Decision fusion scheme implemented in Figure 3b using both irises	0.58 ± 0.32	96.87 ± 1.83

Table 6. Verification Performance and Minimum Total Error Rates of Proposed Multimodal Biometric Systems and Different Levels of Fusions.

Scheme	Minimum TER (%)	GAR (at 0.01% FAR)
Feature fusion scheme implemented in Figure 1b using both irises-with BSA	0.86 ± 0.34	94.91 ± 1.83
Score fusion scheme implemented in Figure 2b using both irises	0.81 ± 0.48	95.00 ± 2.03
Decision fusion scheme implemented in Figure 3b using both irises	0.58 ± 0.32	96.87 ± 1.83
Proposed Scheme	0.27 ± 0.41	98.93 ± 1.11

As the table demonstrates, the best performance is achieved using the proposed scheme since it involves the consideration of each level of fusion advantage for performing the fusion of face and iris biometrics. Specifically, as it is described in [27], the OR Rule threshold-optimized scheme is useful in the presence of outliers. Thus, involving this significant characteristic of the decision level fusion scheme in our proposed scheme, along with consideration of optimized features and weights at feature level and score level fusion, leads to a robust multimodal biometric system. Comparing the results obtained from different levels of fusion with our combined level fusion shows the superiority of the proposed scheme over all unimodal and multimodal schemes implemented in this study. The proposed scheme performance improvement in terms of GAR and TER is obtained as 98.93% and 0.27%, respectively.

On the other hand, in order to compare our proposed scheme with state-of-the-art face-iris fusion methods, we performed the fusion of face and iris using different fusion techniques on the *CASIA-Iris Distance* database in Table 7. The experimental results performed in Table 7 show the superiority of the proposed scheme over other face-iris multimodal biometric systems implemented in this study. Recently, the Support Vector Machine (SVM), mainly as a popular method for classification, is used in the area of statistics learning theory. Generally, SVM is targeted based on structural minimization principle and maps the training data into a higher dimensional feature using the kernel trick to construct an optimal hyperplane with large separating margin between two classes of the labeled data. In this work, the radial basis function (RBF) has been applied as the basic kernel function by iterative trials.

Table 7. Verification Performance and Minimum Total Error Rates of Different Multimodal Biometric Systems and Proposed Scheme. PSO: Particle Swarm Optimization; SVM: Support Vector Machine.

State-of-the-Art Fusion Methods on Face and Iris	Minimum TER (%)	GAR (at 0.01% FAR)
Weighted Sum Rule [15]	2.01 ± 0.77	88.76 ± 2.18
Score concatenation [18]	1.49 ± 0.48	89.53 ± 1.67
SVM [16]	1.56 ± 0.71	91.92 ± 2.03
PSO and SVM [1]	1.06 ± 0.39	94.22 ± 1.48
Proposed scheme	0.27 ± 0.41	98.93 ± 1.11

The ROC analysis of the face unimodal system and multimodal biometric systems, including the proposed scheme, is demonstrated in Figure 6. The ROC analysis covers part (b) of the implemented schemes for each level of fusion.

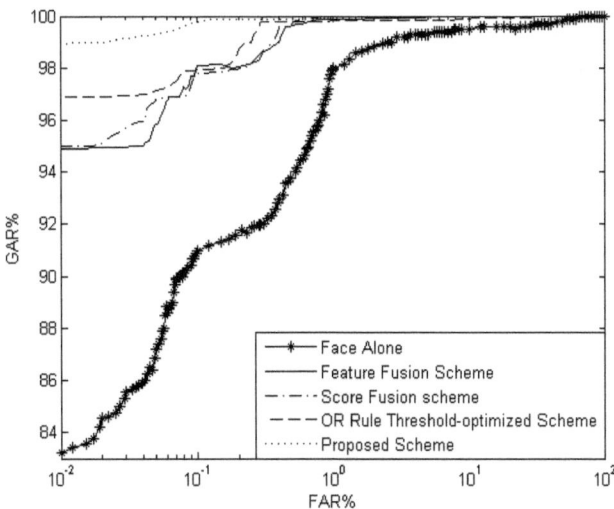

Figure 6. ROC curves of face unimodal and multimodal systems of different schemes.

On the other hand, Figure 7 compares the ROC analysis of proposed scheme and the OR rule threshold-optimized decision level fusion.

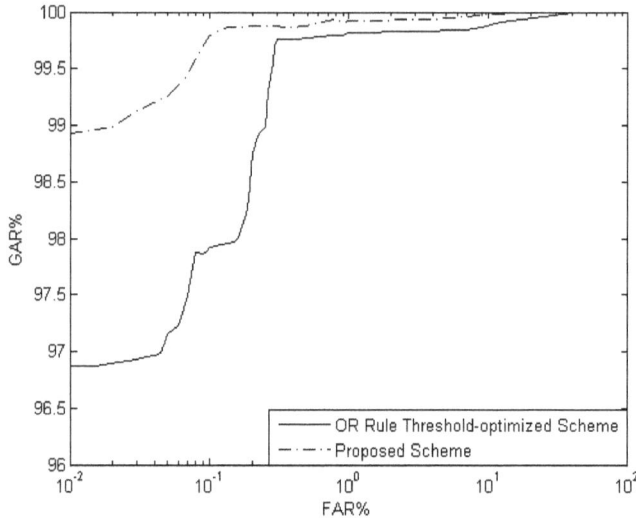

Figure 7. ROC curves of threshold-optimized and proposed schemes.

As observed from the ROC curves, the proposed scheme outperforms unimodal and all multimodal schemes implemented in this study.

4. Conclusions

In this paper, we have investigated the problem of combining different levels of fusion in a face-iris multimodal biometric system framework. Our aim here was to implement different fusion schemes and then compare them with a scheme, including their complementary advantages, in terms of performance. Therefore, we have designed a robust multimodal face-iris biometric system by combining the advantages of score level, feature level and decision level fusion. The proposed scheme has applied Log-Gabor transform as the feature extraction method on face and iris modalities and, subsequently, the corresponding features and scores have been employed to construct different fusion schemes. We specifically have applied a threshold-optimized scheme at the decision level fusion step of the proposed scheme that is useful in the presence of outliers. In addition, BSA as an effective and recent feature selection method has been used with feature and score level fusion of the proposed scheme to construct a more robust biometric system; this has been done by reducing the number of features and improving the performance, and also optimizing the weights. In fact, based on the experimental results provided in this study, we can attract the attention of new perspectives for face-iris multimodal biometric systems that consider the combination of different levels of fusion, in particular decision level fusion, to efficiently represent a robust system.

Author Contributions: The authors performed the experiments and analyzed the results together. Introduction, unimodal and multimodal biometric systems, feature selection algorithm sections were written by Omid Sharifi; while experimental results, discussion and conclusion sections were written by Maryam Eskandari.

Conflicts of Interest: The authors declare no conflict of interest.

References

1. Liau, H.F.; Isa, D. Feature selection for support vector machine-based face-iris multimodal biometric system. *Expert Syst. Appl.* **2011**, *38*, 11105–11111. [CrossRef]
2. Proenca, H.P. Towards Non-Cooperative Biometric Iris Recognition. Ph.D. Thesis, University of Beira Interior Department of Computer Science, Covilhã, Portugal, 2006.
3. Nandakumar, K.; Chen, Y.; Dass, S.C.; Jain, A.K. Likelihood ratio-based biometric score fusion. *IEEE Trans. Pattern Anal. Mach. Intell.* **2008**, *30*, 342–347. [CrossRef] [PubMed]
4. Ross, A.; Nandakumar, K.; Jain, A.K. *Handbook of Multibiometrics*; Springer-Verlag: Berlin, Germany, 2006.
5. Raghavendra, R.; Dorizzi, B.; Rao, A.; Kumar, G.H. Designing efficient fusion schemes for multimodal biometric system using face and palmprint. *Pattern Recognit.* **2011**, *44*, 1076–1088. [CrossRef]
6. Lumini, A.; Nanni, L. Over-complete feature generation and feature selection for biometry. *Expert Syst. Appl.* **2008**, *35*, 2049–2055. [CrossRef]
7. Gökberk, B.; Okan İrfanoğlu, M.; Akarun, L.; Alpaydın, E. Learning the best subset of local features for face recognition. *Pattern Recognit.* **2007**, *40*, 1520–1532. [CrossRef]
8. Eskandari, M.; Toygar, Ö.; Demirel, H. Feature Extractor Selection for Face-Iris Multimodal Recognition. *Signal Image Video Process.* **2014**, *8*, 1189–1198. [CrossRef]
9. Zhang, D.; Jing, X.; Yang, J. *Biometric Image Discrimination (BID) Technologies*; IGI Global: Hershey, PA, USA, 2006.
10. Nandakumar, K. Integration of Multiple Cues in Biometric Systems. Master's Thesis, Michigan State University, East Lansing, MI, USA, 2005.
11. Lam, L.; Suen, C.Y. Application of Majority Voting to Pattern Recognition: An Analysis of Its Behavior and Performance. *IEEE Trans. Syst. Man Cybern. Part A Syst. Hum.* **1997**, *27*, 553–568. [CrossRef]
12. Lam, L.; Suen, C.Y. Optimal Combination of Pattern Classifiers. *Pattern Recognit. Lett.* **1995**, *16*, 945–954. [CrossRef]
13. Xu, L.; Krzyzak, A.; Suen, C.Y. Methods for Combining Multiple Classifiers and their Applications to Handwriting Recognition. *IEEE Trans. Syst. Man Cybernet.* **1992**, *22*, 418–435. [CrossRef]
14. Daugman, J. Combining Multiple Biometrics. Available online: http://www.cl.cam.ac.uk/users/jgd1000/combine/combine.html (accessed on 28 October 2016).
15. Eskandari, M.; Toygar, Ö. Fusion of face and iris biometrics using local and global feature extraction methods. *Signal Image Video Process.* **2014**, *8*, 995–1006. [CrossRef]
16. Wang, F.; Han, J. Multimodal biometric authentication based on score level fusion using support vector machine. *Opto Electron. Rev.* **2009**, *17*, 59–64. [CrossRef]
17. Vasta, M.; Singh, R.; Noore, A. Integrating image quality in 2v-SVM biometric match score fusion. *Int. J. Neural Syst.* **2007**, *17*, 343–351.
18. Eskandari, M.; Toygar, Ö.; Demirel, H. A new approach for Face-Iris multimodal biometric recognition using score fusion. *Int. J. Pattern Recognit. Artif. Intell.* **2013**, *27*. [CrossRef]
19. Wang, Y.; Tan, T.; Wang, Y.; Zhang, D. Combining face and iris biometric for identity verification. In Proceedinmgs of the 4th International Conference on Audio and Video Based Biometric Person Authentication, Guildford, UK, 9–11 June 2003; pp. 805–813.
20. Sim, H.M.; Asmunia, H.; Hassan, R.; Othman, R.M. Multimodal biometrics: Weighted score level fusion based on non-ideal iris and face images. *Expert Syst. Appl.* **2014**, *41*, 5390–5404. [CrossRef]
21. Eskandari, M.; Toygar, Ö. Selection of Optimized features and weights on face-iris fusion using distance images. *Comput. Vis. Image Underst.* **2015**, *137*, 63–75. [CrossRef]
22. Jing, X.Y.; Yao, Y.F.; Yang, J.Y.; Li, M.; Zhang, D. Face and palmprint pixel level fusion and kernel DCV-RBF classifier for small sample biometric recognition. *Pattern Recognit.* **2007**, *40*, 3209–3224. [CrossRef]
23. Yao, Y.; Jing, X.; Wong, H. Face and palmprint feature level fusion for single sample biometric recognition. *Neurocomputing* **2007**, *70*, 1582–1586. [CrossRef]
24. Xiao, Z.; Guo, C.; Yu, M.; Li, Q. Research on log gabor wavelet and its application in image edge detection. In Proceedings of 6th International Conference on Signal Processing (ICSP-2002), Beijing, China, 26–30 August 2002; pp. 592–595.
25. Civicioglu, P. Backtracking Search Optimization Algorithm for numerical optimization problems. *Appl. Math. Comput.* **2013**, *219*, 8121–8144. [CrossRef]

26. Jain, A.K.; Ross, A. Learning User-specific Parameters in a Multibiometric System. In Proceedings of International Conference on Image Processing, New York, NY, USA, 22–25 September 2002; pp. 57–60.

27. Tao, Q.; Veldhuis, R. Threshold-Optimized decision-level fusion and its application to biometrics. *Pattern Recognit.* **2009**, *42*, 823–836. [CrossRef]

28. Biometrics Ideal Test. Available online: http://biometrics.idealtest.org/dbDetailForUser.do?id=4 (assessed on 30 August 2013).

29. Patil, H.; Kothari, A.; Bhurchandi, K. 3-D face recognition: Features, databases, algorithms and challenges. *Artif. Intell. Rev.* **2015**, *44*, 393–441. [CrossRef]

30. Subburaman, V.B.; Marcel, S. Alternative search techniques for face detection using location estimation and binary features. *Comput. Vis. Image Underst.* **2013**, *117*, 551–570. [CrossRef]

31. Gul, G.; Hou, Z.; Chen, C.; Zhao, Y. A dimensionality reduction method based on structured sparse representation for face recognition. *Artif. Intell. Rev.* **2016**. [CrossRef]

32. Bowyer, K.W.; Hollingsworth, K.; Flynn, P.J. Image understanding for iris biometrics: A survey. *Comput. Vis. Image Underst.* **2008**, *110*, 281–307. [CrossRef]

33. Matey, J.R.; Broussard, R.; Kennell, L. Iris image segmentation and sub-optimal images. *Image Vis. Comput.* **2010**, *28*, 215–222. [CrossRef]

34. Galbally, J.; Ross, A.; Gomez-Barrero, M.; Fierrez, J.; Ortega-Garcia, J. Iris image reconstruction from binary templates: An efficient probabilistic approach based on genetic algorithms. *Comput. Vis. Image Underst.* **2013**, *117*, 1512–1525. [CrossRef]

35. Neves, J.; Narducci, F.; Barra, S.; Proença, H. Biometric recognition in surveillance scenarios: A survey. *Artif. Intell. Rev.* **2016**. [CrossRef]

36. Huang, C.; Ding, X.; Fang, C. Pose robust face tracking by combining view-based AAMs and temporal filters. *Comput. Vis. Image Underst.* **2012**, *116*, 777–792. [CrossRef]

37. Active Appearance Modeling. Available online: http://cvsp.cs.ntua.gr/software/AAMtools/ (accessed on 20 April 2013).

38. Pujol, P.; Macho, D.; Nadeu, C. On real-time mean-and- variance normalization of speech recognition features. In Proceedings of the IEEE International Conference on Acoustics, Speech and Signal Processing (ICASSP2006), Toulouse, France, 14–19 May 2006; pp. 773–776.

39. Arora, S.; Londhe, N.D.; Acharya, A.K. Human Identification based on Iris Recognition for Distance Images. *Int. J. Comput. Appl.* **2012**, *45*, 32–39.

40. Masek, L.; Kovesi, P. MATLAB Source Code for a Biometric Identification System Based on Iris Patterns. Bachelor's thesis, the School of Computer Science and Software Engineering, The University of Western Australia, Crawley, Australia, 2003.

41. Tan, C.W.; Kumar, A. A Unified Framework for Automated Iris Segmentation Using Distantly Acquired Face Images. *IEEE Trans. Image Process.* **2012**, *21*, 4068–4079. [CrossRef] [PubMed]

symmetry

MDPI

Article

A Modified GrabCut Using a Clustering Technique to Reduce Image Noise

GangSeong Lee [1], SangHun Lee [1,*], GaOn Kim [2,*], JongHun Park [2] and YoungSoo Park [1]

[1]   Ingenium college of liberal arts, Kwangwoon university, Seoul 01897, Korea; gslee@kw.ac.kr (G.L.); yspark@kw.ac.kr (Y.P.)
[2]   Department of plasmadiodisplay, Kwangwoon university, Seoul 01897, Korea; qwerty@kw.ac.kr
*   Correspondence: leesh58@kw.ac.kr (S.L.); gaon@kw.ac.kr (G.K.); Tel.: +82-2-940-5287 (S.L.)

Academic Editor: Angel Garrido
Received: 31 March 2016; Accepted: 29 June 2016; Published: 14 July 2016

Abstract: In this paper, a modified GrabCut algorithm is proposed using a clustering technique to reduce image noise. GrabCut is an image segmentation method based on GraphCut starting with a user-specified bounding box around the object to be segmented. In the modified version, the original image is filtered using the median filter to reduce noise and then the quantized image using K-means algorithm is used for the normal GrabCut method for object segmentation. This new process showed that it improved the object segmentation performance a lot and the extract segmentation result compared to the standard method.

Keywords: median filter; K-means, image clustering; GraphCut; GrabCut; object segmentation

1. Introduction

Digital image processing deals with a wide variety of applications ranging from biology, military, medical, space science, art, games and movie industries.

The object segmentation is an important step in image processing and analysis [1]. In computer vision, segmentation divides the input image into background and objects. The purpose of the segmentation is to simplify and make it easy to interpret or convert to more meaningful representation of an image. Segmentation is one of the most difficult subjects in an digital image processing, and many studies on this subject have been done to get more accurate results.

GrabCut method is based on object segmentation algorithm called GraphCut [2,3]. While GraphCut algorithm segments an image without user intervention, GrabCut accepts an interest area defined by a user and extracts objects using the clues given to get better results. Many studies have been done to improve performance of GrabCut detecting objects in unknown regions [4,5].

In the proposed method, the image is smoothed using median filter and the quantized using k-means clustering technique. Then, GrabCut extracts objects from the quantized image [6]. In this way, we got improved performance.

2. Related Work

In general, object segmentation is one of the most fundamental tasks in image processing. Image segmentation is to divide an image into a number of pixel sets on the basis of shape or area. In this work, we following image filters and a clustering technique is applied for an efficient object segmentation.

2.1. Image Filter

Filtering is one of the main tasks of signal processing. Filtering is used to remove noise in the image, to extract visual characteristics of interest, and to resample the image. Representative filters are Gaussian filter, Mean, etc.

Gaussian filter is used to remove the noise using the following equations:

$$G(x) = \frac{1}{\sigma\sqrt{2\pi}}\exp\left[\frac{-x^2}{2\sigma}\right] \tag{1}$$

$$G(x,y) = \frac{1}{2\pi\sigma^2}\exp\left[\frac{-(x^2+y^2)}{2\sigma^2}\right] \tag{2}$$

where σ is the standard deviation.

Mean filter is the representative noise removing filter and is defined as follows:

$$O(x,y) = \frac{1}{M \times M}\sum_{i,j}^{M} I(x+i,y+j)$$

$$M \in W$$

$$M = 2N+1 \tag{3}$$

where $I(x+i,y+j)$ is a neighbor pixel and W is a mask of size $(2N+1) \times (2N+1)$.

2.2. Image Segmentation

Segmentation is the process of dividing the digital image into a set of multiple pixels to simplify the image representation. Typical methods of image clustering are Mean Shift (MS), Fuzzy C-Means (FCM), etc. [7].

Mean shift (MS) is a procedure for locating the maxima of a given window area by selecting a pixel (mode) most close to the averaged color, then moving the center of the window to the mode to find local maxima repeatedly until it is converged [8,9]. The result of MS is good in low frequency areas, but it has some difficulties to group high frequency areas. The following Figure 1 show the example.

(a)　　　　　　　　　　　　　(b)

Figure 1. MS clustering. (**a**) Original image; (**b**) Result image.

FCM is to overcome the difficulty of MS by considering belonging or membership degree into the distance. Data-points close to a cluster center have a high belonging degree [10,11].

Let $X = x_1, \cdots, x_N \subseteq R^p$ a data set to be clustered, where p is feature dimension, R^p is a real vector space, and N is the number of pixels. Each pixel of color image is expressed as feature vectors like $x_k = x_{k1}, \cdots, x_{kp}$ and center of cluster is $V = (v_1, \cdots, v_C)$, where p is feature dimension and C is the number of clusters. FCM calculates matrix U minimizing the target function $J_{FCM}(U, V|X)$.

$$J_{FCM}(U, V|X) = \sum_{i=1}^{C}\sum_{k=1}^{N}(u_{ik})^m \|x_k - v_i\|^2 \tag{4}$$

where $\sum_{i=1}^{C} u_{ik} = 1$, and u_{ik} is membership degree that data k belongs to cluster i. V_j is the center of cluster i, defined using Equation (5).

$$v_i = \frac{1}{\sum_{k=1}^{N} u_{ik}} \sum_{k=1}^{N} (u_{ik})^m x_k, m\rangle 1 \tag{5}$$

where u_{ik} is the mean value of x_k using fuzzy constant m. One disadvantage of FCM is that the number of clusters should be provided in advance. Figure 2 is the class membership map applying FCM algorithm and Figure 3 is result of FCM clustering.

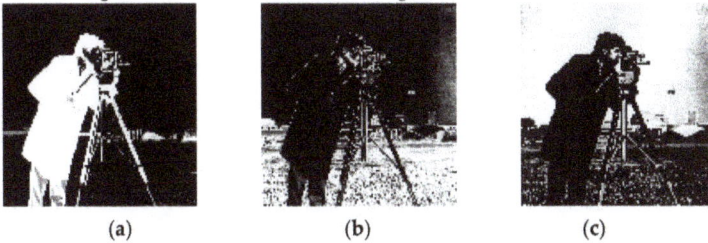

Figure 2. FCM class membership map. (a) Class 1; (b) Class 2; (c) Class 3.

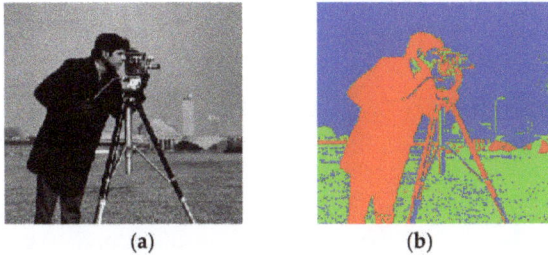

Figure 3. FCM clustering. (a) Original image; (b) Result image.

Image segmentation can be accomplished also by using GraphCut. Boykov proposed GraphCut to get optimized interactive image segmentation:

$$E(\alpha) = U(\alpha) + V(\alpha) \tag{6}$$

where α is a vector of either 0 or 1. 0 means background and 1 means object. $U(\alpha)$ is continuity between adjacent pixels and $V(\alpha)$ is a data term representing how much the data belongs to object or background. Data term requires prior information about object and background, and provided probability density function, data term calculated using Equation (7).

$$V(\alpha) = \begin{cases} \sum_{p \in P} -\log p\left(I_p | H_b\right) & if \quad \alpha_p = 0 \\ \sum_{p \in P} -\log p(I_p | H_o) & otherwise \end{cases} \tag{7}$$

where H_b and H_o are histograms of background and object, p is a pixel and P is a pixel set of an image. I_p is the intensity of pixel P. Figure 4 is result of GraphCut algorithm that is another typical method of image segmentation.

(a)   (b)

Figure 4. GraphCut algorithm. (**a**) Original image; (**b**) Result image.

3. The proposed Method

The flow of proposed method is show in Figure 5.

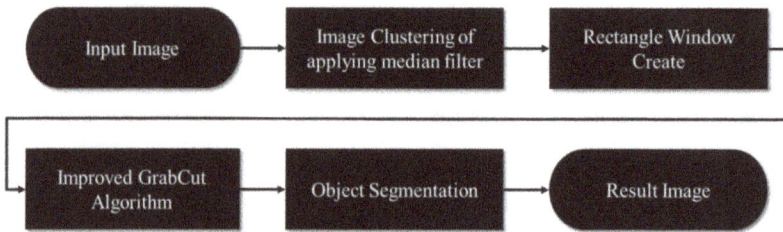

Figure 5. A flowchart of the proposed method.

3.1. Clustering to Reduce Noise

Digital image can be contaminated during data transmission.

$$\sum_{i=1}^{N} |Xmed - E| \leqslant \sum_{i=1}^{N} |Y - Xi| \tag{8}$$

where N is the size of data set. K-means clustering method is applied to the output of the filter. K-means classifies the data set to the predefined number of classes. Let μ_i be the center of i-th cluster and S_i be the set of pixels belongs to cluster i. The variance of all the data set is defined as Equation (9).

$$V = \sum_{i=1}^{k} \sum_{j \in S_i} |x_j - \mu_i|^2 \tag{9}$$

The goal is to find S_i minimizing V. K-means starts with arbitrary initial values μ_i. Allocating pixels to close μ_i and recalculating μ_i is repeated until it is converged.

$$J_{MSE} = \sum_{i=1}^{K} \sum_{x \approx \omega_i} |x - \mu_i|^2$$
$$\text{where} \quad \mu_i = \frac{1}{n} \sum_{x \approx \omega_i} x \tag{10}$$

Equation (10) is the simplest clustering method minimizing J_{MSE} repeatedly. Figure 6 shows applied the median filter in image.

Figure 6. An image applied the median filter image. (**a**) Original image; (**b**) Result image.

3.2. Object Segmentation Using Improved GrabCut

GrabCut accepts an interest area defined by a user and extracts objects using the clue given. Figure 7 shows trimap of foreground of GrabCut algorithm.

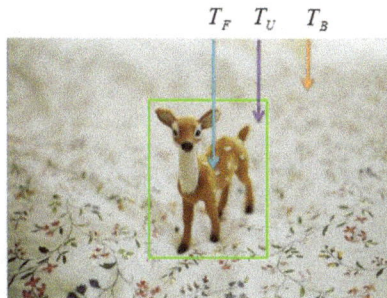

Figure 7. Composition of trimap; shows trimap of foreground (T_F), background (T_B) and unknown region (T_U).

Object and background is mixed in unknown region. Background T_B is defined as Equation (11).

$$T_B = T_F \cap [(T_U \cup \Gamma(z)) \oplus S] \tag{11}$$

where $\Gamma(z) = \left\{ z_n \middle| \sum_{p \in N(z_i)} \nabla g(p) \rangle t \right\}$ is the area greater than gradient. Symbol $\oplus$ is a dilation operator and S is a structure element for it.

A Gaussian mixture model is a probabilistic model that assumes all the data points are generated from a mixture of a finite number of Gaussian distributions with unknown parameters and is given by:

$$p(x|\theta) = \sum_{i=1}^{N} p(x|\omega_i, \theta_i) P(\omega_i) \tag{12}$$

where i-th vector component is characterized by normal distributions with weights α_i and a pair of mean and covariance θ_i. ω_i represents relative importance. α_i is defined as follows:

$$0 \leqslant \alpha_i \leqslant 1 \quad \text{and} \quad \sum_{i=1}^{M} \alpha_i = 1 \tag{13}$$

Parameters for GMM (Gaussian Mixture Model) of M components is expressed as Equation (14):

$$\theta = (\mu_1, \mu_2, \ldots, \mu_M, \ldots, \theta_1^2, \theta_2^2, \ldots, \theta_M^2, \alpha_1, \alpha_2, \ldots, \alpha_M) \tag{14}$$

T_U is defined as follows:

$$T_U = T_F - T_B \tag{15}$$

Figure 8 shows a result of the improved GrabCut algorithm.

| (a) | (b) |

Figure 8. Improved GrabCut algorithm. (a) Original image; (b) Result image.

4. Experiment and Discussion

The experiments were performed using about 400 photos such as figures, plants, food, etc.

Performance of GraphCut, standard GrabCut and proposed method is compared. Figure 9 shows some results. GrabCut is better than GraphCut and, the proposed method shows better results than GrabCut in most cases by detecting background in unknown area [12,13].

Evaluation is performed using precision and recall. Precision is the fraction of retrieved instances that are relevant, while recall (also known as sensitivity) is the fraction of relevant instances that are retrieved:

$$precision = \frac{N(Obj_{EX} \cap Obj_{GT})}{N(Obj_{EX})} \tag{16}$$

$$recall = \frac{N(Obj_{EX} \cap Obj_{GT})}{N(Obj_{GT})} \tag{17}$$

where $N(\cdot)$ is the number of pixels, Obj_{EX} is the object and Obj_{GT} is ground truth objects. Figure 10 shows the precision and recall of three methods. The proposed method gives the best result.

Experiments are performed using *PSNR* (Peak Signal to Noise Ratio). *PSNR* is the ratio between the maximum possible power of a signal and the power of corrupting noise that affects the fidelity of its representation.

$$PSNR = 10 \cdot \log_{10}\left(\frac{MAX^2}{MSE}\right) = 20 \cdot \log_{10}\left(\frac{MAX_I}{\sqrt{MSE}}\right) \tag{18}$$

MSE (Mean Squared Error) is the difference between the estimator and what is estimated. Where *MSE* is defined as follows:

$$MSE = \frac{1}{mn} \sum_{i=0}^{m-1} \sum_{j=0}^{n-1} || I(i,j) - K(i,j) ||^2 \tag{19}$$

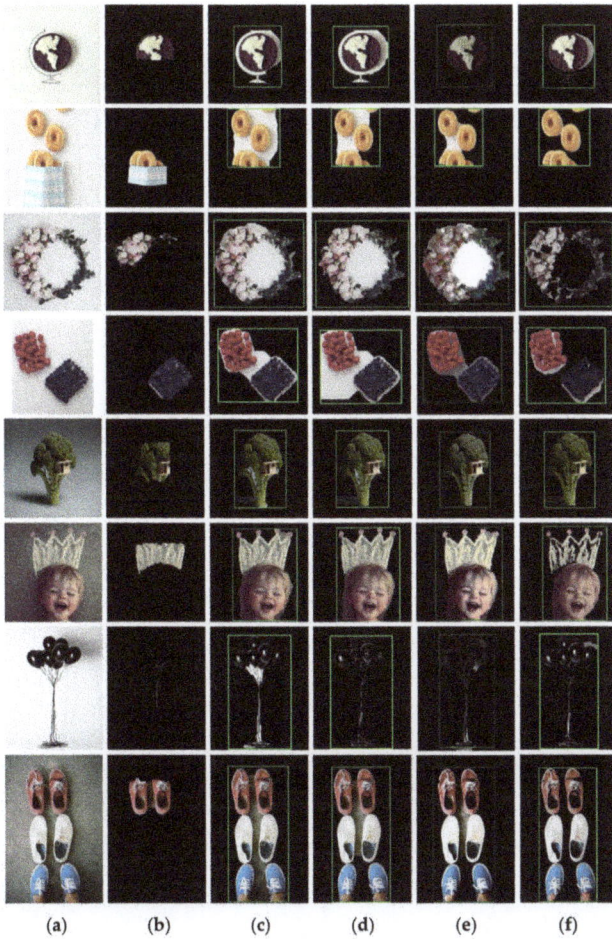

Figure 9. (a) Original image; (b) GraphCut Algorithm; (c) GrabCut Algorithm; (d) Ref. [12]; (e) Ref. [13]; (f) Proposed Method.

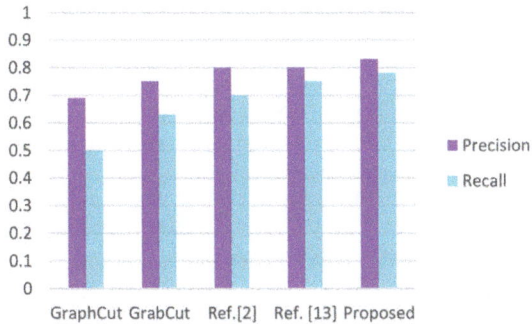

Figure 10. Precision-recall result.

Table 1 shows the quantitative comparison of result experiments.

Table 1. Result of experiment.as quantitative comparison.

Comparision Method	Ref. [2]	Ref. [13]	Proposed Method
MSE	2748.65	2374.43	1308.17
PSNR	29.39 dB	34.17 dB	38.94 dB

5. Conclusions

In this paper, a modified GrabCut method is proposed using median filter and k-means clustering technique to reduce image noise and to extract objects better. An image is preprocessed and then used for the input of standard GrabCut. This method showed better performance than GraphCut or standard GrabCut from the various and complex pictures like medical images, traffic images and people images. This research should be extended further to detect objects in video, and this can be used in many industrial applications.

Acknowledgments: The work reported in this paper was conducted during the sabbatical year of Kwangwoon University in 2013.

Author Contributions: GangSeong Lee provided guidance for this paper; SangHun Lee was the research and academic advisor of editing; GaOn Kim developed and solved the proposed model, carried out this analysis and wrote the manuscript; JongHun Park contributed to the revisions and performed experiments; YoungSoo Park contributed to the revisions and advisor of editing.

Conflicts of Interest: The authors declare no conflict of interest.

References

1. Yoo, T.-H.; Lee, G.-S.; Lee, S.-H. Window Production Method based on Low-Frequency Detection for Automatic Object Extraction of GrabCut. In Proceedings of the Digital Policy & Management, Seoul, Korea, 20 September 2012; pp. 211–217.
2. Tang, M.; Gorelick, L.; Vekler, O.; Boykove, Y. GrabCut in One Cut. In Proceedings of the ICCV Computer Vision, Sydney, Australia, 1–8 December 2013; pp. 1767–1776.
3. Rother, C.; Kolmogorov, V.; Blake, A. GrabCut : Interactive foreground extraction using iterated graph cuts. In Proceedings of the Computer Graphics, Imaging and Visualization (CGIV), Penang, Malaysia, 26–29 July 2004; pp. 309–314.
4. Boykov, Y.Y.; Jolly, M.-P. Interactive graph cut for optimal boundary & region segmentation of object in N-D image. In Proceedings of the IEEE International Conference on Computer Vision, Vancouver, BC, Canada, 7–14 July 2001; pp. 105–112.
5. Hänsch, R.; Hellwich, O.; Wang, X. Graph-cut segmentation of polarimetric SAR image. In Proceedings of the 2014 IEEE Geoscience and Remote Sensing Symposium, Quebec City, QC, Canada, 13–18 July 2014; pp. 1733–1736.
6. Malyszko, D.; Wierzchon, S.T. Standard and Genetic k-means Clustering Techniques in Image Segmentation. In Proceedings of the 6th International Conference on Computer Information Systems and Industrial Management Applications, CISIM '07, Minneapolis, MN, USA, 28–30 June 2007; pp. 299–304.
7. Sulaiman, S.N.; Isa, N.A.M. Adaptive fuzzy-K-means clustering algorithm for image segmentation. *IEEE Trans. Consum. Electron.* **2010**, *56*, 2661–2666. [CrossRef]
8. Ryu, T.; Wang, P.; Lee, S.-H. Image Compression with Meanshift Based Inverse Colorization. In Proceedings of the 2013 IEEE International Conference on Consumer Electronics (ICCE), Las Vegas, NV, USA, 11–14 January 2013; pp. 330–331.
9. Lebourgeols, F.; Drira, F.; Gaceb, D.; Duong, J. Fast Integral MeanShift: Application to Color Segmentation of Document Images. In Proceedings of the ICDAR Document Analysis and Recognition, Washington, DC, USA, 25–28 August 2013; pp. 52–56.
10. Yang, D.-L.; Zhang, J.-W. Study of Image Quantization Technology Based on FCM Clustering Algorithm. In Proceedings of the Intelligent System Design and Engineering Application (ISDEA), Changsha, China, 13–14 October 2010; pp. 421–424.

11. Tian, J.; Huang, Y.; Tian, J. Histogram Constraint Based Fast FCM Cluster Image Segmentation. In Proceedings of the 2007 IEEE International Symposium on Industrial Electronics, Vigo, Spain, 4–7 June 2007; pp. 1623–1627.
12. Tang, M.; Ayed, B.I.; Marin, D.; Boykov, Y. Secrets of GrabCut and Kernel K-means. In Proceedings of the IEEE International Conference on Computer Vision, London, ON, Canada, 7–13 December 2015; pp. 1555–1563.
13. Hua, S.; Shi, P. GrabCut color image segmentation based on region of interest. In Proceedings of the 2014 7th International Congress on Image and Signal Processing (CISP), Dalian, China, 14–16 October 2014; pp. 392–396.

symmetry

MDPI

Article

Computing the Surface Area of Three-Dimensional Scanned Human Data

Seung-Hyun Yoon [1] **and Jieun Lee** [2,†,*]

[1] Department of Multimedia Engineering, Dongguk University, Seoul 04620, Korea; shyun@dongguk.edu
[2] Department of Computer Engineering, Chosun University, Gwangju 61452, Korea
*   Correspondence: jieunjadelee@gmail.com; Tel.: +82-62-230-7473
†   Current address: Department of Computer Engineering, Chosun University, 309 Pimun-daero, Dong-gu, Gwangju 61452, Korea

Academic Editor: Angel Garrido
Received: 16 March 2016; Accepted: 13 July 2016; Published: 20 July 2016

Abstract: An efficient surface area evaluation method is introduced by using smooth surface reconstruction for three-dimensional scanned human body data. Surface area evaluations for various body parts are compared with the results from the traditional alginate-based method, and quite high similarity between the two results is obtained. We expect that our surface area evaluation method can be an alternative to measuring surface area by the cumbersome alginate method.

Keywords: 3D scanner; 3D human model; surface area; surface reconstruction; alginate

1. Introduction

The surface area of human body parts provides important information in medical and medicinal fields, and surface area computation of human body parts is generally a difficult problem. For example, we need to know the accurate surface area when we have to determine the adequate amount of ointment to apply. So far, alginate [1] is generally used to measure surface area. The surface areas of body parts are modeled with alginate, and the models are cut into small pieces. These pieces are spread onto a two-dimensional (2D) plane, and their areas are then measured on the plane, and the total area of the surface is computed by summing the areas of all the pieces. Figure 1 illustrates the overall process for measuring surface area by using alginate. Error is inevitably included in the process of projecting a three-dimensional (3D) surface onto a 2D plane. Moreover, errors by human operators can also accumulate in this process since it requires numerous manual operations.

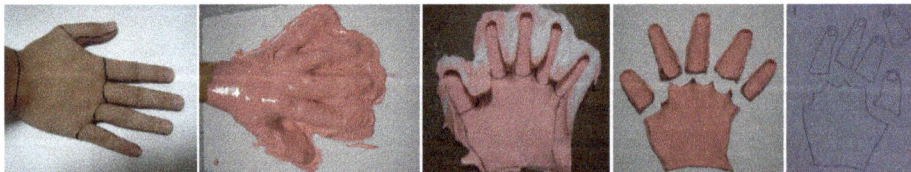

Figure 1. Measuring the surface area of a hand by using alginate [1].

Recently, the rapid advances in 3D shape scanning technology have enabled us to easily obtain geometric information of real 3D models. Three-dimensional shapes from 3D scanners are already used in ergonomic design, e.g., in the garment, furniture, and automobile industries, as well as in the digital content industry such as movies and animations. In this paper, we further extend the usage of 3D scanned human data to medical and medicinal fields. It would be quite useful to utilize 3D scanned human data to avoid the onerousness of the alginate method.

Three-dimensional scanners usually generate polygonal approximation to human model, and its polygon areas are summed to compute the desired surface area. However, this discrete method does not consider the smooth surface property of human skin and the resulting surface area tends to be smaller than the exact one. We prove this fact with some geometric objects whose exact area values are known. Furthermore, we propose an effective area computation method to overcome the limitation of the polygonal approximation. We reconstruct a smooth surface from the polygonal approximation to reflect the smooth surface property of human body parts, thus reducing the error of area measurement. A local part of the scanned data is selected by a user and reconstructed as a smooth surface; the surface area is then accurately computed by using an analytic method.

We compared the surface areas measured by our method with the ones obtained by using alginate. We set up 15 local parts of a human body, and we measured the areas of the local parts of eight people by using alginate. Three-dimensional human models of the same eight people were also generated by 3D scanning. We selected 15 local parts of the 3D models using an intuitive sketch-based user interface that we developed. We reconstructed smooth surfaces for the selected parts and computed the surface areas from the reconstructed surfaces. We analyzes the similarity and the correlation between the area measured by using alginate and the area computed from our reconstruction method, and found a similarity of >95%. Therefore, we expect that our surface area measuring method can be an effective alternative to measuring surface area, replacing the cumbersome alginate method.

The main contributions of this paper can be summarized as follows:

- We propose a simple and effective area computation method based on surface reconstruction for the body parts of 3D scanned human models.
- The area computed using the surface reconstruction method has a 95% similarity with that obtained by using the traditional alginate method.
- Our area computation method proves to be a possible substitute for the cumbersome alginate method.

The rest of this paper is organized as follows. In Section 2 we briefly review some related recent work on scanning technology and surface reconstruction, and in Section 3 we explain how to reconstruct a smooth surface from polygonal meshes and how to compute the surface area from the reconstructed surfaces. In Section 4 we compare the surface areas of various body parts measured by our method with ones obtained by using the traditional alginate method and derive statistical information. In Section 5 we conclude the paper and suggest some future research.

2. Related Work

Recent advances in 3D scanning technology have made it quite easy to achieve 3D shapes of complex objects. Depending on the specific sensors such as lasers, patten lights, optical cameras, and depth cameras, various types of 3D scanners have been developed. In general, 3D scanners can be classified into three types [2]: Contact types, non-contact active types, and non-contact passive types. Contact 3D scanners contact an object with a tiny, thin needle-like sensor and scan the surface of the object. They can scan the front side of the object, but they hardly scan the side portions or concave parts. Non-contact active 3D scanners use a laser to illuminate the object surface to measure the distances or to recognize surficial curves. Non-contact passive 3D scanners use reflective visible light or infrared light from the object to scan the surface of the object, instead of using laser light or sonic waves.

Depending on the specific application, different types of 3D scanners can be used. For example, whole-body 3D scanners [3,4] are widely used for ergonomic design in the garment, furniture, and automobile industries. These whole-body 3D scanners are equipped with four wide-view, high-resolution scanners, which rotate around the person to scan every angle. This high-powered precision scan is able to capture even the smallest details, such as hair, wrinkles on clothes, and buttons. The scanning process generates millions of triangulated surfaces, which are automatically merged and stitched together. A hand-held 3D scanner is similar to a video camera but captures in three dimensions. It is extremely portable and can be used for medical and biomechanical research. For example, portable oral scanners [5,6] are essential for implant surgical guidance and prosthetic design in dentistry.

Even though 3D scanners provide accurate and detailed geometric data from real-world objects, they are restricted to producing a discrete representation such as unorganized point clouds or polygonal meshes. Moreover, these models can have serious problems for many practical applications; these include irregularity, discontinuity, huge dataset size, and missing areas.

Body surface area (BSA) represents the whole area of a human body, and it is an important quantity in the fields of medicine, pharmacy, and ergonomics. Direct BSA measurement uses paper wrapping, bandage, alginate method and so on, but it is very burdensome work. BSA estimation formula is generally determined by one's height and weight, and many efforts have been made to find more accurate estimation. Recently, new BSA estimation formulas have been proposed by using 3D scanned human data [7–9]. Lee and Choi [10] compared alginate method and 3D body scanning in measuring BSA. They reported that BSA measured by the 3D scanning method tended to be smaller than that by the alginate method.

In this paper, we aim to measure the surface area of a selected region of 3D scanned human data. Summing the polygonal area of the selected region can be one of the simplest ways of measuring surface area. However, we take a different approach to obtain a more accurate result than from a polygonal approximation. We reconstruct a smooth surface from the selected region and compute its surface area based on analytic methods rather than on a simple polygonal approximation. Since smooth surface reconstruction is highly important in our method, we briefly review the related techniques for reconstructing a smooth surface from a polygonal mesh.

Vlachos et al. [11] introduced point-normal (PN) triangles for surfacing a triangular mesh. On each triangle of a mesh, they created a cubic Bézier triangle using vertices and normals from the mesh. However, this method is restricted to generating a G^0-continuous surface across the triangle boundaries, which is not suitable for measuring surface area.

Blending techniques are widely used for reconstructing a smooth surface in geometric modeling. Vida et al. [12] surveyed the parametric blending of curves and surfaces. Depending on the number of surfaces to be blended, various approaches have been proposed. Choi and Ju [13] used a rolling ball to generate a tubular surface with G^1-continuous contact to the adjacent surfaces. This technique can be made more flexible by varying the radius of the ball [14]. Hartmann [15] showed how to generate G^n parametric blending surfaces by specifying a blending region on each surface to be blended, and reparameterizing the region with common parameters. A univariate blending function is then defined using one of three common parameters to create a smooth surface. This method was extended to re-parameterize the blending regions automatically in [16].

A more general blending scheme was introduced by Grim and Hughes [17]. They derived manifold structures such as charts and transition functions from a control mesh and reconstructed a smooth surface by blending geometries on overlapping charts using a blending function. Cotrina and Pla [18] generalized this method to construct C^k-continuous surfaces with B-spline boundary curves. This approach was subsequently generalized by Cotrina et al. [19] to produce three different types of surfaces. However, these techniques require complicated transition functions between overlapping charts.

Ying and Zorin [20] created smooth surfaces of arbitrary topology using charts and simple transition functions on the complex plane. This approach provides both C^∞ continuity and local control of the surface. However, the resulting surfaces are not piecewise polynomial or rational. Recently, Yoon [21] extended this technique to reconstruct a smooth surface using displacement functions. Compared to other methods [20,22,23], this method produces a smooth surface that interpolates the vertices of a control mesh, which is an essential condition for measuring the surface area from a smooth surface rather than a polygonal mesh. Therefore, we employ this method to reconstruct a smooth surface and measure its surface area.

3. Computing the Surface Area of 3D Scanned Human Data

In this section we propose a method for computing the surface area of 3D scanned human data. We reconstruct a smooth surface representing the selected region of 3D scanned human data. We then compute the surface area of the selected region from the smooth surface rather than from the triangular mesh, which gives us more accurate results.

3.1. Natural User Interface for Selecting the Region of Interest

Our system provides a user with a sketch-based interface for specifying the region on the 3D scanned human data. A user marks a closed curve on a 2D screen using the sketch interface. We determine the screen coordinates of the vertices of 3D human data using a graphics pipeline and select only vertices with coordinates inside the marked curve [24]. Figure 2 shows a selected region of 3D human data using the sketch-based user interface.

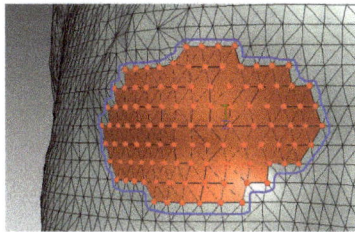

Figure 2. Selected region (in red) from the user's 2D sketch (in blue).

3.2. Smooth Surface Reconstruction

We employ a method proposed by Yoon [21] to reconstruct a smooth surface from the selected region of 3D human data. This section briefly introduces how to reconstruct a smooth surface for the selected region.

Chart and transition function: For each vertex of the selected region, we define a chart in the 2D complex plane. The chart shape is determined by the degree of a vertex. Figure 3 shows the charts U_i and U_j of two vertices with different degrees 6 and 3, respectively. As shown in Figure 3, adjacent charts share two regions and their correspondence is defined by a transition function $\theta_{ij}(z)$ as follows:

$$z' = \theta_{ij}(z) = z^{k_i/k_j}, \tag{1}$$

where k_i and k_j represent the degrees of vertices $\mathbf{v}_i$ and $\mathbf{v}_j$, respectively. For instance, let $z = u + iv = (u, v)$ be the coordinates of z in the chart U_i, then the corresponding coordinates z' in U_j can be computed by $z' = z^{6/3}$ in Figure 3. For more information, refer to [21].

Local Surface Patches: For each chart U_i of a vertex $\mathbf{v}_i$, we construct a 3D surface patch $P_i(u, v)$ approximating the 1-ring neighborhood of $\mathbf{v}_i$. We employ a biquadratic surface patch $P_i(u, v)$ defined as follows:

$$P_i(u, v) = \begin{bmatrix} 1 & u & u^2 \end{bmatrix} \begin{bmatrix} c_1 & c_2 & c_3 \\ c_4 & c_5 & c_6 \\ c_7 & c_8 & c_9 \end{bmatrix} \begin{bmatrix} 1 \\ v \\ v^2 \end{bmatrix}, \tag{2}$$

where c_1 is set to $\mathbf{v}_i$ for $P_i(0,0) = \mathbf{v}_i$ and other coefficient vectors are determined by approximating 1-ring neighboring vertices of $\mathbf{v}_i$ in a least-squares sense. Figure 4 shows a local surface patch $P_i(u, v)$ of $\mathbf{v}_i$ defined on chart U_i.

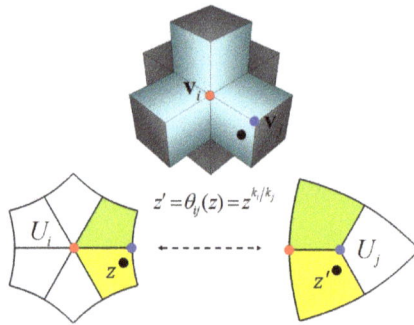

Figure 3. Charts U_i and U_j and their transition function $\theta_{ij}(z)$.

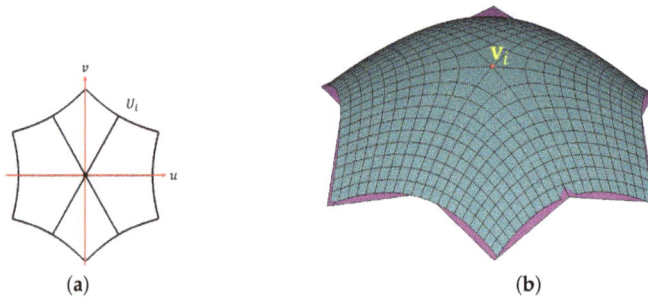

Figure 4. (a) Chart U_i; (b) $P_i(u,v)$ of $\mathbf{v}_i$ defined on U_i.

Blending Surface: We reconstruct a smooth surface by blending the local surface patches. For this, we need a blending function $w_i(u,v)$ on each chart U_i. To construct a blending function $w_i(u,v)$, we first construct a piece of blending function $\eta(u)\eta(v)$ on the unit square $[0,1] \times [0,1]$, where $\eta(t) = 2t^3 - 3t^2 + 1$. We then apply conformal mapping to $\eta(u)\eta(v)$, followed by rotating and copying. Figure 5 shows the example of a blending function $w_i(u,v)$ on a chart of degree $k = 6$. Note that blending functions $w_i(u,v)$ satisfy the partition of unit, $\sum_{\forall i} w_i(u,v) = 1$, on overlapping charts.

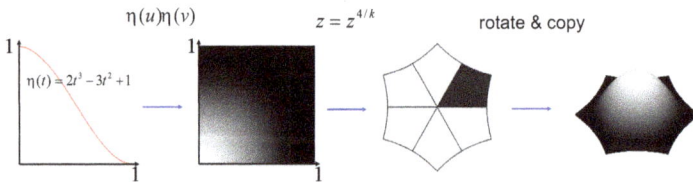

Figure 5. Construction of a blending function.

Finally, our blending surface $S_i(u,v)$ on a chart U_i is defined by a weighted blending of local patches P_j as follows:

$$S_i(u,v) = \sum_{j \in I_z} w_j\left(\theta_{ij}(z)\right) P_j\left(\theta_{ij}(z)\right), \tag{3}$$

where I_z is a set of chart indices containing $z = (u,v)$. Figure 6a shows polygon meshes of different resolutions, generated from a sphere of radius = 5 cm and Figure 6b shows the corresponding blending surfaces generated by using our method.

(a)

(b)

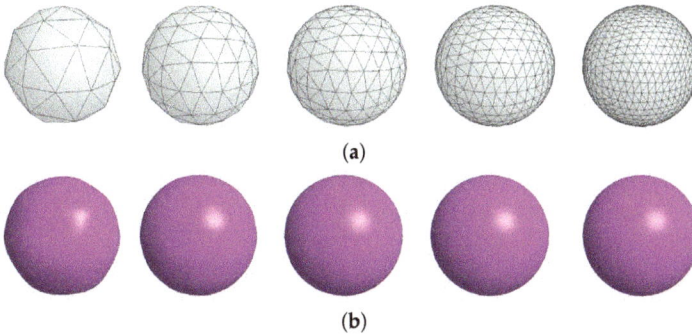

Figure 6. (**a**) Polygon approximations to a sphere of radius = 5 cm; (**b**) blending surfaces reconstructed from (**a**).

Measuring Surface Area: Now we can measure the surface area on a smooth blending surface rather than a polygon mesh as follows:

$$A = \int \int \sqrt{|I|} \, dudv, \tag{4}$$

where $|I|$ is the determinant of the first fundamental form matrix [25]. In general, a polygon mesh generates a surface area smaller than that of a smooth surface. To compare and analyze the accuracy of the proposed method, we measure the surface areas of three geometric objects with different distributions of Gaussian curvature. All 3D shapes, including a human body, can locally be classified into the following cases in terms of Gaussian curvature distributions.

Our first example is a sphere with positive Gaussian curvature ($K > 0$) everywhere. Figure 6a,b show the polygon spheres with the different resolutions and the reconstructed smooth surfaces, respectively. Table 1 compares two surface areas of polygon meshes and reconstructed surfaces in Figure 6. The third column lists the surface areas and computation times measured from polygon meshes and the fourth column lists those from reconstructed surfaces. The next two columns show errors between measured areas and the exact one ($\pi \approx 314.15926535897$), and their ratios are shown in the last column.

Table 1. Comparison of surface areas (in cm^2) and computation time (in ms) in Figure 6.

Cases	# of Triangles	Area (time) (a)	Area (time) (b)	Error (1)	Error (2)	(1)/(2)
1	60	272.46179 (0.03)	293.46164 (3)	41.69747	20.69763	2.01460
2	180	299.35513 (0.05)	308.94577 (9)	14.80413	5.21349	2.83958
3	420	307.64926 (0.06)	312.20694 (22)	6.510004	1.95233	3.33449
4	760	310.52105 (0.11)	313.14139 (40)	3.638208	1.01788	3.57431
5	1740	312.55352 (0.18)	313.73544 (92)	1.605738	0.42382	3.78871

We employ a hyperboloid as the second example, which has negative Gaussian curvature ($K < 0$) everywhere. Figure 7a,b show the polygon approximations to a hyperboloid with different resolutions and the reconstructed smooth surfaces, respectively. Table 2 compares two surface areas of polygon meshes and reconstructed surfaces in Figure 7. The third column lists the surface areas and computation times measured from polygon meshes and the fourth column lists those from the reconstructed surfaces. The next two columns show errors between measured areas and the exact one ($\pi(2\sqrt{6} + \sqrt{2}\sinh^{-1}(\sqrt{2})) \approx 20.01532$), and their ratios are shown in the last column.

(a)

(b)

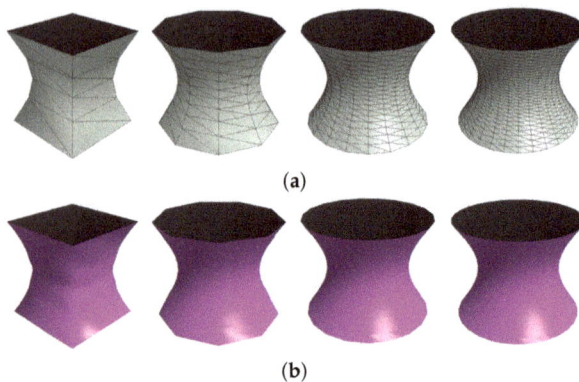

Figure 7. (a) Polygon approximations to a hyperboloid $x^2 + y^2 - z^2 = 1$; (b) blending surfaces reconstructed from (a).

Table 2. Comparison of surface areas (in cm^2) and computation time (in ms) in Figure 7.

Cases	# of Triangles	Area (time) (a)	Area (time) (b)	Error (1)	Error (2)	(1)/(2)
1	32	17.19809 (0.03)	18.06601 (2)	2.81723	1.94931	1.44524
2	162	19.39459 (0.05)	19.68971 (8)	0.62073	0.32561	1.90636
3	722	19.87309 (0.09)	19.95923 (37)	0.14223	0.05609	2.53575
4	1682	19.95392 (0.17)	19.99632 (89)	0.0614	0.019	3.23158

Our last example is a torus which has various distributions of Gaussian curvature as shown in Figure 8a. Table 3 compares two surface areas of polygon meshes and the reconstructed surfaces in Figure 8. The third column lists the surface areas and computation times measured from polygon meshes and the fourth column lists those from the reconstructed surfaces. The next two columns show errors between measured areas and the exact one ($8\pi^2 \approx 78.9568352$), and their ratios are shown in the last column.

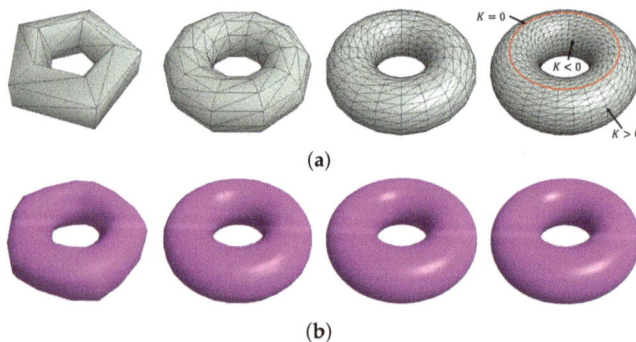

(a)

(b)

Figure 8. (a) Polygon approximations to a torus of radii $r = 1$ cm and $R = 2$ cm; (b) smooth blending surfaces reconstructed from (a).

Table 3. Comparison of surface areas (in cm^2) and computation time (in ms) in Figure 8.

Cases	# of Triangles	Area (time) (a)	Area (time) (b)	Error (1)	Error (2)	(1)/(2)
1	50	62.64104 (0.04)	71.86401 (3)	16.31580	7.09283	2.30032
2	200	74.53550 (0.05)	78.27505 (12)	4.42134	0.68179	6.48495
3	800	77.82805 (0.1)	78.87682 (45)	1.12879	0.08002	14.10562
4	1800	78.45343 (0.18)	78.92898 (98)	0.50341	0.02786	18.06795

Figure 9 shows graphical illustrations of Tables 1–3. Compared with a sphere ($K > 0$) and a hyperboloid ($K < 0$), the surface reconstruction of a torus gives much smaller errors as shown in Figure 9d, which means our method gives more accurate results for the objects with various curvature distributions such as human body skin. Therefore, the surface reconstruction can be an effective method for measuring surface areas on 3D scanned human data.

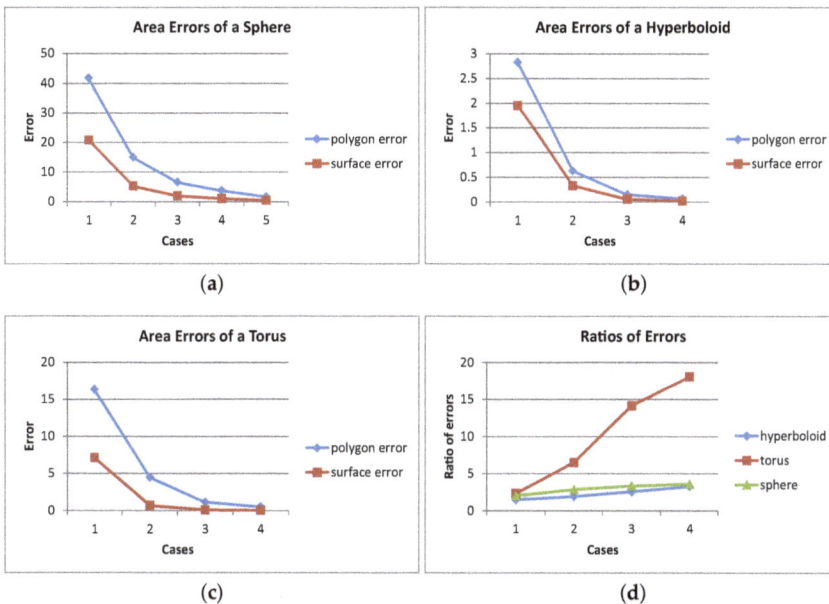

Figure 9. Comparison of errors of (**a**) a sphere; (**b**) a hyperboloid and (**c**) a torus; (**d**) ratios of polygon error to surface error.

4. Experimental Results

We implemented our technique in C++ (Microsoft Visual C++ 2015) on a PC with an Intel Core i7 2.00 GHz CPU with 8GB of main memory and an Intel ® Iris Pro Graphics 5200. In this section, we explain our experiment results of area computation and compare the results with those obtained by using alginate. We measure areas using alginate and compute areas using the proposed method from 8 subjects. Figure 10 shows a 3D scanned human model with different rendering options. We select 15 regions of interest to measure area: upper arms, lower arms, upper legs, lower legs, abdomen, back, pelvis, hips, head, face, and neck. Figure 11 shows examples of the selected regions of interest.

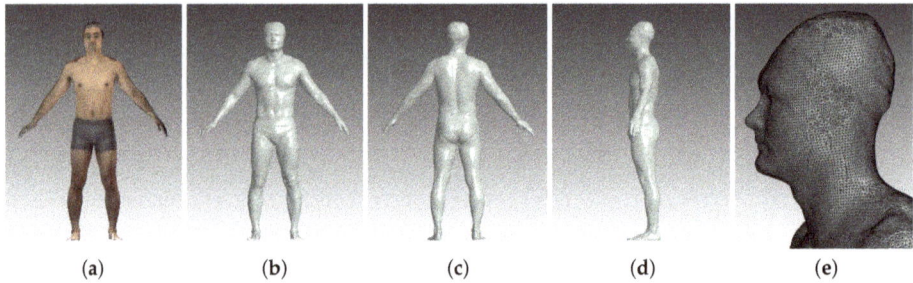

Figure 10. A 3D scanned human model with different rendering options: (**a**) skin texture; (**b**) front view; (**c**) back view; (**d**) side view; (**e**) wireframe.

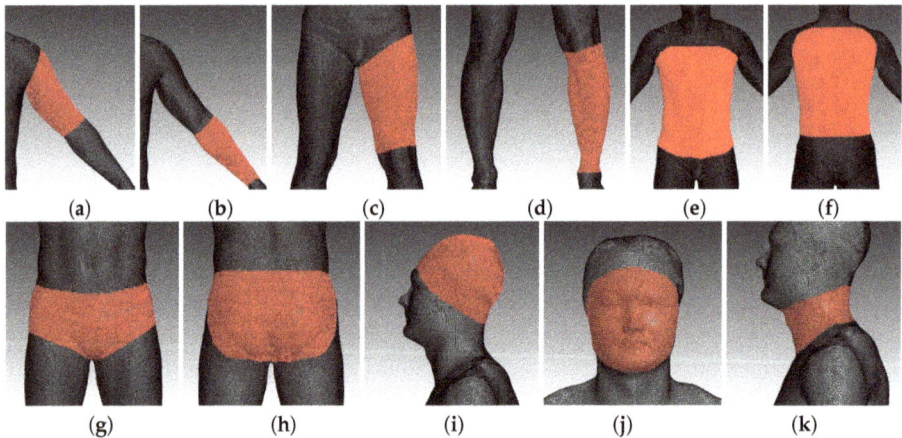

Figure 11. Selected regions of interest: (**a**) left upper arm; (**b**) left lower arm; (**c**) left upper leg; (**d**) left lower leg; (**e**) abdomen; (**f**) back; (**g**) pelvis; (**h**) hips; (**i**) head; (**j**) face; (**k**) neck.

We use the ratio of the difference to the average value to evaluate similarity as follows:

$$similarity = 1 - \frac{A_{df}}{A_{av}}, \tag{5}$$

where A_{ag} is the area value measured by using alginate, A_{sf} is the area value computed by surface reconstruction, and A_{av} is the average value of A_{ag} and A_{sf}. A_{df} is the difference from the average and $A_{df} = |A_{sf} - A_{av}| = |A_{ag} - A_{av}|$. We get the final similarity value for each body part by averaging eight similarity values of eight pairs of area values for each body part.

Figure 12 shows eight pairs of area values of various body parts, which are used in the similarity computation. The similarity values of upper arms, lower arms, upper legs, and lower legs are very high, ranging from 97% to 99% (see Figure 12a–h). The correlations between two area values in those body parts are >0.82. The similarity values in the pelvis and hips are slightly low, being about 95%. Sharp foldings in these parts bring in error in area measurement. Table 4 lists all similarity and correlation values of local body parts.

Table 4. Similarity and correlation between the results of alginate and the proposed surface reconstruction methods.

Region	Similarity	Correlation
left upper arm	0.99232920	0.98651408
right upper arm	0.99050425	0.97836923
left lower arm	0.97442492	0.94239832
right lower arm	0.97565553	0.88847152
left upper leg	0.96904881	0.82351873
right upper leg	0.97294687	0.91311208
left lower leg	0.98809628	0.97643038
right lower leg	0.99031974	0.98423465
abdomen	0.98108957	0.97599424
back	0.97219378	0.89756368
pelvis	0.94844035	0.50870081
hips	0.95367837	0.64129904
head	0.96274736	0.63287971
neck	0.97341437	0.88813431
face	0.97505372	0.87872788
average	0.94925100	0.75430084

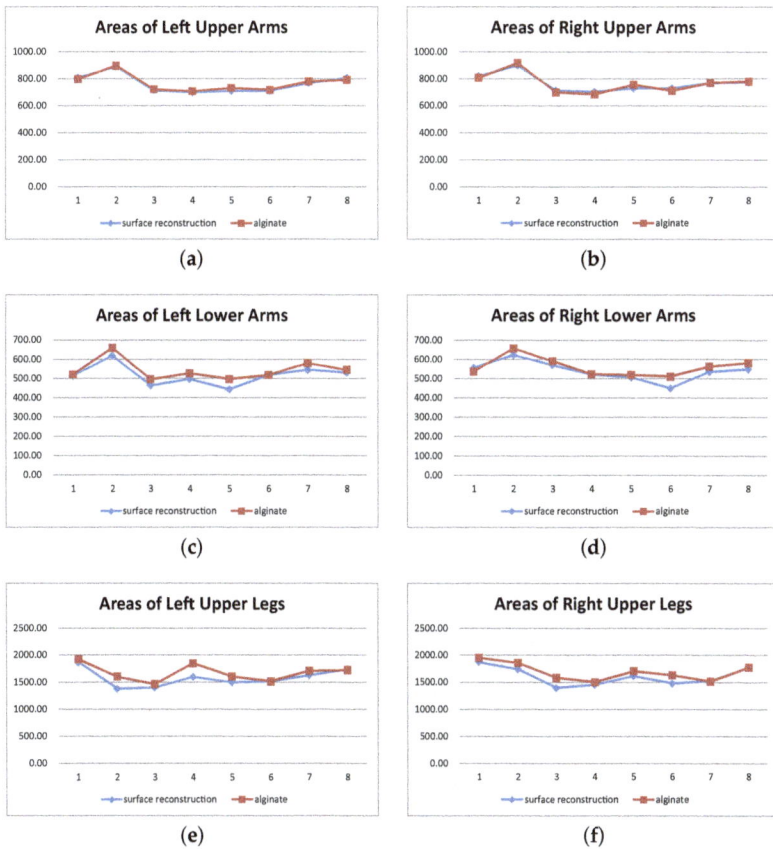

(a)

(b)

(c)

(d)

(e)

(f)

Figure 12. *Cont.*

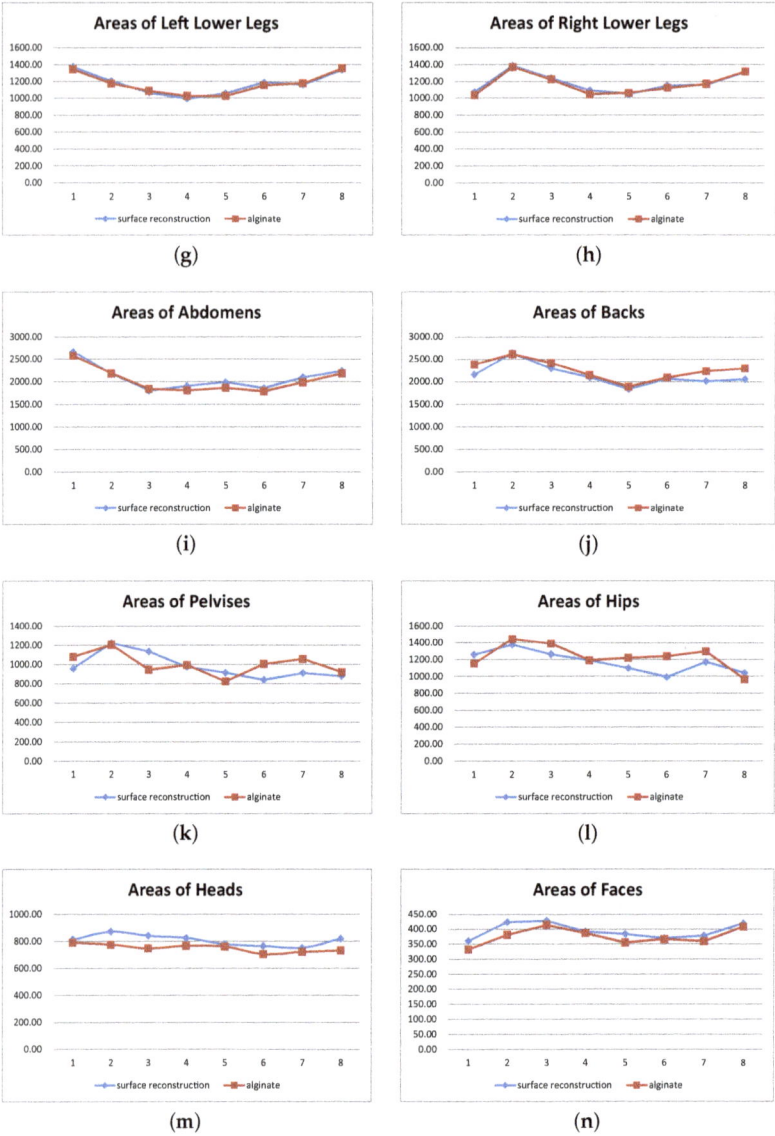

(g)

(h)

(i)

(j)

(k)

(l)

(m)

(n)

Figure 12. *Cont.*

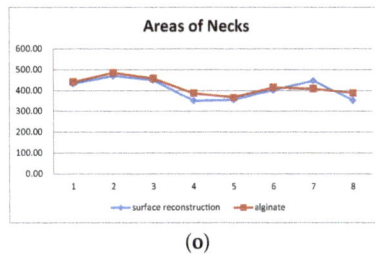

(o)

Figure 12. Areas of body parts of eight people. The red broken line shows eight values of area obtained by using alginate and the blue line shows those obtained by surface reconstruction; (**a**) areas of left upper arms; (**b**) areas of right upper arms; (**c**) areas of left lower arms; (**d**) areas of right lower arms; (**e**) areas of left upper legs; (**f**) areas of right upper legs; (**g**) areas of left lower legs; (**h**) areas of right lower legs; (**i**) areas of abdomens; (**j**) areas of backs; (**k**) areas of pelvises; (**l**) areas of hips; (**m**) areas of heads; (**n**) areas of faces; (**o**) areas of necks.

Finally, we should recall that both area values from alginate and from the proposed surface reconstruction method are not true values. As mentioned before, error is inevitably included in the process of projecting 3D surface onto a 2D plane and it is also attributable to human operators who model surfaces and measure surface area by using alginate. In using surface reconstruction, selected regions are different for different operators. Error is expected to be reduced when expert operators measure the areas with both methods repeatedly. We concentrate on the similarity and correlation between the two results in this work.

We have also measured the computation time of our method that includes surface reconstruction and area computation. Compared to the simplest polygon area computation method, our method takes more time as reported in Section 3. However, the absolute time is sufficiently short to be called real-time. In our work, a 3D scanned human model has 250,000 triangles averagely, a face part with 3000 triangles and a back with 25,000 triangles took 24 ms and 206 ms to compute their surface areas, respectively.

5. Conclusions

In this paper, we developed an analytic area computation method by reconstructing a smooth surface from polygonal meshes. We applied this method to measure the areas of local body parts of 3D scanned human models. We also measured areas of the same body parts using the traditional alginate method to compare area computation results. The results showed 95% similarity between the two methods, and we expect our area computation method can be an efficient alternative to using alginate.

In future work, we plan to extend our technique to measure the volume of volumetric data obtained from computed tomography or magnetic resonance imaging, which can be expected to be a useful diagnostic technique in the medical industry.

Acknowledgments: This research was supported by the Basic Science Research Program through the National Research Foundation of Korea(NRF) funded by Ministry of Education(Grant No. NRF-2013R1A1A4A01011627) and also supported by Broadcasting and Telecommunications Development Fund through the Korea Radio Promotion Association(RAPA) funded by the Ministry of Science, ICT & Future Planning.

Author Contributions: Seung-Hyun Yoon and Jieun Lee conceived and designed the experiments; Jieun Lee performed the experiments; Jieun Lee analyzed the data; Seung-Hyun Yoon contributed analysis tools; Seung-Hyun Yoon and Jieun Lee wrote the paper.

Conflicts of Interest: The authors declare no conflicts of interest.

References

1.   Lee, K.Y.; Mooney, D.J. Alginate: Properties and biomedical applications. *Prog. Polym. Sci.* **2012**, *37*, 106–126.

2. 3D Geomagic. Available online: http://www.geomagic.com/en/products/capture/overview (accessed on 16 March 2016).
3. Whole Body 3D Scanner (WBX). Available online: http://cyberware.com/products/scanners/wbx.html (accessed on 16 March 2016).
4. LaserDesign. Available online: http://www.laserdesign.com/products/category/3d-scanners/ (accessed on 16 March 2016).
5. TRIOS. Available online: http://www.3shape.com/ (accessed on 16 March 2016).
6. Exocad. Available online: http://exocad.com/ (accessed on 16 March 2016).
7. Tikuisis, P.; Meunier, P.; Jubenville, C. Human body surface area: Measurement and prediction using three dimensional body scans. *Eur. J. Appl. Physiol.* **2001**, *85*, 264–271.
8. Yu, C.Y.; Lo, Y.H.; Chiou, W.K. The 3D scanner for measuring body surface area: A simplified calculation in the Chinese adult. *Appl. Ergon.* **2003**, *34*, 273–278.
9. Yu, C.Y.; Lin, C.H.; Yang, Y.H. Human body surface area database and estimation formula. *Burns* **2010**, *36*, 616–629.
10. Lee, J.Y.; Choi, J.W. Comparison between alginate method and 3D whole body scanning in measuring body surface area. *J. Korean Soc. Cloth. Text.* **2005**, *29*, 1507–1519.
11. Vlachos, A.; Peters, J.; Boyd, C.; Mitchell, J.L. Curved PN Triangles. In Proceedings of the 2001 Symposium on Interactive 3D Graphics (I3D '01), Research Triangle Park, NC, USA, 19–21 March 2001; ACM: New York, NY, USA, 2001; pp. 159–166.
12. Vida, J.; Martin, R.; Varady, T. A survey of blending methods that use parametric surfaces. *Computer-Aided Des.* **1994**, *26*, 341–365.
13. Choi, B.; Ju, S. Constant-radius blending in surface modeling. *Computer-Aided Des.* **1989**, *21*, 213–220.
14. Lukács, G. Differential geometry of G^1 variable radius rolling ball blend surfaces. *Computer-Aided Geom. Des.* **1998**, *15*, 585–613.
15. Hartmann, E. Parametric G^n blending of curves and surfaces. *Vis. Comput.* **2001**, *17*, 1–13.
16. Song, Q.; Wang, J. Generating G^n parametric blending surfaces based on partial reparameterization of base surfaces. *Computer-Aided Des.* **2007**, *39*, 953–963.
17. Grimm, C.M.; Hughes, J.F. Modeling Surfaces of Arbitrary Topology Using Manifolds. In Proceedings of the 22nd Annual Conference on Computer Graphics and Interactive Techniques (Siggraph '95), Los Angeles, CA, USA, 6–11 August 1995; ACM: New York, NY, USA, 1995; pp. 359–368.
18. Cotrina-Navau, J.; Pla-Garcia, N. Modeling surfaces from meshes of arbitrary topology. *Computer-Aided Geom. Des.* **2000**, *17*, 643–671.
19. Cotrina-Navau, J.; Pla-Garcia, N.; Vigo-Anglada, M. A Generic Approach to Free Form Surface Generation. In Proceedings of the Seventh ACM Symposium on Solid Modeling and Applications (SMA '02), Saarbrucken, Germany, 17–21 June 2002; ACM: New York, NY, USA, 2002; pp. 35–44.
20. Ying, L.; Zorin, D. A simple manifold-based construction of surfaces of arbitrary smoothness. *ACM Trans. Graph.* **2004**, *23*, 271–275.
21. Yoon, S.H. A Surface Displaced from a Manifold. In Proceedings of Geometric Modeling and Processing (GMP 2006), Pittsburgh, PA, USA, 26–28 July 2006; Springer: New York, NY, USA, 2006; pp. 677–686.
22. Vecchia, G.D.; Jüttler, B.; Kim, M.S. A construction of rational manifold surfaces of arbitrary topology and smoothness from triangular meshes. *Computer-Aided Geom. Des.* **2008**, *29*, 801–815.
23. Vecchia, G.D.; Jüttler, B. Piecewise Rational Manifold Surfaces with Sharp Features. In Proceedings of the 13th IMA International Conference on Mathematics of Surfaces XIII, York, UK, 7–9 September 2009; pp. 90–105.
24. Tomas, A.; Eric, H. *Real-Time Rendering*; AK Peters: Natick, MA, USA, 2002.
25. Farin, G. *Curves and Surfaces for CAGD: A Practical Guide*, 5th ed.; Morgan Kaufmann: Burlington, MA, USA, 2002.

symmetry

MDPI

Article

Fuzzy System-Based Face Detection Robust to In-Plane Rotation Based on Symmetrical Characteristics of a Face

Hyung Gil Hong, Won Oh Lee, Yeong Gon Kim, Ki Wan Kim, Dat Tien Nguyen and Kang Ryoung Park *

Division of Electronics and Electrical Engineering, Dongguk University, 30 Pildong-ro 1-gil, Jung-gu, Seoul 100-715, Korea; hell@dongguk.edu (H.G.H.); 215p8@hanmail.net (W.O.L.); csokyg@dongguk.edu (Y.G.K.); yawara18@hotmail.com (K.W.K.); nguyentiendat@dongguk.edu (D.T.N.)
* Correspondence: parkgr@dongguk.edu; Tel.: +82-10-3111-7022; Fax: +82-2-2277-8735

Academic Editor: Angel Garrido
Received: 15 June 2016; Accepted: 29 July 2016; Published: 3 August 2016

Abstract: As face recognition technology has developed, it has become widely used in various applications such as door access control, intelligent surveillance, and mobile phone security. One of its applications is its adoption in TV environments to supply viewers with intelligent services and high convenience. In a TV environment, the in-plane rotation of a viewer's face frequently occurs because he or she may decide to watch the TV from a lying position, which degrades the accuracy of the face recognition. Nevertheless, there has been little previous research to deal with this problem. Therefore, we propose a new fuzzy system–based face detection algorithm that is robust to in-plane rotation based on the symmetrical characteristics of a face. Experimental results on two databases with one open database show that our method outperforms previous methods.

Keywords: TV environment; Face recognition; In-plane rotation of the face; Fuzzy systems; Symmetrical characteristics of a face

1. Introduction

With the rapid development of face recognition technology, it has been widely used in various applications such as authentication for financial transactions, access control, border control, and intelligent surveillance systems. Many studies on 2 dimensional (2D) face recognition have been performed [1–6] with 2D face detection [7,8], and there have been also previous studies on 3D face recognition [9,10]. They proposed fuzzy system–based facial feature fusion [1], convolutional neural network (CNN)-based face recognition [2,4,6], CNN-based pose-aware face recognition [3], and performance benchmarking of face recognition [5]. In addition, CNN-based face detection [7] with performance benchmarking of face detection [8] was also introduced. Three-dimensional face recognition based on geometrical descriptors and 17 soft-tissue landmarks [9] and the 3D data acquired with structured light [10] were performed as well. However, most of these previous studies were done with face images or data of high pixel resolution which are captured at a close distance from camera.

Along with the recent development of digital TV, studies have analyzed the viewers that use intelligent TV technologies such as smart TV and Internet protocol TV [11–15]. An intelligent TV provides a personalized service to the viewer. It includes a camera to obtain identity information in order to receive consumer feedback [11–15]. In order to obtain the information of the viewer using this camera, a face analysis system is used that includes the functionalities of face detection, recognition, and expression recognition [11–15]. However, different from previous research on face detection and recognition [1–10], because the camera is attached to the TV and the distance between the TV and viewer is far within the environment of watching TV, the input images are usually captured at a far distance from the camera. Consequently, the image pixel resolution of a face area is low quality with blurring of the face image. In addition, it is often the case that people watch TV while lying on their sides. Therefore, the in-plane rotation of a face more frequently happens in images compared to the out-of-plane rotation (yaw and pitch) of a face because the face image is captured while people are watching TV, including the camera.

In previous research, An et al. adopted the methods of face detection and recognition in order to determine the identity of a TV viewer [11]. However, this method is only available for frontal face detection [11,13,16], and cannot be used for face recognition of in-plane or out-of-plane rotated faces [11]. In order to build a smart home environment, Zuo et al. proposed a method for face and facial expression recognition using a smart TV and home server, but this method did not deal with face rotation either [13]. In order to recognize a rotated face, previous methods for multi-view face detection have been based on the adaptive boosting (Adaboost) method [17–19]. However, an intensive training procedure is required to build the multi-view face detector, and these studies did not deal with the face recognition of rotated faces.

There are face detection and recognition studies that consider yaw, pitch, and in-plane face rotations [20–32]. Liu proposed a face recognition method that considers head rotation (yaw and pitch rotation) using Gabor-based kernels and principal component analysis (PCA), but this system does not deal with in-plane rotation [20] although the in-plane rotation of a face frequently occurs when a viewer watches TV while lying on his or her side. Mekuz et al. proposed face recognition that considers in-plane rotation using locally linear embedding (LLE) and PCA [26]. They also proposed face recognition methods that consider the in-plane rotation of a face using complex wavelet transforms [27] and Gabor wavelets [28]. However, they only considered in-plane rotations at small angles [26–28]. Anvar et al. proposed a method for estimating the in-plane rotation angle of a face based on scale invariant feature transforms (SIFTs), but they did not deal with face recognition [30]. In other research [31], Du et al. proposed a face recognition method based on speeded-up robust features (SURF). Their method can cope with in-plane rotated face images because of the characteristics of the scale and the in-plane rotation invariance of SURF. However, they did not show the specific experimental recognition results of in-plane rotated faces. In previous research [32], Lee et al. proposed a method of detecting the correct face box from in-plane rotated faces in a TV environment, but multiple face candidates are obtained by their method. Because all these candidates are used for face recognition, the processing time and recognition error are high.

Recently, there have been studies conducted on keypoint detection of a face image in References [33–35]. Using the results of the keypoint detection of a face image, the compensation of the in-plane rotation of a face can be possible. However, in most previous studies including References [33–35], keypoint detection has been done with face images of high pixel resolution which are captured at a close distance to the camera. In contrast, the input images captured at a far distance from the camera (maximum 2.5 m) are used in our research because our study aims at face recognition at far distances in the environment of watching TV. Consequently, the image pixel resolution of a face area is so low in addition to the blurring of a face image that the previous methods of keypoint detection are difficult to apply to the face images used in our research.

Therefore, in order to address the shortcomings of previous research, we propose a new face recognition algorithm that is robust to in-plane rotation based on symmetrical characteristics of a face in the TV environment. Compared to previous work, our research is novel in the following three ways, which are the main differences between our research and previous research [32].

- Multiple face region candidates for a face are detected by image rotation and an Adaboost face detector in order to cope with the in-plane rotation of a face.
- The credibility scores for each candidate are calculated using a fuzzy system. We use four input features. In general, the more symmetrical the left and right halves of the candidate face box are, the sharper the gray-level difference histogram (GLDH) (which is calculated by the pixel

difference between the symmetrical positions based on the vertical axis that evenly bisects the face box) is. Therefore, we define the degree of sharpness of the GLDH as the Y score in this research. Then, the differences in the Y score, pixels, average, and histogram between the left and right halves of the candidate face box are used as the four features based on the symmetrical characteristics of a face.

- The accuracy of face recognition is increased by selecting the face region whose credibility score is the highest for recognition.

The remainder of this paper is organized as follows. In Section 2, we explain the proposed fuzzy-based face recognition system. The experimental results with discussions and conclusions are described in Sections 3 and 4, respectively.

2. Proposed Face Recognition System

2.1. Overview of the Proposed Method

Figure 1 shows the overall procedure of our face recognition system. Using an image captured by the web camera connected to the set-top box (STB) for the smart TV camera (see the detail explanations in Section 3.1), the region of interest (ROI) of the face is determined by image differences between the captured and (pre-stored) background images, morphological operations, and color filtering [32]. The face region is detected within the face ROI by the Adaboost method and image rotation.

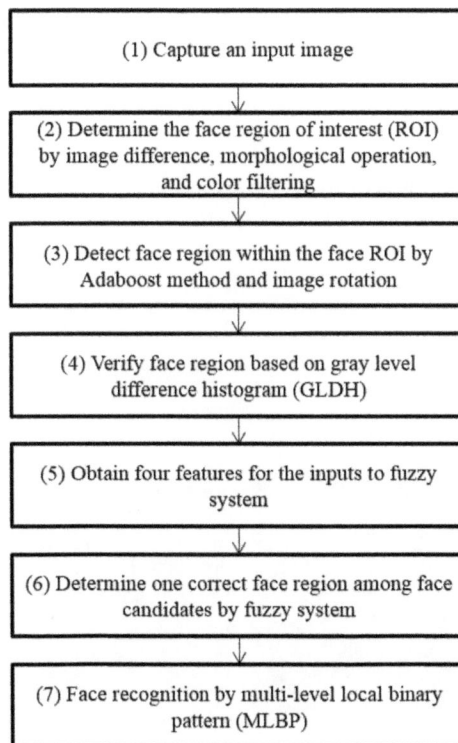

Figure 1. Flowchart of the proposed method.

Incorrect face regions can be removed using verification based on GLDH. With the face candidates, four features are extracted. Using these four features and the fuzzy system, one correct face region is selected from among the candidates. This selected face region is recognized using a multi-level local binary pattern (MLBP). In previous research [32], steps (1)–(4) and (7) of Figure 1 are used, and steps (5) and (6) are newly proposed in our research. Through steps (5) and (6), one correct (upright) face candidate can be selected among multiple candidates, which can reduce the processing time and recognition error.

2.2. Detection and Verification of the Face Region

Using the image captured by the smart TV camera, the face ROIs are detected using image differencing (between the pre-stored background and current captured images), morphological operations, and color filtering [32]. The main goal of our research is face detection robust to in-plane rotation (not facial feature extraction or face recognition). Therefore, we use the simple method of using image differences in order to detect the rough ROI of the human body because this is not the core part of our research. Because the final goal of our research is to detect the correct face region (not the human body) from the roughly detected ROI of the human body, more accurate face ROI can be located by morphological operation, color filtering, and the Adaboost face detector with image rotation, which can reduce the error in the difference image caused by background change. That is, after the difference image is obtained by differencing (between the pre-stored background and current captured images), the area of the human body shows large difference values because the pixels within this area are different between the background and current captured image. Then, the rough area of the human body can be separated from other regions by image binarization. However, small holes inside the area of the human body in the binarized image still exist because some pixel values within this area can be similar between the background and current captured image. These holes give a bad effect on the correct detection of a face, and they can be removed by morphological operation. Because the area of the human body includes the hair, face, and body, the rough candidate region of a face can be separated by color filtering. Then, within the remaining area, more accurate face regions can be detected by the Adaboost face detector. To handle the in-plane rotation of a face, the multiple face regions are located by the face detector according to the in-plane rotation of the image.

The resulting image is shown in Figure 2a. Using the face ROIs, the face regions are detected by Adaboost and image rotation. The Adaboost algorithm is based on a strong classifier that is a combination of weak classifiers [17]. In a TV environment, the in-plane rotation of a viewer's face frequently occurs because he or she can watch the TV from a lying position, which degrades the accuracy of face detection. Therefore, we detected faces using Adaboost with the original image and six (in-plane rotated) images (at $-45°$, $-30°$, $-15°$, $15°$, $30°$ and $45°$). Because Adaboost detection is performed on the original image and six (in-plane rotated) images, multiple face boxes are detected even for areas that contain a single face, as shown in Figure 2b.

(a)   (b)

Figure 2. *Cont.*

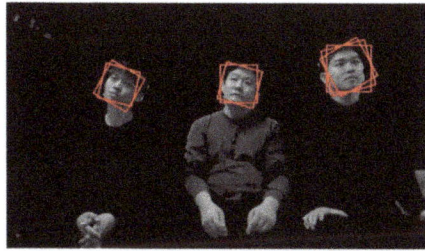

(c)

Figure 2. Detection of the face regions. (**a**) Detected face ROIs; (**b**) Multiple detected face boxes; (**c**) Results of the face detection using GLDH.

From the multiple detected face boxes, as shown in Figure 2b, we select candidates for correct face boxes using GLDH, as shown in Figure 2c. We use the GLDH method to select the correct box because it uses the characteristics of face symmetry to find a vertical axis that optimally bisects the face region [32]. The GLDH is calculated by the pixel difference between the symmetrical positions based on the vertical axis that evenly bisects the face box. Therefore, in general, the more symmetrical the left and right halves of the candidate face box are, the sharper the GLDH is. The GLDHs are shown at the bottom of Figure 3. The horizontal and vertical axes of the graphs respectively represent the gray-level difference (GLD) and number of occurrences [36].

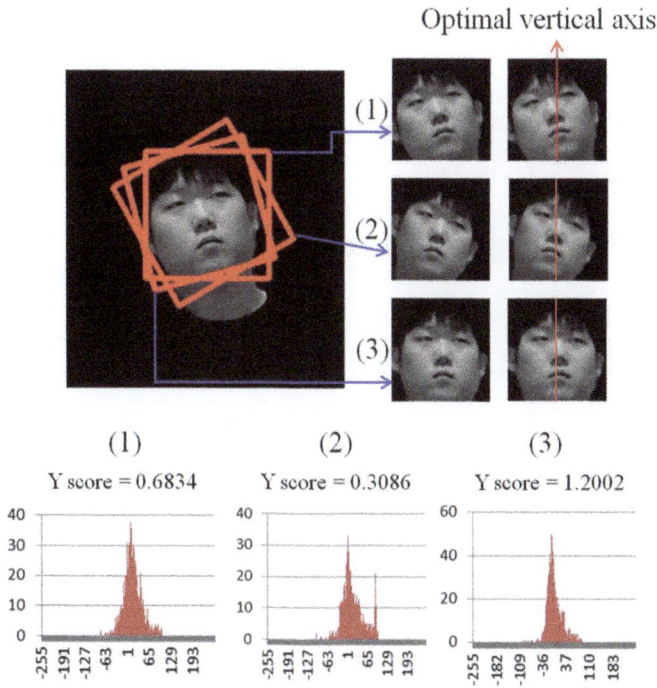

Figure 3. Y scores calculated using GLDH.

It is often the case that the face is originally rotated horizontally (yaw). Therefore, if we vertically bisect the detected face box into two equal areas, the left and right areas are not inevitably symmetrical.

Therefore, if the horizontal position of the vertical axis that evenly bisects the face box is defined as m, our system calculates the GLDHs at five (horizontal) positions ($m - 10$, $m - 5$, m, $m + 5$, and $m + 10$). If one of the five positions is the optimal vertical axis, the GLDH distribution at this position becomes a sharp shape with little variation. In an environment where a user is watching TV, severe rotation (yaw) of the user's head does not occur because he or she is looking at the TV. Therefore, the calculation of GLDH at these five positions can cope with all cases of head rotation (yaw). To measure the sharpness of the GLDH distribution, the Y score is calculated as follows [32,37]:

$$Y \text{ score} = \frac{MEAN}{\sigma^2} \tag{1}$$

where $MEAN$ is the number of pixel pairs whose GLD falls within a specified range (which we set at ± 5) based on the mean of the GLDH distribution. A high $MEAN$ represents a sharp GLDH distribution, which indicates that the corresponding bisected left and right face boxes are symmetrical. In addition, σ is the standard deviation of the distribution. Therefore, the higher the Y score, the more symmetrical the left and right halves of the face box are with respect to the vertical axis (the sharper the GLDH is).

The number of the face candidates is reduced using the Y score, as shown in Figure 2c. However, more than two face boxes still exist, even for a single face area, as shown in Figure 2c. Therefore, if multiple face candidates are used for face recognition, the processing time and face recognition error (false matches) will inevitably increase. In order to solve this problem, we propose a fuzzy-based method to select one correct face candidate. Details are given in Sections 2.3 and 2.4.

2.3. Obtaining Four Features Based on Symmetrical Characteristics of a Face for the Fuzzy System

In previous research [38,39], the characteristics of frontal face symmetry were used for face recognition. We also use four facial symmetry features as inputs for the fuzzy system. The four features (F_1, F_2, F_3, and F_4) are shown below.

$$F_1 = 1 - Y \text{ score} \tag{2}$$

$$F_2 = \sum_{y=0}^{H-1} \sum_{x=0}^{W/2-1} \frac{|I(x, y) - I(W - x - 1, y)|}{W \times H} \tag{3}$$

$$F_3 = |\sum_{y=0}^{H-1} \sum_{x=0}^{W/2-1} \frac{I(x, y) - I(W - x - 1, y)}{W \times H}| \tag{4}$$

$$F_4 = Chi - square \text{ distance between } HistoL \text{ and } HistoR \tag{5}$$

In Equation (2), F_1 is calculated from the Y score of Equation (1) after normalizing it to the range of 0–1. In Equations (3)–(5), $I(x, y)$ is the pixel value at position (x, y), and W and H are the width and height of the detected face box, respectively. Equations (3)–(5) represent the differences between the left and right halves of the candidate face box based on the vertical axis that evenly bisects the face box. Equations (3) and (4) show the exemplary case where the vertical axis is positioned at the half of W.

In Equation (5), $HistoL$ and $HistoR$ respectively represent the histograms of the left-half and right-half regions of a face box.

Features F_2–F_4 are normalized to the range of 0–1. As explained before, the higher the Y score, the more symmetrical the left and right halves of the face box are with respect to the vertical axis. In addition, F_2–F_4 show the dissimilarity between the left and right halves of the face box. Therefore, the more symmetrical the left and right halves of the face box are with respect to the vertical axis, the smaller F_1, F_2, F_3, and F_4 become. To prove this, we show the F_1–F_4 values according to the in-plane rotation of a face as shown in Figure 4. As shown in Figure 4, the greater the amount of in-plane rotation of a face region is, the larger the F_1–F_4 values. That is, the more symmetrical the left and right halves of the face box are with respect to the vertical axis (the smaller the amount of in-plane rotation of a face region is), the smaller F_1, F_2, F_3, and F_4 become. From that, we can confirm that the F_1–F_4

values can be used as inputs for the fuzzy system to select one correct (upright) face candidate among multiple candidates.

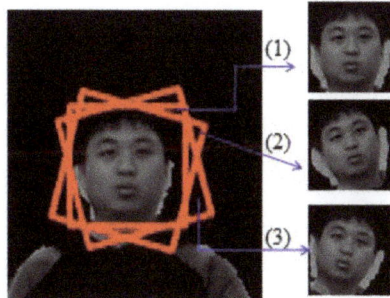

(a)

(1) F_1 (0.000000), F_2 (0.222220), F_3 (0.182540), F_4 (0.104530)

(2) F_1 (0.177360), F_2 (0.416670), F_3 (0.306860), F_4 (0.128390)

(3) F_1 (0.335210), F_2 (0.500000), F_3 (0.362190), F_4 (0.188500)

(b)

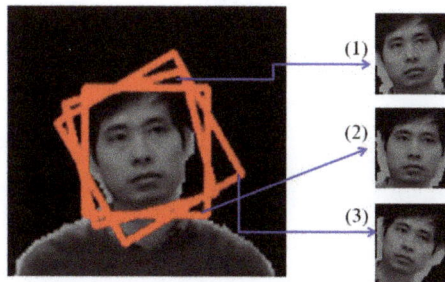

(c)

(1) F_1 (0.103590), F_2 (0.444440), F_3 (0.215660), F_4 (0.107960)

(2) F_1 (0.000000), F_2 (0.305560), F_3 (0.182240), F_4 (0.090740)

(3) F_1 (0.532750), F_2 (0.555560), F_3 (0.382320), F_4 (0.125640)

(d)

Figure 4. F_1–F_4 values according to in-plane rotation of a face. (**a,c**) Examples of detected face regions including in-plane rotation; (**b**) F_1–F_4 values of (1)–(3) face regions of (a); (**d**) F_1–F_4 values of (1)–(3) face regions of (c).

2.4. Determining a Single Correct Face Region Using a Fuzzy System

2.4.1. Definition of Fuzzy Membership Functions and Fuzzy Rule Tables

The four features F_1, F_2, F_3, and F_4 are used as inputs for the fuzzy system, and a single correct face box is its output. To achieve this, we define the input and output membership functions as shown in Figure 5a,b. Two linear functions respectively representing low (L) and high (H) are used as the input membership function. Three linear functions respectively representing low (L), medium (M), and high (H) are used as the output membership function. We acquire fuzzy output values using the input and output membership functions and the defuzzification method [40–44].

(a)

(b)

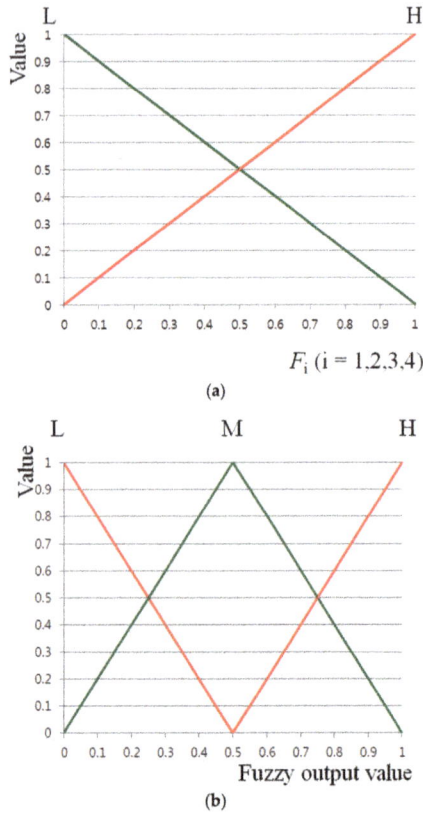

Figure 5. Input (**a**) and output (**b**) fuzzy membership functions.

As explained in Section 2.3, the more symmetrical the left and right halves of the face box are with respect to the vertical axis, the smaller F_1, F_2, F_3, and F_4 become. Based on this fact, we designed the fuzzy rule table shown in Table 1. The fuzzy output values of L and H respectively represent smaller and larger amounts of symmetry of the left and right halves of the face box with respect to the vertical axis.

Table 1. Fuzzy rule table for obtaining the fuzzy system output.

Input 1 (F_1 of Equation (2))	Input 2 (F_2 of Equation (3))	Input 3 (F_3 of Equation (4))	Input 4 (F_4 of Equation (5))	Fuzzy Output Value
L	L	L	L	H
			H	H
		H	L	H
			H	M
	H	L	L	H
			H	M
		H	L	M
			H	L

Table 1. *Cont.*

Input 1 (F_1 of Equation (2))	Input 2 (F_2 of Equation (3))	Input 3 (F_3 of Equation (4))	Input 4 (F_4 of Equation (5))	Fuzzy Output Value
H	L	L	L	H
			H	M
		H	L	M
			H	L
	H	L	L	M
			H	L
		H	L	L
			H	L

2.4.2. Determining a Single Correct Face Region by Defuzzification

In this section, we explain the method for determining a single correct face region based on the output value of the fuzzy system. With one input feature from F_1–F_4 of Equations (2)–(5), we can obtain two outputs using two input membership functions, as shown in Figure 6.

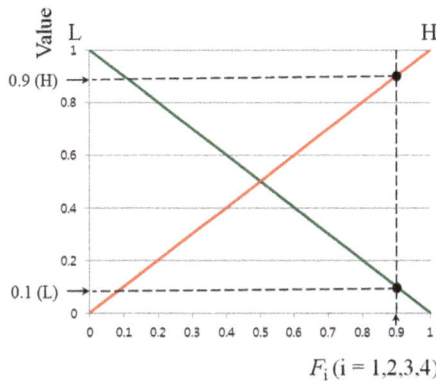

Figure 6. Obtaining two output values from a single input feature (F_i) using two input membership functions.

For example, if we assume that F_1 (of (1) face box of Figure 5a) is 0.9, 0.1 (L) and 0.9 (H) can be obtained from the L and H membership functions, respectively, as shown in Figure 6. Similarly, if F_2, F_3, and F_4 (of (1) face box of Figure 5a) are assumed to be 0.9, three pairs of 0.1 (L) and 0.9 (H) can be obtained. Consequently, the four pairs of 0.1 (L) and 0.9 (H) are obtained from F_1, F_2, F_3, and F_4 using two input membership functions. Based on these four pairs of 0.1 (L) and 0.9 (H), we obtain the combined set as {(0.1 (L), 0.1 (L), 0.1 (L), 0.1 (L)), (0.1 (L), 0.1 (L), 0.1 (L), 0.9 (H)), (0.1 (L), 0.1 (L), 0.9 (H), 0.1 (L)), ..., (0.9 (H), 0.9 (H), 0.9 (H), 0.1 (L) , (0.9 (H), 0.9 (H), 0.9 (H), 0.9 (H))}. With one subset, we can determine a single value (0.1 or 0.9) and a single symbol (L or H) based on the MIN or MAX methods [45,46] and the fuzzy rule table of Table 1.

For example, we can select 0.9 based on the MAX method with one subset (0.1 (L), 0.9 (H), 0.1 (L), 0.1 (L)). In addition, from the input of (L), (H), (L), and (L), we obtain (H) from Table 1. Consequently, we obtain 0.9 (H), which we call the inference value (IV) in this paper. Because the number of components in the combined set of {(0.1 (L), 0.1 (L), 0.1 (L), 0.1 (L)), (0.1 (L), 0.1 (L), 0.1 (L),

0.9 (H)), (0.1 (L), 0.1 (L), 0.9 (H), 0.1 (L)), ..., (0.9 (H), 0.9 (H), 0.9 (H), 0.1 (L) , (0.9 (H), 0.9 (H), 0.9 (H), 0.9 (H))} is 16, we obtain 16 IVs.

We compared the performances of five defuzzification methods, the first of maxima (FOM), last of maxima (LOM), middle of maxima (MOM), mean of maxima (MeOM), and center of gravity (COG) [40–44]. FOM, LOM, MOM, and MeOM select one output value from the outputs determined by the maximum IV (0.9 (M)) of Figure 7a. That is, FOM selects the first output value (S_2 of Figure 7a), and LOM selects the last output value (S_3 of Figure 7a). MOM selects the middle output ((S_2 + S_3)/2). MeOM selects the mean of all the outputs. In Figure 7a, MeOM also selects the (S_2 + S_3)/2.

Different from FOM, LOM, MOM, and MeOM which are based on the maximum IV, COG selects the center for the output based on the weighted average (S_5 of Figure 7c) of all the regions defined by all the IVs (the combined area of three regions R_1, R_2, and R_3 of Figure 7b). The method for calculating the weighted average by COG [42–44] is as follows:

$$S_5 = \frac{\int V \tilde{F}(S) \times S \, dS}{\int V \tilde{F}(S) \, dS} \tag{6}$$

Here, V and S respectively represent the variables for the vertical and horizontal axes of Figure 7b,c and $\tilde{F}$ is the combined area of three regions, R1, R2, and R3, of Figure 7b.

Finally, we select one correct face box whose calculated output value by the defuzzification method is the largest. For example, if the output values of (1), (2), and (3) face boxes of Figure 5a are respectively 0.51, 0.38, and 0.79, the (3) face box is finally selected as the correct one which is used for face recognition.

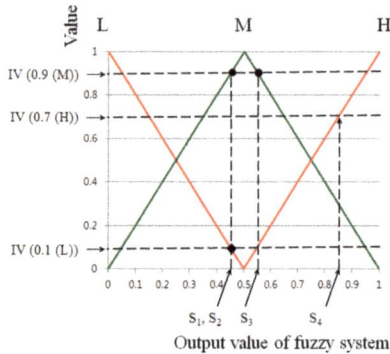

(a)

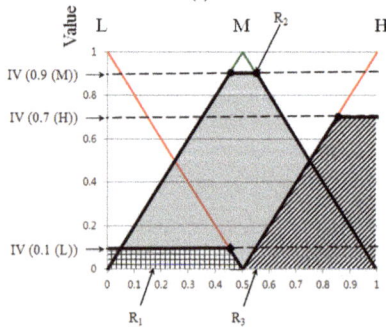

(b)

Figure 7. *Cont.*

Output value of fuzzy system

(c)

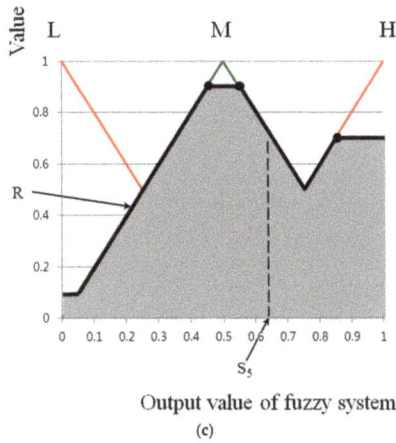

Figure 7. Obtaining the final fuzzy output value by various defuzzification methods: (**a**) by the first of maxima (FOM), last of maxima (LOM), middle of maxima (MOM), and mean of maxima (MeOM); (**b**) by the combined area of three regions of R_1, R_2, and R_3; and (**c**) by center of gravity (COG).

Figure 8 shows an example of the face boxes selected by the previous [32] and proposed methods. As shown in this figure, a more correct face box (where the left and right halves of the face box are more symmetrical) can be obtained using our method. Our system then recognizes faces using MLBP on the selected face box [32]. A more detailed explanation of the face recognition method can be found in [32].

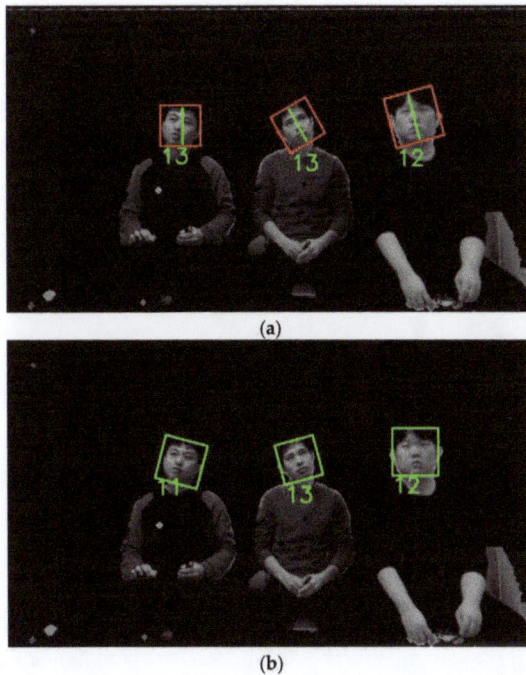

(a)

(b)

Figure 8. Examples of the final selected face boxes by (**a**) the previous method and (**b**) our method.

2.5. Face Recognition Using MLBP

The detected face regions are used for MLBP face recognition. MLBP is based on the local binary pattern (LBP) method, which assigns a binary code to each pixel based on a comparison between the center and its neighboring pixels [47]. MLBP is presented as a histogram-based LBP (concatenation of many histograms), and the LBP is a particular case of MLBP. If the center value is equal to (or greater than) the neighboring pixel, 1 is assigned; if it is less than the neighboring pixel, 0 is assigned. This basic LBP is extended to a multi-resolution method that considers various numbers P of neighboring pixels and distances R between the center and neighboring pixels as follows [32]:

$$LBP_{P,R} = \sum_{p=0}^{P-1} s(g_p - g_c)2^p, \ where \ s(x) = \begin{cases} 1, \ x \geqslant 0 \\ 0, \ x < 0 \end{cases} \tag{7}$$

where g_c is the gray value of the center pixel, g_p ($p = 1, \ldots, P-1$) are the gray values of the p that has equally spaced pixels on a circle of radius R, and $s(x)$ is the threshold function for x. The obtained LBP codes are classified into uniform and non-uniform patterns. Uniform patterns include the number of transitions between 0 and 1 and are 0, 1, or 2. Other patterns are non-uniform patterns. The uniform patterns usually represent edges, corners, and spots, whereas the non-uniform patterns do not contain sufficient texture information. The histograms of uniform and non-uniform patterns are obtained and extracted from various sub-block levels, as shown in Figure 9 [32].

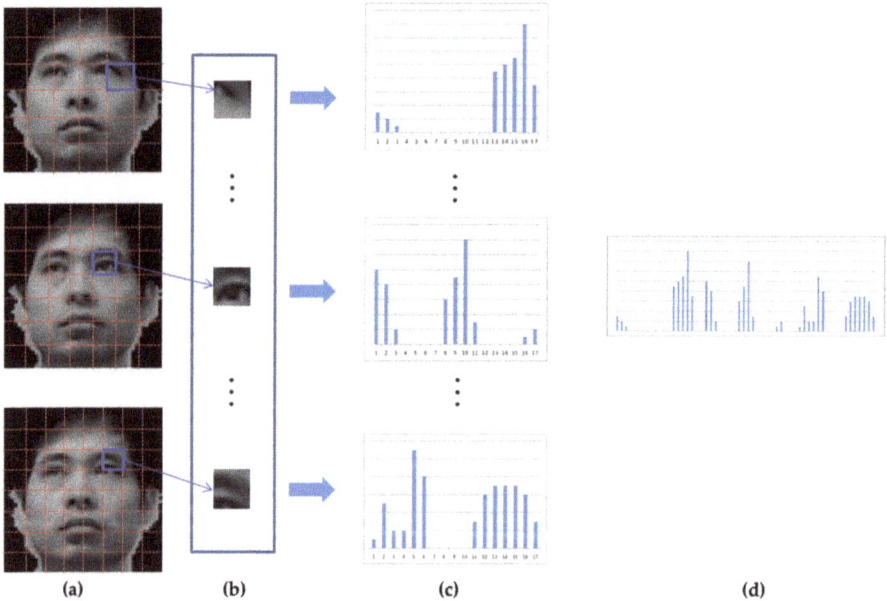

Figure 9. Example of histogram feature extraction using multi-level local binary pattern (MLBP) at three levels. (**a**) Face image divided into various sub-blocks; (**b**) Examples of sub-block regions; (**c**) Histograms for (b) obtained using local binary pattern (LBP); (**d**) Final facial feature histogram obtained by concatenating the histograms of (c).

In order to extract the histogram features globally and locally, sub-blocks of the faces are defined at three levels (the upper (6 × 6), middle (7 × 7), and lower (8 × 8) face of Figure 9). Because the larger-sized sub-blocks are used in the first level (upper face), the global (rough texture) features can be obtained from this sub-block. That is because the histogram information is extracted from the larger area of a face. On the other hand, because the smaller-sized sub-blocks are used in the third level (lower face), the local (fine texture) features can be obtained from this sub-block. That is because the histogram information is extracted from the smaller area of a face.

As shown in Figure 9d, all of the histograms for each sub-block are concatenated in order to form the final feature vector for face recognition. The dissimilarity between the registered and input face histogram features is measured by the chi-square distance

$$\chi^2(E, I) = \sum_i \frac{(E_i - I_i)^2}{E_i + I_i} \tag{8}$$

where E_i is the histogram of the registered face, and I_i is the histogram of the input face. By using the histogram-based distance, a small amount of misalignment between two face images from the same person can be compensated for. In order to deal with faces in various poses (horizontal (yaw) and vertical (pitch) rotation), the histogram feature of the input face is compared with the five registered ones (which were obtained when each user looked at five positions (left-upper, right-upper, center, left-lower, and right-lower positions) on the TV during the initial registration stage) using Equation (8). If the distance calculated by Equation (8) is less than a predetermined threshold, the input face is determined to be a registered person.

3. Experimental Results and Discussions

3.1. Descriptions of Our Databases

Our algorithm is executed in the environment of a server-client-based intelligent TV. We aim to adopt our algorithm into an intelligent TV that can be used in underdeveloped countries where people cannot afford to buy smart TVs with high performance and cost. Therefore, most functionalities of the intelligent TV are provided by a low-cost STB. Additional functionalities requiring a high processing time are provided by a high-performance server, which is connected to the STB by a network. In this environment, our algorithm is executed on a STB (microprocessor without interlocked pipeline stages (MIPS)-based dual core 1.5 GHz, 1 GB double data rate 3 (DDR3) memory, 256/512 MB negative-and (NAND) memory) and server (3.5 GHz CPU and 8 GB of RAM). The STP is attached to a 60 in TV. Steps (1) and (2) of Figure 1 are performed on the STP, and steps (3) to (7) are performed on the server.

There are many face databases, e.g., FEI [48], PAL [49], AR [50], JAFFE [51], YouTube Faces [52], the Honda/UCSD video database [53], and the IIT-NRC facial video database [54]. However, most of them were not collected when a user was watching TV, and face images with in-plane rotation are not included. Therefore, we constructed our own database, which consists of images of users watching TV in natural poses, including face images with in-plane rotation. The database was collected using 15 people by separating them into five groups of three people for the experiments [32]. In order to capture images of users looking at the TV screen naturally, each participant was instructed to watch TV without any restrictions. As a result, we captured a total of 1350 frames (database I) (15 persons × two quantities of participants (one person or three persons) × three seating positions (left, middle, and right) × three Z distances (1.5, 2, and 2.5 m) × five trials (looking naturally)). In addition, a total of 300 images (database II) (five persons × three Z distances (1.5 m, 2 m, and 2.5 m) × two lying directions (left and right) × 10 images) were collected for experiments when each person is lying on his or her side [32]. For face registration for recognition, a total of 75 frames (15 people × five TV gaze points) were obtained at the Z distance of 2 m. Consequently, a total 1725 images were used for the experiments. We make our all databases (used in our research) [55] available for others to use in their own evaluations.

Figure 10 shows examples of the experimental images. For registration, five images were acquired, as shown in Figure 10a, when each user looked at five positions on the TV. Figure 10b shows examples of the images for recognition, which were obtained at various Z-distances, seating positions, and lying directions. Figure 10c shows examples of database II.

(a)

(b)

Figure 10. *Cont.*

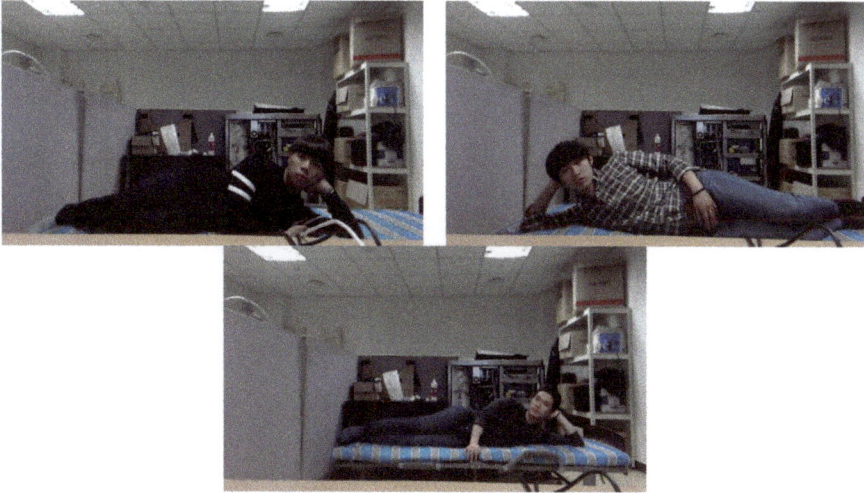

(c)

Figure 10. Examples of experimental images. (**a**) Images for face registration; (**b**) Images for recognition test (database I); (**c**) Images for recognition test (database II).

3.2. Experimental Results of the Face Detection and Recognition with Our Databases I and II

For the first experiment, we measured the accuracy of the face detection using database I. Accuracies were measured based on recall and precision, respectively, calculated as follows [32]:

$$Recall = \frac{\#TP}{m} \tag{9}$$

$$Precision = \frac{\#TP}{\#TP + \#FP} \tag{10}$$

where m is the total number of faces in the images; $\#FP$ and $\#TP$ are the number of false positives and true positives, respectively. False positives are cases where non-faces are incorrectly detected as faces. True positives are faces that are detected correctly. If the recall value is close to 1, the accuracy of the face detection is regarded as high. If the precision value is 1, all of the detected face regions are correct with an $\#FP$ of 0. As explained before, we measured the accuracies of the face detection according to the participant groups as shown in Table 2. In Table 2, recall and precision in the case of equal error rate (EER) are shown in bold type. EER means the error rate when the difference between the recall and precision is minimized in the trade-off relations between recall and precision. The reason why the recall at the EER point for Group 2 was lower than those for the other groups is that the face detection was not successful for the female who had hair occluding part her face and a small face. The reason why the precision at the EER point for Groups 2 and 3 is lower than those for other groups is that the colors of the subjects' clothes were similar to those of the facial skin, which caused false positives.

In Table 3, we measured the face detection accuracies according to the Z distances of the subjects in order to evaluate the effect of the change of image size (resolution). In Table 3, recall and precision in the case of equal error rate (EER) are shown in bold type as well. The recall at the EER point at a Z distance of 2.5 m is lower than for other cases because the face sizes are small, which caused the face detection to fail.

Table 2. Experimental results of the face detection according to participant groups (who have different gaze directions).

Group	Recall (%)	Precision (%)
	94.94	99.91
	96.85	98.87
1	**97.35**	**97.13**
	98.44	96.03
	99.87	94.83
	80.38	99.28
	84.23	95.87
2	**89.44**	**91.15**
	94.67	87.06
	99.45	82.11
	90.38	99.38
	94.27	**95.58**
3	97.78	93.92
	98.38	91.76
	99.89	90.89
	93.09	99.94
	94.18	99.04
4	95.74	98.86
	97.43	**96.24**
	99.91	94.03
	96.87	99.95
	97.05	99.87
5	**98.7**	**99.26**
	99.01	98.07
	99.87	97.16
	91.132	99.692
	93.316	97.846
Average	**95.802**	**96.064**
	97.586	93.832
	99.798	91.804

Table 3. Experimental results of the face detection according to Z distance.

Z Distance (m)	Recall (%)	Precision (%)
	96.54	99.99
	97.42	99.02
1.5	**98.06**	**98.34**
	99.13	97.24
	99.53	96.92
	91.08	99.97
	93.12	97.67
2	**95.78**	**96.59**
	97.59	93.33
	99.93	91.87
	86.98	99.58
	90.98	96.71
2.5	**93.34**	**93.72**
	96.26	90.17
	99.42	86.32
	91.53	99.85
	93.84	97.8
Average	**95.73**	**96.22**
	97.66	93.58
	99.63	91.70

The rows in each group (or Z distance) in Tables 2 and 3 show the changes of recall according to the decreases of precision. Because the recall and precision usually have a trade-off relationship (with a larger recall, a smaller precision is obtained, and vice versa), the changes of recall according to the decrease of precision are presented in our paper in order to show the accuracies of our face detection method more clearly through the various combinations of recall and precision.

In Tables 4 and 5, we respectively measured the accuracies of the face detection according to the seating positions and the number of participants in each image. As shown in Tables 4 and 5, the face detection accuracy is similar, irrespective of the seating position and number of people in each image.

Table 4. Experimental results of the face detection according to seating position.

Seating Position	Recall (%)	Precision (%)
Left	97.11	95.36
Middle	94.22	96.94
Right	95.78	96.15

Table 5. Experimental results of the face detection according to the number of people in each image.

Number of People	Recall (%)	Precision (%)
1	95.79	96.17
3	95.70	96.13

For the second experiment, we measured the accuracy of the face recognition with database I for various defuzzification methods. As explained in Section 2.5, the MLBP histogram of the incoming face is compared (using the chi-squared distance) to the five images of three individuals used to train it and the nearest is chosen as the identity, provided the calculated matching distance is less than the threshold. That is, it is a nearest neighbor classifier and only three identities are included in the tests. We measured the accuracy of the face recognition using the genuine acceptance rate (GAR). As shown in Table 6, the GAR by MOM with the MAX rule is higher than the GARs for other defuzzification methods. Using the MOM with the MAX rule, we compared the GAR of the proposed method to that of the previous one, as shown in Table 7, where it is clear that the GAR of our method is higher than that of the previous method for all cases.

Table 6. Experimental results (genuine acceptance rate (GAR)) of the face recognition using the proposed method and various defuzzification methods (%).

Method		GAR (%)
MIN rule	FOM	83.34
	MOM	90.65
	LOM	90.86
	MEOM	90.78
	COG	90.73
MAX rule	FOM	92.10
	MOM	92.93
	LOM	91.70
	MEOM	91.78
	COG	91.92

Table 7. Comparison of GARs of our method and the previous method according to participant group.

Group	GAR (%)	
	Previous Method [32]	**Proposed Method**
1	90.76	92.02
2	93.2	94.09
3	82.89	90.53
4	96.98	98.08
5	87.33	89.93
Average	90.23	**92.93**

In Tables 8–10, we compared the face recognition accuracy (GAR) of our method to that of the previous method with respect to the Z distance, sitting position, and number of people in each image, respectively. The GAR for various Z distances was measured in order to evaluate the effect of the change of the image size (resolution). The reason why the GAR at a Z distance of 2 m is higher than those at other Z distances is that the registration for face recognition was done with the face images captured at a Z distance of 2 m. The reason why the GAR at a Z distance of 2.5 m is lower than for other cases is that the face sizes in the images are smaller. As shown in Tables 8–10, we confirm that the GARs of our method are higher than those of the previous method in all cases, and the GARs of our method are not affected by the Z distance, sitting position, or the number of people in each image.

Table 8. Comparison of GARs for our method and the previous method for various Z distances.

Z Distance (m)	GAR (%)	
	Previous Method [32]	**Proposed Method**
1.5	89.11	92.22
2	92.97	96.35
2.5	88.61	90.49

Table 9. Comparison of GARs for our method and the previous method for various seating positions.

Seating Position	GAR (%)	
	Previous Method [32]	**Proposed Method**
Left	91.46	94.53
Middle	93.55	94.53
Right	85.64	89.42

Table 10. Comparison GARs for our method and the previous method for various number of people in each image.

Number of People	GAR (%)	
	Previous Method [32]	**Proposed Method**
1	90.12	92.19
3	90.57	93.67

For the next experiments, we compared the GARs of various face recognition methods [47,56–60] with our face detection method. In previous research [47], Ahonen et al. proposed LBP-based feature extraction for face recognition. PCA has been widely used to represent facial features based on eigenfaces [56,57]. Li et al. proposed a local non-negative matrix factorization (LNMF)-based method for the part-based representation of facial features [58]. In a previous study [59], they proposed support vector machine-discriminant analysis (SVM-DA)-based feature extraction for face recognition in order to overcome the limitations of the linear discriminant analysis method that assumes that all classes have Gaussian density functions. Froba et al. proposed the modified census transform (MCT)-based facial feature extraction method which uses the average value of a 3×3 pixel mask, in contrast to the LBP method which uses the center value of a 3×3 pixel neighborhood [60]. As shown in Table 11, the GAR of our MLBP-based recognition method with our face detection method is higher than those of other methods. By using the MLBP histogram features of three levels, as shown in Figure 9, both local and global features can be efficiently used for face recognition, which improves the accuracy of the face recognition.

Table 11. Comparison of GARs of various face recognition methods with our face detection method according to groups in the database.

Group	GAR (%)						
	LBP [47]	PCA [56,57]	LNMF [58]	SVM-D [59]	MCT [60]	Previous Method [32]	MLBP
1	63.03	61.03	50.53	72.44	61.01	90.76	92.02
2	57.02	45.99	42.1	77.59	53.79	93.2	94.09
3	50.47	43.11	48.45	62.61	47.13	82.89	90.53
4	68.08	67.25	61.51	79.63	68.53	96.98	98.08
5	68.4	66.45	65.46	77.76	65.11	87.33	89.93
Average	61.4	56.77	53.61	74.01	59.11	90.23	**92.93**

As shown in Table 12, the GARs of our MLBP-based recognition method with our face detection method are higher than others irrespective of the change of image resolution which is caused by the change of Z distance. As explained before, because the MLBP-based method can use both local and global features for face recognition, the change of image resolution affects the facial features less using MLBP compared to other methods. In Tables 11 and 12, all the methods were applied to the same data of the face ROI detected by our face detection method for fair comparisons.

Table 12. Comparisons of GARs of various face recognition methods with our face detection method for various Z distances.

Z Distance (m)	GAR (%)						
	LBP [47]	PCA [56,57]	LNMF [58]	SVM-DA [59]	MCT [60]	Previous Method [32]	MLBP
1.5	63.06	53.51	52.71	76.55	58.59	89.11	92.22
2	64.96	57.16	56.02	79.29	63.78	92.97	96.35
2.5	56.18	59.4	52.1	66.17	54.98	88.61	90.49

Our research is mainly focused on selecting one correct (upright) face image among multiple (in-plane-rotated) face candidates (without the procedure of detecting eye positions or keypoints) based on a fuzzy system, and on enhancing the performance of face recognition by using only the selected face image. That is, the main goal of our research is face detection robust to in-plane rotation (not facial feature extraction or face recognition). In all the methods of Tables 11 and 12, our face detection method is also commonly used. That is, PCA means PCA-based face recognition with our face detection method. In the same manner, LBP means LBP-based face recognition with our face detection method. Therefore, Tables 11 and 12 just show the accuracies of various face recognition methods with our face detection method. PCA, LBP and MCT are not originally designed to be robust to in-plane rotation. Nevertheless, the reason why we selected PCA, LBP and MCT, etc. (instead of state-of-the-art methods such as deep learning-based face recognition, etc.), for comparisons in Tables 11 and 12 is to show that our face detection method can be used with any kind of traditional or even old-fashioned method whose accuracies are lower than the state-of-the-art methods for face recognition. If we use a recognition method showing high accuracies such as the deep learning-based method in Tables 11 and 12, it is difficult to analyze whether the high accuracies of recognition are caused by our face detection method or the recognition method itself. Therefore, we include only the comparisons with traditional methods in Tables 11 and 12.

For the next test, we performed an additional experiment with database II, which includes extremely rotated faces, as shown in Figure 10c. The recall and precision of the face detection are, respectively, 96.67% and 99.39%, which are similar to those of database I in Tables 2–5. As shown in Table 13, the GAR of our method is 95.15%, which is higher than that of the previous method. In addition, the GAR of our method is similar to those of Tables 6–10. This result confirms that our method can be applied to highly rotated face images.

Table 13. Face recognition accuracy for images of highly rotated faces (database II).

Method	GAR (%)
Previous method [32]	93.10
Proposed method	95.15

Figure 11 shows the examples for which our face recognition method is successful. Figure 12 shows the examples where the face recognition failed. The failures (left person of the left figure of Figure 12 and right person of the right figure of Figure 12) are caused by false matching by the MLBP method, although the correct face boxes are selected by our method.

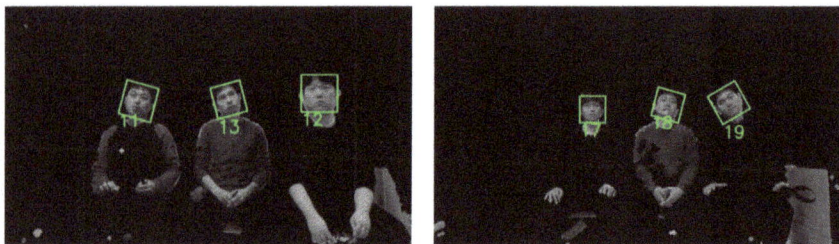

Figure 11. Examples of the success of the face recognition.

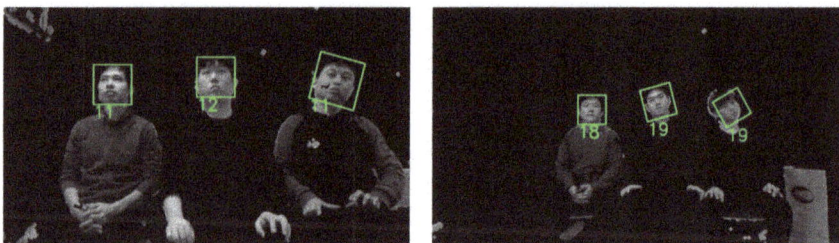

Figure 12. Examples of the failure of the face recognition.

Our method (including fuzzy system–based face detection and MLBP-based face recognition) does not require any training procedure. Even for face candidate detection, we used the original Adaboost face detector provided by the OpenCV library (version 2.4.9 [61]) without additional training. Therefore, all the experimental data were used for testing.

For the next experiment, we measured the processing time of our method. Experimental results show that the processing time per each image is approximately 152 ms. Therefore, our system can be operated at a speed of approximately six or seven frames per second. The processing time of our method is smaller than that of the previous method (185 ms) [32] because only a single face region is selected per individual for recognition. The target applications for TV of our method are the systems for automatic audience rating surveys, program recommendation services, personalized advertising, and TV child locks. Face detection and recognition do not necessary need to be executed at every frame (real-time speed) in these applications. Therefore, our system at the current processing speed of approximately six or seven frames per second can be used for these applications.

Previous research on rotation-invariant face detection exists [62,63]. Their method can detect the correct face region from the face images including various rotations of a face based on the real Adaboost method [62]. However, the processing time of their method is so high (about 250 ms for a 320 × 240 image on a Pentium 4 2.4 GHz PC) that their method cannot be used in our system. In previous research [63], they show that their method can also locate the correct face region from face images including various rotations of a face by a neural network. However, the processing time of their method is so high (about six seconds to process a 160 × 120 pixel image on an SGI O2 workstation (Silicon Graphics Inc., Sunnyvale, CA, USA) with a 174 MHz R10000 processor (Silicon Graphics Inc., Sunnyvale, CA, USA)) that their method cannot be used in our system, either. In our system, the total processing time per one input image (1280 × 720 pixels) by our method is taken as 152 ms on a desktop computer (3.5 GHz CPU and 8 GB of RAM) including the processing time of steps (1) and (2) of Figure 1 on a set-top box (STB) (MIPS-based dual core 1.5 GHz, 1 GB DDR3 memory, 256/512 MB NAND memory). Although the processing time of the previous methods [62,63] includes only the procedure of face detection, our processing time of 152 ms includes both face detection and recognition. In addition, the face images in our research are considerably blurred as shown in Figure 13c,d compared to those in their research because our face images are acquired at far distance of a maximum of 2.5 m (from the camera to the user). Therefore, their methods for face detection based on the training of the real Adaboost or a neural network are difficult to apply to face images in our research.

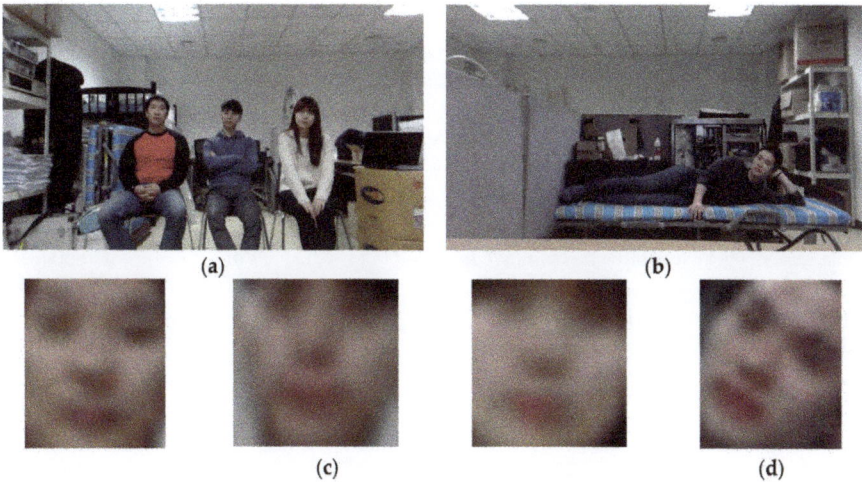

Figure 13. Comparisons of the face images in our research with those in previous studies. (**a**,**b**) Input images in our research; (**c**) Face images of (**a**); (**d**) Face images of (**b**).

In addition, we include the comparative experiments by our method with other rotation-invariant face detection methods [63]. Because our fuzzy-based method is applied to both databases I and II without any parameter tuning or training according to the type of database, the neural network of their method [63] is trained with all the images of databases I and II for fair comparison, and the testing performance are shown with databases I and II, separately.

As shown in Table 14, the accuracy of face detection by our method is higher than that by the previous method with database I. The reason why the accuracy of the previous method is lower than that of our method is that the face images in database I are blurred and the pixel resolution of the face images in database I is very low, as shown in Figure 13c. As shown in Table 15, the accuracy of face detection by our method is also higher than that of the previous method with database II. The reason why the accuracy of the previous method is lower than that of our method is that the pixel resolution of face images in database II is very low and there also exist many variations of in-plane rotation of the face images in addition to the blurring effect as shown in Figure 13d.

Table 14. Comparisons of the face detection accuracy of our method with previous method (database I).

Method	Recall (%)	Precision (%)
Previous method [63]	92.21	92.87
Proposed method	95.80	96.06

Table 15. Comparisons of the face detection accuracy of our method with previous method (database II).

Method	Recall (%)	Precision (%)
Previous method [63]	92.94	93.26
Proposed method	96.67	99.39

3.3. Experimental Results with Labeled Faces in the Wild (LFW) Open Database

As the next experiment, we measured the accuracies of the face detection with the LFW database [64]. Because our research is mainly focused on face detection robust to the in-plane rotation of a face, face images including other factors such as severe out-of-plane rotation and occlusion, etc., are excluded by manual selection for experiments among the images of the LFW database. This manual selection was performed by four people (two males and two females). Two people are in their twenties and the other two people are in their thirties. All four people are not the developers of our system and did not take part in our experiments for unbiased selection. We gave instructions (to the four people) to manually select the face images by comparing the images of the LFW database with those of our databases I and II. Then, only the images (selected by the consensus of all four people) are excluded in our experiments.

In addition, we included the comparative results of our method and the previous method [64]. As shown in Table 16, the accuracies of face detection by our method with the LFW database are similar to those with databases I and II of Tables 14 and 15. In addition, our method outperforms the previous method [63] with the LFW database.

Table 16. Comparisons of the face detection accuracy of our method with the previous method (LFW database).

Method	Recall (%)	Precision (%)
Previous method [63]	91.89	91.92
Proposed method	95.21	95.53

3.4. Discussions

There has been a great deal of previous researches on keypoint detection of a face image in References [33–35]. However, in most previous research including References [33–35], keypoint detection has been done with face images of high pixel resolution which are captured at close distance to the camera. In contrast, the input images captured at a far distance from the camera (maximum 2.5 m) are used in our research because our study aims at face recognition at far distances in the environment of watching TV. Consequently, the image pixel resolution of a face area is so low (less than 40×50 pixels), in addition to the blurring of the face image as shown in Figure 13c,d, that the previous methods of keypoint detection or eye detection are difficult to apply to the face images used in our research.

As an experiment, we measured the accuracies of eye detection by the conventional Adaboost eye detector [17] and subblock-based template matching [65]. Experimental results showed that the recall and precision of eye detection by the Adaboost eye detector within the detected face region were about 10.2% and 12.3%, respectively. In addition, the recall and precision of eye detection by subblock-based template matching within the detected face region were about 12.4% and 13.7%, respectively. These results show that reliable eye positions or keypoints are difficult to detect in our blurred face images of low pixel resolution. Therefore, the procedures of detecting keypoints, alignment (removing in-plane rotation), and face recognition cannot be used in our research.

To overcome these problems, we propose the method of selecting one correct (upright) face image among multiple (in-plane-rotated) face candidates (without the procedure of detecting eye positions or keypoints) based on a fuzzy system, and enhancing the performance of the face recognition by using only the selected face image.

If we synthetically modify (manually rotate) the images of the open dataset, the discontinuous region (between the face and its surrounding areas) occurs in the image as shown in Figure 14b (from the YouTube dataset) and Figure 14e (from the Honda/UCSD dataset), which causes a problem in face detection and the correct accuracy of face detection is difficult to measure with these images. In order to prevent the discontinuous region, we can rotate the whole image. However, the background is also rotated as shown in Figure 14c,f, where an unrealistic background (which does not exist in the real world) is produced in the rotated image, which affects the correct measurement of the face detection accuracy.

As explained before, as shown in Figure 13c,d, the pixel resolution of images used in our research of face recognition is very low in addition to the blurring effect of a face image compared to images in open databases such as the LFPW [33], BioID [34], HELEN [35], YouTube Faces (Figure 14a), and Honda/UCSD (Figure 14d) datasets. These kinds of focused images of high pixel resolution cannot be acquired in our research environment of watching TV where the user's face is captured by a low-cost web camera at the Z distance of a maximum of 2.5 m between the camera and user (as shown in Figure 13c,d). Therefore, the experiments with these open databases cannot reflect the correct measurement of the face recognition accuracy in the environment of watching TV. There is no other open database (acquired at the Z distance of a maximum of 2.5 m) that includes large areas of background and face images of in-plane rotation like our dataset includes, as shown in Figure 13c,d.

Our method cannot deal with occluded or profiled faces. However, the cases of occluded or profiled faces do not occur in our research environment where the use is usually watching TV, as shown in Figure 10. That is because more than two people do not occlude their faces and a profiled face caused by the severe out-of-plane rotation of a face cannot happen when watching TV. Therefore, we do not consider the cases of occluded or profiled faces in our research.

Figure 14. Example of images of YouTube and Honda/UCSD databases. (**a**) Original image of YouTube database [66]; (**b**) Image where cropped face area is rotated and discontinuous region around face area exists with (a); (**c**) Rotated image of YouTube dataset; (**d**) Original image of Honda/UCSD database [67]; (**e**) Image where cropped face is rotated and discontinuous region around face area exists with (d); (**f**) Rotated image of (d).

4. Conclusions

In this paper, we proposed a new fuzzy-based face recognition algorithm that is robust to in-plane rotation. Among the multiple candidate face regions detected by image rotation and the Adaboost face detector, a single correct face region is selected by a fuzzy system and used for recognition. Experimental results using two databases show that our method outperformed previous ones. Furthermore, the performance of our method was not affected by changes in the Z distance, sitting position, or number of people in each image. By using a non-training-based fuzzy system, our method does not require a time-consuming training procedure, and the performance of our method is less affected by the kinds of databases on which it is tested.

As future work, we plan to research a way to combine our fuzzy-based method with a training-based one such as neural networks, SVMs, or deep learning. In addition, we would research a method of enhancing the accuracy of face recognition based on other similarity metrics (such as human vs. machine d-prime) instead of the chi-square distance. In addition, the metric validity would be also checked based on spatial-taxon contours instead of precision and recall when measuring the accuracies of face detection.

Acknowledgments: This research was supported by the MSIP (Ministry of Science, ICT and Future Planning), Korea, under the ITRC (Information Technology Research Center) support program (IITP-2016-H8501-16-1014) supervised by the IITP (Institute for Information & communications Technology Promotion), and in part by the Bio & Medical Technology Development Program of the NRF funded by the Korean government, MSIP (NRF-2016M3A9E1915855).

Author Contributions: Hyung Gil Hong and Kang Ryoung Park designed the overall system for fuzzy-based face detection. In addition, they wrote and revised the paper. Won Oh Lee, Yeong Gon Kim, Ki Wan Kim, and Dat Tien Nguyen helped in the experiments and in the analyses of experimental results.

Conflicts of Interest: The authors declare no conflict of interest.

References

1. Choi, J.Y.; Plataniotis, K.N.; Ro, Y.M. Face Feature Weighted Fusion Based on Fuzzy Membership Degree for Video Face Recognition. *IEEE Trans. Syst. Man Cybern. Part. B Cybern.* **2012**, *42*, 1270–1282.
2. Taigman, Y.; Yang, M.; Ranzato, M.A.; Wolf, L. DeepFace: Closing the Gap to Human-Level Performance in Face Verification. In Proceedings of the IEEE Conference on Computer Vision and Pattern Recognition, Columbus, DC, USA, 23–28 June 2014; pp. 1701–1708.
3. Masi, I.; Rawls, S.; Medioni, G.; Natarajan, P. Pose-Aware Face Recognition in the Wild. In Proceedings of the IEEE Conference on Computer Vision and Pattern Recognition, Las Vegas, CA, USA, 26 June–1 July 2016; pp. 4838–4846.
4. Sun, Y.; Wang, X.; Tang, X. Sparsifying Neural Network Connections for Face Recognition. In Proceedings of the IEEE Conference on Computer Vision and Pattern Recognition, Las Vegas, CA, USA, 26 June–1 July 2016; pp. 4856–4864.
5. Kemelmacher-Shlizerman, I.; Seitz, S.M.; Miller, D.; Brossard, E. The MegaFace Benchmark: 1 Million Faces for Recognition at Scale. In Proceedings of the IEEE Conference on Computer Vision and Pattern Recognition, Las Vegas, CA, USA, 26 June–1 July 2016; pp. 4873–4882.
6. Ghazi, M.M.; Ekenel, H.K. A Comprehensive Analysis of Deep Learning Based Representation for Face Recognition. In Proceedings of the IEEE Conference on Computer Vision and Pattern Recognition, Las Vegas, CA, USA, 26 June–1 July 2016; pp. 102–109.
7. Qin, H.; Yan, J.; Li, X.; Hu, X. Joint Training of Cascaded CNN for Face Detection. In Proceedings of the IEEE Conference on Computer Vision and Pattern Recognition, Las Vegas, CA, USA, 26 June–1 July 2016; pp. 3456–3465.
8. Yang, S.; Luo, P.; Loy, C.C.; Tang, X. WIDER FACE: A Face Detection Benchmark. In Proceedings of the IEEE Conference on Computer Vision and Pattern Recognition, Las Vegas, CA, USA, 26 June–1 July 2016; pp. 5525–5533.
9. Vezzetti, E.; Marcolin, F.; Fracastoro, G. 3D Face Recognition: An Automatic Strategy Based on Geometrical Descriptors and Landmarks. *Robot. Auton. Syst.* **2014**, *62*, 1768–1776. [CrossRef]

10. Beumier, C. 3D Face Recognition. In Proceedings of the IEEE International Conference on Industrial Technology, Mumbai, India, 15–17 December 2006; pp. 369–374.

11. An, K.H.; Chung, M.J. Cognitive Face Analysis System for Future Interactive TV. *IEEE Trans. Consum. Electron.* **2009**, *55*, 2271–2279. [CrossRef]

12. Mlakar, T.; Zaletelj, J.; Tasic, J.F. Viewer Authentication for Personalized iTV Services. In Proceedings of the 8th International Workshop on Image Analysis for Multimedia Interactive Services, Santorini, Greece, 6–8 June 2007; pp. 63–66.

13. Zuo, F.; de With, P.H.N. Real-time Embedded Face Recognition for Smart Home. *IEEE Trans. Consum. Electron.* **2005**, *51*, 183–190.

14. Lee, S.-H.; Sohn, M.-K.; Kim, D.-J.; Kim, B.; Kim, H. Face Recognition of Near-infrared Images for Interactive Smart TV. In Proceedings of the 27th Conference on Image and Vision Computing New Zealand, Dunedin, New Zealand, 26–28 November 2012; pp. 335–339.

15. Lin, K.-H.; Shiue, D.-H.; Chiu, Y.-S.; Tsai, W.-H.; Jang, F.-J.; Chen, J.-S. Design and Implementation of Face Recognition-aided IPTV Adaptive Group Recommendation System Based on NLMS Algorithm. In Proceedings of the International Symposium on Communications and Information Technologies, Gold Coast, Australia, 2–5 October 2012; pp. 626–631.

16. Parveen, P.; Thuraisingham, B. Face Recognition Using Multiple Classifiers. In Proceedings of the 18th IEEE International Conference on Tools with Artificial Intelligence, Arlington, VA, USA, 13–15 November 2006; pp. 179–186.

17. Viola, P.; Jones, M.J. Robust Real-Time Face Detection. *Int. J. Comput. Vis.* **2004**, *57*, 137–154. [CrossRef]

18. Jones, M.; Viola, P. *Fast Multi-View Face Detection*; TR2003–96; Mitsubishi Electric Research Laboratories: Cambridge, MA, USA, August 2003.

19. Xiao, R.; Li, M.-J.; Zhang, H.-J. Robust Multipose Face Detection in Images. *IEEE Trans. Circuits Syst. Video Technol.* **2004**, *14*, 31–41. [CrossRef]

20. Liu, C. Gabor-Based Kernel PCA with Fractional Power Polynomial Models for Face Recognition. *IEEE Trans. Pattern Anal. Mach. Intell.* **2004**, *26*, 572–581. [PubMed]

21. Ban, J.; Féder, M.; Jirka, V.; Loderer, M.; Omelina, L.; Oravec, M.; Pavlovičová, J. An Automatic Training Process Using Clustering Algorithms for Face Recognition System. In Proceedings of the 55th International Symposium ELMAR-2013, Zadar, Croatia, 25–27 September 2013; pp. 15–18.

22. Asthana, A.; Marks, T.K.; Jones, J.M.; Tieu, K.H.; Rohith, M.V. Fully Automatic Pose-Invariant Face Recognition via 3D Pose Normalization. In Proceedings of the IEEE International Conference on Computer Vision, Barcelona, Spain, 6–13 November 2011; pp. 937–944.

23. Joo, Y.H.; An, K.H.; Park, J.W.; Chung, M.J. Real-Time Face Recognition for Mobile Robots. In Proceedings of The 2nd International Conference on Ubiquitous Robots and Ambient Intelligence, Daejeon, Korea, 2–4 November 2005 ; pp. 43–47.

24. Kim, D.H.; An, K.H.; Chung, M.J. Development of a Cognitive Head-Eye System: Real-time Face Detection, Tracking, Recognition and Visual Attention to Human. In Proceedings of the 13th International Conference on Advanced Robotics, Jeju Island, Korea, 21–24 August 2007; pp. 303–307.

25. Ryu, Y.G.; An, K.H.; Kim, S.J.; Chung, M.J. Development of An Active Head Eye System for Human-Robot Interaction. In Proceedings of the 39th International Symposium on Robotics, Seoul, Korea, 15–17 October 2008.

26. Mekuz, N.; Bauckhage, C.; Tsotsos, J.K. Face Recognition with Weighted Locally Linear Embedding. In Proceedings of the 2nd Canadian Conference on Computer and Robot Vision, British Columbia, Canada, 9–11 May 2005; pp. 290–296.

27. Eleyan, A.; Özkaramanli, H.; Demirel, H. Complex Wavelet Transform-Based Face Recognition. *EURASIP J. Adv. Signal. Process.* **2008**, *2008*, 1–13. [CrossRef]

28. Choi, W.-P.; Tse, S.-H.; Wong, K.-W.; Lam, K.-M. Simplified Gabor Wavelets for Human Face Recognition. *Pattern Recognit.* **2008**, *41*, 1186–1199. [CrossRef]

29. Albiol, A.; Monzo, D.; Martin, A.; Sastre, J.; Albiol, A. Face Recognition Using HOG–EBGM. *Pattern Recognit. Lett.* **2008**, *29*, 1537–1543. [CrossRef]

30. Anvar, S.M.H.; Yau, W.-Y.; Nandakumar, K.; Teoh, E.K. Estimating In-Plane Rotation Angle for Face Images from Multi-Poses. In Proceedings of the IEEE Symposium on Computational Intelligence in Biometrics and Identity Management, Singapore, 16–19 April 2013; pp. 52–57.

31. Du, G.; Su, F.; Cai, A. Face Recognition Using SURF Features. In Proceedings of the 6th International Symposium on Multispectral Image Processing and Pattern Recognition, Yichang, China, 29–31 October 2009; pp. 749629-1–749628-7. [CrossRef]

32. Lee, W.O.; Kim, Y.G.; Hong, H.G.; Park, K.R. Face Recognition System for Set-Top Box-Based Intelligent TV. *Sensors* **2014**, *14*, 21726–21749. [CrossRef] [PubMed]

33. Cao, X.; Wei, Y.; Wen, F.; Sun, J. Face Alignment by Explicit Shape Regression. *Int. J. Comput. Vis.* **2014**, *107*, 177–190. [CrossRef]

34. Kazemi, V.; Sullivan, J. One Millisecond Face Alignment with an Ensemble of Regression Trees. In Proceedings of the IEEE Conference on Computer Vision and Pattern Recognition, Columbus, OH, USA, 23–28 June 2014; pp. 1867–1874.

35. Asthana, A.; Zafeiriou, S.; Tzimiropoulos, G.; Cheng, S.; Pantic, M. From Pixels to Response Maps: Discriminative Image Filtering for Face Alignment in the Wild. *IEEE Trans. Pattern Anal. Mach. Intell.* **2015**, *37*, 1312–1320. [CrossRef] [PubMed]

36. Chetverikov, D. GLDH Based Analysis of Texture Anisotropy and Symmetry: An Experimental Study. In Proceedings of the 12th International Conference on Pattern Recognition, Jerusalem, Israel, 9–13 October 1994; pp. 444–448.

37. Chen, X.; Rynn, P.J.; Bowyer, K.W. Fully Automated Facial Symmetry Axis Detection in Frontal Color Images. In Proceedings of the IEEE Workshop on Automatic Identification Advanced Technologies, Buffalo, NY, USA, 17–18 October 2005; pp. 106–111.

38. Song, Y.-J.; Kim, Y.-G.; Chang, U.-D.; Kwon, H.B. Face Recognition Robust to Left/Right Shadows; Facial Symmetry. *Pattern Recognit.* **2006**, *39*, 1542–1545. [CrossRef]

39. Wang, F.; Huang, C.; Liu, X. A Fusion of Face Symmetry of Two-Dimensional Principal Component Analysis and Face Recognition. In Proceedings of International Conference on Computational Intelligence and Security, Beijing, China, 11–14 December 2009; pp. 368–371.

40. Klir, G.J.; Yuan, B. *Fuzzy Sets and Fuzzy Logic*; Prentice-Hall: Upper Saddle River, NJ, USA, 1995.

41. Adeli, H.; Sarma, K.C. *Cost Optimization of Structures: Fuzzy Logic, Genetic Algorithms, and Parallel Computing*; John Wiley and Sons: West Sussex, UK, 2006.

42. Siddique, N.; Adeli, H. *Computational Intelligence: Synergies of Fuzzy Logic, Neural Networks and Evolutionary Computing*; John Wiley and Sons: West Sussex, UK, 2013.

43. Broekhoven, E.V.; Baets, B.D. Fast and Accurate Center of Gravity Defuzzification of Fuzzy System Outputs Defined on Trapezoidal Fuzzy Partitions. *Fuzzy Sets Syst.* **2006**, *157*, 904–918. [CrossRef]

44. Ross, T.M. *Fuzzy Logic. with Engineering Applications*, 3rd ed.; John Wiley and Sons: West Sussex, UK, 2010.

45. Mamdani, E.H. Application of Fuzzy Logic to Approximate Reasoning Using Linguistic Synthesis. *IEEE Trans. Comput.* **1977**, *C-26*, 1182–1191. [CrossRef]

46. Zadeh, L.A. Outline of a New Approach to the Analysis of Complex Systems and Decision Processes. *IEEE Trans. Syst. Man Cybern.* **1973**, *SMC-3*, 28–44. [CrossRef]

47. Ahonen, T.; Hadid, A.; Pietikäinen, M. Face Recognition with Local Binary Patterns. In Proceeding of the 8th European Conference on Computer Vision, Prague, Czech Republic, 11–14 May 2004; pp. 469–481.

48. FEI Face Database. Available online: http://fei.edu.br/~cet/facedatabase.html (accessed on 30 October 2015).

49. Minear, M.; Park, D.C. A Lifespan Database of Adult Facial Stimuli. *Behav. Res. Methods* **2004**, *36*, 630–633. [CrossRef]

50. AR Face Database. Available online: http://www2.ece.ohio-state.edu/~aleix/ARdatabase.html (accessed on 30 October 2015).

51. The Japanese Female Facial Expression (JAFFE) Database. Available online: http://www.kasrl.org/jaffe.html (accessed on 30 October 2015).

52. Wolf, L.; Hassner, T.; Maoz, I. Face Recognition in Unconstrained Videos with Matched Background Similarity. In Proceedings of the IEEE Conference on Computer Vision and Pattern Recognition, Providence, RI, USA, 20–25 June 2011; pp. 529–534.

53. Lee, K.-C.; Ho, J.; Yang, M.-H.; Kriegman, D. Visual Tracking and Recognition Using Probabilistic Appearance Manifolds. *Comput. Vis. Image Underst.* **2005**, *99*, 303–331. [CrossRef]

54. Gorodnichy, D.O. Video-based Framework for Face Recognition in Video. In Proceedings of the 2nd Canadian Conference on Computer and Robot Vision, British Columbia, BC, Canada, 9–11 May 2005; pp. 1–9.

55. Dongguk Face Database (DFace-DB1). Available online: http://dm.dgu.edu/link.html (accessed on 29 April 2016).
56. Belhumeur, P.N.; Hespanha, J.P.; Kriegman, D.J. Eigenfaces vs. Fisherfaces: Recognition Using Class Specific Linear Projection. *IEEE Trans. Pattern Anal. Mach. Intell.* **1997**, *19*, 711–720. [CrossRef]
57. Turk, M.; Pentland, A. Eigenfaces for Recognition. *J. Cogn. Neurosci.* **1991**, *3*, 71–86. [CrossRef] [PubMed]
58. Li, S.Z.; Hou, X.W.; Zhang, H.J.; Cheng, Q.S. Learning Spatially Localized, Parts-based Representation. In Proceedings of the IEEE Conference on Computer Vision and Pattern Recognition, Kauai, HI, USA, 8–14 December 2001; pp. 207–212.
59. Kim, S.-K.; Park, Y.J.; Toh, K.-A.; Lee, S. SVM-based Feature Extraction for Face Recognition. *Pattern Recognit.* **2010**, *43*, 2871–2881. [CrossRef]
60. Froba, B.; Ernst, A. Face Detection with the Modified Census Transform. In Proceedings of the IEEE International Conference on Automatic Face and Gesture Recognition, Seoul, Korea, 17–19 May 2004; pp. 91–96.
61. OpenCV. Available online: http://opencv.org/ (accessed on 11 July 2016).
62. Wu, B.; AI, H.; Huang, C.; Lao, S. Fast Rotation Invariant Multi-View Face Detection Based on Real Adaboost. In Proceedings of the 6th IEEE International Conference on Automatic Face and Gesture Recognition, Seoul, Korea, 17–19 May 2004; pp. 79–84.
63. Rowley, H.A.; Baluja, S.; Kanade, T. Rotation Invariant Neural Network-Based Face Detection. In Proceedings of IEEE Computer Society Conference on Computer Vision and Pattern Recognition, Santa Barbara, CA, USA, 23–25 June 1998; pp. 38–44.
64. Labeled Faces in the Wild. Available online: http://vis-www.cs.umass.edu/lfw/ (accessed on 29 April 2016).
65. Kim, B.-S.; Lee, H.; Kim, W.-Y. Rapid Eye Detection Method for Non-glasses Type 3D Display on Portable Devices. *IEEE Trans. Consum. Electron.* **2010**, *56*, 2498–2505. [CrossRef]
66. YouTube Faces DB. Available online: http://www.cs.tau.ac.il/~wolf/ytfaces/ (accessed on 29 April 2016).
67. The Honda/UCSD Video Database. Available online: http://vision.ucsd.edu/~leekc/HondaUCSDVideoDatabase/HondaUCSD.html (accessed on 29 April 2016).

symmetry

MDPI

Article

M&E-NetPay: A Micropayment System for Mobile and Electronic Commerce

Xiaodi Huang [1,*]**, Jinsong Bao** [2,*]**, Xiaoling Dai** [1]**, Edwin Singh** [3]**, Weidong Huang** [4] **and Changqin Huang** [5]

1   School of Computing and Mathematics, Charles Sturt University, Albury 2640, Australia; sdai@csu.edu.au
2   College of Mechanical Engineering, Donghua University, Shanghai 201620, China
3   The University of the South Pacific, Laucala Campus, Suva, Fiji; singh_edwin@gmail.com
4   School of Engineering and ICT, University of Tasmania, Hobart 7005, Australia; tony.huang@utas.edu.au
5   School of Information Technology in Education, South China Normal University, Guangzhou 510660, China; cqhuang@zju.edu.cn
*   Correspondence: xhuang@csu.edu.au (X.H.); bao@dhu.edu.cn (J.B.);
    Tel.: +61-2-6051-9652 (X.H.); +86-21-6779-2583 (ext. 33) (J.B.)

Academic Editor: Angel Garrido
Received: 17 June 2016; Accepted: 27 July 2016; Published: 3 August 2016

Abstract: As an increasing number of people purchase goods and services online, micropayment systems are becoming particularly important for mobile and electronic commerce. We have designed and developed such a system called M&E-NetPay (Mobile and Electronic NetPay). With open interoperability and mobility, M&E-NetPay uses web services to connect brokers and vendors, providing secure, flexible and reliable credit services over the Internet. In particular, M&E-NetPay makes use of a secure, inexpensive and debit-based off-line protocol that allows vendors to interact only with customers, after validating coins. The design of the architecture and protocol of M&E-NetPay are presented, together with the implementation of its prototype in ringtone and wallpaper sites. To validate our system, we have conducted its evaluations on performance, usability and heuristics. Furthermore, we compare our system to the CORBA-based (Common Object Request Broker Architecture) off-line micro-payment systems. The results have demonstrated that M&E-NetPay outperforms the .NET-based M&E-NetPay system in terms of performance and user satisfaction.

Keywords: mobile and electronic commerce; micro-payment; web services; electronic wallet; mobile networks

1. Introduction

Mobile commerce is concerned with conducting business transactions and providing services on portable, wireless devices over the Internet [1]. Due to the exponential growth of the number of the Internet users and the maturation of wireless communication technologies, mobile commerce has rapidly attained the interest of the business vanguard [2].

M-commerce benefits not only consumers, but also business. It is convenient for consumers to purchase goods and services by using their mobiles. M-commerce enables transactions to be conducted in a high-volume, low-cost per-item way. It is obvious that m-commerce has enormous potentials. However, the current micro-payment systems for m-commerce have the following three main problems.

A desirable protocol of micro-payment should support high-volume, low-cost per-item transactions from vendors [3–10]. Several micro-payment protocols have been proposed for electronic payment in m-commerce recently. The examples of such protocols include MPS (multiparty payment scheme) [11], CMP (chaotic micro-payment protocol) [12], NetPay [13] and the recent ones for wearable devices and clouds [14–16]. Many of the proposed protocols, however, suffer the problems of the dependence on online brokers and a lack of scalability and coin transferability. The interaction between a client and a server in a CORBA-based NetPay system, for example, is mediated by object request brokers (ORBs) on both sides. A problem of this technique is that each node of CORBA has to run ORBs from the same product. In reality, it is difficult for ORBs provided by different vendors to interoperate. In addition, the interoperability does not extend into higher-level services, such as security and transaction management. Furthermore, specific advantages of particular vendors would be lost in this situation. Because this protocol depends on a closely-administered environment, it is unlikely that two random computers can successfully make Distributed Component Object Model (DCOM) or Internet Inter-ORB Protocol (IIOP) calls [17]. As a reasonable protocol for server-to-server communications, CORBA, however, has severe weaknesses in client-to-server communications, especially when client machines are scattered across the Internet.

Middleware interfaces: The recently-developed NetPay makes use of an off-line micro-payment model with a CORBA interface as a middleware that interconnects broker and vendor sites [18]. This prototype is suitable for ecommerce applications. In mobile environments where clients (and possibly servers) keep moving, this requires, however, dealing with the changing network addresses and unreliable connections. As a result, this mobility requirement adds additional constraints to the system. Due to its tight coupling between clients and servers, it is obvious that CORBA is not well suited for this environment. In order to overcome this barrier, in this paper, we present an off-line micropayment model that uses web services rather than CORBA as the middleware. By using the Simple Object Access Protocol (SOAP Web service protocol), the mobility requirement is dealt with by proxies that route messages accordingly. Moreover, the sender of a message and the final recipient do not have to be aware of the proxies. Web services offer greater advantages over CORBA, particularly for developing mobile applications. They cater to a large number of users who use either browsers or mobile devices. Web services add in a new functionality of interoperability, which is independent of the development platforms and programming languages used. In particular, Web Services on the .NET framework are widely available in object-oriented and distributed systems. As such, small and enterprise applications enable connecting to each other over the Internet.

Evaluations: Rowley [19] and Sumak et al. [20] present comprehensive reviews on e-service evaluation frameworks. The evaluations on specific and particular types of e-services, e-shops and e-business include Barnes and Vidgen [21], Behkamal et al. [22], Parasuraman et al. [23], Schubert et al. [24], and Janda et al. [25], In this work, we evaluate our system by using three types of evaluations, which include not only user perceptions, but also system performance.

One of the big challenges for micro-payment systems is that e-coins should be allowed to be spent at a wide range of vendors. Micro-payment systems should enable mobile users to leverage buy-once-spend (almost)-anywhere behaviour. In this work, we extend NetPay into M&E-NetPay. M&E-NetPay uses Web Service interfaces as a middleware for interconnecting the sites of brokers and vendors. Web Service interfaces make it simple to transfer e-coins among vendors. E-coins in M&E-NetPay are easily transferred between multiple vendors, so that M&E users can make multiple purchases. Another challenge in the design of micro-payment systems is the minimization of overheads on the servers of the sites of brokers and vendors. As a fully-distributed multi-tier system deployed over several servers, M&E-NetPay is able to achieve the minimal downtime and maximal competence. As reported in performance evaluations, the .NET framework architecture 4.0 [26] with Web Services in M&E-NetPay improves client-to-server communications. This leads to greatly improving system performance. The architecture with Web Services provides fast, secure and inexpensive communications amongst mobile users and vendor systems. In addition, the M&E-NetPay architecture also supports servers running on different platforms and vendor applications developed by using different programming languages. This allows an M&E-NetPay-enabled vendor to act as a purchasing portal for existing non-M&E-NetPay supporting vendors. In particular, an M&E-NetPay-enabled vendor redirects page accesses to these vendors and manages the debit of user e-coins. As such, existing vendors are encouraged to temporally use M&E-NetPay micropayment services for dynamic registration.

In summary, we design and develop M&E-NetPay in a way that attempts to address the three above-mentioned problems.

The major contributions of this paper are as follows:

- We present a novel micro-payment model of M&E-NetPay and its architecture.
- A new way for the deployment model with a thin-client architecture and Web service interfaces is proposed, i.e., HTML and Wireless Markup Language (WML)-based interfaces for customers.
- We have implemented the prototype of M&E NetPay including one broker and two vendor sites, which are based on the .NET framework using C# and Active Server Pages (ASP.NET). In particular, two vendor sites of ringtones and wallpaper are implemented.
- The three types of evaluations have been performed on the M&E-NetPay prototype. We compare micro-payment with non-micro-payment in terms of usability, performance and heuristic assessment.

The rest of this paper is organized as follows. Section 2 describes the architecture of M&E-NetPay. The protocol and interactions of M&E-NetPay are given in Section 3. Section 4 presents the implementation of the M&E-NetPay prototype. Section 5 reports the evaluations on the system, followed by related work and comparisons in Section 6. We conclude this paper in Section 7.

2. M&E-NetPay Architecture

In this section, we outline the architecture of M&E-NetPay, including the hardware and software architectures.

2.1. M&E-NetPay Software Deployment Architecture

Taking into account the general requirements on performance, security, availability and serviceability, we designed the deployment architecture of M&E-NetPay as shown in Figure 1.

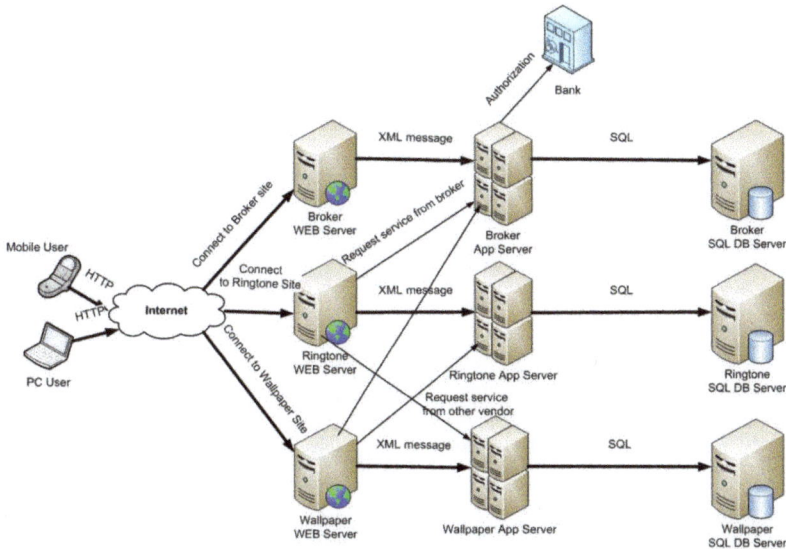

Figure 1. The basic deployment architecture of M&E-NetPay.

As a thin client n-tier application, M&E-NetPay is deployed over three servers: web servers, application servers and database servers. Web servers deploy broker and vendor web/mobile applications. Application servers publish Web services of the broker and vendor. Database servers store required information.

M&E-NetPay is maintainable and serviceable in that any changes result in re-configuring of only part of the application. If the ringtone vendor wants to update its site, for example, then only the web/mobile application on its Web server is re-configured.

2.2. M&E-NetPay Software Architecture

The software architecture of the M&E NetPay micro-payment system is shown in Figure 2. The architecture is designed for Microsoft.NET applications. It consists of the following components.

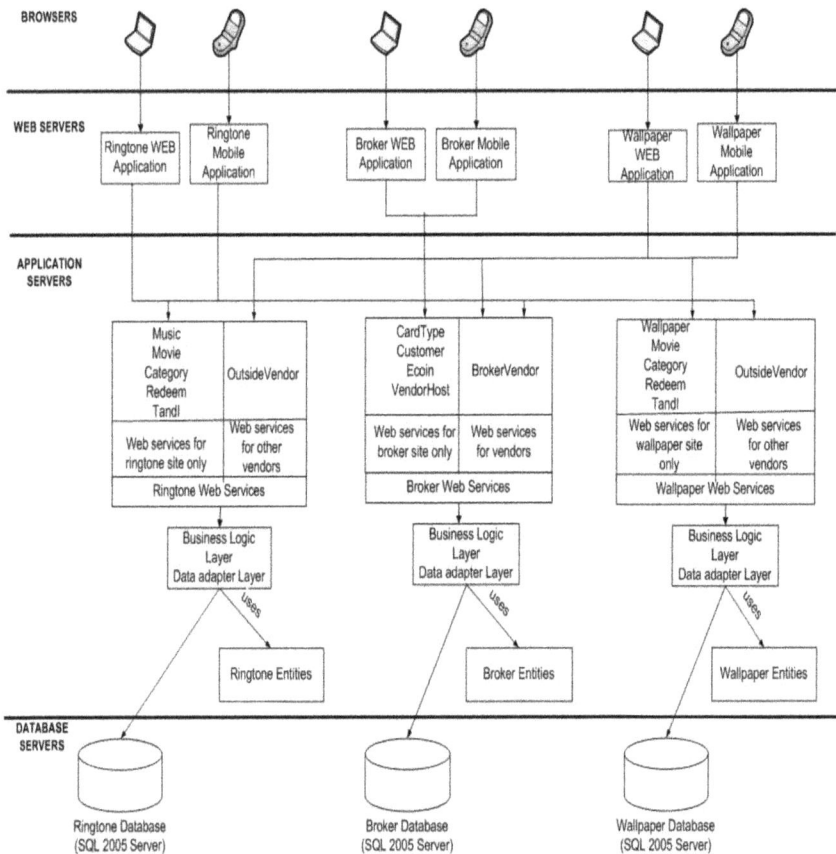

Figure 2. M&E-NetPay basic software application architecture.

Browser: two types of users can access a broker site using their mobile phones or PCs with Internet access. By using a Wireless Markup Language (WML)-based Web browser in their mobile phones, mobile users run the Broker Mobile application with its interface for the small screen of a mobile phone. Internet users can access the Broker Web application through a popular web browser.

Web services: These host the presentation layer. It is much easier to connect remote sites by using web services.

The web service is only available to vendors for accessing certain information from the broker's database. User queries are issued to broker data entities from the client end, and the results are retrieved by data layers. Mobile and Web applications invoke the same Web services hosted on the broker's application server. The broker Web services pass information in an XML-based message to the business logic layer. In our applications, this means that data are retrieved from a database into an entity or entity collection and then updated data are written from an entity back to the database.

Application servers: These mainly accommodate Web services, the business logic layer and the data adapter layer. The business logic layer implements all business rules for the application. The business logic layer passes information to the data adapter layer, the broker database, and executes necessary queries. The data adapters exchange data between a data source and a dataset.

Database servers: These host relational databases, including the ringtone database, the broker database and the wallpaper database. The database in the broker server records account information and transaction histories of all registered users.

The e-wallet of a user resides on the broker's database until she or he logs on to a vendor site using a given e-coin id [18]. Upon login, her or his e-wallet is transferred to the visiting vendor. The broker helps the vendor to verify e-coins, when she or he purchases items from its site. The broker also allows the vendor to redeem e-coins spent on its site and to request touchstones. These functionalities are provided by the "BrokerVendor" Web service of the broker, as shown in Figure 2.

Similarly, vendor sites also provide their interfaces to both mobile and Internet users. The vendor sites allow users to browse their websites and purchase items. When a user logs in to the ringtone site in our system, the ringtone vendor requests her or his e-wallet from the broker. This function is provided by the Web service of "BrokerVendor" of the broker. If the ringtone vendor finds that the e-wallet of this particular user resides on another vendor site, it then requests her or his e-wallet from the vendor that contains e-coin indexes and touchstones. Each vendor has a Web service called "OutsideVendor", which allows other vendors to retrieve e-wallets of their own users. The e-wallet is then stored on the current vendor's site. Once the user purchases an item, her or his e-wallet is debited.

3. Protocol and Interactions of M&E-NetPay

In this section, we describe the protocol and interactions of M&E-NetPay.

3.1. NetPay and M&E-NetPay Micropayment Protocol

M&E-NetPay is evolved from NetPay. Therefore, we start with describing NetPay. It is an off-line micro-payment system by using a secure, inexpensive and debit-based protocol [13]. The NetPay micropayment system has three models of "e-wallets" that manage e-coins. Like other models, e-wallets in the first model are hosted by vendor servers. An e-wallet is passed from one vendor to another, as a customer visits different sites for e-commerce transactions. The second model is a stand-alone client-side application on a client's PC. The third model is a hybrid one that caches E-coins in a Web browser cookie for debiting, if a customer makes a purchase. The NetPay-based system is developed for the CORBA-based broker, vendor and customer networks. By using a set of CORBA interfaces, the broker application server communicates with the vendor application servers for requesting touchstones and redeeming e-coins [18]. CORBA enables clients to invoke methods on remote objects at a server, regardless of by which language objects are programed and where they are located.

M&E-NetPay replaces the CORBA middleware with Web Services, which provide the interoperability (i.e., platform-independent and language-independent). Using a simple XML-based protocol and SOAP, a Web service is an emerging distributed middleware technique that allows applications to exchange data over the Web. It is a new programming model for building distributed applications by open Internet standards. This new technique manoeuvres the openness of specific Internet technologies to address many interoperability issues of CORBA. Web services use Hyper Text Transfer Protocol (HTTP) to transmit messages. This is a major advantage for building an Internet-scale application like the M&E-NetPay system, since most of the Internet proxies and firewalls do not have trouble with HTTP traffic. In contrast, CORBA usually has difficulties with firewalls. Moreover, Web Services are platform-independent and language-independent (i.e., a client program can be programmed in C# and running under Windows, while the Web Service is programmed in Java and running under Linux.). Web Services support different interfaces of client-side application programs. Client code may work by constructing "call" objects that are dispatched to a server or may use a higher level interface that hides the communication level entirely through the use of client-side stub objects with an operational interface that imitates the server [27]. The mechanisms for generating client and server components for Web Services and CORBA are illustrated in Figure 3.

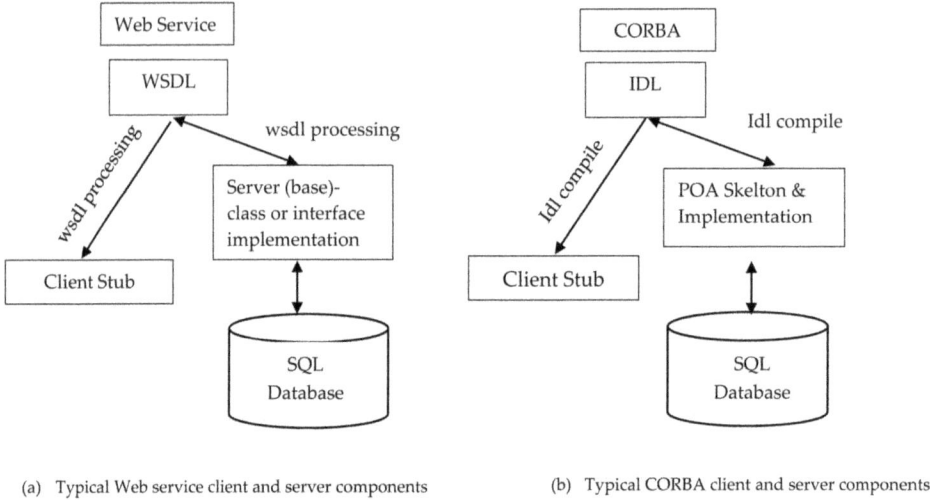

(a) Typical Web service client and server components   (b) Typical CORBA client and server components

Figure 3. Basic client and server components from the interface for Web Services and CORBA.

M&E-NetPay uses a secure, inexpensive widely-available and debit-based protocol. The M&E-NetPay protocol differs from the previous protocols in that running on the .NET platform, it uses Web service interfaces as its middleware.

3.2. M&E-NetPay Micropayment Interaction

Based on the NetPay protocol of the server-side e-wallet [13,28,29], we extend it into the M&E-NetPay protocol in a way that is suitable for mobile and Internet environments. The M&E-NetPay protocol uses touchstones signed by a broker, as well as e-coin indexes signed by requesting peers. The signed touchstone is used by a vendor to verify the electronic currency: paywords. A signed index prevents customers from double spending and resolves disputes between customers and vendors. We assume that an honest broker is trusted by both customers and vendors. The broker manages the bank accounts of all mobile and Internet users. A bank transfers money to a broker on an online request. The mobile/Internet users access the mobile/web application through Web browsers on mobile phones or PCs. In order to purchase items from vendor sites, a mobile or Internet user needs to register with a broker. Upon successful registration, the user's account is created. She or he then needs to buy e-coins from the broker by using her or his credit card. She or he is issued a unique e-coin id each time once having bought e-coins from the broker. She or he can log onto a vendor site using the e-coin id and password. In our system, two vendor sites of ringtones and wallpapers are implemented. A user browses the site and selects the ringtone or wallpaper. A small cost is assigned to each ringtone and wallpaper, depending on their demand and ratings. After the user clicks on the download button in the ringtone site, the broker debits the user account for the cost of the ringtone, e.g., if the user is downloading a ring tone costing "10 c", then the user's account is debited by "10 c".

A user e-wallet is saved on the vendor site last visited. Once the user logs on to the other vendor site for browsing other ringtones, her or his e-wallet is transferred from her or his last visited vendor to the current one. If her or his e-coins are run out, she or he is directed to the broker site to buy more e-coins. At the end of each day, all of the vendors collect the money from the broker in return for e-coins.

As an example, we describe the procedure of macro-payment in M&E-NetPay in the following. Figure 4 also illustrates some key interactions.

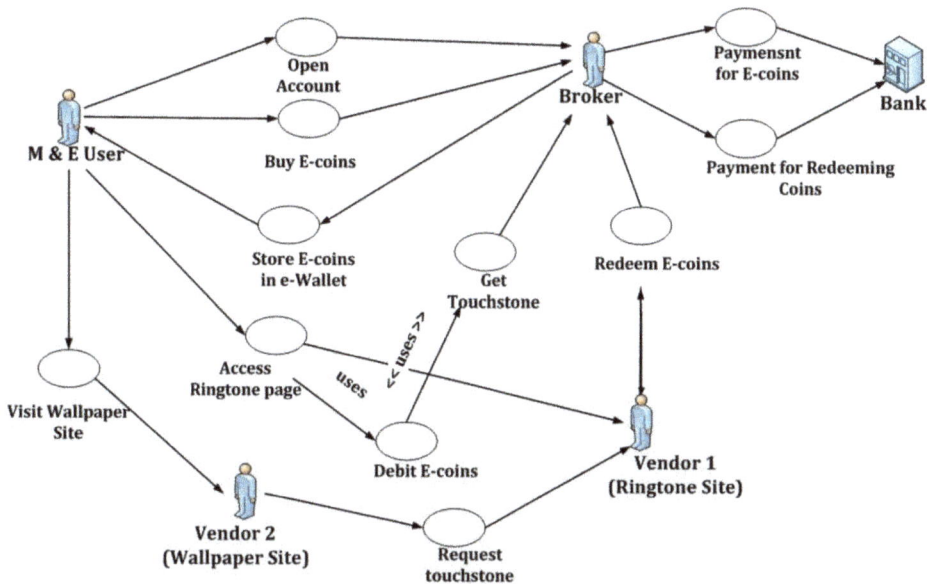

Figure 4. M&E-NetPay component interactions.

Initially, a mobile user registers on a broker's Web site and buys a number of e-coins.

1. The broker may provide credits as "virtual money", which is specific to the network only. The P2P network may require peers to use real money to subscribe and/or to use services. In this case, the broker uses a macro-payment, e.g., credit card transactions with a conventional payment party to buy credits.
2. The broker generates an e-coin chain and stores it in an "e-wallet".
3. When the mobile user selects ringtones to be downloaded from Vendor 1 site, Vendor 1 obtains e-coins from the e-wallet and verifies the e-coins. The mobile user then can download the ringtones.
4. The mobile user may download other ringtones, and her or his coins are debited. If her or his coins run out, she or he is directed to the broker site for buying more. When the mobile user browses Vendor 2, Vendor 2 contacts Vendor 1 in order to obtain the touchstone and index (T & I) and then debits e-coins for this user to download more wallpapers.
5. At the end of each day, the vendors send all of their received e-coins to the broker for redeeming them.
6. For their own credits, vendors may be able to cash them in for real money, again via a conventional macro-payment approach.

As a summary, a mobile user downloads ringtones from Vendor 1. Vendor 1 requests the touchstone, index and e-coins from the broker. After verification, the mobile user is allowed to download ringtones. Vendor 1 sends T & I to the broker. After browsing other vendors, the mobile user wants to download wallpapers from Vendor 2, which contacts Vendor 1 for T & I. If Vendor 1 is off-line, then Vendor 2 requests T & I from the broker.

4. Implementation of M&E-NetPay Prototype

In this section, we present the implementation of the M&E-NetPay prototype.

Our system has implemented one broker and two vendor sites. All applications are developed on the Microsoft.NET platform framework 4.0 [26]. We choose Microsoft Visual Studio 2010 ASP.NET and the C# programming language for frontend implementations and Microsoft SQL Server 2005 for database storage. We use HTML with ASP controls for Web pages and the C# programming language for the back end of the application. The broker and vendors provide access to both mobile and Internet users published on the web servers' IIS. Web service interfaces are implemented on the application servers' IIS, which provides access to the Internet, as well as to remote sites. Vendors and the broker can choose programming languages and operating systems for implementing their systems. A vendor application implemented by the C# programming language on the Windows operating system, for example, can easily communicate with another implemented by the C++ programming language on the UNIX operating system.

To make it more effective and efficient, M&E-NetPay consists of three components: the presentation logic, which presents information to the M&E users; business components, which controls the relationship between inputs and determines business rules; and the data adapter layer, which connects to the database, executes relevant queries and returns the results back to the upper layers. The presentation and business components are communicated only via Web Services, no matter whether they are within a system or between the systems.

Web Services are used as the middleware for M&E-NetPay. Figure 5 shows Web service references on the broker site referenced from the broker Web Service.

```
<applicationSettings>
<BrokerWebApp.Properties.Settings>
<setting name="BrokerWebApp_Customer_CustomerChannel" serializeAs="String">
<value>http://localhost/BrokerWebService/CustomerChannel.asmx</value></setting>
<setting name="BrokerWebApp_CardType_CardTypeChannel" serializeAs="String">
<value>http://localhost/BrokerWebService/CardTypeChannel.asmx</value></setting>
<setting name="BrokerWebApp_Ecoin_EcoinChannel" serializeAs="String">
<value>http://localhost/BrokerWebService/EcoinChannel.asmx</value></setting>
<setting name="BrokerWebApp_VendorHost_VendorHostChannel" serializeAs="String">
<value>http://localhost/BrokerWebService/VendorHostChannel.asmx</value></setting>
</BrokerWebApp.Properties.Settings>
</applicationSettings>
```

Figure 5. Code for Web Service references on the broker Web application.

4.1. Broker

A broker manages customer and vendor accounts, e-coin creation, e-coin redemption, touchstone supply for e-coin verification and macro-payment handling for e-coin purchase and payment to vendors for spent e-coins [13]. The broker database holds user and vendor information. The application server provides business functions. Web service interfaces are for application servers of the broker and vendor. WML interfaces, implemented by using Active Server Pages (ASP.NET) with the ASPX extension, are for mobile users, while HTML interfaces are for Internet users. The Web service interface allows vendors to request e-coin touchstone information, verify e-coins and redeem spent e-coins by other vendors.

Figure 6 shows the screenshots of a customer purchasing e-coins from a broker: (1) registering with the broker to create her or his account; (2) logging in by using the provided customer id and password; (3) authorizing macro-payment by the broker in order to buy e-coins; and (4) debiting the M&E user account for paying e-coins by the bank.

(a)

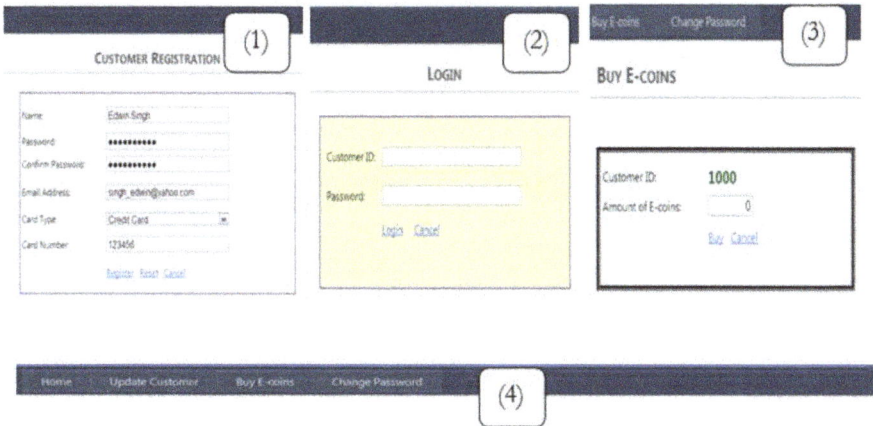

(b)

Figure 6. M&E users purchasing e-coins from a broker. (**a**) Wireless Markup Language (WML) interfaces for mobile phone users; (**b**) HTML interfaces for Internet users.

4.2. Customer

The WML/HTML interfaces of our system are provided for both mobile and Internet users, so that a wide range of customers is allowed to access broker and vendor sites by using a standard Web browser. The use of the thin client technology omits the need to install separate browser software on the client site. The customers use WML/HTML-based ASP.NET pages to browse broker and vendor sites. Being hosted on the server side, the e-wallet of a customer can be transferred from one vendor to another, as the customer makes purchases from those vendors. The e-wallet is held on the vendor server from which the customer is currently buying items.

4.3. Vendor

The site of a vendor displays ASP.NET pages for M&E users to browse. Search functions in sites of the ringtone and wallpaper are provided for users to search for ringtones or wallpapers. The search results are listed as a brief summary of the ringtone or wallpaper with its download cost, as shown in Figure 7a. After downloading an item, the refreshed ASP.NET pages indicate that the amount of e-coins is left with the current vendor in the e-wallet of the user, as shown in Figure 7b.

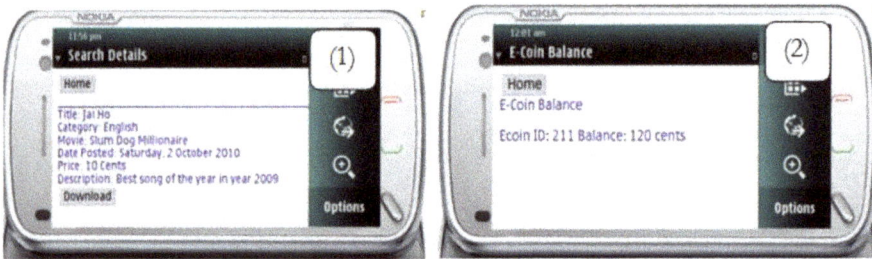

(a)

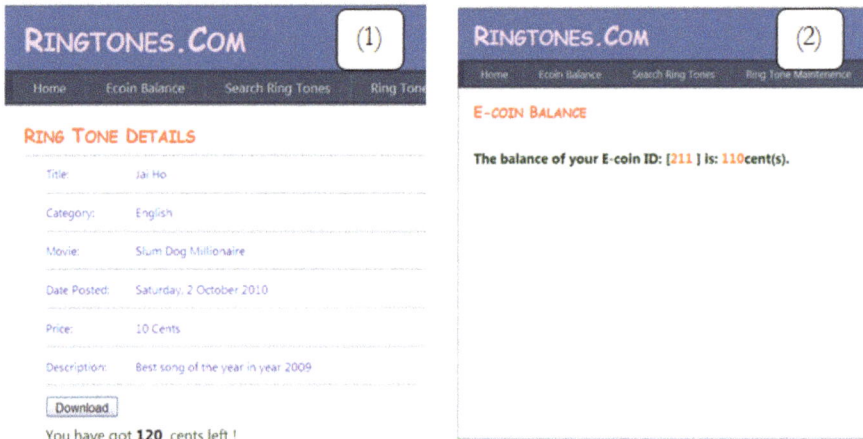

(b)

Figure 7. M&E user spending e-coins at the ringtone site. (**a**) WML interfaces for mobile phone users; (**b**) HTML interfaces for Internet users.

5. Evaluation

In this section, we compare M&E-NetPay to non-micro-payment systems. From different perspectives of end users, we evaluate the M&E-NetPay based micro-payment system by collecting and analysing customer views.

5.1. Experimental Design

Three types of evaluations on the M&E-NetPay micro-payment are carried out:

- Performance evaluation [30], which compares the performance of the M&E-NetPay prototype with that of the CORBA-based NetPay system in terms of response time. This evaluation aims to assess the potential scalability of the system under heavy loading conditions.
- Usability evaluation, which assesses whether M&E-NetPay is useful as far as end users are concerned. Their opinions about our prototype are surveyed, after potential end users purchase items by using the micro-payment, M&E-NetPay and the alternative CORBA-based NetPay system, respectively.
- Heuristic evaluation, which assesses the overall quality of the user interface. Potential design problems of the user interface of the M&E-NetPay prototype are identified by using a range of common HCI design heuristics.

Experiment prototypes and materials: The evaluations are conducted on two prototypes of M&E-NetPay and CORBA-based NetPay. M&E-NetPay is deployed over three servers:

- Web server, which hosts the presentation layer
- Application server, which hosts Web Services, business logic components and data adapter layer
- Database server, which hosts the relational database

The CORBA-based NetPay system is deployed over three servers:

- Web server, which hosts JSP pages as the presentation layer
- CORBA application server, which hosts business logic components
- Database server, which hosts the relational database

A number of PCs connected to the network is used by the participants. Both prototypes are deployed over multiple machines connected via a high speed LAN.

5.2. Performance Evaluation

We carry out experiments on measuring client response time with ten tests. This evaluation aims to compare how long it takes to download wallpapers in the two different payment systems.

Subject: Ten users are a mixture of non-IT specialists, graduate students and college students who volunteer to conduct the evaluation.

Experimental tasks: The users are required to download the same file from both M&E-NetPay and CORBA-based NetPay systems.

The response times of searching for wallpapers, buying e-coins and redeeming e-coins are recorded. They give an indication of the likely scalability of the prototype systems under heavy loading conditions.

Results: As reported in Table 1 and illustrated in Figure 8, we compare M&E-NetPay to CORBA-based NetPay against the response time of downloading wallpapers. The response time delay is the time for downloading a wallpaper. All ten users download the same wallpaper with the size of 38.4 KB.

Table 1. Times for downloading wallpapers.

Test	Response Delay Time with M&E-NetPay (ms)	Response Delay Time with CORBA-Based NetPay (ms)
1	2149	2410
2	2390	2509
3	1734	2294
4	3065	2354
5	2012	2432
6	1976	2091
7	2190	2256
8	1734	2168
9	1637	2005
10	1815	2344
Average	**1976**	**2286**

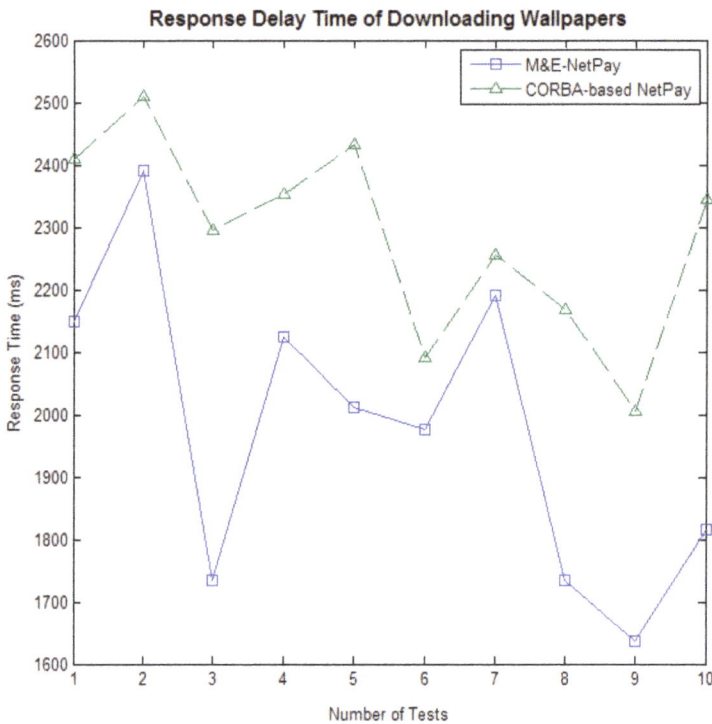

Figure 8. Response delay time of downloading wallpaper.

The result of the *t*-test on the data in the two columns of Table 1 rejects the null hypothesis at the default 5% significance level. That is, the two response delay times of downloading the wallpaper from the two systems have a statistically-significant difference. The test parameters are given below: the *p*-value: 0.0033; confidence interval for the difference in population means of the response time in M&E-NetPay and CORBA-based NetPay: -502.5709 and -117.8291; the test statistic: -3.3878; degrees of freedom (df): 18; and the estimate of the population standard deviation: 204.7455.

It is obvious that the above statistical test result is limited by the size of sample tests. Despite this, the average response delay time for downloading a wallpaper from CORBA-based NetPay is slightly higher than that from M&E-NetPay. On average, the clients take 1976 milliseconds from M&E-NetPay and 2286 milliseconds from CORBA-based NetPay to download the same wallpaper. The time difference is 310 milliseconds. Except for downloading time, we also compare the two systems against the response times of the respective operations of searching for wallpapers, buying and redeeming e-coins in Table 2.

Table 2. Results of searching wallpapers, buying and redeeming e-coins.

Tasks	Average Response Delay Time by M&E-NetPay (ms)	Average Response Delay Time by CORBA-Based NetPay (ms)
Search wallpapers	1501	1703
Buy e-coins	895	920
Redeem e-coins	1990	2110

As listed in Table 2, the searching for wallpapers in M&E-NetPay is 202 milliseconds faster than that in CORBA-based NetPay. Buying and redeeming e-coins also take less time in M&E-NetPay.

There may be other factors that affect the response time of the systems. However, the experiment results still indicate that M&E-NetPay may respond to user interactions faster than NetPay. This observation results from CORBA's limitation in client-to-server communications. In contrast, .NET framework architecture 4.0 with Web Services in M&E-NetPay improves client-to-server communications. It provides relatively fast communications amongst the vendor and broker. In addition, M&E-NetPay, built on a stable, secure and simple architecture, is deployed over multiple servers to share the workload among them.

5.3. Usability Evaluation

We survey the satisfaction levels of the participant users, after they download and purchase items. Furthermore, we ask their preferences for the two systems in general: a CORBA-based NetPay system or M&E-NetPay. As we know, usability evaluation involves testing of the usability of an interface by having a group of individuals performing tasks specific to a system, under the general guidance from a facilitator. It is important to realize that it has multiple components with five attributes associated with an interface [29,31,32]. Specifically, efficiency in our evaluation is measured in terms of how easily one can buy items and the speed of downloading them. Errors are regarded as any actions that prevent the successful occurrence of the expected results. Since some errors escalate the users' transaction time, their effect is measured by the efficiency of use. Learnability and satisfaction are a subjective measure provided by each participant in the experiment. Interface memorability is rarely tested as thoroughly as other attributes. However, it is feasible, to some extent, to conduct comparisons and post-test questionnaires of both systems.

The experiments use pre-test and post-test questionaries. The questions of the pre-test questionnaire are about participants' experience in using mobiles or PCs to download files from the Internet. The post-test questionnaire has the number of questions with scale ratings ranging from one to five, where one is "least favourable" and five "most favourable". The post-test questionnaire also contains open questions for collecting user comments.

Subjects: Fourteen participants are randomly selected with a mixture of non-IT specialists, graduate students and college students. The participants are four non-IT adults, five non-IT graduate students, and the rest are college students. It should be noted that although it is tempting to recruit more participants, it is the general practice to have around 15 participants for usability testing [32].

Experimental tasks: Participants are required to complete the following tasks on M&E-NetPay and CORBA-based NetPay systems, respectively:

- Create an account with the broker;
- Search for a wallpaper on the wallpaper vendor site;
- Download the wallpaper from the wallpaper vendor site;
- Download a ringtone from the ringtone vendor;
- Buy e-coins from the broker; and
- Redeem e-coins with the broker.

Procedure: Before starting the test, participants need to fill out a pre-test questionnaire. Participants are required to carry out all of the tasks listed in a given sheet for the two systems. After finishing the tasks, they then fill out the post-test questionnaire to answer the questions by ticking one level of the rating. One of the questions asks the participants to rank the overall performance of the systems in order of their preference.

Results: From the answers to pre-test questions, we know that all of that participants have used mobiles or PCs to download files from the web weekly or monthly for free. Only two of them use the online credit-card payment systems to purchase goods online. Fortunately, all participants have had such experiences before. This implies that participants' prior knowledge has the least effect on the experiment results.

We survey the participants' satisfaction with buying e-coins, downloading wallpapers and their preference for the two systems. We analyse the post-test questionnaire outcomes and plot the results in Figure 9.

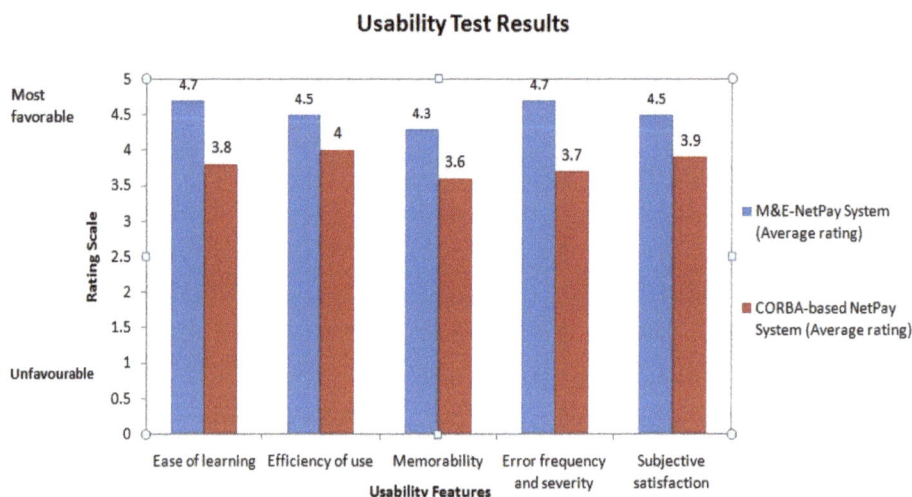

Figure 9. Usability test results with respect to usability features.

Figure 9 shows that the participants significantly favour all of the usability features of M&E-NetPay. With the user friendly interface, M&E-Pay is easy to learn, providing clear instructions on how to accomplish tasks. M&E-NetPay also receives high ratings on its efficiency. The participants comment that the speeds of downloading files (i.e., wallpaper and ringtone) are quite fast with M&E-NetPay in that with a few clicks, they are able to download the file. They also comment that appropriate pop-up error messages prevent them from going off of the right track. The overall average ratings of M&E-NetPay and CORBA-based NetPay are 4.5 and 3.8, respectively. They indicate that the participants prefer M&E-NetPay to CORBA-based NetPay. This fact results from employing new and emerging distributed middleware technique (i.e., Web Services) in M&E-NetPay.

For the open question, some participants write that since M&E-NetPay is available via both mobiles and the web, they will be able to access the system from anywhere at any time with barely any downtime. Twelve participants favour M&E-NetPay.

5.4. Heuristic Evaluation

As the most widely-used inspection method, the heuristic evaluation technique is about identifying usability issues in a user interface by a small number of evaluators (usually one to five) who examine the interface and judge its compliance with usability principles (heuristics) [33–35]. While heuristic reviews are inexpensive and less time consuming, good ideas for improving a user interface may be produced.

Subjects: The evaluators include two IT specialists, one accountant and two graduates. They are experts in either software engineering or applied software fields.

Experimental tasks and procedure: the evaluators are requested to judge the compliance of the M&E-NetPay interface with usability principles ("the heuristic"). Each individual evaluator examines the interface independently. To aid the evaluators in discovering usability problems, a list of heuristics, as shown in Table 3 [35], is provided, which could facilitate the generation of ideas on how to improve the system.

Table 3. Details of the heuristics employed.

Number	Heuristic
1	Visibility
2	Functionality
3	User control and freedom
4	Consistency
5	Help recover from errors
6	Error prevention
7	Memorability
8	Flexibility
9	Aesthetic
10	Help and documentation

With a system checklist provided as a guide, the evaluators are required to first identify the heuristic problems of the interface and then to determine the levels of their seriousness by using the severity ratings as defined in Table 4 [35,36]. The evaluators are also requested to provide recommendations based on their assigned severity ratings.

Table 4. Severity of the heuristic evaluation.

Rank	Interpretation
1	Cosmetic problem only: need not be fixed unless extra time is available on a project
2	Minor usability problem: fixing this should be given a low priority
3	Major usability problem: important to fix, so should be given a high priority
4	Usability catastrophe: imperative to fix this before a product can be released

Results: The five evaluators evaluate M&E-NetPay by relying on the ten heuristics. The results of the heuristic evaluation are given in Table 5.

Table 5. Summary of the findings.

Number	Problem	Heuristic Number	No. of Evaluators	Severity Ranking
1	No sharp colour contrast between product information and its background.	1	2	2
2	No error message is displayed for invalid entries.	5	3	3
3	Multiple options cannot be selected in a menu or dialog box.	2, 8	1	2
4	Insufficient keyboard shortcuts for navigating the activity, function or action.	2	2	2
5	Exit button not provided to exit application from any screen.	2, 3	3	3
6	Not all integers and decimals right-justified.	4	2	1
7	The price associated with the product does not show the currency.	4	4	2
8	No sound used to signal an error.	5	2	1
9	No help topics provided.	10	2	3
10	Borders not used to identify meaningful groups.	1, 7	2	2
11	Titles are not provided on every page.	1, 4	3	2
12	On the login screen, the cursor is not active in the customer id field.	4	2	2

A rating has four levels of severity. The levels of one and two are regarded as minor, which is easily fixed. The levels of three and four should be given high priorities, which have to be fixed. After the evaluation, three major problems have been identified, with each having a severity rating of three. The identified problems, together with their fixing recommendations, are listed in Table 6. We have implemented all recommendations listed in the table.

Table 6. Summary of the findings.

Number	Problem	Recommendation	Severity Ranking	Heuristic Number
1	No error message is displayed for invalid entries	Appropriate error messages should be displayed for invalid entries	3	5
2	An exit button not provided to exit the application from any screen.	An exit button should be implemented on every screen	3	2, 3
3	No help topics provided	Implement help topics, as users may not be aware of the function of the menu or command button	3	10

We have described three kinds of evaluations on the M&E-NetPay prototype to assess performance impact, usability and heuristic evaluations. Usability and performance evaluation have been done on two prototypes of CORBA-based NetPay and M&E-NetPay. Even though heuristic evaluation identifies a few errors, M&E-NetPay is successfully implemented in general. The overall result of the evaluations demonstrates that most participants prefer M&E-NetPay. Participants and evaluators are satisfied very much with M&E-NetPay, recommending the system for wide use.

6. Related Work and Comparisons

In this section, we review related work and compare relevant systems to our system.

As a micro-payment for an ad hoc network, MPS [11] enables a node to join an existing ad hoc network and allows it to pay each node that relays packets on its behalf in real time. Being a lightweight payment scheme based on hash chains, MPS is flexible in route changes without involving a third party (a bank or a broker), in order to pay the nodes in a new path. MPS supports multiple brokers. Off-line verification makes the protocol more efficient and scalable.

Using a micro-payment protocol, CMP [12] is built on symmetry encryption techniques and chaotic double hash chains. The protocol constructs two PayWord chains: one for the merchant and another for users by using the iteration process of the Henon-like chaotic system. The chaotic hash function generates a payment chain. The use of the symmetric algorithm that encrypts transaction information improves the security and efficiency of CMP. CMP is an off-line system with three stakeholders, users, vendors and a broker.

As an off-line micro-payment system, NetPay [13] is a new micro-payment model in e-commerce. It uses CORBA interfaces to support communication between broker and vendor applications. NetPay improves its performance and security by using fast hashing functions. This prototype is quite suitable for e-commerce applications. In a mobile environment where a client (and possibly the server) keeps moving, which results in changing network addresses and unreliable connections, CORBA, however, is not well suited for this scenario. This is because of CORBA's tight coupling between clients and servers.

We compare our M&E-NetPay protocol to other micro-payment protocols. We have analysed the results from the three types of evaluations of M&E-NetPay prototypes to demonstrate their usability, performance and overall satisfaction of the requirements.

It is generally agreed that the key requirements for a mobile micro-payment system are as follows [3,8,11–13,29,30,37–39]:

- Security: The e-coins must be well encrypted to prevent peers from double spending and fraud.
- Anonymity: Peer users and peer vendors should not reveal their identities to each other or to any other third party.
- Ease of use: This is the ability of M&E users who are able to use the system easily without familiarizing themselves with the M&E user interfaces or being involved in any type of authentication at all times.
- Scalability: The load of communication and transaction of any entity must not grow to an unmanageable size. The load should be distributed among the vendors rather than the broker. Payment systems should be able to cater to the rapidly growing number of users without showing a negative impact on the performance.
- Transferability: The e-coins used for payments should be transferable between multiple vendors. This allows the users to use the same e-coins to make payments across multiple vendors.
- Interoperability: This is the ability of a system that operates in conjunction with other supporting protocols, hardware, software, applications and data layers. Interoperability minimizes the complexity of software development by reusing components and performing inter-component communication. Interoperable systems are language and platform independent.

In the following, we compare M&E-NetPay to several well-known micro-payment systems and also to some more recent micro-payment systems in M&E networks. The comparison criteria are the set of the key requirements: the need for an easy-to-use micro-payment system; the need for secure electronic coins and no double spending; ensuring anonymity for customers; supporting transferable e-coins between vendors; a robust, low performance impact, off-line micro-payment-supported, scalable architecture for a very large number of end users; and the ability of the system to be language and platform independent. Table 7 summarises the comparisons of the M&E-NetPay protocol with other systems.

Table 7. Comparison of M&E-NetPay with other micro-payment models. SOAP, Simple Object Access Protocol.

System/Property	MPS	CMP	NetPay	M&E NetPay
Security	Medium (Smart card devices cannot grantee multiple payment protection)	Very High (Uses a chaotic hash function to encrypt messages and services provided)	Very high (prevents users from over spending, prevents vendors from over charging and prevents a third party from forging e-coins)	Very High (Prevents M&E users from over spending, prevents vendors from over charging and prevents a third party from forging e-coins)
Privacy/anonymity	High (The nodes have no information about the user identity)	Very High (The vendor has no information about the user identity; all information services between vendors and users are encrypted)	High (The vendor has no information about the user identity)	High (The vendor has no information about the M&E user identity)
Ease of use	Medium (A lightweight payment scheme; however, too many private, public and sharing secret keys can slow the transaction)	Medium (The use of double hash chains and private and secret keys can slow the transaction)	High (Users need to spend little time in order to buy items from the vendor sites)	High (Provides simple interfaces with easy use; the Web Service that uses the SOAP and simple XML-based protocol makes the payment process faster)
Transferability	Low (Payment chains can only be transferred between the nodes in an ad hoc network; on a new network, a user needs to buy another payment chain from the new broker)	High (The same hash chain can be used on multiple vendor sites)	High (The e-coins can freely be transferred across multiple vendors for users to make multiple purchase)	Very High (The e-coins can freely be transferred across multiple vendors for M&E users to make multiple purchases; Web Service interfaces provide extra simplicity on transferring e-coins between vendors)
Scalability	Low (A change in route does not require contacting the broker; however, the model can only support the nodes in the ad hoc network)	Medium (CMP also requires the vendors to register with the broker if they wish to participate in the M&E commerce, and the use of too many keys can be costly in the future)	Medium (No or less communication with the broker and low volume transactions; however, the CORBA tight coupling between clients and servers in mobile environments makes it less scalable)	High (No or less communication with the broker and low volume information transferred; .NET applications such as a web service are fully accessible at any time on any platform and can support any number of M&E users)
Interoperability	Low (This model is only designed for a fixed network)	Low (Does not support multiple platforms and languages, as no appropriate middleware is specified)	Medium (Supports only a few platforms and languages)	Very High (Supports all platforms and languages)

The above comparisons show that the M&E-NetPay system has advantages over other micro-payment systems.

The security of M&E-NetPay is achieved by using existing security technologies. First, it uses a thin client n-tier architecture. With this deployment architecture, users logging on broker or vendor systems can access only Web servers. From there, transaction information is transferred through a secure channel in an XML message, which cannot be intercepted by a third party. Moreover, Web services on application servers are only available upon the request of mobile/web applications on Web servers. The vendors and broker in M&E-NetPay rely on Web service interfaces of the other party to exchange M&E user information. It is impossible for third parties to log directly or indirectly on to application servers. In addition, application servers are inaccessible from outside the network.

Second, M&E-NetPay relies on the security of Web services. As we know, Web services' security includes three aspects: authentication, which verifies that M&E users and vendors are who they claim to be; confidentiality and privacy, which keep information secret by encrypting the content of a message and obfuscating the sending and receiving parties' identities; and integrity and non-repudiation, which make sure that a message remains unaltered during transit by having the sender digitally sign the message. A digital signature is used to validate the signature and provides non-repudiation.

Finally, M&E-NetPay uses one-way hash functions to generate paywords and prevents M&E users and vendors from over spending and forging paywords from a payword chain. It employs 128-bit encryption of the messages. Since only the broker knows the mapping between the pseudonyms (IDc) and the true identity of an M&E user, M&E user privacy is protected.

In a word, M&E-NetPay has high security features. As an off-line fully-distributed system, the M&E-NetPay is mostly suitable for micro-payments over the WWW. In terms of transferability, e-coins are able to be transferred freely between vendors for multiple purchases. CMP is primarily designed for low value mobile commerce items. The protocol has greater security and faster operation efficiency, but CMP does not support multiple platforms and languages. MPS's design supports multiple brokers. Off-line verification has made the protocol more efficient and scalable. The system, however, still cannot avoid a limited amount of fraud. There may be a wastage of the broker endorsement, which is distributed to the previous path, if the topology of the ad hoc network changes.

M&E-NetPay has greater scalability and performance features, as it supports the rapidly growing number of M&E users. NetPay uses CORBA middleware interfaces to support several programming languages (e.g., Java® and C++®) and platforms (e.g., Windows®, Linux®). Vendor systems have to be "hard-coded" with CORBA by communicating with the NetPay broker to exchange messages. In comparison with M&E-NetPay, NetPay has a lesser ability due to the tight coupling between clients and servers as a result of its use of CORBA. M&E-NetPay is the solution to the above problem, as it can support any languages and platforms. Hence, it has a very high rating of interoperability.

7. Conclusions

In this paper, we have presented the design and implementation of the M&E-NetPay micro-payment system on the .NET platform and its comprehensive evaluations. Interconnecting broker and vendor sites through XML-based interactions, Web services are used to provide greater interoperability than the CORBA middleware. Apart from generating, redeeming and verifying e-coins, the broker in M&E-NetPay provides e-wallets to customers. Through their interfaces, vendor applications allow users to browse their sites, download items and obtain the valid touchstone and index from a broker or the previous vendor. M&E-NetPay uses a secure, inexpensive and debit-based off-line protocol that allows vendors to interact only with customers after an initial validation of coins. M&E-NetPay achieves a secure and high transaction volume per item by using fast hashing functions that validate e-coin unspent indexes. The results of two types of comparison evaluations on the usability and performance of two systems have demonstrated that the users, as their preference, would adopt M&E-NetPay for widespread use. In the future, we will take further advantage of Web services [40] to generalize the proposed architecture as components of a wider range of M&E-commerce applications.

Acknowledgments: This work was partially supported by the National Natural Science Foundation of China under Grant No. 61370229 and the National Key Technology R&D Program of China under Grant No. 2014BAH28F02, the S &T Projects of Guangdong Province under Grant No. 2014B010103004, 2014B010117007, 2015A030401087, and 2016B010109008, and GDUPS (2015).

Author Contributions: X.H., X.D. and J.B. proposed the idea of this work and designed the system. E.S. implemented it. All the authors contributed to evaluating this system and writing this paper.

Conflicts of Interest: The authors declare no conflict of interest.

References

1. Adi, W.; Al-Qayedi, A.; Zarooni, A.A.; Mabrouk, A. Secured multi-identity mobile infrastructure and offline mobile-assisted micro-payment application. In Proceedings of the IEEE International Conference of Wireless Communications and Networking, 21–25 March 2004; pp. 879–882.
2. Bokai, M.; Mohammadi, S. Exploring adoption of NetPay micro-payment: A simulation approach. In Proceedings of the 2010 IEEE International Conference on Educational and Information Technology (ICEIT), Chongqing, China, 17–19 September 2010; pp. 245–250.
3. Furche, A.; Wrightson, G. SubScrip—An efficient protocol for pay-per-view payments on the Internet. In Proceedings of the 5th Annual International Conference on Computer Communications and Networks, Rockville, MD, USA, 16 October 1996.
4. Hwang, M.; Lin, I.; Li, L. A simple micro-payment scheme. *J. Syst. Softw.* **2001**, *55*, 221–229.
5. Isern-Deyà, A.; Payeras-Capellà, M.; Macià Mut-Puigserver, M.; Ferrer-Gomila, J. Untraceable, anonymous and fair micropayment scheme. In Proceedings of the 9th ACM International Conference on Advances in Mobile Computing and Multimedia, Ho Chi Minh City, Vietnam, 5–7 December 2011; pp. 42–48.
6. Park, D.; Boyd, C.; Dawson, E. Micro-Payments for Wireless Communications. In *Security and Cryptology*; Lecture Notes in Computer Science 2015; Springer: Berlin/Heidelberg, Germany, 2001; pp. 192–205.
7. Pedersen, T. Electronic payments of small amounts. In *Security Protocols*; Lomas, M., Ed.; Lecture Notes in Computer Science No. 1189; Springer Verlag: Berlin, Germany, 1997; pp. 59–68.
8. Reddy, S.; Estrin, D.; Hansen, M.; Srivastava, M. Examining micro-payments for participatory sensing data collections. In Proceedings of the 12th ACM International Conference on Ubiquitous Computing, Copenhagen, Denmark, 26–29 September 2010; pp. 33–36.
9. Rivest, R.; Shamir, A. PayWord and MicroMint Two Simple Micropayment Schemes. In *Security Protocols*; Lecture Notes in Computer Science v. 1189; Springer: Berlin/Heidelberg, Germany, 1997; pp. 69–87.
10. Shin, D.; Jun, M. Micro Payment System Using OTP for Customer's Anonymous. In Proceedings of the IEEE 2011 International Conference on Information Science and Applications (ICISA), Jeju Island, Korea, 26–29 April 2011; pp. 1–5.
11. Tewari, H.; O'Mahony, D. Multiparty Micropayment for Ad Hoc Network. In Proceedings of the Wireless Communications and Networking, 2003 (WCNC 2003), New Orleans, LA, USA, 20 March 2003; Volume 3, pp. 2033–2044.
12. Ziang, N.; Liu, X.; Zhao, J.; Yang, D. A Mobile Micropayment Protocol Based on Chaos. In Proceedings of the 2009 Eighth International Conference on Mobile Business, Dalian, China, 27–28 June 2009; pp. 284–289.
13. Dai, X.; Grundy, J. Net Pay: An off-line, decentralized micro-payment system for thin-client applications. *Electron. Commer. Res. Appl.* **2007**, *6*, 91–101. [CrossRef]
14. Chen, L.; Li, X.; Shi, M. A Novel Micro-payment Scheme for M-commerce based on Self-Renewal Hash Chains. In Proceedings of the IEEE International Conference on Communications, Circuits and Systems, Kokura, Japan, 11–13 July 2007; pp. 1343–1346.
15. Yohan, A.; Lo, N.W.; Randy, V.; Chen, S.J.; Hsu, M.Y. A Novel Authentication Protocol for Micropayment with Wearable Devices. In Proceedings of the 10th International Conference on Ubiquitous Information Management and Communication, Danang, Vietnam, 4–6 January 2016; pp. 1–7.
16. Veeraraghavan, P.; Almuairfi, S.; Chilamkurti, N. Anonymous paperless secure payment system using clouds. *J. Supercomput.* **2016**, *72*, 1813–1824. [CrossRef]
17. Gisolfi, D. Web Services Architect, Part 3: Is Web Services the Reincarnation of CORBA? 2010. Available online: http://www.ibm.com/developerworks/webservices/library/ws-arc3 (accessed on 20 May 2016).

18. Dai, X.; Grundy, J. Architecture of a Micro-Payment System for Thin-Client Web Applications. In Proceedings of the 2002 International Conference on Internet Computing, Las Vegas, NV, USA, 24–27 June 2002; pp. 444–450.

19. Rowley, J. An analysis of the e-service literature: Towards a research agenda. *Int. Res.* **2006**, *16*, 339–359. [CrossRef]

20. Sumak, B.; Polancic, G.; Hericko, M. Towards an e-service knowledge system for improving the quality and adoption of e-services. In Proceedings of the 22nd Bled eEnablement: Facilitating an Open, Effective and Representative Society, Bled, Slovenia, 14–17 June 2009.

21. Barnes, S.J.; Vidgen, R.T. An integrative approach to the assessment of ecommerce quality. *J. Electron. Commer. Res.* **2002**, *3*, 114–127.

22. Behkamal, B.; Kahani, M.; Akbari, M.K. Customizing ISO 9126 quality model for evaluation of B2B applications. *Inf. Softw. Technol.* **2009**, *51*, 599–609. [CrossRef]

23. Parasuraman, A.; Zeithaml, V.A.; Malhotra, A. E-S-QUAL: A multiple-item scale for assessing electronic service quality. *J. Serv. Res.* **2005**, *7*, 213–233. [CrossRef]

24. Schubert, P. Extended web assessment method (EWAM)—Evaluation of electronic commerce applications from the customer's viewpoint. *Int. J. Electron. Commer.* **2003**, *7*, 51–80.

25. Janda, S.; Trocchia, P.J.; Gwinner, K.P. Consumer perceptions of Internet retail service quality. *Int. J. Serv. Ind. Manag.* **2002**, *13*, 412–431. [CrossRef]

26. CodeGuru the .NET Architecture, Published by Wiley, September 21, 2004. Available online: http://www.codeguru.com/csharp/sample_chapter/article.php/c8245__1/The-NET-Architecture.htm (accessed on 20 May 2016).

27. Gray, N. Comparison of Web Services, Java-RMI, and CORBA service implementations, Fifth Australasian Workshop on Software and System Architectures. In Proceedings of the Conjunction with ASWEC 2004, Melbourne, Australia, 13 April 2004.

28. Dai, X.; Grundy, J. Three Kinds of E-wallets for a NetPay Micro-payment System. In Proceedings of the Fifth International Conference on Web Information Systems Engineering, Brisbane, Australia, 22–24 November 2004; pp. 66–77.

29. Huang, X.; Dai, X.; Liang, W. BulaPay: A Novel Web Service based Third-Party Payment System for e-Commerce. *Electron. Commer. Res.* **2014**, *14*, 611–633. [CrossRef]

30. Yang, C.; Qi, M. Scheme and applications of mobile payment based on 3-D security protocol. In Proceedings of the 3rd ACM International Conference on Mobile Technology, Applications & Systems, Bangkok, Thailand, 25–27 October 2006; pp. 1–4.

31. Dumas, S.J.; Redish, J.C. *A Practical Guide to Usability Testing*; Ablex Publishing Corporation: Norwood, NJ, USA, 1993.

32. Hwang, W.; Salvendy, G. Number of people required for usability evaluation: The 10 ± 2 rule. *Commun. ACM* **2010**, *53*, 130–133. [CrossRef]

33. Nielsen, J. Heuristic evaluation. In *Usability Inspection Methods*; Nielsen, J., Mack, R.L., Eds.; John Wiley & Sons: New York, NY, USA, 1994.

34. Cao, H.; Wang, L.; Zhu, J. A Trust-Aware Mobile Multiple Micro-Payment Mechanism Based on Smart Agent in Distributed Environment. In Proceedings of the 2nd IEEE International Conference on Pervasive Computing and Applications, ICPCA 2007, Birmingham, UK, 26–27 July 2007; pp. 314–318.

35. Pilla, P. Heuristic Evaluation Report, Heuristic Evaluation of HOMIE, 2009. Available online: http://www.ccs.neu.edu/home/prasu14/hci/i8/HomieReport.pdf (accessed on 12 May 2015).

36. Nielsen, J. Severity Ratings for Usability Problems, Papers and Essays, 1995. Available online: http://www.useit.com/papers/heuristic/severityrating.html (accessed on 21 April 2016).

37. Mallon, D. *mCommerce Security White Paper Key Security Techniques—mCommerce Security White Paper Version 1.1*; Sybase: Dublin, CA, USA, 2010.

38. Williams, L.; Smith, C. Performance Evaluation of Software Architectures. In Proceedings of the Performance Engineering Services and Software Engineering Research, Buenes Aires, Argentina, 5–8 February 2007.

39. Yang, C.; Wu, C.; Chiu, C.; Chiou, S.; Liao, W. Micropayment schemes with ability to return changes. In Proceedings of the 11th ACM International Conference on Information Integration and Web-based Applications & Services, Kuala Lumpur, Malaysia, 14–16 December 2009; pp. 356–363.

40. Huang, X.D. UsageQoS: Estimating the QoS of Web Services through Online User Communities. *ACM Trans. Web* **2013**, *8*, 1–31. [CrossRef]

symmetry

MDPI

Article

Social Content Recommendation Based on Spatial-Temporal Aware Diffusion Modeling in Social Networks

Farman Ullah and Sungchang Lee *

School of Electronics and Information Engineering, Korea Aerospace University, Deogyang-gu, Goyang-si, Gyeonggi-do 412-791, Korea; farman@kau.ac.kr
* Correspondence: sclee@kau.ac.kr; Tel.: +82-2-300-0127

Academic Editor: Angel Garrido
Received: 8 June 2016; Accepted: 23 August 2016; Published: 1 September 2016

Abstract: User interactions in online social networks (OSNs) enable the spread of information and enhance the information dissemination process, but at the same time they exacerbate the information overload problem. In this paper, we propose a social content recommendation method based on spatial-temporal aware controlled information diffusion modeling in OSNs. Users interact more frequently when they are close to each other geographically, have similar behaviors, and fall into similar demographic categories. Considering these facts, we propose multicriteria-based social ties relationship and temporal-aware probabilistic information diffusion modeling for controlled information spread maximization in OSNs. The proposed social ties relationship modeling takes into account user spatial information, content trust, opinion similarity, and demographics. We suggest a ranking algorithm that considers the user ties strength with friends and friends-of-friends to rank users in OSNs and select highly influential injection nodes. These nodes are able to improve social content recommendations, minimize information diffusion time, and maximize information spread. Furthermore, the proposed temporal-aware probabilistic diffusion process categorizes the nodes and diffuses the recommended content to only those users who are highly influential and can enhance information dissemination. The experimental results show the effectiveness of the proposed scheme.

Keywords: spatial; temporal; information diffusion; probabilistic diffusion model; recommender system; online social networks

1. Introduction

Recommender systems are web-based applications, tools, techniques, and programs that are used to provide suggestions for items and products of interest; they do this by analyzing user interactions and consumed content histories [1,2]. Social content recommendations use social networks and user interactions to model recommendation processes. At present, online social networks (OSNs) are increasing in importance and have become a fundamental medium to diffuse information to a large number of people. The surge of social networking sites (SNSs) has enabled user interactions from anywhere and has opened a new era of social interaction, collaboration, preference collection, and tagging for personalization. Information exchange is the keystone of a structured society, and OSNs play important roles in propagating information and enabling users to receive information of interest across many areas, including interest-based community detection [3], political influence [4,5], and economic networks [5–7]. SNSs such as Facebook, Twitter, and LinkedIn enable users to share opinions and status updates (newsfeeds) effortlessly with masses of people on any topic. A recent Facebook study [8] showed that the average separation between users was 4.7 hops. Increasing friends, followers, and acquaintances provides constant information updates, which increases the information overload problem. Koroleva et al. regarded constant information updates as a double-edged sword [9]. On the one hand, the increasing number of users helped to maximize the contagion process to spread information to a large number of OSN users. On the other hand, a user receives hundreds or even thousands of newsfeeds, and most of them are non-newsworthy. User frustration increases when receiving the same newsfeed with non-newsworthy updates at the top of their SNSs walls. Figure 1 shows a newsfeed with updates at different times on a Facebook wall. Figure 1a shows the updated newsfeed after 2 h, and Figure 1b is the same newsfeed with updates after 8 h on the top of the wall. However, there is not much newsworthy information. As a user's network of friends expands, their continuous newsfeed updates increase the probability that a user will miss important newsfeeds. This paper proposes a selective diffusion based recommendation technique that incorporates spatial-temporal information and user specific information to identify whether or not to recommend specific information to a user.

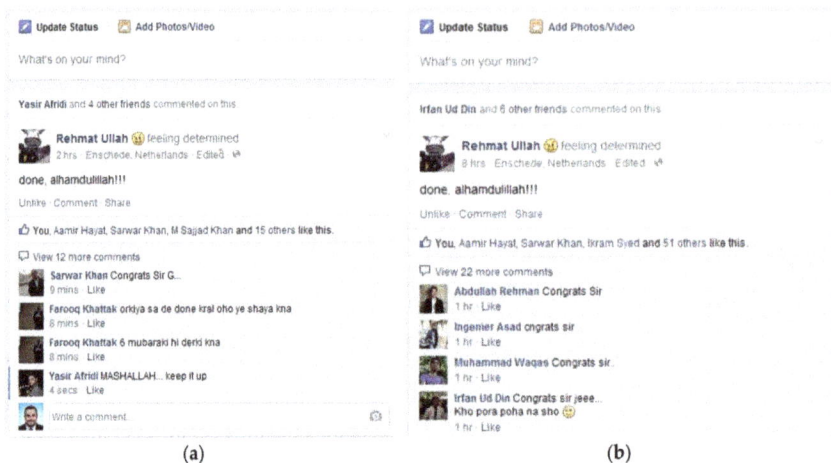

(a) (b)

Figure 1. The same newsfeed with updates at the top of a Facebook wall at different times: (**a**) status of newsfeed after 2 h; and (**b**) same newsfeed after 8 h with some non-newsworthy updates.

A social network (SN) is modeled as a graph of people (nodes) connected by friendship or mutual interests (links/edges). OSNs not only provide a meeting point and facilitate the building of social relations among a large number of people, they also play important roles in spreading information, news, ideas, and innovations. The link between nodes may be directed or undirected and provides a "word-of-mouth" communication channel [10,11]. The Oxford dictionary [12] defines diffusion as the spreading of something more widely. Connected nodes spread information in OSNs, and a network with high connectivity maximizes the information spread. In 1969, Milgram [13] requested 240 people to write a letter to a stockbroker in Boston. The participants did not know the stockbroker or his address. Nevertheless, 60 letters reached the target destination. It took an average of 6.2 hops for this to happen, creating the six-degrees-of-separation adage. Cheng [14] analyzed 5.2 billion Twitter friendships and discovered five degrees of separation within the Twitter network. These studies only provided a degree of separation for the diffusion of information. They did not determine which pieces of information diffused more frequently, or how and through which path they diffused. Typically, it is unnecessary to diffuse messages to an entire network. Groups such as service providers, politicians, security analysts, and crime prevention services are keenly interested in identifying people in society that are effective at spreading information. This knowledge helps these groups to maximize their own information dissemination.

Influence is the capacity to have an effect on the character, development, or behavior of someone or something, or the effect itself [12]. Social influence is a phenomenon in which an individual impels his connection to behave in a similar way [15]. The flow of information in OSNs is only possible if individuals can influence one another. Teo [16] analyzed demographic and motivation variables associated with Internet usage activities and showed that individuals with similar demographics and behaviors tended to engage in similar activities. Social influence involves a three steps process: (i) a set of nodes that can exert influence; (ii) the method of contagion of the influence; and (iii) the set of nodes that can adapt the influence. In social influence maximization, a small set of nodes is considered that can maximize information diffusion. However, selecting these in a large network is an NP-hard problem [5]. In this work, we model the social ties relationship considering multicriteria information. There are four categories of information—user spatial information, user interactions and activity history for social trust, user opinions on similar items, and user demographics. We rank the nodes considering friends and friends-of-friends to select the most influential nodes for diffusion.

The ubiquity of smart devices, communication technologies, and online social media enables people to be connected all the time, and this in turn makes it possible for information to be diffused at anytime from anywhere. Spatial and temporal circumstances no longer constrain information diffusion. Individuals receive information in a continuous stream over time, merging small pieces of information at different spatial locations and then conveying them to the masses over OSNs. Several mathematical models have been proposed to model the distribution of information in OSNs, but none of them is comprehensive [17]. The heterogeneity of user interactions and mobility concerning links, the dynamic structure of OSNs, and the merging of information over time and space have opened a new era of research on information diffusion and user behaviors in OSNs. This article explores how a piece of information diffuses temporally and provides a mathematical model to control the diffusion process in OSNs.

The rest of the paper is organized as follows. Section 2 explains the related work and background knowledge regarding social networks and information diffusion. Section 3 introduces the proposed scheme, which incorporates a spatial-temporal approach with selective diffusion-based social content recommendations. In Section 4, our approaches are compared to various state-of-the-art ranking and diffusion techniques using simulation results. Finally, the conclusion is presented in Section 5.

2. Background and Related Literature Review

The scope of this paper is closely related to SNS dynamics, social influence, and information diffusion. This section presents the background and related work regarding OSN structures, user interactions and information flows, user influence on direct friends and other users, and information diffusion processes in OSNs.

Web-based SNSs have become a popular socialization-based medium that enables users to provide opinions about various products and items. In sociology, Georg Simmel [18] helped pioneer structural theories such as triad dynamics. Mereno [19] depicted "sociograms" for interpersonal relationships, preferences, and choices within groups. The concept of social networks was first coined by Barnes [20] in his article Class and Committees in a Norwegian Island Parish. In web-based SNSs, individuals: (i) make a public or semi-public profile within a bounded system; (ii) maintain a list of users with whom they want to share a connection; and (iii) view and traverse the shared content by direct connections or by others in the system [21]. OSNs are represented by a graph $G (V, E)$, where V is the set of nodes and E is the set of edges showing a relationship among members of V [22]. Mostly, the large-scale OSNs are scale-free graph means that their degree distribution follows the power law [23]. The main feature of the scale-free network is that they have higher degree node called hub. Haythornthwaite [24] stated that SNs were initially introduced to connect families and friends.

User interest in the SNSs has opened a new wave of SNS development in the last decade. The first recognizable SNS was SixDegrees.com, which was launched in 1997. This was followed by LiveJournal, AsianAvenue, and BlackPlanet in 1999; LunarStorm and MinGente in 2000; Cyworld and Ryze in 2001; Fotolog, Friendster and Skyblog in 2002; LinkedIn and Last.FM in 2003; and Flicker and Facebook in 2004 [21]. OSNs have been studied and utilized in many fields. Jaeger et al. [25] and Zappen et al. [26]

explored the potential use of SN platforms in E-Government to deliver and improve services. Viral marketing is a marketing technique using word-of-mouth effects to achieve marketing objectives and increase brand awareness. SNSs can play a vital role in viral marketing. A mechanism to select a node in viral marketing based on its motivation to forward online content is proposed in [27]. Richardson and Domingos [28] used a probabilistic approach to model the customer's network value, and this value showed the expected sales profit gained from some other customers with whom this customer was directly or indirectly connected to in the network. Recommender systems [1,2,29] also collect user opinions and personalize content for them. Lada Adamic and Glance [4] analyzed political blogs prior to the US Presidential election of 2004 and revealed two well-separated clusters. Coleman et al. [30] modeled a network for physicians and found that physicians with more academic citations had their prescriptions of new drugs accepted more frequently than physicians with fewer academic citations.

In OSNs, information disseminates where users influence one another. Social influence determines in particular: (i) who are the most popular and influential users in the OSN; (ii) which user influences whom; (iii) why they are influenced; and (iv) why users are attracted to particular services [31]. A novel social influence model is proposed based on bounded rationality of agents in SNs [32]. Influence is modeled as an influence game, where the players are, influencers and followers. The follower is following the influencers considering the measure of bounded rationality. Higher the rationality of a follower with respect to a seed, higher the probability that the follower will follow the state of the influencer. Influence has a long history of study in social sciences, marketing, communication, and political sciences. Katona et al. [33] found that a user who is connected to many users (user degree) have a higher adoption probability. In addition, the density of connections in a group who already adopted has the strong influence on the adaptation of individuals connected to this group. Influence is node specific, and local links between nodes are more important than global links. In addition, links between two users may be multi-aspect, i.e., they can work in opposite directions between two users depending on the topics [34]. Identifying important nodes is a key problem when determining social influence in an SN analysis. Degree, betweenness, and closeness centrality are measures of determining the criticality and importance of nodes [7,35,36]. Piraveenan et al. [37] introduced a new centrality measure (percolation centrality) to analyze the importance of nodes during percolation in networks. They found that the average of percolation centrality overall possible single contagion source reduces to betweenness centrality. In addition, the percolation centrality reduces to betweenness centrality if all the nodes are infected or partially percolated to some extent. Node centrality ranks nodes based on their central positions in SNs considering either edges, shortest paths, or the nodes passing through a node. Google introduced PageRank algorithm [38,39] to rank a web page. The PageRank algorithm ranked solely based on their location in the Web's graph structure regardless of the web page content. A Webpage has a higher rank that is linked to important WebPages. Zhu et al. [40] suggested a SpreadRank algorithm based on the random walk theory and the spreading ability of the node. Xiang et al. [41] discussed the understanding of PageRank and the relationship between the PageRank and social influence analysis. They developed a linear influence model by introducing the prior knowledge which generalizes the authority computation of PageRank. In this work, Luarn et al. [42] showed that dissemination information frequency was highly affected by the network degree. The social transmission of information and decision-making is highly influenced by the behavior of others [43]. OSN structures and topology affect information diffusion on a large scale [6,44]. In influence diffusion, the node may be either active or inactive. Kempe et al. [45] derived a formula for influence maximization as a discrete optimization problem and suggested two diffusion models—linear threshold (LT) and independent cascade (IC). Suppose u and v are two nodes in an SNS, and $E(u,v)$ is an edge between these nodes. The LT model [5,40,45–47] assigns random weight $w(u,v)$ from interval $[0,1]$ to edge $E(u,v)$ and represents the influence of user u over v. The LT model then sums the influence weights of active neighbors N of node $(\sum_{i \in N} w_i)$ and uses a threshold value to determine whether the node will be switched from inactive to active. The IC model [5,15,40,45] uses a probabilistic approach, and an inactive node is given one chance with a certain probability to activate its inactive neighbors. An extension of IC model is proposed in [48] called independent cascade model with negative opinion (IC-N). They incorporated users' negativity and stated that negative opinions are more dominant over positive opinion. The IC-N model introduced a quality factor. IC-N first selects a set of nodes and activated them. With certain probability (q) the seed node becomes positive if he experienced good quality and with $1 - q$ becomes negative. Bakshy et al. [49] analyzed 250 million Facebook users and examined the role of information diffusion. The main findings of the analysis were that: (i) an exposed user was more likely to spread information and propagate messages more quickly; and (ii) stronger ties were more influential, but weak ties were of greater help in spreading recent information. Kim et al. [50] introduced a conceptual framework for information diffusion across heterogeneous OSNs and provided a macro-level information diffusion model.

OSN structures and topologies, user positions in an SN, and user interactions play important roles in influence maximization and information diffusion. Kasthurirathna et al. [51] simulated a coordination game on four different classes of complex networks. In the study, they found that in all four types of networks that slightly less connected people first adopt the coordination compared to the highest connected people. However, there are two other important factors with respect to OSNs—the spatial awareness to diffuse information to individuals in the same region who are likely to be more interested in the diffused information, and the temporal awareness to consider the diffused message importance with the passage of time. Temporal- and spatial awareness-based diffusion was studied in [7,52–56], but there is still a great deal of research required in the field of OSNs. Nicosia et al. [53] stated that complex networks were time-varying graphs, and interactions among the users were time varying. In OSNs, most edges are active for a short period. Holme et al. discussed several systems that could benefit from a temporal network infrastructure [54]. Contacts and social networks, telephone networks, and the Internet and mobile networks are all examples where the use of only the network topology does not provide full information. Space (location) is one of the most relevant factors to be considered when attempting to acquire complete information [55]. Spatial information is one of the most important factors for location aware services such as advertisements and entertainment.

The contributions of this paper to the aforementioned literature can be summarized as follows: (i) it proposes social content recommendations based on spatial-temporal aware diffusion in SNs; (ii) it models social ties relationships considering multicriteria such as user spatial information, content trust, opinion similarity and demographics; (iii) it provides a ranking algorithm (SocNodeRank) considering a user's direct friends and friends-of-friends to select the most influential nodes; (iv) it offers a temporal aware probabilistic diffusion model to maximize information diffusion and minimize contagion time; and (vii) the proposed scheme scales well with a large number of nodes and links, minimizes the contagion time needed to diffuse a message, and controls recommendations based on social diffusion.

3. Social Content Recommendations Based on Spatial-Temporal Aware Diffusion in Social Networks

This section presents the proposed spatial-temporal aware selective flooding-based information diffusion in OSNs. Focus is placed on algorithms that are able to model the multicriteria-based social ties relationships between users, rank nodes to find the most influential individuals from whom the users can accept information, and provide temporal aware probabilistic diffusion-based content recommendations. In OSNs, spatial and temporal information are two important factors along with network topology and user interactions when attempting to understand information dissemination in SNs. We model multicriteria-based social ties relationships between users considering each user's spatial information, their interactions with respect to content sharing, the extent to which friends share the same content with other users in the SNS, opinions on various items and products, and user demographics. We rank the nodes using the proposed SocNodeRank ranking algorithm to select the Top K nodes as seeds to maximize the information distribution in the SNs. The diffusion process is modeled as a temporal aware probabilistic diffusion model that maximizes the information distribution (i.e., recommendations).

3.1. Multicriteria-Based Social Ties Relationship (Influence) Modeling

The SNS graph will be either directed or undirected. In an undirected network, edges do not have direction and the relationship from either of the nodes remains the same (symmetric). In reality, the heterogeneous nature of user interactions causes OSNs to be directed networks in nature, and the two neighboring users have different levels of influence (asymmetric) on one another. Figure 2 gives a pictorial representation of the proposed social ties relationship modeling considering two direct neighbor users, u and v.

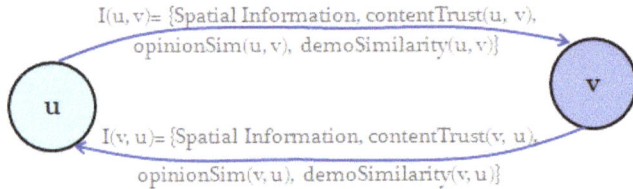

Figure 2. Pictorial overview of multicriteria-based social ties relationship (influence) modeling.

An SNS provides the location of a user, and we use each user's position (GPS coordinates) to determine spatial similarity. Equation (1) finds the spatial similarity of two users using the cosine similarity normalized by the geo-distance between the users.

$$spSim\,(u,v) = \frac{1}{\left[1 + \frac{dist(u,v)}{1000}\right]} \left(\frac{C_u.C_v}{||C_u||\,||C_v||}\right) \tag{1}$$

where $dist\,(u,v)$ is the geo-distance between the users and can be found by Equation (2) using the Haversine formula [57,58].

$$\begin{cases} a = sin^2\left(\frac{\Delta\varnothing}{2}\right) + cos\,(\varnothing_{C_u})\,cos\,(\varnothing_{C_v})\,sin^2\left(\frac{\Delta\varphi}{2}\right); where\,\Delta\varnothing = |\varnothing_{C_u} - \varnothing_{C_v}|,\,\Delta\varphi = |\varphi_{C_u} - \varphi_{C_v}| \\ c = 2arctan\left(\frac{\sqrt{a}}{\sqrt{1-a}}\right) \\ dist\,(u,v) = R \times c; where\,R\,is\,the\,radius\,of\,the\,Earth \end{cases} \tag{2}$$

where C_u and C_v are the GPS coordinates of users u and v, respectively. Each GPS point has latitude ($\varnothing$) and longitude (φ) coordinates. $\varnothing_{C_u}$ and φ_{C_u} are the latitude and longitude of user u GPS coordinate (C_u), and $\varnothing_{C_v}$ and φ_{C_v} are the latitude and longitude of user v GPS coordinate (C_v). R is the radius of earth, 6371 km.

In OSNs, social relationships vary based on the acceptance and adoption of content by other users. Users like to recommend and share message with people who easily accept that information and share it with other users. In this paper, we consider that users only accept and adopt content from individuals who shared that content with other users. Figure 3 shows the overview of content acceptance by a user, and that content being shared to the walls of other users. Figure 3a shows content being shared by user u to user v, and v shares part of that content to his directly neighboring users. Content acceptance (*contentTrust*) between users u and v in Figure 4a will be *contentTrust* $(u,v) = \frac{1}{3}\left(\frac{(2+0+1)}{3}\right) = \frac{1}{3}$. In Figure 3b, it will be *contentTrust* $(u,v) = \frac{1}{3}\left(\frac{3+3+3}{3}\right) = 1$. Mathematically, the *contentTrust* between any two users such as u and v can be found by Equation (3).

$$contentTrust\,(u,v) = \frac{1}{|CSet|} \frac{\sum_{c\in CSet}|n_{v,c}|}{|N_v|} \tag{3}$$

where $|CSet|$ is the set of content user u shared to user v, $|N_v|$ is the number of neighbors of user v and $|n_{v,c}|$ are the users from N_v diffused by content c.

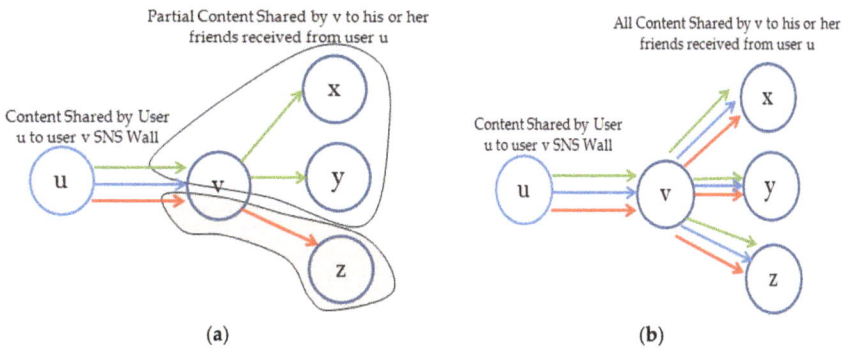

Figure 3. Content sharing by friends and friends-of-friends in an SNS: (**a**) content sharing by user u to v and v share partial content to his friends; and (**b**) content sharing by user u to w and w share all content to his friends.

In OSNs, users easily adopt content from users who experienced content in the past. Recommender systems mostly recommend user content to active users with similar content histories. In this paper, we consider experienced content similarity as a factor to find the influence between users. The content similarity between users u and v can be found by Equation (4)

$$opinionSim\,(u,v) = 1 - \frac{\sum_{i \in (CSet_u \cap CSet_v)} |R_{u,i} - R_{v,i}|}{R_{max}\,|CSet_u \cap CSet_v|} \tag{4}$$

where $CSet_u$ and $CSet_v$ are the sets of content experienced by users u and v, respectively; R_{max} is the maximum rating; and i is the content experienced by both users u and v.

Users with similar demographic characteristics in OSNs are most likely to connect with each other. In reality, most businesses and service providers treat service consumers unequally, and they segment consumers based on demographics such as age, gender, and income for various smart services. This paper considers age, gender, and occupation to find similarities between users using the cosine similarity technique in Equation (5).

$$demoSim\,(u,v) = \frac{\sum_{i=1}^{D} u_i v_i}{\sqrt{\sum_{i=1}^{D} u_i^2}\sqrt{\sum_{i=1}^{D} v_i^2}} \tag{5}$$

where i are the demographic attributes of users u and v, and D is the total of the demographics attributes. The influence, $I\,(u,v)$, of user u on v, is depicted in Figure 3, and it can be calculated using Equation (6).

$$I\,(u,v) = \alpha\,spSim\,(u,v) + \beta\,contentTrust\,(u,\,v) + \gamma\,opinionSim\,(u,\,v) + \omega\,demoSim\,(u,\,v) \tag{6}$$

where α, β, γ, and ω are the real numbers from interval $[0,1]$ such that $\alpha + \beta + \gamma + \omega = 1$. We adjust the coefficients by considering the contributions and importance of the criteria in Equation (6).

3.2. Ranking Algorithm for the Selection of the Most Influential Nodes to Initialize the Diffusion Process in an OSN

In this subsection, we introduce the proposed social node ranking algorithm, which is called SocNodeRank, to select the most influential nodes to initialize the information dissemination process. The proposed algorithm considers the user's direct friends and friends-of-friends to compute the node rank. Numerous ranking algorithms have been proposed, including degree centrality, betweenness centrality and closeness centrality [59,60]. Degree centrality considers the number of edges; a user will be more central if he is more connected. Degree centrality ignores network structure information and instead only considers the ties to direct neighbors. Closeness centrality measures how close a node is to other nodes in the network. The betweenness of a node refers to the number of node pairs passing through a node with the minimum number of edges. Closeness and betweenness consider the network structure, but they lack efficiency in large scale networks.

The SocNodeRank algorithm considers user directed friends and friends-of-friends to determine a ranking value. A network can be directed or undirected, but this paper focuses on directed graph-based social networks. However, we also provide a mechanism to rank the nodes in an undirected network. SocNodeRank considers the direct friends and friends-of-friends. However, if a node is common between the user and his friends, then it will be considered only once, as depicted in Figure 4. Equation (7) finds the rank of the node in an undirected SNS graph.

$$SocNodeRank_{undirected}(w) = \sum_{x \in W} e_{wx} + \sum_{x \in W} \sum_{z \in X \cap z \notin W} e_{xz} \tag{7}$$

where W is a neighbor set of w, x is one of the neighbors of w, and X is a neighbor set of x. e is the edge between two users, which is 1 if present and 0 otherwise. Figure 4 shows the SocNodeRank algorithm, and Figure 4a is the ranking mechanism of an undirected graph (SNS).

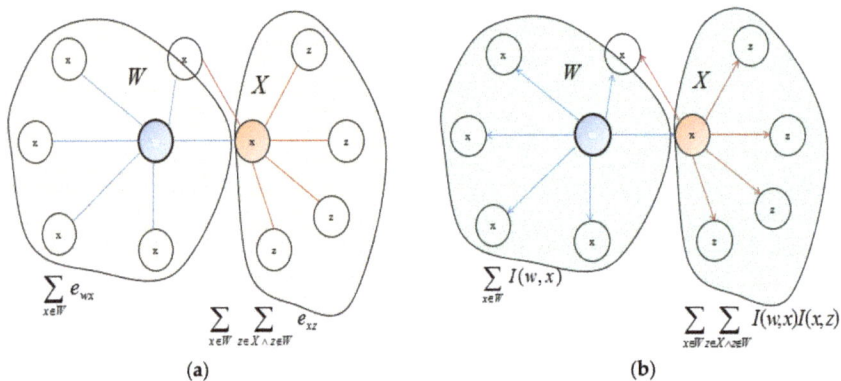

Figure 4. Overview of proposed social node rank (SocNodeRank): (a) SocNodeRank in an undirected network; and (b) SocNodeRank in the directed network.

A social relationship between friends is asymmetric in directed networks. In Section 3.1, we introduced the influence modeling and the multicriteria factors that affect influence. We consider the weighted out degree (outgoing edge) to rank nodes in directed networks. The rank of the node in the directed SNS is given by Equation (8), where $I(w,x)$ is the influence of user w on user x.

$$SocNodeRank_{directed}(w) = \sum_{x \in W} I(w,x) + \sum_{x \in W} \sum_{z \in X \cap z \notin W} I(w,x)\, I(x,z) \tag{8}$$

We consider the direct influence of the user on his friends and the propagation of influence to his friends-of-friends. Figure 4b depicts the ranking mechanism of the proposed scheme in the directed graph. The proposed scheme needs less time to rank the nodes in an SNS because it considers only 2-hops neighbors of user compared to most ranking algorithms such as betweenness, closeness and PageRank except the degree centrality. The proposed algorithm also captures the properties of the betweenness centrality- and degree centrality-based schemes.

3.3. Temporal Aware Probabilistic Diffusion-Based Social Content Recommendations in OSNs

The growth of SNSs and user interactions helps disseminate information. However, the density of users creates an information overload problem. In this subsection, we introduce the probabilistic diffusion model to model the social content recommendation process. This model operates in a controlled manner that considers the recommended message temporal information, network topology, and user multicriteria influence. In SNSs, users can receive the same diffused messages from many users or with minor information updates, which does not happen with biological diffusion. We place diffusing messages into two categories—one in which the user accepts the message and the other in which the user does not accept the message. We model the influence between users based on multicriteria, and then we rank the nodes and select the most influential (initial spreader) K-nodes to initialize the diffusion process. We extend the concept of the continuous-time Markov chain model in [40] and the social influence diffusion model presented in [61] to control the diffusion process and overcome the information overload problem.

In the diffusion process, a node can be inactive or active. At time $t = 1$, K influential nodes are selected to start the diffusion process. At $t = 2$, neighbors of the diffusing nodes only become active when the incoming influence from the diffusing node is greater than the outgoing influence and the recommendation (diffusion acceptance threshold). Suppose u is the diffusing node and v might be the accepting (diffusion adopting) node. The activation state of v at $t = 2$ is given by Equation (9).

$$P_{Diff_{t=2}}(v) = \begin{cases} I(u,v); \text{ where } I(u,v) \geq I(v,u) \text{ and } I(u,v) \geq RAT_1 \\ 0; \text{ otherwise} \end{cases} \quad (9)$$

where $P_{Diff_{t=2}}(v)$ determines whether node v at time $t = 2$ will be diffused or not. RAT_1 is the acceptance threshold of diffusion (recommendation) from direct friends. In this paper, we consider two different thresholds—RAT_1 is the threshold of the content recommendation acceptance from the direct friends at $t = 2$, and RAT_2 is the threshold for friends-of-friends and so on for $t > 2$.

At $t = 1$ in Figure 5a, the highest value of K (influential nodes) is selected using the mechanism of Section 3.2, and for simplicity, $K = 1$ is considered to explain the procedure. At $t = 2$ of Figure 5b, the neighboring nodes {2, 4, 6, 7, 8} of {1} are selected for diffusion, but according to Equation (9), only nodes {2, 6, 8} are diffused because node {1} has a higher influence on them and the diffusion acceptance threshold $RAT_1 = 0.4$. The active node continues to diffuse if the outgoing influence is greater than the incoming influence. The diffusing node accepts only the diffused message if the diffusion contagion probability is greater than the diffusion acceptance threshold. Suppose the threshold of diffusion acceptance is $RAT_2 \geq 0.5$ for $t > 2$ in Figure 5c. Nodes {3} and {5} are diffused because $(0.8 \times 0.74) \geq RAT_2$ and $(0.65 \times 0.8) \geq RAT_2$, while node {9} is not diffused because $(0.55 \times 0.35 + 0.8 \times 0.2) < RAT_2$. Similarly, at $t = 4$ in Figure 5d, node {7} is diffused because $(0.8 \times 0.74 \times 0.85) \geq RAT_2$. Mathematically, the nodes activation probability for $(t + 1) \geq 3$ can be given by Equation (10).

$$P_{Diff_{t+1}}(w; V) = \begin{cases} \sum_{v=1}^{V} I(v,w) \ P_{Diff_t}(v|V) \ ; \ I(v,w) \geq I(w,v) \text{ and } P_{Diff_{t+1}}(w; V) \geq RAT_2 \\ 0; \text{ otherwise} \end{cases} \quad (10)$$

where V is the set of neighboring nodes of node w that can diffuse u at time t where $(t + 1) \geq 3$. The equation shows that users accept recommendations when a large number of higher similarity friends and friends-of-friends recommend the content.

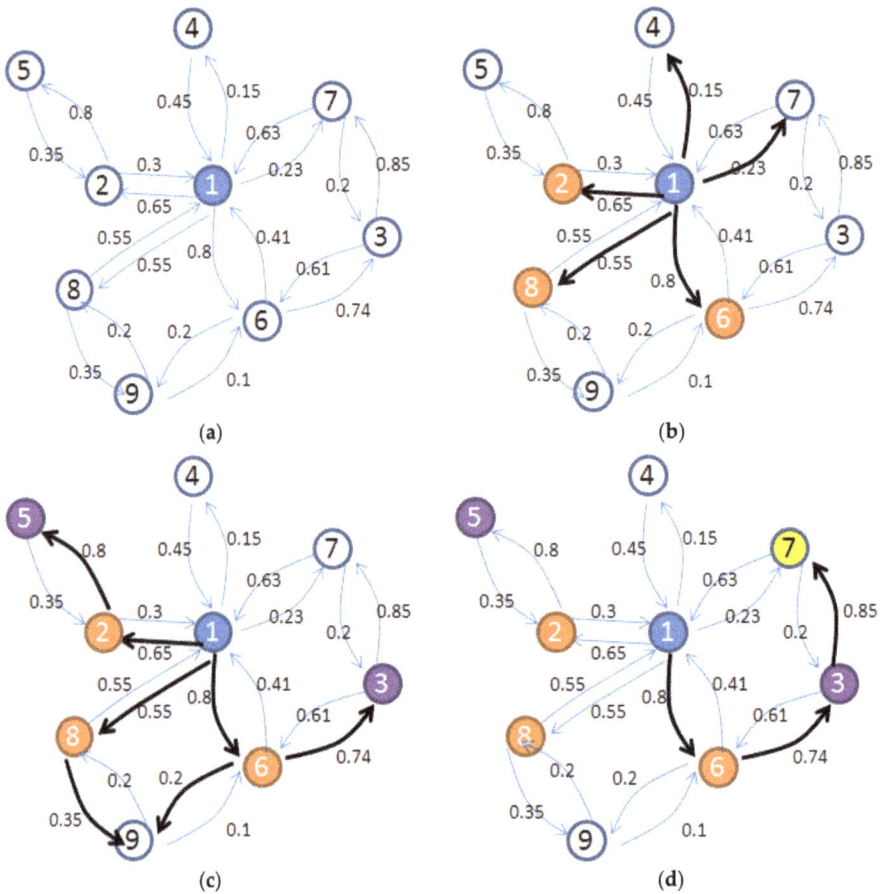

Figure 5. Overview of the proposed temporal aware probabilistic information diffusion process: (a) diffusion process start (*t* = 1); (b) diffused node at (*t* = 2); (c) diffusion status at (*t* = 3); and (d) Information Diffusion at (*t* = 4).

4. Simulation Results and Discussion

In this section, different real-world publicly available datasets used for this paper are introduced, and a comparison of the proposed schemes with various previous rankings and diffusion algorithms is provided.

4.1. Experimental Results for Social Ties Relationship (Influence) Factors

To the best of our knowledge, there is no single dataset network that contains all of the required attributes to determine influence among users and evaluate the proposed influence modeling scheme. To build a single compilation of network data, we extract user spatial information from [62], Advotago user social trust data from [63], and demographics and user ratings from MovieLens [64]. tab:symmetry-08-00089-t001 shows the attributes of the dataset used for the evaluation of influence scheme. Finally, the dataset used for the multicriteria relationship modeling has an average degree of 3.345 and network diameter of 9.

Table 1. Description of dataset for multicriteria-based influence modeling.

Dataset Attributes	Description/Values of Attributes
Number of users	133
Number of edges	450
Trust value between users	{0.6, 0.8, 1.0}
User demographics	Sex, Age
Minimum times same content evaluated by friends	10

In order to evaluate the different influential factors, tab:symmetry-08-00089-t002 shows the Top − 10 most influential nodes in the multicriteria dataset of tab:symmetry-08-00089-t001. Extensive simulations are carried out to determine the optimal values of α, β, γ, and ω to calculate the actual influence between users using the multicriteria. The nodes are ranked using weighted out-degree for each attribute separately and also for combined using the various coefficient criteria in Equation (6). Selecting the coefficient values, the precision is used to compare the node rankings by individual attribute-based influence with node rankings by multicriteria-based influence. Figure 6 shows the precision comparison of the single attribute and the multicriteria influence-based ranked nodes. The coefficients $\{\alpha = 0.20,\ \beta = 0.45,\ \gamma = 0.20,\ \omega = 0.15\}$ are selected because they extract most of the influential nodes of each single attribute.

Table 2. Top-10 influential nodes in the multicriteria influential dataset using weighted out degree.

Spatial Similarity	Content Trust	Opinion Similarity	Demographics Similarity	$\alpha = 0.2,\ \beta = 0.45,$ $\gamma = 0.2,\ \omega = 0.15$	$\alpha = 0.4,\ \beta = 0.4,$ $\gamma = 0.1,\ \omega = 0.1$
70	70	70	70	70	70
29	113	50	109	58	58
58	114	71	58	114	64
27	58	114	85	113	27
64	64	6	114	64	114
102	71	4	61	27	113
78	27	48	32	71	29
32	32	102	55	29	71
4	127	89	1	109	32
99	1	113	127	32	4

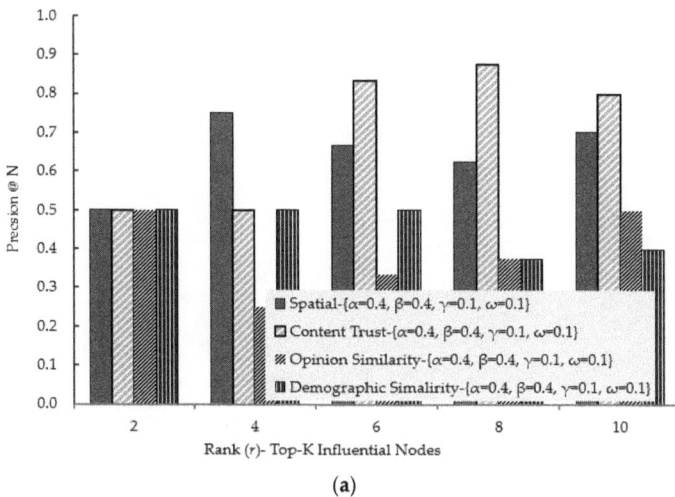

Figure 6. *Cont.*

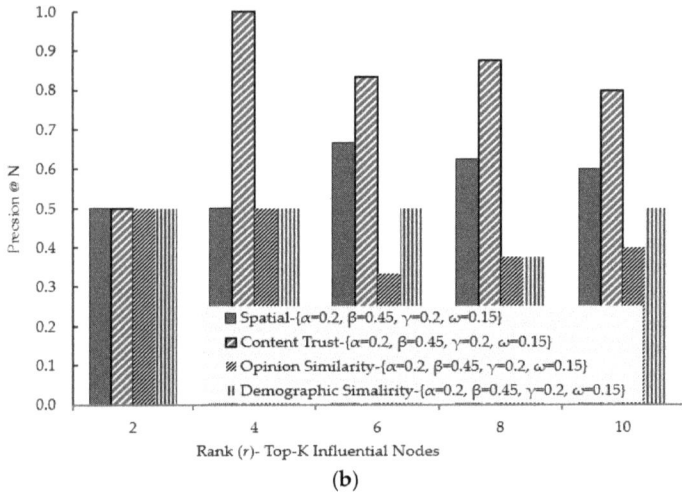

(b)

Figure 6. Precision comparison for the selection of multicriteria influence coefficients: (**a**) precision comparison of the individual attribute based influential node selection with multicriteria consider coefficients $\{\alpha = 0.4,\ \beta = 0.4,\ \gamma = 0.1,\ \omega = 0.1\}$; and (**b**) precision comparison of the individual attribute based influential node selection with multicriteria consider coefficients $\{\alpha = 0.2,\ \beta = 0.45,\ \gamma = 0.2,\ \omega = 0.15\}$.

4.2. Results of Proposed Social Node Ranking Algorithm (Selection of Highest Influential Nodes)

The results of the proposed SocNodeRank are compared with three state-of-art ranking algorithms—weighted out-degree centrality, closeness centrality, and betweenness centrality. We used the directed weighted networks publicly available datasets for results comparison. The details of the datasets are as follows.

Zachary's Karate Network [65]: The small-sized karate network is used as a directed network, but the edge weight (1) is the same between users. The dataset consists of 34 members and 78 friend relationships. The karate club was started at an American university in the late 1970s.

Coauthorships in Network Science [66]: This is a weighted directed coauthors network with 1589 nodes and 2742 edges.

Political Blogs of US Elections [4]: This is a directed network of hyperlinks between 2004 US election weblogs. The dataset consists of 1490 nodes and 19,025 edges.

Coappearance network of characters [67]: The co appearance network of characters in the novel Les Miserables is a weighted network with 77 nodes and 254 edges.

To compare the proposed SocNodeRank with other schemes, the ranking similarity metric from [68] is employed. The ranking similarity F(r) of two schemes at rank *r* can be given by Equation (11).

$$F(r) = \frac{|L(r) \cap L'(r)|}{|r|} \tag{11}$$

where $L(r)$ and $L\prime(r)$ are the two sets of nodes of two different ranking schemes at rank *r*.

Figure 7 shows the simulation and comparison results of ranking similarity $F(r)$ for the above specified networks at various *r* ranking. The figure clearly depicts that the proposed algorithm has higher ranking similarity with the weighted out degree and betweenness. We used Gephi [69] as open source software for most of the ranking algorithm simulation. In the simulation setting for PageRank, we considered (probability = 0.5, Epsilon = 0.1 and use edge weights) and for Eignvector centrality directed and 100 number of iterations. The experimental results show that as the network becomes denser in term of nodes and especially edges (interactions), the proposed SocNodeRank differs from the other schemes. The ranking similarities in Figure 7a (the karate network) and Figure 7b (the coappearance of characters in Les Miserables) are quite alike, whereas the network size increases in Figure 7c (coauthors in Network Science) and Figure 7d (the political blogs of the 2004 US election). SocNodeRank extracts some important potential nodes that can maximize the information diffusion. In addition, it extracts nodes that for the most part have the highest values of betweenness for the 1-edge neighbor friends.

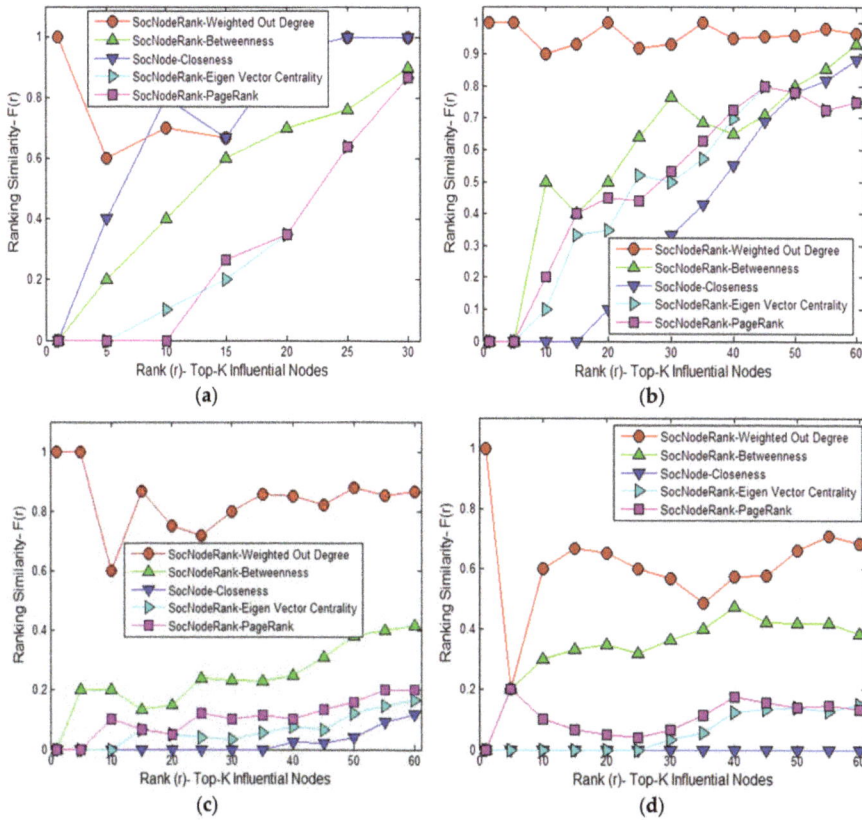

Figure 7. Ranking similarity of the proposed SocNodeRank with weighted out degree, betweenness, and closeness: (**a**) Zachary's Karate Network; (**b**) coappearance network of characters in the novel Les Miserables; (**c**) coauthorships in Network Science; and (**d**) political blogs.

Tables 3 and 4 show the Top-10 highly influential nodes from the coauthorships in Network Science and the 2004 US election political blogs, respectively. In the Top-10 rankings of the coauthorships network, SocNodeRank extracts nodes such as {1087, 132, 133}, which is not extracted by other schemes. Crucially, these nodes for the most part have the highest degree of betweenness and edge centrality for the direct neighbors. Similarly, in the Top-10 rankings of the 2004 US election political blogs, it extracts nodes {1047, 980, 1384, 615}. Node {980} has a high level of betweenness nodes {1051, 1041, 1101, 1479} in the direct friends. Therefore, the proposed SocNodeRank scheme captures the properties of edge centrality and betweenness simultaneously, but it takes less time to rank the nodes as compared to the betweenness-only scheme because it considers only the direct friends and friends-of-friends.

Table 3. Top-10 ranked influential nodes (Nodes IDs) in the directed network of coauthorships in network science.

Rank	Weighted Out Degree	Closeness	Betweenness	PageRank	Eign Vector Centrality	Propose SocNodeRank
1	517	1027	78	33	645	517
2	151	492	150	78	1429	54
3	97	866	301	30	1430	516
4	516	867	516	46	1431	151
5	309	986	281	62	1432	34
6	34	450	46	216	1433	132
7	54	1182	151	294	33	1087
8	744	640	307	150	1434	133
9	377	1026	216	34	30	744
10	655	499	71	69	1435	309

Table 4. Top-10 ranked influential nodes (Nodes IDs) in the political blogs of 2004 US election.

Rank	Weighted Out Degree	Closeness	Betweenness	PageRank	Eign Vector Centrality	Propose SocNodeRank
1	855	1490	855	155	55	855
2	454	1405	55	963	155	1047
3	512	1397	1051	855	641	1000
4	387	1247	155	55	729	980
5	880	230	454	641	1051	524
6	363	216	387	1051	642	880
7	1101	1340	1479	1153	756	387
8	1000	1279	1101	1245	535	1384
9	524	926	1041	729	323	615
10	144	833	729	798	1245	1101

4.3. Results of the Temporal Aware Probabilistic Diffusion-Based Social Content Recommendations in OSNs

In this subsection, the simulation results are introduced for users influenced with the diffusion time and number of Top-K (rank-r) initially diffused nodes for the multicriteria dataset introduced in Section 4.1. The proposed SocNodeRank algorithm is used to select highly influential users that can increase the diffusion process. Figure 8 shows the number of influenced users over the simulation (diffusion) time, and it shows that the diffusion rate is higher at the start, but slows as time progresses. The suggested temporal aware probabilistic diffusion model controls the diffusion process and diffuses only those nodes with high levels of similarity. It is able to diffuse the information further. The recommendation acceptance thresholds (RAT_1 and RAT_2) have a greater impact on the nodes activation process. If the threshold values are higher, then most of the nodes are excluded from diffusion.

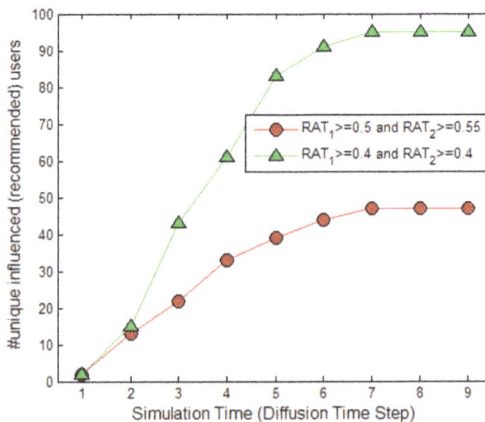

Figure 8. Effect of diffusion time on number of influenced (recommendation accepted) users.

Figure 9 shows the effects of the Top-K influential users and the RAT_1 on the number of influenced (recommended) users. The Top-K influential users help to diffuse and increase the number of diffused nodes. However, after a certain value of K, the influence of the Top-K users decreases, and it does not improve the diffusion process.

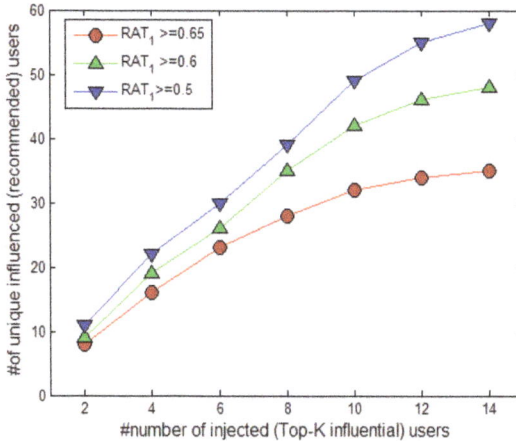

Figure 9. Effect on #number of influenced users by various Top-K influential users and RAT_1.

5. Conclusions

In this paper, we have presented a method of providing social content recommendations using multicriteria-based relationship modeling between social ties and temporal aware probabilistic diffusion model. We modeled the relationship between social ties to identify users with higher levels of relationship similarity based on multicriteria incorporating user spatial information, content trust, opinion similarity, and demographics. We assigned different weights to each attribute of the multicriteria. We suggested a ranking algorithm for selection of the Top-K most influential nodes considering the influence of friends and friends-of-friends to maximize the information spread. The proposed ranking algorithm has higher ranking similarity with the weight out degree and betweenness. However, it needs less processing time to rank the nodes because it only considers the two hops neighbors of a user compared to betweenness which considers the entire network. Furthermore, the suggested temporal-aware probabilistic diffusion process categorized the nodes based on whether or not they were able to recommend and diffuse content. This enabled the process to overcome the information overload problem. We used different publically available datasets to verify the effectiveness of the proposed social content recommendation scheme. We introduced different thresholds in the proposed probabilistic diffusion model, so the inactive node will be activated if it has highest influential diffused nodes and the higher number of diffused nodes. The proposed scheme scaled well with a large number of nodes and links, and it controlled the diffusion process to recommend content to users with high levels of relationship similarity to overcome the information overload problem.

Acknowledgments: This work was supported through Research Program by the National Research Foundation (NRF) of Korea (Grant No. 2014R1A1A2056357).

Author Contributions: Farman Ullah conceived the idea, and performed the experiment and analysis. Sungchang Lee finalized the idea and supervised the research.

Conflicts of Interest: The authors declare no conflict of interest.

References

1. Ricci, F.; Rokach, L.; Shapira, B. *Introduction to Recommender Systems Handbook*; Springer: Berlin, Germany, 2011; pp. 1–35.

2. Ullah, F.; Sarwar, G.; Lee, S. N-screen aware multicriteria hybrid recommender system using weight based subspace clustering. *Sci. World J.* **2014**, *2014*, 679849.

3. Wang, Z.; Zhou, X.; Zhang, D.; Yang, D.; Yu, Z. Cross-domain community detection in heterogeneous social networks. *Pers. Ubiquitous Comput.* **2014**, *18*, 369–383.

4. Adamic, L.A.; Glance, N. The political blogosphere and the 2004 US election: Divided they blog. In Proceedings of the 3rd International Workshop on Link Discovery, Chicago, IL, USA, 21–24 August 2005; pp. 36–43.

5. Doo, M. *Spatial and Social Diffusion of Information and Influence: Models and Algorithms*; Georgia Institute of Technology: Atlanta, GA, USA, 2012.

6. Cowan, R.; Jonard, N. Network structure and the diffusion of knowledge. *J. Econ. Dyn. Control* **2004**, *28*, 1557–1575.

7. Dolde, W.; Tirtiroglu, D. Temporal and spatial information diffusion in real estate price changes and variances. *Real Estate Econ.* **1997**, *25*, 539–565.

8. Ugander, J.; Karrer, B.; Backstrom, L.; Marlow, C. The anatomy of the Facebook Social Graph. 2011. arXiv:1111.4503. arXiv Preprint. Available online: http://arxiv.org/pdf/1111.4503v1.pdf (accessed on 8 June 2016).

9. Koroleva, K.; Krasnova, H.; Günther, O. 'STOP SPAMMING ME!'—Exploring Information Overload on Facebook. In Proceedings of the 16th Americas Conference on Information Systems, Lima, Peru, 12–15 August 2010; p. 447.

10. Goldenberg, J.; Libai, B.; Muller, E. Talk of the network: A complex systems look at the underlying process of word-of-mouth. *Mark. Lett.* **2001**, *12*, 211–223.

11. Jansen, B.J.; Zhang, M.; Sobel, K.; Chowdury, A. Twitter power: Tweets as electronic word of mouth. *J. Am. Soc. Inf. Sci. Technol.* **2009**, *60*, 2169–2188.

12. Oxford dictionaries. Available online: http://www.oxforddictionaries.com/definition/english (accessed on 8 June 2016).

13. Travers, J.; Milgram, S. An experimental study of the small world problem. *Sociometry* **1969**, *32*, 425–443.

14. Cheng, A. Six Degrees of Separation, Twitter Style. 1969, Volume 5. Available online: http://www.sysomos.com/insidetwitter/sixdegrees/ (accessed on 8 June 2016).

15. Guille, A.; Hacid, H.; Favre, C.; Zighed, D.A. Information diffusion in online social networks: A survey. *ACM SIGMOD Rec.* **2013**, *42*, 17–28.

16. Teo, T.S. Demographic and motivation variables associated with Internet usage activities. *Internet Res.* **2001**, *11*, 125–137.

17. Broecheler, M.; Shakarian, P.; Subrahmanian, V.S. A scalable framework for modeling competitive diffusion in social networks. In Proceedings of the IEEE Second International Conference on Social Computing (SocialCom), Minneapolis, MN, USA, 20–22 August 2010; pp. 295–302.

18. Simmel, G. The Metropolis and Mental Life, 1903. Available online: https://books.google.com/books?hl=zh-CN&lr=&id=-mk5rOctXtEC&oi=fnd&pg=PA1999-IA2&dq=The+Metropolis+and+Mental+Life&ots=J_cSXCMc01&sig=qeTE6ihpClC6jjhu5ZkHswrlZ5s#v=onepage&q=The%20Metropolis%20and%20Mental%20Life&f=false (accessed on 8 June 2016).

19. Moreno, J.L. *Who Shall Survive*; ASGPP: Washington, DC, USA, 1934; Volume 58.

20. Barnes, J.A. Class and committees in a Norwegian island parish. *Hum. Relat.* **1954**, *7*, 39–58.

21. Ellison, N.B. Social network sites: Definition, history, and scholarship. *J. Comput. Med. Commun.* **2007**, *13*, 210–230.

22. Garrido, A. Symmetry in complex networks. *Symmetry* **2011**, *3*, 1–15.

23. Chung, K.S.K.; Piraveenan, M.; Hossain, L. Topology of online social networks. In *Encyclopedia of Social Network Analysis and Mining*; Springer: New York, NY, USA, 2014; pp. 2191–2202.

24. Haythornthwaite, C. Social networks and Internet connectivity effects. *Inf. Community Soc.* **2005**, *8*, 125–147.

25. Jaeger, P.T.; Shneiderman, B.; Fleischmann, K.R.; Preece, J.; Qu, Y.; Wu, P.F. Community response grids: E-government, social networks, and effective emergency management. *Telecommun. Policy* **2007**, *31*, 592–604.

26. Zappen, J.P.; Harrison, T.M.; Watson, D. A new paradigm for designing e-government: Web 2.0 and experience design. In Proceedings of the International Conference on Digital Government Research, Montreal, QC, Canada, 18–21 May 2008.

27. Ho, J.Y.; Dempsey, M. Viral marketing: Motivations to forward online content. *J. Bus. Res.* **2010**, *63*, 1000–1006.

28. Richardson, M.; Domingos, P. Mining knowledge-sharing sites for viral marketing. In Proceedings of the Eighth ACM SIGKDD International Conference on Knowledge Discovery and Data Mining, Edmonton, AB, Canada, 23–25 July 2002; pp. 61–70.

29. Pan, Y.; Cong, F.; Chen, K.; Yu, Y. Diffusion-aware personalized social update recommendation. In Proceedings of the 7th ACM Conference on Recommender Systems, Hong Kong, China, 12–16 October 2013; pp. 69–76.

30. Coleman, J.S.; Katz, E.; Menzel, H. *Medical Innovation: A Diffusion Study*; Bobbs-Merrill Co.: Indianapolis, IN, USA, 1966.

31. Saulwick, A.; Trentelman, K. Towards a formal semantics of social influence. *Knowl.-Based Syst.* **2014**, *71*, 52–60.

32. Kasthurirathna, D.; Harre, M.; Piraveenan, M. Influence modelling using bounded rationality in social networks. In Proceedings of the IEEE/ACM International Conference on Advances in Social Networks Analysis and Mining, Paris, France, 25–28 August 2015; pp. 33–40.

33. Katona, Z.; Zubcsek, P.P.; Sarvary, M. Network effects and personal influences: The diffusion of an online social network. *J. Mark. Res.* **2011**, *48*, 425–443.

34. Tang, J.; Sun, J.; Wang, C.; Yang, Z. Social influence analysis in large-scale networks. In Proceedings of the 15th ACM SIGKDD International Conference on Knowledge Discovery and Data mining, Paris, France, 28 June–1 July 2009; pp. 807–816.

35. Bonacich, P.; Lloyd, P. Eigenvector-like measures of centrality for asymmetric relations. *Soc. Netw.* **2001**, *23*, 191–201.

36. Freeman, L.C. A set of measures of centrality based on betweenness. *Sociometry* **1977**, *40*, 35–41.

37. Piraveenan, M.; Prokopenko, M.; Hossain, L. Percolation Centrality: Quantifying Graph-Theoretic Impact of Nodes during Percolation in Networks. *PLoS ONE* **2013**, *8*. [CrossRef]

38. Page, L.; Brin, S.; Motwani, R.; Winograd, T. The PageRank Citation Ranking: Bringing Order to the Web. Available online: http://ilpubs.stanford.edu:8090/422/1/1999-66.pdf (accessed on 8 June 2016).

39. Brin, S.; Page, L. Reprint of: The anatomy of a large-scale hypertextual web search engine. *Comput. Netw.* **2012**, *56*, 3825–3833.

40. Zhu, T.; Wang, B.; Wu, B.; Zhu, C. Maximizing the spread of influence ranking in social networks. *Inf. Sci.* **2014**, *278*, 535–544.

41. Xiang, B.; Liu, Q.; Chen, E.; Xiong, H.; Zheng, Y.; Yang, Y. PageRank with Priors: An Influence Propagation Perspective. In Proceedings of the Twenty-Third International Joint Conference on Artificial Intelligence (IJCAI), Beijing, China, 3–9 August 2013.

42. Luarn, P.; Yang, J.C.; Chiu, Y.P. The network effect on information dissemination on social network sites. *Comput. Hum. Behav.* **2014**, *37*, 1–8.

43. Danchin, É.; Giraldeau, L.A.; Valone, T.J.; Wagner, R.H. Public information: From nosy neighbors to cultural evolution. *Science* **2004**, *305*, 487–491. [PubMed]

44. Delre, S.A.; Jager, W.; Bijmolt, T.H.; Janssen, M.A. Will it spread or not? The effects of social influences and network topology on innovation diffusion. *J. Product Innov. Manag.* **2010**, *27*, 267–282.

45. Kempe, D.; Kleinberg, J.; Tardos, É. Maximizing the spread of influence through a social network. In Proceedings of the Ninth ACM SIGKDD International Conference on Knowledge Discovery and Data Mining, Washington, DC, USA, 24–27 August 2003; pp. 137–146.

46. Kempe, D.; Kleinberg, J.; Tardos, É. Maximizing the spread of influence through a social network. *Theory Comput.* **2015**, *11*, 105–147.

47. Goyal, A.; Lu, W.; Lakshmanan, L.V. Simpath: An efficient algorithm for influence maximization under the linear threshold model. In Proceedings of the 2011 IEEE 11th International Conference on Data Mining, Vancouver, BC, Canada, 11–14 December 2011; pp. 211–220.

48. Chen, W.; Collins, A.; Cummings, R.; Ke, T.; Liu, Z.; Rincon, D.; Yuan, Y. Influence Maximization in Social Networks When Negative Opinions May Emerge and Propagate. In Proceedings of the SIAM International Conference on Data Mining (SDM '2011), Mesa, AZ, USA, 28–30 April 2011; Volume 11, pp. 379–390.

49. Bakshy, E.; Rosenn, I.; Marlow, C.; Adamic, L. The role of social networks in information diffusion. In Proceedings of the 21st International Conference on World Wide Web, Lyon, France, 16—20 April 2012; pp. 519–528.

50. Kim, M.; Newth, D.; Christen, P. Modeling dynamics of diffusion across heterogeneous social networks: News diffusion in social media. *Entropy* **2013**, *15*, 4215–4242.

51. Kasthurirathna, D.; Piraveenan, M.; Harré, M. Influence of topology in the evolution of coordination in complex networks under information diffusion constraints. *Eur. Phys. J. B* **2014**, *87*, 1–15.

52. Ullah, F.; Sarwar, G.; Lee, S.C.; Park, Y.K.; Moon, K.D.; Kim, J.T. Hybrid recommender system with temporal information. In Proceedings of the 26th International Conference on Advanced Information Networking and Applications, Fukuoka, Japan, 26–29 March 2012; pp. 421–425.

53. Nicosia, V.; Tang, J.; Mascolo, C.; Musolesi, M.; Russo, G.; Latora, V. Graph metrics for temporal networks. In *Temporal Networks*; Springer: Berlin/Heidelberg, Germany, 2013; pp. 15–40.

54. Holme, P.; Saramäki, J. Temporal networks. *Phys. Rep.* **2012**, *519*, 97–125.

55. Barthélemy, M. Spatial networks. *Phys. Rep.* **2011**, *499*, 1–101.

56. Liu, S.; Meng, X. A Location-Based Business Information Recommendation Algorithm. *Math. Probl. Eng.* **2015**, *2015*, 345480.

57. Sinnott, R.W. Virtues of the Haversine. *Sky Telesc.* **1984**, *68*, 158.

58. Palmer, M.C. Calculation of distance traveled by fishing vessels using GPS positional data: A theoretical evaluation of the sources of error. *Fish. Res.* **2008**, *89*, 57–64.

59. Borgatti, S.P. Centrality and network flow. *Soc. Netw.* **2005**, *27*, 55–71.

60. Nocera, A.; Ursino, D. PHIS: A system for scouting potential hubs and for favoring their "growth" in a Social Internetworking Scenario. *Knowl.-Based Syst.* **2012**, *36*, 288–299.

61. Doo, M.; Liu, L. Probabilistic Diffusion of Social Influence with Incentives. *IEEE Trans. Serv. Comput.* **2014**, *7*, 387–400.

62. Kurant, M.; Gjoka, M.; Wang, Y.; Almquist, Z.W.; Butts, C.T.; Markopoulou, A. Coarse-grained topology estimation via graph sampling. In Proceedings of the 2012 ACM Workshop on Workshop on Online Social Networks, Helsinki, Finland, 13–17 August 2012; pp. 25–30.

63. Rossi, R.; Ahmed, N. The Network Data Repository with Interactive Graph Analytics and Visualization. In Proceedings of the Twenty-Ninth AAAI Conference on Artificial Intelligence (AAAI-15), Austin, TX, USA, 25–30 January 2015; Volume 15, pp. 4292–4293.

64. GroupLens Dataset (ml-100k). Available online: http://grouplens.org/datasets/movielens/ (accessed on 8 June 2016).

65. Zachary, W.W. An information flow model for conflict and fission in small groups. *J. Anthropol. Res.* **1977**, *33*, 452–473.

66. Newman, M.E. Who is the best connected scientist? A study of scientific coauthorship networks. In *Complex Networks*; Springer: Berlin/Heidelberg, Germany, 2004; pp. 337–370.

67. Knuth, D.E. *The Stanford GraphBase: A Platform for Combinatorial Computing*; Addison-Wesley: Reading, PA, USA, 1993; Volume 37.

68. Kimura, M.; Saito, K. Tractable models for information diffusion in social networks. In Proceedings of the European Conference on Principles of Data Mining and Knowledge Discovery, Berlin, Germany, 18–22 September 2006; Springer: Berlin/Heidelberg, Germany, 2006; pp. 259–271.

69. Bastian, M.; Heymann, S.; Jacomy, M. Gephi: An open source software for exploring and manipulating networks. In Proceedings of the Third International AAAI Conference on Weblogs and Social Media, San Jose, CA, USA, 17–20 May 2009; Volume 8, pp. 361–362.

symmetry

MDPI

Article

A POCS Algorithm Based on Text Features for the Reconstruction of Document Images at Super-Resolution

Fengmei Liang, Yajun Xu *, Mengxia Zhang and Liyuan Zhang

College of Information Engineering, Taiyuan University of Technology, Taiyuan 030024, Shanxi, China; fm_liang@163.com (F.L.); mengxia_zhang@126.com (M.Z.); Lyuan_Zhang@yeah.net (L.Z.)
* Correspondence: yjun_xu@163.com; Tel.: +86-186-0341-0966

Academic Editor: Angel Garrido
Received: 27 June 2016; Accepted: 22 September 2016; Published: 29 September 2016

Abstract: In order to address the problem of the uncertainty of existing noise models and of the complexity and changeability of the edges and textures of low-resolution document images, this paper presents a projection onto convex sets (POCS) algorithm based on text features. The current method preserves the edge details and smooths the noise in text images by adding text features as constraints to original POCS algorithms and converting the fixed threshold to an adaptive one. In this paper, the optimized scale invariant feature transform (SIFT) algorithm was used for the registration of continuous frames, and finally the image was reconstructed under the improved POCS theoretical framework. Experimental results showed that the algorithm can significantly smooth the noise and eliminate noise caused by the shadows of the lines. The lines of the reconstructed text are smoother and the stroke contours of the reconstructed text are clearer, and this largely eliminates the text edge vibration to enhance the resolution of the document image text.

Keywords: super-resolution reconstruction; document image; scale invariant feature transform algorithm; text feature; projection onto convex sets algorithm

1. Introduction

In the early 1980s, Tsai and Huang were the first to use the Fourier-transform method for satellite image reconstruction [1]. Since then, super-resolution image reconstruction has been a hot topic in the field of image processing.

Recently, the use of super-resolution techniques has also drawn many researchers' attention. Domestic and foreign scholars have performed a great deal of research on image super-resolution reconstruction. For instance, Chen et al. used the iterative gradient algorithm combined with the bilateral total variation algorithm to estimate motion for the subpixel level of target images based on the Taylor expansion, and they achieved good results [2]. Kato et al. estimated the relative displacement of the observed image using subpixel block matching and reconstructed super-resolution images based on sparse representation [3]. Panda et al. used the iterative adaptive regularization method and genetic algorithm for image super-resolution [4]. However, these algorithms are not specific to document images, they do not take the characteristics of the document image into account, and the performance of the application on document images is poor. Scholars have also put forward some super-resolution algorithms for document images. For example, Fan et al. found that regions with highly similar characters can self-register and reconstruct themselves using local consistency [5]. Kumar et al., on the basis of sparse representation, pointed out that, although the shapes of the characters were not consistent, their edges and stroke curves were similar [6]. Finally, they found the ideal high-resolution (HR) image block by training the dictionary. Abed et al. proposed a total variation regularization method based on the directions of the text strokes, and it has achieved good results [7]. However, the algorithms mentioned above have several disadvantages. They are highly complex, require a single noise model, and cannot fully reconstruct the text edge information. For document images, the high-frequency details of the edges of the image are very important, and restoring the edge details is a preferred factor in document image reconstruction. The projection onto convex sets (POCS) algorithm is very inclusive of prior information, and it can restore the details of the text boundaries to a considerable extent, but it requires a large amount of computation and the solution is not unique. Consequently, a POCS algorithm is here proposed for super-resolution of text document images based on text features—which improves the original POCS algorithm over the degraded model and threshold—and optimizes the scale invariant feature transform (SIFT) algorithm at the registration stage to enhance the effect of document image reconstruction and ensure its efficiency.

2. Methodology

2.1. Classical Image Super-Resolution Reconstruction Algorithm

2.1.1. Establishment of Degraded Model

The HR image degradation process usually serves as the observation model, and the relationship between HR and low-resolution (LR) is established through the reduced quality model. Finally, the HR equation is generated. The image degradation model is as follows:

$$g_k = D_k B_k W_k f + n_k, \; k = 1, 2, 3 \ldots L \tag{1}$$

Among these, D_k is the geometric distortion matrix, B_k is the fuzzy matrix, W_k is the down-sampling matrix, k is the number of sequence image frames, and the range is from 1 to L. f and g_k refer to the collected HR original image and LR observation sequence after reducing mass. n_k is the additive Gauss noise.

Let $H_k = D_k B_k W_k$; Formula (1) can be written as follows:

$$g_k = H_k f + n_k \tag{2}$$

As shown in Formula (2), the image reconstruction process involves estimating the process of reverse f by regression model. In this paper, the process is divided into two steps: image registration and then super-resolution reconstruction using POCS algorithm.

2.1.2. Image Registration Based on SIFT Algorithm

Image registration involves estimating the location of each pixel using the tiny differences in information between different image sequences, thus providing the registration parameters.

The document image is distinguished from the words and the background, and the texture features, which must be extracted during the feature extraction stage, are very visible. However, image rotation, translation, and scaling deformation can affect text considerably, especially small text. It is therefore necessary to find an algorithm suitable for image deformation in the registration. The SIFT algorithm proposed by Lowe in 1999 can reduce the influence of deformation in the registration of the document image and extract more accurate features [8]. Accordingly, this paper uses the SIFT algorithm for image registration.

Steps in the implementation of the SIFT algorithm:

1. Construct a scale space
2. Detect extreme points in the scale space
3. Locate the extreme points accurately

4.   Determine the main direction of key points
5.   Find the descriptors of key points

Step 5 involves taking the key-point-centered 16×16 pixel neighborhood and then dividing it into 4×4 image sub-blocks. Each pixel defines vector information in 8 directions. Then we generate $4 \times 4 \times 8 = 128$-dimensional feature vectors. After generating the SIFT feature vectors of two images, the Euclidean distance of the feature vectors of key points is used to assess the similarity of the key points in the two images.

2.1.3. Image Reconstruction Based on the POCS Algorithm

As shown above, the image degradation model can be expressed as follows:

$$g_k = H_k f + n_k \tag{3}$$

Let (m_1, m_2) be the arbitrary point of LR image, and the corresponding point in HR image is (n_1, n_2). $g(m_1, m_2, l)$ is a pixel of the LR image of coordinates m_1, m_2, and frame l. $f(n_1, n_2)$ is a pixel of the HR image. $n(m_1, m_2, l)$ is the noise carried by the frame l of LR image, which is generally considered additive noise. The corresponding point of (m_1, m_2) in reference frame is (m'_1, m'_2). $H(n_1, n_2; m'_1, m'_2, l)$ is the point spread function of the observation image at point (m'_1, m'_2). Formula (3) can be expressed as follows:

$$g(m_1, m_2, l) = \sum_{n_1, n_2} H(n_1, n_2; m'_1, m'_2, l) f(n_1, n_2) + n(m_1, m_2, l) \tag{4}$$

According to the data consistency constraint requirements, HR images and any LR image should maintain the same pixel value at the corresponding points under ideal conditions. That is, each observed LR image sequence $g(m_1, m_2, l)$ can be represented by a convex set C [9], such as Formula (5):

$$C_{(m_1, m_2, l)} = \left\{ f(n_1, n_2, k) : \left| r^{(f)}(m_1, m_2, l) \right| \leq \delta_0(m_1, m_2, l) \right\} \tag{5}$$

Among (5), $C_{(m_1, m_2, l)}$ is the collection of pixels in the LR image, $f(n_1, n_2, k)$ is the current frame HR image estimate, $\delta_0(m_1, m_2, l)$ is the threshold, which is determined by noise standard deviation and credibility boundary. $r^{(f)}(m_1, m_2, l)$ is the residual of any point in a convex set.

$$r^{(f)}(m_1, m_2, l) = g(m_1, m_2, l) - \sum_{n_1, n_2} f(n_1, n_2, k) \cdot H(n_1, n_2; m'_1, m'_2, l) \tag{6}$$

Among (6), k is the frame number of the HR image. The operator that the pixel $f(n_1, n_2, k)$ mapped to convex set, with $P_{m_1, m_2, l} [f(n_1, n_2, k)]$.

The basic idea of POCS is to iterate the initial estimation $\hat{f}^{(0)}(n_1, n_2, k)$ of high-resolution images with POCS operator P and finally produce the high-resolution image [10]. Interpolation is usually used to perform the initial estimation $\hat{f}^{(0)}(n_1, n_2, k)$ of high-resolution images, and the constraint set of observation sequences is used to correct the $\hat{f}^{(0)}(n_1, n_2, k)$, until the iteration conditions are met. Assuming that there are t iterations, the basic expression used to reconstruct the high-resolution image f with the POCS algorithm is as follows:

$$\hat{f}^{(t+1)}(n_1, n_2, k) = P_m P_{m-1} \cdots P_1 \hat{f}^{(t)}(n_1, n_2, k) \tag{7}$$

2.2. A POCS Algorithm Based on Text Feature

2.2.1. Features of Document Images

Common document images are based on black and white images. Black represents text and white is background. The gray curve of image is distinguished from text and background and its envelope is a bimodal curve. The bimodal characteristics of document image are shown in Figure 1.

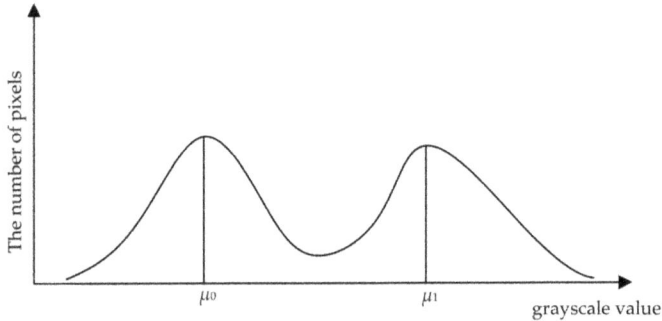

Figure 1. Bimodal distribution of document image.

As shown above, the abscissa represents gray value, the ordinate represents the number of corresponding pixels, and μ_0 and μ_1 represent the positions of the two peaks of the curve. In the ideal gray image, there should be only two gray values, μ_0 and μ_1. If $\mu_0 = 0$, this indicates that the pixel value of black text area is 0; $\mu_1 = 255$ means that the pixel value of the white area is 255. However, the image no longer shows the two value image in the actual document because of the noise. This should be done to reduce the number of pixels between μ_0 and μ_1 infinitely close to zero because they may include noise pixels that can blur the image. Bimodal envelope of document image can be expressed as follows:

$$V_{el}(H') = (H' - \mu_0)^2 \cdot (H' - \mu_1)^2 \qquad (8)$$

Among (8), $V_{el}(H')$ is the number of corresponding pixels and H' is the gray value. The closer the number of pixels near the bimodal comes to μ_0 or μ_1, the closer the envelope comes to the ideal situation, and the better the quality of the document image.

2.2.2. Text Features Based on the POCS Algorithm

This section describes the optimization of the traditional POCS algorithm from the following two aspects.

In POCS steps, building the convex set of low-resolution images is a key step in the introduction of high-frequency information. For document images, because its details are mostly words, the edges are more difficult to discern. The original algorithm is not ideal, especially not for some of the grayer document images. As a consequence, features of the document image were added to the original POCS algorithm in this paper as a priori constraint. These features were then used to repeatedly correct estimation until the results converged.

The original algorithm installs the initial threshold on the basis of noise $n(m_1, m_2, l)$ characteristics in the degradation model. However, in real life, it is difficult to determine the noise characteristics, so a fixed threshold can cause the algorithm to lack flexibility.

In this paper, the fixed threshold was changed to an adaptive threshold, which improved the inclusivity of the algorithm and the quality of the reconstructed image.

1. Formula (8) is added to the constraint conditions as an a priori function. The initial estimation is repeatedly revised based on it until the results meet the reconstruction conditions.
2. In the original POCS algorithm, the image reduction formula is as follows:

$$g(m_1, m_2, l) = \sum_{n_1, n_2} \hat{H}(n_1, n_2; m'_1, m'_2, l) f(n_1, n_2) + n(m_1, m_2, l) \tag{9}$$

There are uncertainty factors in the estimation of the point spread function (PSF) $H(n_1, n_2; m'_1, m'_2, l)$ of the image that collected in reality at point (m'_1, m'_2). In this paper, the error was taken into account. Set $H(n_1, n_2; m'_1, m'_2, l)$ was considered an accurate estimation, $\Delta H(n_1, n_2; m'_1, m'_2, l)$ was considered an error estimation, so the actual estimation $\hat{H}(n_1, n_2; m'_1, m'_2, l)$ was as follows:

$$\hat{H}(n_1, n_2; m'_1, m'_2, l) = H(n_1, n_2; m'_1, m'_2, l) + \Delta H(n_1, n_2; m'_1, m'_2, l) \tag{10}$$

Formula (9) can be written as follows:

$$\begin{aligned} g(m_1, m_2, l) &= \sum_{n_1, n_2} f(n_1, n_2) \hat{H}(n_1, n_2; m'_1, m'_2, l) + n(m_1, m_2, l) \\ &= \sum_{n_1, n_2} f(n_1, n_2) [(H(n_1, n_2; m'_1, m'_2, l) + \Delta H(n_1, n_2; m'_1, m'_2, l)] + n(m_1, m_2, l) \\ &= \sum_{n_1, n_2} f(n_1, n_2) (H + \Delta H)(n_1, n_2; m'_1, m'_2, l) + n(m_1, m_2, l) \end{aligned} \tag{11}$$

This can be deformed into the following:

$$\begin{aligned} & g(m_1, m_2, l) - \sum_{n_1, n_2} f(n_1, n_2) H(n_1, n_2; m'_1, m'_2, l) \\ &= \sum_{n_1, n_2} f(n_1, n_2) \Delta H(n_1, n_2; m'_1, m'_2, l) + n(m_1, m_2, l) \end{aligned} \tag{12}$$

The left side of the formula is the residual $r^{(f)}(m_1, m_2, l)$. At this time, the residual value is determined using the original noise $n(m_1, m_2, l)$ and the added uncertainty factors $f(n_1, n_2) \Delta H(n_1, n_2; m'_1, m'_2, l)$. The threshold value is then determined using the mixed noise, which contains added error factors.

After introducing the mixed noise, the new residual was calculated automatically while correcting the projection. Then, the standard variance is counted as the modified threshold. The threshold does not need to be set in advance at the start of the algorithm. It can be given automatically based on the new residual during revision of the LR image.

2.2.3. Algorithm Implementation

When reconstructing the super-resolution (SR) image in this paper, an LR image of the observed sequences must first be set as a reference frame and interpolated, and a motion estimation must be made between the LR image and the reference frame to produce the image offset. Then, a low-resolution image is projected using the improved degradation model, and the modified residual and the modified threshold value are then calculated. The reference frame is then corrected based on the threshold. Finally, the current high-resolution estimation was iteratively modified according to the prior constraint condition of the document image until the acceptable reconstruction results were achieved. The specific flow chart is shown in Figure 2:

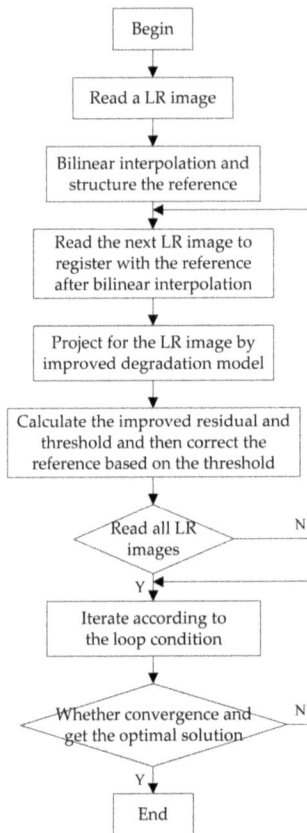

Figure 2. Algorithm flow chart.

As shown in Figure 2, we can see that the implementation of the POCS algorithm can be divided into five steps:

1. Read an LR image and make a bilinear interpolation of it. The interpolation multiplier is a multiple of the desired improved resolution, then this interpolated image was selected as a reference frame. Let the initial estimate be f_0 and the threshold value be δ_0.

2. Read the remaining LR frames to register with the reference frame after bilinear interpolation, and estimate the motion between each frame and the reference frame to produce the registration mapping parameters.

3. Define the convex set C of the sequences and calculate the residual r values to correct the reference frame.

4. According to the data consistency constraint, the operator P was calculated, and the relationship between r and δ_0 was assessed to correct f.

5. The cycle end condition was set to $\frac{\| f^{(t+1)} - f^{(t)} \|}{f^{(t)}} \leq \varepsilon$. If the conditions are met, then $\hat{f}^{(t)} = f^{(t)}$, and the loop ends. Otherwise, let $t = t + 1$, and return to the step 3 until $f^{(t+1)}$ meets the convergence conditions. Then $\hat{f}^{(t)}$ is the eligible solution of the algorithm.

3. Analysis of Experimental Results

In this paper, under the support of a 64 bit Windows 8 operating system and a 2nd generation Intel Core i5-3210M processor, 2.50 GHz CPU, 4 GB RAM, 500 GB HDD, MATLAB R2013b software is used to perform the experiment. In the experiment, there are three sets of LR sequences, one generated by simulation and the others taken with a camera. The reconstruction of image sequences is evaluated using the peak signal to noise ratio (PSNR), the real images using the mean opinion score (MOS), the execution times, and the memory occupations.

3.1. Registration with the Optimized SIFT Operator

Here, text features were added to the document image during the preprocessing stage of registration to improve the matching accuracy, and the feature descriptors were reduced to the fifth step of the original algorithm SIFT, reducing the dimensions from 128 to 48 to improve the matching speed. The results of registration and analysis are as follows.

Figure 3 is the reference image. Figure 4 is the image to be registered. Figure 5 shows the matching results of the original SIFT algorithm, and Figure 6 is the result of the improved SIFT algorithm.

山外青山楼外楼

西湖歌舞几时休

暖风熏得游人醉

直把杭州作汴州

Figure 3. Reference image.

山外青山楼外楼

西湖歌舞几时休

暖风熏得游人醉

直把杭州作汴州

Figure 4. Image to be registered.

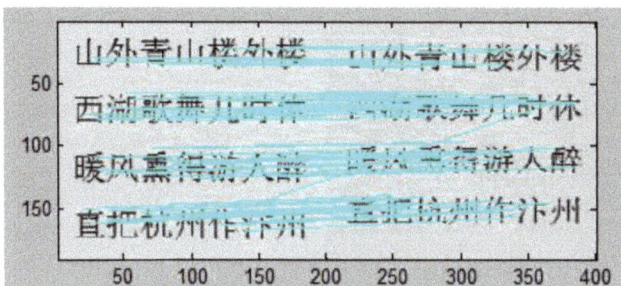

Figure 5. Effective feature point matching results of the original scale invariant feature transform (SIFT) algorithm.

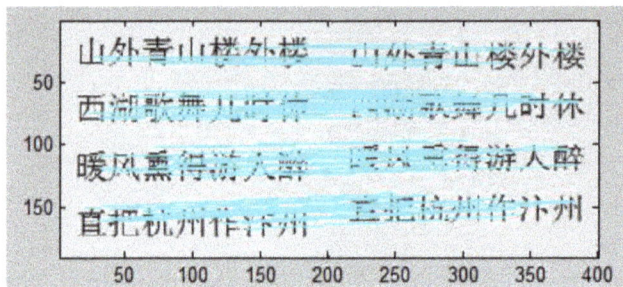

Figure 6. Effective feature point matching results of the improved SIFT algorithm.

The results in Table 1 show that there were fewer matched and mismatched points associated with the improved algorithm, which indicates that the algorithm can quickly produce the best registration parameters with high precision. At the same time, the memory space is released due to the reductions in dimensions.

Table 1. Analysis of the results.

Algorithm Types	Characteristic Points		Matching Points	Matching Points	Time Spent on Matching (s)	The Correct Matching Rate (%)
	n_1	n_2				
Original algorithm	220	227	184	7	1.795	96
Improved algorithm	220	227	180	1	1.763	99

3.2. Reconstructed Results and Analysis

3.2.1. The Reconstructed Result of an LR Image Which Is Generated by Simulation

The original HR image is an image of calligraphy copybook 490 × 391 in size, shown in Figure 7a. The LR image sequences are obtained by down-sampling, panning, blurring, and adding noise to the original HR image, shown in Figure 7b. In these images, down-sampling was performed with a rank sampling of 1/2, and the length of the panning involved selecting parameters randomly between [0, 1]. The size of the blur function window was 3 × 3, and the type of noise added was Gauss noise, of which the standard deviation was 1.

The reconstruction of the original POCS is shown in Figure 7d, and the edges of the words appear ghost in Figure 7d. We compared the performance of the proposed algorithm with the approach in [11], the method in [12], and the improved POCS in this paper. The results are presented in Figure 7e–g, respectively. By comparison, the original POCS algorithm adds high-frequency image details and at the same time introduces a lot of noise. The text edges of the reconstructed document image can gain fuzzy ghosts caused by this noise due to the fixed threshold, which is set by the original algorithm to mix in more noise, so the image quality is affected. There are also some noticeable ringing artifacts in the reconstruction image showed in Figure 7e. The result in Figure 7f shows that the approach in [12] can retain lots of details, but the reconstruction was still unsatisfactory. It can be seen from the Figure 7g that the overall sharpness of the image has been recovered and a significant amount of details has also been restored. Comparison reveals that our approach is superior, as it is able to recover more visual clarity from the LR images, particularly near the textual region.

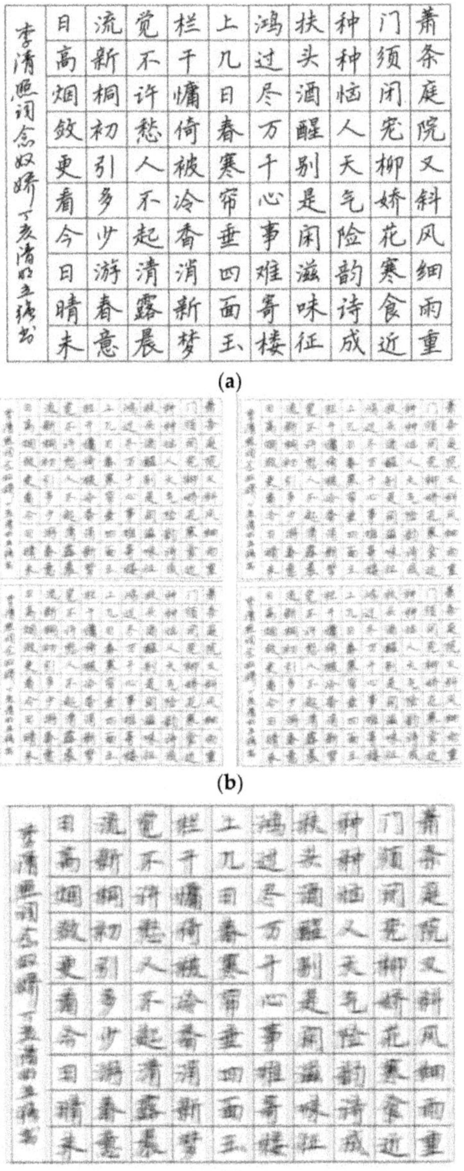

(a) (b) (c)

Figure 7. *Cont.*

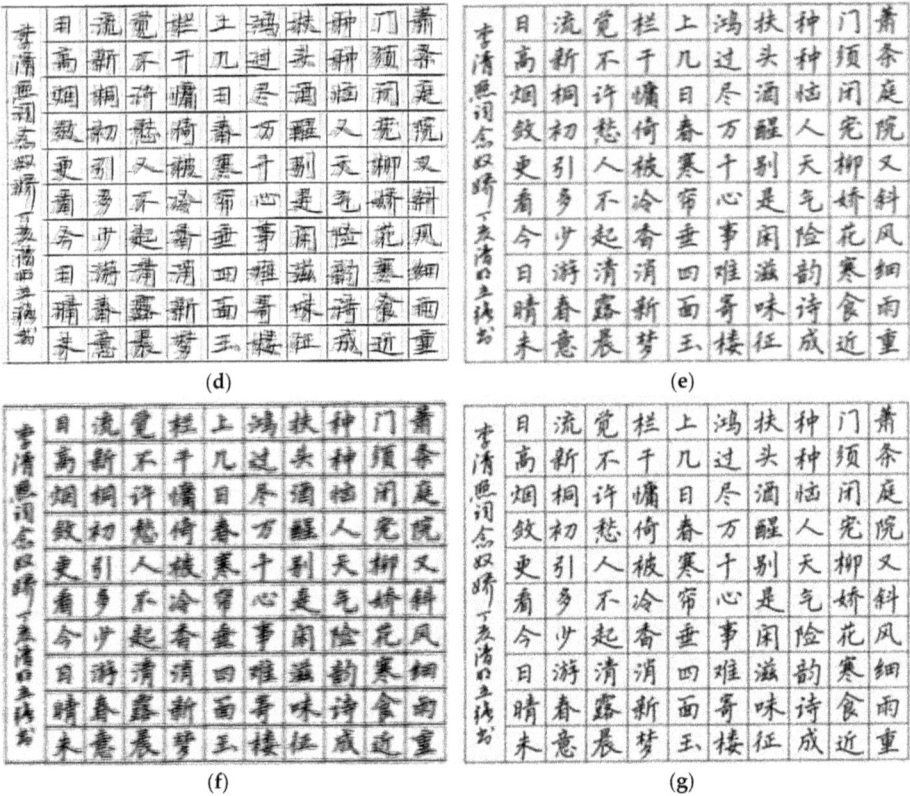

(d)

(e)

(f)

(g)

Figure 7. (**a**) Original high-resolution (HR) image; (**b**) low-resolution (LR) image sequences; (**c**) reference image after registration using the improved SIFT; (**d**) reconstruction of the original projection onto convex sets (POCS); (**e**) reconstruction using the method in [11]; (**f**) reconstruction using the method in [12]; (**g**) reconstruction of the improved POCS.

3.2.2. Reconstruction of Ancient Books and Ancient Inscriptions Taken by Camera

First, the experiment uses a set of LR image sequences of ancient books, four document images with low quality where the words on the images are difficult to identify; the size of the images is 476×356, as shown in Figure 8a. Each image shown in Figure 8a was produced using bilinear interpolation, and an interpolated image was selected to register with other LR images one-by-one to produce the reference frame using the improved SIFT algorithm, as shown in Figure 8b. The result of the reconstruction using the original POCS is shown in Figure 8c. As shown in Figure 8c, the area surrounding the document text in the reconstructed image has an obvious procrastination phenomenon, which can be attributed to the incorporation of noise into the reconstruction process. Next, the methods in [11,12] were performed to obtain HR images. It is observed that the obtained results were unsatisfactory as shown in the Figure 8d,e. The reconstructed HR image using the improved POCS method is given in Figure 8f. The Figure 8f demonstrates that the improved POCS algorithm can rebuild the text clearly and the results are visibly improved. This is because our proposed method utilizes the improved SIFT, hence resulting in better registration. The effectiveness of the registration in turn leads to the satisfactory image quality.

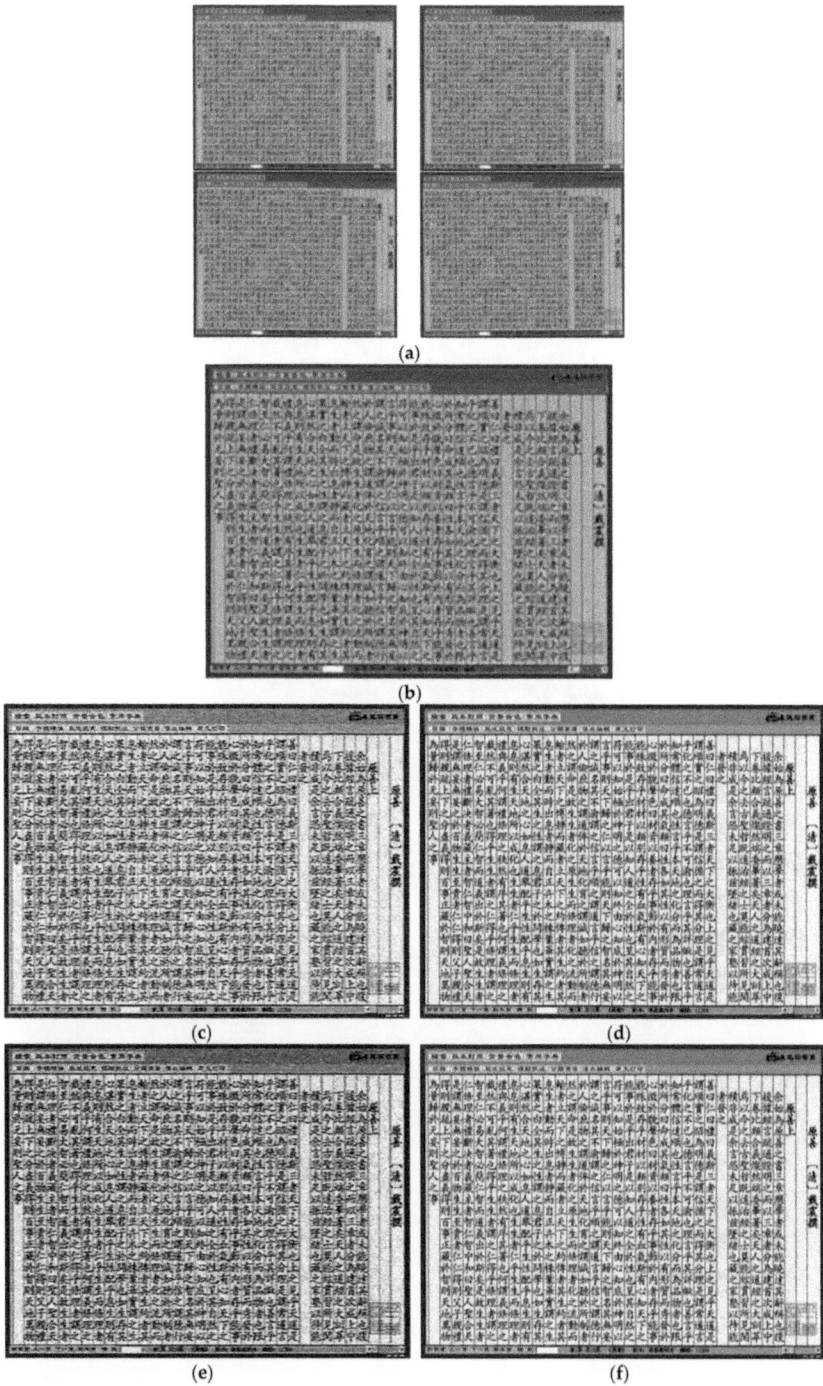

Figure 8. (a) LR image sequences of ancient books; (b) reference image after registration using the improved SIFT; (c) reconstruction of the original POCS; (d) reconstruction using the method in [11]; (e) reconstruction using the method in [12]; (f) reconstruction of the improved POCS.

A set of LR image sequences of three ancient inscriptions was used here. The images are 449×729 in size, as shown in Figure 9a. Because the words were inscribed many years ago, many are illegible in these inscriptions and the defect of the image quality is obvious. Figure 9b shows the reference image. As shown in Figure 9c, the reconstructed image of the inscription as processed using the original algorithm has apparent noise and a ringing effect in edge of the text. The result of implementing the method described in [11] is shown in Figure 9d. The super-resolution method which is proposed in [12] resulted in Figure 9e. Figure 9f shows the implementation of the improved method described in this paper. Comparing Figure 9f to Figure 9c–e, we notice not only has our method removed the outliers more efficiently than other methods, but it has also resulted in sharper edges without any ringing effects.

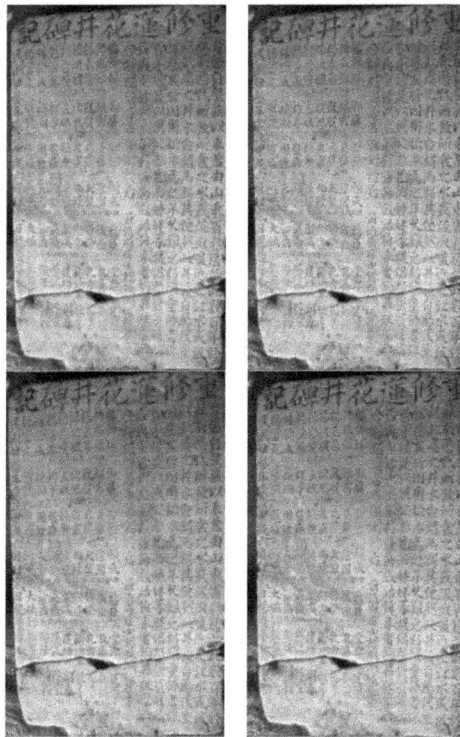

(a)

Figure 9. *Cont.*

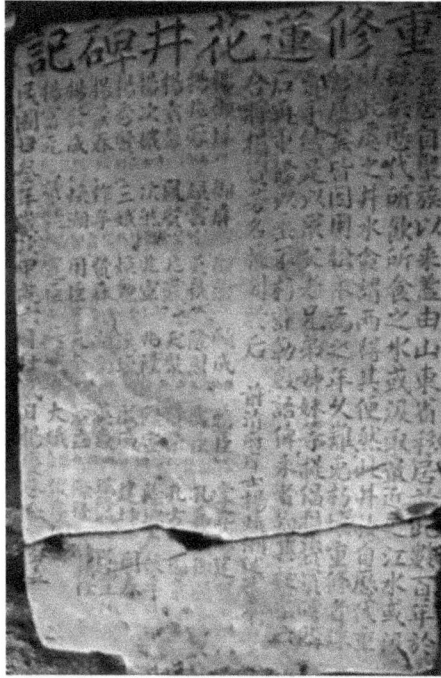

(b)

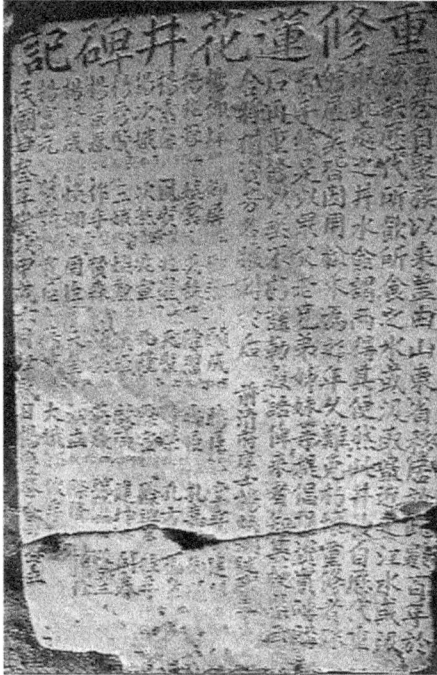

(c)

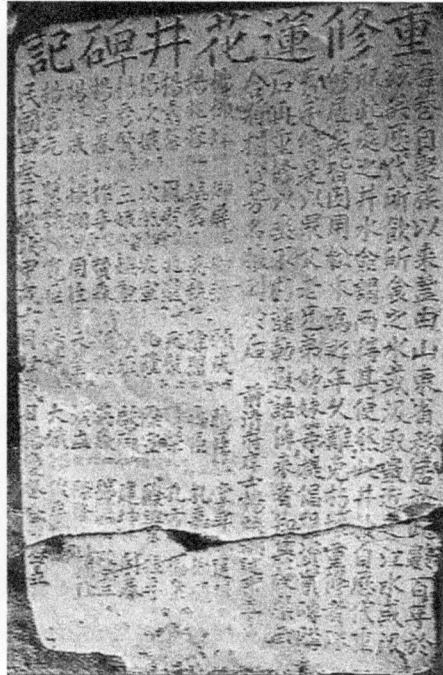

(d)

Figure 9. *Cont.*

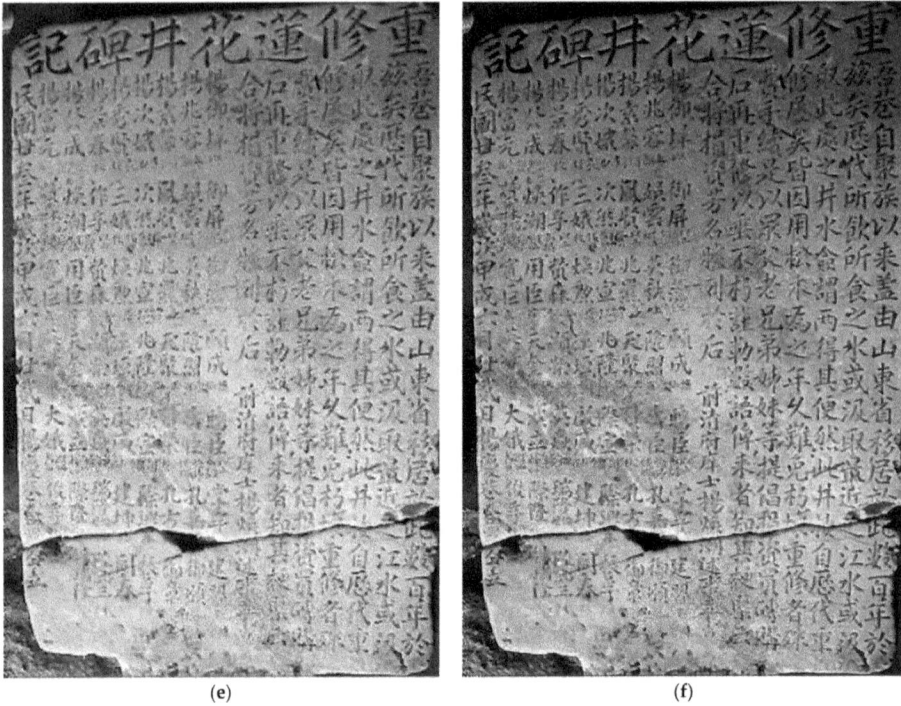

Figure 9. (a) LR image sequences of ancient inscriptions; (b) reference image after registration using the improved SIFT; (c) reconstruction of the original POCS; (d) reconstruction using the method in [11]; (e) reconstruction using the method in [12]; (f) reconstruction of the improved POCS.

3.2.3. Reconstruction Quality Evaluation

In the experiment, there were three sets of low-resolution sequences, the first group was generated by simulation and the others were taken with a camera. To ensure fairness in comparison, the simulation environments are given as follows: Windows 8, MATLAB R2013b, 2.50 GHz CPU, and 4 GB RAM. We used the peak signal to noise ratio (PSNR) and the mean opinion score (MOS) to compare the performance of image sequences of each of these methods. The value of MOS ranges from [0, 5], in which larger value of MOS indicates better image quality and smaller ones indicate poorer quality. The various visual feelings generated by 10 observers for the same image are referenced here. Then a reasonable evaluation score was determined using to the clarity of the image and whether there is any blurring or noise. The MOS value, which is the final result of the assessment, was then calculated. Also, we provide an analysis on the computational complexity of the methods using the execution times and the memory occupations.

(a) Evaluation of the first group

As shown in Table 2, the PSNR and MOS values of the improved POCS are higher than the previous three algorithms, which show the reconstruction of this improved algorithm to be more similar to the original HR image. The evaluation results show that the improved approach gains a better performance both on reconstruction quality and computational time.

Table 2. Quantitative comparison of the performance of the copybook experiments.

Algorithm Types	PSNR/dB	MOS	The Execution Time/s	The Memory Occupation/KB
Original POCS	23.29	3.302	7.319	2464.00
the method in [11]	23.57	3.534	7.434	2934.00
the method in [12]	24.31	3.708	7.327	2585.00
Improved POCS	24.68	4.149	7.323	2324.00

(b) Evaluation of the second group

In order to illustrate the experimental performance of the ancient books using the improved POCS approach, Table 3 shows the quantitative comparison of the four methods. It is clear that the resulting image (Figure 8f) has a better quality than other reconstruction methods.

Table 3. Quantitative comparison of the performance of the ancient books experiments.

Algorithm Types	PSNR/dB	MOS	The Execution Time/s	The Memory Occupation/KB
Original POCS	36.97	3.452	10.371	3173.00
the method in [11]	37.34	3.598	11.453	3528.00
the method in [12]	38.16	3.974	11.185	3259.00
Improved POCS	38.23	4.079	10.548	3194.00

Table 4 shows that the improved POCS method has highest values of PSNR and MOS. The execution time and memory occupation are a little higher than other algorithms, but this is acceptable with respect to its image quality. This means that the algorithm in this paper reduces noise more effectively than other algorithms, and the edges of the reconstructed image are much smoother. These objective performance measures illustrate the improved POCS method has the best visual quality and further reconfirm our subjective evaluation of the reconstructed image.

Table 4. Quantitative comparison of the performance of the ancient inscriptions experiments.

Algorithm Types	PSNR/dB	MOS	The Execution Time/s	The Memory Occupation/KB
Original POCS	37.98	3.548	11.079	3279.00
the method in [11]	38.03	3.746	11.855	3371.00
the method in [12]	38.74	4.032	11.273	3284.00
Improved POCS	38.82	4.214	11.143	3281.00

4. Conclusions

An improved super-resolution algorithm for document images based on the optimization of the classical SIFT algorithm and traditional POCS algorithm is proposed in this paper. By enhancing the contrast and reducing the dimensions, the improved SIFT algorithm improved the efficiency in the registration section. The text feature information was added as a prior condition to correct the estimated value, and an uncertain factor was added to the degenerate formula to calculate residue to produce the threshold automatically in the POCS model.

The improved super-resolution scheme provides better results on a wide class of document images. The results of the simulation showed that the method can be used to suppress a great deal of noise pollution, and properly preserve the edge details with good reconstruction effects. As a result, its recognition rate is higher and the operation time is shorter than in the original algorithm.

There are still many factors that the algorithm does not cover. They provide direction for future work. First, this algorithm is applicable to the document image under simple background. If the background is a mixture image of text and natural scenery, the reconstruction will be of poorer quality. It may be possible to combine some image segmentation technology in future works to render the reconstruction more specific. Second, in the reconstruction phase, all of the initial extracted images are grayscale ones, and this article does not involve color images. The next work will include experiments addressing the reconstruction of color document images to render the algorithm more efficient and more widely applicable.

Acknowledgments: This work was supported by the Project of Natural Science Foundation of Shanxi Province (No. 2013011017-3).

Author Contributions: Fengmei Liang and Yajun Xu conceived and designed the experiments; Yajun Xu performed the experiments; Fengmei Liang and Liyuan Zhang analyzed the data; Mengxia Zhang contributed analysis tools; Fengmei Liang wrote the paper.

Conflicts of Interest: The authors declare no conflict of interest. The founding sponsors had no role in the design of the study; in the collection, analyses, or interpretation of data; in the writing of the manuscript, and in the decision to publish the results.

References

1. Tsai, R.Y.; Huang, T.S. Multi frame image restoration and registration. *Adv. Comput. Vis. Image Process.* **1984**, *1*, 101–106.
2. Chen, Y.; Jin, W.; Wang, L.X.; Liu, C. Robust Multiframe Super-resolution reconstruction based on regularization. *Comput. Symp.* **2010**, *10*, 408–413.
3. Kato, T.; Hino, H.; Murata, N. Multi-frame image super resolution based on sparse coding. *Neural Netw.* **2015**, *66*, 64–78. [CrossRef] [PubMed]
4. Panda, S.S.; Jena, G.; Sahu, S.K. Image Super Resolution Reconstruction Using Iterative Adaptive Regularization Method and Genetic Algorithm. *Indian J. Med. Res.* **2015**, *60*, 19–27.
5. Fan, W.; Sun, J.; Minagawa, A.; Hotta, Y. Local consistency constrained adaptive neighbor embedding for text image super-resolution. In Proceedings of the 10th IAPR International Workshop on Document Analysis System (DAS), Gold Coast, Australia, 27–29 March 2012; pp. 90–94.
6. Kumar, V.; Bansal, A.; Tulsiyan, G.H.; Mishra, A.; Namboodiri, A. Sparse document image coding for restoration. In Proceedings of the 12th International Conference on Document Analysis and Recognition (ICDAR), Washington, DC, USA, 25–28 August 2013; pp. 713–717.
7. Abedi, A.; Kabir, E. Stroke width-based directional total variation regularisation for document image super resolution. *IET Image Process.* **2016**, *10*, 158–166. [CrossRef]
8. Lowe, D.G. Distinctive Image Features from Scale-Invariant Keypoints. *Int. J. Comput. Vis.* **2004**, *60*, 91–110. [CrossRef]
9. Stark, H.; Oskoui, P. High Resolution Image Recovery from Image-plane Arrays Using Convex Projections. *Opt. Soc. Am. J.* **1989**, *6*, 1715–1726. [CrossRef]
10. Xu, H.; Miao, H.; Yang, C.; Xiong, C. Research on super-resolution based on an improved POCS algorithm. In Proceedings of the SPIE—The International Society for Optical Engineering, International Conference on Optical and Photonic Engineering (icOPEN 2015), Singapore, Singapore, 14–16 April 2015; Volume 9524.
11. Al-Anizy, M.H. Super Resolution Image from Low Resolution of Sequenced Frames—Text Image and Image-Based on POCS. *J. Al-Nahrain Univ.* **2012**, *15*, 138–141.
12. He, Y.; Yap, K.H.; Chen, L.; Chau, L.P. A soft map framework for blind super-resolution image reconstruction. *Image Vis. Comput.* **2009**, *27*, 364–373. [CrossRef]

symmetry

MDPI

Article

Uncertain Quality Function Deployment Using a Hybrid Group Decision Making Model

Ze-Ling Wang [1], **Jian-Xin You** [2] **and Hu-Chen Liu** [1,2,*]

1   School of Management, Shanghai University, Shanghai 200444, China; happyking66@126.com
2   School of Economics and Management, Tongji University, Shanghai 200092, China; yjx2256@vip.sina.com
*   Correspondence: huchenliu@shu.edu.cn; Tel.: +86-21-6613-3703; Fax: +86-21-6613-7931

Academic Editor: Angel Garrido
Received: 15 September 2016; Accepted: 28 October 2016; Published: 4 November 2016

Abstract: Quality function deployment (QFD) is a widely used quality system tool for translating customer requirements (CRs) into the engineering design requirements (DRs) of products or services. The conventional QFD analysis, however, has been criticized as having some limitations such as in the assessment of relationships between CRs and DRs, the determination of CR weights and the prioritization of DRs. This paper aims to develop a new hybrid group decision-making model based on hesitant 2-tuple linguistic term sets and an extended QUALIFLEX (qualitative flexible multiple criteria method) approach for handling QFD problems with incomplete weight information. First, hesitant linguistic term sets are combined with interval 2-tuple linguistic variables to express various uncertainties in the assessment information of QFD team members. Borrowing the idea of grey relational analysis (GRA), a multiple objective optimization model is constructed to determine the relative weights of CRs. Then, an extended QUALIFLEX approach with an inclusion comparison method is suggested to determine the ranking of the DRs identified in QFD. Finally, an analysis of a market segment selection problem is conducted to demonstrate and validate the proposed QFD approach.

Keywords: quality function deployment; hesitant 2-tuples; QUALIFLEX approach; multiple criteria decision-making

---

1. Introduction

Quality function deployment (QFD) is a systematic, cross-functional team-based product planning technique used to ensure that customer requirements (CRs) are deployed throughout the research and development (R&D), engineering, and manufacturing stages of products [1]. The QFD methodology originated in the late 1960s and early 1970s in Japan, and first used at the Kobe Shipyards of Mitsubishi Heavy Industries [2]. Later, QFD has become a standard practice in most leading companies such as General Motors, Ford, Xerox, IBM, Procter & Gamble, and Hewlett-Packard [3]. QFD is a customer-driven methodology that supports engineers to efficiently translate CRs into relevant design requirements (DRs) or engineering characteristics (ECs), and subsequently into parts characteristics, process plans and production requirements in the new product development. Generally, the implementation of QFD in an organization can improve engineering knowledge, productivity and quality, and reduce product costs, development cycle time and engineering changes [4]. Due to its visibility and easiness, the QFD method has been successfully introduced in diverse industries as a quality tool to achieve higher product performance and customer satisfaction [5–7].

In the application of QFD, a matrix configuration called house of quality (HoQ) is of fundamental and strategic importance, which is used to translate CRs (WHATs) into appropriate DRs (HOWs) graphically [8]. Generally, constructing the HoQ comprises determining the weights of CRs, the relationship matrix between WHATs and HOWs, the interrelationship matrix among HOWs and the importance of DRs. Besides, QFD is a group decision behavior and often involves a group of cross-functional team members from marketing, design, quality, production, as well as a group of customers [4]. The customers of a product or service in a targeted market are first selected for determining the importance weights of CRs. Then a QFD team is established for assessing the relationships between CRs and DRs and the interrelationships between DRs. The prioritization of DRs is a critical output of the QFD planning process, which guides engineering designers in resource allocation, decision-making and the subsequent QFD analysis [9].

In the real-life world, there are many QFD problems with imperfect, vague and imprecise information due to the existence of conflicting goals, time press, lack of knowledge, etc. It is common for QFD team members to use linguistic terms to express their judgments [4,10]. Moreover, because of uncertainty and incompleteness in the early stage of new product development, a single linguistic term may not suitable or adequate for QFD team members to give their assessments in constructing the HoQ. The experts of a QFD team may prefer to use multiple linguistic terms for expressing their judgment information sufficiently. Therefore, QFD problems could use the linguistic modeling and computational methods in its solving process. Hesitant fuzzy linguistic term sets (HFLTSs) [11] were proposed and used to deal with the situations in which decision makers may hesitate among several possible linguistic values or think of richer expressions for assessing an alternative. Compared with other fuzzy linguistic approaches, HFLTSs are more convenient and flexible to manage the hesitancy and uncertainty of decision makers in practical applications [12]. Recently, HFLTSs have attracted more concerns of researchers and have been widely applied to many fields [13–15]. Additionally, to calculate linguistic information without loss of information, the 2-tuple linguistic representation model was introduced by Herrera and Martínez [16]. A well-known extension of the 2-tuple linguistic model is the interval 2-tuple linguistic model [17], which uses uncertain linguistic variables called interval 2-tuples for computing with words. Due to its characteristics and capabilities, numerous studies have reported decision-making models and methods within the interval 2-tuple linguistic environment [18–20].

On the other hand, prioritizing DRs in QFD can be viewed as a complex multiple criteria decision-making (MCDM) problem and MCDM methods have been found to be a useful tool to solve this kind of problem [21]. The QUALIFLEX (qualitative flexible multiple criteria method), a variation of Jacquet-Lagreze's permutation method, is a very useful outranking method proposed by Paelinck [22] for MCDM in view of its simple logic, full utilization of information contained in the decision analysis, and easy computational procedure [23]. The methodology of QUALIFLEX is based on a metric procedure that evaluates all possible permutations of the considered alternatives and identifies the optimal permutation that exhibits the greatest comprehensive concordance/discordance index [24]. In recent years, the QUALIFLEX has been successfully applied to address different MCDM problems. For example, Chen et al. [25] developed an extended QUALIFLEX method for dealing with MCDM problems in the context of interval type-2 fuzzy sets and applied it for medical decision-making. Chen [26] presented an interval-valued intuitionistic fuzzy QUALIFLEX method with a likelihood-based comparison approach for multiple criteria decision analysis within the environment of interval-valued intuitionistic fuzzy sets. Based on the type-2 fuzzy framework, Wang et al. [23] developed a likelihood-based QUALIFLEX method for multiple criteria group decision-making within the interval type-2 fuzzy decision setting. Zhang and Xu [24] proposed a hesitant fuzzy QUALIFLEX with a signed distance-based comparison method for solving MCDM problems in which the assessments of alternatives and the weights of criteria are both expressed by hesitant fuzzy elements.

Based on the above discussions, this paper attempts to propose an extended QUALIFLEX approach based on hesitant 2-tuple linguistic term sets to capture imprecise or uncertain assessment information in constructing the HoQ and to enhance the analysis capability of the traditional QFD. Although a considerable number of MCDM methods have been adopted by previous QFD models for the prioritization of DRs, to the best of our knowledge, no study has been conducted on applying the QUALIFLEX approach in QFD problems. Therefore, we extend the QUALIFLEX algorithm within the hesitant 2-tuple linguistic environment to determine the priorities of DRs. In addition, the existing approaches proposed to improve QFD analysis only consider the situations where the information about CR weights is completely known; no or little attention has been paid to the QFD problems in which the CR weight information is incompletely known. In response, we propose a linear programming model to determine the weights of CRs for the QFD problem with incomplete weight information. Finally, the computational procedure of the new QFD model is illustrated by an illustrate example concerning market segment evaluation and selection.

The rest of this paper is organized as follows. Section 2 briefly reviews the related work of QFD improvements. In Section 3, we introduce some basic concepts concerning HFLTSs and interval 2-tuples. It is followed by a description of the proposed QFD framework using hesitant 2-tuple linguistic term sets and a modified QUALIFLEX approach. Next, a case study is presented in Section 5 to illustrate the proposed QFD methodology, and a comparative analysis with other relevant QFD methods is also provided in this section. Finally, this article is concluded with discussion of key findings and future research suggestions in Section 6.

2. Literature Review

Over the last two decades, a number of improvements have been developed to eliminate the restrictions and enhance the performance of the traditional QFD. Critical reviews have summarized the concepts and decision methods employed in the QFD process, see, for example, [5,27,28]. In the sequel, we briefly review the existing QFD methodologies from the perspectives of CR weighting, vague assessments, and DR ranking in the HoQ construction.

Determining correct importance weights of CRs is essential in QFD since they significantly affect the target values set for DRs. Therefore, Armacost et al. [29] first integrated analytic hierarchy process (AHP) with QFD to establish a framework for prioritizing CRs and applied it to the manufacture of industrialized housing. Kwong and Bai [1] employed a fuzzy AHP with an extent analysis approach to determine the importance weights for the CRs in QFD. Aye Ho et al. [3] proposed an integrated group decision-making approach to QFD, which first modifies the nominal group technique to obtain CRs and then integrates the agreed and individual criteria approaches to assign customers' importance levels. Lam and Lai [30] proposed an analytical network process (ANP)-QFD model for designing environmental sustainability, which makes use of the ANP to determine the importance degrees of CRs and DRs and to incorporate the inter-dependence between CRs and DRs in the HoQ. Liu et al. [31] developed a fuzzy non-linear regression model using the minimum fuzziness criterion to identify the degree of compensation among CRs in QFD. Ji et al. [32] developed a novel approach that integrates Kano model quantitatively into QFD to optimize product design to maximize customer satisfaction under cost and technical constraints.

In addition, the QFD team members or consumers participating the construction of HoQ may have vague assessments and cannot provide their opinions with exact numerical values. Therefore, to effectively capture inevitable vagueness and uncertainty in the QFD planning process, Chin et al. [4] presented an evidential reasoning (ER)-based methodology for synthesizing various types of assessment information provided by customers and QFD team members. Chan and Wu [8] suggested a systematic approach to QFD on the basis of symmetrical triangular fuzzy numbers (STFNs) to capture the vagueness in linguistic assessments from both customers and technicians. Zhang and Chu [10] proposed a fuzzy group decision-making approach incorporating with two optimization models (i.e., logarithmic least squares model and weighted least squares model) to aggregate the multi-format and multi-granularity linguistic judgments of decision makers for constructing the HoQ. Yan and Ma [33] proposed a two-stage group decision-making approach to tackle with human subjective perception and customer heterogeneity underlying QFD, in which the order-based semantics of linguistic information is used to derive the fuzzy preference relations of different DRs with respect to each customer and the fuzzy majority is used to synthesize all customers' individual fuzzy preference relations to determine the prioritization of DRs.

As another key issue of QFD, the prioritization of DRs have been extensively researched and various methods have been suggested in the QFD literature. For example, Luo et al. [34] proposed a new QFD-based product planning approach to determine the optimal target levels of DRs for a product market with heterogeneous CRs by integrating consumer choice behavior analysis. Zhong et al. [35] constructed a fuzzy chance-constrained programming model to determine the target values of DRs in QFD and designed a hybrid intelligent algorithm by integrating fuzzy simulation and genetic algorithm to solve the proposed model. Jia et al. [9] presented a method for quantifying the importance degree of DRs with a multi-level hierarchical structure in QFD, in which fuzzy ER algorithm is adopted to deal with the fuzziness and incompleteness during the evaluation process and fuzzy discrete Choquet integral is used to characterize the interactions among DRs during the aggregation process. Hosseini Motlagh et al. [36] provided a fuzzy preference ranking organization method for enrichment evaluation (PROMETHEE) approach to rank DRs for the QFD process in multi-criteria product design. Song et al. [21] proposed a group decision approach based on rough set theory and grey relational analysis (GRA) approach for prioritizing DRs in QFD under vague environment.

3. Preliminaries

3.1. Hesitant Fuzzy Linguistic Term Sets

As an extended form of HFSs [37–39], the concept of hesitant fuzzy linguistic term sets (HFLTSs) was introduced by Rodriguez et al. [11] to deal with the linguistic decision-making situations where decision makers hesitate to give appropriate linguistic terms as the assessment expressions. In the following, some basic definitions related to HFLTSs are given [11,40].

Definition 1. *Let $S = \{s_0, s_1, ..., s_g\}$ be a fixed set of linguistic term set. An HFLTS associated with S, H_S is an ordered finite subset of the consecutive linguistic terms of S. The empty and full HFLTSs for a linguistic variable ϑ are defined as $H_S(\vartheta) = \varnothing$ and $H_S(\vartheta) = S$, respectively.*

Definition 2. *Let $S = \{s_0, s_1, ..., s_g\}$ be a linguistic term set, a context-free grammar is a 4-tuple $G_H = (V_N, V_T, I, P)$, where V_N indicates a set of nonterminal symbols, V_T is a set of terminal symbols, I is the starting symbol, and P denotes the production rules. The elements of G_H are defined as follows:*

$$V_N = \{\langle\text{primary term}\rangle, \langle\text{composite term}\rangle, \langle\text{unary relation}\rangle, \langle\text{binary relation}\rangle, \langle\text{conjunction}\rangle\};$$

$$V_T = \{\text{lower than, greater than, at least, at most, between, and}, s_0, s_1, ..., s_g\};$$

$$I \in V_N;$$

$$
\begin{aligned}
P = \{ & I ::= \langle\text{primary term}\rangle \,|\, \langle\text{composite term}\rangle \\
& \langle\text{composite term}\rangle ::= \langle\text{unary relation}\rangle \langle\text{primary term}\rangle \,|\, \langle\text{binary relation}\rangle \\
& \qquad\qquad\qquad\qquad \langle\text{primary term}\rangle \langle\text{conjunction}\rangle \langle\text{primary term}\rangle \\
& \langle\text{primary term}\rangle ::= s_0 \,|\, s_1 \,|\, ... \,|\, s_g \\
& \langle\text{unary relation}\rangle ::= \text{lower than} \,|\, \text{greater than} \,|\, \text{at least} \,|\, \text{at most} \\
& \langle\text{binary relation}\rangle ::= \text{between} \\
& \langle\text{conjunction}\rangle ::= \text{and}\}.
\end{aligned}
$$

Definition 3. *Let E_{GH} be a function that transforms the comparative linguistic expressions obtained by the context-free grammar G_H into an HFLTS H_S of the linguistic term set S. The linguistic expressions generated by G_H using the production rules can be converted into HFLTSs according to the following ways:*

a. E_{GH} (lower than s_i) = $\{s_k | s_k \in S \text{ and } s_k < s_i\}$;
b. E_{GH} (greater than s_i) = $\{s_k | s_k \in S \text{ and } s_k > s_i\}$;
c. E_{GH} (at least s_i) = $\{s_k | s_k \in S \text{ and } s_k \geq s_i\}$;
d. E_{GH} (at most s_i) = $\{s_k | s_k \in S \text{ and } s_k \leq s_i\}$;
e. E_{GH} (between s_i and s_j) = $\{s_k | s_k \in S \text{ and } s_i \leq s_k \leq s_j\}$.

Definition 4. *Let $S = \{s_0, s_1, ..., s_g\}$ be a linguistic term set. The envelope of an HFLTS, denoted by env(H_S), is a linguistic interval whose limits are determined by its upper bound H_S^+ and lower bound H_S^-, shown as follows:*

$$\text{env}(H_S) = [\underline{H_S}, \overline{H_S}], \qquad \underline{H_S} \leq \overline{H_S}, \tag{1}$$

where

$$\overline{H_S} = \max(s_i) = s_j, s_i \leq s_j \text{ and } s_i \in H_S, \forall i, \tag{2}$$

$$\underline{H_S} = \min(s_i) = s_j, s_i \geq s_j \text{ and } s_i \in H_S, \forall i. \tag{3}$$

3.2. Interval 2-Tuple Linguistic Model

The 2-tuple linguistic model was developed by Herrera and Martínez [16] to avoid information loss and to increase the accuracy in carrying out linguistic computing [20,41]. A generalized 2-tuple linguistic model, the interval 2-tuple linguistic model was initiated by Zhang [17] for better expressing uncertain linguistic assessments. It can be defined as follows.

Definition 5. *Let $S = \{s_0, s_1, ..., s_g\}$ be a linguistic term set, an interval 2-tuple linguistic variable is composed of two 2-tuples, denoted by $[(s_k, \alpha_1), (s_l, \alpha_2)]$, where $(s_k, \alpha_1) \leq (s_l, \alpha_2)$, $s_k (s_l)$ and $\alpha_1 (\alpha_2)$ represent the linguistic label of S and symbolic translation, respectively. An interval 2-tuple can be transformed into an interval value $[\beta^L, \beta^U] (\beta^L, \beta^U \in [0,1], \beta^L \leq \beta^U)$ by the following function [17]:*

$$\Delta[\beta^L, \beta^U] = [(s_k, \alpha_1), (s_l, \alpha_2)] \quad \text{with} \quad \begin{cases} s_k, & k = \text{round}(\beta^L \cdot g) \\ s_l, & l = \text{round}(\beta^U \cdot g) \\ \alpha_1 = \beta^L - \frac{k}{g}, & \alpha_1 \in \left[-\frac{1}{2g}, \frac{1}{2g}\right) \\ \alpha_2 = \beta^U - \frac{l}{g}, & \alpha_2 \in \left[-\frac{1}{2g}, \frac{1}{2g}\right). \end{cases} \tag{4}$$

where round $(\cdot)$ is the usual round operation. On the contrary, there exists an inverse function Δ^{-1} that can convert an interval 2-tuple $[(s_k, \alpha_1), (s_l, \alpha_2)]$ into an interval value $[\beta^L, \beta^U] (\beta^L, \beta^U \in [0,1], \beta^L \leq \beta^U)$ as follows:

$$\Delta^{-1}[(s_k, \alpha_1), (s_l, \alpha_2)] = \left[\frac{k}{g} + \alpha_1, \frac{l}{g} + \alpha_2\right] = [\beta^L, \beta^U]. \tag{5}$$

Specifically, the interval 2-tuple linguistic variable reduces to a 2-tuple linguistic variable when $s_k = s_l$ and $\alpha_1 = \alpha_2$. Besides, motivated by [16], a linguistic interval $[s_k, s_l]$ can be converted into an interval 2-tuple $[(s_k, 0), (s_l, 0)]$ by adding a value 0 as symbolic translation.

Definition 6. *Consider any three interval 2-tuples $\tilde{a} = [(r, \alpha), (t, \varepsilon)]$, $\tilde{a}_1 = [(r_1, \alpha_1), (t_1, \varepsilon_1)]$ and $\tilde{a}_2 = [(r_2, \alpha_2), (t_2, \varepsilon_2)]$, and let $\lambda \in [0,1]$, their operational laws are expressed as follows [42,43]:*

a $\tilde{a}_1 \otimes \tilde{a}_2 = [(r_1, \alpha_1), (t_1, \varepsilon_1)] \otimes [(r_2, \alpha_2), (t_2, \varepsilon_2)] = \Delta[\Delta^{-1}(r_1, \alpha_1) \cdot \Delta^{-1}(r_2, \alpha_2), \Delta^{-1}(t_1, \varepsilon_1) \cdot \Delta^{-1}(t_2, \varepsilon_2)]$;
b $\tilde{a}_1 \oplus \tilde{a}_2 = [(r_1, \alpha_1), (t_1, \varepsilon_1)] \oplus [(r_2, \alpha_2), (t_2, \varepsilon_2)] = \Delta[\Delta^{-1}(r_1, \alpha_1) + \Delta^{-1}(r_2, \alpha_2), \Delta^{-1}(t_1, \varepsilon_1) + \Delta^{-1}(t_2, \varepsilon_2)]$;
c $\tilde{a}^\lambda = ([(r, \alpha), (t, \varepsilon)])^\lambda = \Delta[(\Delta^{-1}(r, \alpha))^\lambda, (\Delta^{-1}(t, \varepsilon))^\lambda]$;
d $\lambda \tilde{a} = \lambda[(r, \alpha), (t, \varepsilon)] = \Delta[\lambda \Delta^{-1}(r, \alpha), \lambda \Delta^{-1}(t, \varepsilon)]$.

Definition 7. Let $\tilde{a}_i = [(r_i, \alpha_i), (t_i, \varepsilon_i)]$ $(i = 1, 2, ..., n)$ be a set of interval 2-tuples and $\omega = (\omega_1, \omega_2, ..., \omega_n)^T$ be an associated weight vector, with $\omega_j \in [0, 1]$, $\sum_{j=1}^{n} \omega_j = 1$. The interval 2-tuple ordered weighted average (ITOWA) operator is defined as [17]:

$$\text{ITOWA}_\omega (\tilde{a}_1, \tilde{a}_2, ..., \tilde{a}_n) = \overset{n}{\underset{j=1}{\oplus}} \left(\omega_j \tilde{a}_{\sigma(j)} \right)$$
$$= \Delta \left[\sum_{j=1}^{n} \omega_j \Delta^{-1} \left(r_{\sigma(j)}, \alpha_{\sigma(j)} \right), \sum_{j=1}^{n} \omega_j \Delta^{-1} \left(t_{\sigma(j)}, \varepsilon_{\sigma(j)} \right) \right], \tag{6}$$

where $(\sigma(1), \sigma(2), ..., \sigma(n))$ is a permutation of $(1, 2, ..., n)$, such that $\tilde{a}_{\sigma(j-1)} \geq \tilde{a}_{\sigma(j)}$ for all $j = 2, ..., n$.

Definition 8. Let $\tilde{a}_1 = [(r_1, \alpha_1), (t_1, \varepsilon_1)]$ and $\tilde{a}_2 = [(r_2, \alpha_2), (t_2, \varepsilon_2)]$ be any two interval 2-tuples defined on the linguistic term set $S = \{s_0, s_1, ..., s_g\}$, then the inclusion comparison possibility of $\tilde{a}_1$ and $\tilde{a}_2$ is defined as follows [19]:

$$p(\tilde{a}_1 \supseteq \tilde{a}_2) = \max \left\{ 1 - \max \left(\frac{\delta_2 - \beta_1}{h(\tilde{a}_1) + h(\tilde{a}_2)}, 0 \right), 0 \right\}, \tag{7}$$

where $h(\tilde{a}_1) = \Delta^{-1}(t_1, \varepsilon_1) - \Delta^{-1}(r_1, \alpha_1) = \delta_1 - \beta_1$, $h(\tilde{a}_2) = \Delta^{-1}(t_2, \varepsilon_2) - \Delta^{-1}(r_2, \alpha_2) = \delta_2 - \beta_2$ and $p(\tilde{a}_1 \supseteq \tilde{a}_2)$ is the degree to which $\tilde{a}_1$ is not smaller than $\tilde{a}_2$. The inclusion comparison possibility, $p(\tilde{a}_1 \supseteq \tilde{a}_2)$, satisfies the properties that $0 \leq p(\tilde{a}_1 \supseteq \tilde{a}_2) \leq 1$; $p(\tilde{a}_1 \supseteq \tilde{a}_2) = 0$ if $(t_1, \varepsilon_1) \leq (r_2, \alpha_2)$; $p(\tilde{a}_1 \supseteq \tilde{a}_2) = 1$ if $(r_1, \alpha_1) \geq (t_2, \varepsilon_2)$; $p(\tilde{a}_1 \supseteq \tilde{a}_2) + p(\tilde{a}_2 \supseteq \tilde{a}_1) = 1$, and $p(\tilde{a}_1 \supseteq \tilde{a}_2) = p(\tilde{a}_2 \supseteq \tilde{a}_1) = 0.5$ if $p(\tilde{a}_1 \supseteq \tilde{a}_2) = p(\tilde{a}_2 \supseteq \tilde{a}_1)$.

Based on the inclusion comparison possibility, a comparison between interval 2-tuple linguistic arguments can be obtained. For a set of interval 2-tuples $\tilde{a}_i = [(r_i, \alpha_i), (t_i, \varepsilon_i)]$ $(i = 1, 2, ..., n)$, we first compute the inclusion comparison possibilities of the pairwise interval 2-tuples using Equation (7), and let $p_{ij} = p(\tilde{a}_i \supseteq \tilde{a}_j)$ $(i, j = 1, 2, ..., n)$, we can construct the inclusion comparison matrix $P = [p_{ij}]_{n \times n}$. Then, the optimal degrees of membership for the interval 2-tuples $\tilde{a}_i$ $(i = 1, 2, ..., n)$ are determined by

$$\bar{p}(\tilde{a}_i) = \frac{1}{n(n-1)} \left(\sum_{j=1}^{n} p_{ij} + \frac{n}{2} - 1 \right). \tag{8}$$

As a result, the ranking order of all the interval 2-tuples can be produced in terms of the descending order of the $\bar{p}(\tilde{a}_i)$ values.

Definition 9. Let $\tilde{a}_1 = [(r_1, \alpha_1), (t_1, \varepsilon_1)]$ and $\tilde{a}_2 = [(r_2, \alpha_2), (t_2, \varepsilon_2)]$ be two interval 2-tuples, then

$$D_H(\tilde{a}_1, \tilde{a}_2) = \Delta \left[\frac{1}{2} \left(\left| \Delta^{-1}(r_1, \alpha_1) - \Delta^{-1}(r_2, \alpha_2) \right| + \left| \Delta^{-1}(t_1, \varepsilon_1) - \Delta^{-1}(t_2, \varepsilon_2) \right| \right) \right] \tag{9}$$

is called the Hamming distance between $\tilde{a}_1$ and $\tilde{a}_2$ [42],

$$D_E(\tilde{a}_1, \tilde{a}_2) = \Delta \sqrt{\frac{1}{2} \left((\Delta^{-1}(r_1, \alpha_1) - \Delta^{-1}(r_2, \alpha_2))^2 + (\Delta^{-1}(t_1, \varepsilon_1) - \Delta^{-1}(t_2, \varepsilon_2))^2 \right)} \tag{10}$$

is called the Euclidean distance between $\tilde{a}_1$ and $\tilde{a}_2$ [43].

4. QFD Using Hesitant 2-Tuples and QUALIFLEX Method

In this section, we propose a hybrid analytical model combining hesitant 2-tuple linguistic term sets and an extended QUALIFLEX approach for handling QFD problems with incomplete weight information. The flowchart of the proposed QFD algorithm is depicted in Figure 1. In short, the proposed QFD approach is composed of three key stages: assessing relationships between CRs and

DRs, determining importance weights of CRs, and determining the ranking order of DRs. In the first stage, the relationships between WHATs and HOWs are rated by integrating HFLTSs and interval 2-tuples to form the QFD problem within the hesitant 2-tuple linguistic environment. Then, an optimization model is established based on the GRA method, by which the importance degrees of CRs can be determined. Finally, a modified QUALIFLEX approach is developed for the determination of priority of each DR. In the following subsections, the procedure of the proposed group analytical approach for QFD is described in further detail.

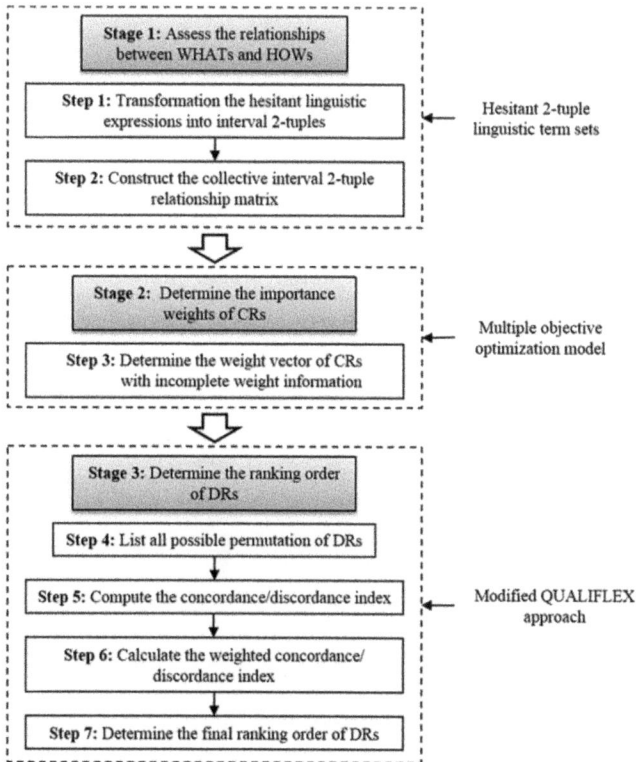

Figure 1. Flowchart of the proposed QFD (Quality function deployment) model.

4.1. Assess the Relationships between WHATs and HOWs

Assume that there are K team members TM_k ($k = 1, 2, ..., K$) in a QFD expert group responsible for the assessment of relationships between a set of WHATs CR_i ($i = 1, 2, ..., m$) and a set of HOWs DR_j ($j = 1, 2, ..., n$) using the linguistic term set $S = \{s_0, s_1, ..., s_g\}$. In our proposal, the QFD team members give their judgments on the relationships between CRs and DRs by means of the context-free grammar approach. Let $H_k = \left[h_{ij}^k \right]_{m \times n}$ be the linguistic assessment matrix of the kth team member, where h_{ij}^k indicates the hesitant linguistic expression provided by TM_k over the relationship between CR_i and DR_j. Based on these assumptions and notations, the steps of dealing with the uncertain CR-DR relationship assessments are presented as follows:

Step 1: Transformation the hesitant linguistic expressions into interval 2-tuples

To homogenize all the judgments for the relationships between WHATs and HOWs, hesitant linguistic assessment matrices are provided by QFD team members by using linguistic expressions based on the context-free grammar G_H. After converting into corresponding HFLTSs according to the transformation function E_{GH}, every QFD team member's hesitant linguistic expressions h_{ij}^k can be transformed into linguistic intervals $env\left(h_{ij}^k\right) = \left[\underline{h_{ij}^k}, \overline{h_{ij}^k}\right]$ by calculating the envelope of each HFLTS (as in Definition 4). Then, the linguistic intervals are represented using the interval 2-tuple linguistic approach and translated into $\left[\left(r_{ij}^k, 0\right), \left(t_{ij}^k, 0\right)\right]$. As a result, the hesitant linguistic assessment information of QFD team members can be expressed by interval 2-tuple assessments as follows:

$$\tilde{R}_k = \begin{bmatrix} \tilde{r}_{11}^k & \tilde{r}_{12}^k & \cdots & \tilde{r}_{1n}^k \\ \tilde{r}_{21}^k & \tilde{r}_{22}^k & \cdots & \tilde{r}_{2n}^k \\ \vdots & \vdots & \cdots & \vdots \\ \tilde{r}_{m1}^k & \tilde{r}_{m2}^k & \cdots & \tilde{r}_{mn}^k \end{bmatrix}, \quad k = 1, 2, ..., K. \tag{11}$$

Step 2: Construct the collective interval 2-tuple relationship matrix

In this step, to relieve the influence of unfair arguments on the QFD results, the ITOWA operator is utilized to aggregate the QFD team members' subjective assessments. That is, the collective interval 2-tuple assessments $\tilde{r}_{ij} = \left[\left(r_{ij}, \alpha_{ij}\right), \left(t_{ij}, \varepsilon_{ij}\right)\right]$ for $i = 1, 2, ..., m$ and $j = 1, 2, ..., n$ are computed as follows:

$$\begin{aligned} \tilde{r}_{ij} &= ITOWA_\omega\left(\tilde{r}_{ij}^1, \tilde{r}_{ij}^2, ..., \tilde{r}_{ij}^K\right) = \overset{K}{\underset{k=1}{\oplus}}\left(\omega_k \tilde{r}_{ij}^{\sigma(k)}\right) \\ &= \Delta\left[\sum_{k=1}^{K} \omega_k \Delta^{-1}\left(r_{ij}^{\sigma(k)}, \alpha_{ij}^{\sigma(k)}\right), \sum_{k=1}^{K} \omega_j \Delta^{-1}\left(t_{ij}^{\sigma(k)}, \varepsilon_{ij}^{\sigma(k)}\right)\right], \end{aligned} \tag{12}$$

where $\tilde{r}_{ij}^{\sigma(k)}$ is the kth largest of the interval 2-tuples $\tilde{r}_{ij}^k$ $(k = 1, 2, ..., L)$, $\omega = (\omega_1, \omega_2, ..., \omega_L)^T$ is the ITOWA weight vector with $\omega_k \in [0,1]$ and $\sum_{k=1}^{K} \omega_k = 1$, which can be obtained via the argument-dependent approach developed by Wu et al. [44].

As a result, a collective interval 2-tuple relationship matrix $\tilde{R}$ can be produced based on the individual assessments of multiple QFD team members.

$$\tilde{R} = \begin{bmatrix} \tilde{r}_{11} & \tilde{r}_{12} & \cdots & \tilde{r}_{1n} \\ \tilde{r}_{21} & \tilde{r}_{22} & \cdots & \tilde{r}_{2n} \\ \vdots & \vdots & \cdots & \vdots \\ \tilde{r}_{m1} & \tilde{r}_{m2} & \cdots & \tilde{r}_{mn} \end{bmatrix}. \tag{13}$$

Note that the HOWs (DRs) are assumed to be independent in the above computations, which may not the case in some circumstances. Thus, for the QFD problems characterized by interdependent HOWs, the relationship matrix $\tilde{R}$ between WHATs and HOWs is adjusted by [33]:

$$\tilde{R}' = \tilde{R} \otimes \tilde{C}, \tag{14}$$

where $\tilde{C} = \left[\tilde{c}_{ij}\right]_{m \times n}$ is the correlation matrix of HOWs. Therefore, the correlations among HOWs can be incorporated into our QFD model with the adjusted collective interval 2-tuple relationship matrix $\tilde{R}'$.

4.2. Determine the Importance Weights of CRs

In practical QFD circumstances, the information concerning relative importance of CRs is usually incompletely known due to time pressure, lack of knowledge or customer's limited expertise. The management of incomplete information has been studied by many researchers [45,46], and lots of methods have been developed for the determination of criteria weights with incomplete information, such as those based on technique for order preference by similarity to an ideal solution (TOPSIS) [19], distance measure [47] and entropy method [48]. In the QFD literature, however, little research has been conducted to estimate the weights of CRs when the weight information is incompletely known. The GRA, proposed by Deng [49], is a kind of method for solving MCDM problems, which aims at choosing the alternative with the highest grey relational grade to the reference sequence. Therefore, in this part, we establish a multiple objective optimization model based on the GRA to determine the relative importance of CRs with partly known weight information.

Let $w = (w_1, w_2, ..., w_m)^T$ be the weight vector of CRs collected from the customer representatives in a targeted market, where $w_i \in [0,1]$ and $\sum_{i=1}^m w_i = 1$, the known weight information on CRs can be usually constructed using the following basic ranking forms [26,50], for $i \neq l$:

a.   A weak ranking: $H_1 = \{w_i \geq w_l\}$;
b.   A strict ranking: $H_2 = \{w_i - w_l \geq \gamma_l\}\,(\gamma_l > 0)$;
c.   A ranking of differences: $H_3 = \{w_i - w_l \geq w_p - w_q\}\,(l \neq p \neq q)$;
d.   A ranking with multiples: $H_4 = \{w_i \geq \gamma_l w_l\}\,(0 \leq \gamma_l \leq 1)$;
e.   An interval form: $H_5 = \{\gamma_i \leq w_i \leq \gamma_i + \varepsilon_i\}\,(0 \leq \gamma_i \leq \gamma_i + \varepsilon_i \leq 1)$.

Let H denote the set of the known weight information of CRs given by a group of customers and $H = H_1 \cup H_2 \cup H_3 \cup H_4 \cup H_5$, the specific steps to compute the CR weights are given below:

Step 3: Determine the weight vector of CRs

According to the basic principle of the GRA method, the most critical DR for customer satisfaction should have the "greatest relation grade" to the reference sequence (the vector of ideal relevance value of each DR with respect to CRs). Under the interval 2-tuple linguistic environment, the reference sequence denoted as $\tilde{r}^* = (\tilde{r}_1^*, \tilde{r}_2^*, ..., \tilde{r}_m^*)^T$ can be defined as follows [43]:

$$\tilde{r}_i^* = [(r_i^*, \alpha_i^*), (t_i^*, \varepsilon_i^*)] = \Delta\,[1.0, 1.0], \quad i = 1, 2, ..., m. \tag{15}$$

For each CR of the associated DRs in QFD, the grey relation coefficient between $\tilde{r}_{ij}$ and $\tilde{r}_i^*$, i.e., $\xi\,(\tilde{r}_{ij}, \tilde{r}_i^*)$, is calculated using the following equation:

$$\xi\,(\tilde{r}_{ij}, \tilde{r}_i^*) = \frac{\delta_{min} + \zeta\delta_{max}}{\delta_{ij} + \zeta\delta_{max}}, \quad i = 1, 2, ..., m, j = 1, 2, ..., n, \tag{16}$$

where $\delta_{ij} = D\,(\tilde{r}_{ij}, \tilde{r}_i^*)$, $\delta_{min} = \min\{\delta_{ij}\}$, $\delta_{max} = \max\{\delta_{ij}\}$ for $i = 1, 2, ..., m; j = 1, 2, ..., n$, ζ is the distinguishing coefficient, $\zeta \in [0,1]$. Normally, the value of ζ is taken as 0.5 since it offers moderate distinguishing effects and good stability. Then the grey relational grade $\xi\,(\tilde{r}_j, \tilde{r}^*)$ between the reference sequence $\tilde{r}^*$ and the comparative sequences $\tilde{r}_j$ corresponding to DR_j can be acquired by

$$\xi\,(\tilde{r}_j, \tilde{r}^*) = \sum_{i=1}^m w_i \xi\,(\tilde{r}_{ij}, \tilde{r}_i^*), \quad j = 1, 2, ..., n. \tag{17}$$

In general, for the given weight vector of CRs, the larger $\xi\,(\tilde{r}_j, \tilde{r}^*)$, the more important the DR_j will be. Thus, a reasonable weight vector of CRs should be determined so as to make all the grey relational grades $\xi\,(\tilde{r}_j, \tilde{r}^*)\,(j = 1, 2, ..., n)$ as larger as possible, which means to maximize the grey relational grade vector $\Gamma\,(w) = (\xi\,(\tilde{r}_1, \tilde{r}^*), \xi\,(\tilde{r}_2, \tilde{r}^*), ..., \xi\,(\tilde{r}_n, \tilde{r}^*))$ under the condition $w \in H$. As a result, we can reasonably form the following multiple objective optimization model:

$$\max\Gamma\,(w) = (\xi\,(\tilde{r}_1, \tilde{r}^*), \xi\,(\tilde{r}_2, \tilde{r}^*), ..., \xi\,(\tilde{r}_n, \tilde{r}^*))$$

$$(M-1) \quad \text{s.t.} \begin{cases} w \in H, \\ \sum_{i=1}^n w_i = 1, w_i \geq 0, i = 1, 2, ..., m. \end{cases}$$

Several approaches have been proposed to solve linear programming problems with multiple objectives. In this paper, the max-min operator [26] is applied to integrate all the grey relational grades $\xi\left(\tilde{r}_j, \tilde{r}^*\right)$ $(j = 1, 2, ..., n)$ into a single objective optimization model:

$$
\max \lambda
$$

$$
(M-2) \quad \text{s.t.} \quad \begin{cases} \xi\left(\tilde{r}_j, \tilde{r}^*\right) \geq \lambda_1, (j = 1, 2, ..., n), \\ w \in H, \sum\limits_{i=1}^{m} w_i = 1, w_i \geq 0, i = 1, 2, ..., m. \end{cases}
$$

By solving Model (M−2), its optimal solution $w^* = \left(w_1^*, w_2^*, ..., w_m^*\right)^T$ can be used as the weight vector of CRs.

4.3. Determine the Ranking Order of DRs

To prioritize DRs, in this subsection we develop a hesitant 2-tuple linguistic QUALIFLEX (HTL-QUALIFLEX) approach with the inclusion comparison method. Since the relationship assessments between WHATs and HOWs are transformed into interval 2-tuples in the first stage, this study utilizes the comparison approach of interval 2-tuples based on the inclusion comparison possibility to recognize the corresponding concordance/discordance index. The best priority order of DRs is generated based on the level of concordance and the most critical DRs can be identified for subsequent QFD analysis. Next, the algorithm of the HTL-QUALIFLEX approach for the ranking of DRs is summarized.

Step 4: List all possible permutation of DRs

Given the set of identified design requirements, i.e., DR_j $(j = 1, 2, ..., n)$, and assume that there exist $n!$ permutations of the ranking of the DRs. Let P_ρ denote the ρth permutation as:

$$
P_\rho = \left(..., \text{DR}_\chi, ..., \text{DR}_\eta, ...\right), \quad \rho = 1, 2, ..., n!, \tag{18}
$$

where DR_ξ and DR_ζ, $\xi, \zeta = 1, 2, ..., n$, are the DRs listed in QFD and DR_ξ is ranked higher than or equal to DR_ζ.

Step 5: Compute the concordance/discordance index

The concordance/discordance index $\phi_i^\rho\left(\text{DR}_\xi, \text{DR}_\zeta\right)$ for each pair of design requirements $\left(\text{DR}_\xi, \text{DR}_\zeta\right)$ at the level of preorder with respect to the ith customer requirement and the ranking corresponding to the ρth permutation is defined as follows:

$$
\phi_i^\rho\left(\text{DR}_\xi, \text{DR}_\zeta\right) = \bar{p}\left(\tilde{r}_{\xi i}\right) - \bar{p}\left(\tilde{r}_{\zeta i}\right), \quad i = 1, 2, ..., m. \tag{19}
$$

Based on the inclusion comparison possibility comparison method of interval 2-tuples, there are concordance, ex aequo and discordance if $\bar{p}\left(\tilde{r}_{\xi i}\right) - \bar{p}\left(\tilde{r}_{\zeta i}\right) > 0$, $\bar{p}\left(\tilde{r}_{\xi i}\right) - \bar{p}\left(\tilde{r}_{\zeta i}\right) = 0$, and $\bar{p}\left(\tilde{r}_{\xi i}\right) - \bar{p}\left(\tilde{r}_{\zeta i}\right) < 0$, respectively.

Step 6: Calculate the weighted concordance/discordance index

By incorporating the weights of CRs $w = \left(w_1, w_2, ..., w_m\right)^T$ derived via Model (M−2), we can calculate the weighted concordance/discordance index $\phi^\rho\left(\text{DR}_\xi, \text{DR}_\zeta\right)$ for each pair of design requirements $\left(\text{DR}_\xi, \text{DR}_\zeta\right)$ at the level of preorder with respect to the m CRs and the ranking corresponding to the permutation P_ρ is determined by

$$
\phi^\rho\left(\text{DR}_\xi, \text{DR}_\zeta\right) = \sum\limits_{i=1}^{m} \phi_i^\rho\left(\text{DR}_\xi, \text{DR}_\zeta\right) w_i. \tag{20}
$$

Step 7: Determine the final ranking order of DRs

Finally, the comprehensive concordance/discordance index ϕ^ρ for the ρth permutation is computed as follows:

$$\phi^\rho = \sum_{\zeta,\zeta=1,2,...,n} \sum_{i=1}^{m} \phi_i^\rho \left(DR_\zeta, DR_\zeta\right) w_i. \tag{21}$$

It is easily seen that the bigger the comprehensive concordance/discordance index value, the better the ranking order of the DRs. Therefore, the final ranking result of DRs should be the permutation with the greatest comprehensive concordance/discordance index ϕ^ρ, i.e., $P^* = \max\limits_{\rho=1,2,...,n!} \{\phi^\rho\}$.

5. Illustrative Example

In this section, we provide a numerical example to illustrate the applicability and implementation process of the proposed QFD approach. This case study involves a QFD analysis for market segment evaluation and selection [51].

5.1. Implementation

Market segment selection is an important marketing activity of a company in the highly competitive market. It can be regarded as a complex decision-making problem because many potential criteria and decision makers must be involved during the selection procedure and the outcomes of any choice are uncertain. QFD provides an effective framework for market segment evaluation and selection due to the multi-dimensional characteristics of market segments. Thuan Yen JSC is a trading service and transportation company located in northern Vietnam, which has more than 50 different sizes of trucks. This company has built a customer network in both domestic and international markets with ten years' experience in providing trading and transportation services. To further expand the company's business in the domestic and international markets, managers of this company have to select the most suitable segment to maximize its profit. Thus, the proposed QFD approach is applied to the first part of the entire market segment selection procedure for this company, i.e., determining the company's business strengths (HOWs) based on market segment features (WHATs).

First, an expert team including five company decision makers, TM_k ($k = 1, 2, ..., 5$), is set up to carry out the QFD analysis. Based on a survey of related literature and interviews with the company's top managers and head of departments, the market segment features (CRs) are determined as segment growth rate (CR_1), expected profit (CR_2), competitive intensity (CR_3), capital required (CR_4), and level of technology utilization (CR_5), and the company business strengths (DRs) are selected as relative cost position (DR_1), delivery reliability (DR_2), technological position (DR_3), and management strength and depth (DR_4). Each member of the QFD team analyzes the match between the market segment features and the company's business strengths (WHATs–HOWs), and judges the relationships between them by means of grammar-free expressions over a seven-point linguistic term set S:

$$S = \left\{ \begin{array}{l} s_0 = \text{Very Low (VL)}, s_1 = \text{Low (L)}, s_2 = \text{Medium Low (ML)}, s_3 = \text{Medium (M)}, \\ s_4 = \text{Medium High (MH)}, s_5 = \text{High (H)}, s_6 = \text{Very High (VH)} \end{array} \right\}$$

Table 1 shows the linguistic relationship assessments of the four DRs with respect to each CR provided by the five QFD team members.

Table 1. Linguistic assessments on relationships between CRs and DRs by the QFD team.

WHATs (CRs)	Team Members	HOWs (DRs)			
		DR$_1$	DR$_2$	DR$_3$	DR$_4$
CR$_1$	TM$_1$	Greater than MH	ML	Between L and M	M
	TM$_2$	H	M	M	At least ML
	TM$_3$	Between H and VH	M	ML	M
	TM$_4$	H	MH	Less than M	M
	TM$_5$	H	At most MH	M	Between ML and MH
CR$_2$	TM$_1$	At least H	Greater than MH	Between MH and VH	H
	TM$_2$	VH	H	H	Greater than H
	TM$_3$	Greater than H	VH	VH	Between MH and VH
	TM$_4$	H	At least H	Greater than MH	H
	TM$_5$	Between MH and VH	H	H	At least H
CR$_3$	TM$_1$	Greater than M	H	At least H	Between MH and VH
	TM$_2$	MH	Between MH and VH	VH	H
	TM$_3$	H	H	Greater than H	H
	TM$_4$	At least H	VH	VH	At most H
	TM$_5$	VH	Greater than MH	VH	VH
CR$_4$	TM$_1$	Less than H	M	Greater than MH	MH
	TM$_2$	H	Between ML and MH	H	H
	TM$_3$	MH	M	H	Less than H
	TM$_4$	M	At most MH	MH	At most MH
	TM$_5$	At most H	Less than MH	Between MH and VH	M
CR$_5$	TM$_1$	Between MH and VH	MH	At most H	Between L and ML
	TM$_2$	H	H	MH	ML
	TM$_3$	H	Less than H	H	L
	TM$_4$	At least MH	M	M	At most M
	TM$_5$	H	MH	Between MH and VH	ML

CR: Customer requirement; DR: Design requirement; QFD: Quality function deployment.

In what follows, the proposed QFD approach is used to help the company obtain the ranking of HOWs for selecting market segments. First, the hesitant linguistic expressions of the QFD team members are converted into HFLTSs by applying the transformation function E_{GH}. Then, the linguistic intervals are yielded by calculating the envelope of each obtained HFLTS and the interval 2-tuple relationship matrix $\widetilde{R}_k$ $(k = 1, 2, ..., 5)$ of every QFD team member is subsequently constructed. For instance, the interval 2-tuple relationship matrix of TM$_1$ $\widetilde{R}_1$ is presented in Table 2. By implementing the ITOWA operator, the collective assessments regarding the relationship judgements between CRs and DRs are taken as the collective interval 2-tuple relationship matrix $\widetilde{R} = \left[\widetilde{r}_{ij}\right]_{5 \times 4}$, as shown in Table 3. Note that the ITOWA operator weights are derived using the argument-dependent approach [44].

Table 2. Interval 2-tuple relationship matrix of TM$_1$.

WHATs	HOWs			
	DR$_1$	DR$_2$	DR$_3$	DR$_4$
CR$_1$	$[(s_5, 0), (s_6, 0)]$	$[(s_2, 0), (s_2, 0)]$	$[(s_1, 0), (s_3, 0)]$	$[(s_3, 0), (s_3, 0)]$
CR$_2$	$[(s_5, 0), (s_6, 0)]$	$[(s_5, 0), (s_6, 0)]$	$[(s_4, 0), (s_6, 0)]$	$[(s_5, 0), (s_5, 0)]$
CR$_3$	$[(s_4, 0), (s_6, 0)]$	$[(s_5, 0), (s_5, 0)]$	$[(s_5, 0), (s_6, 0)]$	$[(s_4, 0), (s_6, 0)]$
CR$_4$	$[(s_0, 0), (s_4, 0)]$	$[(s_3, 0), (s_3, 0)]$	$[(s_5, 0), (s_6, 0)]$	$[(s_4, 0), (s_4, 0)]$
CR$_5$	$[(s_4, 0), (s_6, 0)]$	$[(s_4, 0), (s_4, 0)]$	$[(s_0, 0), (s_5, 0)]$	$[(s_1, 0), (s_2, 0)]$

Table 3. Collective interval 2-tuple relationship matrix.

WHATs	HOWs			
	DR$_1$	DR$_2$	DR$_3$	DR$_4$
CR$_1$	Δ[0.833, 0.884]	Δ[0.884, 0.448]	Δ[0.448, 0.543]	Δ[0.543, 0.322]
CR$_2$	Δ[0.876, 0.994]	Δ[0.994, 0.839]	Δ[0.839, 0.949]	Δ[0.949, 0.833]
CR$_3$	Δ[0.791, 0.935]	Δ[0.935, 0.833]	Δ[0.833, 0.949]	Δ[0.949, 0.994]
CR$_4$	Δ[0.426, 0.709]	Δ[0.709, 0.29]	Δ[0.290, 0.551]	Δ[0.551, 0.782]
CR$_5$	Δ[0.782, 0.884]	Δ[0.884, 0.614]	Δ[0.614, 0.667]	Δ[0.667, 0.614]

In the second stage, it is assumed that the company's managers can only provide their partial information for the CR weights using the basic ranking forms introduced in Section 3.2, and the set of known weight information H is shown as follows:

$$H = \{w_1 \geq 1.20w_4, 0.15 \leq w_2 \leq 0.26, 0.03 \leq w_3 - w_5 \leq 0.10, w_2 - w_3 \geq w_3 - w_4, w_5 \geq 0.17\}.$$

Because the weight information is incompletely known, we employ Model (M−2) to construct the following linear programming model to determine the weights of CRs.

$$\max\lambda$$

$$s.t. \begin{cases} 0.727w_1 + 0.816w_2 + 0.711w_3 + 0.452w_4 + 0.685w_5 \geq \lambda, \\ 0.425w_1 + 0.764w_2 + 0.757w_3 + 0.386w_4 + 0.51w_5 \geq \lambda, \\ 0.377w_1 + 0.757w_2 + w_3 + 0.682w_4 + 0.546w_5 \geq \lambda, \\ 0.431w_1 + 0.757w_2 + 0.683w_3 + 0.444w_4 + 0.337w_5 \geq \lambda, \\ w \in H, \sum\limits_{i=1}^{5} w_i = 1, w_i \geq 0, i = 1, 2, ..., 5. \end{cases}$$

By solving the above linear programming model, the weight vector of the five CRs is derived as $w = (0.196, 0.260, 0.211, 0.163, 0.170)^T$.

In the third stage, there are 24 (=4!) permutations of the rankings for all the DRs that must be tested, which are expressed as follows:

$P_1 = (DR_1, DR_2, DR_3, DR_4)$, $P_2 = (DR_1, DR_2, DR_4, DR_3)$, $P_3 = (DR_1, DR_3, DR_2, DR_4)$,
$P_4 = (DR_1, DR_3, DR_4, DR_2)$, $P_5 = (DR_1, DR_4, DR_2, DR_3)$, $P_6 = (DR_1, DR_4, DR_3, DR_2)$,
$P_7 = (DR_2, DR_1, DR_3, DR_4)$, $P_8 = (DR_2, DR_1, DR_4, DR_3)$, $P_9 = (DR_2, DR_3, DR_1, DR_4)$,
$P_{10} = (DR_2, DR_3, DR_4, DR_1)$, $P_{11} = (DR_2, DR_4, DR_1, DR_3)$, $P_{12} = (DR_2, DR_4, DR_3, DR_1)$,
$P_{13} = (DR_3, DR_1, DR_2, DR_4)$, $P_{14} = (DR_3, DR_1, DR_4, DR_2)$, $P_{15} = (DR_3, DR_2, DR_1, DR_4)$,
$P_{16} = (DR_3, DR_2, DR_4, DR_1)$, $P_{17} = (DR_3, DR_4, DR_1, DR_2)$, $P_{18} = (DR_3, DR_4, DR_2, DR_1)$,
$P_{19} = (DR_4, DR_1, DR_2, DR_3)$, $P_{20} = (DR_4, DR_1, DR_3, DR_2)$, $P_{21} = (DR_4, DR_2, DR_1, DR_3)$,
$P_{22} = (DR_4, DR_2, DR_3, DR_1)$, $P_{23} = (DR_4, DR_3, DR_1, DR_2)$, $P_{24} = (DR_4, DR_3, DR_2, DR_1)$.

In Step 5, we calculated the concordance/discordance index $\phi_i^{\rho}(DR_\phi, DR_\eta)$ using Equation (19) for each pair of DRs (DR_ϕ, DR_η) $(\phi, \eta = 1, 2, 3, 4)$ in the permutation P_ρ in relation with CR$_i$ $(i = 1, 2, ..., 5)$. Considering the first permutation P_1 for example, the results of the concordance/discordance index are shown in Table 4. In Step 6, we utilize Equation (20) to compute the weighted concordance/discordance index $\phi^{\rho}(DR_\xi, DR_\zeta)$ for each pair of (DR_ϕ, DR_η) in the permutation P_ρ, and the results are indicated in

Table 5. In Step 7, the comprehensive concordance/discordance index ϕ^ρ ($\rho = 1, 2, ..., 24$) is calculated by applying Equation (21) for each permutation P_ρ. The computation results are given as follows:

$\phi^1 = 0.2266, \phi^2 = 0.0949, \phi^3 = 0.3183, \phi^4 = 0.2783, \phi^5 = 0.0549, \phi^6 = 0.1465,$
$\phi^7 = 0.0850, \phi^8 = -0.0467, \phi^9 = 0.0351, \phi^{10} = -0.1465, \phi^{11} = -0.2284, \phi^{12} = -0.2783,$
$\phi^{13} = 0.2684, \phi^{14} = 0.2284, \phi^{15} = 0.1268, \phi^{16} = -0.0549, \phi^{17} = 0.0467, \phi^{18} = -0.0949,$
$\phi^{19} = -0.1268, \phi^{20} = -0.0351, \phi^{21} = -0.2684, \phi^{22} = -0.3183, \phi^{23} = -0.0850, \phi^{24} = -0.2266.$

Table 4. Results of the concordance/discordance index for P_1.

P_1	CR_1	CR_2	CR_3	CR_4	CR_5
$\phi_i^1 (FM_1, FM_2)$	0.131	0.059	−0.030	0.069	0.147
$\phi_i^1 (FM_1, FM_3)$	0.250	0.063	−0.165	−0.141	0.103
$\phi_i^1 (FM_1, FM_4)$	0.119	0.063	0.034	0.008	0.250
$\phi_i^1 (FM_2, FM_3)$	0.119	0.004	−0.135	−0.210	−0.044
$\phi_i^1 (FM_2, FM_4)$	−0.012	0.004	0.065	−0.061	0.103
$\phi_i^1 (FM_3, FM_4)$	−0.130	0.000	0.200	0.149	0.147

Table 5. Results of the weighted concordance/discordance index.

P_1	P_2	P_3	P_4	P_5	P_6	P_7	P_8	P_9	P_{10}	P_{11}	P_{12}
0.0708	0.0708	0.0250	0.0250	0.0908	0.0908	−0.0708	−0.0708	−0.0458	−0.0458	0.0200	0.0200
0.0250	0.0908	0.0708	0.0908	0.0708	0.0250	−0.0458	0.0200	−0.0708	0.0200	−0.0708	−0.0458
0.0908	0.0250	0.0908	0.0708	0.0250	0.0708	0.0200	−0.0458	0.0200	−0.0708	−0.0458	−0.0708
−0.0458	0.0200	0.0458	0.0659	−0.0200	−0.0659	0.0250	0.0908	−0.0250	0.0659	−0.0908	−0.0659
0.0200	−0.0458	0.0659	0.0458	−0.0659	−0.0200	0.0908	0.0250	0.0659	−0.0250	−0.0659	−0.0908
0.0659	−0.0659	0.0200	−0.0200	−0.0458	0.0458	0.0659	−0.0659	0.0908	−0.0908	0.0250	−0.0250
0.0708	0.0708	0.0250	0.0250	0.0908	0.0908	−0.0708	−0.0708	−0.0458	−0.0458	0.0200	0.0200
0.0250	0.0908	0.0708	0.0908	0.0708	0.0250	−0.0458	0.0200	−0.0708	0.0200	−0.0708	−0.0458

P_{13}	P_{14}	P_{15}	P_{16}	P_{17}	P_{18}	P_{19}	P_{20}	P_{21}	P_{22}	P_{23}	P_{24}
−0.0250	−0.0250	0.0458	0.0458	0.0659	0.0659	−0.0908	−0.0908	−0.0200	−0.0200	−0.0659	−0.0659
0.0458	0.0659	−0.0250	0.0659	−0.0250	0.0458	−0.0200	−0.0659	−0.0908	−0.0659	−0.0908	−0.0200
0.0659	0.0458	0.0659	−0.0250	0.0458	−0.0250	−0.0659	−0.0200	−0.0659	−0.0908	−0.0200	−0.0908
0.0708	0.0908	−0.0708	0.0200	−0.0908	−0.0200	0.0708	0.0250	−0.0708	−0.0458	−0.0250	0.0458
0.0908	0.0708	0.0200	−0.0708	−0.0200	−0.0908	0.0250	0.0708	−0.0458	−0.0708	0.0458	−0.0250
0.0200	−0.0200	0.0908	−0.0908	0.0708	−0.0708	−0.0458	0.0458	0.0250	−0.0250	0.0708	−0.0708
−0.0250	−0.0250	0.0458	0.0458	0.0659	0.0659	−0.0908	−0.0908	−0.0200	−0.0200	−0.0659	−0.0659
0.0458	0.0659	−0.0250	0.0659	−0.0250	0.0458	−0.0200	−0.0659	−0.0908	−0.0659	−0.0908	−0.0200

Based on the comprehensive concordance/discordance indexes ϕ^ρ ($\rho = 1, 2, ..., 24$) produced, it is easily seen that the best permutation is P_3 because $\phi^3 = 0.3183$ gives the maximum value, and the final priory order of the four DRs is $DR_1 \succ DR_3 \succ DR_2 \succ DR_4$. Therefore, the most important company business strength for the considered case study is "relative cost position (DR1)", which should be given the highest priority for selecting the optimal market segment, followed by DR3, DR2, and DR4.

5.2. Comparisons and Discussions

To validate the effectiveness of the proposed QFD, a comparative analysis with the conventional QFD and the fuzzy QFD [51] methods is conducted on the same problem of market segments evaluation. In addition, an extended linguistic QFD approach based on discrete numbers [52] is chosen to facilitate the comparative analysis. By applying these methods, the ranking results of the four DRs are generated as shown in Table 6.

With respect to the proposed QFD approach, Table 6 shows that our prioritization of the DRs is in accordance with the rankings yielded by the conventional QFD, the fuzzy QFD, and the linguistic QFD methods. Thus, the potential of the proposed QFD is validated through the comparative study. However, compared with the conventional QFD method and its various improvements, the QFD approach here proposed offers some additional advantages as follows:

- Different types of uncertainties in the implementation of QFD, such as imprecision, uncertainty and hesitation, can be well modeled via the hesitant 2-tuple linguistic term sets. The QFD team members can use more flexible and richer expressions to express their subjective judgments.
- By using the ITOWA operator, the proposed method can relieve the influence of unfair judgments concerning the relationships between CRs and DRs on the QFD analysis results, through assigning very low weights to those "false" or "biased" opinions.
- The proposed approach is able to deal with QFD problems in which the information about CR weights is incompletely known. Under the condition of incomplete weight information, a multiple objective programming model can be established to solve the optimal weights of CRs.
- The proposed methodology can get a more reasonable and credible ranking of DRs by using the modified QUALIFLEX approach, which makes the QFD analysis results certain and facilitates product planning decision-making.
- The proposed model is suitable to solve complicated QFD problems with comprehensive CRs and limited DRs, since the number of CRs has little effect upon the implementation efficiency of the proposed method.

Table 6. Ranking results of HOWs by the listed methods.

HOWs	QFD		Fuzzy QFD		Linguistic QFD		Proposed Approach
	W_j	Ranking	$\widetilde{W}_j$	Ranking	$\hat{W}_j$	Ranking	
DR_1	7.221	1	(0.267, 0.475, 0.724)	1	$s_{5.06}$	1	1
DR_2	6.303	3	(0.231, 0.415, 0.659)	3	$s_{4.32}$	3	3
DR_3	7.035	2	(0.253, 0.448, 0.689)	2	$s_{4.60}$	2	2
DR_4	5.919	4	(0.217, 0.400, 0.641)	4	$s_{4.05}$	4	4

6. Conclusions

In this paper, we developed a hybrid group decision-making model using hesitant 2-tuple linguistic term sets and an extended QUALIFLEX method for handling QFD problems with incomplete weight information. The HFLTSs, a new effective tool to express human's hesitancy in decision-making, was used to represent the diversity and uncertainty of subjective assessments given by QFD team members, and the interval 2-tuple linguistic model was employed to process the acquired linguistic assessment information, which can effectively avoid information loss and distortion in the linguistic computing. As a result, the hesitant 2-tuple linguistic approach for the expression of assessment information better reflects the deep-seated uncertainty in the implementation process of QFD. As the weight information of CRs is usually incomplete because additionally complex and abstract, a linear programming model was suggested to determine the optimal weight vector for CRs. Finally, the normal QUALIFLEX method has been modified to obtain the priority order of DRs and to detect the most important ones for the following design stages. The real-world efficacy of the proposed QFD approach was illustrated by using a market segment evaluation and selection problem.

In the future, the following research directions are recommended. First, the linguistic term sets that are uniformly and symmetrically distributed were used in the proposed analytical approach to model and manage QFD team members' linguistic expressions. However, in some situations, the unbalanced linguistic term sets [53] or the linguistic term sets with different granularity of uncertainty [54,55] may be employed by experts to express their opinions. Therefore, in future work, extending the proposed QFD approach to unbalanced linguistic or multi-granular linguistic context should be explored. Second, to obtain a more accurate DR ranking, complex computations are required in applying the QFD model being proposed. Thus, another direction for future research is to develop a computer-based application system using programming languages such as R to facilitate the implementation of the proposed QFD algorithm. Third, a market segment selection example was used in this paper to illustrate the effectiveness of the proposed QFD. In future research, other complex case studies of product development can be applied to further verify the feasibility and practicality of the proposed hybrid group decision-making model.

Acknowledgments: This work was partially supported by the National Natural Science Foundation of China (Nos. 71402090 and 71671125), the Program for Professor of Special Appointment (Young Eastern Scholar) at Shanghai Institutions of Higher Learning (No. QD2015019), and the Shanghai Pujiang Program (No. 14PJC051).

Author Contributions: The individual contribution and responsibilities of the authors were as follows: Hu-Chen Liu and Ze-Ling Wang together designed research; and Jian-Xin You provided extensive advice throughout the study regarding to abstract, introduction, research design, research methodology, findings and revise the manuscript. The discussion was a team task. All authors have read and approved the final manuscript.

Conflicts of Interest: The authors declare no conflict of interest.

References

1. Kwong, C.K.; Bai, H. Determining the importance weights for the customer requirements in QFD using a fuzzy AHP with an extent analysis approach. *IIE Trans.* **2003**, *35*, 619–626. [CrossRef]
2. Akao, Y. *Quality Function Deployment: Integrating Customer Requirements into Product Design*; Productivity Press: Cambridge, MA, USA, 1990.
3. Aye Ho, E.S.S.; Lai, Y.J.; Chang, S.I. An integrated group decision-making approach to quality function deployment. *IIE Trans.* **1999**, *31*, 553–567. [CrossRef]
4. Chin, K.S.; Wang, Y.M.; Yang, J.B.; Poon, K.K.G. An evidential reasoning based approach for quality function deployment under uncertainty. *Expert Syst. Appl.* **2009**, *36*, 5684–5694. [CrossRef]
5. Sivasamy, K.; Arumugam, C.; Devadasan, S.R.; Murugesh, R.; Thilak, V.M.M. Advanced models of quality function deployment: A literature review. *Qual. Quant.* **2016**, *50*, 1399–1414. [CrossRef]
6. Kurtulmuşoğlu, F.B.; Pakdil, F.; Atalay, K.D. Quality improvement strategies of highway bus service based on a fuzzy quality function deployment approach. *Transp. A Trans. Sci.* **2016**, *12*, 175–202. [CrossRef]
7. Onar, S.Ç.; Büyüközkan, G.; Öztayşi, B.; Kahraman, C. A new hesitant fuzzy QFD approach: An application to computer workstation selection. *Appl. Soft Comput.* **2016**, *46*, 1–16. [CrossRef]
8. Chan, L.K.; Wu, M.L. A systematic approach to quality function deployment with a full illustrative example. *Omega* **2005**, *33*, 119–139. [CrossRef]
9. Jia, W.; Liu, Z.; Lin, Z.; Qiu, C.; Tan, J. Quantification for the importance degree of engineering characteristics with a multi-level hierarchical structure in QFD. *Int. J. Prod. Res.* **2016**, *54*, 1627–1649. [CrossRef]
10. Zhang, Z.; Chu, X. Fuzzy group decision-making for multi-format and multi-granularity linguistic judgments in quality function deployment. *Expert Syst. Appl.* **2009**, *36*, 9150–9158. [CrossRef]
11. Rodriguez, R.M.; Martínez, L.; Herrera, F. Hesitant fuzzy linguistic term sets for decision making. *IEEE Trans. Fuzzy Syst.* **2012**, *20*, 109–119. [CrossRef]
12. Rodríguez, R.M.; Labella, Á.; Martínez, L. An overview on fuzzy modelling of complex linguistic preferences in decision making. *Int. J. Comput. Intell. Syst.* **2016**, *9* (Suppl. 1), 81–94. [CrossRef]
13. Zhang, X.; Xu, Z.S.; Wang, H. Heterogeneous multiple criteria group decision making with incomplete weight information: A deviation modeling approach. *Inf. Fusion* **2015**, *25*, 49–62. [CrossRef]
14. Montes, R.; Sanchez, A.M.; Villar, P.; Herrera, F. A web tool to support decision making in the housing market using hesitant fuzzy linguistic term sets. *Appl. Soft Comput.* **2015**, *35*, 949–957. [CrossRef]
15. Liao, H.; Xu, Z.; Zeng, X.J. Hesitant fuzzy linguistic VIKOR method and its application in qualitative multiple criteria decision making. *IEEE Trans. Fuzzy Syst.* **2015**, *23*, 1343–1355. [CrossRef]
16. Herrera, F.; Martínez, L. A 2-tuple fuzzy linguistic representation model for computing with words. *IEEE Trans. Fuzzy Syst.* **2000**, *8*, 746–752.
17. Zhang, H. The multiattribute group decision making method based on aggregation operators with interval-valued 2-tuple linguistic information. *Math. Comput. Model.* **2012**, *56*, 27–35. [CrossRef]
18. Wang, J.Q.; Wang, D.D.; Zhang, H.Y.; Chen, X.H. Multi-criteria group decision making method based on interval 2-tuple linguistic information and Choquet integral aggregation operators. *Soft Comput.* **2015**, *19*, 389–405. [CrossRef]
19. Xue, Y.X.; You, J.X.; Zhao, X.; Liu, H.C. An integrated linguistic MCDM approach for robot evaluation and selection with incomplete weight information. *Int. J. Prod. Res.* **2016**, *54*, 5452–5467. [CrossRef]

20. Martínez, L.; Herrera, F. An overview on the 2-tuple linguistic model for computing with words in decision making: Extensions, applications and challenges. *Inf. Sci.* **2012**, *207*, 1–18. [CrossRef]
21. Song, W.; Ming, X.; Han, Y. Prioritising technical attributes in QFD under vague environment: A rough-grey relational analysis approach. *Int. J. Prod. Res.* **2014**, *52*, 5528–5545. [CrossRef]
22. Paelinck, J.H.P. Qualiflex: A flexible multiple-criteria method. *Econ. Lett.* **1978**, *1*, 193–197. [CrossRef]
23. Wang, J.C.; Tsao, C.Y.; Chen, T.Y. A likelihood-based QUALIFLEX method with interval type-2 fuzzy sets for multiple criteria decision analysis. *Soft Comput.* **2015**, *19*, 2225–2243. [CrossRef]
24. Zhang, X.; Xu, Z.S. Hesitant fuzzy QUALIFLEX approach with a signed distance-based comparison method for multiple criteria decision analysis. *Expert Syst. Appl.* **2015**, *42*, 873–884. [CrossRef]
25. Chen, T.Y.; Chang, C.H.; Rachel Lu, J.F. The extended QUALIFLEX method for multiple criteria decision analysis based on interval type-2 fuzzy sets and applications to medical decision making. *Eur. J. Oper. Res.* **2013**, *226*, 615–625. [CrossRef]
26. Chen, T.Y. Interval-valued intuitionistic fuzzy QUALIFLEX method with a likelihood-based comparison approach for multiple criteria decision analysis. *Inf. Sci.* **2014**, *261*, 149–169. [CrossRef]
27. Carnevalli, J.A.; Miguel, P.C. Review, analysis and classification of the literature on QFD—Types of research, difficulties and benefits. *Int. J. Prod. Econ.* **2008**, *114*, 737–754. [CrossRef]
28. Chan, L.K.; Wu, M.L. Quality function deployment: A comprehensive review of its concepts and methods. *Qual. Eng.* **2002**, *15*, 23–35. [CrossRef]
29. Armacost, R.L.; Componation, P.J.; Mullens, M.A.; Swart, W.W. An AHP framework for prioritizing customer requirements in QFD: An industrialized housing application. *IIE Trans.* **1994**, *26*, 72–79. [CrossRef]
30. Lam, J.S.L.; Lai, K.H. Developing environmental sustainability by ANP-QFD approach: The case of shipping operations. *J. Clean. Prod.* **2015**, *105*, 275–284. [CrossRef]
31. Liu, Y.; Zhou, J.; Chen, Y. Using fuzzy non-linear regression to identify the degree of compensation among customer requirements in QFD. *Neurocomputing* **2014**, *142*, 115–124. [CrossRef]
32. Ji, P.; Jin, J.; Wang, T.; Chen, Y. Quantification and integration of Kano's model into QFD for optimising product design. *Int. J. Prod. Res.* **2014**, *52*, 6335–6348. [CrossRef]
33. Yan, H.B.; Ma, T. A group decision-making approach to uncertain quality function deployment based on fuzzy preference relation and fuzzy majority. *Eur. J. Oper. Res.* **2015**, *241*, 815–829. [CrossRef]
34. Luo, X.G.; Kwong, C.K.; Tang, J.F.; Sun, F.Q. QFD-based product planning with consumer choice analysis. *IEEE Trans. Syst. Man Cybern. Syst.* **2015**, *45*, 454–461. [CrossRef]
35. Zhong, S.; Zhou, J.; Chen, Y. Determination of target values of engineering characteristics in QFD using a fuzzy chance-constrained modelling approach. *Neurocomputing* **2014**, *142*, 125–135. [CrossRef]
36. Hosseini Motlagh, S.; Behzadian, M.; Ignatius, J.; Goh, M.; Sepehri, M.; Hua, T. Fuzzy PROMETHEE GDSS for technical requirements ranking in HOQ. *Int. J. Adv. Manuf. Technol.* **2015**, *76*, 1993–2002. [CrossRef]
37. Torra, V. Hesitant fuzzy sets. *Int. J. Intell. Syst.* **2010**, *25*, 529–539. [CrossRef]
38. Rodríguez, R.; Martínez, L.; Torra, V.; Xu, Z.; Herrera, F. Hesitant fuzzy sets: State of the art and future directions. *Int. J. Intell. Syst.* **2014**, *29*, 495–524. [CrossRef]
39. Rodríguez, R.M.; Bedregal, B.; Bustince, H.; Dong, Y.C.; Farhadinia, B.; Kahraman, C.; Martínez, L.; Torra, V.; Xu, Y.J.; Xu, Z.S.; et al. A position and perspective analysis of hesitant fuzzy sets on information fusion in decision making. Towards high quality progress. *Inf. Fusion* **2016**, *29*, 89–97. [CrossRef]
40. Liu, H.; Rodríguez, R.M. A fuzzy envelope for hesitant fuzzy linguistic term set and its application to multicriteria decision making. *Inf. Sci.* **2014**, *258*, 220–238. [CrossRef]
41. Rodríguez, R.M.; Martínez, L. An analysis of symbolic linguistic computing models in decision making. *Int. J. Gen. Syst.* **2013**, *42*, 121–136. [CrossRef]
42. Liu, H.C.; Lin, Q.L.; Wu, J. Dependent interval 2-tuple linguistic aggregation operators and their application to multiple attribute group decision making. *Int. J. Uncertain. Fuzziness Knowl. Based Syst.* **2014**, *22*, 717–735. [CrossRef]
43. Liu, H.C.; You, J.X.; You, X.Y. Evaluating the risk of healthcare failure modes using interval 2-tuple hybrid weighted distance measure. *Comput. Ind. Eng.* **2014**, *78*, 249–258. [CrossRef]
44. Wu, J.; Liang, C.; Huang, Y. An argument-dependent approach to determining OWA operator weights based on the rule of maximum entropy. *Int. J. Intell. Syst.* **2007**, *22*, 209–221. [CrossRef]
45. Ureña, R.; Chiclana, F.; Morente-Molinera, J.A.; Herrera-Viedma, E. Managing incomplete preference relations in decision making: a review and future trends. *Inf. Sci.* **2015**, *302*, 14–32. [CrossRef]

46. Jiang, Y.; Xu, Z.; Yu, X. Group decision making based on incomplete intuitionistic multiplicative preference relations. *Inf. Sci.* **2015**, *295*, 33–52. [CrossRef]

47. Xue, Y.X.; You, J.X.; Lai, X.D.; Liu, H.C. An interval-valued intuitionistic fuzzy MABAC approach for material selection with incomplete weight information. *Appl. Soft Comput.* **2016**, *38*, 703–713. [CrossRef]

48. Qi, X.; Liang, C.; Zhang, J. Generalized cross-entropy based group decision making with unknown expert and attribute weights under interval-valued intuitionistic fuzzy environment. *Comput. Ind. Eng.* **2015**, *79*, 52–64. [CrossRef]

49. Deng, J.L. Introduction to gray system theory. *J. Grey Syst.* **1989**, *1*, 1–24.

50. Liu, H.C.; You, J.X.; Li, P.; Su, Q. Failure mode and effect analysis under uncertainty: An integrated multiple criteria decision making approach. *IEEE Trans. Reliab.* **2016**, *65*, 1380–1392. [CrossRef]

51. Dat, L.Q.; Phuong, T.T.; Kao, H.P.; Chou, S.Y.; Nghia, P.V. A new integrated fuzzy QFD approach for market segments evaluation and selection. *Appl. Math. Model.* **2015**, *39*, 3653–3665. [CrossRef]

52. Xu, Z.S. A method based on linguistic aggregation operators for group decision making with linguistic preference relations. *Inf. Sci.* **2004**, *166*, 19–30. [CrossRef]

53. Estrella, F.J.; Espinilla, M.; Martínez, L. Fuzzy linguistic olive oil sensory evaluation model based on unbalanced linguistic scales. *J. Mult. Valued Log. Soft Comput.* **2014**, *22*, 1–21.

54. Morente-Molinera, J.A.; Pérez, I.J.; Ureña, M.R.; Herrera-Viedma, E. On multi-granular fuzzy linguistic modeling in group decision making problems: A systematic review and future trends. *Knowl. Based Syst.* **2015**, *74*, 49–60. [CrossRef]

55. Dong, Y.; Wu, Y.; Zhang, H.; Zhang, G. Multi-granular unbalanced linguistic distribution assessments with interval symbolic proportions. *Knowl. Based Syst.* **2015**, *82*, 139–151. [CrossRef]

![symmetry logo] *symmetry*

MDPI

Article

Image Intelligent Detection Based on the Gabor Wavelet and the Neural Network

Yajun Xu [1], Fengmei Liang [1,*], Gang Zhang [1] and Huifang Xu [2]

[1] College of Information Engineering, Taiyuan University of Technology, Taiyuan 030024, China; xuyajun0032@link.tyut.edu.cn (Y.X.); m13840612579@163.com (G.Z.)
[2] Daqin Railway Co. Ltd., Taiyuan Railway Administration, Taiyuan 030013, China; huif_xu@sina.com
[*] Correspondence: yjun_xu@163.com; Tel.: +86-186-0341-0966

Academic Editor: Angel Garrido
Received: 21 September 2016; Accepted: 11 November 2016; Published: 15 November 2016

Abstract: This paper first analyzes the one-dimensional Gabor function and expands it to a two-dimensional one. The two-dimensional Gabor function generates the two-dimensional Gabor wavelet through measure stretching and rotation. At last, the two-dimensional Gabor wavelet transform is employed to extract the image feature information. Based on the back propagation (BP) neural network model, the image intelligent test model based on the Gabor wavelet and the neural network model is built. The human face image detection is adopted as an example. Results suggest that, although there are complex textures and illumination variations on the images of the face database named AT&T, the detection accuracy rate of the proposed method can reach above 0.93. In addition, extensive simulations based on the Yale and extended Yale B datasets further verify the effectiveness of the proposed method.

Keywords: Gabor wavelet; feature information; neural network; face recognition

1. Introduction

Wavelet theory has been increasingly popular and has quickly developed since its appearance in the 1980s [1]. Scholars generally thought that the wavelet transform is a breakthrough of Fourier transform. In addition, the Gabor filter shows a strong robustness towards the luminance and contrast of images and the human facial expressions, and it reflects the most helpful local features for human face recognition [2,3], so the Gabor wavelet has been widely applied to the extraction of human face features. Currently, scholars have deepened their study of neural network theories. The artificial neural network [4,5] is an intelligent system simulated by humans according to information processing of the human brain nerve system, and a new-type structural computing system generated based on a preliminary understanding of the human brain organization structure and activity mechanism. Since it can simulate the human brain nerve system and endow the machine with the perception, learning, and deduction capability of the human brain, it has been widely applied to the model recognition in various fields.

However, how to combine the neural network with the nonlinear theories, such as wavelet theory, fuzzy set, and chaos theory, is a new research direction [6,7]. The neural network boasts a collection of favorable characteristics, including fault tolerance, self-adaption, self-learning and generalization capability, and robustness, and the wavelet transform has the temporal frequency local and zooming characteristics, so the Gabor wavelet transform can be employed to reduce the number of input nodes in the neural network and increase the convergence speed on the one hand, and sufficiently and efficiently express the human face characteristics and improve the neural network recognition capability on the other. However, it has been an issue of great concern to both experts and scholars as to how to combine advantages of the two and apply them to the human face recognition technique. The work [8] proposed a method for detecting facial regions by combining a Gabor filter and a convolutional neural network and obtained a detection rate of 87.5%. Kaushal et al. [9] used the feature vector based on Gabor filters as the input of a feed forward neural network (FFNN). A similar work presented in [10] was implemented using Java environment, aiming to object localization and classification.

Therefore, the research focus of this paper is about the image feature extraction based on the Gabor wavelet transform. Combining the intelligent recognition of the back propagation (BP) neural network, this paper puts forward the image intelligent detection based on the Gabor wavelet transform and the neural network model. The human face image detection is taken as an example. First, the model is tested based on the human face database "Yale" [11] with illumination variation and complex texture. Then, using the human face database "AT&T" [12], the detection accuracy rate of the model is given to evaluate the detection performance. Finally, based on the human face database "extended Yale B" [13], the effectiveness of the method is proved by the performance comparison between our proposal and several state-of-the-art methods.

2. Methodology

2.1. Gabor Wavelet Theory and Feature Transformation

In order to introduce the Gabor wavelet and apply it to the image feature extraction, this paper first introduces the analysis deduction of the one-dimensional Gabor wavelet so as to introduce the two-dimensional Gabor wavelet.

Among them, the one-dimensional Gabor wavelet [14] is constituted by a trigonometric function multiplied by a Gaussian function shown in Equation (1):

$$W(t, t_0, \omega) = e^{-\sigma(t-t_0)^2} e^{i\omega(t-t_0)} \tag{1}$$

Conduct integration of the product of Equation (1) and the signal frequency, and the one-dimensional Gabor wavelet transform can be expressed below:

$$C(x(t))(t_0, \omega) = \int_{-\infty}^{\infty} x(t)W(t, t_0, \omega)dt \tag{2}$$

The left of the equation stands for the frequency information of the signal, $x(t)$, when the frequency is ω and the time is t_0. Put Equation (1) into Equation (2), and expand the mixed one into the following one:

$$C(x(t))(t_0, \omega) = \int_{-\infty}^{\infty} x(t)e^{-\sigma(t-t_0)^2}\cos(\omega(t-t_0))dt + i\int_{-\infty}^{\infty} x(t)e^{-\sigma(t-t_0)^2}\sin(\omega(t-t_0))dt \tag{3}$$

The real and imaginary part of Equation (3) expressed in the form of complex number is shown in Equation (4) below.

The two-dimensional Gabor wavelet can be generated by expanding the one-dimensional Gabor function into a two-dimensional one, and through measure stretching and rotation [15]. The two-dimensional Gabor wavelet can acquire the image information in terms of any measure and any orientation. Through the one-dimensional Gabor wavelet function, it can be seen that the two-dimensional Gabor wavelet function is unique and can be adopted as the primary function for the image extraction and analysis. In other words, the description completeness of images in terms of space and frequency domain can be realized. The wavelet transform reflects a relatively intuitive concept: when the textures are relatively meticulous, the sampling scope of the sample domain is relatively small, while the sampling scope of the opposite frequency domain is relatively large. However, when the textures are relatively coarse, the sampling scope of the space domain is relatively large and the sampling scope of the frequency domain is relatively small. Therefore, the two-dimensional

Gabor wavelet can capture features including the selectivity of space position, orientation and space frequency, and the quadrature phase relationship.

The two-dimensional Gabor wavelet core [16] is defined in Equation (4):

$$G_{u,v} = \frac{||k_{u,v}||^2}{\sigma^2} e^{\left(-\frac{||k_{u,v}||^2||z||^2}{2\sigma^2}\right)} [e^{(ik_{u,v}z)} - e^{\left(-\frac{\sigma^2}{2}\right)}]$$ (4)

where u and v stand for orientation and measure, respectively; z stands for the coordinate point of the fixed position; $\frac{||k_{u,v}||^2}{\sigma^2}$ is used to compensate the weakening of the energy spectrum; $e^{\left(-\frac{||k_{u,v}||^2||z||^2}{2\sigma^2}\right)}$ stands for the Gaussian envelop function; $e^{(ik_{u,v}z)}$ stands for the vibration function, the real part of which is the cosine function and the imaginary part of which is the sine function; $e^{\left(-\frac{\sigma^2}{2}\right)}$ stands for the DC component; σ stands for the size of the two-dimensional Gabor wavelet, namely the radius of the Gaussian function; $k_{u,v}$ stands for the central frequency of the filter, describing the response of the Gabor filter in terms of different orientations and measures. Therefore, when $k_{u,v}$ is different, a group of Gabor filters can be obtained. The real and imaginary part of Gabor filters based on five frequencies (0.2, 0.22, 0.24, 0.26, and 0.28) and eight orientations (0°, 45°, 90°, 135°, 180°, 225°, 270°, and 315°) are shown in Figure 1 below.

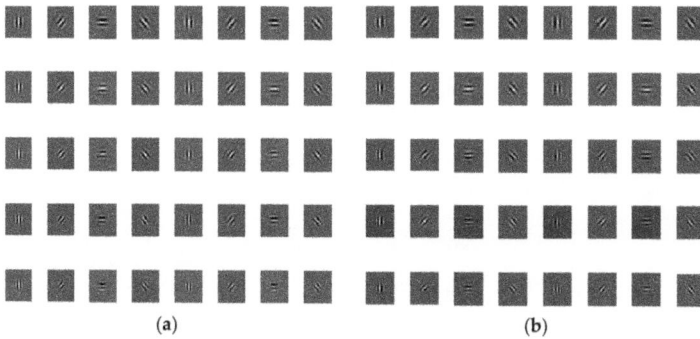

Figure 1. Gabor filter template based on five frequencies and eight orientations: (**a**) the template of the real part; and (**b**) the template of the imaginary part.

Wavelet transform has the following advantages [17] when being applied to image processing: (1) wavelet decomposition can cover the whole frequency domain; (2) by choosing the proper filter, the wavelet filter can largely reduce or even remove the relevance between different characteristics extracted; (3) the wavelet transform has a "zooming" characteristic and can adopt the wide analysis window in the low-frequency section and the narrow analysis window in the high-frequency section.

Therefore, to the image feature extraction process, the Gabor image feature extraction is to conduct the convolution of input images and the Gabor wavelet described in Equation (4). It is assumed that the input image grey scale is $I(x, y)$ and the convolution between I and the Gabor core, $G_{u,v}$, is shown in Equation (5) below:

$$O_{u,v}(x,y) = I(x,y) * G_{u,v}(x,y)$$ (5)

where $*$ stands for the convolution factor; $O_{u,v}(x,y)$ stands for the convolution image in the corresponding measure of u and the corresponding orientation of v.

2.2. Neural Network Model Structure and Its Algorithm

2.2.1. Back Propagation Neural Network Structure

The neural network is a highly nonlinear system. In terms of different functions and research, there are different neural network models. The BP neural network is a feedforward network adopted by the neural network model as the learning algorithm through the error BP algorithm [18]. It is mainly constituted of the input layer, the output layer, and the hidden layer. The nerve cell between layers adopts the fully-interlinked connection style and builds connections through the corresponding network weight coefficient, *w*. In addition, there is no connection between nerve cells within every layer. The basic idea of the BP algorithm is that the learning process is made up of two processes, namely the signal forward-propagation and the error backward-propagation. Figure 2 shows the specific structure.

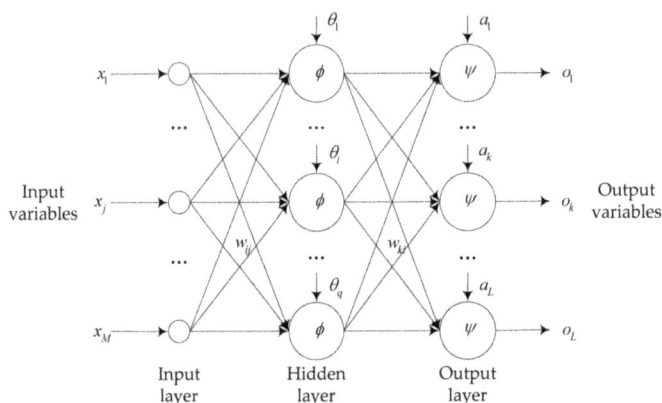

Figure 2. Back propagation (BP) neural network model structure.

x_j stands for the input of the *j* node in the input layer ($j = 1, \ldots , M$); w_{ij} stands for the weight value from the *i* node in the hidden layer to the *j* node in the input layer; θ_i stands for the threshold of the *i* node in the hidden layer; $\phi(x)$ stands for the excitation function of the hidden layer; w_{ki} stands for the weight value from the *k* node in the output layer to the *i* node in the hidden layer ($i = 1, \ldots , q$); a_k stands for the threshold value of the *k* node in the output layer ($k = 1, \ldots , L$); $\psi(x)$ stands for the excitation function of the output layer; o_k stands for the output of the *k* node in the output layer.

2.2.2. Back Propagation Neural Network Model Algorithm Steps

When the signal enters the BP neural network through the signal forward-propagation, the input samples are input through the input layer and are transmitted to the output layer through the processing in the hidden layer. If the practical output of the output layer fails to coincide with the expected output, it will move into the error backward-propagation period. The essence of the above signal forward-propagation and the error backward-propagation is a network iterative process. During the network iterative process, the weight value keeps on adjusting. The process endures until the output network error is reduced below the set error value or until the process reaches the pre-set iterations. Thus, it can be seen that the input and output relationship of the BP neural network is a highly linear system featuring "more input–more output," which is applicable to the prediction and recognition process system.

According to the weight value of the input nodes and the output nodes, the weight value between the input nodes and the hidden nodes, and the weight value between the hidden nodes and the output nodes, the relationship iteration between nodes of various layers is shown below:

(1) Signal forward-propagation process:

The input, net_i, of the i node in the hidden layer:

$$net_i = \sum_{j=1}^{M} w_{ij} x_j + \theta_i \qquad (6)$$

The output, y_i, of the i node in the hidden layer:

$$y_i = \phi(net_i) = \phi(\sum_{j=1}^{M} w_{ij} x_j + \theta_i) \qquad (7)$$

The input, net_k, of the k node in the output layer:

$$net_k = \sum_{i=1}^{q} w_{ki} y_i + a_k = \sum_{i=1}^{q} w_{ki} \phi(\sum_{j=1}^{M} w_{ij} x_j + \theta_i) + a_k \qquad (8)$$

The output, o_k, of the k node in the output layer:

$$o_k = \psi(net_k) = \psi(\sum_{i=1}^{q} w_{ki} y_i + a_k) = \psi(\sum_{i=1}^{q} w_{ki} \phi(\sum_{j=1}^{M} w_{ij} x_j + \theta_i) + a_k) \qquad (9)$$

(2) Error backward-propagation process:

The error backward-propagation first starts from the output layer to calculate the output error of the nerve cells in various layers step by step. Then, the weight value and the threshold of various layers are adjusted according to the error gradient descent to make the final modified network output to approximate the expected value. The quadric form error criterion function of every sample, E_p, is shown in Equation (10):

$$E_p = \frac{1}{2} \sum_{k=1}^{L} (T_k - o_k)^2 \qquad (10)$$

All in all, the major idea of the BP neural network is to modify the threshold and the weight value to make the error function to descend along the gradient orientation. The input layer obtains the practical output by processing the input information in the hidden layer. If the practical output is not in conformity with the sample output, the error will be sent back layer by layer. The weight value of every layer is modified according to the learning rules regulated by the algorithm. Through the repetition of the step, convergence or homeostasis can be achieved. In other words, the step keeps on until the total error between the practical output and the target output reaches the minimum error as required. The BP neural network model structure schematic diagram established is shown in Figure 3 below.

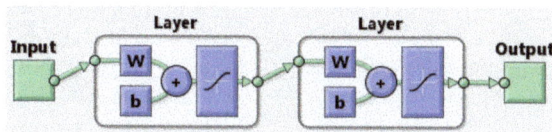

Figure 3. BP neural network structure schematic diagram.

Endow the BP neural network already built with relevant training parameters for model training. The relevant model parameters during network training and network convergence are shown in Table 1 below. The number of neurons takes 100 in the first layer of the network. The number of neurons in output layer, which is also named as the last layer, depends on the need of practical application.

Moreover, we only need to distinguish between faces and non-faces, so one neuron is sufficient to finish the work. Thus, the number of nodes in the output layer takes 1. In this network, we do not have any hidden layers called between the output layer and the input layer. This indicates that the second layer is the output layer. Through experiences, we have found that the training function "trainscg", which is suitable for nonlinear studies, has a good advantage in terms of memory consumption.

Table 1. Key parameters of network training setting and the network convergence results.

Network Training Parameters		Network Convergence Parameters	
Number of nodes in the first layer	100	Network training times	342
Number of nodes in the second layer	1	Network training duration(s)	10
Network target error	1×10^{-5}	Network convergence error	9.86×10^{-6}
Network training function	trainscg	Network goodness-of-fit	0.99999

A sample of dynamic error changes during the network training process is shown in Figure 4 below. Mean squared error (mse) is used as the performance function, namely the network target error. In this network, we set 1×10^{-5} as the target error. It involves that, when the mse value reaches 1×10^{-5}, the training stops. The maximum number of epochs (also called training times) sets 400, after that the training phase stops. Experimentally, we realized that the larger the number, the faster the training will be.

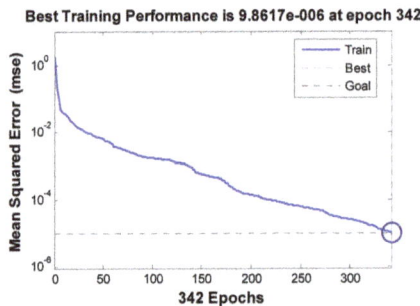

Figure 4. A sample of dynamic error variations during the network training process.

The BP neural network human face detection steps based on the Gabor feature extraction are shown as follows:

(1) Conduct convolution between the image to be recognized and the standard template image and improve the resistance against the image luminosity variation. The standard template image is shown in Figure 5 below:

(2) Generate Gabor filters in terms of n frequencies and m orientations. Here, $n = 5$ and $m = 8$.

(3) Upload the human-face and the non-human-face images as the training samples, extract features of filters generated in Step 1, and adopt the extracted feature data as the input information of the BP neural network model.

(4) Build the BP neural network model and refer to the core network parameters in the following part.

(5) Input the extracted feature data into the BP neural network already built according to Step 3, and train the network. Conduct human face detection of the trained network and draw the human face area on the image.

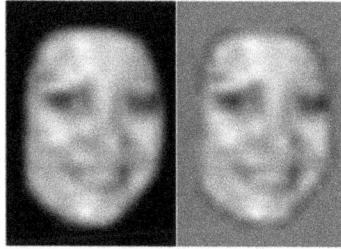

Figure 5. Standard template images.

3. Results and Discussion

Performance evaluation is one of most important aspects in face recognition applications. For the sake of verifying the effectiveness and stability of the proposed method, experiments were conducted on several public face databases, including Yale, AT&T, and extended Yale B, on images in which contain different poses, different expressions, and various illumination conditions. At last, the proposed method is compared with some other state-of-the-art methods.

3.1. Experiments and Analysis on the Yale Face Database

For the purpose of verifying the performance, we experiment on the Yale face database consisted of 165 images (137 × 147) with different variations such as facial expressions, luminance changes, and configuration, which includes 15 individuals and 11 grayscale images per individual. A preview image of the database is shown as Figure 6.

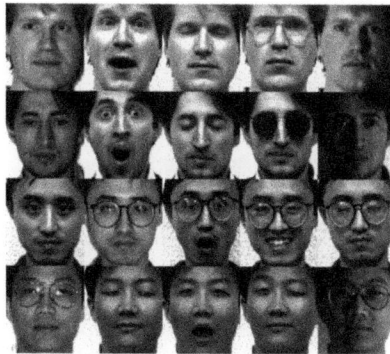

Figure 6. A preview image of the database of faces in the Yale face database.

In the experiments, all images are converted, cropped, and down-sampled to 25 × 30 pixels with grayscale. The training set consisted of partial images per subject from the database, and the rest consisted of the testing set. The images with facial expressions, luminance changes, and configuration each underwent the recognition test. From the experimental results, it can be seen that the neural network based on the Gabor feature extraction can accurately recognize the human face. The test results are shown in Figure 7 below.

(a)

(b)

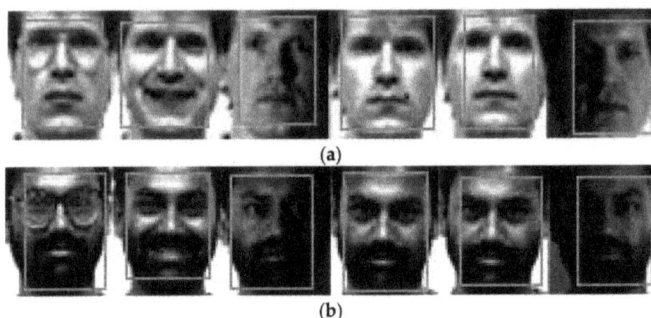

Figure 7. Qualitative results of our method on the Yale face database with illumination variation, complex texture, and facial expressions: (**a**) the recognition results of the face images labeled s1 in the Yale database; and (**b**) the recognition results of the face images labeled s7 in the Yale database.

3.2. Experiments and Analysis on the AT&T Database

Nowadays, the performance of the face recognition system is evaluated by various metrics, in which the recognition rate is commonly used. In order to comprehensively analyze the proposed method's recognition accuracy rate, it was tested in the publicly available AT&T database. This database contains 40 different persons, and every person has 10 different human face images (92 × 112) with 256 grey levels per pixel. For some subjects, the images were taken at different times with illumination variation, different facial expressions, and different facial details. A preview image of the database is shown as Figure 8.

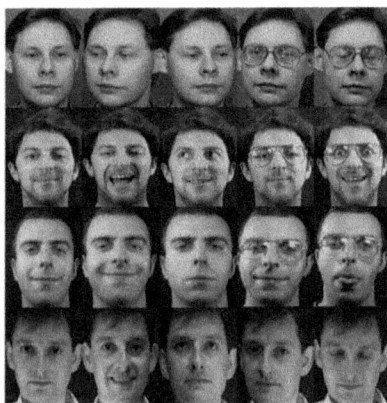

Figure 8. A preview image of the database of faces in the AT&T database.

As in the previous experiment, we converted, cropped, and down-sampled all images to 25 × 30. Then, we selected partial images randomly from each person's images as the training set and others as the testing set. The recognition accuracy results are shown in Table 2 below when we randomly selected four images of each person and performed five experiments. Results suggest that the accuracy rate of human face recognition reaches above 0.93. The recognition accuracy rate is relatively high.

Table 2. Neural network human face recognition accuracy rate based on the Gabor feature extraction.

Test Times	Number of Training Samples	Number of Test Samples	Recognition Accuracy Rate of Training Samples	Recognition Accuracy Rate of Test Samples	Comprehensive Recognition Accuracy Rate
1	160	240	0.965	0.922	0.9392
2	160	240	0.975	0.935	0.9510
3	160	240	0.952	0.922	0.9340
4	160	240	0.961	0.920	0.9364
5	160	240	0.962	0.925	0.9398

3.3. Experiments and Analysis on the Extended Yale B Face Database

In addition to the recognition accuracy rate, there are several other important and critical metrics available for performance evaluation. The following metrics are also introduced: false accept rate (FAR), false reject rate (FRR), and receiver operating characteristics (ROCs). FAR indicates the percent of the individuals that are incorrectly accepted; FRR measures the percent of valid inputs that are incorrectly rejected; ROC graphs are increasingly used in machine learning and data processing research for organizing and visualizing the performance of a system in recent years [19]. It is a graphical representation for visualizing characterization change between FAR and FRR. In the ROC graph, the points on the top left have high FFR and low FAR; thus, the ROC represents smart classifiers.

Furthermore, experiments were carried out using face images from the face database named the extended Yale B face database to analyze the performances of the face recognition using these metrics, and comparative analysis of the experimental results with existing methods is provided precisely in this section. The dataset contains 16,128 grayscale images in GIF format of 28 human subjects under nine poses and 64 illumination conditions. A preview image of the database of faces in the extended Yale B face database is shown as Figure 9. For this database, we simply use the cropped images and resize them to 32 × 32 pixels.

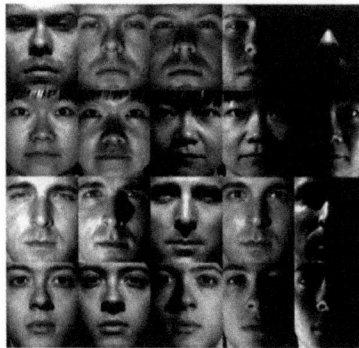

Figure 9. A preview image of the database of faces in the extended Yale B face database.

In this experiment, we selected a total of 28 individuals' facial images in the database, and each individual has 64 images with different poses and different illuminations. Moreover, 2, 4, 8, 16, and 32 images were randomly chosen from each group as training set; meanwhile, the remaining images were selected as a testing set. Comparison results of recognition rates of the proposed method, Local Gabor (LG, method proposed in [20]) and local gabor binary pattern (LGBP, method proposed in [21]) are shown in Table 3, and the ROC graph is shown as Figure 10. From Table 3, it can be seen that, as the training sample numbers increase, recognition rates of all methods also increase. In addition, when the training sample number is 32, the recognition rate of the proposed method outperforms LG and LGBP by an interval of 6.87% and 3.91%, respectively. Therefore, under the environment with different poses and different illuminations, the proposed method is better than the LG and LGBP face recognition method.

Table 3. Recognition rates of methods on the extended Yale B with different training sample numbers.

Methods	Training Sample Numbers					
	2	4	6	8	16	32
LG, Loris et al.'s method in [20]	50.49%	53.29%	54.32%	65.77%	75.32%	79.36%
LGBP, Xie et al.'s method in [21]	53.65%	59.35%	65.89%	74.49%	78.96%	82.32%
Our method	54.76%	60.78%	76.44%	79.53%	83.56%	86.23%

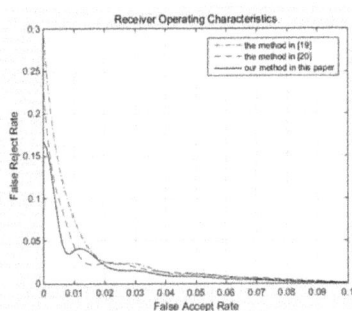

Figure 10. The receiver operating characteristic (ROC) curves of the various face recognition methods.

4. Conclusions

This paper first analyzes the Gabor wavelet theory with relatively strong resistance against image luminance and texture changes and its transform features, puts forward the idea to extract image feature information based on the Gabor wavelet transform, and then builds the image intelligent detection model based on the Gabor wavelet and neural network model. The human face detection experiments based on three datasets are conducted to analyze the validity of the model algorithm. When Gabor wavelet transform and the neural network are combined to test human face, the AT&T human face database is adopted to test the accuracy of the model algorithm, finding out that its accuracy rate is above 0.93, despite the complex texture and luminance changes. Based on the Yale and the extended Yale B face databases, and through the comparison with other state-of-the-art methods, the results illustrate that the method we proposed has an improved performance in face recognition.

Face recognition technology, which attracts an increasing amount of scientific research workers to it, has been vigorously developed and has been applied to many fields. However, there are still some gaps between the actual application and the ideal situation. In future work, we will test our proposed method on real world databases to further validate its effectiveness, such as the face detection data set and benchmark (FDDB) database, the labeled faces in the wild (LFW) database, and so on. At present, the technology has reached a bottleneck, and the research of it has very limited space for improvement. As a result, other techniques based on face recognition have become more challenging and more market-oriented topics, such as age estimation and gender estimation. This provides a good direction for our future research.

Acknowledgments: This work was supported by the Project of Natural Science Foundation of Shanxi Province (No. 2013011017-3).

Author Contributions: Fengmei Liang and Gang Zhang conceived and designed the experiments; Yajun Xu performed the experiments; Fengmei Liang and Yajun Xu analyzed the data; Huifang Xu contributed analysis tools; Yajun Xu wrote the paper.

Conflicts of Interest: The authors declare no conflict of interest.

References

1. Dong, Y.; Jiang, W. Facial expression recognition based on lifting wavelet and FLD. *Opt. Tech.* **2012**, *38*, 579–582.

2. Zhou, L.J.; Ma, Y.-Y.; Sun, J. Face recognition approach based on local medium frequency Gabor filters. *Comput. Eng. Des.* **2013**, *34*, 3635–3638.

3. Zhan, S.; Zhang, Q.X.; Jiang, J.G.; Ando, S. 3D face recognition by kernel collaborative representation based on Gabor feature. *Acta Photonica Sin.* **2013**, *42*, 1448–1453. [CrossRef]

4. Li, H.; Ding, X. Application research on improved fusion algorithm based on BP neural network and POS. *Appl. Mech. Mater.* **2015**, *733*, 898–901. [CrossRef]

5. Govindaraju, R.S. Artificial neural networks in hydrology. I: Preliminary concepts. *J. Hydrol. Eng.* **2000**, *5*, 115–123.

6. Shao, J. Applications of wavelet fuzzy neural network in approximating non-linear functions. *Comput. Dig. Eng.* **2013**, *41*, 4–6.

7. Wu, W.J.; Huang, D.G. Research on fault diagnosis for rotating machinery vibration of aero-engine based on wavelet transformation and probabilistic neural network. *Adv. Mater. Res.* **2011**, *295–297*, 2272–2278. [CrossRef]

8. Kwolek, B. Face Detection Using Convolutional Neural Networks And Gabor Filters. In Proceedings of the International Conference Artificial Neural Networks: Biological Inspirations (ICANN 2005), Warsaw, Poland, 11–15 September 2005; Volume 3696, pp. 551–556.

9. Kaushal, A.; Raina, J.P.S. Face detection using neural network & Gabor wavelet transform. *Int. J. Comput. Sci. Technol.* **2010**, *1*, 58–63.

10. Andrzej, B.; Teresa, N.; Stefan, P. Face detection and recognition using back propagation neural network and fourier gabor filters. *Signal Image Process.* **2011**, *2*, 705–708.

11. Yale Face Database. Available online: http://vision.ucsd.edu/content/yale-face-database (accessed on 14 November 2016).

12. Database of Faces. 2016. Available online: http://www.cl.cam.ac.uk/research/dtg/attarchive/facedatabase.html (accessed on 14 November 2016).

13. Extended Yale Face Database B (B+). Available online: http://vision.ucsd.edu/content/extended-yale-face-database-b-b (accessed on 14 November 2016).

14. Khalil, M.S. Erratum to: Reference point detection for camera-based fingerprint image based on wavelet transformation. *Biomed. Eng. Online* **2016**, *15*, 30. [CrossRef] [PubMed]

15. Yifrach, A.; Novoselsky, E.; Solewicz, Y.A.; Yitzhaky, Y. Improved nuisance attribute projection for face recognition. *Form. Pattern Anal. Appl.* **2016**, *19*, 69–78. [CrossRef]

16. Lades, M.; Vorbruggen, J.C.; Buhmann, J.; Lange, J.; Malsburg, C.V.D.; Wurtz, R.P. Distortion invariant object recognition in the dynamic link architecture. *IEEE Trans. Comput.* **1993**, *42*, 300–311. [CrossRef]

17. Gao, Y.; Zhang, R. Analysis of house price prediction based on genetic algorithm and BP neural network. *Comput. Eng.* **2014**, *40*, 187–191.

18. Lee, K.C.; Ho, J.; Kriegman, D. Acquiring Linear Subspaces for Face Recognition under Variable Lighting. *IEEE Trans. Pattern Anal. Mach. Intell.* **2005**, *27*, 684–698. [PubMed]

19. Ramakrishnan, S. *Introductory Chapter: Face Recognition-Overview, Dimensionality Reduction, and Evaluation Methods. Face Recognition-Semisupervised Classification, Subspace Projection and Evaluation Methods*; InTech: Vienna, Austria, 2016.

20. Nanni, L.; Brahnam, S.; Ghidoni, S.; Menegatti, E. Region based approaches and descriptors extracted from the cooccurrence matrix. *Int. J. Latest Res. Sci. Technol.* **2014**, *3*, 192–200.

21. Xie, S.; Shan, S.; Chen, X.; Chen, J. Fusing local patterns of Gabor magnitude and phase for face recognition. *IEEE Trans. Image Process.* **2010**, *19*, 1349–1361. [PubMed]

symmetry

MDPI

Article

Some Invariants of Circulant Graphs

Mobeen Munir [1], Waqas Nazeer [1], Zakia Shahzadi [1] and Shin Min Kang [2,3,*]

[1] Division of Science and Technology, University of Education, Lahore 54000, Pakistan;
    mmunir@ue.edu.pk (M.M.); nazeer.waqas@ue.edu.pk (W.N.); zakiashahzadi@yahoo.com (Z.S.)
[2] Department of Mathematics and Research Institute of Natural Science, Gyeongsang National University,
    Jinju 52828, Korea
[3] Center for General Education, China Medical University, Taichung 40402, Taiwan
*   Correspondence: smkang@gnu.ac.kr; Tel.: +82-55-772-1420

Academic Editor: Angel Garrido
Received: 9 October 2016; Accepted: 11 November 2016; Published: 18 November 2016

Abstract: Topological indices and polynomials are predicting properties like boiling points, fracture toughness, heat of formation, etc., of different materials, and thus save us from extra experimental burden. In this article we compute many topological indices for the family of circulant graphs. At first, we give a general closed form of M-polynomial of this family and recover many degree-based topological indices out of it. We also compute Zagreb indices and Zagreb polynomials of this family. Our results extend many existing results.

Keywords: circulant graphs; topological indices; polynomials

1. Introduction

A number, polynomial or a matrix can uniquely identify a graph. A topological index is a numeric number associated to a graph which completely describes the topology of the graph, and this quantity is invariant under the isomorphism of graphs. The degree-based topological indices are derived from degrees of vertices in the graph. These indices have many correlations to chemical properties. In other words, a topological index remains invariant under graph isomorphism.

The study of topological indices, based on distance in a graph, was effectively employed in 1947 in chemistry by Weiner [1]. He introduced a distance-based topological index called the "Wiener index" to correlate properties of alkenes and the structures of their molecular graphs. Recent progress in nano-technology is attracting attention to the topological indices of molecular graphs, such as nanotubes, nanocones, and fullerenes to cut short experimental labor. Since their introduction, more than 140 topological indices have been developed, and experiments reveal that these indices, in combination, determine the material properties such as melting point, boiling point, heat of formation, toxicity, toughness, and stability [2]. These indices play a vital role in computational and theoretical aspects of chemistry in predicting material properties [3–8].

Several algebraic polynomials have useful applications in chemistry, such as the Hosoya Polynomial (also called the Wiener polynomial) [9]. It plays a vital role in determining distance-based topological indices. Among other algebraic polynomials, the M-polynomial—introduced recently in 2015 [10]—plays the same role in determining the closed form of many degree-based topological indices. Other famous polynomials are the first Zagreb polynomial and the second Zagreb polynomial.

A graph G is an ordered pair (V, E), where V is the set of vertices and E is the set of edges. A path from a vertex v to a vertex w is a sequence of vertices and edges that starts from v and stops at w. The number of edges in a path is called the length of that path. A graph is said to be connected if there is a path between any two of its vertices. The distance $d(u, v)$ between two vertices u, v of a connected graph G is the length of a shortest path between them. Graph theory is contributing a lion's share in many areas such as chemistry, physics, pharmacy, as well as in industry [11]. We will start with some preliminary facts.

Let G be a simple connected graph and let uv represent the edge between the vertices u and v. The number of vertices of G, adjacent to a given vertex v, is the "degree" of this vertex, and will be denoted by d_v. We define $V_k = \{v \in V(G) | d_v = k\}$, $E_{i,j} = \{uv \in E(G) | d_u = j \text{ and } d_v = i\}$, $\delta = Min\{d_v | v \in V(G)\}$, $\triangle = Max\{d_v | v \in V(G)\}$, and m_{ij} as the number of edges uv of G such that $\{d_v, d_u\} = \{i, j\}$. The M-polynomial of G is defined as:

$$M(G, x, y) = \sum_{\delta \leq i \leq j \leq \triangle} m_{ij} x^i y^j \tag{1}$$

Active research is in progress, and many authors computed M-polynomials for different types of nonmaterial, for example see [12–16] and the references therein.

The Wiener index of G is defined as:

$$W(G) = \frac{1}{2} \sum_{(u,v)} d(u,v) \tag{2}$$

where (u,v) is any ordered pair of vertices in G. Gutman and Trinajstić [11] introduces important topological index called first Zagreb index, denoted by $M_1(G)$, and is defined as:

$$M_1(G) = \sum_{uv \in E(G)} (d_u + d_v) \tag{3}$$

The second Zagreb index $M_2(G)$ and the second modified Zagreb index $^mM_2(G)$ are defined as:

$$M_2(G) = \sum_{uv \in E(G)} (d_u \times d_v) \tag{4}$$

and:

$$^mM_2(G) = \sum_{uv \in E(G)} \frac{1}{d_u \cdot d_v} \tag{5}$$

Results obtained in the theory of Zagreb indices are summarized in the review [17].

In 1998, working independently, Bollobas and Erdos [18] and Amic et al. [19] proposed general Randić index. It has been extensively studied by both mathematicians and theoretical chemists (see, for example, [20,21]). The Randić index denoted by $R_\alpha(G)$ is the sum of $(d_u d_v)^\alpha$; i.e.:

$$R_\alpha(G) = \sum_{uv \in E(G)} (d_u d_v)^\alpha \tag{6}$$

where α is any constant.

The symmetric division index is defined by:

$$SDD(G) = \sum_{uv \in E(G)} \left(\frac{min\{d_u, d_v\}}{max\{d_u, d_v\}} + \frac{max\{d_u, d_v\}}{min\{d_u, d_v\}} \right) \tag{7}$$

These indices can help to characterize the chemical and physical properties of molecules (see [9]).

Table 1 enlists some standard degree-based topological indices and their derivation from M-polynomial [10].

Table 1. Derivation of topological indices from M-polynomial.

Topological Index	$f(x,y)$	Derivation from $M(G,x,y)$
First Zagreb	$x+y$	$(D_x+D_y)(M(G;x,y))\|_{x=y=1}$
Second Zagreb	xy	$(D_xD_y)(M(G;x,y))\|_{x=y=1}$
$^mM_2(G)$	$\frac{1}{xy}$	$(S_xD_y)(M(G;x,y))\|_{x=y=1}$
General Randić $\alpha \epsilon N$	$(xy)^\alpha$	$(D_x^\alpha D_y^\alpha)(M(G;x,y))\|_{x=y=1}$
General Randić $\alpha \epsilon N$	$\frac{1}{xy}^\alpha$	$(S_x^\alpha S_y^\alpha)(M(G;x,y))\|_{x=y=1}$
Symmetric Division Index	$\frac{x^2+y^2}{xy}$	$(D_xS_y+D_yS_x)(M(G;x,y))\|_{x=y=1}$

Where $D_x(f(x,y)) = x\frac{\partial f(x,y)}{\partial x}, D_y(f(x,y)) = y\frac{\partial f(x,y)}{\partial y}, S_x(f(x,y)) = \int_0^x \frac{f(t,y)}{t}dt, S_y(f(x,y)) = \int_0^y \frac{f(x,t)}{t}dt$.
For a simple connected graph, the first Zagreb polynomial is defined as:

$$M_1(G,x) = \sum_{uc\in E(G)} x^{[d_u+d_v]} \tag{8}$$

and the second Zagreb polynomial is defined as:

$$M_2(G,x) = \sum_{uc\in E(G)} x^{[d_u\times d_v]} \tag{9}$$

In 2013, Shirdel et al. in [22] proposed the hyper-Zagreb index, which is also degree-based, given as:

$$HM(G) = \sum_{uc\in E(G)} [d_u+d_v]^2 \tag{10}$$

In 2012, Ghorbani and Azimi [23] proposed two new variants of Zagreb indices; namely, the first multiple Zagreb index $PM_1(G)$ and the second multiple Zagreb index $PM_2(G)$, which are defined as:

$$PM_1(G) = \prod_{uv\in E(G)} [d_u+d_v] \tag{11}$$

$$PM_2(G) = \prod_{uv\in E(G)} [d_u \times d_v] \tag{12}$$

In this paper, we address the family of circulant graphs. We give closed forms of M-polynomial and Zagreb Polynomials for this family. We also compute many degree-based topological indices.

Definition 1. *Let n, m, and $a_1,\ldots,a_m$ be positive integers, where $1 \leq a_i \leq \lfloor\frac{n}{2}\rfloor$ and $a_i \neq a_j$ for all $1 \leq i < j \leq m$. An undirected graph with the set of vertices $V = \{v_1,\ldots,v_n\}$ and the set of edges $E = \{v_iv_{i+a_j} : 1 \leq i \leq n, 1 \leq j \leq m\}$, where the indices being taken modulo n, is called the circulant graph, and is denoted by $C_n(a_1,\ldots,a_m)$.*

The graph of $C_{11}(1,2,3)$ is shown in Figure 1.

Figure 1. $C_{11}(1,2,3)$.

This is one of the most comprehensive families, as its specializations give some important families. Classes of graphs that are circulant include the Andrásfai graphs, antiprism graphs, cocktail party graphs, complete graphs, complete bipartite graphs, crown graphs, empty graphs, rook graphs, Möbius ladders, Paley graphs of prime order, prism graphs, and torus grid graphs. Special cases are summarized in the Table 2.

Table 2. Special cases of circulant graphs.

Graph	Symbol	Graph	Symbol
2-path graph	$C_{i_2}(1)$	triangle graph	$C_{i_3}(1)$
square graph	$C_{i_4}(1)$	tetrahedral graph	$C_{i_4}(1,2)$
5-cycle graph	$C_{i_5}(1)$	pentatope graph	$C_{i_5}(1,2)$
6-cycle graph	$C_{i_6}(1)$	octahedral graph	$C_{i_6}(1,2)$
utility graph	$C_{i_6}(1,3)$	3-prism graph	$C_{i_6}(2,3)$
6-complete graph	$C_{i_6}(1,2,3)$	7-cycle graph	$C_{i_7}(1)$
7-complete graph	$C_{i_7}(1,2,3)$	8-cycle graph	$C_{i_8}(1)$
4-antiprism graph	$C_{i_8}(1,2)$	(4,4)-complete bipartite graph	$C_{i_8}(1,3)$
4-Möbius ladder graph	$C_{i_8}(1,4)$	16-cell graph	$C_{i_8}(1,2,3)$
8-complete graph	$C_{i_8}(1,2,3,4)$	9-cycle graph	$C_{i_9}(1)$
9-complete graph	$C_{i_9}(1,2,3,4)$	10-cycle graph	$C_{i_{10}}(1)$
5-antiprism graph	$C_{i_{10}}(1,2)$	5-crown graph	$C_{i_{10}}(1,3)$
5-Möbius ladder graph	$C_{i_{10}}(1,5)$	5-prism graph	$C_{i_{10}}(2,5)$
5-cocktail party graph	$C_{i_{10}}(1,2,3,4)$	(5,5)-complete bipartite graph	$C_{i_{10}}(1,3,5)$

Because of this somewhat universality, circulant graphs have been the subject of much investigation; for example, the chromatic index for circulant graphs is computed in [24]. Connectivity is discussed in [25], and the Weiner index is computed in [26]. Exact values of the domination number of some families of circulant graphs are given in [27]. Habibi et. al. computed the revised Szeged spectrum of circulant graphs [28]. Multi-level and antipodal labelings for circulant graphs is discussed in [29,30].

2. Main Theorem

We divided our main results into two parts.

2.1. Polynomials

In this section, we computed the closed forms of some polynomials.

Theorem 1. *Let $C_n(a_1, a_2, ..., a_m)$ be a circulant graph. Then, the M-Polynomial is:*

$$M((C_n(a_1, a_2, ..., a_m), x, y) = nx^{n-1}y^{n-1}$$

Proof. Let $C_n(a_1, a_2, ..., a_m)$, where $n = 3, 4...n$. and $1 \le a_i \le \lfloor \frac{n}{2} \rfloor$ and $a_i \ne a_j$ when $n =$ even and when $1 \le a_i \le \lfloor \frac{n}{2} \rfloor$ and $a_i < a_j$ when $n =$ odd be the circulant graph. From the structure of $C_n(a_1, a_2, ..., a_m)$, we can see that there is one partition $V_{\{1\}} = \{v \in V(C_n(a_1, a_2, ..., a_m)) | d_v = n\}$. We see that the edge set of $C_n(a_1, a_2, ..., a_m)$ partitions as follows:

$$E_{\{n-1,n-1\}} = \{e = uv \in E(C_n(a_1, a_2, ..., a_m)) | d_u = n - 1 \& d_v = n - 1\} \to |E_{\{n-1,n-1\}}| = n$$

Thus, the M-Polynomial of $(C_n(a_1, a_2, ..., a_m), x, y)$ is:

$$
\begin{aligned}
M(C_n(a_1, a_2, ..., a_m), x, y) &= \sum_{i \le j} m_{ij}(C_n(a_1, a_2, ..., a_m)) x^i y^j \\
&= \sum_{n-1 \le n-1} m_{n-1 \times n-1}(C_n(a_1, a_2, ..., a_m)) x^{n-1} y^{n-1} \\
&= \sum_{uv \in E_{\{n-1, n-1\}}} m_{n-1 \times n-1}(C_n(a_1, a_2, ..., a_m)) x^{n-1} y^{n-1} \\
&= |E_{\{n-1, n-1\}}| x^{n-1} y^{n-1} \\
&= n x^{n-1} y^{n-1}
\end{aligned}
$$

□

In the following theorem, we computed first and second Zagreb polynomials.

Theorem 2. *Let $C_n(a_1, a_2, ..., a_m)$ be a circulant graph. Then:*

(1) $M_1(C_n(a_1, a_2, ..., a_m), x) = n x^{2(n-1)}$
(2) $M_2(C_n(a_1, a_2, ..., a_m), x) = n x^{(n-1)^2}$

Proof. Let $C_n(a_1, a_2, ..., a_m)$ be a complete circulant graph. The edge set of $C_n(a_1, a_2, ..., a_m)$ has one partition based on degree of vertices. The edge partition has n edges uv, where $d_u = d_v = n - 1$. It is easy to see that $|E_1(C_n(a_1, a_2, ..., a_m))| = d_{n-1 \times n-1}$. Now we have:

(1)

$$
\begin{aligned}
M_1(C_n(a_1, a_2, ..., a_m)) &= \sum_{uv \in E(C_n(a_1, a_2, ..., a_m))} x^{[d_u + d_v]}, \\
&= \sum_{uv \in E_1(C_n(a_1, a_2, ..., a_m))} x^{[d_u + d_v]} \\
&= |E_1(C_n(a_1, a_2, ..., a_m))| x^{2(n-1)} \\
&= n x^{2(n-1)}
\end{aligned}
$$

(2)

$$
\begin{aligned}
M_2(C_n(a_1, a_2, ..., a_m)) &= \sum_{uv \in E(C_n(a_1, a_2, ..., a_m))} x^{[d_u \times d_v]} \\
&= \sum_{uv \in E_1(C_n(a_1, a_2, ..., a_m))} x^{[d_u \times d_v]} \\
&= |E_1(C_n(a_1, a_2, ..., a_m))| x^{(n-1)^2} \\
&= n x^{(n-1)^2}
\end{aligned}
$$

□

2.2. Topological Indices

In this section, we will recover some topological indices from polynomials computed in the above section.

Theorem 3. *For the circulant graph $C_n(a_1, a_2, ..., a_m)$, we have:*

(1) $M_1(C_n(a_1, a_2, ..., a_m)) = 2n(n - 1)$
(2) $M_2(C_n(a_1, a_2, ..., a_m)) = n(n - 1)^2$
(3) $^m M_2(C_n(a_1, a_2, ..., a_m)) = \frac{n}{(n-1)^2}$

(4) $R_\alpha(C_n(a_1, a_2, ..., a_m)) = n\{(n-1)^2\}^\alpha$
(5) $R_\alpha(C_n(a_1, a_2, ..., a_m)) = \frac{n}{(n-1)^\alpha}$
(6) $SDD(C_n(a_1, a_2, ..., a_m)) = 2n$

Proof. Let $f(x, y) = M((C_n(a_1, a_2, ..., a_m), x, y) = nx^{n-1}y^{n-1}$. Then:

$$D_x(f(x, y)) = n(n-1)x^{n-1}y^{n-1}$$

$$(D_x f(x, y))(M(C_n(a_1, a_2, ..., a_m); x, y))|_{x=y=1} = n(n-1)$$

$$D_y(f(x, y)) = n(n-1)x^{n-1}y^{n-1}$$

$$(D_x f(x, y))(M(C_n(a_1, a_2, ..., a_m); x, y))|_{x=y=1} = n(n-1)$$

$$S_x f(x, y) = \frac{n}{n-1}x^{n-1}y^{n-1}$$

$$(S_x f(x, y))(M(C_n(a_1, a_2, ..., a_m); x, y))|_{x=y=1} = \frac{n}{n-1}$$

$$S_y f(x, y) = \frac{n}{n-1}x^{n-1}y^{n-1}$$

$$(S_y f(x, y))(M(C_n(a_1, a_2, ..., a_m); x, y))|_{x=y=1} = \frac{n}{n-1}$$

(1) $M_1(C_n(a_1, a_2, ..., a_m))$:

$$(D_x + D_y)f(x, y))(M(C_n(a_1, a_2, ..., a_m); x, y))|_{x=y=1} = 2n(n-1)$$

(2) $M_2(C_n(a_1, a_2, ..., a_m))$:

$$(D_x D_y)f(x, y))(M(C_n(a_1, a_2, ..., a_m); x, y))|_{x=y=1} = n(n-1)^2$$

(3) $^m M_2(C_n(a_1, a_2, ..., a_m))$:

$$(S_x S_y)f(x, y))(M(C_n(a_1, a_2, ..., a_m); x, y))|_{x=y=1} = \frac{n}{(n-1)^2}$$

(4) $R_\alpha(C_n(a_1, a_2, ..., a_m))$:

$$(D_x^\alpha D_y^\alpha)f(x, y))(M(C_n(a_1, a_2, ..., a_m); x, y))|_{x=y=1} = n(n-1)^{2\alpha}$$

(5) $R_\alpha(C_n(a_1, a_2, ..., a_m))$:

$$(S_x^\alpha S_y^\alpha)f(x, y))(M(C_n(a_1, a_2, ..., a_m); x, y))|_{x=y=1} = \frac{n}{(n-1)^\alpha}$$

(6) $SDD(C_n(a_1, a_2, ..., a_m))$:

$$(D_x S_y + D_y S_x)(M(C_n(a_1, a_2, ..., a_m); x, y))|_{x=y=1} = 2n$$

□

Theorem 4. *Let* $(C_n(a_1, a_2, ..., a_m)$ *be a circulant graph. Then:*

(1) $PM_1(M_n) = 2(n-1)^n$
(2) $PM_2(M_n) = \{(n-1)^2\}^n$
(3) $HM(M_n) = \{2(n-1)\}^2(n)$

Proof. Let $C_n(a_1, a_2, ..., a_m)$ be a complete circulant graph. The edge set of $C_n(a_1, a_2, ..., a_m)$ has one partition based on degree of vertices. The edge partition has n edges uv, where $d_u = d_v = n - 1$. It is easy to see that $|E_1(M_n)| = d_{(n-1) \times (n-1)}$. Now, we have:

(1)

$$
\begin{aligned}
PM_1(C_n(a_1, a_2, ..., a_m)) &= \prod_{uv \in E(C_n(a_1, a_2, ..., a_m))} [d_u + d_v] \\
&= \prod_{uv \in E_1(C_n(S_i}))[d_u + d_v] \\
&= \{2(n-1)\}^{|E_1(C_n(a_1, a_2, ..., a_m))|} \\
&= 2(n-1)^n
\end{aligned}
$$

(2)

$$
\begin{aligned}
PM_2(C_n(a_1, a_2, ..., a_m)) &= \prod_{uv \in E(C_n(S_i}} [d_u \times d_v] \\
&= \prod_{uv \in E_1(C_n(a_1, a_2, ..., a_m))} [d_u \times d_v] \\
&= \{(n-1)^2\}^{|E_1(C_n(a_1, a_2, ..., a_m))|} \\
&= \{(n-1)^2\}^n
\end{aligned}
$$

(3)

$$
\begin{aligned}
HM(C_n(a_1, a_2, ..., a_m)) &= \sum_{uv \in E(C_n(a_1, a_2, ..., a_m))} [d_u + d_v]^2 \\
&= \sum_{uv \in E_1(C_n(a_1, a_2, ..., a_m))} [d_u + d_v]^2 \\
&= \{2(n-1)\}^2 |E_1(C_n(a_1, a_2, ..., a_m))| \\
&= \{2(n-1)\}^2 (n)
\end{aligned}
$$

□

3. Conclusions

In this article, we computed many topological indices for the family of circulant graphs. At first we give a general closed form of M-polynomial of this family and recover many degree-based topological indices out of it. We also compute Zagreb indices and Zagreb polynomials of this family. Our results actually extend many existing results about crown graphs, Paley graphs, complete bipartite, Möbius Ladders, any many other families; see Table 2.

Acknowledgments: This research is supported by Gyeongsang National University, Jinju 52828, Korea. We are thankful to the reviewers for suggestions that really improve this paper.

Author Contributions: All authors contributed equally to the writing of this paper. All authors read and approved the final manuscript.

Conflicts of Interest: The authors declare no conflict of interest.

References

1. Wiener, H. Structural determination of paraffin boiling points. *J. Am. Chem. Soc.* **1947**, *69*, 17–20.
2. Katritzky, A.R.; Jain, R.; Lomaka, A.; Petrukhin, R.; Maran, U.; Karelson, M. Perspective on the Relationship between Melting Points and Chemical Structure. *Cryst. Growth Design* **2001**, *1*, 261–265.
3. Rucker, G.; Rucker, C. On topological indices, boiling points, and cycloalkanes. *J. Chem. Inf. Comput. Sci.* **1991**, *39*, 788.

4. Dobrynin, A.A.; Entringer, R.; Gutman, I. Wiener index of trees: Theory and applications. *Acta Appl. Math.* **2001**, *66*, 211–249.

5. Du, W.; Li, X.; Shi, Y. Algorithms and extremal problem on Wiener polarity index. *MATCH Commun. Math. Comput. Chem.* **2009**, *62*, 235–244.

6. Gutman, I.; Polansky, O.E. *Mathematical Concepts in Organic Chemistry*; Springer: New York, NY, USA, 1986.

7. Ma, J.; Shi, Y.; Yue, J. The wiener polarity index of graph products. *Ars Combin.* **2014**, *116*, 235–244.

8. Ma, J.; Shi, Y.; Wang, Z.; Yue, J. On wiener polarity index of bicyclic networks. *Sci. Rep.* **2016**, *6*, 19066.

9. Gutman, I. *Some Properties of the Wiener Polynomial*; Graph Theory Notes: New York, NY, USA, 1993; Volume 125, pp. 13–18.

10. Klavzar, S.; Deutsch, E. M-Polynomial and Degree-Based Topological Indices. *Iran. J. Math. Chem.* **2015**, *6*, 93–102.

11. Gutman, I.; Trinajstic, N. Graph theory and molecular orbitals total ϕ-electron energy of alternant hydrocarbons. *Chem. Phys. Lett.* **1972**, *17*, 535–538.

12. Munir, M.; Nazeer, W.; Rafique, S.; Kang, S.M. M-polynomial and degree-based topological indices of Nano star dendrimers. *Symmetry* **2016**, *8*, 97.

13. Munir, M.; Nazeer, W.; Rafique, S.; Nizami, A.R.; Kang, S.M. M-polynomial and degree-based topological indices of Titania Nanotubes. *Symmetry* **2016**, *8*, 117.

14. Munir, M.; Nazeer, W.; Rafique, S.; Kang, S.M. M-polynomial and degree-based topological indices of Buckytubes. *Symmetry* **2016**, submitted.

15. Kang, S.; Munir, M.; Nizami, A.; Shahzadi, Z.; Nazeer, W. Some Topological Invariants of the Möbius Ladder. Preprints 2016, 2016110040, doi:10.20944/preprints201611.0040.v1.

16. Munir, M.; Nazeer, W.; Rafique, S.; Nizami, A.; Kang, S.M. Some Computational Aspects of Triangular Boron Nanotubes. *Symmetry* **2016**, doi:10.20944/preprints201611.0041.v1.

17. Gutman, I.; Das, K.C. The first Zagreb indices 30 years after. *MATCH Commun. Math. Comput. Chem.* **2004**, *50*, 83–92.

18. Bollobas, B.; Erdös, P. Graphs of extremal weights. *Ars Combin.* **1998**, *50*, 225–233.

19. Amic, D.; Beslo, D.; Lucic, B.; Nikolic, S.; Trinajstić, N. The Vertex-Connectivity Index Revisited. *J. Chem. Inf Comput. Sci.* **1998**, *38*, 819–822.

20. Kier, L.B.; Hall, L.H. *Molecular Connectivity in Structure-Activity Analysis*; John Wiley & Sons: New York, NY, USA, 1986.

21. Li, X.; Gutman, I. *Mathematical Aspects of Randic-Type Molecular Structure Descriptors*; Mathematical Chemistry Monographs, No. 1; University of Kragujevac: Kragujevac, Serbia, 2006.

22. Shirdel, G.H.; Pour, H.R.; Sayadi, A.M. The hyper-Zagreb index of graph operations. *Iran. J. Math. Chem.* **2013**, *4*, 213–220

23. Ghorbani, M.; Azimi, N. Note on multiple Zagreb indices. *Iran. J. Math. Chem.* **2012**, *3*, 137–143.

24. Voigt, M.; Walther, H. On the chromatic number of special distance graphs. *Discrete Math.* **1991**, *97*, 395–397.

25. Boesch, F.; Tindell, R. Circulants and their connectivity. *J. Graph Theory* **1984**, *8*, 487–499.

26. Zhou, H. The Wiener Index of Circulant Graphs. *J. Chem.* **2014**, doi:10.1155/2014/742121.

27. Fu, X.L.; Yang, Y.S.; Jiang, B.Q. On the domination number of the circulant graphs $C(n; 1, 2)$, $C(n; 1, 3)$ and $C(n; 1, 4)$. *Ars Comb.* **2011**, *102*, 173–182.

28. Habibi, N.; Ashrafi, A.R. On revised szeged spectrum of a graph. *TAMKANG J. Math.* **2014**, *45*, 375–387.

29. Kang, S.M.; Nazeer, S.; Kousar, I.; Nazeer, W.; Kwun, Y.C. Multi-level and antipodal labelings for certain classes of circulant graphs. *J. Nonlinear Sci. Appl.* **2016**, *9*, 2832–2845.

30. Nazeer, S.; Kousar, I.; Nazeer, W. Radio and radio antipodal labelings for circulant graphs $G(4k + 2; 1; 2)$. *J. Appl. Math. Inf.* **2015**, *33*, 173–183.

symmetry

MDPI

Article

Segmentation of Brain Tumors in MRI Images Using Three-Dimensional Active Contour without Edge

Ali M. Hasan [1,2], Farid Meziane [1], Rob Aspin [1] and Hamid A. Jalab [3,*]

[1] School of Computing, Science and Engineering, University of Salford, Manchester M5 4WT, UK;
   a.hasan4@edu.salford.ac.uk (A.M.H.); f.meziane@salford.ac.uk (F.M.); r.aspin@salford.ac.uk (R.A.)
[2] Computers Unit, College of Medicine, Al-Nahrain University, Baghdad 64074, Iraq
[3] Faculty of Computer Science & Information Technology, University of Malaya,
   Kuala Lumpur 50603, Malaysia
*   Correspondence: hamidjalab@um.edu.my; Tel.: +60-3-7967-2503

Academic Editor: Angel Garrido
Received: 13 October 2016; Accepted: 14 November 2016; Published: 18 November 2016

Abstract: Brain tumor segmentation in magnetic resonance imaging (MRI) is considered a complex procedure because of the variability of tumor shapes and the complexity of determining the tumor location, size, and texture. Manual tumor segmentation is a time-consuming task highly prone to human error. Hence, this study proposes an automated method that can identify tumor slices and segment the tumor across all image slices in volumetric MRI brain scans. First, a set of algorithms in the pre-processing stage is used to clean and standardize the collected data. A modified gray-level co-occurrence matrix and Analysis of Variance (ANOVA) are employed for feature extraction and feature selection, respectively. A multi-layer perceptron neural network is adopted as a classifier, and a bounding 3D-box-based genetic algorithm is used to identify the location of pathological tissues in the MRI slices. Finally, the 3D active contour without edge is applied to segment the brain tumors in volumetric MRI scans. The experimental dataset consists of 165 patient images collected from the MRI Unit of Al-Kadhimiya Teaching Hospital in Iraq. Results of the tumor segmentation achieved an accuracy of 89% ± 4.7% compared with manual processes.

Keywords: magnetic resonance imaging; modified gray level co-occurrence matrix; three-dimensional active contour without edge; two-dimensional active contour without edge

1. Introduction

Brain tumors are relatively less common than other neoplasms, such as those of the lung and breast, but are considered highly important because of prognostic effects and high morbidity [1]. Clinical diagnosis, predicted prognosis, and treatment are significantly affected by the accurate detection and segmentation of brain tumors and stroke lesions [2].

The Iraqi Ministry of Health reported that the use of depleted uranium and other toxic substances in the first and second Gulf Wars had increased the average annual number of registered cancerous brain tumor cases and birth defects since 1990 [3]. This study was conducted in collaboration with MRI units in Iraqi hospitals that have witnessed the high numbers of these cases. The role of image processing in medicine has expanded with the progress of medical imaging technologies, and additional images are obtained using an increased number of acquisition modalities. Therefore, image processing was embedded in medical systems and used widely in medicine, from diagnosis to therapy. To date, diagnostic imaging is an invaluable tool in medicine. Standard medical imaging techniques, such as ultrasonography, computed tomography, and magnetic resonance imaging (MRI), have significantly increased knowledge on anatomy and disease diagnosis in medical research. Among these medical technologies, MRI is considered a more useful and appropriate imaging technique for brain tumors than other modalities. MRI presents detailed information on the type, position, and size of tumors in a noninvasive manner. Additionally, MRI is more sensitive to local changes in tissue density. Spatial resolution, which represents the digitization process of assigning a number to each pixel in the original image, has increased significantly in recent years. Standard MRI protocols are commonly used to produce multiple images of the same tissue with different contrast after the administration of parametric agents, including T1-weighted (T1-w), T2-weighted (T2-w), fluid-attenuated inversion recovery (FLAIR), and T1-weighted images with contrast enhancement (T1c-w). T1-w images are obtained during the T1 relaxation time of the excited net magnetization or protons to recover 63% of the original net magnetization after the radiofrequency pulse of the MRI scanner is switched off. By contrast, T2-w images are obtained during T2 relaxation time, which represents the required time for the decline of net magnetization to 37% of the original net magnetization [4]. FLAIR is a special protocol in MRI scanners and produces adaptive T2-w images by removing the signal of brain edema and other structures with high water content, such as cerebrospinal fluid (CSF) [5].

Most brain tumors appear as hypo-intense relative to normal brain tissue on T1-w images and hyper-intense on T2-w images. Therefore, T2-w images are commonly used for providing an initial assessment, identifying tumor types, and distinguishing tumors from non-tumor tissues [6]. A contrast material is commonly used to enhance the tumor boundary against the surrounding normal brain tissue on T1-w images. This technique enables tumor detection that cannot be distinguished and recognized from T2-w and T1-w images because of similarity with adjacent normal brain tissue [7]. In clinical routine, a T2-w scan is performed immediately after patient positioning to identify the tumor location. T1-w scan is used before and after contrast administration for tumors showing contrast enhancement. The T2-w scan in axial viewing with FLAIR is used to show non-enhanced tumors [6].

As scanner resolutions improved and slice thickness decreased, an increasing number of slices were produced and clinicians required increasing time to diagnose each patient from image sets. Therefore, automated tumor detection and segmentation have attracted considerable attention in the past two decades [8].

One particular challenge in imaging features is the similarity between tumors located inside the brain white matter and those that overlap intensity distributions with the gray matter. This pattern is particularly evident at the boundary between a tumor and the surrounding tissue. Partial volumes (PVs) are considered as boundary features containing a mixture of different tissue types [9]. The thicknesses of the image slices (5–7 mm) produce significant PV effects, in which individual image pixels describe more than one tissue type. As a result, peripheral tumor regions are misclassified. This occurrence is common in T2-w images. A similar problem occurs toward the outer brain edge, where the CSF and gray matter overlap with the image sample. This circumstance may generate image intensities that erroneously indicate tumor presence.

In the past few decades, the number of studies devoted to automated brain tumor segmentation has grown rapidly because of the progress in the medical imaging field [8]. Active contour models, or snakes, are highly important applications for brain tumor segmentation. These tools are strongly suitable for determining the boundary between the tumor and the surrounding tissue [10]. This approach enables segmentation, matching, and tracking of anatomical areas by exploiting conditions derived from the anatomical and biological knowledge regarding location, size, and shape of anatomical areas [11]. Active contour models are defined as curves or surfaces that move under the influence of weighted internal and external forces. Internal forces are responsible for curve smoothness, whereas external forces are responsible for the pushing and pulling of curves toward the anatomical area boundaries.

Generally, the active contour models suffer from the problem of initial contour determination and leakage in imprecise edges. The majority of the proposed approaches in brain abnormality detection and segmentation are limited by (i) computational complexity; the (ii) absence of full automation because of brain tumor diversity; and (iii) the problem of contour initialization and imprecise edges.

To overcome these problems, we developed a fully automated method for locating the initial contour and segmentation of brain tumors by using a three-dimensional active contour without edge (3DACWE). Moreover, we compared the resulting accuracies of 2D and 3D segmentations.

Our system is based on the use of a single MRI modality (T2-w images) in axial viewing for detecting brain abnormality instead of multi-modal MRI (e.g., sagittal and coronal images). The system searches in parallel for dissimilar regions corresponding to its reflection on the opposite hemisphere of the brain by exploiting normal brain structural symmetry. This method would help commence the segmentation process automatically. Consequently, the proposed system becomes fully automated and is independent from atlas registration to avoid any inaccurate registration process that may directly affect the precision of tumor segmentation. Such a strategy also does not require a prior skull-removing step.

The remaining sections of this paper are organized as follows: in Section 2, the proposed method is explained; in Section 3, experimental results are discussed while describing how to locate and identify the tumor; and in Section 3, the conclusions are given.

2. Proposed Method

This research aimed to develop an automated method that can locate the initial contour of brain tumor segmentation across all axial slices of volumetric MRI brain scans. The overall flow chart of the proposed method is shown in Figure 1.

2.1. Data Collection

The clinical image dataset consists of 165 MRI brain scans acquired during routine diagnostic procedure at the MRI Unit in Al-Kadhimiya Teaching Hospital in Baghdad, Iraq. This dataset was diagnosed and classified into normal and abnormal by the clinicians of this unit. The MRI slice sets were obtained using a SIEMENS MAGNETOM Avanto 1.5 Tesla scanner (Malvern, PA, USA) and PHILIPS Achieva 1.5 Tesla scanner (Best, Netherlands). The provided dataset consisted of tumors with different sizes, shapes, locations, orientations, and types. A total of 88 patients in this dataset exhibited different brain abnormalities with tumor sizes, shapes, locations, orientations, and types. The remaining patient images exhibited no detectable pathology. The dataset included the four MRI image modalities, namely, T2-w, T1-w, T1c-w, and FLAIR images, under axial viewing and 3–5 mm slice thickness. An additional enhanced dataset of 50 pathological patients was prepared, although the brain tumors were manually segmented and labeled by an expert in this unit who evaluates segmentation algorithm accuracy.

The standard benchmark Multimodal Brain Tumor Segmentation dataset (BRATS 2013) obtained from the International Conference on Medical Image Computing and Computer-Assisted Interventions [8] was adopted to evaluate the proposed method.

2.2. Image Preprocessing

The preprocessing step involved the performance of a set of algorithms on MRI brain scan slices as a preparation for the feature extraction step. This step included dimension resizing of the MRI slices, image enhancement by Gaussian filter, and normalization of MRI image intensity because of image intensity variation. Finally, mid-sagittal plane (MSP) detection and correction algorithm were implemented.

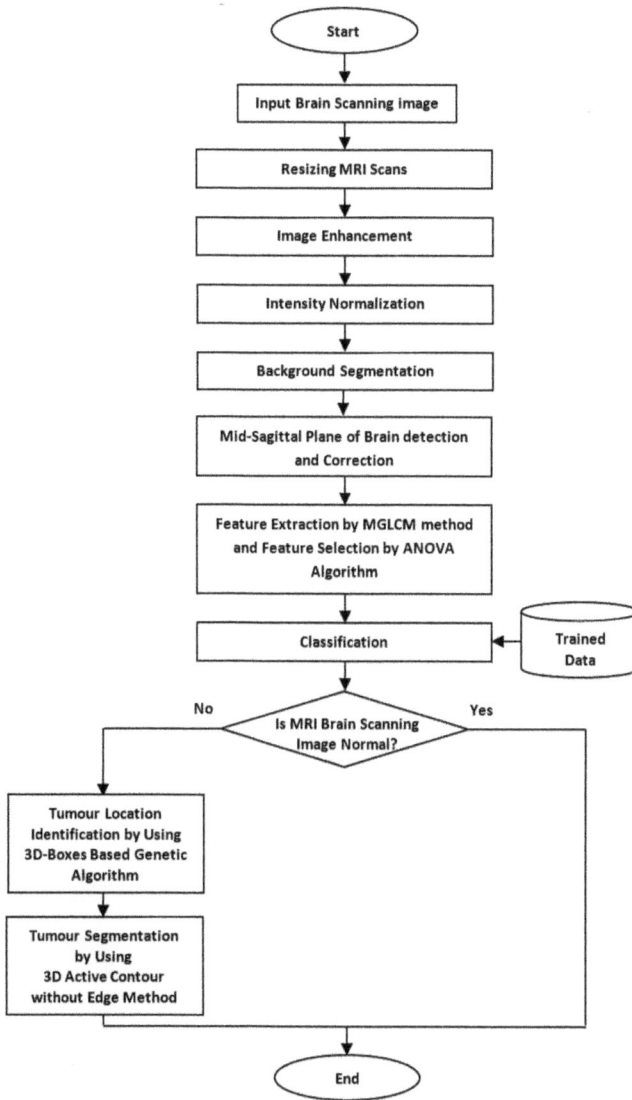

Figure 1. Flowchart of the proposed algorithm.

2.2.1. Resizing the Dimensions of MRI (Magnetic Resonance Imaging) Slices

The provided MRI brain slices were collected from two scanners with different spatial resolutions. To enable the use of the full set without bias, the MRI scans were resized to 512 × 512 pixels. All algorithms developed in this study were implemented on squared slices. When the dimensions of the given MRI slices were changed to a square ratio, care was taken to maintain the ratio of voxels to pixels (e.g., pixel spacing). The MRI slices were then resized by adding additional columns from the left and right and additional rows from the top and bottom portions of the MRI slice until the slice size became 512 × 512 pixels in resolution (Figure 2).

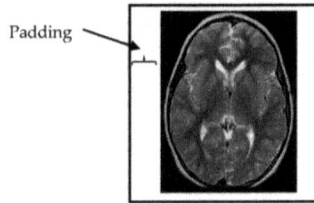

Figure 2. Padding of MRI brain image margins by zeros.

2.2.2. MRI Enhancement Algorithm

The typical noise in MRI images appeared as a small random modification of the intensity in an individual or a small groups of pixels. These differences can be sufficiently large to lead to erroneous segmentation. A spatial domain low-pass filter (Gaussian filter, The Math Works, Natick, MA, USA) was used and contributed a negative effect to the responses of noise smoothing linear image enhancement. Consequently, the performance of the Gaussian filter was evaluated visually because of its preferably low value for σ [11,12].

2.2.3. Intensity Normalization

The pixel intensity values of each MRI slice were normalized to the same intensity interval to achieve dynamic range consistency. Histogram normalization was then applied to stretch and shift the original histogram of the image and cover all the grayscale levels in the image. The resulting normalized image achieved a higher contrast than that of the original image because the histogram normalization method enhanced image contrast and provided a wider range of intensity transformation. This approach demonstrated an enhanced classification of pathological tissues that can be achieved using the unmodified image [13].

2.2.4. Background Segmentation

Prior knowledge suggests that the background intensity values of MRI brain slices often approaches zero to enable background segmentation. The ability to eliminate and exclude the background from the region of interest is important because the background normally contains a much higher number of pixels than that of the brain region but without meaningful information [13]. In this study, histogram thresholding was used as a segmentation method to isolate the background. This approach is based on the thresholding of intensity values by a specific T value. Subsequently, the application employs a set of morphological operators to remove any hole appearing in the region. Notably, the T2-w MR image histograms attained almost identical distribution shapes [14]. Therefore, the T value was selected experimentally and set to 0.1 after the effects of a range of threshold values (0.05, 0.1, 0.2, and 0.3) were manually observed. Hence, if an intensity value of a pixel is less than 0.1, the pixel is considered as a background.

2.2.5. Mid-Sagittal Plane Detection and Correction

Mid-sagittal plane identification is an important initial step in brain image analysis because this method provides an initial estimation of the brain's pathology assessment and tumor detection. The human brain is divided into two hemispheres with an approximately bilateral symmetry around the MSP. The two hemispheres are separated by the longitudinal fissure, which represents a membrane between the left and right hemisphere. MSP extraction methods can be divided into two groups as follows. Content-based methods find a plane that maximizes a symmetrical measure between both sides of the brain. By contrast, shape-based methods use the inter-hemispheric fissure as a simple landmark to extract and detect the MSP. In this study, we focused on determining the orientation of the patient's head instead of measuring the symmetry to identify the brain MSP [3].

2.3. Feature Extraction

The fundamental objective of any diagnostic medical imaging investigation is tissue characterization. Texture analysis is commonly used to provide unique information on the intensity variation of spatially related pixels in medical images [2]. The choice of an appropriate technique for feature extraction depends on the particular image and application [4]. Texture features are extracted from MRI brain slices to encode clinically valuable information by using modified gray-level co-occurrence matrix (MGLCM). This method is a second-order statistical method proposed by Hasan and Meziane [3] to generate textural features and provide information about the patterning of MRI brain scan textures. These features are used to measure statistically the degree of symmetry between the two brain hemispheres. Symmetry is an important indicator that can be used to detect the normality and abnormality of the human brain. MGLCM generates texture features by computing the spatial relationship of the joint frequencies of all pairwise combinations of gray-level configuration of each pixel in the left hemisphere. These pixels are considered as reference pixels, with one of nine opposite pixels existing in the right hemisphere under nine offsets and one distance. Therefore, nine co-occurrence matrices are generated for each MRI brain scanning image.

To reduce the dimensionality of the feature space, we added the resultant MGLCM matrices of all the MRI slices at all orientations. The maximum number of gray levels considered for each image was typically scaled down to 256 gray levels (8 bits/pixel), rather than using the full dynamic range of 65,536 gray levels (16 bits/pixel) before computing the MGLCM. This quantization step was essential to reduce a large number of zero-valued entries in the co-occurrence matrix [15,16]. The computing time for implementing MGLCM for each slice was about 2.3 min by using an HP workstation Z820 (Natick, MA, USA) with Xeon E5-3.8 GHz (Quad-Core) and 16 GB of RAM (random access memory).

2.3.1. Feature Aggregation

The MGLCM method determines nine co-occurrence matrices. For each matrix, 21 statistical descriptors are determined, generating 189 descriptors for each MRI brain scan [3]. The cross correlation descriptor is also determined for the original MRI brain scan. Accordingly, 190 descriptors are attained for each MRI brain scan image. These features are used by the subsequent classification to differentiate between normal and abnormal brain images.

2.3.2. Feature Selection

High-dimensional feature sets can negatively affect the classification results because high numbers of features may reduce the classification accuracy owing to the redundancy or irrelevance of some features. Feature-selection techniques aim to identify a small subset of features that minimizes redundancy and maximizes relevancy. Therefore, feature selection is an important step in exposing the most informative features and for optimally tuning the classifier's performance to reliably classify unknown data. In this study, ANOVA was employed to measure feature significance and relevance [3].

2.4. Classification

Classification is the process of sorting objects in images into separate classes and plays an important role in medical imaging, especially in tumor detection and classification. This step is also a common process employed in many other applications, such as robotic and speech recognition [4]. In the present study, a multi-layer perceptron neural network (MLP) was adopted to classify MRI brain scans into normal and abnormal images. MLP is used in different applications, such as optimization, classification, and feature extraction [3].

2.5. Brain Tumors Location Identification

Many tumor-segmentation methods are not fully automated. These approaches require user involvement in selecting a seed point. Usually, the MRI slices of a patient are interpreted visually and subjectively by radiologists, in which tumors are segmented by hand or by semi-automatic tools. Both manual and semiautomatic approaches are considered as tedious, time-consuming, error-prone processes. Tumors are more condensed than the surrounding material and present as brighter pixels than the surrounding brain tissue. Therefore, the basic concept of brain tumor detection algorithms is finding pixel clusters with a different or higher intensity than that of their surroundings. In this study, a bounding 3D-box-based genetic algorithm (BBBGA) method was proposed by Hasan [17] to search and identify the location of most dissimilar regions between the left and right hemispheres of the brain automatically without the need for user interaction. The input was a set of MR slices belonging to the scans of a single patient, and its output was a subset of slices covering and circumscribing the tumor with a 3D box. The BBBGA method exploits the symmetry feature of axial viewing of MRI brain slices to search for the most dissimilar region between the left and right brain hemispheres. This dissimilarity is detected using genetic algorithm (GA) and an objective-function-based mean intensity computation. The process involves randomly generating hundreds of 3D boxes with different sizes and locations in the left brain hemisphere. Such boxes are then compared with the corresponding 3D boxes in the right brain hemisphere through the objective function. These 3D boxes are moved and updated during the iterations of the GA toward the region that maximized the objective function value. An advantage of the BBBGA method is its lack of necessity for image registration or intensity standardization in MR slices. The approach is an unsupervised method; hence, the problems on observer variability in supervised techniques are ignored.

Prior to BBBGA, exponential transformation is implemented to compress the low-contrast regions in MRI brain images and expand the high-contrast regions in a nonlinear manner. This action would increase the intensity difference between the brain tumor and the surrounding soft tissue [17,18]. Figure 3 illustrates the pseudo-code for BBBGA.

1. read **Img** ← **MRI slices**
2. **Img** ← **exp(Img)**
3. // Initialization
4. generate *N* feasible individuals *ind* randomly in current population , which is set experimentally to 100.
5. compute *mean*(Img(*ind* $_i$)) for each $i \in N$
6. // loop until termination condition is achieved
7. for $i = 1$ to N
8. // Selection
9. select the best two individual from current population (*ind* $_1$, *ind* $_2$)
10. // Crossover
11. *newind* $_1$, *newind* $_2$ ← with crossover-probability crossover *ind* $_1$, *ind* $_2$
12. // Mutation
13. *newind* $_1$ ← with mutation-probability mutate *ind* $_1$
14. *newind* $_2$ ← with mutation-probability mutate *ind* $_2$
15. // Evaluation *newind* $_1$, *newind* $_2$
16. compute *mean*(Img(*newind* $_1$))
17. compute *mean*(Img(*newind* $_2$))
18. new population ← newind$_1$, newind$_2$
19. endfor

Figure 3. Pseudo-code for BBBGA.

For additional details on how each individual in the GA population is mapped into binary form, we use the following scenario. Suppose we have a MRI brain scan (dimensions $512 \times 512 \times 32$ pixels) of a pathological patient, each individual in the GA population is denoted by the binary representation of the coordinates of one 3D box $(x_1, x_2, y_1, y_2, z_1,$ and $z_2)$. In this case, x_1 and x_2 represent the height of the 3D box and are subject to the constraints $1 \leq x_1 < 512$ and $x_1 < x_2 \leq 512$. Meanwhile, y_1 and y_2 signify the width of the 3D box and are subject to the constraints $1 \leq y_1 < 256$ and $y_1 < y_2 \leq 256$. Finally, z_1 and z_2 represent the depth of the 3D box and are subject to the constraints $1 \leq z_1 < 32$ and $z_1 < z_2 \leq 32$. Herein, we assume that the maximum number of MRI slices is 32. Figure 4 shows an example of how the coordinates of 3D box $(x_1, x_2, y_1, y_2, z_1, z_2)$ are mapped to the individual of GA in a binary form.

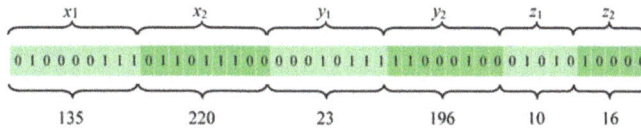

Figure 4. Individual structure.

2.6. Three-Dimensional Brain Tumors Segmentation

The principal goal of the segmentation process is partitioning an image into meaningful and homogeneous regions with respect to one or more characteristics [4]. Subdivision levels of segmentation depend on the problem being solved. Segmentation in medical imaging is normally used to classify pixels to different anatomical regions, such as bones, muscles, and blood vessels. Furthermore, this function is used to classify the pixels of pathological regions, such as malignancies and necrotic and fibrotic areas.

Brain tumor segmentation is difficult to achieve using conventional methods (e.g., pixel-based, region-based, and edge-based methods) [10,19]. Additionally, given the appearance of volumetric 3D medical imaging data, the segmentation of these data for extracting boundary elements belonging to the same structure offers an additional challenge. Therefore, the deformable model was proposed to improve this concern by combining constraints derived from the image and a priori knowledge of the object, such as location, shape, and orientation. The deformable model was originally developed to solve a set of problems in computer vision and medical image analysis. Both 2D and 3D deformable model variants have been applied to segment, visualize, track, and quantify various anatomic structures, such as brain tumors, heart, face, cerebral, coronary and retinal arteries, kidney, and lungs.

2.6.1. Level Set Method

The level set method is a powerful tool for implementing contour evolution and managing topology changes. This approach simply defines an evolving contour C implicitly. This contour is represented by the zero-level set of a Lipschitz continuous function as given in Equation (1):

$$\varnothing : \Omega \rightarrow \mathbb{R}$$

where

$$C = [(x,y) \in \Omega; \varnothing(x,y) = 0]$$
$$C_{inside} = [(x,y) \in \Omega; \varnothing(x,y) > 0] \tag{1}$$
$$C_{outside} = [(x,y) \in \Omega; \varnothing(x,y) < 0]$$

where x and y are coordinates in the image plane. The level set function simultaneously defines both an edge contour and a segmentation of the image.

A crucial step is determining the set function level that segments the image to different important regions. This step is achieved by defining the level set function through subtracting the threshold from each pixel gray-level value. This action results in a level set function positive in regions where the gray level exceeds the threshold (Figure 5) [20].

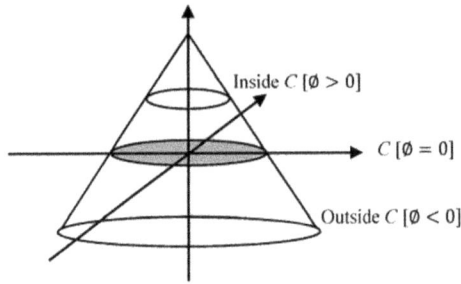

Figure 5. Evolving of contour C.

2.6.2. 3DACWE (Three-Dimensional Active Contour without Edge)

The 3DACWE method, also known as Chan–Vese model, is an example of a geometrically active contour model [21]. The initial contour is evolved using a level set method and does not rely on the gradient of the image for the stopping process. The stopping process of the contour is based on minimizing the Mumford–Shah function [20,21]. Therefore, the 3DACWE method can detect object boundaries not necessarily defined by the gradient, even if the object boundaries are highly smooth or discontinuous.

The 3DACWE algorithm evolves the 3D level set function and minimizes the Mumford–Shah function by setting the value of piecewise smooth function u. Moreover, the function u approximates the original image I besides smoothing the connected components in the image domain Ω.

Mumford and Shah proposed the energy function given in Equation (2), which can be used for segmenting an image I into non-overlapping regions [21–24]:

$$\mathcal{F}^{MS}(u,C) = \int_{\Omega} (u - I)^2 \, dxdydz + \mu \int_{\Omega/C} |\nabla u|^2 \, dxdydz + v\,|C| \tag{2}$$

where the first term encourages u to be close to I, the second term ensures that u is differentiable on Ω/C, and the third term ensures regularity on C. To overcome the time complexity of solving the general Mumford and Shah function, it is required to suppose u to be constant on each connected component . An active contour approach was proposed by Chan and Vese [21] based on minimizing Mumford and Shah functional by penalizing the enclosed area assuming that u is supposed to have only the two values which are given in Equation (3) [24]:

$$u(x,y) = \begin{cases} c_1 \text{ where } x,\,y,\,z \text{ are inside } C \\ c_2 \text{ where } x,\,y,\,z \text{ are outside } C \end{cases} \tag{3}$$

where C is the boundary of a contour, and c_1 and c_2 are the values of u inside and outside the contour, respectively. The Chan–Vese energy function is given by Equation (4) [21,22,24]:

$$\mathcal{F}^{CV}(C,c_1,c_2) = \mu\,Length\,(C) + v\,Area\,(inside\,(C)) + \lambda_1 \int_{inside(C)} |I(x,y,z) - c_1|^2 \, dxdydz$$

$$+ \lambda_2 \int_{outside(C)} |I(x,y,z) - c_2|^2 \, dxdydz \tag{4}$$

The regularity is controlled by penalizing the length in the first term, and the size is controlled by penalizing the enclosed area of C.

These terms are called regularizing terms and are given in Equations (5) and (6), and encourage the contour C to be smooth and short, and can be written by using 3D-level set form $\emptyset$ as [21,25]

$$length\left(\varnothing\left(x,y,z\right)\right)=\int_{\Omega}\delta_0\left(\varnothing\left(x,y,z\right)\right)\left|\nabla\varnothing\left(x,y,z\right)\right|dxdydz \qquad (5)$$

$$area\left(\varnothing\left(x,y,z\right)\right)=\int_{\Omega}H\left(\varnothing\left(x,y,z\right)\right)dxdydz \qquad (6)$$

$\lambda_1,\lambda_2,\mu\geq0$, and $v\geq0$ are fixed parameters controlling selectivity, where the energy function is minimized by fixing these parameters optimally. Meanwhile, λ_1, λ_2 control the internal and external forces, respectively. These terms usually hold the same constant and hence cause a fair competition between these two regions [24]. Generally, $\lambda_1=\lambda_2=1$ [21,26]. Meanwhile, μ controls the smoothness of contour C and assumes a scaling role. However, the parameter is not constant across all experiments. If μ is large, only larger objects with smooth boundaries are segmented. If μ is small, objects of smaller size are segmented accurately [21,23,27]. Typically, μ depends on image resolution, where $\mu=0.1\times255^2$ [26]. Meanwhile, v sets the penalty for the area inside the contour C. This parameter is essential only when two sides of boundaries (internal and external boundaries) are present in the desired object [23]. δ_0 is a 2D Dirac function that represents $\frac{d}{d\varnothing}H\left(\varnothing\left(x,y\right)\right)$. ∇ (Equation (5)) is the gradient operator, and H is the Heaviside function [20–22,25]. Accordingly, by using the 3D-level set function, we can rewrite the Chan–Vese energy function in Equation (7) [28] as follows:

$$\begin{aligned}\mathcal{F}^{CV}\left(\varnothing\left(x,y,z\right)\right)=\mu\int_{\Omega}\delta_0\left(\varnothing\left(x,y,z\right)\right)\left|\nabla\varnothing\left(x,y,z\right)\right|dxdydz+\\v\int_{\Omega}H\left(\varnothing\left(x,y,z\right)\right)dxdydz+\lambda_1\int_{inside(C)}\left|I\left(x,y,z\right)-c_1\right|^2H(\varnothing\left(x,y,z\right)dxdydz+\\\lambda_2\int_{outside(C)}\left|I\left(x,y,z\right)-c_2\right|^2\left(1-H\left(\varnothing\left(x,y,z\right)\right)dxdydz\right.\end{aligned} \qquad (7)$$

The minimization is solved by alternatively updating c_1, c_2, and $\varnothing$, keeping $\varnothing$ fixed, and minimizing the energy function $\mathcal{F}^{CV}$ with respect to the optimal values c_1 and c_2. Consequently, Equations (8) and (9) are attained for c_1 and c_2 as functions of $\varnothing$:

$$c_1\left(\varnothing\left(x,y,z\right)\right)=\frac{\int_{\Omega}I\left(x,y,z\right).H\left(\varnothing\left(x,y,z\right)\right)dx\,dy\,dz}{\int_{\Omega}H\left(\varnothing\left(x,y,z\right)\right)dx\,dy\,dz} \qquad (8)$$

$$c_2\left(\varnothing\left(x,y,z\right)\right)=\frac{\int_{\Omega}I\left(x,y,z\right).\left(1-H\left(\varnothing\left(x,y,z\right)\right)\right)dx\,dy\,dz}{\int_{\Omega}\left(1-H\left(\varnothing\left(x,y,z\right)\right)\right)dx\,dy\,dz} \qquad (9)$$

To minimize the energy function $\mathcal{F}^{CV}$ with respect to $\varnothing$ and fix the c_1 and c_2, a gradient descent method is adopted and has yielded the associated Euler-Lagrange equation for $\varnothing$, which is given by Equation (10) (parameterizing the descent direction by an artificial time) [21,22,27,29,30]:

$$\begin{cases}\frac{\partial\varnothing}{\partial t}=\delta\left(\varnothing\right)\left[\mu\,div\left(\frac{\nabla\varnothing}{\left|\nabla\varnothing\right|}\right)-v-\lambda_1\left(I\left(x,y\right)-c_1\right)^2+\lambda_2\left(I\left(x,y\right)-c_2\right)^2\right]\;in\;\Omega\\\frac{\delta(\varnothing)}{\left|\nabla\varnothing\right|}\frac{\partial\varnothing}{\partial\vec{n}}=0\;on\;\partial\end{cases} \qquad (10)$$

where, $\vec{n}$ represents the exterior normal to the boundary of $\partial\Omega$, and $\frac{\partial\varnothing}{\partial\vec{n}}$ represents the normal derivative of $\varnothing$ at the boundary.

2.6.3. Evaluation of the Segmentation

Image segmentation evaluation can be categorized into subjective and objective evaluation. The subjective evaluation method requires to compare visually the result of the image [31]. While the objective evaluation is divided into supervised and unsupervised techniques. Supervised evaluation methods evaluate segmentation algorithms by comparing segmentation results with manually-segmented reference images which are segmented by experts. It is also known as ground truth reference images or gold standard. While unsupervised evaluation methods do not require to compare with any additional reference images but it just relies on the degree of matching among the characteristics of segmented images as desired by humans. The main advantage of unsupervised evaluation method is that it does not need to compare against a manual segmented reference image. This merit makes it more suitable for real-time application where a large number of images with unknown content and no ground truth need.

In this study supervised evaluation is preferred because of the complexity of the brain tissue and variety of brain tumors. It measures the degree of similarity between the segmented MRI brain tumors and those that are segmented manually or with ground truth dataset. A set of statistical measures have been used to evaluate the segmentation outcomes. They are True Positive (TP), False Positive (FP), True Negative (TN), and False Negative (FN). Such that the TP denotes number of pixels that are correctly segmented as part of a tumor, FP denotes number of pixels that are incorrectly segmented as part of a tumor, FN denotes number of pixels that are incorrectly segmented as healthy pixels, and TN denotes number of pixels that are correctly segmented as healthy pixels. These measures are used to evaluate the segmentation process; accuracy, sensitivity, and specificity. Accuracy is defined as the ratio of numbers of pixels that are correctly segmented to the total number of pixels in MRI slices (Equation (11)). Sensitivity considers the proportion of the tumor that is correctly segmented (Equation (12)). Specificity refers to the proportion of the non-tumor region that is correctly segmented and is given in Equation (13) [8].

$$Accuracy = \frac{TP + TN}{TP + TN + FP + FN} \times 100\% \tag{11}$$

$$Sensitivity = \frac{TP}{TP + FN} \times 100\% \tag{12}$$

$$Specificity = \frac{TN}{TN + FP} \times 100\% \tag{13}$$

In the evaluation of segmentation, both accuracy and specificity are not highly relevant because these two measures adopt the TN, which depends on the relative size of the MRI image and the object of interest. Therefore, the following additional metrics are used to evaluate the segmentation results: Dice, Jaccard, and matching coefficients [32,33].

3. Experimental Results

Experiments were performed using MATLAB R2015a (The Math Works, Natick, MA, USA), on Windows 10. The collected dataset is initially classified into normal and pathological cases to train the algorithms and evaluate the classification and tumor identification accuracy by cross-validation. The standard (BRATS 2013) dataset is then used to evaluate the accuracy of the proposed 3D segmentation system.

3.1. Classification Results

For each MRI image, 190 features were extracted using the MGLCM method. The highest classification accuracy with the optimum performance was achieved by the MLP network at 91% accuracy for correctly classifying the collected dataset by cross validation.

In this study, ANOVA was used for relevance analysis. The critical value α was set to 0.001 to obtain highly significant features [34]. The assessment of predictors depends on both F-statistic value and p-value because a p-value less than 0.001 is insufficient for a predictor. Instead, the predictor must also hold a high F-statistic value. The high F-statistic value indicates that the classes significantly separated from one another [17]. The differences between the features of normal and abnormal MRI brain scan groups of the co-occurrence matrix at $\theta_1 = 0$ and $\theta_2 = 0$ is shown in Table 1. All features seemed acceptable except the weighted mean predictor. Nevertheless, significant variation existed in the F-statistic values between features, indicating a degree of significant difference between the selected features. The *p*-value does not actually signify the degree of separation of each group from others and ignores feature redundancy [35]. This drawback is overcome by the *F*-statistic to determine the power of feature discrimination through thresholding, in which different threshold values are taken to ignore the redundant features and evaluate the selected features at each time by observing the performance of the classifier. When the *F*-statistic threshold value increases, the numbers of selected predictors and the vector of the features decrease. The optimal threshold value that can provide the highest accuracy is 35 (Figure 6). Under this condition, only one normal patient was classified incorrectly as pathological, and three pathological patients were classified incorrectly as normal. Thus, some patients may be misclassified in both ways, but the high classification accuracy (97.8%) reduces these cases to a very small number. These cases are not treated and are passed to the segmentation phase as "erroneous cases."

Consequently, 11 relevant and significant features for each angle of the MGLCM are selected by ANOVA, namely, contrast, correlation, dissimilarity, sum of square variance, sum average, sum variance, difference entropy, information measure of correlation I, inverse difference normalized (IDN), inverse difference moment normalized, and weighted distance, in addition to the cross correlation.

Table 1. Comparison of MRI brain scan features (mean ± standard deviation (SD)) between normal and abnormal patients.

Features	Abnormal MRI Scans	Normal MRI Scans	*F*-Statistic	*p*-Value
Auto correlation ($\times 10^3$)	5.62 ± 1.2	4.92 ± 1.24	13.67	<0.001
Contrast ($\times 10^3$)	1.89 ± 0.618	0.918 ± 0.22	166.2	<0.001
Correlation ($\div 10$)	7.1 ± 0.91	8.07 ± 0.72	291.5	<0.001
Cluster Prominence ($\times 10^8$)	3.6 ± 1.87	2.7 ± 1.09	14.62	<0.001
Cluster Shade ($\times 10^5$)	7.6 ± 4.26	5.5 ± 2.9	13.14	<0.001
Dissimilarity ($\times 10$)	2.42 ± 0.47	1.58 ± 0.21	209	<0.001
Energy ($\div 10$)	1.02 ± 0.2	1.05 ± 0.18	368.15	<0.001
Entropy	7.07 ± 0.336	6.87 ± 0.25	15.21	<0.001
Homogeneity ($\div 10$)	3.55 ± 0.34	3.76 ± 0.26	451.3	<0.001
Max. Probability ($\div 10$)	3.17 ± 0.33	3.23 ± 0.28	444.96	<0.001
Sum of Square Variance ($\times 10^3$)	6.5 ± 1.6	5.38 ± 1.23	24.36	<0.001
Sum Average ($\times 10^2$)	1.15 ± 0.112	1.06 ± 0.15	20.84	<0.001
Sum Variance ($\times 10^4$)	2.33 ± 0.47	1.97 ± 0.48	24.25	<0.001
Sum Entropy	4.46 ± 0.177	4.16 ± 0.147	35.98	<0.001
Difference Entropy	3.64 ± 0.2	3.34 ± 0.124	132.2	<0.001
Information Measure of Correlation I ($\div 10$)	−2.24 ± 0.3	−2.53 ± 0.26	430.15	<0.001
Information Measure of Correlation II ($\div 10$)	9.11 ± 0.2	9.26 ± 0.18	355.48	<0.001
Inverse difference Normalized ($\div 10$)	9.25 ± 0.12	9.48 ± 0.06	407.8	<0.001
Inverse difference Moment Normalized ($\div 10$)	9.78 ± 0.07	9.87 ± 0.028	316.89	<0.001
Weighted Mean ($\div 10$)	−8.73 ± 84	0.53 ± 18.7	0.92	0.339
Weighted Distance	3.05 ± 2.91	0.77 ± 0.52	46.1	<0.001
Cross Correlation ($\div 10$)	7.1 ± 0.91	8.07 ± 0.72	291.5	<0.001

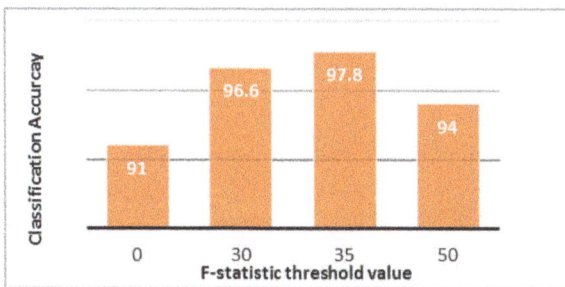

Figure 6. Optimal *F*-statistic threshold value. The achievable accuracy was 97.8% ± 0.1% for the classification of the collected dataset into normal and abnormal brain scans with sensitivity and specificity rates of 98.1% ± 0.3% and 97.6% ± 0.4%, respectively.

3.2. Tumor Identification Results

The main factor that differentiates tumor from healthy tissue is tumor brightness relative to the surrounding brain tissue. Therefore, brain tumor detection algorithms are based mainly on finding pixel clusters with a different intensity from that of their surroundings on the basis of brain symmetry.

After the MRI brain scans are classified into normal and pathological images, the BBBGA method was applied on those identified as pathological cases as shown in the pathological patient in Figure 7. The red rectangles denoting the optimized 3D box refer to the pathological area in slices 6–9 where the tumor appears.

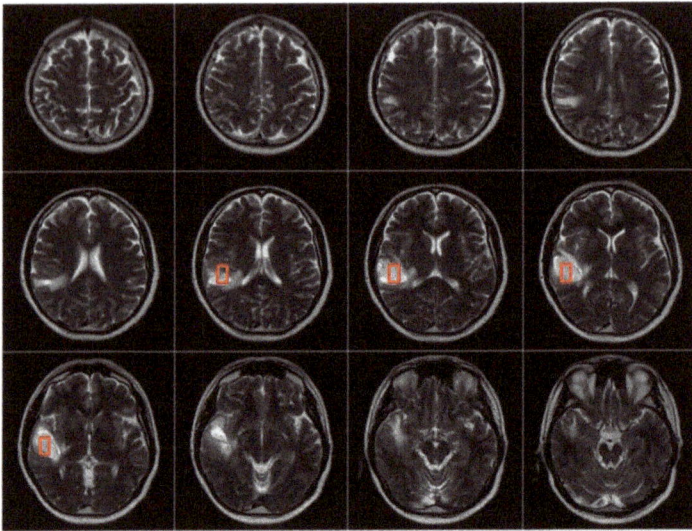

Figure 7. MRI brain scanning slices, the red rectangles denote the optimized location of 3D-box.

The BBBGA method was implemented on MRI brain slices of pathological patients with population size (N) equal to 100. The individuals were selected using the roulette wheel selection method because this approach is more popular and efficient in different applications [36]. The selected individuals were then mated using a multi-point crossover with probability of 0.5 [37]. Finally, a single-point mutation was implemented with a probability of 0.05.

Evidence extracted from previous studies [18,38] indicates the lack of a standard method for evaluating the BBBGA approach. Saha, et al. [38] used an example to observe and measure the noise sensitivity of this approach by adding Gaussian noise with different amounts of $\sigma = 0$, 0.1, 0.2, 0.3, and 0.4 although this addition is not important and irrelevant to evaluation. We measure the noise sensitivity of our approach after the addition of Gaussian noise of the same noise amounts. Figure 8 shows that FP is proportional to the amount of noise in the MRI scan. Hence, our approach was evaluated using the collected dataset that included 88 pathological cases. Among 84 pathological cases, an abnormality was successfully located. Only four cases remained undetected because of the method's inability to detect hardly visible tumors of size less than 1 cm^3. Moreover, tiny tumors hold a spatial scale relatively similar to normal anatomic variability [39].

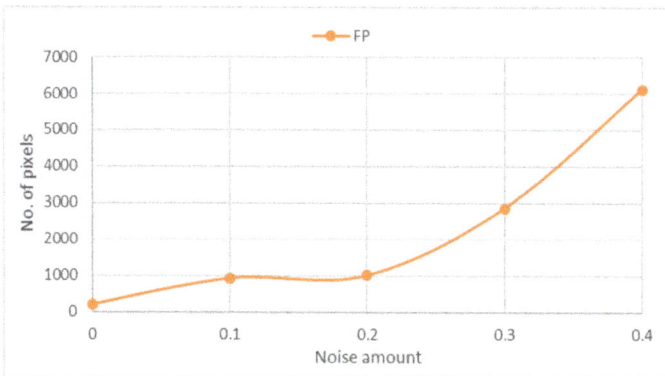

Figure 8. FP increase with increasing noise amount in the MRI scan.

3.3. Tumor Segmentation Results

After the brain tumor location was recognized and identified by BBBDG, the 3DACWE approach was initialized and applied to the T2-w MRI brain scan images of 12 MRI slices from the collected dataset (Figure 9). The ground truth provided by the clinician are marked in green, and the tumor boundaries extracted by 3DACWE are marked in red. This patient holds a brain tumor in the left brain hemisphere. The 3DACWE was initialized by the following initial parameters: $\lambda_1 = \lambda_2 = 1$ and length penalty $\mu = 106$.

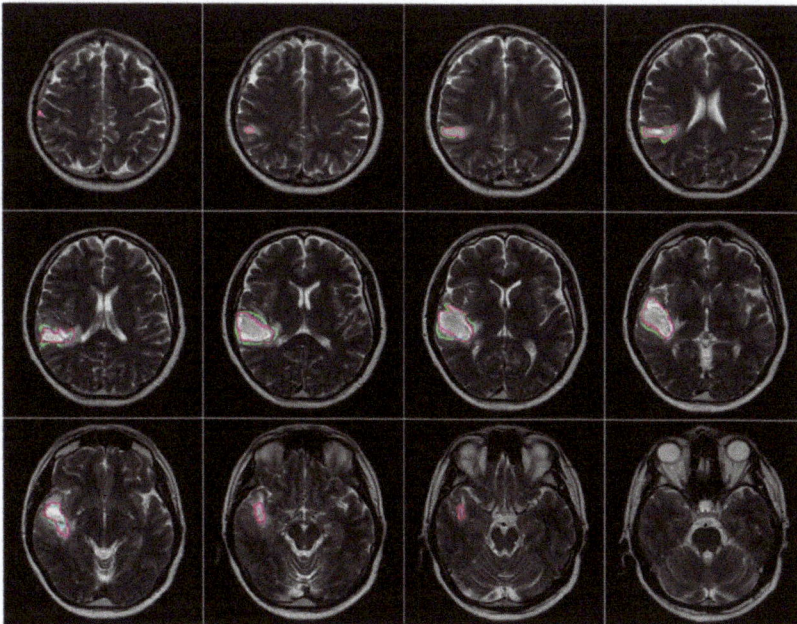

Figure 9. Comparative segmentation results on MRI T2-w (normalized) scan (matching images in Figure 7) by 3DACWE. The ground truth is marked in green, and the output of 3DACWE is marked in red.

The Chan–Vese energy function was minimized within the iterations of 3DACWE and reached a steady state in 1250 iterations (Figure 10). This patient's MRI scan was segmented with a Dice score of 88.4% by comparing with manual segmentation.

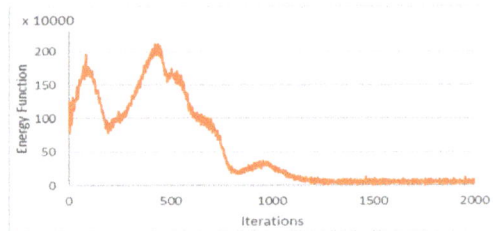

Figure 10. The Chan–Vese energy function convergence to steady state.

Under this segmentation process, some brain tumor parts in the MRI slices were incorrectly segmented as healthy tissues, and some healthy brain tissues were incorrectly segmented as pathological tissues. To eliminate these ambiguities and reduce the FPs and FNs, a consistency verification algorithm was applied for post-processing [11,40]. The majority filter was adopted to remove the FPs and FNs by replacing the segmented pixels inconsistent with their neighboring segmented pixels in a certain neighborhood. For example, if the center pixel of a window is segmented as a tumor but the majority of the neighboring pixels were segmented as healthy, then the center pixel is changed to healthy. Otherwise, if a pixel within the tumor area is segmented incorrectly as healthy and is surrounded by pixels that are segmented as tumor, then the pixel is changed to a pathological pixel. If the window size of the applied filter is increased, the quality of the output image significantly augments at the expense of complexity increase [40]. Herein, consistency verification was applied in a 5×5 neighborhood window [11].

For instance, the MRI brain slices in Figure 11A include some MRI slices with incorrectly segmented pixels (Figure 11B). The consistency verification algorithm results are shown in Figure 11C. Tables 2 and 3 demonstrate the average results of segmentation for both collected and BRATS 2013 datasets, respectively, in which the collected dataset was manually segmented by experts.

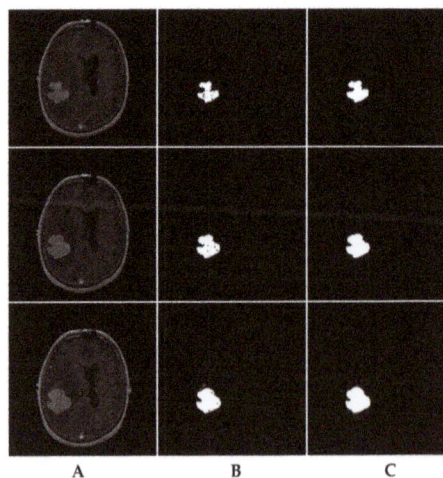

Figure 11. Results after applying the consistency verification algorithm. (**A**) original MRI slices; (**B**) segmented MRI brain slices by 3DACWE; (**C**) output of consistency verification algorithm.

Table 2. Segmentation results for each patient in the collected dataset.

Results	Sensitivity	Specificity	Accuracy	Dice Index	Jaccard	Matching
Average	0.854	0.999	0.998	0.890	0.804	0.854
STD	0.069	0.001	0.002	0.047	0.075	0.069
Min	0.690	0.996	0.994	0.764	0.619	0.690
Max	0.971	1.000	1.000	0.956	0.915	0.971

Table 3. Segmentation results for each patient in the BRATS 2013 dataset.

Results	Sensitivity	Specificity	Accuracy	Dice Index	Jaccard	Matching
Average	0.909	1.000	0.999	0.893	0.809	0.909
STD	0.076	0.000	0.000	0.043	0.068	0.076
Min	0.736	0.998	0.998	0.768	0.624	0.736
Max	0.999	1.000	1.000	0.947	0.899	0.999

The overall results of the segmentation of the four MRI modalities (T1-w, T2-w, T1c-w, and FLAIR) for the collected dataset with the four most common metrics reported in previous studies (accuracy, Dice, Jaccard, and matching methods) are summarized in Figure 12. The T1c-w-based segmentation attained the highest average metric rates because of contrast enhancement of the pathological tissues. The T2-w-based segmentation was rated as the lowest among all metrics because of highly inhomogeneous intensity distribution despite the sharp edges and high intensity of brain tumor with respect to the surrounding tissues.

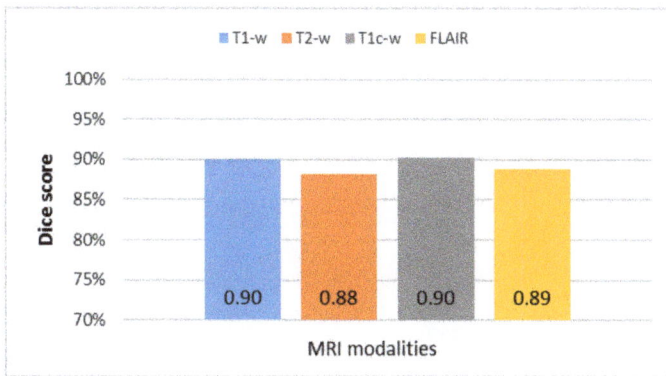

Figure 12. Average score with SD obtained from segmentation score rates by the four metrics.

Table 4 shows a comparison between the achieved mean Dice score and SD of 2DACWE and 3DACWE methods in segmenting images obtained under the four MRI modalities T2-w, T1-w, T1c-w, and FLAIR (Figure 13).

Table 4. Comparison of tumor segmentation accuracy (mean ± SD) using 2DACWE and 3DACWE.

Method	T2-w	T1-w	T1c-w	FLAIR
2DACWE	83.73% ± 4.6%	84.43% ± 5.3%	86.6% ± 3.4%	82% ± 4.5%
3DACWE	88.11% ± 4.4%	89.92% ± 4.9%	90.3% ± 3.6%	88.8% ± 6.9%

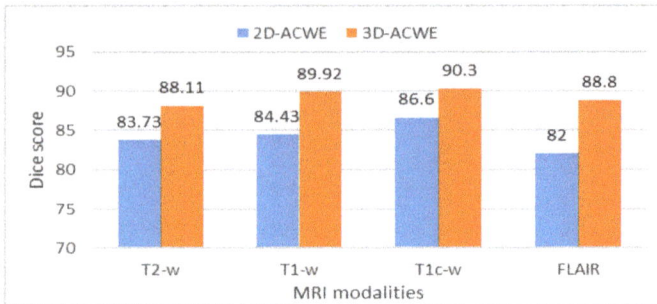

Figure 13. Comparison of tumor segmentation of the average Dice scores of four MRI modalities under 2DACWE and 3DACWE.

Figure 14 compares between the 3DACWE and 2DACWE segmentation results of the collected dataset. Notably, 3DACWE outweighs the 2DACWE method for all patients in the given dataset.

Figure 15 shows a comparison of the clinical and experimental identifications of the most relevant slices for the 20 pathological patients by measuring the means and SD between the two groups. Notably, the means and SD of the clinical and experimental tests are similar.

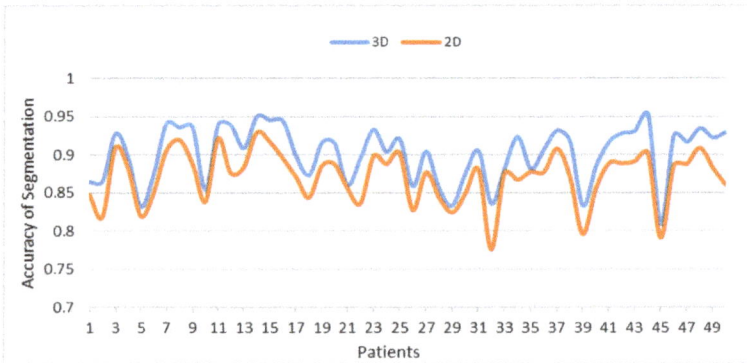

Figure 14. Comparison between 3DACWE and 2DACWE segmentation results for the given dataset.

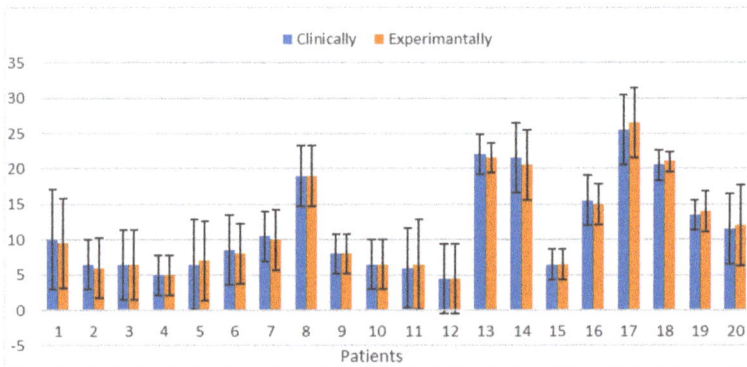

Figure 15. Comparison between clinical and experimental MRI slice identification features (mean ± SD).

3.4. Discussion

The 3DACWE method was successfully applied to both the collected and standard datasets of BRATS 2013. The achieved Dice score of the collected dataset was 89% ± 4.7%, whereas the achieved Dice score of BRATS 2013 was 89.3% ± 4.2%. A major difficulty was encountered during white matter tumor segmentation because of the overlapping white and gray matter intensity distributions in such case. Some parts of the tumors in the gray matter were not distinguished because of restricted image resolution and the complex network of brain tissue with various shapes and sizes. These factors significantly affected a large number of voxels located on the tissue borders. Moreover, central tumor image intensity slightly differed from the peripheral tumor intensity. As such, the intensity near the borders can be considered similar to that of the gray matter. Consequently, the tumor and gray matter may be confused, and the peripheral tumor regions may be misclassified.

Tumor size affects segmentation accuracy, and errors usually occur at the tumor boundaries. Large brain tumors contain a high number of image pixels that can be misclassified. Moreover, large tumors likely ingress into the brain boundary and CSF fluid and render the precise determination of tumor boundary challenging. With regard to the overall executive processing time, the proposed system handled volumetric MRI data with different characteristics, such as number of slices and tumor size, type, boundary, and location. These characteristics make the overall segmentation process time consuming. Hence, the processing time of the proposed system was measured by second per MRI slice. Our proposed system required 243 s/MRI slice to run the segmentation.

Segmentation accuracy decreased significantly with increasing slice thickness as shown in the scatter plot of the segmentation accuracy to the summation of slice thickness and space between slices (Figure 16).

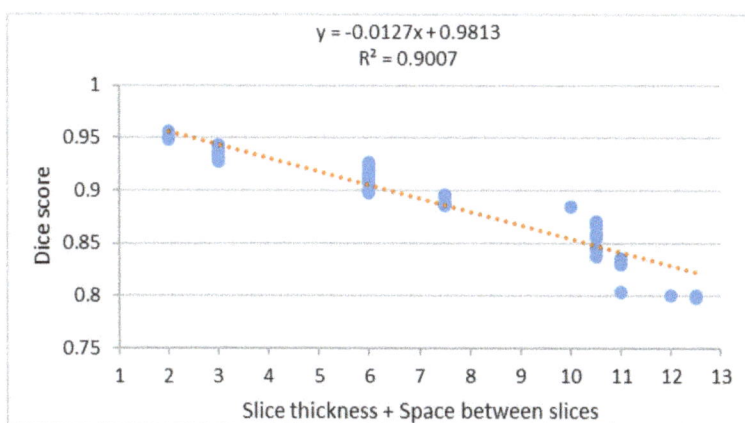

Figure 16. Scatterplot of segmentation accuracy to the summation of slice thickness and space between slices, showing the mean accuracy (*R*-squared) as the dotted brown line.

The scatter plot in Figure 16 shows a negative correlation between the Dice score and the summation of the slice thickness and space between slices. Therefore, to achieve high segmentation accuracy, a reduction of the slice thickness and space between slices to a minimum is essential and diminishes the PV effect.

Table 5 contains an overview of the segmentation methods demonstrated in [8].

Table 5. Overview of the segmentation methods compared with the proposed system.

Reference	MRI Modalities	Approach	No. of Patients	Accuracy (100%)	Match (100%)	Jaccard (100%)	Dice (100%)
[33]	T_1, T_2, and PD	Fuzzy clustering	6	-	53–91	-	-
[41]	T_1	Template-moderated classification	20	95	-	-	-
[42]	T_1, and T_1-c	Level-sets	5	-	-	85–93	-
[43]	T_2	Generative model	3	-	-	59–89	-
[44]	T_1, T_1-C, T_2, and FLAIR	Weighted aggregation	20	-	-	62–69	-
[12]	T_1, T_1-c, T_2, and FLAIR	SVM	14	34–93	-	-	-
[45]	T_1, T_1-c, T_2, and FLAIR	Generative model	25	-	-	-	40–70
Proposed System	T_1, T_1-c, T_2, and FLAIR	3D-ACWE	50 (collected)	99.8 ± 0.2	85.4 ± 6.9	80.4 ± 7.4	89 ± 4.7
			25 (BRATS 2013)	99.9	91 ± 7.6	81 ± 6.8	89.3 ± 4.2

4. Conclusions

Visual diagnosis of MRI scan images is subjective and highly dependent on clinician expertise. The proposed method offers a reduction of clinician evaluation time from 3–5 h to 5–10 min without significant reduction in the accuracy of the diagnosis. Indeed, the proposed method can recognize and segment MRI brain abnormality (tumor) on T2-w, T1-w, T1c-w, and FLAIR images. The 3DACWE segmentation technique reduces manual input, offers a rapid operation, and exhibits high accuracy compared with manual segmentation as evaluated using both the Al-Kadhimiya and BRATS 2013 datasets. We conclude that the 3DACWE method is effective in brain tumor segmentation because the approach does not only consider local tumor properties, such as gradients, but also relies on global properties, such as intensity, contour length, and region length. Although the achieved accuracy was high relative to those of other segmentation techniques, the 3DACWE was relatively slow for brain tumor segmentation. Such a slow pace was ascribed to the processing of a massive number of MRI slices of 512×512 pixel resolution with a high number of iterations used to attain the required accuracy.

Acknowledgments: The authors would like to thank the MRI Unit of Al Kadhimiya Teaching Hospital in Baghdad, Iraq for providing us with MRI brain scanning images dataset. They would also like to thank the reviewers for their comments and suggestions for improving the paper. This research is funded by the Ministry of Higher Education Malaysia under the Fundamental Research Grant Scheme (FRGS), Project No. FP073-2015A.

Author Contributions: Ali M. Hasan developed the algorithm; Farid Meziane performed the experiments; Rob Aspin analyzed the data; Ali M. Hasan and Hamid A. Jalab wrote the paper.

Conflicts of Interest: The authors declare no conflict of interest.

References

1. Karkavelas, G.; Tascos, N. Epidemiology, histologic classification and clinical course of brain tumors. In *Imaging of Brain Tumors with Histological Correlations*; Springer: Heidelberg, Germany, 2002; pp. 1–10.
2. Nabizadeh, N.; Kubat, M. Brain tumors detection and segmentation in mr images: Gabor wavelet vs. Statistical features. *Comput. Electr. Eng.* **2015**, *45*, 286–301. [CrossRef]
3. Hasan, A.; Meziane, F. Automated screening of mri brain scanning using grey level statistics. *Comput. Electr. Eng.* **2016**, *53*, 276–291. [CrossRef]
4. Dougherty, G. *Digital Image Processing for Medical Applications*; Cambridge University Press: Cambridge, UK, 2009.
5. Mechtler, L. Neuroimaging in neuro-oncology. *Neurol. Clin.* **2009**, *27*, 171–201. [CrossRef] [PubMed]
6. Tonarelli, L. Magnetic resonance imaging of brain tumor. *Enterp. Contin. Educ.* **2013**, *300*, 48116–40300.
7. Belkic, D.; Belkic, K. *Signal Processing in Magnetic Resonance Spectroscopy with Biomedical Applications*; CRC Press: Boca Raton, FL, USA, 2010.

8. Menze, H.; Jakab, A.; Bauer, S.; Kalpathy-Cramer, J.; Farahani, K.; Kirby, J.; Burren, Y.; Porz, N.; Slotboom, J.; Wiest, R.; et al. The multimodal brain tumor image segmentation benchmark (brats). *IEEE Trans. Med. Imaging* **2015**, *34*, 1993–2024. [CrossRef] [PubMed]
9. Tohka, J. Partial volume effect modeling for segmentation and tissue classification of brain magnetic resonance images: A review. *World J. Radiol.* **2014**, *6*, 855–864. [CrossRef] [PubMed]
10. Gordillo, N.; Montseny, E.; Sobrevilla, P. State of the art survey on MRI brain tumor segmentation. *Magn. Resonan. Imaging* **2013**, *31*, 1426–1438. [CrossRef] [PubMed]
11. Nabizadeh, N. Automated Brain Lesion Detection and Segmentation Using Magnetic Resonance Images. Ph.D. Thesis, University of Miami, Coral Gables, FL, USA, April 2015.
12. Verma, R.; Zacharaki, E.I.; Ou, Y.; Cai, H.; Chawla, S.; Lee, S.-K.; Melhem, E.R.; Wolf, R.; Davatzikos, C. Multiparametric tissue characterization of brain neoplasms and their recurrence using pattern classification of mr images. *Acad. Radiol.* **2008**, *15*, 966–977. [CrossRef] [PubMed]
13. Tantisatirapong, S. Texture Analysis of Multimodal Magnetic Resonance Images in Support of Diagnostic Classification of Childhood Brain Tumours. Ph.D. Thesis, University of Birmingham, Birmingham, UK, 2015.
14. Udomchaiporn, A.; Coenen, F.; García-Fiñana, M.; Sluming, V. 3-D MRI Brain Scan Feature Classification Using an Oct-Tree Representation. In *Advanced Data Mining and Applications*; Motoda, H., Wu, Z., Cao, L., Zaiane, O., Yao, M., Wang, W., Eds.; Springer: Heidelberg, Germany, 2013; pp. 229–240.
15. Kassner, A.; Thornhill, R. Texture analysis: A review of neurologic mr imaging applications. *Am. J. Neuroradiol.* **2010**, *31*, 809–816. [CrossRef] [PubMed]
16. Gomez, W.; Pereira, W.; Infantosi, A. Analysis of co-occurrence texture statistics as a function of gray-level quantization for classifying breast ultrasound. *IEEE Trans. Med. Imaging* **2012**, *31*, 1889–1899. [CrossRef] [PubMed]
17. Hasan, A.; Meziane, F.; Abd Kadhim, M. Automated segmentation of tumours in mri brain scans. In Proceedings of the 9th International Joint Conference on Biomedical Engineering Systems and Technologies (BIOSTEC 2016), Rome, Italy, 21–23 February 2016; SCITEPRESS: Rome, Italy, 2016; Volume 2, pp. 55–62.
18. Khandani, M.; Bajcsy, R.; Fallah, Y. Automated segmentation of brain tumors in mri using force data clustering algorithm. In Proceedings of the International Symposium on Visual Computing, Las Vegas, NV, USA, 30 November–2 December 2009; Springer: Heidelberg, Germany, 2009; Volume 5875, pp. 317–326.
19. Liu, J.; Li, M.; Wang, J.; Wu, F.; Liu, T.; Pan, Y. A survey of MRI-based brain tumor segmentation methods. *Tsinghua Sci. Technol.* **2014**, *19*, 578–595.
20. Rousseau, O. *Geometrical Modeling of the Heart*; University of Ottawa: Ottawa, ON, Canada, 2009.
21. Chan, F.; Vese, A. Active contours without edges. *IEEE Trans. Image Process.* **2001**, *10*, 266–277. [CrossRef] [PubMed]
22. Chan, F.; Vese, A. An active contour model without edges. In Proceedings of the Scale-Space'99, Corfu, Greece, 26–27 September 1999; Nielsen, M., Ed.; Springer: Heidelberg, Germany, 1999; pp. 141–151.
23. Getreuer, P. Chan–Vese segmentation. *Image Process. Line* **2012**, *2*, 214–224. [CrossRef]
24. Li, C.; Kao, C.-Y.; Gore, J.C.; Ding, Z. Minimization of region-scalable fitting energy for image segmentation. *IEEE Trans. Image Process.* **2008**, *17*, 1940–1949. [PubMed]
25. Klotz, A. 2D and 3D Multiphase Active Contours without Edges Based Algorithms for Simultaneous Segmentation of Retinal Layers from Oct Images. Master's Thesis, University of Texas, Austin, TX, USA, 2013.
26. Nixon, M.; Aguado, A. *Feature Extraction Image Processing*, 2nd ed.; Academic Press: Oxford, UK, 2008; p. 424.
27. Tai, X.-C.; Lie, K.-A.; Chan, T.F.; Osher, S. Image processing based on partial differential equations. In Proceedings of the International Conference on PDE-Based Image Processing and Related Inverse Problems, Oslo, Norway, 8–12 August 2005.
28. Crandall, R. *Image Segmentation Using the Chan-Vese Algorithm*; Project Report; Technion–Israel Institute of Technology: Haifa, Israel, 2009.
29. Thapaliya, K.; Pyun, J.-Y.; Park, C.-S.; Kwon, G.-R. Level set method with automatic selective local statistics for brain tumor segmentation in MR images. *Comput. Med. Imaging Graph.* **2013**, *37*, 522–537. [CrossRef] [PubMed]
30. Chan, F.; Sandberg, B.; Vese, A. Active contours without edges for vector-valued images. *J. Vis. Commun. Image Represent.* **2000**, *11*, 130–141. [CrossRef]
31. Zhang, H.; Fritts, J.E.; Goldman, S.A. Image segmentation evaluation: A survey of unsupervised methods. *Comput. Vis. Image Underst.* **2008**, *110*, 260–280. [CrossRef]

32. Agrawal, R.; Sharma, M. Review of segmentation methods for brain tissue with magnetic resonance images. *Int. J. Comput. Netw. Inf. Secur.* **2014**, *6*, 55–62. [CrossRef]
33. Fletcher-Heath, L.M.; Hall, L.O.; Goldgof, D.B.; Murtagh, F.R. Automatic segmentation of non-enhancing brain tumors in magnetic resonance images. *Artif. Intell. Med.* **2001**, *21*, 43–63. [CrossRef]
34. Quinn, G.; Keough, M. *Experimental Design and Data Analysis for Biologists*; Cambridge University Press: Cambridge, UK, 2002.
35. Johnson, K.; Synovec, R. Pattern recognition of jet fuels: Comprehensive GC × GC with ANOVA-based feature selection and principal component analysis. *Chemom. Intell. Lab. Syst.* **2002**, *60*, 225–237. [CrossRef]
36. Talebi, M.; Ayatollahi, A.; Kermani, A. Medical ultrasound image segmentation using genetic active contour. *Biomed. Sci. Eng.* **2010**, *4*, 105–109. [CrossRef]
37. Chipperfield, A.; Fleming, P.; Pohlheim, H.; Fonseca, C. *Genetic Algorithm Toolbox for Use with Matlab*; Techical Report; University of Sheffield: Sheffield, UK, 1994.
38. Saha, B.; Ray, N.; Greiner, R.; Murtha, A.; Zhang, H. Quick detection of brain tumors and edemas: A bounding box method using symmetry. *Comput. Med. Imaging Graph.* **2012**, *36*, 95–107. [CrossRef] [PubMed]
39. Sanjuán, A.; Price, C.J.; Mancini, L.; Josse, G.; Grogan, A.; Yamamoto, A.K.; Geva, S.; Leff, A.P.; Yousry, T.A.; Seghier, M.L. Automated identification of brain tumors from single MR images based on segmentation with refined patient-specific priors. *Front. Neurosci.* **2013**, *7*, 241. [CrossRef] [PubMed]
40. Haghighat, M.B.A.; Aghagolzadeh, A.; Seyedarabi, H. Multi-focus image fusion for visual sensor networks in dct domain. *Comput. Electr. Eng.* **2011**, *37*, 789–797. [CrossRef]
41. Kaus, M.R.; Warfield, S.K.; Nabavi, A.; Black, P.M.; Jolesz, F.A.; Kikinis, R. Automated segmentation of mr images of brain tumors 1. *Radiology* **2001**, *218*, 586–591. [CrossRef] [PubMed]
42. Ho, S.; Bullitt, L.; Gerig, G. Level-set evolution with region competition: Automatic 3-d segmentation of brain tumors. In Proceedings of the 16th International Conference on Pattern Recognition, Trois-Rivieres, PQ, Canada, 1–15 August 2002; IEEE: New York, NY, USA, 2002; pp. 532–535.
43. Prastawa, M.; Bullitt, E.; Ho, S.; Gerig, G. A brain tumor segmentation framework based on outlier detection. *Med. Image Anal.* **2004**, *8*, 275–283. [CrossRef] [PubMed]
44. Corso, J.J.; Sharon, E.; Dube, S.; El-Saden, S.; Sinha, U.; Yuille, A. Efficient multilevel brain tumor segmentation with integrated bayesian model classification. *IEEE Trans. Med. Imaging* **2008**, *27*, 629–640. [CrossRef] [PubMed]
45. Menze, B.H.; Van Leemput, K.; Lashkari, D.; Weber, M.-A.; Ayache, N.; Golland, P. A generative model for brain tumor segmentation in multi-modal images. In Proceedings of the International Conference on Medical Image Computing and Computer-Assisted Intervention, Beijing, China, 20–24 September 2010; Springer: Heidelberg, Germany, 2010; pp. 151–159.

symmetry

MDPI

Article

Attribute Control Chart Construction Based on Fuzzy Score Number

Shiwang Hou [1,*], Hui Wang [2] and Shunxiao Feng [2]

[1] School of Business, Huaihua University, Huaihua 418000, Hunan, China
[2] School of Mechanical and Power Engineering, North University of China, Taiyuan 030051, Shanxi, China; 15135172544@163.com (H.W.); fengshunxiao@sina.com (S.F.)
* Correspondence: houshiwan@163.com; Tel.: +86-182-7455-3180

Academic Editor: Angel Garrido
Received: 12 October 2016; Accepted: 22 November 2016; Published: 26 November 2016

Abstract: There is much uncertainty and fuzziness in product quality attributes or quality parameters of a manufacturing process, so the traditional quality control chart can be difficult to apply. This paper proposes a fuzzy control chart. The plotted data was obtained by transforming expert scores into fuzzy numbers. Two types of nonconformity judgment rules—necessity and possibility measurement rules—are proposed. Through graphical analysis, the nonconformity judging method (i.e., assessing directly based on the shape feature of a fuzzy control chart) is proposed. For four different widely used membership functions, control levels were analyzed and compared by observing gaps between the upper and lower control limits. The result of the case study validates the feasibility and reliability of the proposed approach.

Keywords: fuzzy scores; fuzzy attribute control chart; membership function

PACS: 0510-*a*

1. Introduction

International quality management expert Dr. Juran pointed out that, for users, quality is the fitness for use, and not conformance to specification. End users rarely know what the specifications are—they evaluate the product mostly based on the applicability and the durability of its applicability. From the quality viewpoint of fitness for use, the conformance to specifications is downplayed and the users' evaluation is strengthened. Due to the emphasis on the user's feelings and psychological factors, there are many fuzzy attributes of quality from the fitness-for-use viewpoint. As a result, there are not only two distinct judgments (applicative or inapplicable) when evaluating the quality from a fitness point of view. In this sense, considering the fuzzy property of fitness-for-use quality is more practical.

As one of the main tools of SPC (statistical process control), the control chart is widely used for monitoring the state of a process. However, for the aforementioned fitness quality, the conventional continuous control chart or attributes control charts cannot be applied directly.

In recent years, many researchers have applied fuzzy set theory to solve the problems that arise during the construction of uncertain quality control charts. Gülbay and Kahraman put forward a direct fuzzy approach in [1–3]. They represented the linguistic variables and control limits of sample quality evaluation with a fuzzy set, without any defuzzification operation, and judged the process control state by the degree of overlap of α-cut set of sample fuzzy set and the control limits' fuzzy set. In their approach, the control level can be adjusted by the parameter α. Taleb and Sorooshian used fuzzy set to depict linguistic data, defuzzified the fuzzy set to crisp values by use of the weighted average method, and then built the control chart in [4,5]. By describing the magnitude of the process shift and the occurrence rate of an assignable cause as fuzzy numbers, the fuzziness was modeled using both minimax and maximin approaches, and a control chart suitable for processes with fuzzy parameters was presented by Morabi et al. in [6]. Grzegorzewski et al. [7] proposed a fuzzy control chart based on a necessity coefficient. Hsieh et al. constructed a c-chart of wafer defects in an integrated circuit manufacturing process by using fuzzy theory in [8]. Tannock [9] proposed an approach to construct a fuzzy individual control chart.

Some indirect approaches were also proposed to build different types of control charts, for example, Shu et al. [10] constructed fuzzy and R control charts based on the fuzzy dominance between the fuzzy averages and variances, and Gildeh et al. [11] used Dp,q-distance between fuzzy numbers to calculate their variance, covariance, and autocorrelation coefficient, and then used the autocorrelation coefficient to modify the limits of the control chart. Zarandi et al. [12] constructed a fuzzy control chart for different process shifts by adjusting the membership function parameters. In order to detect small shifts represented by fuzzy numbers, fuzzy exponentially weighted moving average (EWMA) control charts for univariate variable was developed by Sentürk et al. in [13]. Faraz et al. [14] constructed a fuzzy control chart to treat two uncertainties of fuzziness and randomness in data.

Product quality is often measured by various characteristics that are generally correlated. Multivariate control charts are necessary for quality control in such situations. Attribute quality characteristics are sometimes defined by linguistic variables, or product units are classified into several categories with linguistic forms. Fernandez et al. developed a method to control these fuzzy quality evaluations with fuzzy multivariate control charts in [15]. In order to simultaneously monitor the quality characteristics of a product or process measured by linguistic or fuzzy data, Ghobadi et al. [16] developed a fuzzy multivariate cumulative sum (F-MCUSUM) control chart by means of fuzzy set theory. In [17], Alipour et al. combined multivariate statistical quality control with fuzzy set theory to develop a fuzzy multivariate exponentially weighted moving average (F-MEWMA) control chart.

The aforementioned literatures have provided good ideas to deal with the fitness quality attributes. However, their common feature is that they constructed fuzzy control charts by fuzzy operation and defuzzification based on fuzzy set membership functions given in advance. When there is some deviation of a priori information, the control chart based on such a priori information can produce distorted signals, and the application effect can be discounted greatly. Literature that describes the construction of the fuzzy membership functions by use of rating scores, and subsequent creation of a corresponding fuzzy-number-based control chart, is still rare.

This paper proposes a method to build a control chart based on fuzzy score number, and describes the design of nonconformity judging criteria and analysis of the type selection of fuzzy numbers. The rest of this paper is organized as follows. Section 2 analyzes the data acquisition and transformation method. Possibility and necessity rules for nonconformity judgment are presented in Section 3. The basic form of a control chart based on fuzzy score number is described in Section 4, and an application case study is given in Section 5. Section 6 quantitatively analyzes the influence of different types of fuzzy numbers to the control chart. Section 7 ends the paper with a summary and conclusions.

2. The Plotted Data of a Control Chart Based on Fuzzy Number

Fuzzy number-based control chart has a similar working principle of Shewhart control chart. It plots the fuzzy number data of sample mean, and takes fuzzy number data of process mean to construct the control limits, including central line and upper and lower limit. The state of the control chart is judged according the relationship between the fuzzy number characteristic value of sample mean and control limits.

In order to acquire the process information used to construct a fuzzy control chart, this paper adopted sample mean to evaluate the process mean when the process was in steady state. Suppose a triangular fuzzy number sample, whose size is n, $S_1(L_1, m_1, R_1)$, $S_2(L_2, m_2, R_2)$, ... $S_n(L_n, m_n, R_n)$, and the fuzzy number estimator of process mean $\hat{S}(\overline{L}, \overline{m}, \overline{R})$ is as follows:

$$\hat{S} = \frac{\sum_{i=1}^n S_i}{n} \tag{1}$$

The parameters of process fuzzy number take the mean of corresponding parameters of sample fuzzy number (i.e., $\hat{S}(\hat{L}, \hat{m}, \hat{R})$):

$$\hat{L} = \frac{\sum_{i=1}^n L_i}{n}, \ \hat{m} = \frac{\sum_{i=1}^n m_i}{n}, \ \hat{R} = \frac{\sum_{i=1}^n R_i}{n} \tag{2}$$

After determining the fuzzy number of process mean, plotted data of the control chart is obtained by transforming the expert score of sample evaluation into a fuzzy number.

3. Control Limits of a Fuzzy Number-Based Control Chart and Its Nonconformity Judgment

3.1. Possibility and Necessity Measures

A fuzzy score number-based control chart indicates the status of a process according to the matching degree between the fuzzy number of sample mean and the fuzzy number of process mean. A degree of matching higher than a predefined value between the two means indicates that the process is in a better control state. This is on the basis of two measures of fuzzy events: possibility (*Pos*) and necessity (*Nec*) measures,.

Suppose fuzzy set A, $B \in F(X)$, then we have the following definition (as mentioned by Jamison et al. in [18] and Zadeh in [19]):

$$Pos(B|A) = \sup_{x \in X} \min \{\mu_A(x), \mu_B(x)\} \tag{3}$$

$$Nec(B|A) = \inf_{x \in X} \max \{1 - \mu_A(x), \mu_B(x)\} \tag{4}$$

The above two formulas will be called possibility measure and necessity measure of fuzzy set B under the given fuzzy set A, respectively, where $x \in A$, B, and μ_A and μ_B are the membership functions of fuzzy sets A and B, respectively. The former is an optimistic estimation of the possibility of fuzzy events, and the latter is a conservative estimation of the possibility of fuzzy event, satisfying the following relations:

$$\begin{aligned} Nec(B|A) &= 1 - Pos(\overline{B}|A) \\ &= 1 - Pos(\overline{B}|A) = 1 - \sup_{x \in X} \min \{\mu_A(x), 1 - \mu_B(x)\} = \inf_{x \in X} \max \{1 - \mu_A(x), \mu_B(x)\} \end{aligned} \tag{5}$$

Formula (5) shows that the necessity measure of some fuzzy event is the possibility measure of its opposite event.

Taking triangular fuzzy number as an example, the possibility measure is shown in Figure 1a; its value is the highest point of the shadowed area and denotes the matching degree of possibility between fuzzy numbers A and B. However, for a given *Pos*, there are no constraints on the vertex of B, and in Figure 1a, a and a' all satisfy the given *Pos*.

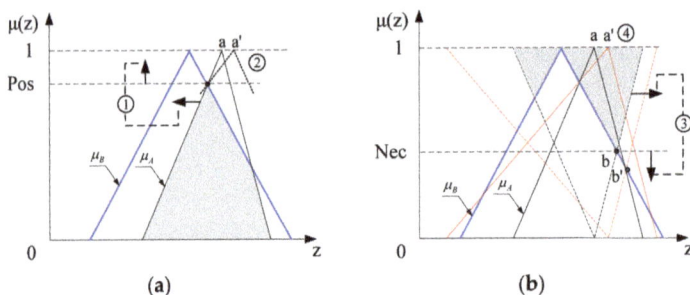

Figure 1. Schematic diagram of two fuzzy measures. (a) Possibility measure (*Pos*); (b) necessity measure (*Nec*).

The necessity measure is shown in Figure 1b, whose value is the lowest point of the shadowed area, and denotes the matching degree of necessity between fuzzy numbers *A* and *B*. The given *Nec* can limit the value range of elements in *B* for known *A*. As shown by the symbol ④ in Figure 1b, when the vertex of *B* moves from *a* to *a'*, the value of *Nec* decease from *b* to *b'*, although the value of *Pos* remains the same, shown by the symbol ② in Figure 1a. So, the matching degree between fuzzy numbers *A* and *B* can be judged synthetically by selecting the appropriate combination of *Pos* values and *Nec* values.

Let fuzzy number $\overline{S}$ be the estimation of current process level. For known fuzzy number of sample mean $\widetilde{S}$ and its possibility distribution, the matching degree between $\widetilde{S}$ and $\overline{S}$ can be decided by the possibility measure and necessity measure as follows:

$$Pos(\overline{S}|\widetilde{S}) = \sup_{z \in U}[\min\{\mu_{\overline{S}}(z), \mu_{\widetilde{S}}(z)\}] \tag{6}$$

$$Nec(\overline{S}|\widetilde{S}) = \inf_{z \in U}[\max\{\mu_{\overline{S}}(z), 1 - \mu_{\widetilde{S}}(z)\}] \tag{7}$$

The above two formulas can be used to judge the abnormal status of control chart.

3.2. Control Chart Nonconformity Judgment Rules

Suppose $\overline{S}(\overline{L}, \overline{m}, \overline{R})$ and $\widetilde{S}(\widetilde{L}, \widetilde{m}, \widetilde{R})$ are the triangular fuzzy numbers of process mean and sample mean, respectively. The in-control fuzzy control chart must satisfy the following two conditions:

(1) The possibility measure of $\widetilde{S}$ under known $\overline{S}$ must be no less than the preset $\alpha(0 < \alpha \leq 1)$ (i.e., $Pos(\overline{S}|\widetilde{S}) \geq \alpha$ (See [20]));

(2) The necessity measure of $\widetilde{S}$ under known $\overline{S}$ must be no less than the preset $\beta(0 < \beta \leq 1)$ (i.e., $Nec(\overline{S}|\widetilde{S}) \geq \beta$).

By analyzing the graphical characteristics of these two measures, it was found that the fuzzy set $\overline{S}$ and $\widetilde{S}$, meeting conditions of $Pos(\overline{S}|\widetilde{S}) \geq \alpha$ and $Nec(\overline{S}|\widetilde{S}) \geq \beta$, has the following features, as shown in Figures 2 and 3:

$$\text{If } Pos(\overline{S}|\widetilde{S}) \geq \alpha, \text{ then } \overline{S}_\alpha \cap \widetilde{S}_\alpha \neq \varnothing; \tag{8}$$

$$\text{If } Nec(\overline{S}|\widetilde{S}) \geq \beta, \text{ then } \overline{S}_\beta \supset \widetilde{S}_{1-\beta} \tag{9}$$

For trapezoidal fuzzy numbers, graphical analysis results show that the above two features are still true (as shown in Figures 4 and 5). So, the above two features regarding probability measure and necessity measure can be used as the criteria for judging the fuzzy control chart's abnormal state.

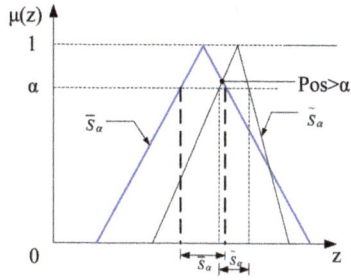

Figure 2. Probability measure (*Pos*) of triangular fuzzy number.

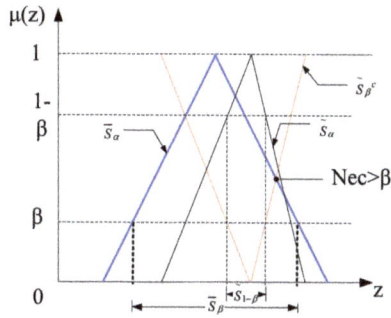

Figure 3. Necessity measure (*Nec*) of triangular fuzzy number.

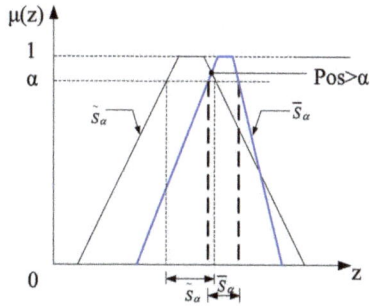

Figure 4. *Pos* of trapezoidal fuzzy number.

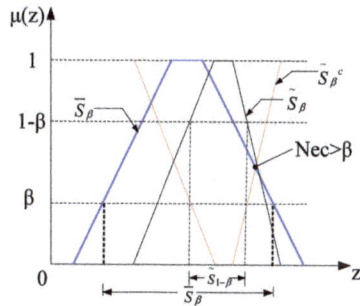

Figure 5. *Nec* of trapezoidal fuzzy number.

3.3. Parameters of Threshold for Nonconformity Judgment

In the aforementioned two rules, the value of parameters α and β are not uniformly set and depend on the specific process mean fuzzy number and sample mean fuzzy number.

For example, let $\overline{\hat{S}}(\hat{L}, \hat{m}, \hat{R})$ and $\widetilde{S}_i(L_i, m_i, R_i)$ be the fuzzy numbers of process mean and sample mean, respectively. If there is no abnormity in the control chart during a certain sampling period, the values of parameters α and β can be taken as the minimum of *Pos* and *Nec* of $\overline{\hat{S}}$ under different $\widetilde{S}$ as follows:

$$\alpha = \min(Pos(\overset{\wedge}{\overline{S}}|\widetilde{S}_i)) = \min_{z \in U} \sup[\min\{\mu_{\hat{S}}(z), \mu_{\widetilde{S}_i}(z)\}] \tag{10}$$

$$\beta = \min(Nec(\overset{\wedge}{\overline{S}}|\widetilde{S}_i)) = \min_{z \in U} \inf[\max\{\mu_{\hat{S}}(z), 1 - \mu_{\widetilde{S}_i}(z)\}] \tag{11}$$

Figure 6 illustrates the above process.

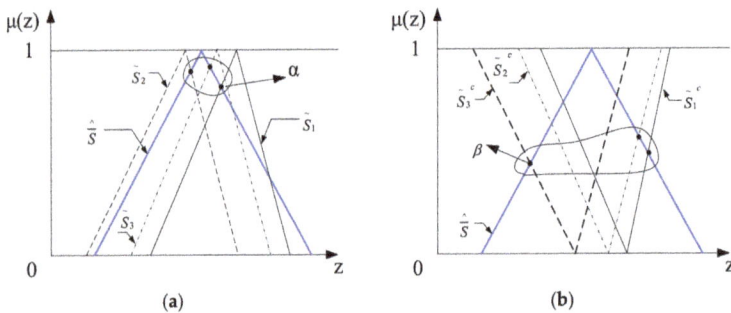

Figure 6. Determination of parameters. (**a**) Determination of parameter α; (**b**) determination of parameter β.

4. Basic Form of a Fuzzy Control Chart

The plotted data of fuzzy control chart is the supremum $Sup(\widetilde{S}_i)_\alpha$ and infimum $Inf(\widetilde{S}_i)_\alpha$ of α-cut of quality characteristic-scoring sample fuzzy number, which is denoted as a line segment connecting $Sup(\widetilde{S}_i)_\alpha$ and $Inf(\widetilde{S}_i)_\alpha$. The upper and lower control limits are supremum $Sup(\overline{S}_\alpha)$ and infimum $Inf(\overline{S}_\alpha)$ of mean scoring fuzzy number, respectively. The fuzzy control chart is judged abnormal when the α-cut set of sample fuzzy number and process mean fuzzy number do not meet the two criteria.

Taking triangular fuzzy number as example, $Sup(\widetilde{S}_i)_\alpha$ and $Inf(\widetilde{S}_i)_\alpha$ can be defined as follows, also shown in Figure 7.

$$Sup(\widetilde{S}_i)_\alpha = \max(\widetilde{S}_i)_\alpha = \max\{z | \mu(z) > \alpha\} \tag{12}$$

$$Inf(\widetilde{S}_i)_\alpha = \min(\widetilde{S}_i)_\alpha = \min\{z | \mu(z) > \alpha\} \tag{13}$$

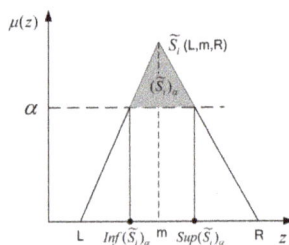

Figure 7. Supremum ($Sup(S_i)_a$) and infimum ($Inf(S_i)_a$) of fuzzy number α-cut.

Figure 8 shows the basic form of fuzzy score number-based control chat.

Fuzzy quality control chart

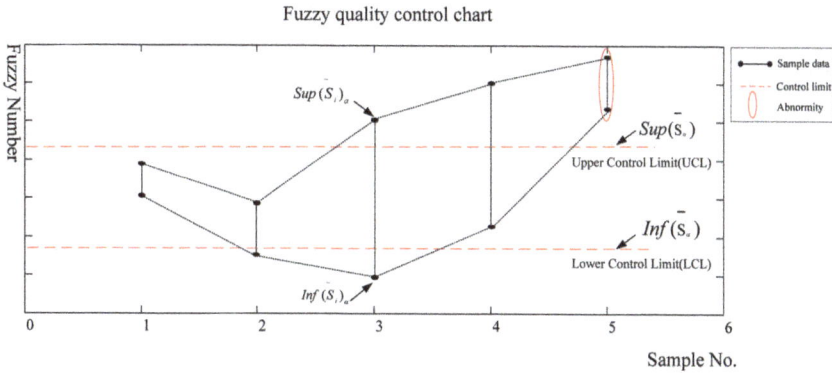

Figure 8. Basic form of a fuzzy score number-based fuzzy control chart.

5. Case Study

5.1. Construction of a Fuzzy Control Chart Based on Triangular Fuzzy Number

In the outer packaging inspection station of a certain workshop, the outer packaging quality was evaluated by a professional quality inspector. The process of monitoring packaging quality using a fuzzy control chart was as follows.

Step 1: Obtain the statistics of process mean. When process was in steady status, the quality inspector evaluated the cases by random sampling, and the triangular fuzzy number of the process mean was calculated as $\overline{S}$ (5, 6, 7).

Step 2: Determine the parameters of nonconformity judgment. In this case study, the parameters were set as $\alpha = 0.8$ and $\beta = 0.3$ in terms of Formulas (10) and (11). The upper and lower control limits, 6.2, 5.8 and 6.7, 5.3, were the right and left endpoints of α-cut and β-cut of process mean fuzzy number $\overline{S}$ (5, 6, 7), respectively, according to Formulas (12) and (13).

Step 3: Prepare the plotted data. A sample size of 5 was taken randomly and the scoring triangular fuzzy numbers were $\widetilde{S}_1(5.5, 6.3, 6.8)$, $\widetilde{S}_2$ (4.6, 5.6, 6.6), $\widetilde{S}_3$ (3, 6, 9), $\widetilde{S}_4$ (3, 7, 8.5), and $\widetilde{S}_5$ (6.5, 7.5, 8.5). According to Formulas (6), (7), (11), and (12), we obtain the results shown in Table 1.

Table 1. The calculation results of sample fuzzy number parameters.

Sample No.	Pos	$Sup(\widetilde{S}_i)_{ff}$	$Inf(\widetilde{S}_i)_{ff}$	Nec	$Sup(\widetilde{S}_i)_{1-fi}$	$Inf(\widetilde{S}_i)_{1-fi}$
1	0.83	6.4	6.14	0.47	6.45	6.06
2	0.8	5.8	5.4	0.3	5.9	5.3
3	1	6.6	5.4	0.25	6.9	5.1
4	0.8	7.3	6.2	0	7.45	5.8
5	0.25	7.7	7.3	0	7.8	7.2

Step 4: Build a fuzzy control chart for control. A fuzzy control chart can be built by using the data in Table 1. In Figures 9 and 10, the vertical segments were drawn in the chart by connecting the left and right endpoints of a complementary set of sample fuzzy number α-cut $\widetilde{S}_\alpha$, and the control chart state can be determined by the cross-relationship between the limit lines and vertical segments according to Formulas (8) and (9).

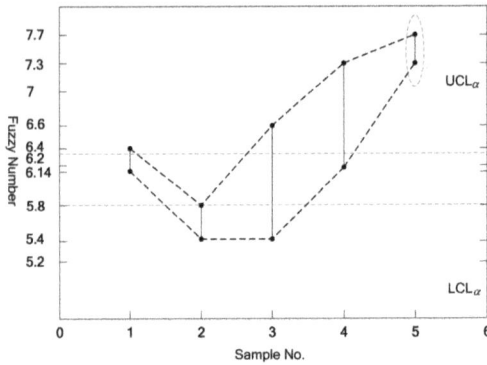

Figure 9. Possibility measure of control chart.

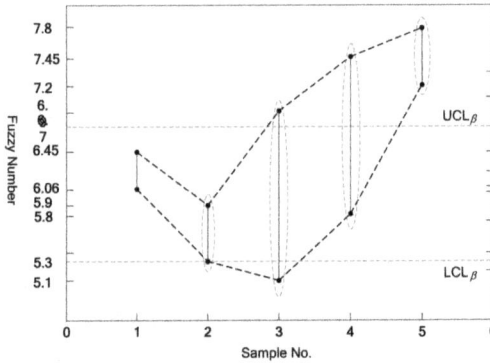

Figure 10. Necessity measure of control chart.

5.2. Analysis of Control Chart

For the fifth sample in Figure 9, the intersection of $\widetilde{S}_\alpha$ and $\overline{S}_\alpha$ was empty. It did not satisfy the criterion of possibility measure, thus the process was judged as abnormal. For the second, third, fourth, and fifth samples in Figure 10, $\widetilde{S}_\beta$ did not include $\overline{S}_{1-\beta}$. This did not satisfy the criterion of necessity measure, and the process was judged as abnormal. So, the process was in control only for Sample 1. The control level can be adjusted by parameters α and β.

For the same sample data, the following control chart (Figure 11) can be built by the commonly used method. The plotted data was the α-level fuzzy median of fuzzy number of each sample. The α-level fuzzy median of ith sample was obtained by $\frac{Sup(\widetilde{S}_i)_\alpha + Inf(\widetilde{S}_i)_\alpha}{2}$. If the fuzzy median is between the upper limit $Sup(\overline{S}_\alpha)$ and the lower limit $Inf(\overline{S}_\alpha)$, the process can be judged as being in a controlled state. We can see from Figure 11 that, except for the third sample, the other four samples were all nonconformities. This method has higher sensitivity than the control chart proposed by this paper, but this also limits actual application for its excessive alarms. In contrast, the proposed control chart is more suitable for practical application.

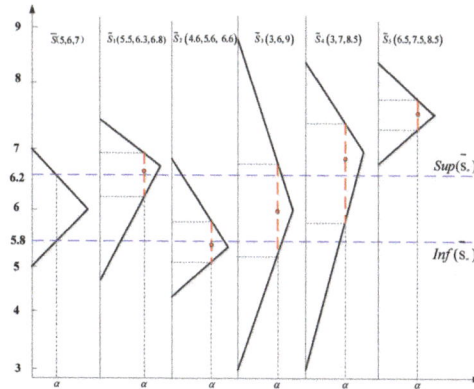

Figure 11. Fuzzy control chart based on α-level fuzzy median.

6. Influence of the of Membership Function Type on a Fuzzy Control Chart

The membership function of a fuzzy number plotted in a control chart can be any curve with a domain of $[0,1]$. In addition to the triangular membership function, trapezoidal-type, Gauss-type, and π-type functions are also used in fuzzy system.

In order to compare the impact of different types of membership functions on a fuzzy control chart, we designed the corresponding algorithm to calculate the parameters of other membership function by use of the parameters R, L, and m of the triangular membership function.

Suppose a triangular membership function $Tri\,(x, L, m, R)$ is:

$$
Tri\,(x, L, m, R) = \begin{cases} 0 & x \leq L \\ \frac{x-L}{m-L} & L < x \leq m \\ \frac{R-x}{R-m} & m < x \leq c \\ 0 & x > cc \end{cases}
$$

(1) For π-type membership function,

$$
Pi\,(x, a, b, c, d) = \begin{cases} 0, & x \leq a \\ 2\left(\frac{x-a}{b-a}\right)^2, & a \leq x \leq \frac{a+b}{2} \\ 1 - 2\left(\frac{x-a}{b-a}\right)^2, & \frac{a+b}{2} \leq x \leq b \\ 1 & b \leq x \leq c \\ 1 - 2\left(\frac{x-c}{d-c}\right)^2, & c \leq x \leq \frac{c+d}{2} \\ 2\left(d - \frac{x}{d-c}\right)^2, & \frac{c+d}{2} \leq x \leq d \\ 0, & d \leq x \end{cases} \tag{14}
$$

and trapezoidal-type membership function,

$$
Trap\,(x, a, b, c, d) = \begin{cases} 0 & x \leq a \\ \frac{x-a}{b-a} & a \leq x \leq b \\ 1 & b \leq x \leq c, \\ \frac{d-x}{d-c} & c \leq x \leq d \\ 0 & x \geq d \end{cases} \tag{15}
$$

291

then the parameters a, b, c, and d can be obtained by the following formulas:

$$a = 2 \times [k \times (m - L) + L] - [p \times (m - L) + L] \tag{16}$$

$$b = p \times (m - L) + L \tag{17}$$

$$c = (1 - p) \times (R - m) + m \tag{18}$$

$$d = 2 \times [(1 - k) \times (R - m) + m] - [(1 - p) \times (R - m) + m] \tag{19}$$

where k, p are shape parameters valued between 0 and 1.

(2) For a Gauss-type membership function,

$$G(x, \sigma, n) = e^{\frac{-(x-n)^2}{2\sigma^2}}, \tag{20}$$

where the parameters n and σ can be obtained by the following formulas:

$$n = k \times m + \frac{1}{2} \times p \times (L + R) \tag{21}$$

$$\sigma = [p \times (R - m) + (m - c)] / \sqrt{2\ln 2} \tag{22}$$

Taking the triangular fuzzy number data set shown in Table 1 as an example, the respective parameters of a fuzzy control chart for trapezoidal-type, Gauss-type, and π-type functions are shown in Table 2.

Table 2. Calculated results of fuzzy number parameters for different membership functions.

Type	Sample Fuzzy Number	Pos	$Sup(\widetilde{S}_i)_a$	$Inf(\widetilde{S}_i)_{ff}$	Nec	$Sup(\widetilde{S}_i)_{1-fi}$	$Inf(\widetilde{S}_i)_{1-fi}$
T	$\widetilde{S}_1$ (5.58, 6.22, 6.35, 6.75)	0.9162	6.43	6.092	0.4588	6.47	6.03
	$\widetilde{S}_2$ (4.70, 5.50, 5.70, 6.50)	0.8750	5.86	5.34	0.2500	5.94	5.26
	$\widetilde{S}_3$ (3.30, 5.70, 6.30, 8.70)	1.0000	6.78	5.22	0.1875	7.02	4.98
	$\widetilde{S}_4$ (3.40, 6.60, 7.15, 8.35)	0.8750	7.39	5.96	0	7.51	5.64
	$\widetilde{S}_5$ (6.6, 7.4, 7.6, 8.4)	0.1875	7.76	7.24	0	7.84	7.16
	$\overline{S}$ (5.10, 5.90, 6.10, 6.90)	-	6.26	5.74	-	6.66	5.34
G	$\widetilde{S}_1$ (0.2251, 6.285)	0.9078	6.44	6.13	0.4550	6.48	6.09
	$\widetilde{S}_2$ (0.4247, 5.6)	0.895	5.88	5.32	0.2418	5.96	5.24
	$\widetilde{S}_3$ (1.274, 6)	1	6.85	5.15	0.1757	7.08	4.93
	$\widetilde{S}_4$ (0.7432, 6.875)	0.7551	7.37	6.38	0.0387	7.50	6.25
	$\widetilde{S}_5$ (0.4247, 7.5)	0.2103	7.78	7.22	0.0016	7.86	7.14
	$\overline{S}$ (5.10, 5.90, 6.10, 6.90)	-	6.28	5.72	-	6.66	5.34
P	$\widetilde{S}_1$ (5.58, 6.22, 6.35, 6.75)	0.986	6.47	6.02	0.4209	6.51	5.97
	$\widetilde{S}_2$ (4.7, 5.5, 5.7, 6.5)	0.9687	5.96	5.25	0.1250	6.01	5.19
	$\widetilde{S}_3$ (3.3, 5.7, 6.3, 8.7)	1	7.05	4.94	0.0703	7.23	4.77
	$\widetilde{S}_4$ (3.4, 6.6, 7.15, 8.35)	0.9688	7.53	5.59	0	7.62	5.36
	$\widetilde{S}_5$ (6.6, 7.4, 7.6, 8.4)	0.0703	7.85	7.15	0	7.91	7.09
	$\overline{S}$ (5.10, 5.90, 6.10, 6.90)	-	6.36	5.65	-	6.59	5.41

Note: T—trapezoidal, G—Gauss, P—π.

Figures 12 and 13 show the difference between possibility and necessity measures of a fuzzy control chart constructed by four types of fuzzy numbers.

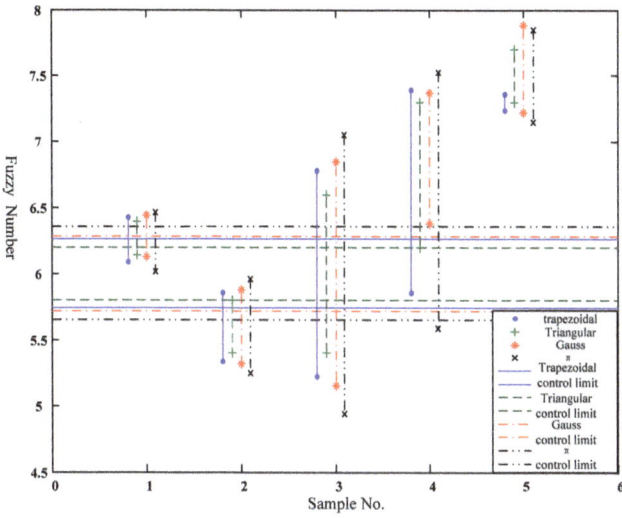

Figure 12. Possibility measure of a control chart built by different fuzzy numbers.

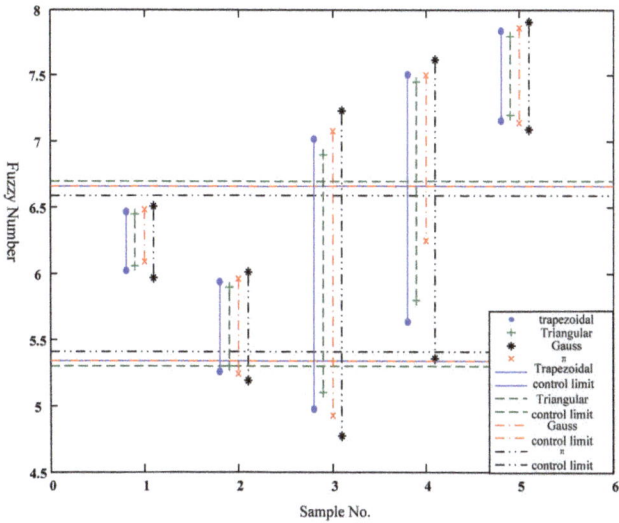

Figure 13. Necessity measure of a control chart built by different fuzzy numbers.

The following conclusions can be drawn:

(1) As for the possibility measure, triangular membership function has the narrowest control limit (i.e., the highest control level), trapezoid type takes second place, and π-type function is last.

(2) As for the necessity measure, π-type function has the narrowest control limit (i.e., the highest control level), trapezoid type takes second place, and triangular-type function is last; there is no distinct difference between the trapezoid-type and Gauss-type function, and their control limits are almost overlapping. Furthermore, the control interval of each sample has little difference.

Considering the product features and quality control level, we can adopt different membership functions to construct the fuzzy number and build the control chart. Correspondingly, different combinations of functions can be used to carry out the possibility and necessity measures in order to satisfy the quality control requirement of given applications.

7. Conclusions

Focusing on the fuzzy uncertain and immeasurable product attributes, this paper proposed an approach to construct a control chart based on fuzzy score number obtained from experts' quality scores. This approach uses actual score values to calculate the statistical parameters of a fuzzy number, and can effectively avoid the influence of a prior distortion of a predefined membership function.

Two kinds of nonconformity judging rules were proposed, and their mathematical and graphical features were also analyzed. By use of the graphical features, the process state can be judged by the distribution of the plotted data in the control chart directly, and this facilitates the popularization and application of the proposed control chart.

The influence of membership function on the control chart was analyzed. According to the results of this analysis, for quality attributes needing strict control, the triangular membership function should be selected; however, for quality attributes more loosely controlled, π-type function should be selected.

The parameters have a great influence on the effectiveness of proposed control chart. Therefore, determining the appropriate values of these two parameters for different process conditions is worth studying further.

Acknowledgments: This work was supported in part by Scientific Research Fund of Hunan Provincial Education Department (Grant No. 16B208); MOE (Ministry of Education in China) Project of Humanities and Social Sciences (Grant No. 13YJC630049), China; Natural Science Foundation of Shanxi Province (Grant No. 2013021021-2), China.

Author Contributions: Shiwang Hou, Hui Wang and Shunxiao Feng conceived and worked together to achieve this work, Shiwang Hou wrote the paper, Hui Wang and Shunxiao Feng made contribution to the case study.

Conflicts of Interest: The authors declare no conflict of interest.

References

1. Gülbay, M.; Kahraman, C. An alternative approach to fuzzy control charts: Direct fuzzy approach. *Inf. Sci.* **2007**, *177*, 1463–1480. [CrossRef]
2. Gülbay, M.; Kahraman, C. Development of fuzzy process control charts and fuzzy unnatural pattern analyses. *Comput. Stat. Data Anal.* **2006**, *51*, 434–451. [CrossRef]
3. Gülbay, M.; Kahraman, C.; Ruan, D. α-Cut fuzzy control charts for linguistic data. *Int. J. Intell. Syst.* **2004**, *19*, 1173–1195. [CrossRef]
4. Taleb, H.; Limam, M. On fuzzy and probabilistic control charts. *Int. J. Prod. Res.* **2002**, *40*, 2849–2863. [CrossRef]
5. Sorooshian, S. Fuzzy Approach to Statistical Control Charts. *J. Appl. Math.* **2013**, *19*, 4499–4588. [CrossRef]
6. Morabi, Z.S.; Owlia, M.S.; Bashiri, M.; Doroudyan, M.H. Multi-objective design of control charts with fuzzy process parameters using the hybrid epsilon constraint PSO. *Appl. Soft Comput.* **2015**, *30*, 390–399. [CrossRef]
7. Grzegorzewski, P.; Hryniewicz, O. Soft methods in statistical quality control. *Control Cybern.* **2000**, *29*, 119–140.
8. Hsieh, K.L.; Tong, L.I.; Wang, M.C. The application of control chart for defects and defect clustering in IC manufacturing based on fuzzy theory. *Expert Syst. Appl.* **2007**, *32*, 765–776. [CrossRef]
9. Tannock, J.D.T. A fuzzy control charting method for individuals. *Int. J. Prod. Res.* **2003**, *41*, 1017–1032. [CrossRef]
10. Shu, M.H.; Wu, H.C. Fuzzy and R control charts: Fuzzy dominance approach. *Comput. Ind. Eng.* **2011**, *61*, 676–685. [CrossRef]
11. Gildeh, B.S.; Shafiee, N. X-MR control chart for autocorrelated fuzzy data using D$_{p,q}$-distance. *Int. J. Adv. Manuf. Technol.* **2015**, *5*, 1047–1054. [CrossRef]
12. Zarandi, M.H.F.; Alaeddini, A.; Turksen, I.B. A hybrid fuzzy adaptive sampling—Run rules for Shewhart control charts. *Inf. Sci.* **2008**, *178*, 1152–1170. [CrossRef]
13. Sentürk, S.; Erginel, N.; Kaya, I.; Kahraman, C. Fuzzy exponentially weighted moving average control chart for univariate data with a real case application. *Appl. Soft Comput.* **2014**, *22*, 1–10. [CrossRef]

14. Faraz, A.; Shapiro, A.F. An application of fuzzy random variables to control charts. *Fuzzy Sets Syst.* **2010**, *16*, 2684–2694. [CrossRef]

15. Fernández, M.N.P.; García, A.C.; Barzola, O.R. Multivariate multinomial control chart using fuzzy approach. *Int. J. Prod. Res.* **2015**, *53*, 2225–2238. [CrossRef]

16. Ghobadi, S.; Noghondarian, K.; Noorossana, R. Developing a fuzzy multivariate CUSUM control chart to monitor multinomial linguistic quality characteristics. *Int. J. Adv. Manuf. Technol.* **2015**, *79*, 1893–1903. [CrossRef]

17. Alipour, H.; Noorossana, R. Fuzzy multivariate exponentially weighted moving average control chart. *Int. J. Adv. Manuf. Technol.* **2010**, *48*, 1001–1007. [CrossRef]

18. Jamison, K.D.; Lodwick, W.A. The construction of consistent possibility and necessity measures. *Fuzzy Sets Syst.* **2002**, *132*, 1–10. [CrossRef]

19. Zadeh, L.A. Fuzzy sets as a basis for a theory of possibility. *Fuzzy Sets Syst.* **2002**, *100*, 9–34. [CrossRef]

20. Dubois, D.; Prade, H. Ranking fuzzy numbers in the setting of possibility theory. *Inf. Sci.* **1983**, *30*, 183–224. [CrossRef]

symmetry

MDPI

Article

Comparing Bayesian and Maximum Likelihood Predictors in Structural Equation Modeling of Children's Lifestyle Index

Che Wan Jasimah bt Wan Mohamed Radzi *, Huang Hui and Hashem Salarzadeh Jenatabadi

Department of Science and Technology Studies, Faculty of Science, University of Malaya, Kuala Lumpur 50603, Malaysia; huanghui@siswa.um.edu.my (H.H.); jenatabadi@um.edu.my (H.S.J.)
* Correspondence: jasimah@um.edu.my; Tel.: +603-7967-5182

Academic Editor: Angel Garrido
Received: 17 July 2016; Accepted: 18 November 2016; Published: 28 November 2016

Abstract: Several factors may influence children's lifestyle. The main purpose of this study is to introduce a children's lifestyle index framework and model it based on structural equation modeling (SEM) with Maximum likelihood (ML) and Bayesian predictors. This framework includes parental socioeconomic status, household food security, parental lifestyle, and children's lifestyle. The sample for this study involves 452 volunteer Chinese families with children 7–12 years old. The experimental results are compared in terms of root mean square error, coefficient of determination, mean absolute error, and mean absolute percentage error metrics. An analysis of the proposed causal model suggests there are multiple significant interconnections among the variables of interest. According to both Bayesian and ML techniques, the proposed framework illustrates that parental socioeconomic status and parental lifestyle strongly impact children's lifestyle. The impact of household food security on children's lifestyle is rejected. However, there is a strong relationship between household food security and both parental socioeconomic status and parental lifestyle. Moreover, the outputs illustrate that the Bayesian prediction model has a good fit with the data, unlike the ML approach. The reasons for this discrepancy between ML and Bayesian prediction are debated and potential advantages and caveats with the application of the Bayesian approach in future studies are discussed.

Keywords: Bayesian structural equation modeling; public health; maximum likelihood structural equation modeling; Gibbs sampler algorithm

---

1. Introduction

Children's lifestyle behaviors, such as technology usage time, home studying, physical activity, and sleep duration tend to change in non-favorable directions. Some studies indicate that the family environment is an important determinant of children's lifestyle [1]. Therefore, information on children's lifestyle is often gathered based on household environment surveys. Decision-makers can use such data to allocate resources prudently when planning activities aimed at improving the overall lifestyle of children in a particular community. For ease of interpretation, this type of information is summarized in a single value called the children's lifestyle index. It is also important to identify factors potentially affecting this index. Various studies have indicated that many factors are related to the lifestyle index of children, including parental socioeconomic status [2–4] and parental lifestyle [5]. However, there are insufficient studies on the impact of household food security on children's lifestyle. Moreover, there are links between parental socioeconomic situation and household food security [6]. Nevertheless, research on the simultaneous integration of the interrelationships among the four well-known concepts into one model remains scarce. These influential factors are interrelated and latent because they cannot be measured directly, and it is thus quite complicated to determine the lifestyle index.

Figure 1 shows the hypothesized model involving measurement and structural components used to illustrate the children's lifestyle index. Six important relationships are examined in the current research framework. The influence of socioeconomic status as an independent variable of behavior further complicates our understanding of children's lifestyle and related behaviors.

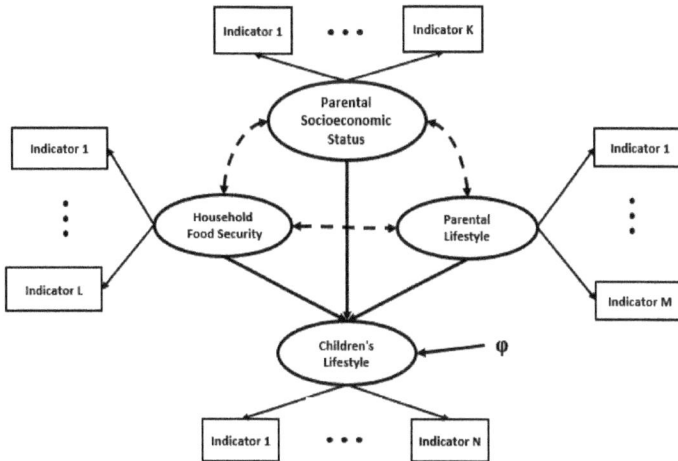

Figure 1. Research framework.

First, the direct relationships between parental socioeconomic status and children's lifestyle, as well between parental socioeconomic status and both parental lifestyle and household food security, are considered. Second, as Ishida [7] confirmed, there is an interconnection between household food security and lifestyle behavior. Therefore, these two latent variables with their relationships are included in the research model as mediators. The third relationship considers the direct impact of both household food security and parental lifestyle on children's lifestyle. The goal is to distinguish how these two family environment indicators influence children's lifestyle.

Since it is also reasonable to hypothesize that parental socioeconomic status, household food security, and parental lifestyle are correlated, the interrelationships among these three latent variables are indicated in Figure 1 by dashed lines with double-headed arrows connecting the latent variables.

The six hypotheses considered in the research model are:

H_1: Parental socioeconomic status has a significant impact on children's lifestyle.

H_2: Household food security has a significant impact on children's lifestyle.

H_3: Parental lifestyle has a significant impact on children's lifestyle.

H_4: There is a significant relationship between parental socioeconomic status and parental lifestyle in the research model.

H_5: There is a significant relationship between parental socioeconomic status and household food security in the research model.

H_6: There is a significant relationship between household food security and parental lifestyle in the research model.

Linear and nonlinear regression analyses have become bases of modeling techniques in statistics. However, individual regression analysis for each dependent variable is hardly challenged as a realistic approach in situations where the outcomes are naturally related. Additionally, it is difficult to analyze some research frameworks using regression models when an outcome is determined not only by the direct impacts of the predictor variables but also by their unobserved common causes. Structural equation modeling (SEM) is a suitable technique that can address the above limitations by providing a robust means of studying interdependencies among a set of correlated variables. Maximum likelihood SEM (ML-SEM) has been used by many researchers to analyze a complex phenomenon involving hypothesized relationships between independent and dependent latent variables. Classical methods based on the covariance structure approach encounter serious difficulties when dealing with complicated models and/or data structures. ML-SEM is applied to analyze the proper amount of hidden indicators (constructs or latent variable) to determine the observed indicators. ML-SEM is capable of performing concurrent analysis to illustrate the connections among observed indicators and corresponding latent variables as well as the connection among latent variables (Ullman [8]).

A computational algorithm in ML-SEM is developed based on the sample covariance matrix. ML-SEM employs the assumption that the observations are independent and identically distributed according to multivariate normal distribution [9]. If this assumption is not fulfilled, the sample covariance matrix cannot be determined in the usual way and is difficult to obtain [10]. Therefore, a number of researchers, such as Bashir and Schilizzi [11], Radzi and Jenatabadi [12], and Scheines and Hoijtink [13] have proposed using the Bayesian approach in SEM to overcome these problems. The Bayesian approach is attractive as users are able to employ prior information to update current information pertaining to the parameters of interest.

In his book *Structural Equation Modeling: A Bayesian Approach*, Lee [9] presents some advantages of Bayesian SEM (B-SEM) prediction:

- First moment properties of raw individual observations are mainly used in statistical techniques, thus making the techniques much simpler than second moment properties of the sample covariance matrix. Hence, B-SEM is easier to apply in more complex states.
- Direct latent variable estimation is possible, which simplifies the process of obtaining factor score estimates compared to classical regression methods.
- As manifest variables are directly modeled with their latent variables using familiar regression functions, B-SEM provides a more direct interpretation. It can also use common methods of regression modeling, such as residual and outlier analyses in conducting statistical analysis.

As pointed out by Scheines and Hoijtink [13], Lee and Song [14], and Dunson [15], the Bayesian predictor technique allows researchers to use prior experts' theories in addition to the sample information to produce better outputs and deliver valuable statistics and indices, including the mean and percentiles of the posterior distribution of unknown parameters. In conclusion, more reliable results can be achieved for small samples. In contrast, the Bayesian approach has much more flexibility in handling complex situations. Even though many studies have been done on determining the lifestyle index, not much has been done on modeling this index using SEM, particularly when considering information on parental socioeconomic status, household food security, and parental lifestyle. Therefore, the main purpose of this study is to illustrate the worth of ML-SEM and Bayesian SEM (B-SEM) in developing a model that describes the lifestyle index of children.

2. Theoretical Background of Maximum Likelihood-Structural Equation Modeling (ML-SEM) and Bayesian-SEM (B-SEM)

In the field of SEM, new techniques and statistical prediction analyses have been developed to better evaluate more complex data structures. These contain but are not limited to: linear/nonlinear SEM with covariates [16,17], SEM with multilevel dimensions [18,19], SEM with multi-samples [20,21], SEM analysis with categorical data [22,23], SEM with exponential indicators [24], and SEM with nonlinear correlations [25,26]. The above research works endeavor not only to prepare theoretical results but also to produce significant practical values. Indeed, the B-SEM technique is developed based on a Bayesian approach as the second generation of ML-SEM, which involves a much wider class of models [9].

2.1. ML-SEM

SEM is strongly capable of hypothesizing any types of relations and interactions among research variables in a single causal framework. This technique is helpful for researchers to better understand the concept of latent variables and their action within the model. Based on Bollen 's [27] study, "latent variables provide a degree of abstraction that permits us to describe relations among a class of events or variables that share something in common". For instance, with this ability of latent variables of SEM, we were able to combine indicators that are related to children's behavior in a household environment and named it the "children's lifestyle" latent variable. Another capability of SEM is determining the interconnection between three predictors (parental socioeconomic status, household food security, and parental lifestyle) and the impact of them on children's lifestyle (see Figure 2).

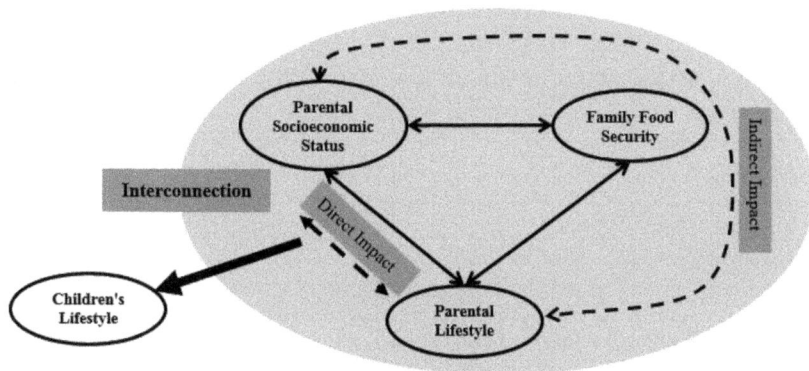

Figure 2. Interconnection among family environment variables and impact on children's lifestyle.

For predicting and estimating the research parameters in ML-SEM, measurements and structural models are the main procedures. The measurement model is defined by a $p \times 1$ vector y_i that is given by:

$$y_i = \mu \Omega_i + \varepsilon_i; \ i = 1, 2, \ldots, n \tag{1}$$

where

$$Y = \begin{bmatrix} y_1 \\ y_2 \\ \vdots \\ y_n \end{bmatrix}; \Omega = \begin{bmatrix} \Omega_1 \\ \Omega_2 \\ \vdots \\ \Omega_n \end{bmatrix}; \mu = \begin{bmatrix} \mu_{11} & \mu_{12} & \cdots & \mu_{1n} \\ \mu_{21} & \mu_{22} & \cdots & \mu_{2n} \\ \vdots & \vdots & \ddots & \vdots \\ \mu_{m1} & \mu_{m2} & \cdots & \mu_{mn} \end{bmatrix}; \varepsilon = \begin{bmatrix} \varepsilon_1 \\ \varepsilon_2 \\ \vdots \\ \varepsilon_n \end{bmatrix}.$$

In the research model Y includes five indicators: technology use, hours of study at home, child's physical exercise, child's sleep amount, and school grade (see Section 3.1).

In Equation (1):

(a) μ is a $(m \times n)$ matrix that represents factor loadings from modeling the regressions of y_i on Ω_i.

(b) Ω_i is a $(n \times 1)$ vector with normal distribution $N(0, \Phi)$ and is representative of the constructs (latent variables). $\Omega_i \ i = 1, \ldots, n$, are identically independent, have no correlation with ε_i, and have normal distribution $N(0, \Phi)$. To modify the exogenous and endogenous latent variables' association, Ω_i is partitioned into (λ_i, ω_i), where λ_i and ω_i are $r \times 1$ and $s \times 1$ vector variables, respectively, with latent structures.

(c) ε_i is a $(m \times 1)$ random vector with $N(0, \psi_\varepsilon)$ distribution that represents the error measurement.

Equation (2) presents the structural function elements:

$$\lambda_i = \Sigma \lambda_i + \gamma \omega_i + \pi_i; \ i = 1, 2, \ldots, n, \tag{2}$$

$$
\Sigma = \begin{bmatrix} \Sigma_{11} & \Sigma_{12} & \cdots & \Sigma_{1r} \\ \Sigma_{21} & \Sigma_{22} & \cdots & \Sigma_{2r} \\ \vdots & \vdots & \ddots & \vdots \\ \Sigma_{r1} & \Sigma_{r2} & \cdots & \Sigma_{rr} \end{bmatrix} ; \gamma = \begin{bmatrix} \gamma_{11} & \gamma_{12} & \cdots & \gamma_{1s} \\ \gamma_{21} & \gamma_{22} & \cdots & \gamma_{2s} \\ \vdots & \vdots & \ddots & \vdots \\ \gamma_{r1} & \gamma_{r2} & \cdots & \gamma_{rs} \end{bmatrix},
$$

where

(a) Σ is an $r \times r$ matrix of structural parameters representing the relationships among endogenous latent variables. This matrix is assumed to have zeroes in the diagonal elements.

(b) γ is an $r \times s$ matrix of regression parameters relating both exogenous and endogenous latent variables, and π_i is a $r \times 1$ vector of disturbances.

(c) π_i is an error term presumed to have $N(0, \psi_\pi)$ distribution, where ψ_π is a diagonal covariance matrix and this vector is uncorrelated with ω_i.

In this paper children's lifestyle is the endogenous latent variable and parental socioeconomic status, household food security, and parental lifestyle are exogenous latent variables for dependent variable. However, parental lifestyle and household food security are endogenous latent variables for parental socioeconomic status. Therefore, in the research model parental socioeconomic status acts as exogenous, children's lifestyle acts as endogenous, and household food security and parental lifestyle act both endogenous and exogenous.

To estimate the research parameters with ML-SEM, the robust-weighted least-squares (RWLS) procedure is used. RWLS incurs standard errors, estimates the research parameter coefficients, calculates χ^2 and fit indices created by applying the diagonal weight matrix components produced based on the thresholds' asymptotic variances, and estimates the latent correlation [28]. Model evaluation is the next step in ML-SEM. In this respect, the model goodness-of-fit can be checked through the related Chi-square statistic (CMIN), normed fit index (NFI), comparative fit index (CFI), Tucker Lewis index (TLI), incremental fit index (IFI), relative fit index (RFI), goodness-of-fit index (GFI), and root mean square error of approximation (RMSEA) [8].

2.2. B-SEM

The ML method finds estimates by maximizing the likelihood function, assuming observed data. Specifically, if $x = (x_1, \ldots, x_n)$ is the observed value of a random sample $X = (X_1, \ldots, X_n)$ from distribution $f(\cdot)$, $f \in \mathcal{F} = \{f(x|\theta) : x \in \chi, \theta \in \Omega\}$, then the likelihood function of θ has the form

$$
\mathcal{L}(\theta) = f(x_1, \ldots, x_n|\theta) = \prod_{i=1}^{n} f(x_i|\theta). \tag{3}
$$

The ML estimate of θ is given by

$$
\hat{\theta} = \operatorname*{argmin}_{\theta} \mathcal{L}(\theta) \tag{4}
$$

Before observing the data in Bayesian analysis, the practitioner/expert has an idea/belief/ information about the unknown parameter $\theta \in \Omega$. This prior information is updated with information obtained from the sample, forming the posterior distribution of θ, which will be used to estimate θ. This procedure is shown in Figure 3, where the distributions for a prior and its respective posterior for a given parameter, together with the likelihood, are illustrated. Note that the likelihood can be considered the distribution of the data given the parameter values. Based on Figure 3, the major portion of the prior distribution has lower parameter values than the likelihood distribution. The posterior is obtained as a compromise between the prior and the likelihood.

From Figure 3, it is apparent that the prior does not allocate sufficient probability where the likelihood is high, and there exists prior-data conflict. See Evans and Moshonov [29] for more details.

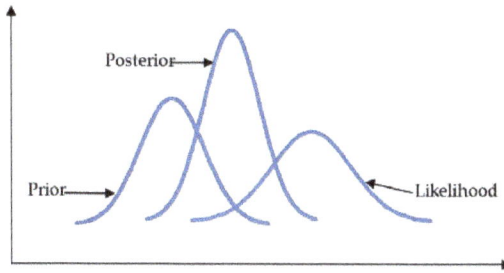

Figure 3. Likelihood, posterior, and prior for a parameter (source: [30]).

Priors can be non-informative or informative. A non-informative prior, also named a diffuse prior, can, for instance, have a normal or uniform distribution with large variance. In statistical modeling, a large uncertainty in the parameter value is reflected by a large variance. Consequently, with a large prior variance, the likelihood contributes relatively more information to the construction of the posterior, and the estimate is closer to an ML estimate. Evans [31] cautioned that using a large variance prior may lead to the Jeffreys-Lindley paradox.

Formally, the formation of a posterior draws on Bayes's theorem. Consider the probabilities of events C and D, $P_r(C)$ and $P_r(D)$. Based on probability theory, the joint event C and D can be expressed in terms of conditional and marginal probabilities:

$$P_r(C, D) = P_r(C|D) \, P_r(D) = P_r(D|C) \, P_r(C) \tag{5}$$

In Equation (5), if we divide every side by $P_r(C)$ then we get:

$$P_r(D|C) = \frac{P_r(C|D) \, P_r(D)}{P_r(C)} \tag{6}$$

which is known as Bayes's theorem. By applying this theorem in modeling, it lets the data x take the role of C and the parameter value takes the role of D. Thus, the posterior can be symbolically illustrated as

$$posterior = parameter \ given \ data = \tfrac{data \ | \ parameters \ \times parameters}{data} = \tfrac{likelihood \ \times prior}{data} \propto likelihood \times prior \tag{7}$$

In the above formula "$\propto$" means "proportional to". More specifically, we have

$$P(\theta|x) = \frac{\mathcal{L}(\theta) \, \pi(\theta)}{m(x)} \tag{8}$$

where $\pi(\theta)$ is the prior distribution (probability) of $\theta \in \Omega$ and $m(x)$ is called the prior predictive distribution of x obtained as (for a continuous case):

$$m(x) = \int_\Omega \mathcal{L}(\theta) \, \pi(\theta) \, d\theta \tag{9}$$

In this study, the variables gathered are in the form of ordered categories. Yanuar and Ibrahim [32] believe that, before conducting Bayesian analysis, a threshold specification must be identified in order to treat the ordered categorical data as manifestations of a hidden continuous normal distribution. A brief explanation of the threshold specification is given below.

Suppose X and Y are defined as:

$$X = \begin{bmatrix} x_1 \\ x_2 \\ \vdots \\ x_n \end{bmatrix} ; Y = \begin{bmatrix} y_1 \\ y_2 \\ \vdots \\ y_n \end{bmatrix},$$

which can be considered as the ordered categorical data matrix and latent continuous variables, respectively. Moreover, the relationship between X and Y is described by applying the threshold specification. The procedure for x_1 is described as an instant. More precisely, let

$$x_1 = c \; if \; \tau_c - 1 < y_1 < \tau_c \tag{10}$$

- c is the number of categories for x_1;
- $\tau_c - 1$ and τ_c represent the threshold levels associated with y_1.

For example, in the current study we assumed $c = 3$, which leads to $\tau_0 = -\infty$ and $\tau_3 = \infty$. Meanwhile, the values of τ_1 and τ_2 are calculated based on the proportion of cases in each category of x_1 using

$$\tau_k = \Phi^{-1} \left(\sum_{r=1}^{2} \frac{N_r}{N} \right), \; k = 1, 2, \tag{11}$$

We assumed that Y is distributed as a multivariate normal. Therefore, in Equation (10) we have:

- $\Phi^{-1}(\cdot)$ is the inverse standardized normal distribution;
- N is the total number of cases;
- N_r is the number of cases in the rth category.

Under the Bayesian SEM, $X = (x_1, x_2, \ldots, x_n)$ and $Y = (y_1, y_2, \ldots, y_n)$ are the ordered categorical data matrix and latent continuous variables, respectively, and $\Omega = (\omega_1, \omega_2, \ldots, \omega_n)$ is the matrix of latent variables. The observed data X are augmented with the latent data (Y, Ω) in the posterior analysis. The parameter space is denoted by $\Theta = (\tau, \theta, \Omega)$, where $\theta = (\Phi, \Lambda, \Lambda_\omega, \Psi_\delta, \Psi_\varepsilon)$ is the structural parameter. In line with Lee (2007), the prior model is given by

$$\pi(\Theta) = \pi(\tau) \pi(\theta) \pi(\Omega | \tau, \theta) \tag{12}$$

where, due to the ordinal nature of thresholds, a diffuse prior can be adopted. Specifically, for some constant c,

$$\pi(\tau) = c \tag{13}$$

Further, to accommodate a subjective viewpoint, a natural conjugate prior can be adopted for θ with the conditional representation $\pi(\theta) = \pi(\Lambda | \Psi_\varepsilon) \pi(\Psi_\varepsilon)$. More specifically, let

$$\psi_{\varepsilon k}^{-1} \sim \Gamma(\alpha_{0\varepsilon k}, \beta_{0\varepsilon k}) \tag{14}$$

$$\left(\Lambda_k | \psi_{\varepsilon k}^{-1} \right) \sim N(\Lambda_{0k}, \psi_{\varepsilon k} H_{0yk}) \tag{15}$$

where $\psi_{\varepsilon k}$ is the kth diagonal element of ψ_ε, Λ_k is the kth row of Λ, and Γ denotes the gamma distribution. Finally, an inverse-Wishart distribution is adopted for Φ as follows:

$$\Phi^{-1} \sim W_q(R_0, \rho_0) \tag{16}$$

It is further supposed that all hyperparameters are known. Posterior distribution can be found by normalizing the product $L(\Theta | X = x) \pi(\Theta)$.

In order to sample from the posterior distribution $\Theta|X = x$, the Markov Chain Monte Carlo (MCMC) technique is used to handle the computational complexity.

2.3. Modeling Description

The model hypothesized in this study consists of 16 indicator variables with one exogenous latent variable and three endogenous latent variables. The following measurement model is then formulated:

$$y_i = \Lambda \omega_i + \varepsilon_i, \ i = 1, 2, \ldots, n \tag{17}$$

where $\omega_i = (\eta_i, \xi_{i1}, \xi_{i2}, \xi_{i3})^T$. The structural part of the current SEM model has the form

$$\eta_i = \gamma_1 \xi_{i1} + \gamma_2 \xi_{i2} + \gamma_3 \xi_{i3} + \delta_i \tag{18}$$

where $(\xi_{i1}, \xi_{i2}, \xi_{i3})^T$ is distributed as $N(0, \Phi)$ and independent of δ_i, which is distributed as $N(0, \psi_\delta)$.

In data analysis, we applied AMOS 18 to estimate research parameters for ML-SEM, while WinBUGS 1.4 was used for B-SEM analysis. The hierarchical structure is employed by choosing the prior information for parameters involved in the hypothesized model, as defined in Equations (14)–(16).

3. Materials and Methods

3.1. Data Structure

The information gathered in this survey includes information about parental socioeconomic status, household food security, parental lifestyle, and children's lifestyle of individuals living in Urumqi, Xinjiang, China. The parental socioeconomic situation was measured as the initial independent variable, including nine indicators. Eight of the indicators are the mother's age, father's age, mother's education level, father's education level, mother's income level, father's income level, mother's work experience, and father's work experience. The added question is "How long have the parents been married? The parents' ages were classified into four groups, namely "30 years old or below", "31 to 40 years old", "41 to 50 years old" and "over 50 years old", which were coded as 1, 2, 3, and 4, respectively. With respect to education level, the responses obtained were coded as 1 for "Less than High School", 2 for "High school", 3 for "Diploma", 4 for "Bachelor's", and 5 for "Master's or PhD". The respondents were asked about the parental income status, and the responses were denoted by 1, 2, 3, 4, and 5 for "less than RMB2,000 per month", "RMB2,001–RMB3,000 per month", "RMB3,001–RMB4,000 per month", "RMB4,001–RMB5,000 per month", and "more than RMB5,000 per month", respectively. The respondents were asked about the parental work experience and the responses were coded as 1, 2, 3, 4, and 5, denoting "less than 5 years", "5–10 years", "11–15 years", "16–20 years", and "more than 20 years", respectively. The last question in the socioeconomic part is related to the duration of the parents' marriage, and responses were labeled 1, 2, 3, 4, and 5 for "less than 2 years", "2–4 years", "5–7 years", "8–10 years", and "more than 10 years". Family food security status was the first mediator, based on a study by Bickel, Nord [33], which included 18 standard questions. We extracted nine questions that are representative of the food security indicators, which were measured on a Likert scale from 1 to 9. The third variable is parental lifestyle and in the research model it acts as the second mediator. Several authors have proposed lists of health-related behaviors for measuring parental lifestyle. Nakayama and Yamaguchi [34] suggested a list of health-related behaviors including physical exercise, smoking habits, average sleeping hours, and average working hours per day. In our study, we added drinking alcohol to Nakayama's list and measured all factors for fathers and mothers separately. Therefore, parental lifestyle was measured based on 10 indicators, namely alcohol drinking, smoking habits, physical exercise, working hours, and average sleeping hours per day, for the mother and father. The respondents were asked about their alcohol drinking habits and the responses were coded as 1, 2, 3, 4, 5, 6, and 7, denoting "less than 1 time per month", "1 time per month", "2 to 3 times per month", "1 time per week", "2 to 3 times per week", "4 to

6 times per week", and "every day". The respondents were asked about their smoking habits and the responses were coded as 1, 2, and 3, denoting "smoker", "quit", and "non-smoker", respectively. Regarding the frequency of physical exercise, respondents were asked "how many times a week on average do you do physical exercise?" The responses for this question were placed into four categories coded as 1, 2, 3, and 4, indicating "none", "1 or 2 times per week", "3 or 4 times per week", and "more than 4 times per week", respectively. Working hours per day were coded as 1 for "more than 14 h", 2 for "9–14 h", and 3 for "less than 9 h". The average sleeping hours per day were grouped as 1 referring to "less than 7 h per day", 2 for "more than 8 h per day", and 3 for "between 7 and 8 h per day". Children's lifestyle was the fourth latent variable and acted as the dependent variable. Five indicators were considered in measuring the children's lifestyle index. These are hours of study at home, child's sleep amount, technology use, school grade, and child's physical exercise. The average hours of study at home per day were grouped into four categories: 1 referring to "less than 1 h per day", 2 for "1 to 2 h per day", 3 for "3 to 4 h per day", and 4 for "more than 4 h per day". The child's average sleeping hours per day were grouped into 1 referring to "less than 7 h per day", 2 for "between 7 and 8 h per day", 3 for "between 8 and 9 h per day", and 4 for "more than 9 h per day". Respondents were asked "how many hours on average does your child use technology per day?" The responses for this question consist of four categories that were coded as 1, 2, 3, and 4, indicating "less than 1 h per day", "1 to 2 h per day", "3 to 4 h per day", and "more than 4 h per day", respectively. Children's school levels were coded from 1 to 6, denoting grades 1 to 6. Children's physical activity per week was coded as 1 for "none", 2 for "1 or 2 times per week", 3 for "3 or 4 times per week", and 4 for "more than 4 times per week".

3.2. Ethics Statement

For this research, the questionnaires were self-administered/reported. These surveys were collected anonymously, with no way of identifying the participants. Therefore, based on the Health Research Ethics Authority [35], the research does not require an ethics review "based solely on the researcher's personal reflections and self-observation".

3.3. Sampling

Five primary schools were selected from Urumqi City, Xinjiang Province, China and 120 questionnaires were delivered to every school. Every primary school includes six grades, and 20 questionnaires were distributed to each grade. Therefore, $5 \times 6 \times 20 = 600$ questionnaires were distributed in 2014 to five schools. Every questionnaire was for one family including a father, a mother, and a child between seven and 12 years old. The sample comprised parents who joined school parent meetings that take place four times per year. For each of the six grades, 20 volunteers were selected and trained on filling out the questionnaire. The survey was conducted with University of Malaya funding. A parent was retained in the sample if they had a child between seven (grade 1) and 12 (grade 6) years of age.

Of 600 distributed questionnaires, 483 were returned. The rest of the families refused to continue their cooperation. Among 483 questionnaires, 22 were eliminated based on missing data. Mahalanobis distance is an extremely general measure that is utilized for the measurement of multivariate outliers [36]. Based on Mahalanobis Distance testing, nine observations (observation number; 36, 88, 92, 134, 228, 256, 372, 411, and 444) were eliminated from the list because they were considered outliers that could affect the model fit, R^2, and the size and direction of parameter estimates (see Table 1). Therefore, $(483 - 22 - 9 = 452)$ 452 observations were considered as the final data of the study.

Table 1. Mahalanobis distance.

Observation Number	Mahalanobis D-Squared	p1	p2
36	22.56	0.0016	0.0084
88	20.31	0.0067	0.0091
92	18.92	0.0092	0.0104
134	36.58	0.0116	0.0124
228	32.71	0.0231	0.0178
256	30.08	0.0854	0.0364
372	28.19	0.0932	0.0392
411	25.44	0.1589	0.0421
444	19.76	0.2876	0.0482

If $p1$ or $p2$ is less than 0.05 then the observation is an outlier.

4. Results

Tables 2 and 3 show a descriptive analysis of the child and parental characteristics.

Table 2. Descriptive analysis of child characteristics.

Characteristics	Percentage	Characteristics	Percentage
Gender:		*Average hours per day of using technology:*	
Boy	45.60%	Less than one hour per day	13.20%
Girl	54.40%	1 to 2 h per day	15.70%
School grades:		3 to 4 h per day	40.70%
Grade 1	14.20%	More than 4 h per day	30.70%
Grade 2	16.80%	*Physical activities in a week:*	
Grade 3	16.60%	None	44.20%
Grade 4	16.30%	1 or 2 times per week	28.40%
Grade 5	17.30%	3 or 4 times per week	19.70%
Grade 6	18.80%	More than 4 times per week	7.70%
Study at home:		*Average sleeping hours in a day:*	
Less than one hour per day	21.10%	Less than 7 h per day	5.80%
1 to 2 h per day	29.40%	Between 7 and 8 h per day	22.20%
3 to 4 h per day	33.10%	Between 8 and 9 h per day	56.30%
More than 4 h per day	16.40%	More than 9 h per day	15.70%

Only the essential factors of each latent variable were sustained in the research model by applying factor loading. Table 4 presents the indicators' factor loadings on three latent variables. According to Argyris and Schön [37], the standardized factor loading must be over 0.5. As illustrated in Table 4, some factor loadings of four latent variables are below 0.5; therefore, these indicators must be excluded from the measurement model. For the parental socioeconomic latent variable, six indicators were excluded from the research model. These are the mother's age, father's age, father's education, mother's income, mother's work experience, and father's work experience. For the parental lifestyle latent variable, five indicators were excluded from the research model. These are the mother's alcohol drinking habit, mother's smoking habit, father's smoking habit, mother's physical exercise habit, and father's physical exercise habit. Among nine indicators of household food security, four were excluded from the research model. Finally, two indicators of children's lifestyle were excluded from the research model, the child's physical exercise habit and school grade.

Table 3. Descriptive analysis of parental characteristics

Characteristics	Father (%)	Mother (%)	Characteristics	Father (%)	Mother (%)
Age:			*Smoking Habit:*		
Less than or equal 30 years old	18.5%	21.6%	Smoker	66.6%	23.8%
Between 31and 40 years old	36.2%	25.1%	Quitted	15.7%	13.7%
Between 41 and 50 years old	22.1%	28.4%	Non-smoker	17.7%	62.5%
More than 50 years old	23.2%	24.9%	*Physical exercise:*		
Education:			None	54.4%	33.6%
Less than High School	11.3%	9.5%	1 or 2 times in a week	27.7%	38.7%
High school	19.8%	6.7%	3 or 4 times per week	14.8%	11.2%
Diploma	37.7%	41.9%	More than 4 times in a week	3.1%	16.5%
Bachelor	29.1%	33.1%	*Working hours in a day:*		
Master or PhD	2.1%	8.8%	More than 14 hours per day	26.7%	8.2%
Income:			9–14 hours per day	62.8%	73.5%
Less than RMB2000 per month	11.7%	20.6%	Less than 9 hours per day	10.5%	18.3%
RMB2001-RMB3000 per month	22.6%	24.5%	*Average sleeping hours in a day:*		
RMB3001-RMB4000 per month	33.9%	22.1%	Less than 7 hours per day	55.4%	61.9%
RMB4001-RMB5000 per month	19.9%	17.3%	Between 7 to 8 hours per day	27.9%	30.0%
More than RMB5000 per month	11.9%	15.5%	More than 8 hours per day	16.7%	8.1%
Work experience:			*Drinking Alcohol Habit:*		
No work experience	0.00%	0.00%	Less than one time per month	3.2%	10.6%
Less than 5 years	7.4%	19.2%	1 time per month	4.5%	22.7%
5-10 years	12.9%	21.7%	2 to 3 times per month	16.1%	32.1%
11-15 years	36.6%	26.6%	1 time per week	16.7%	28.2%
16-20 years	32.8%	23.6%	2 to 3 times per week	39.5%	6.4%
More than 20 years	10.3%	8.9%	4 to 6 times per week	18.7%	0.00%
			Every day	1.3%	0.00%

Table 4. Factor loading analysis of research latent variables.

Parameter Description	Factor Loading
Parental Socioeconomic	
Mother's age	0.43
Father's age	0.38
Mother's education	0.74
Father's education	0.39
Mother's income	0.43
Father's income	0.68
Mother's work experience	0.06
Father's work experience	0.05
Parents' marriage length	0.82
Parental Lifestyle	
Mother's drinking alcohol	0.36
Father's drinking alcohol	0.73
Mother's smoking habit	0.48
Father's smoking habit	0.41
Mother's physical exercises	0.21
Father's physical exercises	0.09
Mother's working hours	0.76
Father's working hours	0.88
Mother's average sleeping hours	0.83
Father's average sleeping hours	0.71

Table 4. *Cont.*

Parameter Description	Factor Loading
Household Food Security	
Worry about running out of food	0.73
Do not have money: household	0.82
Cannot afford to eat balanced meals: household	0.93
Cut down food portions: household	0.12
Do not eat the whole day: adults	0.98
Do not have money: children	0.04
Cannot afford to eat balanced meals: children	0.25
Cannot afford enough food: children	0.82
Skip a meal: children	0.24
Children's Lifestyle	
Technology use	0.92
Hours of study at home	0.73
Child's physical exercise	0.49
Child's sleep amount	0.68
School grade	0.46

Figure 4 represents the results of model fitting based on the SEM approach. In this respect, the model's goodness-of-fit can be checked with normed fit index (NFI), comparative fit index (CFI), Tucker Lewis index (TLI), incremental fit index (IFI), relative fit index (RFI), and goodness-of-fit index (GFI). The values of GFI, IFI, RFI, TLI, and NFI are within the acceptable range. Therefore, the current model is fitted for our data at the 5% significance level.

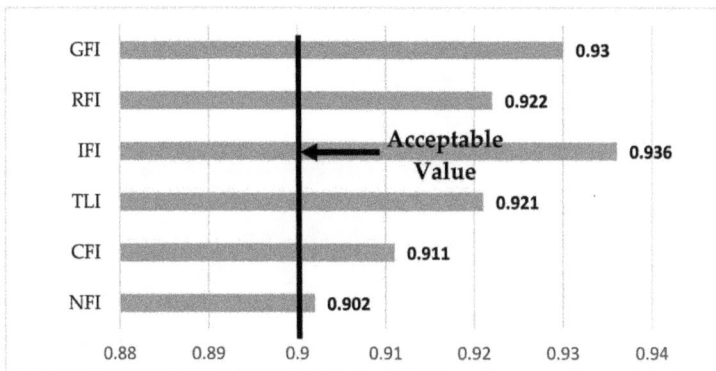

Figure 4. Model fitting analysis.

For some particular Bayesians, priors can come from any source, objective or otherwise [38]. The issue just described is referred to as the "elicitation problem" and has been discussed by Van Wesel [39] and Rietbergen and Klugkist [40]. Moreover, elicitation procedure is a time-consuming task, and even experts are often mistaken and prone to overstating their certainty [41]. Therefore, instead of depending fully on expert decisions, research scholars engaging Bayesian analysis often attempt to select the priors such that they are informative enough to yield B-SEM's advantages, while not being so informative as to bias the results [42]. By the end if one is unsure about the prior distribution, a sensitivity analysis is suggested [43]. In such an analysis, the outcomes of different prior specifications are compared to inspect the influence of the prior. To achieve this goal, models with four types of prior inputs were compared. In assigning hyperparameter values, a small variance was taken for each parameter. Fixed values of $\alpha = 5$ and $\beta = 5$ were evaluated for four inputs. Furthermore, values corresponding to Φ were measured with $\rho_0 = 25$ and $R_0^{-1} = 6.5I$.

Accordingly, the four prior inputs calculated are:

1. Prior I: Unknown loadings in Λ are all made equal to 0.35, and the measures corresponding to $\{\theta_1, \theta_2, \theta_3\}$ are $\{0.6, 0.5, 0.2\}$.
2. Prior II: The hyperparameter values are considered half of the values in prior I.
3. Prior III: The hyperparameter values are considered a quarter of the values in prior I.
4. Prior IV: The hyperparameter values are considered double the values in prior I.

Table 5 presents the outputs based on four types of prior inputs. This table indicates that the parameter estimates and standard errors obtained for various prior inputs are reasonably close. It can be concluded that the statistics found based on B-SEM are not sensitive to these four prior inputs. Therefore, our approach is only valid with the adopted prior and B-SEM applied here is quite robust against the different prior inputs. Accordingly, for the purpose of discussing the results obtained using B-SEM, the results obtained using type I prior are used.

Table 5. Parameter estimation and standard error for four types of prior in B-SEM analysis.

Parameter	Prior I Estimate	STD	Prior II Estimate	STD	Prior III Estimate	STD	Prior IV Estimate	STD
θ_1	0.561	0.021	0.555	0.033	0.549	0.069	0.584	0.121
θ_2	0.493	0.088	0.461	0.097	0.452	0.102	0.503	0.201
θ_3	0.203	0.096	0.192	0.051	0.180	0.091	0.221	0.138
θ_{13}	0.739	0.108	0.721	0.101	0.598	0.027	0.751	0.102
θ_{16}	0.683	0.112	0.677	0.109	0.655	0.111	0.686	0.138
θ_{19}	0.822	0.087	0.816	0.078	0.801	0.098	0.852	0.203
θ_{22}	0.733	0.039	0.730	0.035	0.722	0.069	0.763	0.093
θ_{27}	0.763	0.109	0.755	0.099	0.743	0.106	0.771	0.126
θ_{28}	0.883	0.119	0.844	0.081	0.822	0.077	0.896	0.119
θ_{29}	0.827	0.044	0.814	0.041	0.759	0.036	0.834	0.66
θ_{210}	0.711	0.066	0.697	0.057	0.666	0.051	0.723	0.107
θ_{31}	0.734	0.029	0.726	0.026	0.669	0.039	0.742	0.127
θ_{32}	0.822	0.071	0.816	0.064	0.798	0.061	0.831	0.104
θ_{33}	0.928	0.191	0.909	0.161	0.852	0.170	0.832	0.206
θ_{35}	0.981	0.058	0.921	0.052	0.832	0.048	0.883	0.067
θ_{38}	0.816	0.161	0.799	0.152	0.764	0.143	0.802	0.188

Based on Figures 5 and 6, the estimated structural equations that address the relationships between the children's lifestyle index and parental socioeconomic status, household food security, and parental lifestyle for ML-SEM and B-SEM are given by:

$$\hat{\phi} (ML - SEM) = 0.549\theta_1 + 0.198\,\theta_2 + 0.488\,\theta_3 \qquad (19)$$

and

$$\hat{\phi} (B - SEM) = 0.561\theta_1 + 0.203\,\theta_2 + 0.493\,\theta_3 \qquad (20)$$

respectively, where

θ_1 is the coefficient of parental socioeconomic status indicator;
θ_2 is the coefficient of household food security indicator;
θ_3 is the coefficient of parental lifestyle indicator.

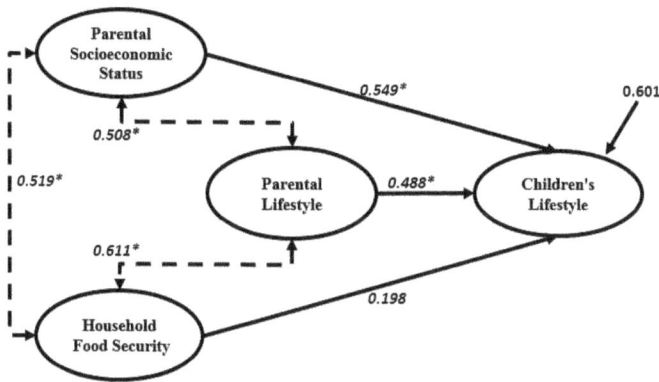

Figure 5. ML-SEM output.

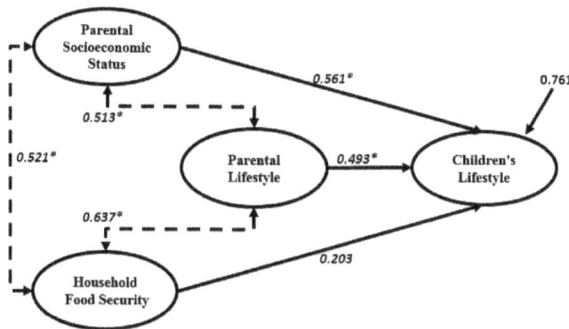

Figure 6. B-SEM output.

Table 6 presents the outputs of the research hypotheses regarding the relationships among variables in this study. In both models, the impact of parental socioeconomic status and parental lifestyle on children's lifestyle is significant. However, the impact of household food security on children's lifestyle is not significant. Moreover, the relationships between parental socioeconomic status and parental lifestyle, parental socioeconomic status and household food security, and household food security and parental lifestyle are significant and positive.

Table 6. Estimated parameters estimation of SEM using ML and Bayesian predictors.

Relation	Estimated Coefficients	
	ML-SEM	B-SEM
Parental socioeconomic → Children's life style	0.549 *	0.561 *
Household food security → Children's life style	0.198	0.203
Parental lifestyle → Children's life style	0.488 *	0.493 *
Parental socioeconomic ↔ Parental lifestyle	0.508 *	0.513 *
Parental socioeconomic ↔ Household food security	0.519 *	0.521 *
Household food security ↔ Parental lifestyle	0.611 *	0.637 *

* Presents a significant relationship with 95% confidence.

This part of the study presents an analysis of the comparison between the ML-SEM and B-SEM techniques in predicting the children's lifestyle index. Two main stages were considered in the comparison. In the first stage, four indices were used to compare ML-SEM with B-SEM, which are representative of the strength and correctness of the predictions. Among various prediction techniques, the root mean square error (RMSE), coefficient of determination (R^2), mean absolute error (MSE), and mean absolute percentage error (MAPE) are the most familiar statistical indices for comparison purposes. Table 7 presents the formulas of these indices and outputs of the ML and Bayesian approaches.

Table 7. Comparative outputs of ML-SEM and B-SEM.

Name of Index	Formula	ML-SEM Value	B-SEM Value
MAPE	$MAPE = \frac{1}{n} \sum_{i=1}^{n} \left\| \frac{y_i' - y_i}{y_i} \right\|$	0.094	0.088
RMSE	$RMSE = \sqrt[2]{\frac{\sum_{i=1}^{n} (y_i' - y_i)^2}{n}}$	0.091	0.051
MSE	$MSE = \frac{\sum_{i=1}^{n} \|y_i' - y_i\|}{n}$	0.128	0.105
R^2	$R^2 = \frac{[\sum_{i=1}^{n} (y_i' - \bar{y}_i') \cdot (y_i - \bar{y}_i)]^2}{\sum_{i=1}^{n} (y_i' - \bar{y}_i') \cdot \sum_{i=1}^{n} (y_i - \bar{y}_i)}$	0.601	0.761

In the above indices, y_i is the ith actual value of the dependent variable and y_i' is the ith predicted value. The R^2 value for the B-SEM model was greater than for the ML-SEM model, and the RMSE, MSE, and MAPE values for the B-SEM model were lower than for ML-SEM. Therefore, according to the performance indices, B-SEM predicted children's lifestyle better than the ML-SEM model. The main reason B-SEM performed better is the ML framework defined, which permits simultaneous self-adjustment of parameters and effective learning of the association between inputs and outputs in causal and complex models.

The present comparative analysis illustrates that the B-SEM has superior evaluation capability over ML-SEM in children's lifestyle index prediction. This conclusion is only made for this empirical analysis and it does not prove that B-SEM is always superior to ML-SEM.

5. Discussion

The main purpose of the present study was to demonstrate the potential of the maximum likelihood SEM and Bayesian SEM approaches in modeling the children's lifestyle index. The strength of SEM is its ability to perform a simultaneous test to describe the relationship between the observed variables and the respective unobserved variables as well as the connection among the unobserved variables. AMOS version 18 was used to analyze the data in this study, which is a flexible tool that enables researchers to examine relationships that violate the normal assumption of the variables considered in a model. Additionally, the outputs were compared by applying Bayesian SEM using winBUGS version 1.4.

In the current study, ML-SEM served as a representative parametric analysis method and B-SEM as a representative semi-parametric technique for predicting the children's lifestyle index. Based on the R^2, RMSE, MSE, and MPEA indices, SEM with the Bayesian approach was more effective at predicting children's lifestyle with the dataset obtained from Urumqi, Xinjiang, China.

Although much work has focused on determining the children's lifestyle index, not much has been done on modeling this index using SEM, particularly with Bayesian approaches. This is especially true when information on parental socioeconomic status, household food security, and parental lifestyle is concerned. The indicators that were found to be significant in explaining the latent factors considered in this study are as follows. The socioeconomic indicators are age, income, work experience, education level, and length of parents' marriage. Parental lifestyle is explained by smoking habit, frequency of engaging in physical exercise, alcohol drinking habit, number of working hours, and number of sleeping hours per day. Worry about running out of food, not having money (household), inability to afford eating balanced meals (household), cutting down food portions (household), not eating the

Symmetry **2016**, *8*, 141

whole day (adults), not having money (children), inability to afford to eat balanced meals (children), inability to afford enough food (children), and skipping a meal (children) served as indicators to measure household food security.

Therefore, the research model includes three predictors, which are parental socioeconomic status, parental lifestyle, and household food security. Among these, parental socioeconomic status and parental lifestyle have a significant impact on predicting children's lifestyle. However, household food security does not have a direct impact on children's lifestyle. Parental socioeconomic status was the main and first predictor in the study, as it has the highest impact on children's lifestyle. Parental socioeconomic status is a combination of nine indicators (Table 4), only three of which have acceptable factor loadings in the research model. These are the mother's education, father's income, and parents' marriage length. This means that helping a family with a longer marriage by improving the mother's education and father's income can provide better children's lifestyle quality. The second predictor was parental lifestyle, which was measured based on 10 indicators. Among these indicators, five had significant factor loadings and remained in the final research model. The five indicators are father's alcohol drinking habit, mother's working hours, father's working hours, mother's average sleeping hours, and father's average sleeping hours. This means that, in Urumqi, Xinjiang, China, controlling the father's alcohol drinking habit and optimizing both parents' working hours and average sleeping hours can lead to higher children's lifestyle quality. The third predictor was household food security, which was measured with nine indicators. Among these, five indicators had significant factor loadings (Table 4) and were considered in the final research model. Household food security does not have a direct impact on children's lifestyle. However, it has a strong significant relationship with parental socioeconomic status and parental lifestyle. Therefore, this predictor cannot be eliminated from the research model. In other words, household food security has an indirect impact on children's lifestyle with relations to parental socioeconomic status and parental lifestyle.

We proposed a Bayesian approach to analyze useful structural equations for children's lifestyle index modeling. In formulating ML-SEM and developing the Bayesian method, emphasis was placed on raw individual random observations rather than on the sample covariance matrix.

Through this study it was found that parental socioeconomic status and parental lifestyle have a significant effect on the children's lifestyle index, but household food security does not. The concept of modeling the children's lifestyle index by considering various indicators that describe latent factors can be explored further by incorporating new survey data. This idea is particularly suitable with the sequential Bayesian approach by considering the results from this study as prior input for future studies. The research framework introduced (Figure 1) can be used in any area. Hence, another suggestion for future studies is a comparison analysis modeling children's lifestyle in China and Malaysia. It is worth noting that there is a lack of evidence to indicate a connection between children's calorie intake and energy expenditure and overall lifestyle. Clinical causes and effects were not examined in the present research, so it is recommended to study them in future investigations.

Acknowledgments: This work was fully supported by University of Malaya project numbers RP026-2012F and PS013-2012A. The authors would like to thank the anonymous reviewers for their constructive comments.

Author Contributions: Conceived and designed the experiments: Che Wan Jasimah bt Wan Mohamed Radzi, Huang Hui, and Hashem Salarzadeh Jenatabadi; performed the experiments: Huang Hui; analyzed the data: Hashem Salarzadeh Jenatabadi; contributed reagents and materials: Huang Hui, Che Wan Jasimah bt Wan Mohamed Radzi, and Hashem Salarzadeh Jenatabadi; wrote the paper: Huang Hui, Che Wan Jasimah bt Wan Mohamed Radzi, and Hashem Salarzadeh Jenatabadi.

Conflicts of Interest: The authors declare no competing financial interests.

References

1. Brown, J.E.; Broom, D.H.; Nicholson, J.M.; Bittman, M. Do working mothers raise couch potato kids? Maternal employment and children's lifestyle behaviours and weight in early childhood. *Soc. Sci. Med.* **2010**, *70*, 1816–1824. [CrossRef] [PubMed]

2. Okubo, H.; Miyake, Y.; Sasaki, S.; Tanaka, K.; Murakami, K.; Hirota, Y. Dietary patterns in infancy and their associations with maternal socio-economic and lifestyle factors among 758 Japanese mother-child pairs: The Osaka maternal and child health study. *Matern. Child Nutr.* **2014**, *10*, 213–225. [CrossRef] [PubMed]

3. Veldhuis, L.; Vogel, I.; van Rossem, L.; Renders, C.M.; HiraSing, R.A.; Mackenbach, J.P.; Raat, H. Influence of maternal and child lifestyle-related characteristics on the socioeconomic inequality in overweight and obesity among 5-year-old children; the "Be Active, Eat Right" Study. *Int. J. Environ. Res. Public Health* **2013**, *10*, 2336–2347. [CrossRef] [PubMed]

4. Mangrio, E.; Wremp, A.; Moghaddassi, M.; Merlo, J.; Bramhagen, A.C.; Rosvall, M. Antibiotic use among 8-month-old children in Malmö, Sweden-in relation to child characteristics and parental sociodemographic, psychosocial and lifestyle factors. *BMC Pediatr.* **2009**, *9*, 1–6. [CrossRef] [PubMed]

5. Ek, A.; Sorjonen, K.; Nyman, J.; Marcus, C.; Nowicka, P. Child behaviors associated with childhood obesity and parents' self-efficacy to handle them: Confirmatory factor analysis of the lifestyle behavior checklist. *Int. J. Behav. Nutr. Phys. Acta* **2015**, *12*, 1–13. [CrossRef] [PubMed]

6. Chi, D.L.; Masterson, E.E.; Carle, A.C.; Mancl, L.A.; Coldwell, S.E. Socioeconomic status, food security, and dental caries in US children: Mediation analyses of data from the National Health and Nutrition Examination Survey, 2007–2008. *Am. J. Public Health* **2014**, *104*, 860–864. [CrossRef] [PubMed]

7. Ishida, R. Impacts of beautiful natural surroundings on happiness: Issues of environmental disruption, food, water security and lifestyle in modern times. *Br. J. Med. Med. Res.* **2015**, *9*, 1–6. [CrossRef]

8. Ullman, J.B. Structural equation modeling: Reviewing the basics and moving forward. *J. Personal. Assess.* **2006**, *87*, 35–50. [CrossRef]

9. Lee, S.Y. *Structural Equation Modeling: A Bayesian Approach 2007*, 1st ed.; John Wiley & Sons: West Sussex, UK, 2007; pp. 24–25.

10. Olsson, U.H.; Foss, T.; Troye, S.V.; Howell, R.D. The performance of ML, GLS, and WLS estimation in structural equation modeling under conditions of misspecification and nonnormality. *Struct. Equ. Model.* **2000**, *7*, 557–595. [CrossRef]

11. Bashir, M.K.; Schilizzi, S. Food security policy assessment in the Punjab, Pakistan: Effectiveness, distortions and their perceptions. *Food Secur.* **2015**, *7*, 1071–1089. [CrossRef]

12. Wan Mohamed Radzi, C.W.J.B.; Salarzadeh Jenatabadi, H.; Hasbullah, M.B. Firm Sustainability Performance Index Modeling. *Sustainability* **2015**, *7*, 16196–16212. [CrossRef]

13. Scheines, R.; Hoijtink, H.; Boomsma, A. Bayesian estimation and testing of structural equation models. *Psychometrika* **1999**, *64*, 37–52. [CrossRef]

14. Lee, S.Y.; Song, X.Y. Evaluation of the Bayesian and maximum likelihood approaches in analyzing structural equation models with small sample sizes. *Multivar. Behav. Res.* **2004**, *39*, 653–686. [CrossRef] [PubMed]

15. Dunson, D.B. Bayesian latent variable models for clustered mixed outcomes. *J. R. Stat. Soc. Ser. B Stat. Methodol.* **2000**, *62*, 355–366. [CrossRef]

16. Lovaglio, P.G.; Boselli, R. Simulation studies of structural equation models with covariates in a redundancy analysis framework. *Qual. Quant.* **2015**, *49*, 881–890. [CrossRef]

17. Lovaglio, P.G.; Vittadini, G. Structural equation models in a redundancy analysis framework with covariates. *Multivar. Behav. Res.* **2014**, *49*, 486–501. [CrossRef] [PubMed]

18. Cho, S.J.; Preacher, K.J.; Bottge, B.A. Detecting intervention effects in a cluster-randomized design using multilevel structural equation modeling for binary responses. *Appl. Psychol. Meas.* **2015**, *39*, 627–642. [CrossRef]

19. Scherer, R.; Gustafsson, J.E. Student assessment of teaching as a source of information about aspects of teaching quality in multiple subject domains: An application of multilevel bifactor structural equation modeling. *Front. Psychol.* **2015**, *6*, 1–15. [CrossRef] [PubMed]

20. Song, X.Y.; Lee, S.Y. A maximum likelihood approach for multisample nonlinear structural equation models with missing continuous and dichotomous data. *Struct. Equ. Model.* **2006**, *13*, 325–351. [CrossRef]

21. Lee, S.Y.; Ho, W.T. Analysis of multisample identified and non-identified structural equation models with stochastic constraints. *Comput. Stat. Data Anal.* **1993**, *16*, 441–453. [CrossRef]

22. Rabe-Hesketh, S.; Skrondal, A.; Pickles, A. Generalized multilevel structural equation modeling. *Psychometrika* **2014**, *69*, 167–190. [CrossRef]

23. Yang, Y.; Green, S.B. Evaluation of structural equation modeling estimates of reliability for scales with ordered categorical items. *Methodology* **2015**, *11*, 23–34. [CrossRef]

24. Wedel, M.; Kamakura, W.A. Factor analysis with (mixed) observed and latent variables in the exponential family. *Psychometrika* **2011**, *66*, 515–530. [CrossRef]

25. Finch, W.H. Modeling nonlinear structural equation models: A comparison of the two-stage generalized additive models and the finite mixture structural equation model. *Struct. Equ. Model.* **2015**, *22*, 60–75. [CrossRef]

26. Wall, M.M.; Amemiya, Y. Estimation for polynomial structural equation models. *J. Am. Stat. Assoc.* **2000**, *95*, 929–940. [CrossRef]

27. Bollen, K.A. Latent variables in psychology and the social sciences. *Annu. Rev. Psychol.* **2002**, *53*, 605–634. [CrossRef] [PubMed]

28. Flora, D.B.; Curran, P.J. An empirical evaluation of alternative methods of estimation for confirmatory factor analysis with ordinal data. *Psychol. Methods* **2004**, *9*, 466–491. [CrossRef] [PubMed]

29. Evans, M.; Moshonov, H. Checking for prior-data conflict. *Bayesian Anal.* **2006**, *1*, 893–914. [CrossRef]

30. Muthén, B.; Asparouhov, T. Bayesian structural equation modeling: A more flexible representation of substantive theory. *Psychol. Methods* **2012**, *17*, 313–335. [CrossRef] [PubMed]

31. Evans, M. Measuring statistical evidence using relative belief. *Comput. Struct. Biotechnol. J.* **2016**, *14*, 91–96. [CrossRef] [PubMed]

32. Yanuar, F.; Ibrahim, K.; Jemain, A.A. Bayesian structural equation modeling for the health index. *J. Appl. Stat.* **2013**, *40*, 1254–1269. [CrossRef]

33. Bickel, G.; Nord, M.; Price, C.; Hamilton, W.; Cook, J. *Guide to Measuring Household Food Security in the United States*; USDA, Food and Nutrition Services: Alexandria, VA, USA, 2000.

34. Nakayama, K.; Yamaguchi, K.; Maruyama, S.; Morimoto, K. The relationship of lifestyle factors, personal character, and mental health status of employees of a major Japanese electrical manufacturer. *Environ. Health Prev.* **2001**, *5*, 144–149. [CrossRef] [PubMed]

35. Health Research Ethics Authority (HREA). Available online: http://www.hrea.ca/Ethics-Review-Required.aspx (accessed on 22 January 2016).

36. Mullen, M.R.; Milne, G.R.; Doney, P.M. An international marketing application of outlier analysis for structural equations: A methodological note. *J. Int. Market.* **1995**, *3*, 45–62.

37. Argyris, C.; Schön, D.A. Organizational learning: A theory of action perspective. *Reis* **1997**, *77*, 345–348. [CrossRef]

38. Schoot, R.; Kaplan, D.; Denissen, J.; Asendorpf, J.B.; Neyer, F.J.; Aken, M.A. A gentle introduction to Bayesian analysis: Applications to developmental research. *Child Dev.* **2014**, *85*, 842–860. [CrossRef] [PubMed]

39. Van Wesel, F. Priors & Prejudice: Using Existing Knowledge in Social Science Research. Ph.D. Thesis, Utrecht University, Utrecht, The Netherlands, January 2011.

40. Rietbergen, C.; Klugkist, I.; Janssen, K.J.; Moons, K.G.; Hoijtink, H.J. Incorporation of historical data in the analysis of randomized therapeutic trials. *Contemp. Clin. Trials* **2011**, *32*, 848–855. [CrossRef] [PubMed]

41. Garthwaite, P.H.; Kadane, J.B.; O'Hagan, A. Statistical methods for eliciting probability distributions. *J. Am. Stat. Assoc.* **2005**, *100*, 680–701. [CrossRef]

42. Van Erp, S.; Mulder, J.; Oberski, D.L. Prior sensitivity analysis in default Bayesian structural equation modeling. In Proceedings of the 31th International Organisation of Pension Supervisors (IOPS) Summer Conference, Enschede, The Netherlands, 9–10 June 2016.

43. Gelman, A.; Carlin, J.B.; Stern, H.S.; Rubin, D.B. *Bayesian Data Analysis*, 2nd ed.; Chapman & Hall/Chemical Rubber Company (CRC): Boca Raton, FL, USA, 2004; p. 189.

symmetry

MDPI

Article

Comprehensive Reputation-Based Security Mechanism against Dynamic SSDF Attack in Cognitive Radio Networks

Fang Ye, Xun Zhang and Yibing Li *

College of Information and Communication Engineering, Harbin Engineering University, Harbin 150001, China; yefang0923@126.com (F.Y.); zhangxun0611@gmail.com (X.Z.)
* Correspondence: liyibing0920@126.com; Tel.: +86-133-0460-5678

Academic Editor: Angel Garrido
Received: 8 November 2016; Accepted: 30 November 2016; Published: 3 December 2016

Abstract: Collaborative spectrum sensing (CSS) was envisioned to improve the reliability of spectrum sensing in centralized cognitive radio networks (CRNs). However, secondary users (SUs)' changeable environment and ease of compromise make CSS vulnerable to security threats, which further mislead the global decision making and degrade the overall performance. A popular attack in CSS is the called spectrum sensing data falsification (SSDF) attack. In the SSDF attack, malicious cognitive users (MUs) send false sensing results to the fusion center, which significantly degrades detection accuracy. In this paper, a comprehensive reputation-based security mechanism against dynamic SSDF attack for CRNs is proposed. In the mechanism, the reliability of SUs in collaborative sensing is measured with comprehensive reputation values in accordance with the SUs' current and historical sensing behaviors. Meanwhile a punishment strategy is presented to revise the reputation, in which a reward factor and a penalty factor are introduced to encourage SUs to engage in positive and honest sensing activities. The whole mechanism focuses on ensuring the correctness of the global decision continuously. Specifically, the proposed security scheme can effectively alleviate the effect of users' malicious behaviors on network decision making, which contributes greatly to enhancing the fairness and robustness of CRNs. Considering that the attack strategy adopted by MUs has been gradually transforming from simplicity, fixedness and singleness into complexity, dynamic and crypticity, we introduce two dynamic behavior patterns (true to false and then to true (TFT) and false to true and then to false (FTF)) to further validate the effectiveness of our proposed defense mechanism. Abundant simulation results verify the rationality and validity of our proposed mechanism.

Keywords: cognitive radio networks; collaborative spectrum sensing; dynamic spectrum sensing data falsification attack; comprehensive reputation

---

1. Introduction

With the rapid development of wireless services and applications, the conventional static spectrum management policy inevitably causes scarcity in specific spectrum bands. Moreover, a large portion of the allocated spectrum is unused occasionally, leading to underutilization and wastage of valuable spectrum resources [1]. As the most promising solution to the spectrum scarcity problem, cognitive radio networks (CRNs) have attracted widespread attention recently. With this new communication paradigm, unlicensed users (also referred to as secondary users, SUs) can opportunistically utilize the spectrum for licensed users (also referred to as primary or incumbent users, PUs). When the primary user is detected back to the band, SUs in the band must forsake the spectrum immediately. Therefore, as an initial step, SUs must accurately sense the spectrum occupancy conditions for available opportunities to avoid any interference with the licensed users [2].

However, due to SUs' changeable environment and ease of compromise, the open characteristic of CRNs produces various security threats in the reliability of sensing data in CRNs [3]. For example, channel impairment, such as shadowing and multipath fading, lead to the fact that local spectrum sensing conducted by the individual user is often incorrect. Although the participation of multiple SUs in collaborative spectrum sensing (CSS) contributes to the improvement of detection accuracy, the global decision making may be misguided when SUs intentionally or unintentionally send falsified sensing information to the fusion center (FC) during coopesration. This sort of attack in CSS, called the spectrum sensing data falsification (SSDF) attack (also referred to as Byzantine attacks), significantly degrades collaborative detection correctness [4].

To address the above issues, various secure CSS schemes have been proposed [5–13]. When simple attack patterns are adopted by only a few malicious users (MUs) in CRNs, the schemes presented in [5–7] can work well enough. The concept of applying the trust and reputation model in CRNs has also attracted interest recently [8–12]. A reputation-based secure CSS algorithm with trusted node assistance based on [5–7] was proposed in [8], which started with trusted SUs merely to assure the inerrability of global decision making. In [9], a soft reputation-based sensing scheme was presented by modeling the operative mode of PU as a renewal process. These two schemes can still work availably even in the presence of a large number of malicious users. Qingqi Pei exploited the cognitive cycle to build the trust model, thus ensuring the security of CSS [10]. The authors in [11] considered the number of false sensing as the attenuation factor of trust to punish MUs; however, they ignored the dynamic characteristics of SSDF attack behavior. In [12], the OGKmethod was employed to mitigate the effect of MUs and improve sensing robustness. A novel trust scheme called SensingGuardis proposed in [13] to mitigate the harmful effect of SSDF attackers and enhance the performance of CSS.

Nevertheless, all of the existing methods mentioned above possess respective limitations. The schemes proposed in [5–7] become defective either by the increasing proportion of malicious users or in the face of complex attack strategies. Others in [8–10] have no ability to maintain robustness under the presence of a high proportion of MUs, while the rest of the methods, as in [11–13], cannot cope with complicated attack patterns.

To target the aforementioned problems in the CSS, this paper establishes a comprehensive reputation-based security mechanism against dynamic SSDF attack patterns for CRNs. Specifically, each SU is assigned one comprehensive reputation by the FC, and the reliability of SUs in collaborative sensing is measured with comprehensive reputation in accordance with SUs' historical sensing behaviors. Meanwhile, a punishment strategy is presented to revise the reputation, among which a reward factor and a penalty factor are introduced to encourage SUs to engage in positive and honest sensing activities. The whole mechanism focuses on mitigating the threat of dynamic malicious behaviors on network decisions and ensuring the correctness of the global decision continuously. Simulation results verify the robustness and effectiveness of the proposed security mechanism. Our scheme maintains a satisfactory sensing performance even under the circumstance that a large portion of malicious cognitive users exists in the network and employs complex attack behavior patterns. The improvements and novelties of our proposed scheme are presented clearly in Table 1, which concerns the main characteristics of the above defense schemes and compares them with the characteristics of our proposed scheme.

The rest of the paper is organized as follows. The system model and dynamic malicious attack behavior patterns are described in detail in Section 2. The proposed comprehensive reputation-based security mechanism and data fusion solution are respectively discussed in Sections 3 and 4. The simulation results are presented in Section 5. Finally, Section 6 concludes this paper.

Table 1. Performance enhancements achieved by the advanced SSDF defense mechanisms in CRNs.

Performance Enhancement Compared with Existing Approaches	Counter a Small Number of Attackers	Counter a High Proportion of Attackers	Counter Simple Attack Patterns	Counter Complex Attack Patterns
Chen et al. [5]	×		×	
Zhao and Zhao [6]	×		×	
Kaligineedi et al. [7]	×		×	
Zeng et al. [8]	×		×	×
Du [9]	×		×	×
Pei et al. [10]	×		×	×
Feng et al. [11]	×	×	×	
Lu et al. [12]	×	×	×	
FENG et al. [13]	×	×	×	
Our's	×	×	×	×

2. Application Scenario

In this section, we give a brief introduction of the cognitive radio network model adopted in this paper and establish the security problems against Byzantine attacks.

2.1. Network Architecture

The problem of spectrum sensing is to decide whether a particular slice of the spectrum is available or not. Consider a cognitive radio network where K secondary users are collaborating in the spectrum sensing process in the presence of one primary user, as shown in Figure 1. Without loss of generality, the energy detection [14–16] method is applied by each SU for individual spectrum sensing. Based on its observations, each SU solves a hypothesis testing problem and discriminates between the two hypotheses during the t-th sensing slot.

$$\begin{aligned} \mathcal{H}_0: \quad & x_i(t) = v_i(t) \\ \mathcal{H}_1: \quad & x_i(t) = h_i s_i(t) + v_i(t) \end{aligned} \tag{1}$$

where $x_i(t)$ represents the received signal at the i-th SU, h_i is the complex channel gain between the PU and SU$_i$, and the sensing channel is assumed to be time-invariant during the sensing process. The PU's transmitted signal, $s_i(t)$, is assumed to be a BPSK modulated signal. The noise $v_i(t)$ is additive white Gaussian noise (AWGN) with zero mean and variance σ_v^2. $s_i(t)$ and $v_i(t)$ are mutually independent. The hypotheses $\mathcal{H}_0$ and $\mathcal{H}_1$ represent the absence and presence of the PU, respectively.

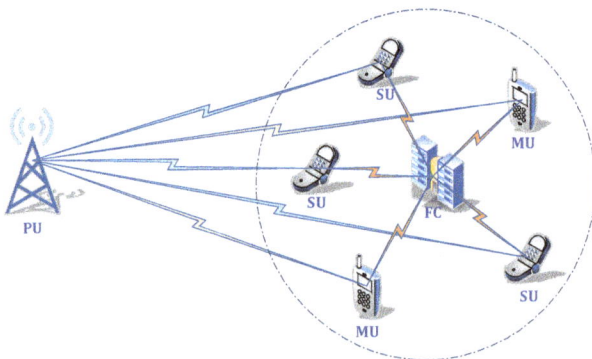

Figure 1. The sketch map of the network scene model.

In our system model, the test statistics for the energy detector for the i-th cognitive user is computed as the sum of the received signal energy over an interval of N samples and is given by:

$$Y_i \sim \begin{cases} \chi^2_{2m}, & \mathcal{H}_0 \\ \chi^2_{2m}(2\gamma_i), & \mathcal{H}_1 \end{cases} \tag{2}$$

Under hypothesis $\mathcal{H}_0$, the test statistic Y_i is a random variable whose probability density function is a chi-square distribution χ^2_{2m} with $N = 2m$ degrees of freedom, and $m = TW$ is the time-bandwidth product; otherwise, Y_i follows a non-central chi-square distribution $\chi^2_{2m}(2\gamma_i)$ with N degrees of freedom and non-central parameter $2\gamma_i$. The instantaneous signal-to-noise ratio (SNR) at the i-th SU is γ_i.

We consider the case in which each individual SU makes a one-bit hard decision, $d_i(t)$, on the absence or presence of the PU based on the sensing information, such that:

$$d_i(t) = \begin{cases} 0, & decision\ \mathcal{H}_0 \quad if\ Y_i < \lambda_i \\ 1, & decision\ \mathcal{H}_1 \quad if\ Y_i \geq \lambda_i \end{cases} \tag{3}$$

where λ_i is the decision threshold of SU_i. Then, the detection probability and false alarm probability of SU_i can be respectively expressed as:

$$\begin{aligned} P_d^{(i)} &= P\{d_i(t) = 1 \mid \mathcal{H}_1\} \\ P_f^{(i)} &= P\{d_i(t) = 1 \mid \mathcal{H}_0\} \end{aligned} \tag{4}$$

In this model, each SU in the network forwards its processed binary local decision $u_i(t)$ to the central entity, then the fusion center makes the final decision $u_0(t)$ about the state of the spectrum based on all of the information received from the participating SUs. The communication channels between SUs and the FC are assumed to be error-free in this paper.

In collaborative spectrum sensing, the global probabilities of false alarm, detection and misdetection for evaluating the performance of final joint decisions are expressed as Q_f, Q_d and Q_m, respectively, which can be written as follows [1]:

$$Q_f = \sum_{l=n}^{K} \binom{K}{l} \prod_{j=1}^{l} P_f^{(j)} \prod_{i=l+1}^{K} (1 - P_f^{(i)}) \tag{5}$$

$$Q_d = \sum_{l=n}^{K} \binom{K}{l} \prod_{j=1}^{l} P_d^{(j)} \prod_{i=l+1}^{K} (1 - P_d^{(i)}) \tag{6}$$

$$Q_m = 1 - Q_d \tag{7}$$

where $P_d^{(i)}$ and $P_f^{(i)}$ respectively denote the local detection probability and false alarm probability of the i-th SU. It can be seen that the OR fusion rule corresponds to the case of $n = 1$; the AND fusion rule corresponds to the case of $n = K$; and the majority fusion rule corresponds to the case of $n \geq K/2$.

2.2. Dynamic Attack Behavior Patterns

As the input data of the cognition cycle's follow-up processes, sensing information sent by SUs is essential to network decision. Therefore, at the sensing information reporting stage, the system's expectation is that SUs can actively report true sensing data to the FC. However, the FC may receive wrong or dishonest sensing data due to channel shading, shadowing and SSDF attack [17,18]. In particular, the SSDF attack is caused by two reasons: (1) SUs with cognition ability are compromised, and their reports are falsified; (2) the malfunction or fault of SUs leads to sensing reports contrary to the fact.

Although dishonest manners appear in different patterns, their common goal is to mislead the FC to make wrong decisions on current channel states. To be specific, the common attack models include [19,20]:

- Always present (AP): the attacker asserts the channel is busy in any case, i.e., $u_i(t) = 1$;
- Always absent (AA): the attacker asserts the channel is idle in any case, i.e., $u_i(t) = 0$;
- Always opposite (AO): the attacker with strong sensing ability always sends sensing reports contrary to its local spectrum sensing results, i.e., $u_i(t) = 1 - d_i(t)$.

With the booming growth of artificial intelligence, the attack patterns adopted by malicious users become increasingly complicated, which should be carefully taken into consideration. The attack strategy adopted by malicious cognitive users has been gradually transforming from simplicity, fixedness and singleness into complexity, dynamic and crypticity. They are inclined to achieve the goal of gaining additional spectrum access opportunities by cheating, undermining the licensed user and cognitive systems or other purposes to a larger extent. Besides, in an actual network, an individual SU can change between the true and false state back and forth due to both objective and artificial causes. Particularly, in a hostile environment, an honest SU may be manipulated by its adversary, thus suffering severe performance degradation (even turning into a malicious user) during a certain period. However, the adversary may evacuate from the battlefield after a while, and the behavior of the SU will transform again. Still another condition is, when the block that shelters an SU no longer exists, or the SU leaves the shadow zone, a better performance may be achieved.

Based on the above argumentation, we introduce two dynamic behavior patterns in this paper, which are the behavior of changing from true to false and then to true (TFT) and the behavior of changing from false to true and then to false (FTF); the details of the patterns are as follows.

2.2.1. TFT Behavior Pattern

Specifically, in the TFT behavior pattern, cognitive users with a virtuous nature report correct and true sensing results to the fusion center according to its normal working condition in the first period of time; due to some uncontrollable factors, such as being controlled by the enemy or sheltered from the shadow block, the SU reports false decisions for the next period; when it gets rid of the enemy or leaves the shadow zone, the user resubmits normal sensing data. The "true" and "false" respectively denote the normal working status (NWS) and temporary working status (TWS) of cognitive users. This kind of malicious dynamic behavior occurs under the unconsciousness and passiveness of cognitive users.

In the TFT behavior pattern, the detection probability and false alarm probability of SU_i in NWS are denoted by $P_{d1}^{(i)}(TFT)$, $P_{d3}^{(i)}(TFT)$ and $P_{f1}^{(i)}(TFT)$, $P_{f3}^{(i)}(TFT)$, respectively.

$$P_{d1}^{(i)}(TFT) = P_{d3}^{(i)}(TFT) = P\{u_i(t) = 1 \mid \mathcal{H}_1\} = P\{d_i(t) = 1 \mid \mathcal{H}_1\} = P_d^{(i)} \tag{8}$$

$$P_{f1}^{(i)}(TFT) = P_{f3}^{(i)}(TFT) = P\{u_i(t) = 1 \mid \mathcal{H}_0\} = P\{d_i(t) = 1 \mid \mathcal{H}_0\} = P_f^{(i)} \tag{9}$$

We use $P_{d2}^{(i)}(TFT)$ and $P_{f2}^{(i)}(TFT)$ to indicate the detection and false alarm probability of SU_i in TWS:

$$P_{d2}^{(i)}(TFT) = P\{u_i(t) = 1 \mid \mathcal{H}_1\} = P\{d_i(t) = 0 \mid \mathcal{H}_1\} = 1 - P_d^{(i)} \tag{10}$$

$$P_{f2}^{(i)}(TFT) = P\{u_i(t) = 1 \mid \mathcal{H}_0\} = P\{d_i(t) = 0 \mid \mathcal{H}_0\} = 1 - P_f^{(i)} \tag{11}$$

2.2.2. FTF Behavior Pattern

In the FTF behavior pattern, cognitive users with a vicious nature report incorrect and false sensing results to the fusion center according to its normal attacking condition in the first period of time; in order to avoid exposing their own malicious identity, malicious users will temporarily disguise themselves as normal SUs and submit true local decision results within the next period of time; after successfully achieving the purpose of deception, malicious users immediately expose their harsh nature and resubmit the reversed local decision results. "False" and "true" denote the NWS and TWS of SUs, respectively. This kind of malicious dynamic behavior arises when cognitive users possess deliberate and proactive motivation.

In the FTF behavior pattern, the detection and false alarm probability of SU_i in NWS are denoted by $P_{d1}^{(i)}(\text{FTF})$, $P_{d3}^{(i)}(\text{FTF})$ and $P_{f1}^{(i)}(\text{FTF})$, $P_{f3}^{(i)}(\text{FTF})$, respectively.

$$P_{d1}^{(i)}(\text{FTF}) = P_{d3}^{(i)}(\text{FTF}) = P\{u_i(t) = 1 \mid \mathcal{H}_1\} = P\{d_i(t) = 0 \mid \mathcal{H}_1\} = 1 - P_d^{(i)} \tag{12}$$

$$P_{f1}^{(i)}(\text{FTF}) = P_{f3}^{(i)}(\text{FTF}) = P\{u_i(t) = 1 \mid \mathcal{H}_0\} = P\{d_i(t) = 0 \mid \mathcal{H}_0\} = 1 - P_f^{(i)} \tag{13}$$

$P_{d2}^{(i)}(\text{FTF})$ and $P_{f2}^{(i)}(\text{FTF})$ are employed to indicate the detection and false alarm probability of SU_i in TWS.

$$P_{d2}^{(i)}(\text{FTF}) = P\{u_i(t) = 1 \mid \mathcal{H}_1\} = P\{d_i(t) = 1 \mid \mathcal{H}_1\} = P_d^{(i)} \tag{14}$$

$$P_{f2}^{(i)}(\text{FTF}) = P\{u_i(t) = 1 \mid \mathcal{H}_0\} = P\{d_i(t) = 1 \mid \mathcal{H}_0\} = P_f^{(i)} \tag{15}$$

Precisely speaking, both behavior patterns possess a sensing performance similar to normal SUs within a certain period and the performance similar to AO attackers in the other period. It can be seen from the difference of their respective detection and false alarm probability that these two generalized behavior patterns exert distinct effects on CSS, which will be shown later. For the convenience of expression, the cognitive users described by both of these behavior patterns are referred to as malicious secondary users in this paper.

3. Comprehensive Reputation-Based Security Mechanism

In order to identify and defend against the complicated attack behavior of malicious users more effectively and rapidly, this paper proposes a novel reputation-based security mechanism. In the mechanism, each SU is allocated a continuously updated comprehensive reputation (CR) value by the FC in accordance with its reported sensing data. The CR value evaluates the reliability and correctness of the individual user's sensing data sent to the FC. Higher reputation means that the user's sensing data in the past are more beneficial for the FC to make the right global decisions. The CR value is an important reference in the next sensing round.

The comprehensive reputation integrally considers four influencing factors of user reliability, including current reliability, historical reputation, reward factor and punishment factor. A malicious user obtains low reputation and fusion weight due to submitting falsified sensing data, and the FC weakens its harmful effect in the process of data fusion or directly ignores its sensing results. The comprehensive reputation adequately measures and reflects the reliability of individual sensing results for cognitive users in an appropriate time scale and is constantly changed and updated.

3.1. Current and Historical Reputation

3.1.1. Current Reliability

In CSS, the global decision is usually more reliable than local decisions [21,22]. Therefore, the global decision can be treated as a reference to determine whether the sensing result of a single user is errorless or not at one certain slot.

The current reliability is the consistency check between the local decision of SUs and the final decision of the FC. The setting principle is to slow down the ascending rate and speed up the descending rate to improve the reliability of reputation; SSDF attackers can be availably restrained in this way. Considering that the comprehensive reputation will be updated at the end of each sensing round with higher calculation frequency, thus this requires the reputation quantization algorithm to be simple and efficient. In view of the above analysis, the current reliability value of the i-th SU at the t-th sensing slot is as follows:

$$CurR_i(t) = (-1)^{u_i(t)+u_0(t)} \times \tau^\theta, \quad t = 1, 2, \cdots \tag{16}$$

where $u_i(t)$ and $u_0(t)$ respectively represent the local report and global decision made by the i-th SU and FC. The current reputation will be incremented by one if $u_i(t)$ is consistent with $u_0(t)$; otherwise, it will be decremented by τ^θ. The constant τ acts on accelerating descent velocity and decelerating increased velocity, $\tau > 1$. The specific size of τ can be adjusted according to the actual situation to achieve the compromise of weighted efficiency and correctness. When $\tau = 2$, the cumulative rate of consistency accuracy is only half of the decay rate. The calculation method of θ is as follows:

$$\theta = \begin{cases} 1, & u_i(t) \neq u_0(t) \\ 0, & u_i(t) = u_0(t) \end{cases} \tag{17}$$

3.1.2. Historical Reputation

The reputation is the subjective probability prediction of the subject concerning whether the object can complete a certain collaborative activity correctly and non-devastatingly, and historical sensing behavior reflects the reliability variation of cognitive users. In order to highlight the historical behavior of SUs in the role of reputation evaluation, we introduce the historical reputation variable denoted as $HisR_i(t)$ to describe and evaluate the reliability of the i-th SU at the t-th slot.

If all of the historical reputation of cognitive users is taken into account, that would require much storage space occupation and high computing complexity. Hence, we consider employing an observing window to assess the detection stability of SUs in the most recent period. The observation window calculates the weighted sum of the corresponding CR value in up-to-date L sensing events and moves forward along with the occurrence of a new sensing event.

Historical sensing information has a near-far effect on the update process of reputation; in reality, recent sensing events in historical sensing behaviors play a more significant role than long-term sensing events in real-time reputation calculation. Therefore, the reputation of different slots should be endowed with distinct time weights, called the time attenuation factor (TAF) in this paper. The TAF of the comprehensive reputation for the i-th SU at the $(t - k)$-th time slot is represented as $\alpha_{i,k}$.

$$\alpha_{i,k} = \frac{L-k}{\frac{L(L+1)}{2}} = \frac{2(L-k)}{L(L+1)} \tag{18}$$

With sensing time increasing, even if a misbehaving user wins high trust in a certain slot, during the period of its opportunistic attack, the reputation of the attacker will gradually decay over time. Time attenuation factor contributes greatly to supervising and urging cognitive users to submit genuine sensing results continuously.

The historical reputation $HisR_i(t)$ of the i-th SU at the t-th sensing slot is evaluated as the following rule:

$$HisR_i(t) = \sum_{k=1}^{L} \alpha_{i,k} ComUpR_i(t-k), \quad k = 1, 2, \cdots, L \quad t = 1, 2, \cdots \tag{19}$$

where $ComUpR_i(t - k)$ denotes the CR value of SU_i at the $(t - k)$-th slot; L is the length of the observation window.

In the calculation of historical reputation, the observation window length should not be set too small; otherwise, the decay rate of the reputation value is too fast to fully assess the reliability of the cognitive users, and the historical behavior information cannot be brought into sufficient usage; on the other hand, remaining sensitive to the potential behavior change of SUs requires that L should not be set too large, either. In actual spectrum sensing, the length of the observation window can be reasonably selected according to the computation time length of the reputation value and the change of the sensing performance of the cognitive users.

Instead of only considering the influence of SUs afterone sensing round is exerted on the current CR value, the design regulation of historical reputation conducts distributed processing for the instantaneous growth or decline of the reputation value via choosing appropriate observation lengths of the window according to specific demand; thus, the adverse effects of the burst fluctuation of the reputation value on the reliability of the cognitive users can be avoided.

3.2. Punishment Strategy

Since security has played a major role in CRNs, numerous research works have mainly focused on attack detection based on detection probability, but few of them took the penalty of attacks into consideration and neglected how to implement effective punitive strategies against attackers. In addition, in the dynamic SSDF attacks; behavior pattern, malicious users alternately submit authentic and spurious sensing data; general reputation mechanisms cannot effectively identify this sort of attack, and MUs may always be in a believable state, while a well-built reputation update mechanism should be sensitive to changes in users' behaviors and able to punish their villainy.

Aiming at this issue, this paper introduces a reward and punishment strategy to modify the CR value in line with the behavior characteristics of cognitive users, in which a reward factor and a penalty factor are introduced to encourage SUs to engage in positive and honest sensing activities. On the one hand, the reputation of users who continuously report false sensing results ought to be attenuated in a timely manner, making them unable to participate in cooperative sensing; on the other, users who conduct persistent honest sensing are supposed to be rewarded appropriately, thus encouraging them to continue to submit real detection outcomes.

3.2.1. Reward Factor

Assuming the i-th SU performed true sensing at the $(t - k)$-th round and continuous honest sensing behaviors occur in the next $(t - h + 1, t - h + 2, \cdots, t - 1)$-th sensing round, then the reward factor has a positive effect on modifying the CR value of the cognitive user. The calculation of the reward factor is according to the following method:

$$RewF_i(t) = |\frac{1}{h-1}(\sum_{l=t-h}^{t-1} ComUpR_i(l) - maxComUpR_i(l))| \qquad (20)$$

where $ComUpR_i(t)$ is the CR value of SU_i at the t-th sensing round and h denotes the times of continuous honest sensing events.

h sensing reputation values are employed during the computational process of the reward factor. The reward dynamics is constantly adjusted with the cumulative reputation. However, MUs may accumulate relatively high reputation through continuously providing honest decision results inside a shorter time; in view of this kind of speculation, the calculation of the reward factor removes the maximum reputation value $maxComUpR_i(l)$ of SU_i in h successive true sensing slots, thus reducing the reward intensity for honest sensing efforts. Only by ceaselessly submitting real local results can SUs establish a favorable credit status for themselves; thus the reward factor can motivate cognitive users to make a positive contribution to collaborative sensing.

3.2.2. Penalty Factor

Supposing SU_i conducted false sensing at the $(t - g)$-th slot and continuous false sensing behaviors emerge in the following $(t - g + 1, t - g + 2, \cdots, t - 1)$ sensing round, then the penalty factor has influence on inhibiting malicious attack behavior. The calculation method of penalty factor is as follows:

$$PenF_i(t) = |\frac{1}{g-1}(\sum_{l=t-g}^{t-1} ComUpR_i(l) - minComUpR_i(l))| \tag{21}$$

where $ComUpR_i(t)$ is the CR value of SU_i at the t-th sensing round and g denotes the times of continuous false sensing events.

The punishment scheme follows a habit of human society, that is the initial criminal punishment is light, and the cumulative crime will be punished heavily. Therefore, the greater the threat is, the more serious of a punishment should be imposed. The penalty factor removes the minimum value in g; CR values are removed in the penalty factor computing; in this way, the influence of accidental behavior on reputation in spectrum sensing is weakened; moreover, cognitive users will pay a great price for short-term opportunistic behavior caused by their unlikely mind, so as to achieve the purpose of restraining malicious attacks.

3.3. Calculation of the CR Value

In the proposed security mechanism, four influential elements for evaluating sensing reliability are generally considered, including current reliability, historical reputation, reward factor and punishment factor. We utilize $ComUpR_i(t)$ to represent the comprehensive reputation value of the i-th SU at the t-th sensing slot:

$$ComUpR_i(t) = \rho_0 \cdot HisR_i(t) + \rho_1 \cdot CurR_i(t) + \beta \cdot RewF_i(t) + \gamma \cdot PenF_i(t) \tag{22}$$

where ρ_0 and ρ_1 respectively are the proportion coefficients of historical reputation and current reliability, $0 < \rho_0, \rho_1 < 1$ and $\rho_0 + \rho_1 = 1$. Their values can be appropriately adjusted according to the demand of network security. When demand for the sensitivity of the security mechanism is higher, increase ρ_1, which means raising the weight of current trust evidence; then, it can be detected immediately once any untrustworthy behavior appears; when the long-term influence of reputation plays an important role, increase ρ_0, which signifies raising the weight of historical reputation to encourage the SUs to be legitimate in the long run. In fact, ρ_1 is a kind of response speed; a high speed of response means that cognitive users can make more effective and rapid response to changes in their CR value. The determination method of β and γ is as follows:

$$\beta = \begin{cases} 1, & continuous\ honest\ sensing\ events\ exist \\ 0, & else \end{cases} \tag{23}$$

$$\gamma = \begin{cases} -1, & continuous\ false\ sensing\ events\ exist \\ 0, & else \end{cases} \tag{24}$$

The literature [23] has pointed out that data fusion schemes become completely incapable, and no reputation-based fusion scheme can achieve any performance gain when the number of attackers exceeds a certain fraction in the CRN. If the number of independent attackers is greater than half of the total users, the FC will be rendered "blind". To tackle this problem and ensure the correctness of the global decision, we assume only some reliable nodes (RN), instead of all SUs, are trustworthy initially. In reality, the RNs can be a base station, access point, cluster head, etc. Since they share the generality as foundations of the cognitive system, it is reasonable to grant the position of these RNs exceeding that of the remaining SUs.

Symmetry **2016**, *8*, 147

In the first instance, only RNs participate in the deciding procedure, meaning the global decision is made merely based on their sensing results. Though the remaining SUs are not contained in the step of cooperative sensing, their CRs are accumulated continuously. A SU can be considered as a reliable one only when its CR value exceeds the predetermined reputation threshold η_r.

Employ C to describe the set of RNs, and $A(t)$ represents the set of cognitive users that can participate in the fusion decision, which are given by:

$$C = \{i | SU_i \text{ is a CN}\} \tag{25}$$

$$A(t) = \{j | ComUpR_i(t) \geq \eta_r, \quad j \in \{1, 2, \cdots, N\}\}, \quad t = 1, 2, \cdots \tag{26}$$

where C is determined on the basis of the specific circumstance, while $A(t)$ varies with the results of identifying procedure each sensing round.

The initial CR values are $ComUpR_{i \in C}(0) = \eta_r + \Delta$ and $ComUpR_{i \notin C}(0) = \eta_r - \Delta$ for RNs and remaining nodes, respectively. The setting of margin Δ is to distinguish RNs from other SUs, namely the degree of tolerance for potentially sensing errors. Consequently, only SUs belonging to C make contributions to the global decision making for the first round, then the range enlarges to $A(t)$.

Unlike the existing mechanisms, the proposed security mechanism does not abandon any user, and their identification is conducted all the way. This is more equitable and reasonable particularly when the complicated behavior patterns are taken into account, noticing that one FTF user that behaves poorly at the intermediate stage may obtain a better performance eventually.

3.4. Reliable Nodes' Credibility Verification

This step is conducted within the RNs by inspecting the variances of their CR values. After the identifying step completed at each round, we compare the real-time CR value of each RN $ComUpR_i(t)$, $i \in C$ with its highest CR value in previous sensing slots, which is denoted by $ComUpR_{imax}(t-1)$. Initially, $ComUpR_{imax}(0) = \eta_r + \Delta$, $i \in C$. If the real-time CR value $ComUpR_i(t)$ is higher than $ComUpR_{imax}(t-1)$, then the new highest CR value is updated as the current one, otherwise $ComUpR_{imax}(t)$ remains unchanged. Accordingly, the highest CR value update mode can be presented as:

$$ComUpR_{imax}(t) = \begin{cases} ComUpR_i(t), & ComUpR_i(t) > ComUpR_{imax}(t-1) \\ ComUpR_{imax}(t-1), & ComUpR_i(t) < ComUpR_{imax}(t-1) \end{cases} \quad i \in C, t = 1, 2, \cdots \tag{27}$$

Then, the following inequality set is verified immediately:

$$ComUpR_{imax}(t) - ComUpR_i(t) < \Delta \quad i \in C, t = 1, 2, \cdots \tag{28}$$

where Δ denotes the degree of tolerance for potentially sensing errors as discussed earlier. If all RNs satisfy the above inequalities, the deciding procedure can be performed directly; otherwise, it means that the local results sent by the corresponding RN have been inconsistent with the global decision many times. Under such a circumstance, we conclude that the global decision is incorrect (which may be caused by various reasons), considering that these RNs are trusted all of the time. This verification process is called sustained credible node assistance (SCNA). In order to ensure the correctness of the final decision in the future, resetting is performed before center fusion, which is to clear all of the accumulated CR and weight values via setting them to the initial state. The accumulation restarts hereafter. In this step, the inequality set Equation (25) serves as a trigger and decides whether the resetting is required.

4. Comprehensive Reputation-Based Data Fusion Solution

4.1. Weight Allocation

The center data fusion is ultimately implemented after the above steps are accomplished, in which all of the elected trusted users will participate. Distinct fusion weights are allocated to SUs corresponding to their comprehensive reputation values. Users with greater reputation have stronger impact on the final decision making; hence, the sensing accuracy of CSS can be improved.

The fusion weight value for the i-th SU at the t-th sensing slot can be calculated as:

$$w_i(t) = \begin{cases} 0, & i \notin A(t) \\ \frac{ComUpR_i(t-1)}{\overline{ComUpR_i(t-1)}}, & i \in A(t) \end{cases} \quad t = 1, 2, \cdots \tag{29}$$

where:

$$\overline{ComUpR_i(t-1)} = \frac{ComUpR_i(t-1)}{\sum_{i \in A(t)} ComUpR_i(t-1)}, \quad t = 1, 2, \cdots \tag{30}$$

denotes the average CR value of reliable nodes. The initial weight is $w_{i \in C}(0) = 1$, $w_{i \notin C}(0) = 0$.

4.2. Measurement Combining Stage

Evidently, the idea of comprehensive reputation updating and sustained credible node assistance are not restricted to specific designated fusion techniques and can be widely applied. For simplicity, we employ the majority fusion rule as an example in this paper, which is proven to be relatively ideal in both detection accuracy and energy efficiency [23].

Majority rule implies that the final decision is in accord with the decision of the majority of the received local decisions. Assuming M SUs are qualified to participate in the collaboration, mathematically, the final decision is made according to the majority rule as follows:

$$Final\ Decision \begin{cases} 1 \equiv occupied, & if \sum_{i=1}^{M} w_i u_i \geq \frac{M}{2} \\ 0 \equiv unoccupied, & else \end{cases} \tag{31}$$

Similar to the local decision, the accuracy and reliability of the final decision is measured and evaluated by two acknowledged metrics, the global false alarm probability (Q_f) and the global misdetection probability (Q_m). Both depend on the final decision rather than the local decision.

4.3. The Mechanism Flow

Based on the above discussions, the operation process of the comprehensive reputation-based security mechanism is shown in Figure 2. The CSS system starts working with the step of reputation initialization; all SUs conduct individual sensing to obtain the one-bit decision result. If the CR value $ComUpR_i(t)$ of SU_i exceeds reputation threshold η_r, then FC would allow this user to join the cooperation. Different fusion weights are assigned to qualified cognitive users for center decision fusion. After obtaining the global decision $u_0(t)$, credibility verification is performed for reliable users to ensure that the whole CSS system has not been held hostage by malicious users. If all RNs pass the verification, then the CR values can be updated in accordance with users' sensing behaviors, which comprises the current reliability, historical reputation, reward factor and penalty factor. Consequently, the proposed security mechanism gives a system-wide view of the satisfaction of a cognitive user.

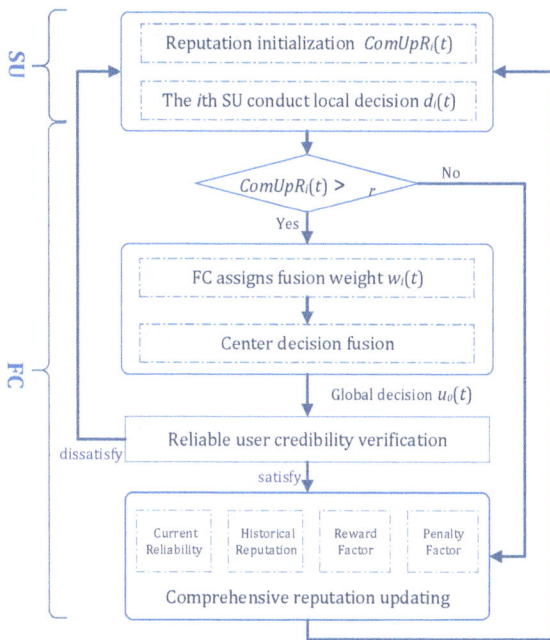

Figure 2. The flow chart of the proposed security mechanism.

5. Numerical and Simulation Results

In this section, we present the numerical results for the proposed reputation mechanism. The simulations are conducted with $K = 50$ cognitive users in a centralized CRN, among which $N_1 = 10$ reliable nodes exist. In the numerical simulation, the sensing performance is given as the reference curve when malicious users employ the AO attack strategy. We investigate the impact of malicious users exerted on the collaborative sensing when the two dynamic behavior patterns introduced in Section 2.2 are used. The number of MUs is expressed as N_0, and the proportion of MUs is set to $[0, 0.8]$. That is, from no malicious users exist in the network (the proportion is 0%), till 40 cognitive nodes, all are misbehaving users except the 10 reliable nodes (the proportion is 80%).

Without loss of generality, the primary signal is assumed to be the BPSK signal with $P(H_1) = 0.3$, and all SUs experience independently and identically distributed (i.i.d.) fading or shadowing with the same average SNR $\gamma = -10$dB to simplify the implementation. The time-band product m is five, and the same energy detection threshold $\lambda = 12$ is utilized. The entire simulation runs 100,000 rounds, and we assume that the moment of behavior changes respectively occurs at the 40,000 and 70,000 round in both the TFT and FTF patterns. For the sake of taking advantage of the user's historical sensing results, the observation window length L is set to three. The margin $\Delta = 50$ is set to evidently distinguish RNs from other SUs in the initial sensing stage, namely the degree of tolerance for potentially sensing errors. To avoid mistaking normal SUs as MUs, h and g should not be too small, meanwhile the proposed punishment strategy should be sensitive enough to punish misbehaving users or reward honest users; thus, h and g should not be too large. Hence, we set the times of continuous/false sensing events $h = g = 3$. The reputation threshold $\eta_r = 100$ is set to effectively identify MUs and prevent them from participating in the collaboration in each sensing slot. We set the variable τ to two to ensure that the cumulative consistency accuracy rate of current reliability in the comprehensive reputation value is only half of the decay rate, which is also a compromise of weighted efficiency and correctness. The proportion coefficients of historical reputation and current reliability in the CR value calculation ρ_0 and ρ_1 are set to 0.5 to balance current trust evidence and the long-term influence of reputation. All simulations are conducted in the MATLAB R2015a environment.

The following six scenarios are carefully considered in this section:

- Scenario 1   there are N_1 RNs in the CRN, performing CSS with no reputation mechanism;
- Scenario 2   there are K RNs in the CRN, performing CSS with no reputation mechanism;
- Scenario 3   there are N_0 misbehaving SUs and $K - N_0$ RNs in the CRN, performing CSS with the proposed scheme in this paper;
- Scenario 4   there are N_0 MUs in the CRN, performing CSS with the security scheme in [13];
- Scenario 5   there are N_0 MUs in the CRN, performing CSS with the security scheme in [12];
- Scenario 6   there are N_0 MUs and N_1 RNs in the CRN, performing CSS with the proposed security scheme in this paper to counter diverse SSDF attacks.

The purpose of considering Scenarios 1–3 is to provide the simulation experiments with clear contrast reference curves. Specifically, we consider Scenario 1 to explore when all SUs are reliable cognitive nodes; what the performance of the non-reputation-based sensing scheme is like under diverse SSDF attacks. Scenario 2 is set to experiment on the performance of the non-reputation-based method in the presence of partial reliable users. In Scenario 3, there are only two kinds of cognitive users, i.e., RNs and MUs, and we test the detection performance under this circumstance when suffering different types of SSDF attack. Scenarios 4 and 5 are two contrast algorithms to further verify and evaluate the effectiveness of the proposed mechanism in this paper.

5.1. The Sensing Performance under AO Attack

As mentioned above, the always opposite attack strategy refers to the attack mode that MUs report after reversing the local decision result. Figures 3 and 4 present the cooperative sensing performance under AO attack. The horizontal axis accounts for the proportion of malicious users. The vertical axis in Figure 3 represents the global false alarm probability Q_f, and the vertical axis in Figure 4 represents the global misdetection probability Q_m.

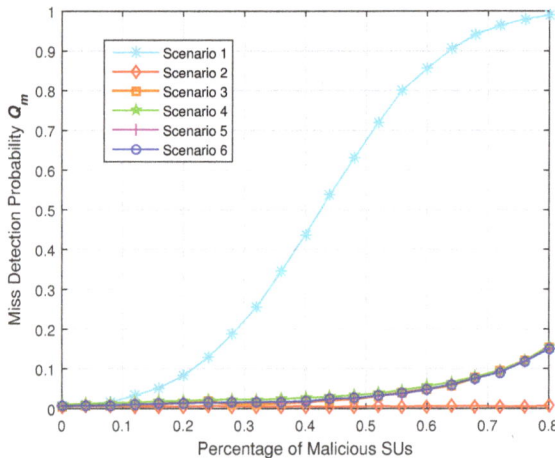

Figure 3. The misdetection probability under the always opposite (AO) attack.

It can be seen from the Figures 3 and 4, when there is no malicious users in the network (Scenario 2), the optimal sensing performance can be achieved if K cognitive users are reliable nodes. Additionally, the detection performance under Scenario 3, in which $K - N_0$ RNs participate in the collaboration, is inferior to that in Scenario 2. The detection performance of the scheme with no reputation mechanism in Scenario 1 dropped dramatically under the AO attack pattern, which means it is indispensable for CRNs to adopt a necessary and effective security mechanism to defend against various types of spiteful attack behaviors. When the number of MUs exceeds half of all cognitive users, the sensing performance is even worse than that of random guessing. Besides the proposed reputation mechanism (Scenario 6), the scheme in [12] (Scenario 5) and [13] (Scenario 4) can achieve the equivalent performance of $K - N_0$ reliable nodes, meaning that they can availably identify the malicious SUs and eliminate their harmful effects via only using reliable reported results for fusion decision making.

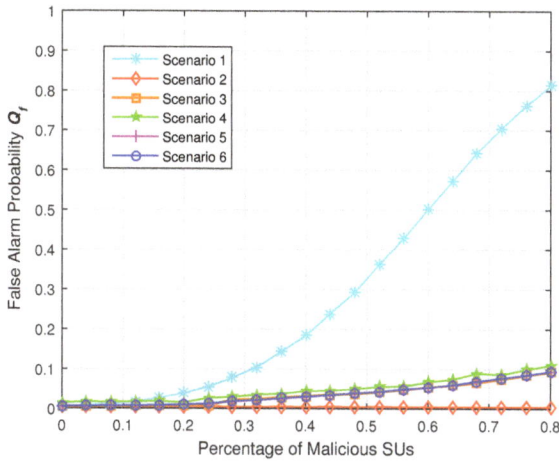

Figure 4. The false alarm probability under AO attack.

5.2. The Sensing Performance under TFT Attack

As introduced in Section 2.2, in the TFT behavior pattern, cognitive users report true sensing results in the first period of time; the SU reports false decisions for the next period; the user finally resubmits the normal sensing data. Figures 5 and 6 show the collaborative sensing performance of the above several scenarios under the TFT attack.

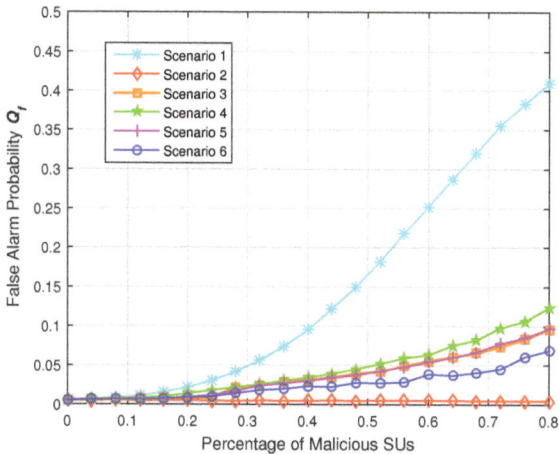

Figure 5. The false alarm probability under the true to false and then to true (TFT) attack.

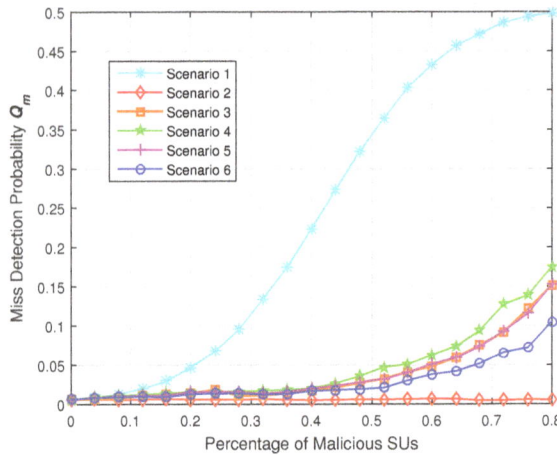

Figure 6. The misdetection probability under the TFT attack.

Through the observation, we can learn that the sensing performance of Scenarios 1 and 2 are identical to that in Section 5.1, which is attributed to the non-participation of misbehaving cognitive users. Scenario 5 still can achieve the sensing performance when $K - N_0$ reliable nodes are collaborating; the performance in Scenario 4. It is worth noting that the proposed mechanism in this paper can achieve better performance than the schemes in [12,13]. When MUs occupy 80% of all of the SUs, specifically, the false alarm probability of the proposed mechanism in this paper, [12,13] respectively, is 0.0690, 0.0965 and 0.1233, and the misdetection probability, respectively, is 0.1020, 0.1506 and 0.1748. The proposed mechanism in this paper possesses an obvious performance advantage both in false alarm probability and misdetection probability compared to the contrasted algorithms.

The reason for the performance advantage is that the cognitive users with poor performance at the initial stage will be permanently abandoned in the literature [12,13], which does not consider that the SUs' behavior may be dynamically changed, and a better individual user's sensing performance may be obtained after a period of time. In this paper, the mechanism is proposed to continuously evaluate the reliability of each cognitive user through the calculation of the comprehensive reputation. Our scheme forgives the repentance behavior (change from poor performance to good performance) of cognitive users, that is continually mitigating the effect of the correctness of reported results in earlier time slots exerted on assessing the reliable degree of cognitive users. SUs are allowed to continue to participate in the fusion decision of cooperative spectrum sensing when the comprehensive reputation value exceeds the reputation threshold. This way is equivalent to increasing the user number of participation cooperative sensing, thus obtaining an obvious gain of the sensing performance.

5.3. The Sensing Performance under FTF Attack

As presented in Section 2.2, in the FTF behavior pattern, SUs report false sensing results to the FC in the first period of time and then report true decisions for the next period; the user finally resubmits reversed local decision results. Figures 7 and 8 show the collaborative sensing performance of the above several scenarios under the FTF attack.

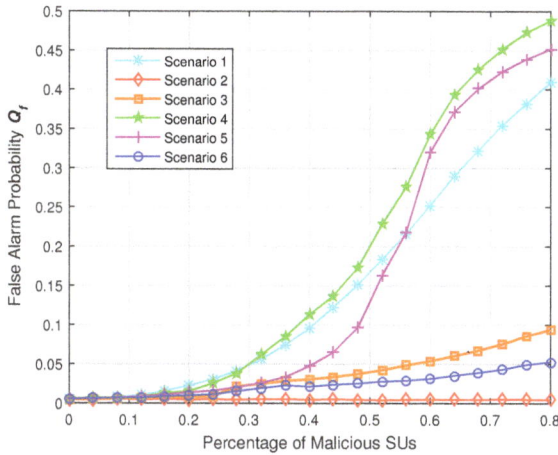

Figure 7. The false alarm probability under the false to true and then to false (FTF) attack.

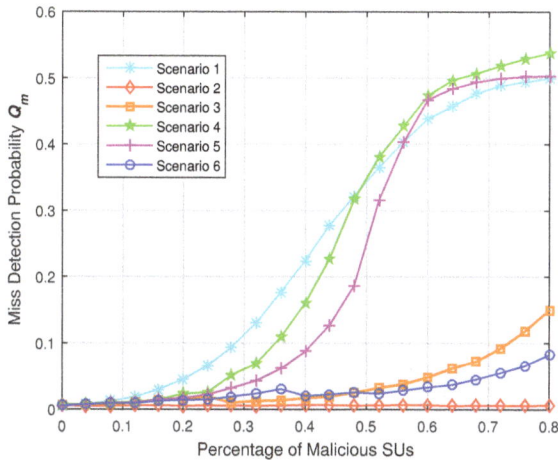

Figure 8. The misdetection probability under the FTF attack.

Similarly, the sensing performances of Scenarios 1 and 2 are identical to that in Section 5.1. However, the performance of Scenarios 4 and 5 deteriorated significantly when malicious SUs adopt the FTF attack mode. In Figure 7, when the proportion of MUs is greater than 40%, the performance of the algorithms in Scenarios 4 and 5 sharply declines. The false alarm probability of the proposed mechanism in this paper, [12,13] respectively, is 0.0524, 0.4516 and 0.4876 when the percentage of MUs is 80%, meaning that while the ratio of MUs continues to increase (account for the majority), the algorithms in [12,13] become completely ineffective.

This phenomenon can be explained as follows: in the stage of temporary working status in FTF attack, malicious cognitive users, together with normal SUs, obtain a higher level of reputation through accumulation and are identified as cognitive users that can participate in the fusion decision. When the working state of MU changes from TWS to NWS and it resubmits reversed local decision results, it affects the fusion decision process, which makes the false alarm probability and detection probability of the global decision increase simultaneously because of its higher reputation level. Especially when they occupy a higher proportion, these malicious users are enough to control the global decision making process of the fusion center; at this moment, the entire collaborative sensing system is hijacked by malicious users.

In Figure 8, the misdetection probability of the proposed mechanism in this paper, [12,13] respectively, is 0.0138, 0.1693 and 0.2291 when MUs occupy 80% of all SUs, which means that the proposed method can effectively reduce the misdetection probability of CSS, meanwhile protecting the system from complex SSDF attacks. In other words, even if the malicious cognitive user accounts for a high proportion, as 80%, the CSS algorithm based on the reputation mechanism in this paper still possesses higher robustness. The obvious performance advantage profits from reliable nodes' credibility verification. If the declining range of some reliable user's CR value is greater than Δ, the global decision fusion of the FC is identified as occurring persistent errors and triggers the resetting mechanism to clean the comprehensive reputation value for each cognitive user. This method avoids the global decision of the FC being controlled by MUs and reduces the adverse effects of MUs on the global decision results, ultimately achieving better performance than $K - N_0$ users cooperating.

6. Conclusions

In order to effectively resist malicious cognitive users' attack behaviors in cognitive radio networks, the security mechanism for CSS is studied in this paper. We first introduce two new cognitive user dynamic behavior patterns to describe the changing behavior strategies of SUs. On this basis, a comprehensive reputation-based security mechanism against dynamic SSDF attack is proposed. In the mechanism, current and historical sensing behaviors of cognitive users are utilized to integrally evaluate sensing reliability; moreover, a punishment strategy is presented to encourage SUs to engage in positive and honest sensing activities. In addition, the sustained verification of reliable nodes ensures the correctness of the global decision of the fusion center and prevents collaborative sensing from being hijacked by misbehaving cognitive users. Simulation results verify that the proposed security mechanism can effectively alleviate the effect of SUs' malicious behaviors, which guarantees the effectiveness and robustness of CRNs.

Acknowledgments: This paper is funded by the National Natural Science Foundation of China (Grant No. 51509049), the Natural Science Foundation of Heilongjiang Province, China (Grant No. F201345), the Fundamental Research Funds for the Central Universities of China (Grant No. GK2080260140) and the National key research and development program (2016YFF0102806).

Author Contributions: Fang Ye conceived of the concept and performed the research. Xun Zhang conducted the experiments to evaluate the performance of the proposed security mechanism and wrote the manuscript. Yibing Li reviewed the manuscript. All authors have read and approved the final manuscript.

Conflicts of Interest: The authors declare no conflict of interest.

References

1. Axell, E.; Leus, G.; Larsson, E.G.; Poor, H.V. Spectrum sensing for cognitive radio: State-of-the-art and recent advances. *IEEE Signal Proc. Mag.* **2012**, *29*, 101–116.
2. Haykin, S. Cognitive radio: Brain-empowered wireless communications. *IEEE J. Sel. Areas Commun.* **2005**, *23*, 201–220.
3. Althunibat, S.; Denise, B.J.; Granelli, F. Identification and punishment policies for spectrum sensing data falsification attackers using delivery-based assessment. *IEEE Trans. Veh. Technol.* **2016**, *65*, 7308–7321.
4. Zhang, L.Y.; Ding, G.R.; Wu, Q.H.; Zou, Y.L.; Han, Z.; Wang, J.L. Byzantine attack and defense in cognitive radio networks: A Survey. *IEEE Commun. Surv. Tutor.* **2015**, *17*, 1342–1363.
5. Chen, R.L.; Park, J.M.; Bian, K. Robust distributed spectrum sensing in cognitive radio networks. In Proceedings of the 27th Conference on Computer Communications (INFOCOM 2008), Phoenix, AZ, USA, 13–18 April 2008.
6. Zhao, T.; Zhao, Y. A new cooperative detection technique with malicious user suppression. In Proceedings of the IEEE International Conference on Communications, Dresden, Germany, 14–18 June 2009.
7. Kaligineedi, P.; Khabbazian, M.; Bhargava, V. K. Malicious user detection in a cognitive radio cooperative sensing system. *IEEE Trans. Wirel. Commun.* **2010**, *9*, 2488–2497.

8. Kun, Z.; Paweczak, P.; Cabric, D. Reputation-based cooperative spectrum sensing with trusted nodes assistance. *IEEE Commun. Lett.* **2010**, *14*, 226–228.
9. Du, D. Soft reputation-based secure cooperative spectrum sensing. In Proceedings of the 2012 International Conference on Computer Science and Electronics Engineering (ICCSEE), Hangzhou, China, 23–25 March 2012.
10. Pei, Q.Q.; Yuan, B.B.; Li, L.; Li, H.N. A sensing and etiquette reputation-based trust management for centralized cognitive radio networks. *Neurocomputing* **2013**, *101*, 129–138.
11. Feng, J.Y.; Lu, G.Y.; Bao, Z.Q. Supporting trustworthy cooperative spectrum sensing in cognitive radio networks. *J. Comput. Inf. Syst.* **2014**, *10*, 1–12.
12. Lu, J.Q.; Wei, P. Improved cooperative spectrum sensing based on the reputation in cognitive radio networks. *Int. J. Electron.* **2015**, *102*, 855–863.
13. Feng, J.Y.; Lu, G.Y.; Bao, Z.Q.; Zhang, L. Securing cooperative spectrum sensing against rational SSDF attack in cognitive radio networks. *KSII Trans. Internet Inf. Syst.* **2014**, *8*, 1–17.
14. Digham, F.F.; Alouini, M.S.; Simon, M.K. On the energy detection of unknown signals over fading channels. *IEEE Trans. Commun.* **2007**, *55*, 21–24.
15. Zhang, R.; Zhang, J.; Zhang, Y.; Zhang, C. Secure crowdsourcing-based cooperative pectrum sensing. In Proceedings of the 2013 Proceedings of IEEE INFOCOM, Turin, Italy, 14–19 April 2013; pp. 2526–2534.
16. Qin, Z.; Li, Q.; Hsieh, G. Defending against cooperative attacks in cooperative spectrum sensing. *IEEE Trans. Wirel. Commun.* **2013**, *12*, 2680–2687.
17. Bhattacharjee, S.; Debroy, S.; Chatterjee, M.; Kwiat, K. Trust based fusion over noisy channels through anomaly detection in cognitive radio networks. In Proceedings of the 4th International Conference on Security of Information and Networks, Sydney, Australia, 14–19 November 2011; pp. 73–80.
18. Hu, Z.; Ranganathan, R.; Zhang, C.; Qiu, R.C.; Bryant, M.; Wicks, M.C.; Li, L. Robust non-negative matrix factorization for joint spectrum sensing and primary user localization in cognitive radio networks. In Proceedings of the 2012 International Waveform Diversity & Design Conference (WDD), Kauai, HI, USA, 22–27 Janaury 2012; pp. 303–307.
19. Fragkiadakis, A.G.; Tragos, E.Z.; Askoxylakis, I.G. A survey on security threats and detection techniques in cognitive radio networks. *IEEE Commun. Surv. Tutor.* **2013**, *15*, 428–445.
20. Khan, A.A.; Rehmani, M.H.; Reisslein, M. Cognitive radio for smart grids: Survey of architectures, spectrum sensing mechanisms, and networking protocols. *IEEE Commun. Surv. Tutor.* **2016**, *18*, 860–898.
21. Rawat, A.S.; Anand, P.; Chen, H.; Varshney, P.K. Collaborative spectrum sensing in the presence of Byzantine attacks in cognitive radio networks. *IEEE Trans Signal Proc.* **2011**, *59*, 774–786.
22. Zina, C.; Hasna, M.; Hamila, R.; Hamdi, N. Location privacy preservation in secure crowdsourcing-based cooperative spectrum sensing. *EURASIP J. Wirel. Commun. Netw.* **2016**, *85*, 1–11.
23. Althunibat, S.; Sucasas, V.; Marques, H.; Rodriguez, J.; Tafazolli, R.; Granelli, F. On the trade-off eetween security and energy efficiency in cooperative spectrum sensing for cognitive radio. *IEEE Commun. Lett.* **2013**, *17*, 1564–1567.

symmetry

MDPI

Article

Accurate Dense Stereo Matching Based on Image Segmentation Using an Adaptive Multi-Cost Approach

Ning Ma [1,2], Yubo Men [1], Chaoguang Men [1,*] and Xiang Li [1]

[1] College of Computer Science and Technology, Harbin Engineering University, Nantong Street 145, Harbin 150001, China; maning@hrbeu.edu.cn (N.M.); menyubo@hrbeu.edu.cn (Y.M.); leexiang@hrbeu.edu.cn (X.L.)

[2] College of Computer Science and Information Engineering, Harbin Normal University, Normal University Road 1, Harbin 150001, China

* Correspondence: menchaoguang@hrbeu.edu.cn; Tel.: +86-451-86207610

Academic Editor: Angel Garrido
Received: 23 August 2016; Accepted: 12 December 2016; Published: 21 December 2016

Abstract: This paper presents a segmentation-based stereo matching algorithm using an adaptive multi-cost approach, which is exploited for obtaining accuracy disparity maps. The main contribution is to integrate the appealing properties of multi-cost approach into the segmentation-based framework. Firstly, the reference image is segmented by using the mean-shift algorithm. Secondly, the initial disparity of each segment is estimated by an adaptive multi-cost method, which consists of a novel multi-cost function and an adaptive support window cost aggregation strategy. The multi-cost function increases the robustness of the initial raw matching costs calculation and the adaptive window reduces the matching ambiguity effectively. Thirdly, an iterative outlier suppression and disparity plane parameters fitting algorithm is designed to estimate the disparity plane parameters. Lastly, an energy function is formulated in segment domain, and the optimal plane label is approximated by belief propagation. The experimental results with the Middlebury stereo datasets, along with synthesized and real-world stereo images, demonstrate the effectiveness of the proposed approach.

Keywords: stereo matching; multi-cost; image segmentation; disparity plane fitting; belief propagation

1. Introduction

Stereo matching is one of the most widely studied topics in computer vision. The aim of stereo matching is to estimate the disparity map between two or more images taken from different views for the same scene, and then extract the 3D information from the estimated disparity [1]. Intuitively, the disparity represents the displacement vectors between corresponding pixels that horizontally shift from the left image to the right image [2]. Stereo matching serves an important role in a wide range of applications, such as robot navigation, virtual reality, photogrammetry, people/object tracking, autonomous vehicles, and free-view video [3]. A large number of techniques have been invented for stereo matching, and a valuable taxonomy and categorization scheme of dense stereo matching algorithms can be found in the Middlebury stereo evaluation [1,4,5]. According to the taxonomy, most dense stereo algorithms perform the following four steps: (1) initial raw matching cost calculation; (2) cost aggregation; (3) disparity computation/optimization; and (4) disparity refinement. Due to the ill-posed nature of the stereo matching problem, the recovery of accurate disparity still remains challenging due to textureless areas, occlusion, perspective distortion, repetitive patterns, reflections, shadows, illumination variations and poor image quality, sensory noise, and high computing load. Thus, the robust stereo matching algorithm has become a research hotspot recently [6].

By using a combination of multiple single similarity measures into composite similarity measure, itmulti-cost has been proven to be an effective method for calculating the matching cost [7–10]. Stentoumis et al. proposed a multi-cost approach and obtained excellent results for disparity estimation [7]. This is the most well-known multi-cost approach method and represents a state-of-the-art multi-cost algorithm. On the other hand, segmentation-based approaches have attracted attention due to their excellent performance for occlusion, textureless areas in stereo matching [3,11–16]. Our work is directly motivated by the multi-cost approach and the segmentation-based framework therefore, the image segmentation-based framework and an adaptive multi-cost approach are both utilized in our algorithm. The stereo matching problem can be formalized as an energy minimization problem in the segment domain, which ensures our method will correctly estimate large textureless areas and precisely localize depth boundaries. For each segment region, the initial disparity is estimated using an adaptive multi-cost approach, which consists of a multi-cost function and an adaptive support window cost aggregation strategy. An improved census transformation and illumination normal vector are utilized for the multi-cost function, which increases the robustness of the initial raw matching cost calculation. The shape and size of the adaptive support window based on the cross-shaped skeleton can be adjusted according to the color information of the image, which ensures that all pixels belonging to the same support window have the same disparity. In order to estimate the disparity plane parameters precisely, an iterative outlier suppression and disparity plane parameters fitting algorithm is designed after the initial disparity estimation. The main contribution of this work is to integrate the appealing properties of multi-cost approach into the segmentation-based framework. The adaptive multi-cost approach, which consists of a multi-cost function and an adaptive support window, improves the accuracy of the disparity map. This ensures our algorithm works well with the Middlebury stereo datasets, as well as synthesized and real-world stereo image pairs. This paper is organized as follows: In Section 2, related works are reviewed. In Section 3, the proposed approach is described in detail. In Section 4, experimental results and analysis are given using an extensive evaluation dataset, which includes Middlebury standard data, synthesized images, and real-world images. Finally, the paper is concluded in Section 5.

2. Related Works

The stereo matching technique is widely used in computer vision for 3D reconstruction. A large number of algorithms have been developed for estimating disparity maps from stereo image pairs. According to the analysis and taxonomy scheme, stereo algorithms can be categorized into two groups: local algorithms and global algorithms [1].

Local algorithms utilize a finite neighboring support window that surrounds the given pixel to aggregate the cost volume and generate the disparity by winner takes all (WTA) optimization. It implicitly models the assumption that the scene is piecewise smooth and all the pixels of the support window have similar disparities. These methods have simple structure and high efficiency, and could easily capture accurate disparity in ideal conditions. However, local algorithms cannot work well due to the image noise and local ambiguities like occlusion or textureless areas. In general, there are two major research topics for local methods: similarity measure function and cost aggregation [17]. Typical functions are color- or intensity-based (such as sum of absolute difference, sum of squared difference, normalized cross-correlation) and non-parametric transform-based (such as rank and census). The non-parametric transform-based similarity measure function is more robust to radiometric distortion and noise than the intensity based. For cost aggregation aspect, the adaptive window [18–20] and adaptive weight [17,21,22] are two principal methods. Adaptive window methods try to assign an appropriate size and shape support region for the given pixel to aggregate the raw costs. However, adaptive weight methods inspired by the Gestalt principles adopt the fixed-size square window and assign appropriate weights to all pixels within the support window of the given pixel.

Global algorithms are formulated in an energy minimization framework, which makes explicit smoothness assumptions and solves global optimization by minimizing the energy function. This kind of method has achieved excellent results, with examples such as dynamic programming (DP), belief propagation (BP), graph cuts (GC), and simulated annealing (SA). The DP approach is an efficient solution since the global optimization can be performed in one dimension [23]. Generally, DP is the first choice for numerous real-time stereo applications. Due to smoothness consistency, inter-scanlines cannot be well enforced; the major problem of computed disparity maps-based DP presents the well-known horizontal "streaks" artifacts. The BP and GC approaches are formulated in a two-dimensional Markov random field energy function, which consists of a data term and a smoothness term [24,25]. The data term measures the dissimilarity of correspondence pixels in stereo image pairs, and the smoothness term penalizes adjacent pixels that are assigned to different disparities. The optimization of the energy function is considered to be NP-complete problem. Although a number of excellent results have been obtained, both the BP and the GC approaches are typically expensive in terms of computation and storage. Another disadvantage of these approaches is that there are so many parameters that need to be determined. The semi-global method proposed by Hirschmüller is a compromise between one-dimensional optimization and two-dimensional optimization. It employs the "mutual information" cost function in a semi-global context [26]. While this strategy allows higher execution efficiency, it sacrifices some disparity accuracy.

Recently, segmentation-based approaches have attracted attention due to their excellent performance for stereo correspondence [3,11–16]. This kind of method performs well in reducing the ambiguity associated with textureless or depth discontinuity areas, and enhancing noise tolerance. It is based on two assumptions: The scene structure of the image captured can be approximated by a group of non-overlapping planes in the disparity space, and each plane is coincident with at least one homogeneous color segment region in the reference image. Generally, a segmentation-based stereo matching algorithm can be concluded in four steps as follows: (1) segment the reference image into regions of homogeneous color by applying a robust segmentation method (usually the mean-shift image segmentation technique); (2) estimate initial disparities of reliable pixels using the local matching approach; (3) a plane fitting technique is employed to obtain disparity plane parameters, which are considered as a label set; and (4) an optimal disparity plane assignment is approximated utilizing a global optimization approach.

We mainly contribute to Steps (2) and (3) in this work, and Steps (1) and (4) are commonly used techniques in the context of stereo matching. The key idea behind our disparity estimation scheme is utilizing the multi-cost approach that is usually adopted in local methods to achieve a more accurate initial disparity map, and then utilizing the iterative outlier suppression and disparity plane parameters fitting approach to achieve a more reliable disparity plane. For Step (2), the accurate and reliable initial disparity map can improve the accuracy of the final result; however, this step is usually performed utilizing some simple local algorithm [11,13]. A lot of false matching exists, and these matching errors will reduce the accuracy of the final result. Stentoumis et al. have demonstrated that the multi-cost approach can effectively improve the accuracy of the disparity [7]. In order to estimate an accurate initial disparity map, an adaptive multi-cost approach that consists of a multi-cost function and an adaptive support window cost aggregation strategy is employed. For Step (3), for most segmentation-based algorithms, the RANDom Sample Consensus (RANSAC) algorithm is usually used to filter out outliers and fit the disparity plane. RANSAC algorithm is a classical efficient algorithm; the principle of RANSAC is used to estimate the optimal parameter model in a set of data that contains "outliers" using the iteration method. However, the result of the RANSAC algorithm relies on the selection of initial points. Since the selection is random, the result obtained is not satisfying in some cases [13]. Furthermore, in a disparity estimation scheme, the outliers could be determined by a variety of criteria, e.g., mutual consistency criterion, correlation confidence criterion, disparity distance criterion, and convergence criterion. The different outliers will be obtained from different criteria. In order to combine multiple outlier filtering criteria to filter out the outliers and obtain accurate plane fitting parameters, an iterative outlier suppression and disparity plane parameters fitting algorithm is developed.

3. Stereo Matching Algorithm

In this section, the proposed stereo matching algorithm is described in detail. The entire algorithm is shown in the block diagram representation in Figure 1, which involves four steps: image segmentation, initial disparity estimation, disparity plane fitting, and disparity plane optimization.

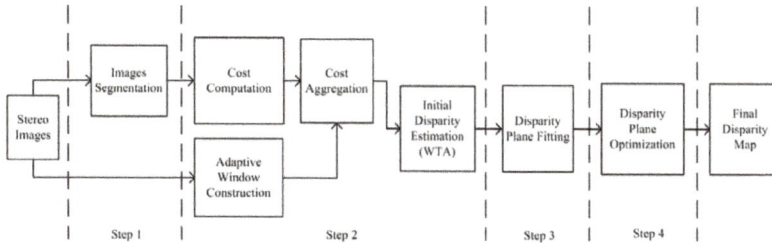

Figure 1. Block diagram representation of the proposed stereo algorithm.

3.1. Image Segmentation

Due to the proposed algorithm being based on the segmentation framework, the first step is that the reference image is divided into a group of non-overlapping, homogeneous color segment regions. The segmentation-based framework implicitly assumes that the disparity varies smoothly in the same segment region, and depth discontinuities coincide with the boundaries of those segment regions. Generally, over-segmentation of the image is preferred, which ensures the above assumptions can be met for most natural scenes. The mean-shift color segmentation algorithm is employed to decompose the reference image into different regions [27]. The mean-shift algorithm is based on the kernel density estimation theory, and takes account of the relationship between color information and distribution characteristics of the pixels. The main advantage of the mean-shift technique is that edge information is incorporated as well, which ensures our approach will obtain disparity in textureless regions and depth discontinuities precisely. The segmentation results of partial images in the Middlebury stereo datasets are shown in Figure 2, and pixels belonging to the same segment region are assigned the same color.

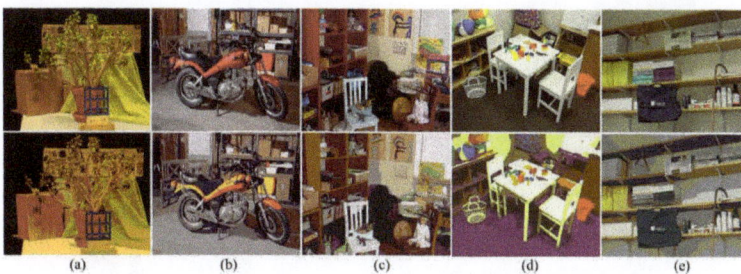

Figure 2. The image segmentation results. (**a**) Jade plant image and corresponding segmentation results; (**b**) motorcycle image and corresponding segmentation results; (**c**) playroom image and corresponding segmentation results; (**d**) play table image and corresponding segmentation results; and (**e**) shelves image and corresponding segmentation results.

3.2. Initial Disparity Map Estimation

The initial disparity map is estimated by an adaptive multi-cost approach, which is shown in the block diagram representation in Figure 3. By using the combination of multiple single similarity measures into a composite similarity measure, it has been proven to be an effective method to calculate the matching cost [7–10]. The adaptive multi-cost approach proposed in this work defines a novel multi-cost function to calculate the raw matching score and employs an adaptive window aggregation strategy to filter the cost volume. The main advantage of the adaptive multi-cost approach is that it improves the robustness of raw initial matching costs calculation and reduces the matching ambiguity, thus the matching accuracy is enhanced.

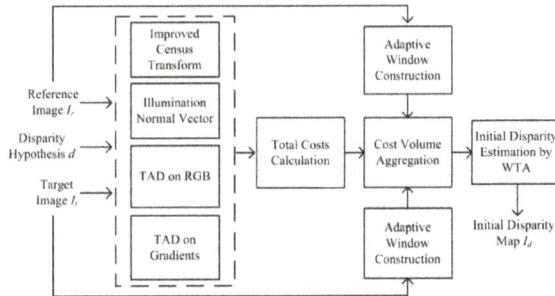

Figure 3. Block diagram representation of the initial disparity map estimation.

The multi-cost function is formulated by combining four individual similarity functions. Two of them are traditional similarity functions, which are absolute difference similarity functions that take into account information from RGB (Red, Green, Blue) channels, and the similarity function based on the principal image gradients. The other two similarity functions are improved census transform [7] and illumination normal vector [22]. An efficient adaptive method of aggregating initial matching cost for each pixel is then applied, which relies on a linearly expanded cross skeleton support window. Some similarity cost functions used here and the shape of the adaptive support window are shown in Figure 4. Finally, the initial disparity map of each segment region is estimated by the "winner takes all" (WTA) strategy.

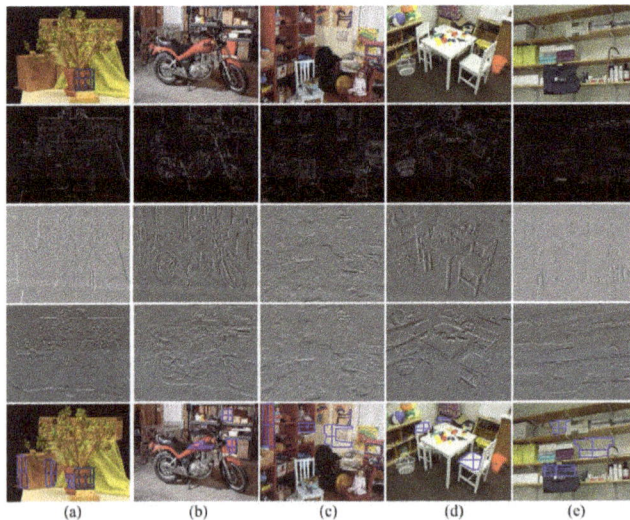

Figure 4. The similarity cost functions and the shape of the adaptive support window. (**a**) Jade plant; (**b**) motorcycle; (**c**) playroom; (**d**) play table; and (**e**) shelves. From top to bottom: the reference images; the modulus images of the illumination normal vector; the gradient maps along horizontal direction; the gradient maps along vertical direction; and the examples of the adaptive support window.

3.2.1. The Multi-Cost Function

The multi-cost function is formulated by combining four individual similarity functions. The improved census transform (ICT) is the first similarity function of multi-cost function. It extends the original census transform approaches; the transform is performed here not only on grayscale image intensity, but also on its gradients in the horizontal and vertical directions. The census transform is a high robustness stereo measure to illuminate variations or noise, and the image gradients have a close relationship with characteristic image features, i.e., edges or corners. The similarity function based on improved census transform exploits the abovementioned advantages. In the preparation phase, we use the mean value over the census block instead of the center pixel value, and calculate the gradient images in the x and y directions using the Sobel operator. Consequently, the ICT over the intensity images I as well as the gradient images I_x (x directions) and I_y (y directions) are shown as:

$$T_{ICT}(x,y) = \underset{[m,n]\in W}{\otimes} \mathcal{E}[\overline{I(x,y)}, I(x+m,y+n)] \underset{[m,n]\in W}{\otimes} \mathcal{E}[\overline{I_x(x,y)}, I_x(x+m,y+n)] \\ \underset{[m,n]\in W}{\otimes} \mathcal{E}[\overline{I_y(x,y)}, I_y(x+m,y+n)] \tag{1}$$

where the operator $\otimes$ denotes a bit-wise catenation, and the auxiliary function $\mathcal{E}$ is defined as:

$$\mathcal{E}(x,y) = \begin{cases} 0 & if\ x \leq y \\ 1 & if\ x > y \end{cases}, \tag{2}$$

The matching cost between two pixels by applying ICT are calculated via the Hamming distance of the two bit strings in Equation (3):

$$C_{ICT}(x,y,d) = Hamming(T_{GCT}^{reference}(x,y), T_{GCT}^{target}(x+d,y)), \tag{3}$$

Illumination normal vector (INV) is the second similarity function of multi-cost function. INV reflects the high-frequency information of the image, which generally exists at the boundaries of objects and fine texture area. Consequently, the high-frequency information reflects some small-scale details of the image, which is very useful for stereo correspondence [22]. Denote a pixel of the image as a point in 3D space $P[x, y, f(x,y)]$, where x and y are the horizontal and vertical coordinates, and $f(x,y)$ is the intensity value of position (x,y). The INV of point P is calculated by the cross-product of its horizontal vector $V_{horizonal}$ and vertical vector $V_{vertical}$. Define $V(P)$ as the INV of point P.

$$V(P) = V_{horizonal} \times V_{vertical} = [V_i(P), V_j(P), V_k(P)], \tag{4}$$

where the horizontal vector $V_{horizonal}$ and vertical vector $V_{vertical}$ are defined as follows:

$$\begin{cases} V_{horizonal} = P[x+1,y,f(x+1,y)] - P[x,y,f(x,y)] \\ V_{vertical} = P[x,y+1,f(x,y+1)] - P[x,y,f(x,y)] \end{cases}, \tag{5}$$

Consequently, Equation (4) can be rewritten as:

$$\begin{aligned} V(P) &= V_{horizonal} \times V_{vertical} \\ &= \begin{vmatrix} i & j & k \\ 1 & 0 & f(x+1,y) - f(x,y) \\ 0 & 1 & f(x,y+1) - f(x,y) \end{vmatrix} \\ &= (f(x,y) - f(x+1,y))i + (f(x,y) - f(x,y+1))j + k \end{aligned} \tag{6}$$

The modulus images of the illumination normal vector of images are shown in the second line of Figure 4. The matching cost between two pixels based on INV measure is calculated via the Euclidean distance of the two vectors as:

$$C_{INV}(x,y,d) = \|V_{reference}(x,y) - V_{target}(x+d,y)\|_2, \tag{7}$$

The next two similarity functions are the traditional similarity functions, truncated absolute difference on RGB color channels (TADc) and truncated absolute difference on the image principal gradient (TADg). TADc is a simple and easily implementable measure, widely used in image matching. Although sensitive to radiometric differences, it has been proven to be an effective measure when flexible aggregation areas and multiple color layers are involved. For each pixel, the cost term is intuitively computed as the minimum value between the absolute difference from RGB vector space and the user-defined truncation value T. It is formally expressed as:

$$C_{TADc}(x,y,d) = \frac{1}{3} \sum_{i \in (r,g,b)} \min \left| I_i^{reference}(x,y) - I_i^{target}(x+d,y), \, T \right|, \tag{8}$$

In the TADg, the gradients of image in the two principal directions are extracted, and the sum of absolute differences of each gradient value in the x and y directions are used as a cost measure. The use of directional gradients separately, i.e., before summing them up to the single measure, introduces the directional information for each gradient into the cost measure. The gradients in the horizontal and vertical directions are shown in the third and fourth lines of Figure 4, respectively. The cost based on TADg can be expressed as Equation (9) with a truncated value T:

$$C_{TADg}(x,y,d) = \min \left| \nabla_x I_{reference}(x,y) - \nabla_x I_{target}(x+d,y), T \right| \\ + \min \left| \nabla_y I_{reference}(x,y) - \nabla_y I_{target}(x+d,y), T \right|, \tag{9}$$

Total matching cost $C_{RAW}(x,y,d)$ is derived by merging the four individual similarity functions. A robust exponential function that resembles a Laplacian kernel is employed for cost combination:

$$C_{RAW}(x,y,d) = \exp\left(-\frac{C_{INV}(x,y,d)}{\gamma_{INV}}\right) + \exp\left(-\frac{C_{GCT}(x,y,d)}{\gamma_{ICT}}\right) \\ + \exp\left(-\frac{C_{TADC}(x,y,d)}{\gamma_{TADC}}\right) + \exp\left(-\frac{C_{TADG}(x,y,d)}{\gamma_{TADG}}\right), \tag{10}$$

Each individual matching cost score is normalized by its corresponding constant γ_{INV}, γ_{ICT}, γ_{TADC}, and γ_{TADG}, to ensure equal contribution to the final cost score, or tuned differently to adjust their impact on the matching cost accordingly. Tests of multi-cost function performed on the Middlebury stereo datasets for stereo matching are presented in Figures 5 and 6. The test results show that the matching precision is increased by combining the individual similarity functions. In Figure 5, disparity maps are estimated with different combinations of similarity functions after the aggregation step. From top to bottom: the reference images; the ground truth; the disparity maps estimated by ICT; ICT+TADc; ICT+TADc+TADg; ICT+TADc+TADg+INV; the corresponding bad 2.0 error maps for ICT; ICT+TADc; ICT+TADc+TADg; and ICT+TADc+TADg+INV. The same region of the error maps is marked by red rectangles. The marked regions show that the error is reduced through combining the individual similarity functions. The disparity plane fitting and optimization steps described in Sections 3.3 and 3.4 have not been used here, in order to illustrate individual results and the improvement achieved by fusing the four similarity functions. Figure 6 shows the visualized quantitative performance of similarity functions (in % of erroneous disparities at 2 error threshold) by comparing different combinations of similarity functions against the ground truth. From left to right, the charts correspond to the error matching rate of (a) non-occluded pixels and (b) all image pixels. On the horizontal axis, A: ICT; B: ICT+TADc; C: ICT+TADc+TADg; and D: ICT+TADc+TADg+INV. Following that, the matching cost $C_{RAW}(x,y,d)$ is stored in a 3D matrix known as the disparity space image (DSI).

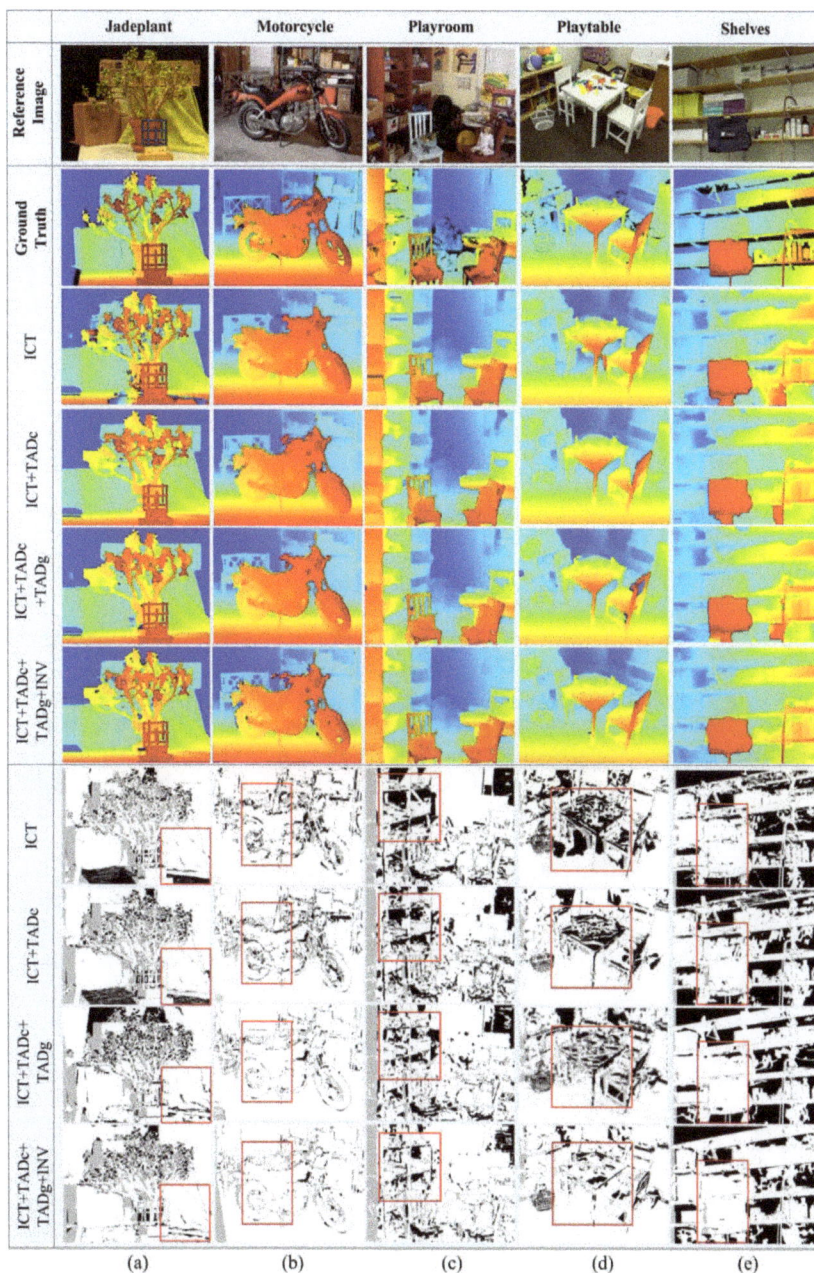

Figure 5. Comparison of different ways of similarity functions combination for Middlebury stereo datasets. (**a**) Jade plant; (**b**) motorcycle; (**c**) playroom; (**d**) play table; and (**e**) shelves. Disparity maps are estimated by different combinations of similarity functions after the aggregation step. From top to bottom: the reference images; the ground truth; the disparity maps estimated by ICT; ICT+TADc; ICT+TADc+TADg; ICT+TADc+TADg+INV; the corresponding bad 2.0 error maps for ICT; ICT+TADc; ICT+TADc+TADg; and ICT+TADc+TADg+INV. The same region of the error maps is marked by red rectangles.

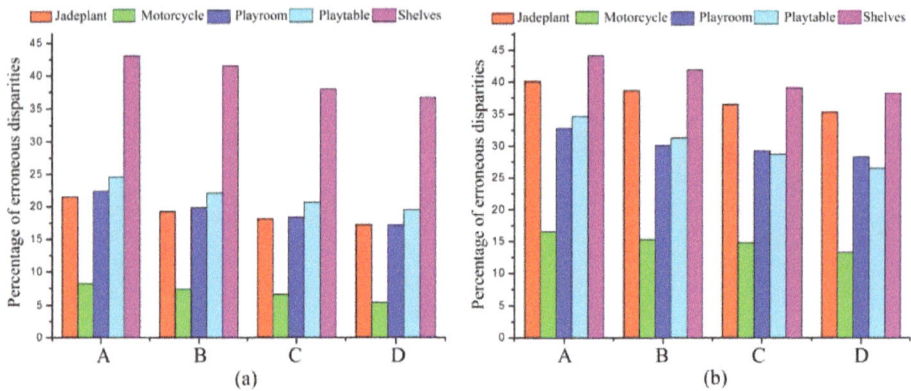

Figure 6. The visualized quantitative performance of similarity functions (in % of erroneous disparities at 2 error threshold) by comparing different combinations of similarity functions against the ground truth. From left to right, the charts correspond to the error matching rate of (**a**) non-occluded pixels and (**b**) all image pixels. On the horizontal axis, A: ICT; B: ICT+TADc; C: ICT+TADc+TADg; and D: ICT+TADc+TADg+INV.

3.2.2. Cost Aggregation

As mentioned above, matching cost $C_{RAW}(x, y, d)$ is called raw DSI since it is always accompanied with aliasing and noise. Cost aggregation can decrease the aliasing and noise by averaging or summing up the DSI over a support window. This implicitly assumes that the support window is a front parallel surface and all pixels in the window have similar disparities. In order to obtain accurate disparity results at near depth discontinuities, an appropriate support window should be constructed. An adaptive cross-based window that relies on a linearly expanded cross skeleton support region for cost aggregation is adopted [7,10,18,28]. The shape of the adaptive support window is visually presented in the fifth line of Figure 4. The cross-based region consists of multiple horizontal line segments spanning several neighboring rows. This aggregation strategy has two main advantages: firstly, the support window can vary adaptively, with arbitrary size and shape according to the scene color similarity; secondly, the aggregation over irregularly shaped support windows can be performed quickly by utilizing the integral image technique.

The construction of cross-based support regions is achieved by expanding around each pixel a cross-shaped skeleton to create four segments $\left\{ h_p^-, h_p^+, v_p^-, v_p^+ \right\}$ defining the corresponding sets of pixels $H(p)$ and $V(p)$ in the horizontal and vertical directions, as seen in Figure 7a [7].

$$\left\{ \begin{array}{l} H(p) = \left\{ (x, y) \middle| x \in [x_p - h_p^-, x_p + h_p^+], y = y_p \right\} \\ V(p) = \left\{ (x, y) \middle| x = x_p, y \in [y_p - v_p^-, y_p + v_p^+] \right\} \end{array} \right. , \tag{11}$$

In our approach, the linear threshold proposed in [7] is used to expand the skeleton around each pixel: $T(L_q) = -(T_{max} / L_{max}) \times L_q + T_{max}$. This linear threshold $T(L_q)$ in color similarity involves the maximum semi-dimension L_{max} of the support window size, the maximum color dissimilarity T_{max} between pixels p and q, and the spatial closeness L_q. According to [7], the values of T_{max} and L_{max} are 20 and 35, respectively. The final support window $U(p)$ for p is formulated as a union of horizontal segment $H(q)$, in which q traverses the vertical segment $V(p)$. A symmetric support window is also adopted to avoid distortion by the outliers in the reference image [7]. This is shown in Figure 7b.

$$U(p) = \bigcup_{q \in V(p)} H(q), \tag{12}$$

The final aggregation cost for each pixel is calculated by aggregating the matching cost over the support window. This process can be quickly realized by integrating image technology, as shown in Figure 7c.

$$C_{aggregation}(x_p, y_p, d) = \frac{1}{|U(p)|} \sum_{(x_i, y_i) \in U(p)} C_{RAW}(x_i, y_i, d), \tag{13}$$

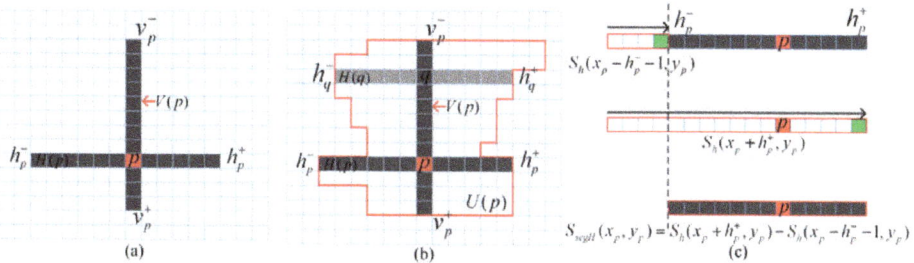

Figure 7. The illustration of the adaptive cross-based aggregation algorithm. (**a**) The upright cross skeleton. The upright cross consists of a horizontal segment $H(p) = \left\{(x,y) \middle| x \in [x_p - h_p^-, x_p + h_p^+], y = y_p\right\}$ and a vertical segment $V(p) = \left\{(x,y) \middle| x = x_p, y \in [y_p - v_p^-, y_p + v_p^+]\right\}$; (**b**) the support region $U(P)$ is a combination of each horizontal segment $H(q)$, where q traverses the vertical segment $V(p)$ of p; (**c**) a schematic of a 1D integral image technique.

Subsequently, the initial disparity $d_{x,y}$ at coordinates (x,y) is estimated by using the WTA strategy where the lowest matching cost is selected:

$$d_{x,y} = \underset{D_{min} \le d \le D_{max}}{\operatorname{argmin}} C_{aggregation}(x, y, d), \tag{14}$$

3.3. Disparity Plane Fitting

Although the RANSAC algorithm has been widely used for rejecting outliers fitting data, it is usually not suitable for the segmentation-based framework of stereo matching [13]. That is because the outliers are caused by many different factors, like textureless areas, occlusion, etc. If the filtering criteria are different, that produces different outliers. In this section, an iterative outlier suppression and disparity plane parameters fitting algorithm is designed for plane fitting. The disparity of each segment region can be modeled as:

$$d(x, y) = ax + by + c, \tag{15}$$

where d is the corresponding disparity of pixel (x,y), and a, b, c are the plane parameters of the arbitrary segment region. In order to solve the plane parameters, a linear system for the arbitrary segment region can be formulated as follows:

$$A[a, b, c]^T = B, \tag{16}$$

where the i'th row of the matrix A is $[x_i, y_i, 1]$, and the i'th element of the vector B is $d(x_i, y_i)$. Then the linear system can be transformed into the form of $A^T A[a, b, c]^T = A^T B$; the detailed function is expressed as follows:

$$\begin{bmatrix} \sum_{i=1}^{m} x_i^2 & \sum_{i=1}^{m} x_i y_i & \sum_{i=1}^{m} x_i \\ \sum_{i=1}^{m} x_i y_i & \sum_{i=1}^{m} y_i^2 & \sum_{i=1}^{m} y_i \\ \sum_{i=1}^{m} x_i & \sum_{i=1}^{m} y_i & 1 \end{bmatrix} \begin{bmatrix} a \\ b \\ c \end{bmatrix} = \begin{bmatrix} \sum_{i=1}^{m} x_i d_i \\ \sum_{i=1}^{m} y_i d_i \\ \sum_{i=1}^{m} d_i \end{bmatrix}, \tag{17}$$

where m is the number of pixels inside the corresponding segment region. After that, the Singular Value Decomposition (SVD) approach is employed to solve the least square equation to obtain the disparity plane parameters:

$$[a, b, c]^T = (A^T A)^+ A^T B, \tag{18}$$

where $(A^T A)^+$ is the pseudo-inverse of $A^T A$ and can be solved through SVD.

However, as is well known, the least square solution is extremely sensitive to outliers. The outliers in this stage are usually generated at the last stage due to the matching error inevitable in initial disparity map estimation. In order to filter out these outliers and obtain accurate plane parameters, four filters are combined.

The first filter is mutual consistency check (often called left-right check). The principle of mutual consistency check is that the same point of the stereo image pair should have the same disparity. Thus, the occluded pixels in the scene can be filtered out. Let $D_{reference}$ be the disparity map from reference image to target image, and D_{target} be the disparity map from target image to reference image. The mutual consistency check is formulated as:

$$\left| D_{reference}(x, y) - D_{target}(x - D_{reference}(x, y), y) \right| \leq t_{consistency}, \tag{19}$$

where $t_{consistency}$ is a constant threshold (typically 1). If the pixels of the reference image satisfy Equation (19), these pixels are marked as non-occluded pixels; otherwise these pixels are marked as occluded pixels, which should be filtered out as outliers.

Afterwards, correlation confidence filter is established to judge whether the non-occluded pixels are reliable. Generally, some of the disparity in the textureless areas may be incorrect but will be consistent for both views. Thus, the correlation confidence filter is adopted to overcome this difficulty and obtain reliable pixels. Let $C_{aggregation}^{first}(x, y)$ be the best cost score of a pixel in the non-occluded pixels set, and $C_{aggregation}^{second}(x, y)$ be the second best cost score of this pixel. The correlation confidence filter is formulated as:

$$\left| \frac{C_{aggregation}^{first}(x, y) - C_{aggregation}^{second}(x, y)}{C_{aggregation}^{second}(x, y)} \right| \geq t_{confidence}, \tag{20}$$

where $t_{confidence}$ is a threshold to adjust the confidence level. If the cost score of the pixels in the reference image satisfies Equation (20), these pixels are considered reliable. If the ratio between the number of the reliable pixels and the total number of the pixels in arbitrary segment region is equal to or greater than 0.5, this segment region is considered a reliable segment region. Otherwise segment regions are marked as unreliable regions, which lack sufficient data to provide reliable plane estimations. The disparity plane of the unreliable region is stuffed through its nearest reliable segment region.

Followed by the above filters, the initial disparity plane parameters of each reliable segment region can be estimated through the reliable pixels. The disparity distance filter is adopted to measure the Euclidean distance between initial disparity and the estimated disparity plane:

$$|d(x, y) - (ax + by + c)| \leq t_{outlier}, \tag{21}$$

where $t_{outlier}$ is a constant threshold (typically 1). If the pixel does not satisfy Equation (21), it would be an outlier. Then we can exclude the outliers, update the reliable pixels of the segment region, and re-estimate the disparity plane parameters of the segment region.

After the abovementioned three filters, the convergence filter is utilized to judge whether disparity plane is convergent. The new disparity plane parameters will be estimated until:

$$|a' - a| + |b' - b| + |c' - c| \leq t_{convergence}, \tag{22}$$

where (a',b',c') are the parameters of the new disparity plane, (a,b,c) are the parameters of the plane obtained in the previous iteration, and $t_{convergence}$ is the convergence threshold of the iterative and is usually set as (typically 10^{-6}).

The flow chart of the iterative outlier suppression and disparity plane parameters fitting algorithm is shown in Figure 8. The detailed implementation of the algorithm is presented as follows, from Step (1) to Step (6):

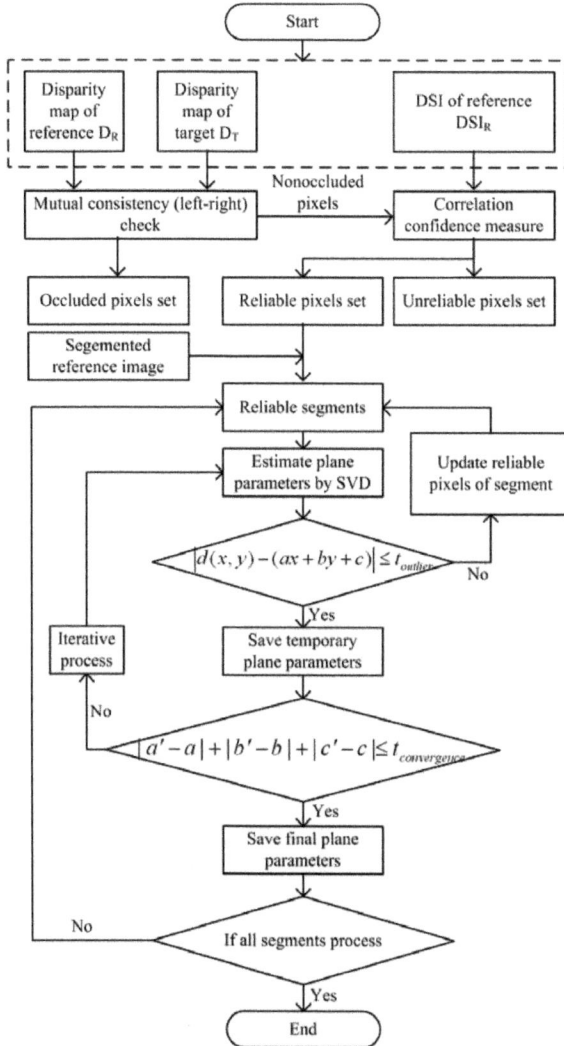

Figure 8. The flow chart of the iterative outlier suppression and disparity plane parameters fitting algorithm.

Step (1): Input segmented reference image, disparity map of stereo image pair, and the DSI of the reference image.

Step (2): Mutual consistency filter is utilized to check the initial disparity of each pixel as in Equation (19); the pixels are detected as non-occluded or occluded pixels.

Step (3): The reliable pixels and reliable segment region are determined by a correlation confidence filter, as in Equation (20).

Step (4): The initial disparity plane parameters of each reliable segment region are estimated through the reliable pixels, and the disparity distance filter described in Equation (21) is utilized to update the reliable pixels.

Step (5): Iterate Step (4) until the convergence filter is satisfied.

Step (6): The algorithm will be terminated when the disparity plane parameters of all segment regions have been estimated. Otherwise, return to Step (3) to process the remainder of the reliable segment regions.

3.4. Disparity Plane Optimization by Belief Propagation

The last step of the segmentation-based stereo matching algorithm is usually global optimization. The stereo matching is formulated as an energy minimization problem in the segment domain. We label each segment region with its corresponding disparity plane by using the BP algorithm [25]. Assume that each segment region $s \in R$, R is the reference image, its corresponding plane $f(s) \in D$, and D is the disparity plane set. The energy function for labeling f can be formulated as:

$$E_{TOTAL}(f) = E_{DATA}(f) + E_{SMOOTH}(f) + E_{OCCLUSION}(f), \tag{23}$$

where $E_{TOTAL}(f)$ is the whole energy function, $E_{DATA}(f)$ is the data term, $E_{SMOOTH}(f)$ is the smoothness penalty term, and $E_{OCCLUSION}(f)$ is the occlusion penalty term.

The data term $E_{DATA}(f)$ is formulated for each segment region and its corresponding disparity plane assignment. It is calculated by summing up the matching cost of each segment region:

$$E_{DATA}(f) = \sum_{s \in R} C_{SEG}(s, f(s)), \tag{24}$$

where $C_{SEG}(s, f(s))$ is the summation of matching cost, which is defined in Section 3.2 for all the reliable pixels inside the segment:

$$C_{SEG}(s, f(s)) = \sum_{(x,y) \in s} C(x, y, d), \tag{25}$$

The smoothness penalty term $E_{SMOOTH}(f)$ is used to punish the adjacent segment regions with different disparity plane:

$$E_{SMOOTH}(f) = \sum_{(\forall (s_i, s_j) \in S_N | f(s_i) \neq f(s_j))} \lambda_{disc}(s_i, s_j), \tag{26}$$

where S_N is a set of all adjacent segment regions, S_i, S_j are neighboring segment regions, and $\lambda_{disc}(x, y)$ is a discontinuity penalty function.

The occlusion penalty term $E_{OCCLUSION}(f)$ is used to punish the occlusion pixels of each segment region:

$$E_{OCCLUSION}(f) = \sum_{s \in R} \omega_{occ} N_{occ}, \tag{27}$$

where ω_{occ} is a coefficient for occlusion penalty and N_{occ} is the number of occluded pixels of the segment region. The energy function $E_{TOTAL}(f)$ is minimized by a BP algorithm, and the final disparity map can be obtained.

4. Experimental Results

The proposed stereo matching algorithm has been implemented by VS2010, and the performance of the algorithm is evaluated using the 2014 Middlebury stereo datasets [29], 2006 [30], 2005 [4], the synthesized stereo image pairs [31], and the real-world stereo image pairs. The set of parameter values used in this paper are shown in Table 1, and the results are shown in Figure 9. The average error rate of the stereo pairs for each evaluation area (all, non-occlusion) are displayed. The percentage of erroneous pixels in the complete image (all) and no-nocclusion areas (nonocc) for the 2 pixels threshold is counted. Figure 9 illustrates the stability of the algorithm to parameter tuning. The test results show that the algorithm is stable within a wide range of values for each parameter. We choose the parameters corresponding to the minimum error rate for all the tested stereo image datasets.

Table 1. Parameters values used for all stereo image pairs.

Parameter Name	Purpose	Algorithm Steps	Parameter Value
Spatial bandwidth hs Spectral bandwidth hr	Image segmentation	Step (1)	10 7
Gamma γ_{INV} Gamma γ_{ICT} Gamma γ_{TADC} Gamma γ_{TADG}	Matching cost computation	Step (2)	40 20 40 20
Threshold $t_{consistency}$ Threshold $t_{confidence}$ Threshold $t_{outlier}$ Threshold $t_{convergence}$	Outliers filter and disparity plane parameters fitting	Step (3)	1 0.04 1 10^{-6}
Smoothness penalty λ_{disc} Occlusion penalty ω_{occ}	Smoothness and occlusion penalty	Step (4)	5 5

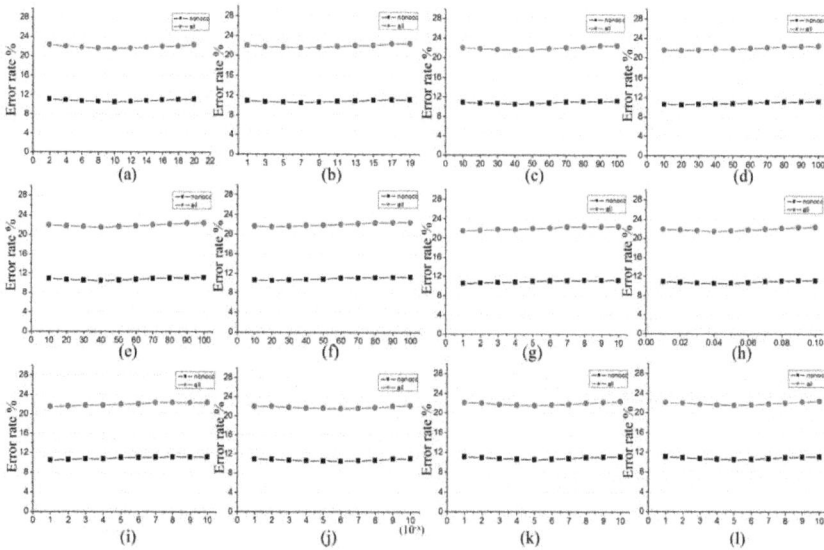

Figure 9. Diagrams presenting the response of the algorithm to the tuning parameters with the rest of the parameter set remaining constant. The average error rate of the stereo pairs for each evaluation area (all, nonocc) is displayed. (**a**) Spatial bandwidth and (**b**) spectral bandwidth used for image segmentation in Step 1 of the algorithm; (**c**) Gamma INV(Illumination Normal Vector); (**d**) Gamma ICT(Improved Census Transform); (**e**) Gamma TADC(Truncated Absolute Difference on Color); and (**f**) Gamma TADG(Truncated Absolute Difference on Gradient) used for matching cost computation in Step (2) of the algorithm. (**g**) Threshold consistency; (**h**) threshold confidence; (**i**) threshold outlier; and (**j**) threshold convergence used for outliers filter and disparity plane parameters fitting in Step (3) of the algorithm; (**k**) smoothness penalty and (**l**) occlusion penalty used for smoothness and occlusion penalty in Step (4) of the algorithm.

Table 2. Quantitative evaluation based on the training set of the 2014 Middlebury stereo datasets at 2 Error Threshold. The best results for each test column are highlighted in bold. Res and Avg represent resolution scale and average error respectively. Adiron, ArtL, Jadepl, Motor, MotorE, Piano, PianoL, Pipes, Playrm, Playt, PlaytP, Recyc, Shelvs, Teddy, and Vintge are the names of experimental data in the training set.

Name	Res	Avg	Adiron	ArtL	Jadepl	Motor	MotorE	Piano	PianoL	Pipes	Playrm	Playt	PlaytP	Recyc	Shelvs	Teddy	Vintge
APAP-Stereo [32]	H	**7.78**	3.04	7.22	13.5	4.39	4.68	10.7	16.1	5.35	**10.1**	8.60	8.11	7.70	**12.2**	5.16	**7.97**
PMSC [33]	H	8.35	**1.46**	**4.44**	**11.2**	3.68	4.07	11.9	18.2	5.25	12.6	**8.03**	6.89	7.58	31.6	3.77	17.9
MeshStereoExt [34]	H	9.51	3.53	6.76	18.1	5.30	5.88	**8.80**	**13.8**	8.10	11.1	8.87	8.33	10.5	31.2	4.96	12.2
MCCNN_Layout	H	9.54	3.49	7.97	14.0	3.91	4.23	12.6	15.6	4.56	12.3	14.9	12.9	7.79	24.9	5.20	17.6
NTDE [35]	H	10.1	4.54	7.00	15.7	3.97	4.37	13.3	19.3	5.12	14.4	12.1	11.7	8.35	33.5	**3.75**	17.8
MC-CNN-acrt [36]	H	10.3	3.33	8.04	16.1	**3.66**	**3.76**	12.5	18.5	**4.22**	14.6	15.1	13.3	**6.92**	30.5	4.65	24.8
LPU	H	10.4	3.17	6.83	11.5	5.8	6.35	13.5	26	7.4	15.3	9.63	**6.48**	10.7	35.9	4.19	21.6
MC-CNN+RBS [37]	H	10.9	3.85	10	18.6	4.17	4.31	12.6	17.6	7.33	14.8	15.6	13.3	7.32	30.1	5.02	22.2
SGM [26]	F	22.1	28.4	6.52	20.1	13.9	11.7	19.7	33.2	15.5	30	58.3	18.5	23.8	49.5	7.38	49.9
TSGO [38]	F	31.3	27.3	12.3	53.1	23.5	25.7	33.4	54.5	22.5	49.6	45	27	24.2	52.2	13.3	57.5
Our method	Q	33.9	36.5	20.6	35.7	27.6	30.5	38.8	59	26.6	46.8	56.9	31.8	29.6	53.3	12.2	52.8

Table 3. Quantitative evaluation based on the test set of the 2014 Middlebury stereo datasets at 2 Error Threshold. The best results for each test column are highlighted in bold. Austr, AustrP, Bicyc2, Class, ClassE, Compu, Crusa, CrusaP, Djemb, DjembL, Hoops, Livgrm, Nkuba, Plants and Stairs are the names of experimental data in the test set.

Name	Res	Avg	Austr	AustrP	Bicyc2	Class	ClassE	Compu	Crusa	CrusaP	Djemb	DjembL	Hoops	Livgrm	Nkuba	Plants	Stairs
PMSC [33]	H	**6.87**	**3.46**	**2.68**	6.19	**2.54**	**6.92**	**6.54**	**3.96**	**4.04**	**2.37**	13.1	**12.3**	12.2	16.2	5.88	10.8
MeshStereoExt [34]	H	7.29	4.41	3.98	5.4	3.17	10	8.89	4.62	4.77	3.49	**12.7**	12.4	10.4	14.5	7.8	8.85
APAP-Stereo [32]	H	7.46	5.43	4.91	5.11	5.17	21.6	9.5	4.31	4.23	3.24	14.3	9.78	**7.32**	13.4	6.3	**8.46**
NTDE [35]	H	7.62	5.72	4.36	5.92	2.83	10.4	8.02	5.3	5.54	2.4	13.5	14.1	12.6	13.9	6.39	12.2
MC-CNN-acrt [36]	H	8.29	5.59	4.55	5.96	2.83	11.4	8.44	8.32	8.89	2.71	16.3	14.1	13.2	**13**	6.4	11.1
MC-CNN+RBS [37]	H	8.62	6.05	5.16	6.24	3.27	11.1	8.91	8.87	9.83	3.21	15.1	15.9	12.8	13.5	7.04	9.99
MCCNN_Layout	H	9.16	5.53	5.63	**5.06**	3.59	12.6	9.97	7.53	8.86	5.79	23	13.6	15	14.7	**5.85**	10.4
LPU	H	10.5	11.4	3.18	8.1	6.08	20.9	9.84	6.94	4	4.04	33.9	16.9	15.2	17.8	9.12	11.6
SGM [26]	F	25.3	45.1	4.33	6.87	32.2	50	13	48.1	18.3	7.66	29.6	36.1	31.2	24.2	24.5	50.2
Our method	Q	38.7	40.4	20.3	27.3	35.1	55.9	22.3	56.1	50.9	24.2	58	56.3	36.5	32.1	38.7	69.7
TSGO [38]	F	39.1	34.1	16.9	20	43.3	55.4	14.3	54.1	49.2	33.9	66.2	45.9	39.8	42.6	47.2	52.6

Tables 2 and 3 show the performance evaluation on the training set and test set of the Middlebury stereo datasets from 2014. Error rates in the table are calculated by setting the threshold value to a two-pixel disparity. The best results for each test column are highlighted in bold. In Table 2, APAP-Stereo [32], PMSC [33], MeshStereoExt [34], NTDE [35], MC-CNN-acrt [36], and MC-CNN+RBS [37] are the state-of-the-art stereo matching methods of Middlebury Stereo Evaluation Version 3, and the MCCNN_Layout and LPU methods are anonymously published. SGM is a classical algorithm based on semi-global matching and mutual information [26]. TSGO is an accurate global stereo matching algorithm based on energy minimization [38]. The results of the 2014 Middlebury stereo datasets show that our method is comparable to these excellent algorithms. Some disparity maps of these stereo pairs are presented in Figure 10. The reference images and ground truth maps are shown in Figure 10a,b, respectively; the final disparity maps are given in Figure 10c; and the bad matching pixels are marked in Figure 10d, where a disparity absolute difference greater than 2 is counted as error. Figure 10d indicates that our proposed approach has excellent performance, especially in textureless regions, disparity discontinuous boundaries, and occluded regions.

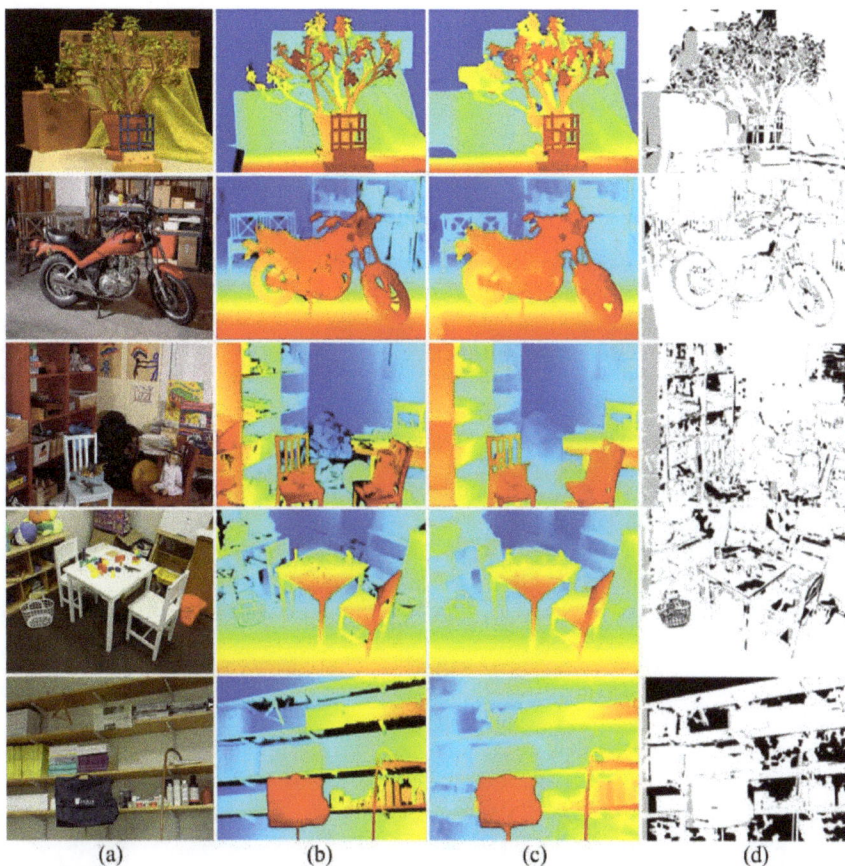

| (a) | (b) | (c) | (d) |

Figure 10. Results of Middlebury stereo datasets "Jade plant", "Motorcycle", "Playroom", "Play table" and "Shelves" (from top to bottom). (**a**) Reference images; (**b**) ground truth images; (**c**) results of the proposed method; and (**d**) error maps (bad estimates with absolute disparity error >2.0 are marked in black).

In order to verify the effect and importance of the four similarity functions during the minimization stage, different combinations of similarity functions and different fitting algorithms are utilized to evaluate the 2014 Middlebury stereo datasets. The statistical results are shown in Figure 11. Firstly, the initial disparity is estimated by ICT, ICT+TADc, ICT+TADc+TADg, and ICT+TADc+TADg+INV, respectively. Secondly, the disparity plane fitting for initial disparity is performed by RANSAC and our fitting algorithm, respectively. Finally, the corresponding disparity plane is optimized by BP. The effect and importance of the four similarity functions during the disparity plane fitting stage and minimization stage can be observed through the histogram. The results illustrate that the most accurate initial disparity can be estimated by ICT+TADc+TADg+INV, and the most accurate final disparity map can be obtained by optimizing the most accurate initial disparity.

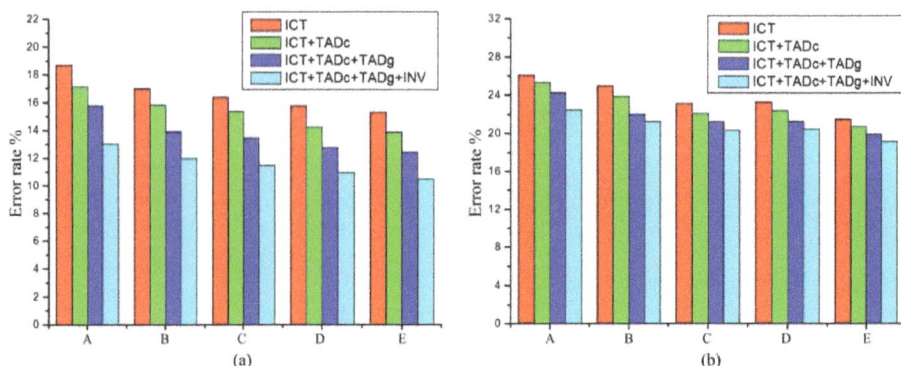

Figure 11. The statistical results of different combinations of similarity functions and different fitting algorithms. (**a**) The average error rates of the non-occlusion areas (nonocc) and (**b**) of the complete image (all). A: initial disparity; B: disparity plane fitting by RANSAC; C: disparity plane optimization of B; D: disparity plane fitting by our iterative outlier suppression and disparity plane parameters fitting algorithm; E: disparity plane optimization of D.

The degree to which each step of the algorithm contributes to the reduction of the disparity error with respect to ground truth is shown in Figure 12. The disparity results are evaluated on the 2014 Middlebury stereo datasets. The charts in Figure 12 present the improvement obtained at each step for the 0.5, 1.0, 2.0, and 4.0 pixel thresholds. The errors refer to non-occlusion areas (nonocc) and to the whole image (all). The contribution of each step to disparity improvement is seen at the nonocc and all curves. One may observe that the error rate is reduced by adding the algorithm steps.

The results of some representative data of the Middlebury stereo data are presented in Figure 13. They are: Moebius and Laundry choose from 2005 Middlebury stereo datasets [4]; Bowling 2 and Plastic choose from 2006 Middlebury stereo datasets [30]. These stereo pairs are captured by high-end cameras in a controlled laboratory environment. The produced disparity maps are accurate, and the error rates of the four Middlebury stereo data with reference to the whole image are given as follows: Moebius, 8.28%; Laundry, 12.65%; Bowling 2, 8.5%; and Plastic, 13.49%. Data Moebius presents an indoor scene with many stacking objects. Our method can generate accurate disparity for most parts of the scene, and the disparity of small toys on the floor is correctly recovered. For data Laundry, a relatively good disparity map is generated for a laundry basket with repeated textures. In Bowling 2, objects with curved surfaces are presented, e.g., ball and Bowling. Disparities of these objects are both accurate and smooth. The disparity of the background (map) is also obtained with few mismatches. In Plastic, the texture information is much weaker; nevertheless, our method can still generate an accurate and smooth disparity map that is close to the ground truth. These examples demonstrate the ability of our approach to produce promising results.

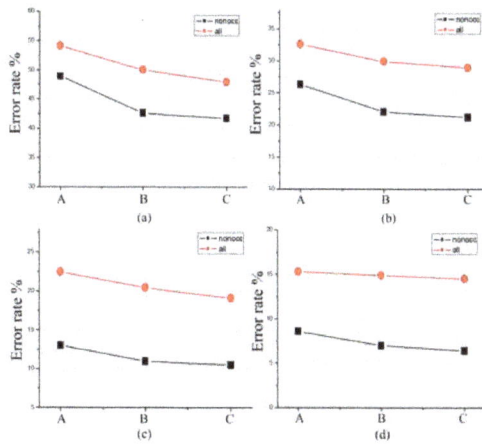

Figure 12. Performance of each step of the algorithm regarding disparity map accuracy. (**a**) The average error rates of the complete image (all) and non-occlusion areas (nonocc) for the 0.75 pixel threshold; (**b**) the average error rates of the complete image (all) and non-occlusion areas (nonocc) for the 1.0 pixel threshold; (**c**) the average error rates of the complete image (all) and non-occlusion areas (nonocc) for the 2.0 pixel threshold; and (**d**) the average error rates of the complete image (all) and non-occlusion areas (nonocc) for the 4.0 pixel threshold. A: initial disparity estimation, B: disparity plane fitting, and C: disparity plane optimization.

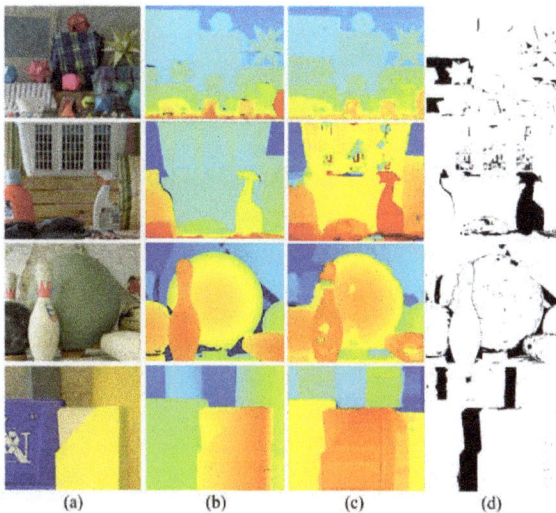

Figure 13. Results of representative data on the Middlebury website. From top to bottom: Moebius, Laundry, Bowling 2 and Plastic. (**a**) Reference image; (**b**) ground truth images; (**c**) results of the proposed method; and (**d**) error maps (bad estimates with absolute disparity error >1.0 are marked in black).

Apart from the Middlebury benchmark images, we also tested the proposed method on both synthesized [31] and real-world stereo pairs. Figure 14 presents the results of the proposed algorithm on three synthesized stereo pairs: Tanks, Temple, and Street. High-quality disparity maps are generated and compared with the ground truth in the second column. The produced disparity maps are accurate, and the error rates of the three synthesized stereo pairs with reference to the whole image are given as follows: Tanks, 4.42%; Temple, 2.66%; and Street, 7.93%. It is clear that our algorithm performs well for details, e.g., the gun barrels of the tanks, as well as for large textureless background regions and repetitive patterns.

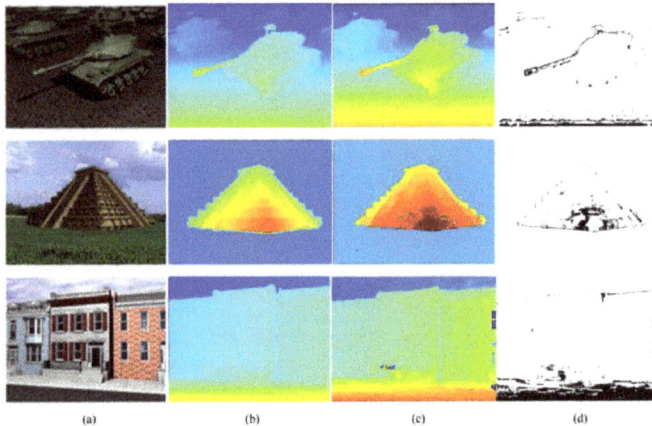

Figure 14. Results of synthesized stereo pairs. From top to bottom: Tanks, Temple, and Street. (a) Reference image; (b) ground truth images; (c) results of the proposed method; and (d) error maps (bad estimates with absolute disparity error >1.0 are marked in black).

The proposed algorithm also performs well on publicly available, real-world stereo video datasets: a "Book Arrival" sequence from FhG-HHI database and an "Ilkay" sequence from Microsoft i2i database. The snapshots for the two video sequences and corresponding disparity maps are presented in Figure 15. For both examples, our system performs reasonably well. In this experiment, we did not give the error maps and the error rate, due to there being no ground truth disparity map for the real-world stereo video datasets. However, in terms of the visual effect, the proposed algorithm can be applied to this dataset very well.

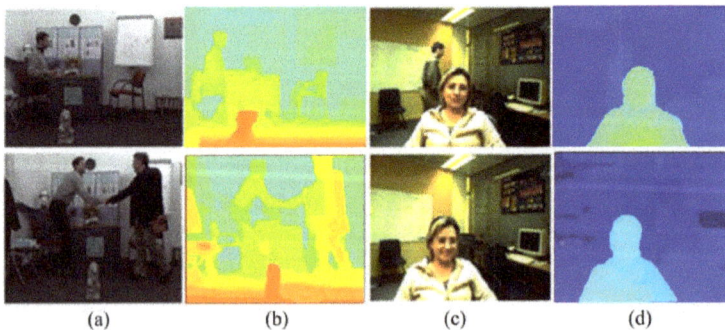

Figure 15. Results of real-world stereo data. (a) Frames of "Book Arrival" stereo video sequence; (b) estimated disparity maps of "Book Arrival"; (c) frames of "Ilkay" stereo video sequence; and (d) estimated disparity maps of "Ilkay".

Runtime. The algorithm implementation is written in VS2010 and uses the OpenCV core library for basic matrix operations. The runtime is measured on a desktop with Core i7-6700HQ 2.60 GHz CPU and 16 GB RAM, and no parallelism technique is utilized. All operations are carried out with floating point precision. Our algorithm require 0.59 s/megapixels (s/mp) for image segmentation, 3.4 s/mp for initial disparity estimation, 15.8 s/mp for disparity plane fitting, and 7.6 s/mp for disparity plane optimization.

5. Discussion and Conclusions

In summary, we present a highly accurate solution to the stereo correspondence problem. The main contribution of this work is to integrate the appealing properties of the multi-cost approach into the segmentation-based framework. Our algorithm has two advantages. Firstly, an adaptive multi-cost method for disparity evaluation is designed to ensure the accuracy of the initial disparity map. The combined similarity function increases the robustness of the initial raw matching costs calculation and the adaptive support window effectively reduces the matching ambiguity. Secondly, an iterative outlier suppression and disparity plane parameters fitting algorithm is developed to ensure a reliable pixel set for each segment region and obtain accurate disparity plane parameters. The ability to deal with textureless areas and occlusion is enhanced by segment constraint. The experimental results demonstrated that the proposed algorithm can generate state-of-the-art disparity results. The ideas introduced in this paper could be used or extended in future stereo algorithms in order to boost their accuracy.

Acknowledgments: This work was supported by the National Natural Science Foundation of China (Grant No. 11547157 and 61100004), the Natural Science Foundation of Heilongjiang Province of China (Grant No. F201320), and Harbin Municipal Science and Technology Bureau (Grant No. 2014RFQXJ073). Sincere thanks are given for the comments and contributions of anonymous reviewers and members of the editorial team.

Author Contributions: Ning Ma proposed the idea for the method. Ning Ma and Yubo Men performed the validation experiment. Ning Ma, Chaoguang Men, and Xiang Li wrote the paper. All the authors read and approved the final paper.

Conflicts of Interest: The authors declare no conflict of interest.

References

1. Scharstein, D.; Szeliski, R. A taxonomy and evaluation of dense two-frame stereo correspondence algorithms. *Int. J. Comput. Vis.* **2002**, *47*, 7–42. [CrossRef]
2. Pham, C.C.; Jeon, J.W. Domain transformation-based efficient cost aggregation for local stereo matching. *IEEE Trans. Circuits Syst. Video Technol.* **2013**, *23*, 1119–1130. [CrossRef]
3. Wang, D.; Lim, K.B. Obtaining depth map from segment-based stereo matching using graph cuts. *J. Vis. Commun. Image Represent.* **2011**, *22*, 325–331. [CrossRef]
4. Scharstein, D.; Pal, C. Learning conditional random fields for stereo. In Proceedings of the IEEE Conference on Computer Vision and Pattern Recognition, Minneapolis, MN, USA, 17–22 June 2007; pp. 1–8.
5. Scharstein, D.; Szeliski, R. High-accuracy stereo depth maps using structured light. In Proceedings of the Computer Vision and Pattern Recognition, Madison, WI, USA, 18–20 June 2003.
6. Shan, Y.; Hao, Y.; Wang, W.; Wang, Y.; Chen, X.; Yang, H.; Luk, W. Hardware acceleration for an accurate stereo vision system using mini-census adaptive support region. *ACM Trans. Embed. Comput. Syst.* **2014**, *13*, 132. [CrossRef]
7. Stentoumis, C.; Grammatikopoulos, L.; Kalisperakis, I.; Karras, G. On accurate dense stereo-matching using a local adaptive multi-cost approach. *ISPRS J. Photogramm. Remote Sens.* **2014**, *91*, 29–49. [CrossRef]
8. Miron, A.; Ainouz, S.; Rogozan, A.; Bensrhair, A. A robust cost function for stereo matching of road scenes. *Pattern Recognit. Lett.* **2014**, *38*, 70–77. [CrossRef]
9. Saygili, G.; van der Maaten, L.; Hendriks, E.A. Adaptive stereo similarity fusion using confidence measures. *Comput. Vis. Image Underst.* **2015**, *135*, 95–108. [CrossRef]
10. Stentoumis, C.; Grammatikopoulos, L.; Kalisperakis, I.; Karras, G.; Petsa, E. Stereo matching based on census transformation of image gradients. In Proceedings of the SPIE Optical Metrology, International Society for Optics and Photonics, Munich, Germany, 21 June 2015.
11. Klaus, A.; Sormann, M.; Karner, K. Segment-based stereo matching using belief propagation and a self-adapting dissimilarity measure. In Proceedings of the IEEE 18th International Conference on Pattern Recognition (ICPR'06), Hong Kong, China, 20–24 August 2006; pp. 15–18.

12. Kordelas, G.A.; Alexiadis, D.S.; Daras, P.; Izquierdo, E. Enhanced disparity estimation in stereo images. *Image Vis. Comput.* **2015**, *35*, 31–49. [CrossRef]

13. Wang, Z.F.; Zheng, Z.G. A region based stereo matching algorithm using cooperative optimization. In Proceedings of the IEEE Conference on Computer Vision and Pattern Recognition, Anchorage, AK, USA, 23–28 June 2008; pp. 1–8.

14. Xu, S.; Zhang, F.; He, X.; Zhang, X. PM-PM: PatchMatch with Potts model for object segmentation and stereo matching. *IEEE Trans. Image Process.* **2015**, *24*, 2182–2196. [PubMed]

15. Taguchi, Y.; Wilburn, B.; Zitnick, C.L. Stereo reconstruction with mixed pixels using adaptive over-segmentation. In Proceedings of the IEEE Conference on Computer Vision and Pattern Recognition, Anchorage, AK, USA, 23–28 June 2008; pp. 1–8.

16. Damjanović, S.; van der Heijden, F.; Spreeuwers, L.J. Local stereo matching using adaptive local segmentation. *ISRN Mach. Vis.* **2012**, *2012*, 163285. [CrossRef]

17. Yoon, K.J.; Kweon, I.S. Adaptive support-weight approach for correspondence search. *IEEE Trans. Pattern Anal.* **2006**, *28*, 650–656. [CrossRef] [PubMed]

18. Zhang, K.; Lu, J.; Lafruit, G. Cross-based local stereo matching using orthogonal integral images. *IEEE Trans. Circuits Syst. Video Technol.* **2009**, *19*, 1073–1079. [CrossRef]

19. Veksler, O. Fast variable window for stereo correspondence using integral images. In Proceedings of the 2003 IEEE Computer Society Conference on Computer Vision and Pattern Recognition, Madison, WI, USA, 18–20 June 2003.

20. Okutomi, M.; Katayama, Y.; Oka, S. A simple stereo algorithm to recover precise object boundaries and smooth surfaces. *Int. J. Comput. Vis.* **2002**, *47*, 261–273. [CrossRef]

21. Hosni, A.; Rhemann, C.; Bleyer, M.; Rother, C.; Gelautz, M. Fast cost-volume filtering for visual correspondence and beyond. *IEEE Trans. Pattern Anal.* **2013**, *35*, 504–511. [CrossRef] [PubMed]

22. Gao, K.; Chen, H.; Zhao, Y.; Geng, Y.N.; Wang, G. Stereo matching algorithm based on illumination normal similarity and adaptive support weight. *Opt. Eng.* **2013**, *52*, 027201. [CrossRef]

23. Wang, L.; Yang, R.; Gong, M.; Liao, M. Real-time stereo using approximated joint bilateral filtering and dynamic programming. *J. Real-Time Image Process.* **2014**, *9*, 447–461. [CrossRef]

24. Taniai, T.; Matsushita, Y.; Naemura, T. Graph cut based continuous stereo matching using locally shared labels. In Proceedings of the IEEE Conference on Computer Vision and Pattern Recognition, Columbus, OH, USA, 23–28 June 2014; pp. 1613–1620.

25. Besse, F.; Rother, C.; Fitzgibbon, A.; Kautz, J. Pmbp: Patchmatch belief propagation for correspondence field estimation. *Int. J. Comput. Vis.* **2014**, *110*, 2–13. [CrossRef]

26. Hirschmuller, H. Stereo Processing by Semiglobal Matching and Mutual Information. *IEEE Trans. Pattern Anal.* **2008**, *30*, 328–341. [CrossRef] [PubMed]

27. Comaniciu, D.; Meer, P. Mean shift: A robust approach toward feature space analysis. *IEEE Trans. Pattern Anal.* **2002**, *24*, 603–619. [CrossRef]

28. Yao, L.; Li, D.; Zhang, J.; Wang, L.H.; Zhang, M. Accurate real-time stereo correspondence using intra-and inter-scanline optimization. *J. Zhejiang Univ. Sci. C* **2012**, *13*, 472–482. [CrossRef]

29. Scharstein, D.; Hirschmüller, H.; Kitajima, Y.; Krathwohl, G.; Nešić, N.; Wang, X.; Westling, P. High-resolution stereo datasets with subpixel-accurate ground truth. In Proceedings of the German Conference on Pattern Recognition, Münster, Germany, 2–5 September 2014; pp. 31–42.

30. Hirschmuller, H.; Scharstein, D. Evaluation of cost functions for stereo matching. In Proceedings of the IEEE Conference on Computer Vision and Pattern Recognition, Minneapolis, MN, USA, 17–22 June 2007; pp. 1–8.

31. Richardt, C.; Orr, D.; Davies, I.; Criminisi, A.; Dodgson, N.A. Real-time spatiotemporal stereo matching using the dual-cross-bilateral grid. In Proceedings of the European Conference on Computer Vision, Crete, Greece, 5–11 September 2010; pp. 510–523.

32. Park, M.G.; Yoon, K.J. As-planar-as-possible depth map estimation. *IEEE Trans. Pattern Anal.* **2016**, submitted.

33. Li, L.; Zhang, S.; Yu, X.; Zhang, L. PMSC: PatchMatch-based superpixel cut for accurate stereo matching. *IEEE T. Circ. Syst. Vid.* **2016**. [CrossRef]

34. Zhang, C.; Li, Z.; Cheng, Y.; Cai, R.; Chao, H.; Rui, Y. Meshstereo: A global stereo model with mesh alignment regularization for view interpolation. In Proceedings of the IEEE International Conference on Computer Vision, Santiago, Chile, 13–16 December 2015; pp. 2057–2065.

35. Kim, K.R.; Kim, C.S. Adaptive smoothness constraints for efficient stereo matching using texture and edge information. In Proceedings of the Image Processing (ICIP), Phoenix, AZ, USA, 25–28 September 2016; pp. 3429–3433.

36. Zbontar, J.; LeCun, Y. Stereo matching by training a convolutional neural network to compare image patches. *J. Mach. Learn. Res.* **2016**, *17*, 1–32.

37. Barron, J.T.; Poole, B. The Fast Bilateral Solver. In Proceedings of the European Conference on Computer Vision, Amsterdam, The Netherlands, 11–14 October 2016; pp. 617–632.

38. Mozerov, M.G.; van de Weijer, J. Accurate stereo matching by two-step energy minimization. *IEEE Trans. Image Process.* **2015**, *24*, 1153–1163. [CrossRef] [PubMed]

![symmetry logo] *symmetry*

MDPI

Article

First and Second Zagreb Eccentricity Indices of Thorny Graphs

Nazeran Idrees [1,*], Muhammad Jawwad Saif [2], Asia Rauf [3] and Saba Mustafa [1]

[1] Department of Mathematics, Government College University Faisalabad, 38000 Faisalabad, Pakistan; sabamustafa48@gmail.com

[2] Department of Applied Chemistry, Government College University Faisalabad, 38000 Faisalabad, Pakistan; jawwadsaif@gmail.com

[3] Department of Mathematics, Government College Women University Faisalabad, 38000 Faisalabad, Pakistan; Asia.rauf@gmail.com

* Correspondence: nazeranjawwad@gmail.com

Academic Editor: Angel Garrido

Received: 21 November 2016; Accepted: 22 December 2016; Published: 6 January 2017

Abstract: The Zagreb eccentricity indices are the eccentricity reformulation of the Zagreb indices. Let H be a simple graph. The first Zagreb eccentricity index $(E_1(H))$ is defined to be the summation of squares of the eccentricity of vertices, i.e., $E_1(H) = \sum_{u \in V(H)} \varepsilon_H^2(u)$. The second Zagreb eccentricity index $(E_2(H))$ is the summation of product of the eccentricities of the adjacent vertices, i.e., $E_2(H) = \sum_{uv \in E(H)} \varepsilon_H(u)\varepsilon_H(v)$. We obtain the thorny graph of a graph H by attaching thorns i.e., vertices of degree one to every vertex of H. In this paper, we will find closed formulation for the first Zagreb eccentricity index and second Zagreb eccentricity index of different well known classes of thorny graphs.

Keywords: graphs; vertices; complete graph; path; star; cycle

1. Introduction

In theoretical chemistry, molecular descriptors or topological indices are utilized to configure properties of chemical compounds. A topological index is a real number connected with chemical structure indicating relationships of chemical configuration with different physical properties, chemical reactivity or biological activity, which is utilized to understand properties of chemical compounds in theoretical chemistry. Topological indices have been observed to be helpful in chemical documentation, isomer discrimination, structure-property relations, structure-activity (SAR) relations and pharmaceutical medication plans. All through the paper, all graphs are considered to be simple and connected.

Let $H = (V, E)$ be a simple graph with $m = |V|$ vertices and $n = |E|$ edges. For $u \in V$, degree of u, denoted by $d(u)$, is number of vertices attached to u in the graph. The maximum distance from a vertex to any other vertex in the graph H is called eccentricity of the vertex and is denoted by $\varepsilon_H(u)$ i.e., $\varepsilon_H(u) = \max\{d(u,v)|v \in V\}$, where $d(u,v)$ denotes the distance between u and v in H. The first Zagreb index (M_1) and second Zagreb index (M_2) are the oldest known indices introduced by Gutman and Trinajstić [1] defined as

$$M_1 = M_1(H) = \sum_{u \in V(H)} d_u^2,$$

$$M_2 = M_2(H) = \sum_{uv \in E(H)} d_u d_v.$$

Several topological indices depend upon the eccentricity of the vertices and are very effective in drug design. Sharma, Goswami and Madan [2] proposed the eccentric connectivity index of the graph H, which is defined as

$$C^{\xi}(H) = \sum_{w \in V(H)} \frac{d_H(w)}{\varepsilon_H(w)}.$$

In 2000, Gupta, Singh and Madan [3] introduced another distance-cum-degree based topological descriptor termed the connective eccentricity index:

$$\xi^C(H) = \sum_{w \in V(H)} d_H(w)\varepsilon_H(w).$$

Other eccentricity related indices include the eccentric distance sum [4], augmented and super augmented eccentric connectivity indices [5–7], and adjacent eccentric distance sum index [8,9].

Recently, the first Zagreb eccentricity index and second Zagreb eccentricity index E_1 and E_2 have been proposed as the revised versions of the Zagreb indices M_1 and M_2, respectively, by Ghorbani and Hosseinzadeh [10]. The first Zagreb eccentricity index (E_1) and the second Zagreb eccentricity index (E_2) of a graph H are defined as

$$E_1 = E_1(H) = \sum_{u \in V(H)} \varepsilon_H^2(u),$$

$$E_2 = E_2(H) = \sum_{uv \in E(H)} \varepsilon_H(u)\varepsilon_H(v),$$

respectively. Das et al. [11] gave a few lower and upper bounds on the first Zagreb eccentricity index and the second Zagreb eccentricity index of trees and graphs, and also characterized the extremal graphs. Nilanjan [12] computed a few new lower and upper bounds on the first Zagreb eccentricity index and the second Zagreb eccentricity index. Zhaoyang and Jianliang [13] computed Zagreb eccentricity indices under different graph operations. Farahani [14] computed precise equations for the First Zagreb Eccentricity index of Polycyclic Aromatic Hydrocarbons. Evidently, Zagreb indices and the family of all connectivity indices express mathematically attractive invariants. In this manner, we expect numerous more studies on these indices and anticipate further development of this area of mathematical chemistry.

2. Results and Discussion

Consider a graph H with vertex set $\{u_1, u_2, \ldots, u_m\}$ and a set of positive integers $\{p_1, p_2, \ldots, p_m\}$. The thorn graph of H, denoted by $H^*(p_1, p_2, \ldots, p_n)$, is obtained by attaching p_j pendant vertices to u_j for each j. The idea of a thorn graph was presented by Gutman [15], and various studies on thorn graphs and different topological indices have been conducted by some researchers in the recent past [16–19]. In this paper, we will derive explicit expressions for computing the first Zagreb eccentricity index and the second Zagreb eccentricity index of thorny graphs of some well-known classes of graphs like complete graphs, complete bipartite graphs, star graphs, cycles and paths.

2.1. The Thorny Complete Graph

Suppose that we take the complete graph K_m with m vertices. Obviously, $E_1(K_m) = m$ and $E_2(K_m) = \frac{m(m-1)}{2}$. The thorny complete graph K_m^* is obtained from K_m by attaching p_j thorns at each vertex of K_m, $j = 1, 2, \ldots, m$. Suppose that the total number of thorns attached to K_m are denoted by T.

Theorem 1. *The first Zagreb eccentricity index and the second Zagreb eccentricity index of K_m^* are given by:*

$$E_1(K_m^*) = 4E_1(K_m) + 9T \text{ and } E_2(K_m^*) = 4E_2(K_m) + 6T, \text{ respectively.}$$

Proof. Let K_m be a compete graph. Suppose that v_j, $j = 1, 2, \ldots, m$ are the vertices of K_m, and v_{jk}, $j = 1, 2, \ldots, m$; $k = 1, 2, \ldots, p_j$ are the newly attached pendant vertices. Then, $\varepsilon_{K_m^*}(v_j) = 2$, $\varepsilon_{K_m^*}(v_{jk}) = 3$ for $j = 1, 2, \ldots, m$; $k = 1, 2, \ldots, p_j$ are the eccentricities of the vertices of K_m^*. Thus, the first Zagreb eccentricity index and the second Zagreb eccentricity index of K_m^* are given by

$$
E_1(K_m^*) = \sum_{j=1}^{m} \varepsilon_{K_m^*}^2(v_j) + \sum_{j=1}^{m} \sum_{k=1}^{p_j} \varepsilon_{K_m^*}^2(v_{jk})
$$

$$
= \sum_{j=1}^{m} (2)^2 + \sum_{j=1}^{m} \sum_{k=1}^{p_j} (3)^2
$$

$$
= 4m + 9T = 4E_1(K_m) + 9T, \text{ and}
$$

$$
M_2^*(K_m^*) = \sum_{v_j v_k \in E(K_m^*)} \varepsilon_{K_m^*}(v_j)\varepsilon_{K_m^*}(v_k) + \sum_{k=1}^{p_j} \sum_{v_j v_{jk} \in E(K_m^*)} \varepsilon_{K_m^*}(v_j)\varepsilon_{K_m^*}(v_{jk})
$$

$$
= \sum_{v_j v_k \in E(K_m^*)} 4 + \sum_{k=1}^{p_j} \sum_{v_j v_{jk} \in E(K_m^*)} 6
$$

$$
= 4\frac{m(m-1)}{2} + 6T
$$

$$
= 4E_2(K_m^*) + 6T.
$$

2.2. The Thorny Complete Bipartite Graph

Assume that we take the complete bipartite graph $K_{n,m}$ having $(n + m)$ vertices. Obviously, the eccentricities are equal to two for all the vertices of $K_{n,m}$. Then, $E_1(K_{n,m}) = 4(n + m)$ and $E_2(K_{n,m}) = 4nm$. The thorny complete bipartite graph $K_{n,m}^*$ is attained by attaching pendant vertices to each vertex of $K_{n,m}$. Let T be the total number of pendant vertices.

Theorem 2. *The first Zagreb eccentricity index and the second Zagreb eccentricity index of $K_{n,m}^*$ are given by:*

$$
E_1(K_{n,m}^*) = 9(n + m) + 16T \text{ and } E_2(K_{n,m}^*) = 9nm + 12T, \text{ respectively.}
$$

Proof. Suppose that $\{v_1, v_2, \ldots, v_n, u_1, u_2, \ldots, u_m\}$ is the vertex set of $K_{n,m}$, and let v_{ik} be the newly attached pendant vertices to v_i, $i = 1, 2, \ldots, n$; $k = 1, 2, \ldots, p_i$ and u_{jl} be the pendant vertices of u_j, $j = 1, 2, \ldots, m$; $k = 1, 2, \ldots, p_l$. Then, the eccentricity of the vertices of $K_{n,m}^*$ is given by $\varepsilon_{K_{n,m}^*}(v_i) = 3$, $\varepsilon_{K_{n,m}^*}(v_{ik}) = 4$, for $i = 1, 2, \ldots n$; $k = 1, 2, \ldots, p_i$ and $\varepsilon_{K_{n,m}^*}(u_j) = 3$, $\varepsilon_{K_{n,m}^*}(u_{jl}) = 4$, for $j = 1, 2, \ldots, m$; $l = 1, 2, \ldots, p_j'$. Thus, the Zagreb eccentricity indices of $K_{n,m}^*$ are given by:

$$
E_1(K_{n,m}^*) = \sum_{i=1}^{n} \varepsilon_{K_{n,m}^*}^2(v_i) + \sum_{j=1}^{m} \varepsilon_{K_{n,m}^*}^2(u_j) + \sum_{i=1}^{n} \sum_{k=1}^{p_i} \varepsilon_{K_{n,m}^*}^2(v_{ik}) + \sum_{j=1}^{m} \sum_{l=1}^{p_j} \varepsilon_{K_{n,m}^*}^2(u_{jl})
$$

$$
= \sum_{i=1}^{n} (3)^2 + \sum_{j=1}^{m} (3)^2 + \sum_{i=1}^{n} \sum_{k=1}^{p_i} (4)^2 + \sum_{j=1}^{m} \sum_{l=1}^{p_j} (4)^2
$$

$$
= 9n + 9m + 16 \sum_{i=1}^{n} p_i + 16 \sum_{j=1}^{m} p_j
$$

$$
= 9(n + m) + 16\left(\sum_{i=1}^{n} p_i + \sum_{j=1}^{m} p_j \right).
$$

The second Zagreb eccentricity index is computed as:

$$E_2(K^*_{n,m}) = \sum_{u_i v_j \in E(K^*_{n,m})} \varepsilon_{K^*_{n,m}}(u_i)\varepsilon_{K^*_{n,m}}(v_j) + \sum_{i=1}^{n}\sum_{k=1}^{p_i} \varepsilon_{K^*_{n,m}}(v_i)\varepsilon_{K^*_{n,m}}(v_{ik}) + \sum_{j=1}^{m}\sum_{l=1}^{p_j} \varepsilon_{K^*_{n,m}}(u_j)\varepsilon_{K^*_{n,m}}(u_{jl})$$

$$= \sum_{u_i v_j \in E(K^*_{n,m})} 9 + \sum_{i=1}^{n}\sum_{k=1}^{p_i} 12 + \sum_{j=1}^{m}\sum_{l=1}^{p_j} 12$$

$$= 9|E(K_{n,m})| + 12\sum_{i=1}^{n} p_i + 12\sum_{j=1}^{m} p_j.$$

2.3. The Thorny Star Graph

Suppose that we have the star graph $S_m = K_{1,(m-1)}$ of m vertices. Obviously, $E_1(S_m) = 4m - 3$ and $E_2(S_m) = 2(m-1)$. Let the thorny star graph S^*_m be obtained by joining p_j pendant vertices to every vertex v_j, $j = 2, 3, \ldots, m$ and p_1 pendant vertices to the central vertex v_1 of S_m.

Theorem 3. *The first Zagreb eccentricity index and the second Zagreb eccentricity index of S^*_m are given by:*

$$E_1(S^*_m) = 9m - 5 + 16T - 7p_1 \text{ and } E_2(S^*_m) = 6(m-1) - 6p_1 + 12T, \text{ respectively.}$$

Proof. Assume v_{1k}, $k = 1, 2, \ldots, p_1$ and v_{jk}, for $j = 2, 3, \ldots, m$; $k = 1, 2, \ldots, p_j$ are the newly attached pendant vertices. Then, the eccentricities of the vertices of S^*_m are given by $\varepsilon_{S^*_m}(v_1) = 2$, $\varepsilon_{S^*_m}(v_j) = 3$, for $j = 2, 3, \ldots, m$, $\varepsilon_{S^*_m}\left(v_{jk}\right) = 4$, for $j = 2, 3, \ldots, m$; $k = 1, 2, \ldots, p_j$, $\varepsilon_{S^*_m}(v_{1k}) = 3$, for $k = 1, 2, \ldots, p_1$. Thus, the Zagreb eccentricity indices of S^*_m are

$$E_1(S^*_m) = \sum_{j=1}^{m} \varepsilon^2_{S^*_m}(v_j) + \sum_{j=1}^{m}\sum_{k=1}^{p_j} \varepsilon^2_{S^*_m}(v_{jk})$$

$$= \varepsilon^2_{S^*_m}(v_1) + \sum_{j=2}^{m} \varepsilon^2_{S^*_m}(v_j) + \sum_{k=1}^{p_1} \varepsilon^2_{S^*_m}(v_{1k}) + \sum_{j=2}^{m}\sum_{k=1}^{p_j} \varepsilon^2_{S^*_m}(v_{jk})$$

$$= (2)^2 + \sum_{j=2}^{m} (3)^2 + \sum_{k=1}^{p_1} (3)^2 + \sum_{j=2}^{m}\sum_{k=1}^{p_j} \varepsilon^2_{S^*_m}(4)^2$$

$$= 4 + 9(m-1) + 9p_1 + 16\sum_{j=2}^{m} p_j$$

$$= 9m - 5 + 9p_1 + 16\sum_{j=1}^{m} p_j - 16p_1,$$

from which we get the desired result. Now,

$$E_2(S^*_m) = \sum_{j=2}^{m} \varepsilon_{S^*_m}(v_1)\varepsilon_{S^*_m}(v_j) + \sum_{j=1}^{m}\sum_{k=1}^{p_j} \varepsilon_{S^*_m}(v_j)\varepsilon_{S^*_m}(v_{jk})$$

$$= \sum_{j=2}^{m} 6 + \sum_{k=1}^{p_1} 6 + \sum_{j=2}^{m}\sum_{k=1}^{p_j} 12$$

$$= 6(m-1) + 6p_1 + 12\sum_{j=2}^{m} p_j$$

$$= 6(m-1) + 6p_1 + 12\sum_{j=1}^{m} p_j - 12p_1,$$

and the result follows.

2.4. The Thorny Cycle

Let C_m be a cycle having m vertices and m edges. Clearly, $E_1(C_m) = E_2(C_m) = \frac{m(m-1)^2}{4}$, if m is odd and $E_1(C_m) = E_2(C_m) = \frac{m^3}{4}$, if m is even. Let C_m^* be the thorny cycle of C_m obtained by joining p_j thorns v_{jk} to each vertex v_j, $j = 1, 2, \ldots, m$ of C_m.

Theorem 4. *The first Zagreb eccentricity index and the second Zagreb eccentricity index of C_m^* are given by*

$$E_1(C_m^*) = \begin{cases} \frac{m(m+1)^2 + T(m+3)^2}{4}, & \text{if } m \text{ is odd} \\ \frac{m(m+2)^2 + T(m+4)^2}{4}, & \text{if } m \text{ is even} \end{cases}$$

and

$$E_2(C_m^*) = \begin{cases} \frac{(m+1)[m(m+1) + (m+3)T]}{4}, & \text{if } m \text{ is odd} \\ \frac{(m+2)[m(m+2) + (m+4)T]}{4}, & \text{if } m \text{ is even} \end{cases},$$

respectively.

Proof. The vertex eccentricities of C_m^* are given as $\varepsilon_{C_m^*}(v_j) = \frac{m+1}{2}$ and $\varepsilon_{C_m^*}(v_{jk}) = \frac{m+3}{2}$, if m is odd; $\varepsilon_{C_m^*}(v_j) = \frac{m+2}{2}$ and $\varepsilon_{C_m^*}(v_{jk}) = \frac{m+4}{2}$, if m is even; for $j = 1, 2, \ldots, m$; $k = 1, 2, \ldots, p_j$.

Thus, when m is an odd number, the first Zagreb eccentricity index of C_m^* is

$$E_1(C_m^*) = \sum_{j=1}^{m} \varepsilon_{C_m^*}^2(v_j) + \sum_{j=1}^{m} \sum_{k=1}^{p_j} \varepsilon_{C_m^*}^2(v_{jk})$$

$$= \sum_{j=1}^{m} \left(\frac{m+1}{2}\right)^2 + \sum_{j=1}^{m} \sum_{k=1}^{p_j} \left(\frac{m+3}{2}\right)^2$$

$$= \frac{m(m+1)^2}{4} + \frac{(m+3)^2}{4} \sum_{j=1}^{m} p_j,$$

and the second Zagreb eccentricity index of C_m^* is

$$E_2(C_m^*) = \sum_{v_j v_k \in E(C_m^*)} \varepsilon_{C_m^*}(v_j) \varepsilon_{C_m^*}(v_k) + \sum_{j=1}^{m} \sum_{k=1}^{p_j} \varepsilon_{C_m^*}(v_j) \varepsilon_{C_m^*}(v_{jk})$$

$$= \sum_{v_j v_k \in E(C_m^*)} \left(\frac{m+1}{2}\right)^2 + \sum_{j=1}^{m} \sum_{k=1}^{p_j} \left(\frac{m+1}{2}\right)\left(\frac{m+3}{2}\right)$$

$$= \left(\frac{m+1}{2}\right)^2 |E(C_m)| + \left(\frac{m+1}{2}\right)\left(\frac{m+3}{2}\right) \sum_{j=1}^{m} p_j$$

$$= \frac{m(m+1)^2}{4} + \frac{(m+1)(m+3)}{4} T.$$

Now, when m is an even number, the first Zagreb eccentricity index of C_m^* is given by

$$E_1(C_m^*) = \sum_{j=1}^{m} \varepsilon_{C_m^*}^2(v_j) + \sum_{j=1}^{m} \sum_{k=1}^{p_j} \varepsilon_{C_m^*}^2(v_{jk})$$

$$= \sum_{j=1}^{m} \left(\frac{m+2}{2}\right)^2 + \sum_{j=1}^{m} \sum_{k=1}^{p_j} \left(\frac{m+4}{2}\right)^2$$

$$= \frac{m(m+2)^2}{4} + \frac{(m+4)^2}{4} \sum_{j=1}^{m} p_j.$$

Next, we proceed for the second Zagreb eccentricity index as

$$
\begin{aligned}
E_2(C_m^*) &= \sum_{v_j v_k \in E(C_m^*)} \varepsilon_{C_m^*}(v_j)\varepsilon_{C_m^*}(v_k) + \sum_{j=1}^{m}\sum_{k=1}^{p_j} \varepsilon_{C_m^*}(v_j)\varepsilon_{C_m^*}(v_{jk}) \\
&= \sum_{v_j v_k \in E(C_m^*)} \left(\tfrac{m+2}{2}\right)^2 + \sum_{j=1}^{m}\sum_{k=1}^{p_j} \left(\tfrac{m+2}{2}\right)\left(\tfrac{m+4}{2}\right) \\
&= \left(\tfrac{m+2}{2}\right)^2 |E(C_m)| + \left(\tfrac{m+2}{2}\right)\left(\tfrac{m+4}{2}\right)\sum_{j=1}^{m} p_j \\
&= \frac{m(m+2)^2}{4} + \frac{(m+2)(m+4)}{4} T.
\end{aligned}
$$

2.5. The Thorny Path Graph

Consider the path graph P_m with m vertices. If m is even, then we write $m = 2n + 2$, and suppose that the vertices of P_m are serially indicated by $v_n', v_{n-1}', \ldots, v_2', v_1', v_0', v_0, v_1, v_2, \ldots, v_{n-1}, v_n$, where the centers of the path P_{2n+2} are v_0' and v_0 having eccentricity $n + 1$. If m is odd, then we write $m = 2n + 1$, and we suppose that we have $v_n', v_{n-1}', \ldots, v_2', v_1', v_0, v_1, v_2, \ldots, v_{n-1}, v_n$ as the consecutive vertices of P_m, where the center of the path P_{2n+1} is v_0 having the eccentricity n. Then, the thorny path graph P_m^* is obtained from P_m by attaching p_j and p_j' pendant vertices to each v_j and v_j' ($j = 1, 2, \ldots, n$), respectively. We define $p_0' = 0$. Now, we will find the first Zagreb eccentricity index and the second Zagreb eccentricity index of P_m^*.

Theorem 5. *The first Zagreb eccentricity index and the second Zagreb eccentricity index of P_m^* are given by*

$$
E_1(P_m^*) = \begin{cases} 2\sum\limits_{j=0}^{n}(n+j+2)^2 + \sum\limits_{j=0}^{n}(p_j + p_j')(n+j+3)^2, & \text{if } m \text{ is even} \\[2ex] 2\sum\limits_{j=0}^{n}(n+j+1)^2 + \sum\limits_{j=0}^{n}(p_j + p_j')(n+j+2)^2, & \text{if } m \text{ is odd} \end{cases},
$$

and

$$
E_2(P_m^*) = \begin{cases} 2\sum\limits_{j=0}^{n-1}(n+j+2)(n+j+3) + \sum\limits_{j=0}^{n}(p_j + p_j')(n+j+2)(n+j+3)(n+2)^2, & \text{if } m \text{ is even} \\[2ex] 2\sum\limits_{j=0}^{n-1}(n+j+1)(n+j+2) + \sum\limits_{j=0}^{n}(p_j + p_j')(n+j+1)(n+j+2), & \text{if } m \text{ is odd} \end{cases},
$$

respectively.

Proof. If $m = 2n + 2$, then all the vertices of P_m^* have eccentricities $\varepsilon_{P_m^*}(v_j) = n+j+2 = \varepsilon_{P_m^*}(v_j')$, for $j = 0, 1, \ldots, n$; $\varepsilon_{P_m^*}(v_{jk}) = n+j+3 = \varepsilon_{P_m^*}(v_{jk}')$, for $j = 0, 1, \ldots, n$; $k = 1, 2, \ldots, p_j$. Thus, the Zagreb eccentricity indices of P_m^* are given by

$$
\begin{aligned}
E_1(P_m^*) &= \sum_{j=0}^{n} \varepsilon_{P_m^*}^2(v_j) + \sum_{j=0}^{n} \varepsilon_{P_m^*}^2(v_j') + \sum_{j=0}^{n}\sum_{k=1}^{p_j} \varepsilon_{P_m^*}^2(v_{jk}) + \sum_{j=0}^{n}\sum_{k=1}^{p_j'} \varepsilon_{P_m^*}^2(v_{jk}') \\
&= \sum_{j=0}^{n} 2(n+j+2)^2 + \sum_{j=0}^{n} p_j(n+j+3)^2 + \sum_{j=0}^{n} p_j'(n+j+3)^2 \\
&= 2\sum_{j=0}^{n}(n+j+2)^2 + \sum_{j=0}^{n}(p_j + p_j')(n+j+3)^2,
\end{aligned}
$$

and

$$E_2(P_m^*) = E_2'(P_m^*) + E_2''(P_m^*)$$

$$E_2'(P_m^*) = \sum_{j=0}^{n-1} \varepsilon_{P_m^*}(v_j)\varepsilon_{P_m^*}(v_{j+1}) + \varepsilon_{P_m^*}(v_0)\varepsilon_{P_m^*}(v_0') + \sum_{j=0}^{n-1} \varepsilon_{P_m^*}(v_j')\varepsilon_{P_m^*}(v_{j+1}')$$

$$= \sum_{j=0}^{n-1}(n+j+2)(n+j+3) + (n+2)^2 + \sum_{j=0}^{n-1}(n+j+2)(n+j+3)$$

$$= 2\sum_{j=0}^{n-1}(n+j+2)(n+j+3) + (n+2)^2.$$

In addition,

$$E_2''(P_m^*) = \sum_{j=0}^{n}\sum_{k=1}^{p_j} \varepsilon_{P_m^*}(v_j)\varepsilon_{P_m^*}(v_{jk}) + \sum_{j=0}^{n}\sum_{k=1}^{p_j'} \varepsilon_{P_m^*}(v_j')\varepsilon_{P_m^*}(v_{jk}')$$

$$= \sum_{j=0}^{n}\sum_{k=1}^{p_j}(n+j+2)(n+j+3) + \sum_{j=0}^{n}\sum_{k=1}^{p_j'}(n+j+2)(n+j+3)$$

$$= \sum_{j=0}^{n} p_j(n+j+2)(n+j+3) + \sum_{j=0}^{n} p_j'(n+j+2)(n+j+3)$$

$$= \sum_{j=0}^{n}(p_j + p_j')(n+j+2)(n+j+3),$$

and the result follows.

If m is odd, then the vertices of P_m^* have the eccentricities, $\varepsilon_{P_m^*}(v_j) = n+j+1 = \varepsilon_{P_m^*}(v_j')$, for $j = 0, 1, \ldots, n$; $\varepsilon_{P_m^*}(v_0) = n+1$, $\varepsilon_{P_m^*}(v_{jk}) = n+j+2 = \varepsilon_{P_m^*}(v_{jk}')$, for $j = 0, 1, \ldots, n$; $k = 1, 2, \ldots, p_j$ (the equalities do not apply for v_0' and v_{0j}'). Now, the Zagreb eccentricity indices of P_{2n+1}^* are given as

$$E_1(P_m^*) = E_1'(P_m^*) + E_1''(P_m^*)$$

$$E_1'(P_m^*) = \sum_{j=1}^{n-1} \varepsilon_{P_m^*}^2(v_j) + \varepsilon_{P_m^*}^2(v_n) + \varepsilon_{P_m^*}^2(v_0) + \sum_{j=1}^{n-1} \varepsilon_{P_m^*}^2(v_j') + \varepsilon_{P_m^*}^2(v_n')$$

$$= \sum_{j=1}^{n-1}(n+j+1)^2 + (2n+1)^2 + (n+1)^2 + \sum_{j=1}^{n-1}(n+j+1)^2 + (2n+1)^2$$

$$= 2\sum_{j=1}^{n}(n+j+1)^2 + (n+1)^2.$$

In addition,

$$E_1''(P_m^*) = \sum_{j=1}^{n}\sum_{k=1}^{p_j} \varepsilon_{P_m^*}^2(v_{jk}) + \sum_{j=1}^{n}\sum_{k=1}^{p_j'} \varepsilon_{P_m^*}^2(v_{jk}') + \sum_{k=1}^{p_0} \varepsilon_{P_m^*}^2(v_{0k})$$

$$= \sum_{j=1}^{n}\sum_{k=1}^{p_j}(n+j+2)^2 + \sum_{j=1}^{n}\sum_{k=1}^{p_j'}(n+j+2)^2 + \sum_{k=1}^{p_0}(n+2)^2$$

$$= \sum_{j=1}^{n}(p_j + p_j')(n+j+2)^2 + p_0(n+2)^2$$

$$= \sum_{j=0}^{n}(p_j + p_j')(n+j+2)^2,$$

and we get the desired result.

Now, $E_2(P_m^*) = E_2'(P_m^*) + E_2''(P_m^*)$

$$E_2'(P_m^*) = \sum_{j=0}^{n-1} \varepsilon_{P_m^*}(v_j)\varepsilon_{P_m^*}(v_{j+1}) + \varepsilon_{P_m^*}(v_0)\varepsilon_{P_m^*}(v_1') + \sum_{j=1}^{n-1} \varepsilon_{P_m^*}(v_j')\varepsilon_{P_m^*}(v_{j+1}')$$

$$= \sum_{j=1}^{n-1}(n+j+1)(n+j+2) + (n+1)(n+2) + \sum_{j=1}^{n-1}(n+j+1)(n+j+2)$$

$$= 2\sum_{j=0}^{n-1}(n+j+1)(n+j+2).$$

In addition,

$$E_2''(P_m^*) = \sum_{j=0}^{n}\sum_{k=1}^{p_j} \varepsilon_{P_m^*}(v_j)\varepsilon_{P_m^*}(v_{jk}) + \sum_{j=1}^{n}\sum_{k=1}^{p_j'} \varepsilon_{P_m^*}(v_j')\varepsilon_{P_m^*}(v_{jk}')$$

$$= \sum_{k=1}^{p_0} \varepsilon_{P_m^*}(v_0)\varepsilon_{P_m^*}(v_{0k}) + \sum_{j=1}^{n}\sum_{k=1}^{p_j} \varepsilon_{P_m^*}(v_j)\varepsilon_{P_m^*}(v_{jk}) + \sum_{j=1}^{n}\sum_{k=1}^{p_j'} \varepsilon_{P_m^*}(v_j')\varepsilon_{P_m^*}(v_{jk}')$$

$$= \sum_{k=1}^{p_0}(n+1)(n+2) + \sum_{j=1}^{n}\sum_{k=1}^{p_j}(n+j+1)(n+j+2) + \sum_{j=1}^{n}\sum_{k=1}^{p_j'}(n+j+1)(n+j+2)$$

$$= p_0(n+1)(n+2) + \sum_{j=1}^{n}(p_j+p_j')(n+j+1)(n+j+2)$$

$$= \sum_{j=0}^{n}(p_j+p_j')(n+j+1)(n+j+2),$$

and we obtain the equality.

3. Conclusions

In this article we computed closed formulas for computing first Zagreb eccentricity index as well as second Zagreb eccentricity index for thorny graphs of important families of graphs like complete graph, complete bipartite graph, cycle, star and path. These relations are given in Theorems 1–4. Moreover, it can be observed from these formulas that values of these indices increase by increasing the number of vertices and number of thorns attached to graphs. These invariants have applications in computational chemistry.

Acknowledgments: The authors are highly grateful to the referees for their valuable comments, which led to great improvement of the original manuscript.

Author Contributions: Nazeran Idrees, Muhammad Jawwad Saif, Asia Rauf and Saba Mustafa contributed equally in computation of results, writing the manuscript and proofreading. All authors have read and approved the final manuscript.

Conflicts of Interest: The authors declare no conflict of interest.

References

1. Gutman, I.; Trinajstić, N. Graph theory and molecular orbitals. Total pi-electron energy of alternant hydrocarbons. *Chem. Phys. Lett.* **1972**, *17*, 535–538. [CrossRef]
2. Sharma, V.; Goswami, R.; Madan, A.K. Eccentric Connectivity Index: A Novel Highly Discriminating Topological Descriptor for Structure-Property and Structure-Activity Studies. *J. Chem. Inf. Comput. Sci.* **1997**, *37*, 273–282. [CrossRef]
3. Gupta, S.; Singh, M.; Madan, A.K. Connective eccentricity index: A novel topological descriptor for predicting biological activity. *J. Mol. Graph. Model.* **2000**, *18*, 18–25. [CrossRef]
4. Gupta, S.; Singh, M.; Madan, A.K. Eccentric distance sum: A novel graph invariant for predicting biological and physical properties. *J. Math. Anal. Appl.* **2002**, *275*, 386–401. [CrossRef]
5. Bajaj, S.; Sambi, S.S.; Gupta, S.; Madan, A.K. Model for prediction of anti-HIV activity of 2-pyridinone derivatives using novel topological descriptor. *QSAR Comb. Sci.* **2006**, *25*, 813–823. [CrossRef]

6. Dureja, H.; Gupta, S.; Madan, A.K. Predicting anti-HIV-1 activity of 6-arylbenzonitriles: Computational approach using superaugmented eccentric connectivity topochemical indices. *J. Mol. Graph. Model.* **2008**, *26*, 1020–1029. [CrossRef] [PubMed]
7. Sedlar, J. On augmented eccentric connectivity index of graphs and trees. *MATCH Commun. Math. Comput. Chem.* **2012**, *68*, 325–342.
8. Sardana, S.; Madan, A.K. Predicting anti-HIV activity of TIBO derivatives: A computational approach using a novel topological descriptor. *J. Mol. Model.* **2002**, *8*, 258–265. [CrossRef] [PubMed]
9. Sardana, S.; Madan, A.K. Relationship of Wiener's index and adjacent eccentric distance sum index with nitroxide free radicals and their precursors as modifiers against oxidative damage. *J. Mol. Struct. THEOCHEM* **2003**, *624*, 53–59. [CrossRef]
10. Ghorbani, M.; Hosseinzade, M.A. A new version of Zagreb indices. *Filomat* **2012**, *26*, 93–100. [CrossRef]
11. Das, K.C.; Lee, D.W.; Graovac, A. Some properties of the Zagreb eccentricity indices. *ARS Math. Contemp.* **2013**, *6*, 117–125.
12. De, N. New Bounds for Zagreb Eccentricity Indices. *Sci. Res.* **2013**, *3*, 70–74. [CrossRef]
13. Luo, Z.; Wu, J. Zagreb eccentricity indices of the generalized hierarchical product graphs and their applications. *J. Appl. Math.* **2014**, *2014*, 241712. [CrossRef]
14. Farahani, M. Exact Formulas for the First Zagreb Eccentricity Index of Polycyclic Aromatic Hydrocarbons (PAHs). *J. Appl. Phys. Sci. Int.* **2015**, *4*, 185–190.
15. Gutman, I. Distance of thorny graphs. *Publ. Inst. Math., Nouv. Sér.* **1998**, *63*, 31–36.
16. De, N. On eccentric connectivity index and polynomial of thorn graph. *Appl. Math.* **2012**, *3*, 931–934. [CrossRef]
17. De, N. Augmented eccentric connectivity index of some thorn graphs. *Int. J. Appl. Math.* **2012**, *1*, 4. [CrossRef]
18. Hernández, J.; Rodríguez, J.; Sigarreta, J. On the geometric–arithmetic index by decompositions-CMMSE. *J. Math. Chem.* **2016**, 1–16. [CrossRef]
19. Rodrıguez, J.; Sigarreta, J. Spectral study of the geometric-arithmetic index. *MATCH Commun. Math. Comput.* **2015**, *74*, 121–135.

symmetry

MDPI

Article

A Study on Immersion of Hand Interaction for Mobile Platform Virtual Reality Contents

Seunghun Han and Jinmo Kim *

Department of Software, Catholic University of Pusan, Busan 46252, Korea; tmd37star@naver.com
* Correspondence: jmkim11@cup.ac.kr; Tel.: +82-51-510-0645

Academic Editor: Angel Garrido
Received: 11 October 2016; Accepted: 30 January 2017; Published: 5 February 2017

Abstract: This study proposes gaze-based hand interaction, which is helpful for improving the user's immersion in the production process of virtual reality content for the mobile platform, and analyzes efficiency through an experiment using a questionnaire. First, three-dimensional interactive content is produced for use in the proposed interaction experiment while presenting an experiential environment that gives users a high sense of immersion in the mobile virtual reality environment. This is designed to induce the tension and concentration of users in line with the immersive virtual reality environment. Additionally, a hand interaction method based on gaze—which is mainly used for the entry of mobile virtual reality content—is proposed as a design method for immersive mobile virtual reality environment. The user satisfaction level of the immersive environment provided by the proposed gaze-based hand interaction is analyzed through experiments in comparison with the general method that uses gaze only. Furthermore, detailed analysis is conducted by dividing the effects of the proposed interaction method on user's psychology into positive factors such as immersion and interest and negative factors such as virtual reality (VR) sickness and dizziness. In this process, a new direction is proposed for improving the immersion of users in the production of mobile platform virtual reality content.

Keywords: mobile virtual reality; hand interface; interaction; immersion; VR sickness; Leap Motion

1. Introduction

With the development of various virtual reality devices and related technologies, an environment where general users can easily enjoy virtual reality content is being formed. As a result, content that enables users to feel an experience that is similar to reality is continuously needed, and various research and technical development related to virtual reality are being carried out to satisfy these needs. To provide a visual experience with three-dimensional (3D) effects, Sutherland [1] studied the HMD (Head Mounted Display) system in the 1960s. Since then, input processing techniques based on virtual reality began to be researched and developed to control physical events in a virtual space while satisfying the users' five senses, including auditory and tactile senses.

As the hardware performance of smart phones is increasing and low-priced mobile virtual reality HMDs are being propagated, a wide variety of mobile platform virtual reality content is being produced, and many related studies are being conducted. The popularization of mobile HMD is especially providing an environment where anyone can experience immersive virtual reality content anywhere. However, mobile HMD requires the attachment of the mobile device inside the HMD, unlike personal computer (PC) or console platforms such as Oculus Rift.

For this reason, the touch input method of mobile devices cannot be used. Because of this limitation, mobile virtual reality content generally uses simple input methods using gaze or connects a game pad for controlling the virtual reality. Recently, dedicated controllers interconnected with mobile HMDs such as Samsung Rink are being developed, and studies [2] are being conducted to develop hardware systems that can process inputs by direct touch of the mobile screen. To provide a virtual reality environment with a greater sense of immersion for users, hardware devices such as Leap Motion and data glove have been developed, which reflect the finger movements and motions of users in real time in the virtual reality environment. As suggested by the above discussion, which virtual reality technology and devices are used to experience immersive content based on virtual reality can be an important factor. However, it is also important to check the generation of motion sickness due to dizziness by the input processing technique while the user is experiencing the content. Nevertheless, there are few studies that analyze the effects of the interaction methods on the physical and psychological responses of users in comparison to the studies on the input processing techniques and interaction methods to improve user immersion.

This study was conducted to design user-oriented interaction to improve immersion in the production of interactive 3D content for the virtual reality environment and to systematically analyze the effects of the interaction on the actual immersion of users. For this purpose, interactive 3D content was designed for a comparative experiment on the suitability and immersion of the proposed input processing technique and interaction. Furthermore, as the core contribution of this study, an interaction method that can increase immersion more than gaze—which is the main input method of the existing mobile virtual reality content—is proposed. To this end, hand interaction that combines Leap Motion with gaze—which is the beginning of users' immersion for interaction with objects in the virtual space—has been included. Lastly, the physical and psychological factors of users that are influenced by the input processing through the proposed hand interaction are evaluated through a questionnaire, and input processing techniques and interactions that can increase users' immersion are analyzed.

Section 2 analyzes various input processing techniques and the psychological factors of users that are required to develop contents based on virtual reality. Section 3 describes the production process for immersive 3D interactive contents proposed in this study. Section 4 describes the core technique of hand interaction using gaze and Leap Motion in this study. Section 5 describes the experiment and analysis process for the proposed method. Finally, Section 6 outlines the conclusion and presents future research directions.

2. Related Works

In the early 1900s, studies on virtual reality were conducted to satisfy the visual sense of users through such devices as head-mounted virtual reality systems [1]. Since then, many researchers have tried to improve the realism of the virtual reality environment and the immersion of user, which has led to studies on haptic system and other devices to satisfy various senses such as tactile sense by improving the physical responses of the virtual world [3,4]. With the development of mobile devices, many application studies using mobile devices were conducted in the virtual reality arena. Lopes et al. [5] proposed a mobile force feedback system through muscle simulation using electricity. Yano et al. [6] conducted research on a handheld haptic device that can touch objects with its fingers. In addition, GyroTab, which gives feedback of a mobile torque based on a gyroscope, was proposed [7]. Another example is POKE [8], a mobile haptic device that interacts through an air pump and silicon membranes. These studies were conducted to provide tactile sense as well as vision in mobile virtual reality, but they were not developed into systems that can be easily accessed and used by anyone.

How to provide input processing for users in a limited environment is as important for virtual reality content in a mobile environment as the design of hardware devices to satisfy the five senses of users. Unlike the PC platform, the mobile virtual reality embedded in the HMD has limitations in users' input environments due to the impossibility of touch, which is the only input function. For this reason, many researchers designed interfaces that can process magnetic input for mobile HMD, such as Google Cardboard. Representative magnetic input devices include Abracadabra [9], Nenya [10], MagiTact [11], and MagGetz [12]. They processed interactions by wearing or holding magnetic objects. Later, Smus et al. [13] proposed a wireless, unpowered, and inexpensive mobile virtual reality (VR) magnetic input processing method that provides physical feedback with the smart phone, only without the calibration process. Gugenheimer et al. [2] proposed the facetouch interface, which processes the interaction of the virtual reality environment by attaching a touch-sensitive surface on the mobile HMD. This was also difficult to use because a separate magnetic device must be attached to process input.

Hands are body parts that are often used to interact with objects in both virtual and real environments. For this reason, controllers are frequently used to indirectly replace the movement of hands in the interaction process of virtual reality content. For more direct control, however, studies are being conducted to accurately capture the movements of hands, including joints, and use them for interaction. For instance, Metcalf et al. [14] conducted a study to capture and control the movements of hands and fingers through optical motion capture using a surface marker. Zhao et al. [15] also proposed a high-fidelity measurement method of 3D hand joint information by combining a motion capture system based on an optical marker and a Kinect camera from Microsoft.

Stollenwerk et al. [16] proposed an optical hand motion capture method based on a marker and tested whether colored marker detection is correctly performed under various lighting conditions and applied the detected hand and finger movements to keyboard performance. Oikonomidis et al. [17] proposed a method of tracking the 3D movements of hands based on the depth information detected by the Kinect. Arkenbout et al. [18] researched an immersive hand motion control method incorporating the Kinect-based Nimble VR system using a fifth dimension technologies (5DT) data glove and a Kalman filter. Furthermore, studies [19–22] on various approaches for analyzing motion by capturing the human hand have been carried out, including a study on articulated hand motion and graphic presentation of data generated from the interaction between objects in certain time intervals [23]. These studies enable users to interact more directly in a virtual environment, but research has not yet been developed into a VR system. In particular, in the case of mobile platform VR, if VR sickness is considered, many factors other than hand motion detection, such as frames per second (FPS) and the refresh rate, should be considered together. Therefore, in order for the hand motion capture research to be used as a VR application, these various technical factors and compatibility with other VR systems such as HMD should be considered in a comprehensive way.

Recently, studies are being conducted using Leap Motion as a technology for expressing free motions in 3D space by capturing the finger movement and motions of user. A method of receiving signature or certificate was researched by detecting hand gestures using Leap Motion, recognizing the tip of the detected finger and writing along the movement of the fingertip point [24]. In another study, hand gestures were divided into four motions of circle, swipe, screen tap, and key tap, and the possibility of accurate perception through matching with predefined templates was tested [25]. Hand gestures were also used with Leap Motions for the training of surgical experiments [26]. Recently, an interface using Leap Motion was designed, and user reactions were analyzed to use hand motions as interaction for playing a virtual reality game [27]. However, there are still few cases of applying Leap Motion to virtual reality content, and in particular, almost no studies have been conducted to design an interaction applied to mobile virtual reality. More importantly, research on user psychology is also required to analyze whether or not the proposed interaction method improves user immersion or causes VR sickness. In relation to this, studies were conducted to analyze whether or not the cue conflict of the head-mounted virtual reality display causes motion sickness [28,29] or to analyze the effect of unstable positions on motion sickness [30,31]. However, few studies have been conducted on the effects of the input processing technique and interaction method of virtual reality on the psychology of users.

Considering this situation, this study designs interactive content based on mobile platform virtual reality, and proposes a hand interaction method using gaze and Leap Motion to improve user immersion. Furthermore, experiments evaluating the suitability of the designed content and proposed interaction method for the virtual reality environment are conducted, and the results analyzed in terms of various factors, such as immersion and VR sickness.

3. Immersive Mobile Virtual Reality Content

The goal of this study is to design an interaction method that can improve immersion through a convenient control process by a user who experiences virtual reality content on a mobile platform. To achieve this, immersive mobile virtual reality content must be provided which enables the experience of a virtual environment through an interaction that is directly proposed by the user. In this study, new interactive 3D content is produced to objectively analyze the immersion and efficiency of the proposed hand interaction in a virtual reality environment that can induce tension and concentration in the user. The interactive structure is designed using gaze, sound, and gesture evenly to provide various experiences with a high level of user satisfaction.

A method often used in board games is used for the proposed flow of content. The goal of this content is to change the random five cards given to us by exchanging them for the same cards with characters beside us, and anyone who collects five identical cards touches the screen quickly. To explain the detailed flow of the content, four characters are deployed in the virtual space, including the user. Then, a set of cards with four patterns are randomly mixed, and five cards are distributed to each character. Users who received the cards take action to collect five cards of one type from the four types of cards. What they do is to select one of their five cards and give it to the person on the left. The user also receives one card from the character on the right. The game progresses in this way until one character has five cards of one type. Then, a 3D virtual object appears at the center of the screen and anyone who selects it quickly wins the game. Figure 1 shows the flow of the proposed content using actual cards.

Figure 1. The flow of the proposed interactive content: (**a**) start the content; (**b**) select cards; (**c**) result of card exchange; (**d**) card game finish condition and event after finish.

The card selection and delivery of three virtual characters (excluding the user) are implemented through a simple exchange behavior pattern. Algorithm 1 defines the exchange behavior pattern representing the card selection and delivery of virtual characters.

The user's card selection time needs to be limited for the proposed content process to cause tension and concentration in the user. In this study, sound is used to raise tension as an element of communicating the limited time to the user. When the user must select a card, the sound "ready-go" is played. The user must quickly select a card in line with the "go" sound. If the user fails to meet the timing, he or she loses his/her right of selection, and one of the cards he or she has is randomly delivered to the character. Interactive content containing sound elements is appropriate for mobile virtual reality content, and whether or not it gives an experience of high satisfaction level to users is analyzed through an experiment in Section 5.

Algorithm 1 Exchange Behavior Pattern of Virtual Characters.

1: Array_Card[3][5] ← Array of five cards that three characters have.
2: l ← Index of the left character from the current character.
3: i ← Index of the current character.
4: **procedure** SELECT CARD(Array_Card)
5:　　Analyze the pattern of five cards (Array_Card[i][5]) of the ith character.
6:　　ptrn_card ← The card that the ith character has the largest number in possession.
7:　　sel_card ← One card among the cards of different patterns from ptrn_card.
8:　　Remove sel_card from Array_Card[i][5].
9:　　return sel_card ← Return one of the cards that has the lowest number.
10: **end procedure**

11: **procedure** DELIVERY CARD(Array_Card, sel_card, l, i)
12:　　Deliver the selected sel_card to the character on the left.
13:　　Remove sel_card from Array_Card[i][5].
14:　　Save sel_card in the card array of l character (Array_Card[l][5]).
15:　　Analyze the card pattern of Array_Card[i][5]
16:　　cnt_card ← Number of the cards of the highest number in possession.
17:　　return cnt_card.
18: **end procedure**

19: **procedure** FINISH GAME(cnt_card)
20:　　**if** cnt_card = 5 **then**
21:　　　　Generate a 3D virtual object at the center of screen.
22:　　**end if**
23: **end procedure**

---

4. Gaze-Based Hand Interaction

　　User interaction is required for the smooth progress of the immersive mobile virtual reality content. The interaction elements of the proposed content consist of a process of selecting one among the five received cards, and a process of selecting the virtual object that is created when five cards of the same type are collected. An interaction method that can enhance immersion while not interfering with the object control must be designed because the content progresses quickly within a limited time. Gaze-based hand interaction is proposed for this purpose in this study.

　　Hayhoe et al. [32] proved that people focus on gaze first when controlling virtual objects in a virtual space. Therefore, the user's gaze must be considered before designing the interaction using hands. Then, the hand motion and gesture are recognized based on gaze to design input processing. Then, input processing is designed by recognizing the hand motions and gestures of users. Figure 2 shows an overview of the gaze-based hand interaction.

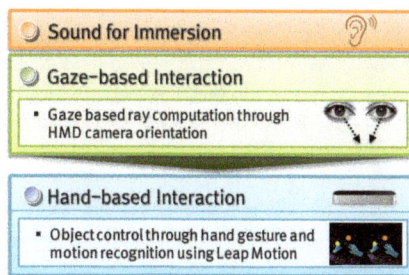

Figure 2. Overview of the proposed gaze-based hand interaction. HMD: head-mounted display.

　　For users who are wearing a mobile HMD, the viewpoint and direction of the camera in the virtual space correspond to the user's gaze. When the user's head moves, the mobile sensor and HMD track it and update the changed information on the screen. In this study, the interactive content is configured in such a way that the user's gaze is not distributed, and the user can concentrate on the screen where their card is displayed because, if you miss the screen, you can also miss the fast moving flow with sound. At this time, the raycasting method is used so that the gaze of the user can accurately select the desired card. The ray is calculated in the camera direction corresponding to the gaze. Then, the object selection or not of virtual space is calculated through the collision detection with the calculated ray and with the virtual object. Figure 3 shows the card selection process of the content proposed through the gaze of user.

Figure 3. Interaction process of the proposed content using gaze: (**a**) card browsing; (**b**) card selection using gaze.

The user is induced to concentrate on a specific area (location of the arrayed card) on the screen using gaze. Then, the hand interaction structure is designed to reflect a similar behavior as card selection in the virtual environment. In this study, Leap Motion is used as an input processing technique to increase the user's immersion in the virtual reality content of the mobile environment. The Leap Motion sensor is a hand motion detection sensor, which consists of two infrared cameras and one infrared ray (light-emitting diode (LED)). This sensor is a small USB peripheral device with a height of 12.7 mm and a width of 80 mm. It can be attached to the HMD device, and the hand gestures and motions are recognized and processed by the infrared sensor. Figure 4 shows the configuration result of input environment consisting of a Leap Motion device used in this study attached to a mobile HMD. Leap Motion is not providing software development kit (SDK) for mobile HMD. Therefore, a mobile virtual reality experiment environment to which hand interaction is applied is constructed in this study by using Unity 3D (Unity Technologies, San Francisco, CA, USA) to produce mobile virtual reality content and integrating it with the Leap Motion development tool and remotely sending divided virtual reality scenes to the mobile phone.

Two hand interactions are proposed: the first is a card selection interaction which checks if the user's gaze is looking at the card to select. Then, the user's finger is perceived at the go timing of the "ready-go" sound. Next, the gesture is set by a clicking motion with the index finger. The second hand interaction is the process of selecting a virtual object generated at a random location in the virtual space when five identical cards are collected and the content finishes. During the progress of the content, the user's gaze is concentrated on his/her cards. In this situation, when a card combination is completed by another character, a virtual object is created instantly, and the user perceives them in the order of gaze to gesture. In other words, the behavior is recognized when the user first looks at the generated virtual object, instantly stretches his/her hand and makes a gesture of holding the object. Algorithm 2 represents the process of these two hand interactions.

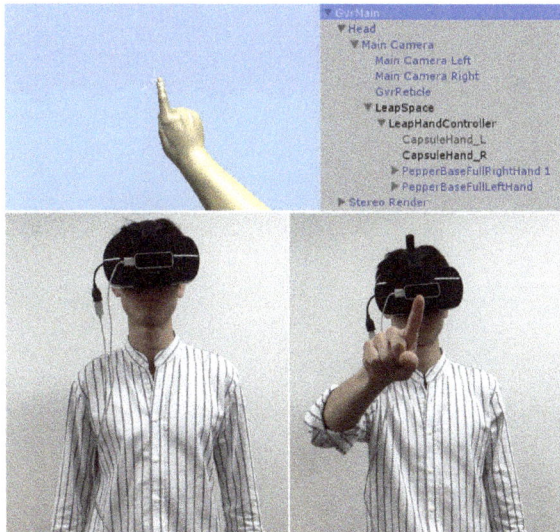

Figure 4. Construction of mobile virtual reality input environment using a Leap Motion device.

Algorithm 2 Hand Interaction Process.

1: Array_UserCard[5] ← Array of the user's five cards.
2: **procedure** HAND-BASED CARD SELECTION(Array_UserCard)
3:     range_sound ← A certain time range is saved around the time when "go" sound is played.
4:     check_gaze ← Checks if the user's gaze is directed to one element of the Array_UserCard.
5:     **if** check_gaze = true **then**
6:         Save recognized finger information of 0th hand.
7:         FingerList OneHand = frame.Hands[0].Fingers .
8:         OneHand [1].IsExtended ← Activation of the index finger is tested.
9:         time_hand ← Time when the index finger is detected.
10:         **if** OneHand [1].IsExtended = true **And** time_hand < range_sound **then**
11:             Card selection finished.
12:         **end if**
13:     **end if**
14: **end procedure**

15: Obj_Finish ← 3D virtual object that is randomly generated when the content finishes.
16: **procedure** HAND-BASED TOUCH OBJECT(Obj_Finish)
17:     check_gaze ← Checks if the user's gaze is directed to the Obj_Finish.
18:     **if** check_gaze = true **then**
19:         cnt_hands ← Number of activated fingers.
20:         **for** i=0,4 **do**
21:             **if** frame.Hands[0].Fingers[i].IsExtended = true **then**
22:                 cnt_hands++
23:             **end if**
24:         **end for**
25:         **if** cnt_hands = 5 **then**
26:             Perceive object touch.
27:             Record the time from object generation to touch.
28:         **end if**
29:     **end if**
30: **end procedure**

Figure 5 shows the process of perceiving hand motions and controlling the content through the proposed hand interactions.

Figure 5. Interaction of content through the proposed gaze-based hand interaction: (**a**) card selection process; (**b**) action event process when the game finishes.

5. Experimental Results and Analysis

The proposed mobile virtual reality content production and virtual reality technique used Unity 3D 5.3.4f1 (Unity Technologies, San Francisco, CA, USA) and a Google virtual reality development tool (gvr-unity-sdk, Google, Mountain View, CA, USA). The hand interface—which is the core input technology of this study—was implemented using Leap Motion SDK v4.1.4 (Leap Motion, Inc., San Francisco, CA, USA). The PC environment used in this environment was IntelR CoreTM i7-4790 (Intel Corporation, Santa Clara, CA, USA), 8 GB random access memory (RAM), and Geforce GTX 960 GPU (NVIDIA, Santa Clara, CA, USA). Furthermore, a Samsung Galaxy S5 (Suwan, Korea) was used as the mobile phone for this experiment, and the Baofeng Mojing 4 (Beijing, China) HMD was used.

The experimental process consists of checking the production result of the proposed content and the analysis of the physical and psychological effects of the gaze-based hand interaction of this study on users in the virtual reality environment. First, the virtual reality content of the mobile platform was produced in accordance with the plan, and the accurate operation of the interaction process was verified based on the proposed input processing. When the content is started, the main screen is switched to the content screen. On the content screen, four characters (including the user) are deployed in the virtual space, five cards are randomly distributed to each character, and the process of selecting the last finish object by matching the cards of one type is implemented. In this process, an interaction method was designed by which the user selects cards using his/her hands based on his/her gaze and touch of the last virtual object. Figure 6 shows this process, and the progress of accurate interaction can be checked in the smooth flow of the proposed interactive content.

Next, the effect of the proposed hand interaction on the psychological elements of the actual result are tested and analyzed. For this experiment, 50 participants in their 20 and 40s were randomly chosen, and they were allowed to experience the produced content before analyzing the results through a questionnaire. The first part of this experiment is the result for the suitability of the proposed virtual reality content. If the card-type content proposed in this study is not suitable for virtual reality content, the later interface has no significance. Therefore, the questionnaire was collected to check this.

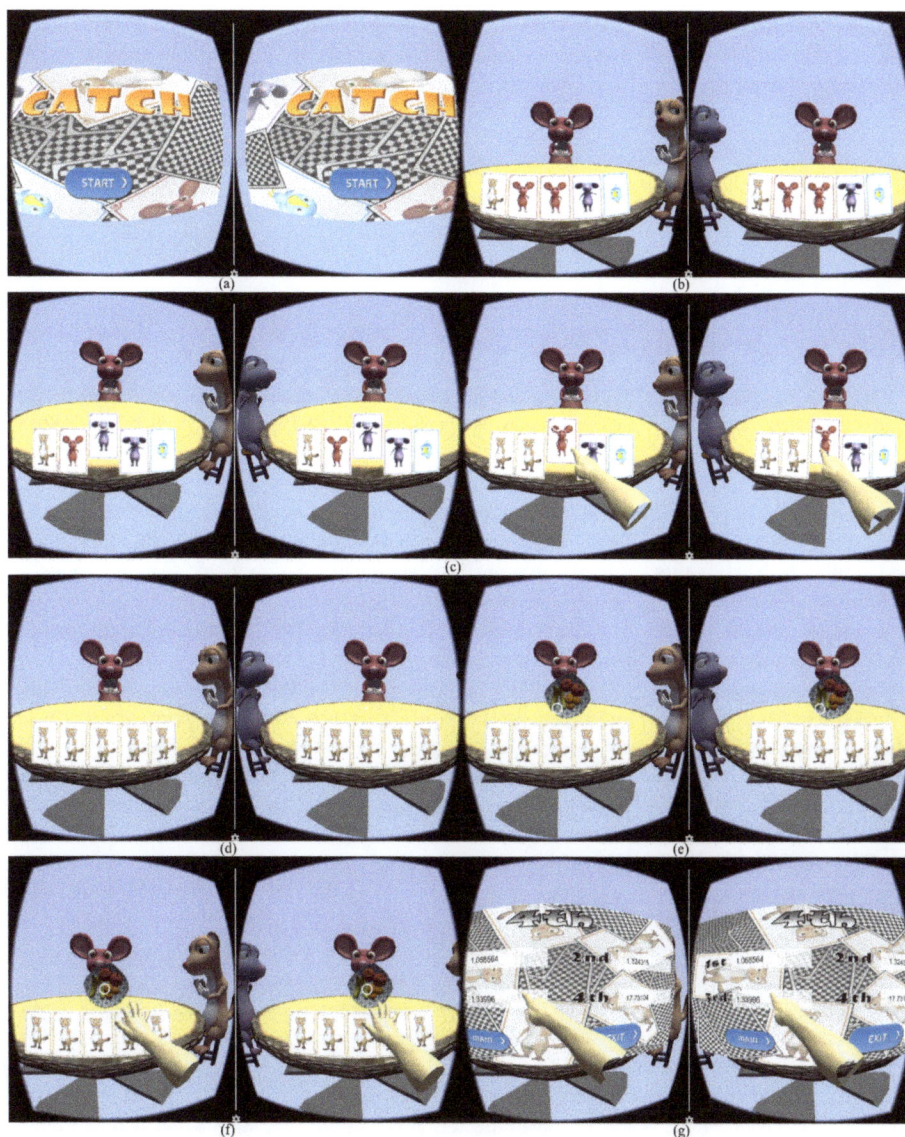

Figure 6. Implementation result of the proposed mobile virtual reality content including our hand interaction: (**a**) content starting screen; (**b**) content initial card setting screen; (**c**) card selection interaction using gaze and hand; (**d**) content finish condition; (**e**) generation of a virtual object for event in the event of content's finish; (**f**) event object control through hand interaction; (**g**) delivery of information by converting the reaction speed of users into their scores.

Figure 7 shows the result: 86% of all participants replied that the proposed content was suitable for testing the virtual reality environment. Furthermore, the participants were asked to write the satisfaction score between 1 and 5 to accurately analyze the numerical data. The respondents gave a high satisfaction score of 3.74 (standard deviation (SD): 0.68) for the experience of the virtual reality environment of the proposed content. In particular, when the reasons for the positive evaluation of the proposed content were analyzed, it was selected because it is an interesting topic for virtual reality content and provides a high level of immersion and a new experience. The satisfaction score was also high at approximately 3.7 out of 5.0. Thus, the proposed content was suitable for experimenting and analyzing mobile virtual reality content in line with the intended purpose.

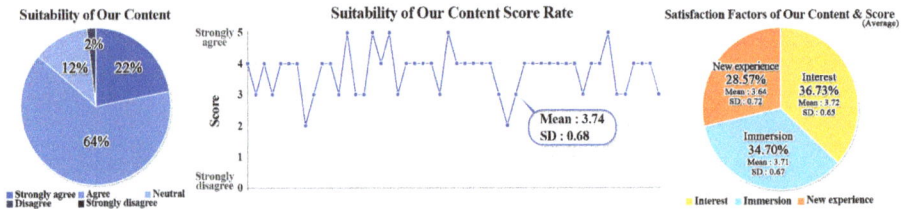

Figure 7. Analysis results for the suitability and satisfaction of the proposed virtual reality content (left to right: suitability level of five items, score distribution between 1 and 5, satisfaction factors consisting of three items).

The second is a comparative experiment for the proposed interaction. This study proposed an interaction that combines hand gesture and motion with the gaze interaction method that is mainly used in mobile virtual reality content. Therefore, an experiment was conducted to analyze whether the proposed interaction can give high immersion and satisfaction to users in comparison with the conventional general interaction method, which uses gaze only. Four experimental groups were constructed for this experiment because the experience of users may vary according to the order of the interaction experiences. In the description of the content, the method of using the gaze only is defined as "G", and gaze-based hand interaction is defined as "H". The first experimental group experienced G first, followed by H, and the second experimental group experienced H first, followed by G, in the reverse order. The third and fourth experimental groups experienced only G or H, respectively, to obtain objective data. The participants in all experimental groups were asked to evaluate their interaction experiences on a scale of 1 to 5.

Figure 8 shows the results of the four experimental groups. First, 80% of the participants who experienced the interactions in the order of G and H replied that the hand interaction was more satisfactory. Their satisfaction scores showed an average difference of 0.85 between G and H (Figure 8a). Figure 8b shows the results of the experimental group who experienced the interactions in the reverse order, and 83.34% of the experimental group was more satisfied with the gaze-based hand interaction. Their satisfaction score difference between G and H was 1.25. There was a slight difference in satisfaction depending on which interaction was experienced first between the gaze and the hand. In particular, participants who experienced the hand interaction first were more satisfied with the gaze-based hand interaction than with the gaze interaction only.

We conducted a Wilcoxon test to prove the alternative hypothesis that assumes that H provides more advanced interaction than G. First, significance probability (p-value) was approximately 0.0016162 in the case of participants who experienced Leap Motion first (Figure 8a). Since this is smaller than significance level (0.05), the null hypothesis was rejected. Next, significance probability was approximately 0.0111225 in the case of participants who experienced the proposed gaze based hand interaction first (Figure 8b). This is also smaller than the significance level, showing consistent results of rejecting the null hypothesis. That is, both of the two tests rejected the null hypothesis, proving that the alternative hypothesis is correct. Finally, computing the significance probability by combining these two, the statistical test results also proved that the proposed hand interaction gave stronger satisfaction and immersion to users compared to the method that uses gaze only, with p-value at 6.4767297×10^{-5}.

Next, the responses of the experimental groups who used only one interaction were analyzed. As shown in Figure 8c, more than 90% of the participants using the gaze-only method were satisfied with the interaction and recorded an average score of 3.8 (SD: 0.60). Participants using the gaze-based hand interaction only scored an average of 4.0 (SD: 0.91) points and more than 83.34% of them gave positive responses (Figure 8d). Although the difference is small, the proposed hand interaction was found to result in better satisfaction and immersion for participants in the same conditions and situations.

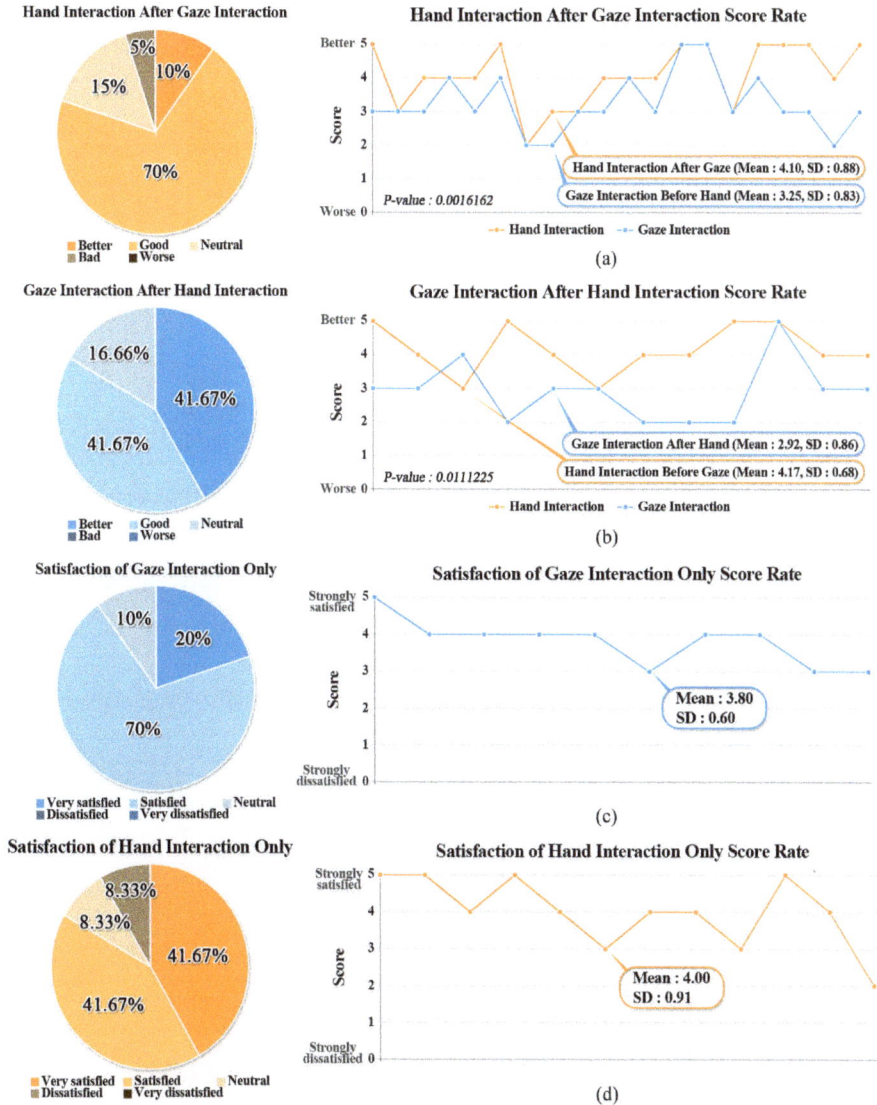

Figure 8. Comparison experiment result of the gaze-based hand interaction and the method using gaze only: (**a**) results of the experimental group who experienced H followed by G; (**b**) results of the experimental group who experienced G followed by H; (**c**) results of the experimental group who experienced G only; and (**d**) results of the experimental group who experienced H only.

The overall analysis results showed that users are more familiar and satisfied with the gaze interaction method, which is mainly used in the existing mobile platform virtual reality content. Thus, we can see that hand interaction can provide greater immersion and satisfaction if it is combined with the appropriate context for the purpose of the content.

The last is a detailed analytical experiment for psychological factors. The psychological factors of users for the proposed gaze-based hand interaction were subdivided into the four items of improved immersion, inducement of interest, provision of new experience, and convenient control for the analysis of positive factors. Furthermore, the four items of VR sickness, fatigue, difficult operation, and inconvenience were analyzed as negative factors. The details of the eight psychological factors presented here referred to the existing studies on immersion in VR [33,34].

The results of the aforementioned experiment showed that the gaze-based hand interaction was more satisfactory. Therefore, the experiment involving the detailed psychological factors was conducted with the proposed hand interaction. As shown in Figure 8a,b,d, among the participants in the experimental groups experiencing the proposed hand interaction, those who gave relatively or objectively positive responses were asked to select one of the four positive factors and record a score. In addition, participants who gave negative responses were also asked to select one of the negative factors and record a score. However, only 4.55% of the respondents gave negative responses below average, so accurate analysis results could not be derived from them.

Figure 9 shows that 45.24% of the participants who gave positive responses, which constituted 81.28% of all respondents, replied that hand interaction improves their immersion and helps them to accurately control the virtual objects. The scores were also generally high, at 3.8 or higher out of 5.0. Therefore, the proposed hand interaction was found to have the greatest influence on the provision of an experiential environment with high immersion in the virtual reality environment, although it is also helpful for the inducement of interest and convenient control.

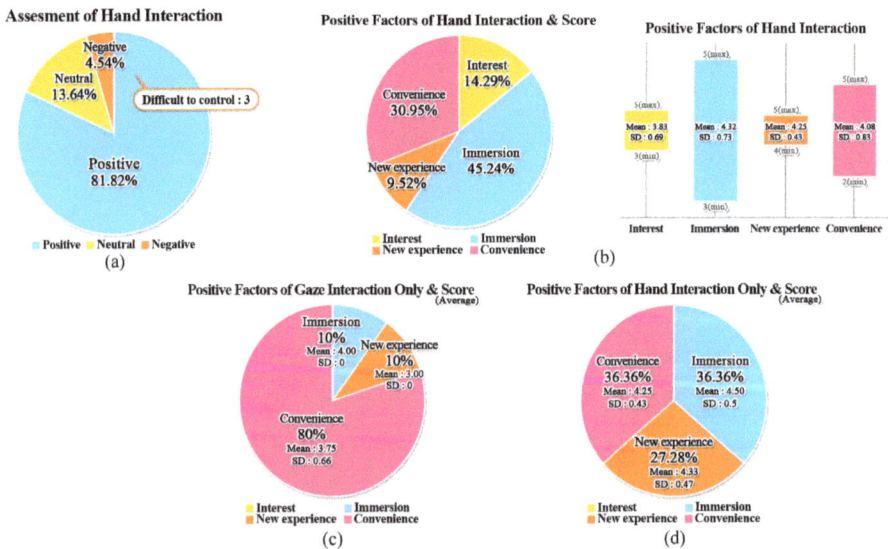

Figure 9. Detailed analysis results for psychological factors of the proposed gaze-based hand interaction: (**a**) satisfaction distribution of the proposed hand interaction; (**b**) distribution of positive factors and score analysis results; (**c**) distribution of the positive psychological factors of participants who experienced the gaze only; (**d**) distribution of the positive psychological factors of participants who experienced the gaze-based hand interaction only.

The analysis of the negative factors showed that many participants selected the difficulty of manipulation. The information on the participants who gave negative answers revealed that they had either considerable or no experience with virtual reality content. Their negative responses seem to be caused by the inconvenience of the new operation method due to their familiarity with the existing gaze method or their lack of experience with interaction. For reference, the questionnaire results of participants who only experienced G or H were analyzed (Figure 8c,d). Most participants experienced only gaze selected convenience, as expected. Participants who experienced the hand interaction only selected various items, such as immersion, convenience, and novelty. Their satisfaction score was also found to be higher by at least 0.5. This suggests that, if the gaze and the hand are combined appropriately in line with the situation, it can provide users with various experiences and high satisfaction.

In this study, it was analyzed whether the proposed hand interaction can lead to higher immersion compared to merely gaze, which is mainly used in mobile platform virtual reality content, and also whether it can cause VR sickness. As a result of various experiments, VR sickness was not found to be a problem. For the virtual reality content of the mobile platform in general, the FPS and the polygon count of all objects comprising the content should be considered carefully. In mobile VR, the recommended number of polygons is 50 to 100 kilobyte (k) and the recommended FPS is 75 or more. If these recommendations are not followed, the user may experience sickness due to the delay and distortion of the screen. In the proposed virtual reality content, the number of polygons ranged from 41.4 to 80.0 k with an average of 56.2 k depending on the camera position, and the FPS ranged from 82.1 to 84.0 with an average of 82.8. Thus, users experienced no hardware problems. Therefore, if the technical performance condition is satisfied in the VR environment, the proposed hand interaction will not cause VR sickness (Table 1).

Table 1. Technical performance analysis results of the proposed mobile virtual reality content.

	Minimum	Maximum	Mean	Recommended
Polygon count	41.4 k	80.0 k	56.2 k	50–100 k
Frame per second (FPS)	82.1	84.0	82.8	≥75

The simulator sickness questionnaire (SSQ) experiment was conducted to analyze the VR sickness of the proposed hand interaction more systematically and statistically [35,36]. In SSQ, the sickness that users can feel from the simulator was deduced in 16 items through various experiments. Participants were asked to select one of four severity levels between none and severe for such items as general discomfort and fatigue. In this study, the sickness was analyzed using raw data excluding weight for the absolute analysis of the proposed interaction [37]. Based on the aforementioned four experimental groups, the questions were designed to compare the values when only gaze was used and after experiencing the gaze based hand interaction. Table 2 lists the results of the SSQ experiment. The result of hand interaction increased slightly compared to the method using gaze only. However, most participants felt almost no VR sickness for both interactions (total mean (mean for each of the 16 items) : 1.05 (0.07), 1.57 (0.1), slight or less). For the detailed factors consisting of nausea, oculomotor, and disorientation, the highest values (0.67, 1.0) were obtained for oculomotor in both interactions, but they were not a level that can cause problems in user's interaction.

Consequently, various new interactions that combine hand motion and movement with gaze should be researched to provide users with satisfying and diverse experiences as well as immersive interactions in the virtual reality content of the mobile environment. Therefore, the current environment where the virtual reality input processing techniques are mostly limited to PCs, consoles, and platforms, and cannot be extended to the mobile environment must be improved so that users can conveniently and easily experience virtual reality content. More specifically, we need to think about how to combine hand gestures and motions with the interaction that handles the movements and controls of objects based on the user's gaze in order to increase immersion while reducing the VR sickness of users.

Table 2. Results of an simulator sickness questionnaire (SSQ) experiment for analysis of the virtual reality (VR) sickness of the proposed interaction.

		Mean	Standard Deviation (SD)	Minimum	Maximum
Total	G	1.05 (0.07)	1.43	0.0	5.0
	H	1.57 (0.1)	1.68	0.0	6.0
Nausea	G	0.33 (0.02)	0.56	0.0	2.0
	H	0.52 (0.03)	0.66	0.0	2.0
Oculomotor	G	0.67 (0.04)	0.89	0.0	3.0
	H	1.0 (0.06)	1.07	0.0	3.0
Disorientation	G	0.43 (0.03)	0.73	0.0	3.0
	H	0.62 (0.04)	1.0	0.0	4.0

6. Conclusions

This study proposed gaze-based hand interaction considering various psychological factors, such as immersion and VR sickness in the process of producing mobile platform virtual reality content. To this end, a card-type interactive content was produced which induces tension and concentration in users in order to plan an interactive content that is suitable for the virtual reality environment of the mobile platform. Then, an environment where users can receive and concentrate on scenes with 3D effects using mobile HMD regardless of the place was provided. A gaze-based hand interaction method was designed to improve the immersion of users in the process of controlling 3D objects and taking actions in the virtual space. Furthermore, the interaction method using gaze only and the interaction method to which hand interface was added were applied with the same content separately, and the experimental results were derived after asking general users to experience them. In a situation where the proposed interactive content was suitable for virtual reality content, the method of combining gaze with hand interaction improved user satisfaction and immersion compared to the interaction method using gaze only. When the detailed psychological factors were analyzed, a high percentage or 45.24% of respondents answered that hand interaction provided interaction with high immersion among the items of improved immersion, inducement of interest, convenient control, and new experience. Thus, it was found that presenting an environment that enables more direct control in the 3D virtual space is helpful for improving the immersion of content when producing virtual reality content of the mobile platform. Finally, it was verified that VR sickness due to hand interaction will not occur if the system requirements, such as FPS and the polygon count, are observed when producing virtual reality content in the mobile environment. The results of the survey experiment through SSQ showed that all respondents felt almost no VR sickness.

The conventional input processing of the mobile platform is providing an interface that mainly uses gaze, due to the limited environment. In this study, interaction combining the conventional gaze method with hand gestures and motions was proposed to improve the immersion of content, and the performance of the proposed method was evaluated. Even though hand interaction is very helpful when analyzed from the aspect of immersion improvement, a perfect development environment for the mobile platform has not yet been provided, due to the problem of installing separate devices. Therefore, it is important to design an interaction method in line with the degree of immersion that the content to be produced desires. In the future, the efficiency of various input processing techniques will be analyzed by experimenting with immersion, motion sickness, etc. through the interaction design as well as using various input devices that support the production of mobile platform virtual reality content.

Supplementary Materials: The following are available online at www.mdpi.com/2073-8994/9/2/22/s1, Video S1: The proposed mobile virtual content including our hand interaction demo video.

Acknowledgments: This research was supported by the Basic Science Research Program through the National Research Foundation of Korea (NRF) funded by the Ministry of Education (No. NRF-2014R1A1A2055834).

Author Contributions: Seunghun Han and Jinmo Kim conceived and designed the experiments; Seunghun Han performed the experiments; Seunghun Han and Jinmo Kim analyzed the data; and Jinmo Kim wrote the paper.

Conflicts of Interest: The authors declare no conflict of interest.

References

1. Sutherland, I.E. A Head-mounted Three Dimensional Display. In Proceedings of the Fall Joint Computer Conference, Part I (AFIPS '68 (Fall, Part I)), San Francisco, CA, USA, 9–11 December 1968; ACM: New York, NY, USA, 1968; pp. 757–764.
2. Gugenheimer, J.; Dobbelstein, D.; Winkler, C.; Haas, G.; Rukzio, E. FaceTouch: Touch Interaction for Mobile Virtual Reality. In Proceedings of the 2016 CHI Conference Extended Abstracts on Human Factors in Computing Systems (CHI EA '16), Santa Clara, CA, USA, 7–12 May 2016; ACM: New York, NY, USA, 2016; pp. 3679–3682.
3. Bergamasco, M. The GLAD-IN-ART Project. In *Virtual Reality: Anwendungen und Trends*; Warnecke, H.J., Bullinger, H.J., Eds.; Springer: Berlin/Heidelberg, Germany, 1993; pp. 251–258.
4. Brooks, F.P. What's Real about Virtual Reality? *IEEE Comput. Graph. Appl.* **1999**, *19*, 16–27.
5. Lopes, P.; Baudisch, P. Muscle-Propelled Force Feedback: Bringing Force Feedback to Mobile Devices. In Proceedings of the SIGCHI Conference on Human Factors in Computing Systems (CHI '13), Paris, France, 27 April–2 May 2013; ACM: New York, NY, USA, 2013; pp. 2577–2580.
6. Yano, H.; Miyamoto, Y.; Iwata, H. Haptic Interface for Perceiving Remote Object Using a Laser Range Finder. In Proceedings of the World Haptics 2009—Third Joint EuroHaptics Conference and Symposium on Haptic Interfaces for Virtual Environment and Teleoperator Systems (WHC '09), Salt Lake City, UT, USA, 18–20 March 2009; IEEE Computer Society: Washington, DC, USA, 2009; pp. 196–201.
7. Badshah, A.; Gupta, S.; Morris, D.; Patel, S.; Tan, D. GyroTab: A Handheld Device That Provides Reactive Torque Feedback. In Proceedings of the SIGCHI Conference on Human Factors in Computing Systems (CHI '12), Austin, TX, USA, 5–10 May 2012; ACM: New York, NY, USA, 2012; pp. 3153–3156.
8. Park, Y.W.; Baek, K.M.; Nam, T.J. The Roles of Touch During Phone Conversations: Long-distance Couples' Use of POKE in Their Homes. In Proceedings of the SIGCHI Conference on Human Factors in Computing Systems (CHI '13), Paris, France, 27 April–2 May 2013; ACM: New York, NY, USA, 2013; pp. 1679–1688.
9. Harrison, C.; Hudson, S.E. Abracadabra: Wireless, High-precision, and Unpowered Finger Input for Very Small Mobile Devices. In Proceedings of the 22nd Annual ACM Symposium on User Interface Software and Technology (UIST '09), Victoria, BC, Canada, 4–7 October 2009; ACM: New York, NY, USA, 2009; pp. 121–124.
10. Ashbrook, D.; Baudisch, P.; White, S. Nenya: Subtle and Eyes-Free Mobile Input with a Magnetically-Tracked Finger Ring. In Proceedings of the SIGCHI Conference on Human Factors in Computing Systems (CHI '11), Vancouver, BC, Canada, 7–12 May 2011; ACM: New York, NY, USA, 2011; pp. 2043–2046.
11. Ketabdar, H.; Yüksel, K.A.; Roshandel, M. MagiTact: Interaction with Mobile Devices Based on Compass (Magnetic) Sensor. In Proceedings of the 15th International Conference on Intelligent User Interfaces (IUI '10), Hong Kong, China, 7–10 February 2010; ACM: New York, NY, USA, 2010; pp. 413–414.
12. Hwang, S.; Ahn, M.; Wohn, K.Y. MagGetz: Customizable Passive Tangible Controllers on and Around Conventional Mobile Devices. In Proceedings of the 26th Annual ACM Symposium on User Interface Software and Technology (UIST '13), St. Andrews, UK, 8–11 October 2013; ACM: New York, NY, USA, 2013; pp. 411–416.
13. Smus, B.; Riederer, C. Magnetic Input for Mobile Virtual Reality. In Proceedings of the 2015 ACM International Symposium on Wearable Computers (ISWC '15), Osaka, Japan, 7–11 September 2015; ACM: New York, NY, USA, 2015; pp. 43–44.
14. Metcalf, C.D.; Notley, S.V.; Chappell, P.H.; Burridge, J.H.; Yule, V.T. Validation and Application of a Computational Model for Wrist and Hand Movements Using Surface Markers. *IEEE Trans. Biomed. Eng.* **2008**, *55*, 1199–1210.

15. Zhao, W.; Chai, J.; Xu, Y.Q. Combining Marker-Based Mocap and RGB-D Camera for Acquiring High-Fidelity Hand Motion Data. In Proceedings of the ACM SIGGRAPH/Eurographics Symposium on Computer Animation (SCA '12), Lausanne, Switzerland, 29–31 July 2012; Eurographics Association: Aire-la-Ville, Switzerland, 2012; pp. 33–42.

16. Stollenwerk, K.; Vögele, A.; Sarlette, R.; Krüger, B.; Hinkenjann, A.; Klein, R. Evaluation of Markers for Optical Hand Motion Capture. In Proceedings of Workshop Virtuelle Realität und Augmented Reality der GI-Fachgruppe VR/AR; Bielefeld, Germany, 8–9 September 2016; Shaker Verlag: Aachen, Deutschland, 2016.

17. Iason Oikonomidis, N.K.; Argyros, A. Efficient model-based 3D tracking of hand articulations using Kinect. In Proceedings of the British Machine Vision Conference (BMVC '12), Dundee, UK, 29 August–2 September 2011. BMVA Press: Durham, UK, 2011; pp. 101.1–101.11.

18. Arkenbout, E.A.; de Winter, J.C.F.; Breedveld, P. Robust Hand Motion Tracking through Data Fusion of 5DT Data Glove and Nimble VR Kinect Camera Measurements. *Sensors* **2015**, *15*, 31644–31671.

19. Ju, Z.; Liu, H. Human Hand Motion Analysis with Multisensory Information. *IEEE/ASME Trans. Mechatron.* **2014**, *19*, 456–466.

20. Mousas, C.; Anagnostopoulos, C.N.; Newbury, P. Finger Motion Estimation and Synthesis for Gesturing Characters. In Proceedings of the 31st Spring Conference on Computer Graphics (SCCG '15), Smolenice, Slovakia, 22–24 April 2015; ACM: New York, NY, USA, 2015; pp. 97–104.

21. Wheatland, N.; Wang, Y.; Song, H.; Neff, M.; Zordan, V.; Jörg, S. State of the Art in Hand and Finger Modeling and Animation. *Comput. Graph. Forum* **2015**, *34*, 735–760.

22. MacRitchie, J.; Bailey, N.J. Efficient Tracking of Pianists' Finger Movements. *J. New Music Res.* **2013**, *42*, 79–95.

23. Stollenwerk, K.; Vögele, A.; Krüger, B.; Hinkenjann, A.; Klein, R. Automatic Temporal Segmentation of Articulated Hand Motion. In *Computational Science and Its Applications—ICCSA 2016: 16th International Conference, Beijing, China, July 4–7, 2016, Proceedings, Part II*; Springer: Berlin, Germany 2016; pp. 433–449.

24. Kamaishi, S.; Uda, R. Biometric Authentication by Handwriting Using Leap Motion. In Proceedings of the 10th International Conference on Ubiquitous Information Management and Communication (IMCOM '16), Danang, Viet Nam, 4–6 January 2016; ACM: New York, NY, USA, 2016; pp. 36:1–36:5.

25. Chan, A.; Halevi, T.; Memon, N. Leap motion controller for authentication via hand geometry and gestures. In *Human Aspects of Information Security, Privacy and Trust, Proceedings of Third International Conference (HAS 2015 Held as Part of HCI International), Los Angeles, CA, USA, 2–7 August 2015*; Tryfonas, T., Askoxylakis, I., Eds.; Springer: Berlin, Germany, 2015; Volume 9190, pp. 13–22.

26. Juanes, J.A.; Gómez, J.J.; Peguero, P.D.; Ruisoto, P. Practical Applications of Movement Control Technology in the Acquisition of Clinical Skills. In Proceedings of the 3rd International Conference on Technological Ecosystems for Enhancing Multiculturality (TEEM '15), Porto, Portugal, 7–9 October 2015; ACM: New York, NY, USA, 2015; pp. 13–17.

27. Lee, P.W.; Wang, H.Y.; Tung, Y.C.; Lin, J.W.; Valstar, A. TranSection: Hand-Based Interaction for Playing a Game within a Virtual Reality Game. In Proceedings of the 33rd Annual ACM Conference Extended Abstracts on Human Factors in Computing Systems (CHI EA '15), Seoul, Korea, 18–23 April 2015; ACM: New York, NY, USA, 2015; pp. 73–76.

28. Reason, J. Motion sickness adaptation: A neural mismatch model. *J. R. Soc. Med.* **1978**, *71*, 819–829.

29. Duh, H.B.L.; Parker, D.E.; Philips, J.O.; Furness, T.A. Conflicting motion cues to the visual and vestibular self-motion systems around 0.06 Hz evoke simulator sickness. *Hum. Factors* **2004**, *46*, 142–154.

30. Riccio, G.; Stoffregen, T. An Ecological Theory of Motion Sickness and Postural Instability. *Ecol. Psychol.* **1991**, *3*, 195–240.

31. Stoffregen, T.A., Jr.; Smart, L.S. Postural instability precedes motion sickness. *Brain Res. Bull.* **1998**, *47*, 437–448.

32. Hayhoe, M.M.; Shrivastava, A.; Mruczek, R.; Pelz, J.B. Visual memory and motor planning in a natural task. *J. Vis.* **2003**, *3*, 49–63.

33. Budhiraja, P.; Sodhi, R.; Jones, B.R.; Karsch, K.; Bailey, B.P.; Forsyth, D.A. Where's My Drink? Enabling Peripheral Real World Interactions While Using HMDs. *CoRR* **2015**, arXiv:1502.04744.

34. Lee, Y.S.; Seo, W.D.; Lee, J.; Sohn, B.S. Immersive Gesture Interface Design for HMD Based Virtual World Navigation. In Proceedings of the Extended Abstracts of HCI Korea 2016 (HCI Korea '16), Kangwon, Korea, 27–29 January 2016; pp. 9–14.

35. Kennedy, R.S.; Lane, N.E.; Berbaum, K.S.; Lilienthal, M.G. Simulator Sickness Questionnaire: An Enhanced Method for Quantifying Simulator Sickness. *Int. J. Aviat. Psychol.* **1993**, *3*, 203–220.

36. Rebenitsch, L.; Owen, C. Review on Cybersickness in Applications and Visual Displays. *Virtual Real.* **2016**, *20*, 101–125.
37. Bouchard, S.; St-Jacques, J.; Renaud, P.; Wiederhold, B.K. Side effects of immersions in virtual reality for people suffering from anxiety disorders. *J. Cyberther. Rehabil.* **2009**, *2*, 127–137.

symmetry

MDPI

Article

Deformable Object Matching Algorithm Using Fast Agglomerative Binary Search Tree Clustering

Jaehyup Jeong, Insu Won, Hunjun Yang, Bowon Lee * and Dongseok Jeong *

Department of Electronic Engineering, Inha University, Incheon 22212, Korea; jaehyup@inha.edu (J.J.);
is.won@inha.ac.kr (I.W.); pinion@inha.edu (H.Y.)
* Correspondence: bowon.lee@inha.ac.kr (B.L.); dsjeong@inha.ac.kr (D.J.); Tel.: +82-32-860-7415 (B.L. & D.J.)

Academic Editor: Angel Garrido
Received: 7 November 2016; Accepted: 4 February 2017; Published: 10 February 2017

Abstract: Deformable objects have changeable shapes and they require a different method of matching algorithm compared to rigid objects. This paper proposes a fast and robust deformable object matching algorithm. First, robust feature points are selected using a statistical characteristic to obtain the feature points with the extraction method. Next, matching pairs are composed by the feature point matching of two images using the matching method. Rapid clustering is performed using the BST (Binary Search Tree) method by obtaining the geometric similarity between the matching pairs. Finally, the matching of the two images is determined after verifying the suitability of the composed cluster. An experiment with five different image sets with deformable objects confirmed the superior robustness and independence of the proposed algorithm while demonstrating up to 60 times faster matching speed compared to the conventional deformable object matching algorithms.

Keywords: content-based image retrieval; image matching; deformable object; clustering

1. Introduction

Humans can recognize and determine objects through vision. Human vision is fast and robust, and it is the most powerful perceptual function to acquire information. Vision is an ability that humans have from birth, and the human performance is far better than that of a computer. Computers may have better performance in fields that are difficult to work with human eyes, such as precision measurements. In a field of recognizing and determining objects, however, their ability is still worse than that of humans. Therefore, research to provide computers with the visual ability at the human level is currently active. Such research is called computer vision. Studies of computer vision are being performed for the recognition of face, object, gesture, from videos or images.

In image recognition, computer vision is divided into the extraction method, which belongs to low-level vision, and the matching method, which belongs to high-level vision. The typical algorithms of the extraction method include D. Lowe's SIFT (Scale-Invariant Feature Transform) [1], which is robust to size and angle change, H. Bay's SURF (Speeded Up Robust Features) [2], which is faster than SIFT, J. Matas's region-based MSER (Maximally Stable Extremal Regions) [3], and K. Mikolajczyk's Harris affine detector [4], which is robust to affine changes. The matching method is divided into a step for composing matching pairs between all the feature points of two images, and a step for performing geometric verification between the matching pairs. In particular, the geometric verification step is the final step in image recognition, and it is very important because, even if many matching pairs are composed, two images may be determined to be mutually different images if geometric verification fails. A typical algorithm for geometric verification is RANSAC [5].

In recent years, image recognition using deep learning has become popular [6]. Deep learning is different from conventional computer vision algorithms (divided into low-and high-level vision). It enables a computer to learn by itself using neural networks, without image feature extraction and matching method, and it is leading to unparalleled levels of accuracy in image recognition. However, deep learning has not yet been used in various object matching due to the requirement for a large amount of data. With a small amount of data in a database, it is still difficult to achieve reasonably good performance for image recognition using deep learning. In addition, to detect unique objects, neural networks have to become much deeper and deeper networks require high computational power. Thus, we still need computer vision technology that uses low-level and high-level vision for image recognition.

A representative technology that uses image recognition is content-based image retrieval, which was established as the MPEG-7 standard. Recently, at MPEG-7, by constructing the CDVS (Compact Descriptor Visual Search) [7], a study was performed for content-based image retrieval, which retrieves an image fast for mobile devices. Content-based image retrieval is a technology that retrieves an image by extracting robust features even if various deformations in brightness, rotation, affine, and size, occur in the image. On the other hand, most matching algorithms perform retrieval by targeting images with rigid objects [8–10]. The object types also include deformable objects; typical examples include clothes, packs, and bags. For rigid objects, the object shapes do not change, but for deformable objects, the object shapes can change in various ways. Because of this difference, the conventional rigid object matching algorithms that are robust to images with rigid objects are not suitable for matching images with deformable objects. Therefore, developing a matching algorithm that is robust to images that contain deformable objects has become an important issue.

The three aspects of excellent matching algorithm are robustness, independence, and fast matching [11]. Robustness is a characteristic that determines that two images with the same object, even if deformation occurs in the object, must be determined to be identical. Independence is a characteristic that determines that two images with mutually different objects are different. Finally, matching is done rapidly in fast matching. If fast matching does not occur, an algorithm may not be appropriate for applications that require fast image retrieval. The most significant weakness of conventional deformable object matching algorithms is slow matching.

In this paper, these three aspects are considered to propose an optimal algorithm for the matching of two images with deformable objects. The remainder of this paper is organized as follows. Section 2 introduces the related works about image matching. In Section 3, the proposed algorithm is described by dividing it into extraction and matching methods. In Section 4, the experiment is described and its results are confirmed and analyzed from five image sets with various deformable objects. Section 5 evaluates the proposed algorithm and reports the conclusion.

2. Related Works

This section introduces well-known feature descriptors developed recently. In the past few years, a number of feature descriptors using binary features were developed. These feature descriptors which have fast feature extraction and less computational complexity are suitable for real-time image matching. This section also introduces the conventional deformable object matching algorithms. Deformable object matching algorithms use different matching methods from rigid object matching algorithms.

2.1. Recent Feature Descriptors

In recent years, binary feature descriptors such as BRIEF (Binary Robust Independent Elementary Features) [12], BRISK (Binary Robust Invariant Scalable Keypoints) [13], FREAK (Fast Retina Keypoint) [14], SYBA (Synthetic Basis) [15], and TreeBASIS [16] have been reported. BRIEF uses a binary string, which results in intensity comparisons at random pre-determined pixel locations. The descriptor similarity is evaluated using the Hamming distance. It trades robustness and independence for fast processing speed, but it is sensitive to image distortions and transformations. BRISK is a 512 bit binary descriptor using a FAST-based detector. It relies on easily configurable circular sampling patterns from which it computes a binary descriptor. It uses the distance ratio of the two nearest neighbors to improve the accuracy of the detection of corresponding keypoint pairs. BRISK requires more computational complexity and more storage space than BRIEF. FREAK improves upon the sampling pattern and method of pair selection that BRISK uses. The features are much more concentrated near the keypoint.

SYBA uses a number of synthetic basis images to measure the similarity between a small image region surrounding a detected feature point and the randomly generated synthetic basis images. The TreeBASIS descriptor uses a binary vocabulary tree that is computed using basis dictionary images and a test set of feature region images. It provides improvements in descriptor size, computation time, matching speed, and accuracy.

2.2. The Conventional Deformable Object Matching Algorithms

The feature-based deformable object matching algorithms include transformation model-based [17], mesh-based [18], cluster-based [19] and graph-based [20] algorithms. The transformation model-based and mesh-based algorithms require high complexity and are not suitable for various deformations of objects. The graph-based algorithms have fast processing speed but relatively poor performance. The conventional deformable object matching algorithm is the ACC (Agglomerative Correspondence Clustering) algorithm [21], which uses the clustering method. This algorithm calculates the dissimilarity between clusters using the adaptive partial linkage model in the framework of hierarchical agglomerative clustering. The IACC (Improved ACC) algorithm [22] includes the feature selection method for selecting robust features. These two algorithms show good performance for deformable objects, but high complexity in the clustering process. The matching speed becomes slower with higher complexity, and it cannot be called a good matching algorithm with slow matching speed.

3. Proposed Algorithm

This section discusses the proposed algorithm. This section is divided into two subsections: the first discusses the extraction method, and the second discusses the matching method. Figure 1 shows the flow chart of the proposed algorithm, consisting of the extraction part (feature extraction and feature selection) and the matching part (the rest).

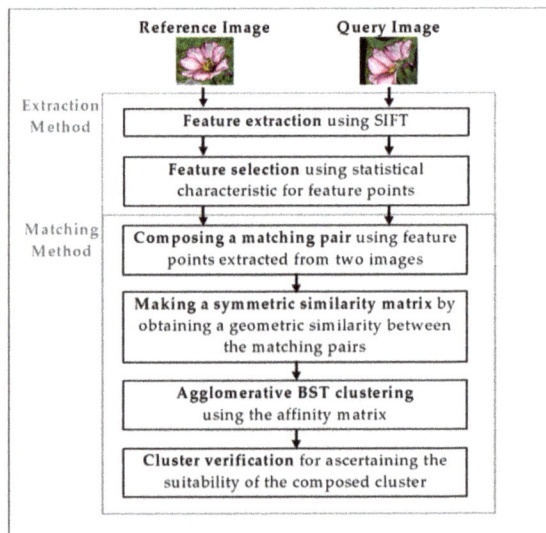

Figure 1. Flowchart of the proposed algorithm.

3.1. Extraction Method

3.1.1. Feature Extraction

There exist methods for extracting the global features and local features from images. A global feature is unsuitable for an image with deformable objects because such features are extracted from the entire image. This is because the various deformations of deformable objects cannot be defined with a single feature. On the other hand, a local feature is suitable for an image with deformable objects because the features are defined for each local region. Furthermore, a local feature is suitable for applying clustering because additional information in terms of position, scale, and orientation is stored. In this study, a typical algorithm for local features, SIFT [1], was used. The feature $F(\cdot)$ stored through SIFT is expressed as (1).

$$F(i) = \{\, p_i \,,\, s_i \,,\, o_i \,,\, f_i \,\},\, (1 \le I \le N) \tag{1}$$

where N is the number of extracted feature points, and every feature point has four components. Here, p_i is the feature point's position, s_i is the scale, o_i is the orientation, and f_i is a feature vector with 128 dimensions.

3.1.2. Feature Selection

Non-matching and higher complexity can occur if the extracted features just use matching. This is because some of the feature points could be the outliers. Therefore, it requires a process that selects the robust feature points included in the inliers. The feature selection is a process for selecting robust feature points in composing matching pairs with the extracted features. In general, when the feature points matched in two images are compared, the statistical characteristic is different between the feature points included in the outliers and those included in the inliers [23]. Therefore, the use of the inlier's statistical characteristic can distinguish the points of the inlier from the outlier. To obtain the inlier's statistical characteristics, the position (p_i), scale (s_i), orientation (o_i), and distance of the center (c_i) components are learned from various image sets [24,25]. When a large value (e_i) is produced by substituting p_i, s_i, o_i, and c_i in the learned inlier's statistical characteristic $ISC(\cdot)$, the probability of belonging to the inlier region is high. The following pseudocode shows a process for selecting N_S feature points from a total of N feature points using $ISC(\cdot)$. If N_S is bigger than N, N_S become N. We use $N_S = 300$. Figure 2b gives an example of using feature selection, and when compared with Figure 2a, where this is not used, some of the outlier points are removed. When the feature points of the outlier are removed because the complexity becomes lower, the features become more robust and the matching speed becomes faster.

```
Feature selection
 E = {ø}, i = 0
 repeat
  i = i + 1
  eᵢ = ISC(pᵢ, sᵢ, oᵢ, cᵢ)
  Insert eᵢ into E
  E, ranked in descending order
 until i = = N
 E = {e₁, e₂, e₃, ... , e_Nₛ ... , e_N}
 Selecting Nₛ feature points from N feature points.
```

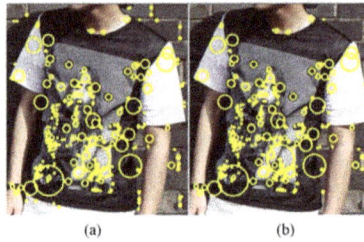

Figure 2. Example of the feature points in an image: (**a**) feature points using only SIFT; and (**b**) the feature points using feature selection.

3.2. Matching Method

3.2.1. Composing a Matching Pair

To compose a matching pair, the feature points extracted from two images are compared [26]. The formula used here is the Euclidean distance, as expressed in Equation (2).

$$Euclid\left(F_{R(i)}, F_{Q(j)}\right) = \sqrt{\sum_{k=1}^{128} \left(F_{R(i)}^k - F_{Q(j)}^k\right)^2} \tag{2}$$

Equation (2) is an equation for finding the Euclidean distance of $F_{R(i)}$, which is the ith feature vector of the reference image, and $F_{Q(j)}$, which is the jth feature vector of the query image. If $Euclid(\cdot)$ is smaller than an arbitrary threshold, the feature points $R(i)$ and $Q(j)$ are composed as a matching pair. One feature point can compose up to the maximum of k matching pairs using the *knn* method. N_M matching pairs composed in this manner undergo the overlap checking process expressed as Equation (3).

$$ovlp[i,j] = \begin{cases} 1, & \text{if } m_i \text{ and } m_j \text{ are overlapping,} \\ 0, & \text{otherwise.} \end{cases} \quad (1 \le i, j \le N_M) \tag{3}$$

A matching pair (m_k) is composed with two feature points matched in two images. In other words, m_k consists of the respective feature points from the reference and query images. In Equation (3), m_k represents the respective positions of two feature points. Here, $m_k = \left(p_k^R, p_k^Q\right)$, where p_k^R is the position of the feature point extracted from the reference image, and p_k^Q is the position of the feature point extracted from the query image. When the ith matching pair (m_i) and jth matching pair (m_j) are compared, if p_i^R matches p_j^R, or p_i^Q matches p_j^Q, they are determined to be overlapped, and the number one is assigned to $ovlp[i,j]$. With this equation, one or zero is assigned to every $ovlp[i,j]$, and finally, an overlap matrix of size $N_M \times N_M$ with $ovlp[i,j]$ for all i, j as its elements is generated. In Figure 3, the circles mean the feature points and lines mean the matching pairs. In addition, dotted lines are overlapped matching pairs and the solid-lines are non-overlapped matching pairs. The generated overlap matrix is used in the clustering process.

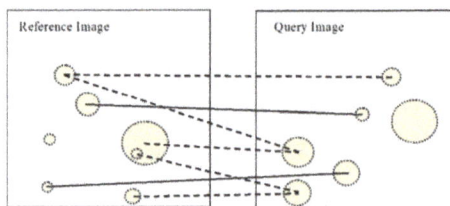

Figure 3. Example of matching pairs that overlap or not.

3.2.2. Making a Symmetric Similarity Matrix

With a deformable object, various deformations may occur because its shape can change. Therefore, it is difficult to evaluate image matching with deformable objects using conventional geometric verification. From the matching pairs composed of the typical conventional geometric verification RANSAC [5], a transform matrix is generated and inliers and outliers are distinguished. On the other hand, a deformable object cannot be defined with a single transform matrix.

Figure 4a presents two images with rigid objects, one of which has one transform matrix (T_1). The reference image's rigid object is transformed geometrically to T_1 in the query image. On the other hand, Figure 4b shows two images with deformable objects, and has many transform matrices (T_2, T_3, and T_4). In this case, a deformable object of the reference image is transformed geometrically to T_2, T_3, and T_4, in the query image. Therefore, because a deformable object cannot be defined with one transform matrix, a new method is required for the approach by generating many transform matrices in a small region. One method used here is to make a symmetric similarity matrix. The symmetric similarity matrix consists of the similarity between transform matrices composed in a point unit. In other words, a symmetric similarity matrix is composed of geometric similarity between all matching pairs.

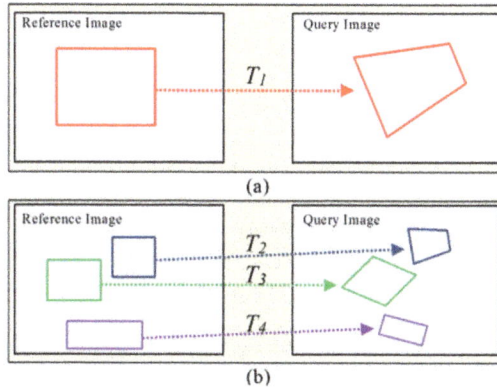

(a)

(b)

Figure 4. Comparison example of a transform matrix (T_i): (**a**) rigid object in the images; and (**b**) deformable object in the images.

To find the geometric similarity between a matching pair, first, a transform matrix is obtained between a matching pair. The transform matrix used here is a homography matrix [27]. Because a homography matrix uses the projective transform method among various transform matrices, it is suitable for obtaining geometric similarity. To compose a homography matrix, the position (p_i), scale (s_i), and orientation (o_i) of a feature point are used, and the matrix is composed using the WGC (Weak Geometric Consistency) [28] method. Using the homography matrix (H_k) composed this way, the geometric similarity (d_{gs}) between a matching pair is found using the Pairwise-WGC [29] method, as expressed in (4).

$$d_{gs}(m_i, m_j) = \frac{1}{2}\left(\left|p^Q_{\ j} - H_i p^R_{\ j}\right| + \left|p^Q_{\ i} - H_j p^R_{\ i}\right|\right), (1 \leq i, j \leq N_M) \tag{4}$$

The two matching pairs to be compared are given as $m_i = (p^R_{\ i}, p^Q_{\ i}, H_i)$ and $m_j = (p^R_{\ j}, p^Q_{\ j}, H_j)$. $|\cdot|$ denotes the Euclidean distance, and $d_{gs}(m_i, m_j)$ is small if H_i and H_j are similar. If geometric similarity is obtained between every matching pair, a symmetric similarity matrix of size $N_M \times N_M$ with $d_{gs}(m_i, m_j)$ as the element is composed, as shown in Figure 5. The symmetric similarity matrix has zero diagonal elements.

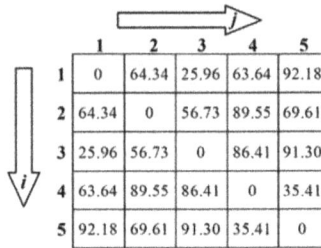

Figure 5. Example of a symmetric similarity matrix (N_M = 5).

$$d_{gs}(m_i, m_j) = sim(i, j) \tag{5}$$

As written in Equation (5), each element of a symmetric similarity matrix represents geometric similarity (d_{gs}) between a matching pair m_i and m_j, and means the similarity (*sim*) between *i* and *j*. Here, the *i* and *j* indices become the minimum units for clustering.

Simply composing a symmetric similarity matrix does not mean a new geometric verification. The new geometric verification intended here refers to everything, from using the composed symmetric similarity matrix, to finally performing the cluster verification after undergoing the clustering process.

3.2.3. Agglomerative BST (Binary Search Tree) Clustering

For clustering, agglomerating clusters by identifying the similarities between the cluster hierarchically is common. The methods for identifying the similarity between clusters include AGNES using the single-link, complete-link, and average-link methods [30]. In the ACC and IACC algorithm [21,22], clustering is performed adaptively using the adaptive partial link method. These clustering methods, however, have a large limitation in that the speed decreases with increasing number of clusters. In general, when the number of initial clusters is n, the hierarchical clustering method has a complexity of $O(n^3)$ because the similarity between clusters needs to be calculated and updated. Here, updating means obtaining a new similarity between an agglomerated cluster and the remaining clusters. The complexity of the similarity calculation between clusters can be reduced using the symmetric similarity matrix obtained earlier, but an additional calculation is essential in the case of an update. In this paper, an algorithm is proposed to reduce the complexity by simplifying the conventional agglomerative hierarchical clustering. The update process that comprises a large proportion of the complexity is omitted, and clustering is performed by constructing a BST (Binary Search Tree) [31] with the basic clusters obtained from symmetric similarity matrix.

The pseudocode presented earlier shows the BST clustering process in detail. In the initialization part, N_{tree} is the number of binary trees (BT_t) generated, and BT_t represents the *t*th binary tree. The BST clustering process that appears hereon is performed the maximum of N_{bc} times. N_{bc} is the number of $sim(i, j)$ in the upper triangular part, excluding the diagonal elements in the symmetric similarity matrix, and $N_{bc} = \frac{N_M \times N_M - N_M}{2}$. When the BST clustering process is examined, first, *i* and *j* with minimum similarity are found in the symmetric similarity matrix (because the symmetric similarity matrix is a symmetrical matrix, they are found only when $i > j$). Here, BST clustering is terminated if the similarity is larger than the given threshold δ_s (similarity threshold). Next, an element of the overlap matrix with *i* and *j* as the index is confirmed. If the value for $ovlp[i, j]$ is one, clustering is not formed because the feature point with an overlap between positions cannot be considered as a robust feature.

Agglomerative BST Clustering

$N_{tree} = 0, k = 0, BT_t = \{\emptyset\}, sumS = 0$ // Initialization

/* BST clustering */

repeat

  $k = k + 1$

  // Find i,j $\{i, j\} = \underset{i>j}{\text{argmin}}$ (symmetric similarity matrix)

  if $sim(i,j) > \delta_s$ then {**break**}

  // overlap check

  if $ovlp[i,j]$ then {$sim(i,j) = \infty$, **continue**}

  // Using BST, Searching & Inserting

  $chk = 0, t = 0$

  repeat

    if $\{i,j\} \in BT_t$ then {$chk = 1$, **break**}

    else if $i \in BT_t$ then {Insert j into BT_t , $chk = 1$, **break**}

    else if $j \in BT_t$ then {Insert i into BT_t , $chk = 1$, **break**}

    else {$t = t + 1$}

  until $t == N_{tree}$

  // make new BT_t

  if $chk == 0$ and $sim(i,j) < thres(\delta_s, sumS)$ then {

    Make BT_t and Insert i,j into BT_t

    $N_{tree} = N_{tree} + 1$

  $sumS \mathrel{+}= sim(i,j)$ }

    $sim(i,j) = \infty$

until $k == N_{bc}$

if any one of the nodes in BT_t ($0 \le t \le N_{tree}$) is the same, merges them.

The rest of BT_t is cluster C_t ($0 \le t \le N_{cluster}$)

In the next part, searching and inserting i and j is performed using BST. This process is performed the maximum of N_{tree} times, and if a node is searched at least once in BT_t, it is terminated. In total, there are three cases of nodes searched from BT_t. The first is the case where both i and j are searched. Here, because all pertinent nodes exist, the process is terminated without insertion. Next is a case where only i is searched. Here, j is inserted as a new leaf node in BT_t, and the process is terminated. Finally, in the case where only j is searched, i is inserted as a new leaf node, and the process is terminated. Figure 6 gives an example of the searching and inserting process of BT_t. For example, when the $i = 8$ and $j = 35$, Figure 6a shows that the node 8 of BT_0 is searched. This is the case where i is searched. As shown in Figure 6b, $j = 35$ is inserted as a new leaf node in BT_0 because j is not searched in BT_0.

$$thres(\delta_s, sumS) = \frac{\delta_s}{sumS/(N_{tree} + 1)} \tag{6}$$

A new BT_t is generated when $t = 0$ or searching is not done. To generate a new BT_t, an additional threshold is required. The root node (first node) is important for generating binary trees. If the root node is incorrect, binary tree generated from the root node can generate large errors. The additional threshold makes the root node more robust. As written in Equation (6), it is an adaptive threshold. Because $sim(i, j)$ increases as BT_t is generated, threshold must also increase. The adaptive threshold is the value that divides similarity threshold (δ_s) by the mean of the sum of root node's similarities. In the BT_t generated here, i and j are inserted as new nodes. Next, it finds new i and j with the minimum similarity value again by providing $sim(i, j) = \infty$ and clustering is repeated the maximum of N_{bc} times. Finally, it checks whether to merge between the generated binary trees. If any one of the nodes in the generated binary trees is the same, they are merged. To merge or not, all the rest of BT_t generated this way become cluster C_t with the basic clusters. For example, in BT_5 of Figure 6, because all nodes form a basic cluster, $C_5 = \{7,6,60,42,28,44\}$. The clusters C_t generated this way finally undergo cluster verification.

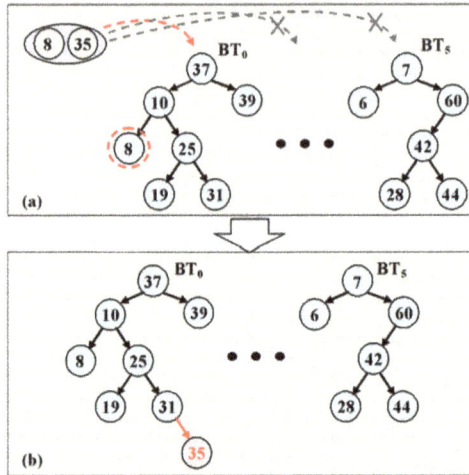

Figure 6. Example of binary search tree ($t = 5$). The circles in blue indicate the nodes in BT_t and the oval in purple indicate two candidate node $\{i = 8, j = 35\}$. (**a**) Node 8 is searched in BT_0 (red dotted arrow and circle); (**b**) Node 35 is inserted as a new leaf node in BT_0 (red solid arrow and red number in the circle).

3.2.4. Cluster Verification

Finally, in the matching method, the cluster verification step determines the suitability of the clusters C_t obtained as described earlier. This step is required because even if a cluster is agglomerated by the geometric similarity between the basic clusters, there is still the possibility of error. In particular, this must be considered when the cluster area is too small when the possibility of error is high. Figure 7 gives examples of mismatching results without using cluster verification, where the cluster area is too small compared to the entire image area.

Figure 7. Examples of mismatching results without using cluster verification.

Cluster Verification

cluster C_t, $t = 0$

$area_{img1}$ = entire reference image(=img1) area

$area_{img2}$ = entire query image(=img2) area

repeat

$\{cv_{img1}, cv_{img2}\}$ = find each convex-hull in C_t

$ratio_{img1}$ = (calculate area of cv_{img1})/$area_{img1}$

$ratio_{img2}$ = (calculate area of cv_{img2})/$area_{img2}$

q_{min} = min($ratio_{img1}$, $ratio_{img2}$)

q_{ratio} = q_{min}/max($ratio_{img1}$, $ratio_{img2}$)

q_{size} = the number of elements in C_t

if $q_{min} > \tau_{min}$ and $q_{ratio} > \tau_{ratio}$

and $q_{size} > \tau_{size}$ then $\{C_t$ is TRUE$\}$

$t = t + 1$

until $t == N_{cluster}$

The previous pseudocode shows the proposed cluster verification step. Cluster verification obtains the determination criteria based on the ratio between the entire image area and the cluster area. The cluster area is calculated by obtaining a convex hull from the positions of the feature points. Here, the feature points can be obtained from the indices that correspond to each element of cluster C_t. Using the ratio that can be obtained from both the reference and query images, the minimum value q_{min} and ratio q_{ratio} of the minimum and maximum values are obtained. As another criterion, q_{size}, the number of elements of C_t, is obtained. These three determination criteria and respective thresholds, τ_{min}, τ_{ratio}, and τ_{size}, are compared, and when they are all larger than the respective thresholds, the pertinent cluster C_t is determined to be suitable. If at least one is determined to be suitable from the clusters, C_t, two images are finally determined to be matching.

4. Experiment

4.1. Experiment Conditions

To evaluate the matching performance, an experiment was performed with five types of image sets. Among these, two types were image sets that contain actual deformable objects, and the other three types were image sets where the images become artificially deformable using TPS (Thin-Plate-Spline). As shown in Figure 8, the image sets that contain actual deformable objects were composed of clothes and snack packs, which are commonly encountered in real life. For the image sets that uses TPS, Stanford University's SMVS standard images [32] and some of the ImageNet's Natural images (flowers, trees, leaves,) [33] and Oxford University's buildings images [34] were used. In the image set, the reference images were constructed with those images where a feature that could represent an object appears at the front. In the case of query images, they were constructed with the images of clothes where a person wears the clothes in various poses; images of snack packs, where various deformations are applied due to the contents in the snack packs; and SMVS and IN-N (ImageNet's Natural), and Oxbuild (Oxford building images), where warping is applied based on several arbitrary points using TPS. Table 1 lists the composition of the five types of image sets. The annotations consist of images, matching pairs of images, and non-matching pairs of images.

	Reference Images	Query Images
(a)		
(b)		
(c)		
(d)		
(e)		

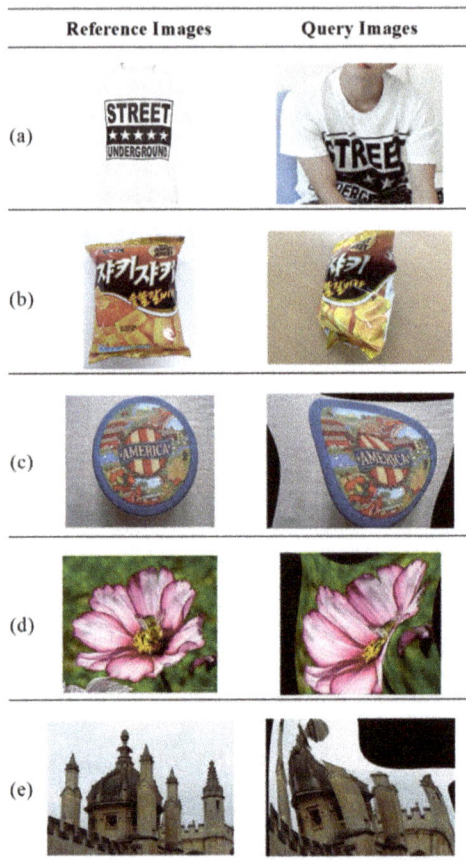

Figure 8. Examples of reference and query (deformable) images: (**a**) clothes; (**b**) snack packs; (**c**) SMVS (using TPS); (**d**) IN-Natural (using TPS); and (**e**) Oxbulid (using TPS).

Table 1. Configuration of image set.

Image Set	Annotations
Clothes	1250 images 996 matching pairs of images 4233 non-matching pairs of images
Snack packs	400 images 300 matching pairs of images 3000 non-matching pairs of images
SMVS (using TPS)	20,400 images 6576 matching pairs of images 7805 non-matching pairs of images
IN-N (using TPS)	1246 images 623 matching pairs of images 5598 non-matching pairs of images
Oxbuild (using TPS)	5063 images 5063 matching pairs of images 20,252 non-matching pairs of images

To measure the proposed algorithm performance, TPR (True Positive Rate) in Equation (7) and FPR (False Positive Rate) in Equation (8) were used. TPR is an equation for finding the robustness among the algorithm characteristics; a larger value indicates better performance. On the other hand, FPR is an equation for finding the independence among the algorithm characteristics; a smaller value indicates better performance. TPR was obtained from the matching pairs of images in Table 1, and FPR is obtained from the non-matching pairs of images in Table 1. The accuracy defined in Equation (9) represents the relationship between TPR and FPR for an objective comparison. Finally, the matching time was measured to determine the fast matching speed.

The proposed algorithm use SIFT [1] for feature extraction like the common comparison algorithms such as ACC [21], IACC [22], and RANSAC [5]. By doing this, we can compare the performance of matching method under the same conditions. In addition, SIFT showed better performance compared with the other feature descriptors such as SURF and BRISK in our experiment which is consistent with other findings [35,36] for images with various deformations. Although SIFT has slower speed for extracting features, it was determined to be an appropriate choice for the feature descriptor.

Here, the experiment was performed by applying all the major parameters required for feature extraction in SIFT. The thresholds for cluster verification were fixed as τ_{min}= 0.001, τ_{ratio}= 0.5, τ_{size}= 3.

$$TPR = \frac{TP}{TP + FN} = \frac{TP}{P} \tag{7}$$

$$FPR = \frac{FP}{FP + TN} = \frac{FP}{N} \tag{8}$$

$$Accuracy = \frac{TP + TN}{P + N} \tag{9}$$

For performance test, we used an Intel Core i5-2500 (quad core) CPU with the clock speed of 3.3 GHz and 8 GB RAM running the Windows 7(64-bit). In addition, all algorithms are implemented in the C ++ environment.

4.2. Experiment Results

Table 2 presents the average computational time and memory storage required to build and use binary trees. Compared with non-binary tree case, when δ_s increases, the algorithm runs faster; when δ_s is above 30, it is faster than non-binary tree case. Since average memory storage required to build binary trees occupies a small part of the whole memory, it is determined to be better to use binary trees.

Table 2. Requirements of the computational time and memory storage about binary tress.

δ_s	Non-Binary Tree	Use of Binary Trees	
	Average Time (ms)	Average Time (ms)	Average Memory (MB)
1	0.004	0.005	0.257
10	0.236	0.266	3.876
20	0.753	0.776	5.935
30	1.501	1.439	7.472
40	2.548	2.366	8.747
50	3.784	3.366	9.784

Figure 9 presents the top three values of accuracy (A1, A2, A3) for each algorithm using Equation (9). These are the results of experimenting with the image set of clothes, snack packs, SMVS (using TPS), IN-N (using TPS) and Oxbuild (using TPS). In the case of RANSAC, the accuracies were very low because it is not an algorithm suitable for images with deformable objects. The other algorithms showed better performance with the proposed algorithm showing the best performance.

Figure 10 presents the recall vs. precision curve using similarity threshold (δ_s) in each image set. The proposed algorithm outperformed the other algorithms, especially for high recall values.

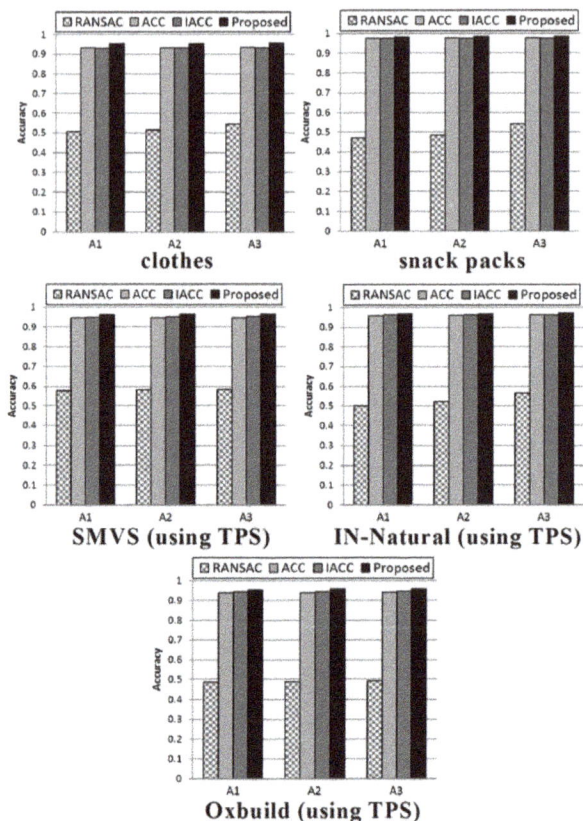

Figure 9. Accuracy of the proposed and other algorithms.

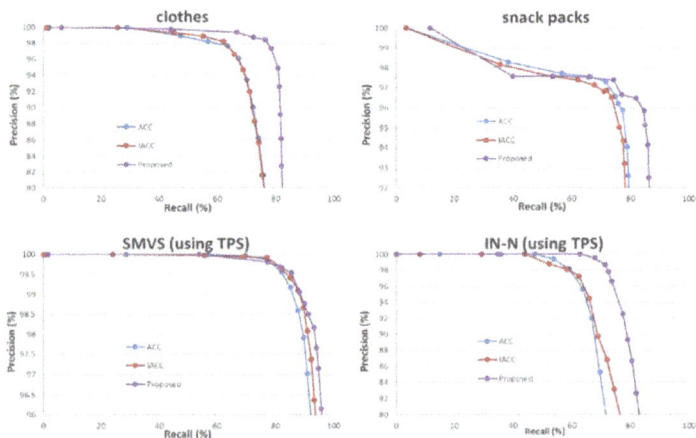

Figure 10. *Cont.*

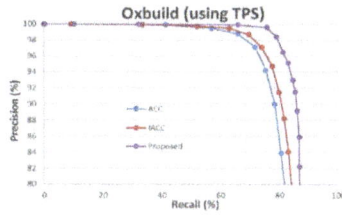

Figure 10. Recall vs. Precision curve of the proposed and other algorithms.

Tables 3–7 list the matching times for each image set. Here, the matching time means the average matching time between two images, and the unit is ms (milliseconds). The matching time was obtained by changing the value of the threshold δ_s, which is a common parameter of the three algorithms (δ_s = 1, 10, 20, 30, 40, and 50). When δ_s decreases, TPR and FPR become lower. On the other hand, when δ_s becomes larger, TPR and FPR become higher. For each algorithm, "match" and "n_match" are obtained. Here, "match" is the average matching time for the matching pairs of images, and "n_match" is the average matching time for the non-matching pairs of images. As δ_s becomes relatively large, the matching time increases, and the matching time for "match" takes longer than for "n_match". "n_match" is faster because there are relatively fewer matching pairs composed from the feature points, and there are little or no clusters composed. A comparison of the algorithms showed that the matching time of the proposed algorithm was faster than the other algorithms. In particular, for "match", it was approximately 10–70 times faster than the ACC algorithm, and approximately 2–10 times faster than the IACC algorithm. Although there was some difference depending on the image set, the proposed algorithm's matching time was the fastest.

Table 8 is a summary of the final results. The values from the table pertain to TPR (Equation (7)), FPR (Equation (8)), Accuracy (Equation (9)) and time (=matching time) in the case of δ_s where the accuracy of each algorithm is highest. Here, "time" is the total average matching time of adding "match" and "n_match" from Tables 3–7. Comprehensive examination of the results confirms that the proposed algorithm is superior to the other algorithms.

Table 3. Matching time (ms) on the "clothes" image set.

δ_s	ACC		IACC		Proposed	
	Match	n_Match	Match	n_Match	Match	n_Match
1	269.60	31.90	57.31	4.39	10.21	4.11
10	777.11	41.78	284.79	6.17	13.08	4.16
20	1113.03	64.06	436.48	8.53	18.80	4.20
30	1227.30	81.64	514.15	10.64	26.52	4.27
40	1334.33	100.00	561.21	12.36	29.65	4.34
50	1365.15	121.29	584.29	13.61	35.32	4.48

Table 4. Matching time (ms) on the "snack packs" image set.

δ_s	ACC		IACC		Proposed	
	Match	n_Match	Match	n_Match	Match	n_Match
1	62.61	6.03	10.27	5.01	7.64	4.98
10	204.05	6.43	21.05	5.08	8.11	4.97
20	231.66	6.64	23.86	5.17	8.78	4.98
30	244.62	6.80	25.71	5.24	9.38	5.03
40	252.61	6.98	26.74	5.29	10.08	5.01
50	257.75	7.09	27.59	5.32	10.49	4.99

Table 5. Matching time (ms) on the "SMVS (using TPS)" image set.

δ_s	ACC		IACC		Proposed	
	Match	n_Match	Match	n_Match	Match	n_Match
1	127.33	10.16	14.43	3.86	6.50	3.38
10	843.84	42.93	105.06	7.53	8.98	3.59
20	1063.17	69.73	142.06	10.68	10.88	3.63
30	1155.25	87.01	157.00	12.99	13.07	3.81
40	1189.76	98.73	164.42	13.53	15.42	3.95
50	1212.29	105.59	170.44	14.31	18.57	4.25

Table 6. Matching time (ms) on the "IN-N (using TPS)" image set.

δ_s	ACC		IACC		Proposed	
	Match	n_Match	match	n_Match	Match	n_Match
1	62.16	9.26	10.47	3.55	7.77	3.61
10	671.48	31.29	121.12	5.94	10.86	3.66
20	938.00	62.39	198.80	8.75	13.72	3.71
30	1072.40	85.72	240.55	10.60	16.87	3.76
40	1158.80	102.34	261.71	11.59	20.04	3.88
50	1208.94	113.75	280.67	12.63	22.93	3.87

Table 7. Matching time (ms) on the "Oxbuild (using TPS)" image set.

δ_s	ACC		IACC		Proposed	
	Match	n_Match	match	n_Match	Match	n_Match
1	115.44	22.67	34.61	9.68	21.09	7.14
10	1102.45	96.69	283.84	16.29	28.76	7.37
20	1518.45	177.52	405.74	23.83	32.11	7.42
30	1740.47	241.87	455.43	29.62	37.32	7.66
40	1826.40	278.72	486.04	32.35	44.66	7.85
50	1907.35	309.01	501.31	35.28	52.92	8.09

Table 8. Experiment results (TPR, FPR, Accuracy, and time (ms)).

Image Set	Result	RANSAC	ACC	IACC	Proposed
clothes	TPR	0.319	0.701	0.689	0.807
	FPR	0.401	0.012	0.009	0.010
	Accuracy	0.546	0.934	0.933	0.955
	time (ms)	71.98	358.22	126.21	14.77
Snack packs	TPR	0.317	0.773	0.777	0.847
	FPR	0.436	0.003	0.005	0.004
	Accuracy	0.541	0.976	0.975	0.983
	time (ms)	28.77	28.89	7.31	5.47
SMVS (using TPS)	TPR	0.983	0.923	0.923	0.948
	FPR	0.750	0.034	0.021	0.023
	Accuracy	0.585	0.946	0.954	0.963
	time (ms)	39.23	611.61	85.70	10.80
IN-N (using TPS)	TPR	0.852	0.669	0.659	0.775
	FPR	0.740	0.006	0.004	0.007
	Accuracy	0.566	0.961	0.962	0.971
	time (ms)	37.06	198.18	34.95	4.71
Oxbuild (using TPS)	TPR	0.832	0.753	0.775	0.830
	FPR	0.858	0.012	0.011	0.011
	Accuracy	0.494	0.941	0.946	0.957
	time (ms)	69.45	539.59	114.79	13.59

Figure 11 presents examples that show the matching results using the proposed algorithm, where red convex hull indicates a suitable cluster.

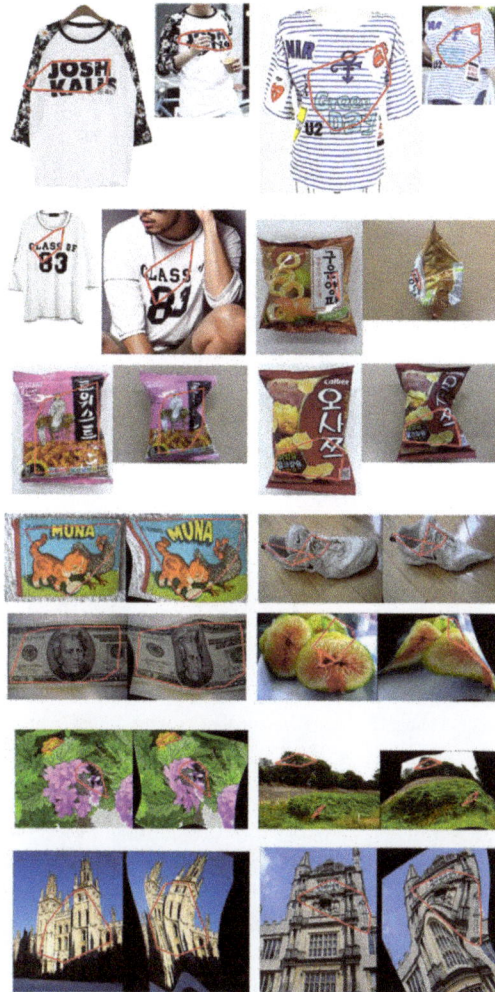

Figure 11. Examples of matching results using proposed algorithm.

5. Conclusions

In this paper, a new matching algorithm between images with deformable objects was proposed. A matching algorithm can be called a good algorithm if three aspects, i.e., robustness, independence, and fast matching, are excellent. Among these aspects, slow matching is the most significant weakness of conventional deformable object matching algorithms. To resolve this weakness, the speed was dramatically improved by reducing the complexity using the feature selection and BST (Binary Search Tree) clustering. The matching results were reliable because the suitability of the composed clusters is determined by the cluster verification step.

The experiment was performed using image sets with various deformable characteristics. As a result, while showing better TPR and FPR performance, compared to conventional algorithms, the proposed algorithm achieves 2–60 times faster matching speed than the conventional algorithms. Fast matching is a very important characteristic because image matching is used for content-based image retrieval. Therefore, the algorithm proposed in this paper can be used more effectively than the conventional algorithms in deformable object-contained image retrieval.

Acknowledgments: We would like to thank the anonymous reviewers for their generous review. This research was supported by Basic Science Research Program through the National Research Foundation of Korea (NRF) funded by the Ministry of Education (2010-0020163) and the Ministry of Science, ICT & Future Planning (2015R1C1A1A01055914).

Author Contributions: Jaehyup Jeong and Insu Won provided the main idea of this paper, designed the overall architecture of the proposed algorithm and wrote the paper; Jaehyup Jeong and HunJun Yang conducted the test data collection and designed the experiments; and Bowon Lee and Dongseork Jeong supervised the work and revised the paper.

Conflicts of Interest: The authors declare no conflict of interest.

References

1. Lowe, D.G. Distinctive image features from scale-invariant keypoints. *Int. J. Comput. Vis.* **2004**, *60*, 91–110. [CrossRef]
2. Bay, H.; Tuytelaars, T.; Van Gool, L. Surf: Speeded up robust features. In Proceedings of the European Conference on Computer Vision (ECCV), Graz, Austria, 7–13 May 2006; pp. 404–417.
3. Matas, J.; Chum, O.; Urban, M.; Pajdla, T. Robust wide-baseline stereo from maximally stable extremal regions. *Image Vis. Comput.* **2004**, *22*, 761–767. [CrossRef]
4. Mikolajczyk, M.; Schmid, C. Scale & affine invariant interest point detectors. *Int. J. Comput. Vis.* **2004**, *60*, 63–86.
5. Fischler, M.A.; Martin, R.C. Random sample consensus: A paradigm for model fitting with applications to image analysis and automated cartography. *ACM Proc. Commun.* **1981**, *24*, 381–395. [CrossRef]
6. Krizhevsky, A.; Sutskever, I.; Hinton, G.E. Imagenet classification with deep convolutional neural networks. In Proceedings of the Advances in Neural Information Processing Systems (NIPS), Stateline, NV, USA, 3–8 December 2012; pp. 1097–1105.
7. Duan, L.Y.; Lin, J.; Chen, J.; Huang, T.; Gao, W. Compact Descriptors for Visual Search. *IEEE Multimed.* **2014**, *21*, 30–41. [CrossRef]
8. Chen, D.M.; Tsai, S.S.; Chandrasekhar, V.; Takacs, G. Inverted Index Compression for Scalable Image Matching. In Proceedings of the IEEE 2010 Data Compression Conference, Snowbird, UT, USA, 24–26 March 2010; p. 525.
9. Chum, O.; Matas, J.; Kittler, J. Locally optimized RANSAC. *Pattern Recognit.* **2003**, *2781*, 236–243.
10. Li, Y.; Snavely, N.; Huttenlocher, D.P. Location recognition using prioritized feature matching. In Proceedings of the European Conference on Computer Vision, Heraklion, Greece, 5–11 September 2010; pp. 791–804.
11. Na, S.; Oh, W.; Jeong, D. A Frame-Based Video Signature Method for Very Quick Video Identification and Location. *ETRI J.* **2013**, *35*, 281–291. [CrossRef]
12. Calonder, M.; Lepetit, V.; Strecha, C.; Fua, P. BRIEF: Binary Robust Independent Elementary Features. In Proceedings of the European Conference on Computer Vision (ECCV), Heraklion, Greece, 5–11 September 2010; pp. 778–792.
13. Leutenegger, S.; Chli, M.; Siegwart, R. BRISK: Binary Robust Invariant Scalable Keypoints. In Proceedings of the IEEE International Conference on Computer Vision, Barcelona, Spain, 6–13 November 2011; pp. 2548–2555.
14. Alahi, A.; Ortiz, R.; Vandergheynst, P. FREAK: Fast Retina Keypoint. In Proceedings of the IEEE Conference on Computer Vision and Pattern Recognition, Providence, RI, USA, 16–21 June 2012; pp. 510–517.
15. Desai, A.; Lee, D.J.; Ventura, D. Matching Affine Features with the SYBA Feature Descriptor. In Proceedings of the Advances in Visual Computing, Las Vegas, NV, USA, 8–10 December 2014; pp. 448–457.
16. Fowers, S.G.; Desai, A.; Lee, D.J.; Ventura, D.; Wilde, D.K. An efficient tree-based feature descriptor and matching algorithm. *AIAA J. Aerosp. Inf. Syst.* **2014**, *11*, 596–606. [CrossRef]
17. Tran, Q.H.; Chin, T.J.; Carneiro, G.; Brown, M.S.; Suter, D. In defence of RANSAC for outlier rejection in deformable registration. In Proceedings of the European Conference on Computer Vision (ECCV), Firenze, Italy, 7–13 October 2012; pp. 274–287.
18. Pilet, J.; Lepetit, V.; Fua, P. Real-time nonrigid surface detection. In Proceedings of the IEEE Conference on Computer Vision and Pattern Recognition (CVPR), Miami, FL, USA, 20–25 June 2009; pp. 822–828.

19. Kettani, O.; Ramdani, F.; Tadili, B. An Agglomerative Clustering Method for Large Data Sets. *Int. J. Comput. Appl.* **2014**, *92*, 1–7. [CrossRef]
20. Zhou, F.; Torre, F.D. Factorized graph matching. In Proceedings of the IEEE Conference on Computer Vision and Pattern Recognition (CVPR), Providence, RI, USA, 16–21 June 2012; pp. 127–134.
21. Cho, M.; Lee, J.; Lee, K.M. Feature correspondence and deformable object matching via agglomerative correspondence clustering. In Proceedings of the IEEE International Conference on Computer Vision, Kyoto, Japan, 29 September–2 October 2009; pp. 1280–1287.
22. Yang, H.; Won, I.; Jeong, D. On the Improvement of Deformable Object Matching. In Proceedings of the Korea-Japan Joint Workshop on Frontiers of Computer Vision (FCV), Okinawa, Japan, 2–5 February 2014; pp. 279–282.
23. Francini, G.; Lepsøy, S.; Balestri, M. Selection of local features for visual search. *Signal Process. Image Commun.* **2013**, *28*, 311–322. [CrossRef]
24. Tsai, S.S.; Chen, D.; Takacs, G.; Chandrasekhar, V.; Vedantham, R.; Grzeszczuk, R.; Girod, B. Fast geometric re-ranking for image-based retrieval. In Proceedings of the IEEE International Conference on Image Processing, Hong Kong, China, 26–29 September 2010; pp. 1029–1032.
25. Lepsøy, S.; Francini, G.; Cordara, G.; Gusmão, P.P.B. Statistical modelling of outliers for fast visual search. In Proceedings of the IEEE International Conference on Multimedia and Expo, Barcelona, Spain, 11–15 July 2011; pp. 1–6.
26. Won, I.; Jeong, J.; Yang, H.; Kwon, J.; Jeong, D. Adaptive Image Matching Using Discrimination of Deformable Objects. *Symmetry* **2016**, *8*, 68. [CrossRef]
27. Chum, O.; Pajdla, T.; Sturm, P. The Geometric Error for Homographies. *Comput. Vis. Image Underst.* **2005**, *97*, 86–102. [CrossRef]
28. Jegou, H.; Douze, M.; Schmid, C. Hamming embedding and weak geometric consistency for large scale image search. In Proceedings of the European Conference on Computer Vision, Marseille, France, 12–18 October 2008; pp. 304–317.
29. Xie, H.; Gao, K.; Zhang, Y.; Li, J.; Liu, Y. Pairwise weak geometric consistency for large scale image search. In Proceedings of the ACM International Conference on Multimedia Retrieval, Trento, Italy, 18–20 April 2011; pp. 42–50.
30. Theodoridis, S.; Koutroumbas, K. *Pattern Recognition*, 3rd ed.; Academic Press: Cambridge, MA, USA, 2006; pp. 541–587.
31. Cormen, T.H.; Leiscrson, C.E.; Rivers, R.L.; Stein, C. *Introduction to Algorithms*, 3rd ed.; MIT Press: Cambridge, MA, USA; McGraw-Hill: New York, NY, USA, 2009; pp. 286–307.
32. Chandrasekhar, V.R.; Chen, D.M.; Tsai, S.S.; Cheung, N.; Chen, H.; Takacs, G.; Reznik, Y.; Vedantham, R.; Grzeszczuk, R.; Bach, J. The stanford mobile visual search data set. In Proceedings of the ACM Conference on Multimedia Systems, San Jose, CA, USA, 23–25 February 2011; pp. 117–122.
33. Deng, J.; Dong, W.; Socher, R. ImageNet: A Large-Scale Hierarchical Image Database. In Proceedings of the IEEE Conference on Computer Vision and Pattern Recognition, Miami, FL, USA, 20–25 June 2009; pp. 248–255.
34. Philbin, J.; Chum, O.; Isard, M.; Sivic, J.; Zisserman, A. Object retrieval with large vocabularies and fast spatial matching. In Proceedings of the IEEE Conference on Computer Vision and Pattern Recognition (CVPR), Minneapolis, MN, USA, 17–22 June 2007.
35. Khan, N.; McCane, B.; Mills, S. Better than SIFT? *Mach. Vis. Appl.* **2015**, *26*, 819–836. [CrossRef]
36. Kashif, M.; Deserno, T.M.; Haak, D.; Jonas, S. Feature description with SIFT, SURF, BRIEF, BRISK, or FREAK? A general question answered for bone age assessment. *Comput. Biol. Med.* **2016**, *68*, 67–75. [CrossRef] [PubMed]

MDPI

Article

Generalized Degree-Based Graph Entropies

Guoxiang Lu

School of Statistics and Mathematics, Zhongnan University of Economics and Law, No. 182 Nanhu Avenue, Wuhan 430073, China; lgxmath@zuel.edu.cn; Tel.: +86-27-8838-5335

Academic Editor: Angel Garrido
Received: 23 November 2016; Accepted: 20 February 2017; Published: 28 February 2017

Abstract: Inspired by the generalized entropies for graphs, a class of generalized degree-based graph entropies is proposed using the known information-theoretic measures to characterize the structure of complex networks. The new entropies depend on assigning a probability distribution about the degrees to a network. In this paper, some extremal properties of the generalized degree-based graph entropies by using the degree powers are proved. Moreover, the relationships among the entropies are studied. Finally, numerical results are presented to illustrate the features of the new entropies.

Keywords: network; information theory; entropy measure; graph entropy; generalized degree-based graph entropy; degree powers

1. Introduction

Nowadays, the research of complex networks has attracted many researchers. One interesting and important problem is to study the network structure by using different graph and network measures. Meanwhile, these graph and network measures have been widely applied in many different fields, such as chemistry, biology, ecology, sociology and computer science [1–7]. From the viewpoint of information theory, the entropy of graphs was initiated to be applied by Mowshowitz [8] and Trucco [9]. Afterwards, Dehmer introduced graph entropies based on information functionals which capture structural information and studied their properties [10–12]. The graph entropies have been used as the complexity measures of networks and measures for symmetry analysis. Recently, so-called generalized graph entropies have been investigated by Dehmer and Mowshowitz [13] for better analysis and applications such as machine learning. The generalized graph entropies can characterize the topology of complex networks more effectively [14].

The degree powers are extremely considerable invariants and studied extensively in graph theory and network science, so they are commonly used as the information functionals to explore the networks [15,16]. To study more properties of graph entropies based on the degree powers, Lu et al. obtained some upper and lower bounds which have different performances to bound the graph entropies in different kinds of graphs and showed their applications in structural complexity analysis [17,18]. Inspired by Dehmer and Mowshowitz [13], we focus on the relationships between degree powers and the parametric complexity measures and then we construct generalized degree-based graph entropies by using the concept of the mentioned generalized graph entropies. The structure of this paper is as follows: In Section 2, some definitions and notations of graph theory and the graph entropies we are going to study are reviewed. In Section 3, we describe the definition of generalized degree-based graph entropies which are motivated by Dehmer and Mowshowitz [13]. In Section 4, we present some extremal properties of such entropies related to the degree powers. Moreover, we give some inequalities among the generalized degree-based graph entropies. In Section 5, numerical results of an exemplary network are shown to demonstrate the new entropies. Finally, a short summary and conclusion are drawn in the last section.

2. Preliminaries to Degree-Based Graph Entropy

A *graph* or *network* G is an ordered pair (V, E) comprising a set V of *vertices* together with a set E of *edges*. In network science, vertices are called *nodes* sometimes. The *order* of a graph means the number of vertices. The *size* of a graph means the number of edges. A graph of order n and size m is recorded as an (n, m)-graph. The *degree* of a vertex v denoted by $d(v)$ or in short d_v means the number of edges that connect to it, where an edge that connects a vertex to itself (a loop) is counted twice. The maximum and minimum degree in a graph are often denoted by $\Delta(G)$ and $\delta(G)$. If every vertex has the same degree $(\Delta(G) = \delta(G))$, then G is called a *regular* graph, or is called a *d-regular* graph with vertices of degree d. An unordered pair of vertices $\{u, v\}$ is called connected if a path leads from u to v. A *connected* graph is a graph in which every unordered pair of vertices is connected. Otherwise, it is called a *disconnected* graph. Obviously, in a connected graph, $1 \leq \delta(G) \leq \Delta(G) \leq n - 1$. A *tree* is a connected graph in which any two vertices are connected by exactly only one path. So tree is a connected $(n, n - 1)$-graph. P_n is denoted a *path graph* characterized as a tree in which the degree of all but two vertices is 2 and the degree of the two remaining vertices is 1. S_n is denoted a *star graph* characterized as a tree in which the degree of all but one vertex is 1. More details can be seen in [17,18].

Next, we describe the concept of (Shannon's) entropy [19,20]. The notation "log" means the logarithm is based 2, and the notation "ln" means the logarithm is based e.

Definition 1. *Let* $\mathbf{p} = (p_1, p_2, \cdots, p_n)$ *be a probability distribution, namely,* $0 \leq p_i \leq 1$ *and* $\sum_{i=1}^{n} p_i = 1$. *The (Shannon's) entropy of the probability distribution is defined by*

$$H(\mathbf{p}) := -\sum_{i=1}^{n} p_i \log p_i.$$

In the above definition, we use $0 \log 0 = 0$ *for continuity of corresponding function.*

Definition 2. *Let* $G = (V, E)$ *be a graph of order n. For* $v_i \in V$, *we define*

$$p(v_i) := \frac{f(v_i)}{\sum_{j=1}^{n} f(v_j)},$$

where f is a meaningful information functional. According to the information functional f, the vertices are mapped to the non-negative real numbers.

Because $\sum_{i=1}^{n} p(v_i) = 1$, the quantities $p(v_i)$ can be seen as probability values. Then the graph entropy of G has been defined as follows [10,12,17,18].

Definition 3. *Let* $G = (V, E)$ *be a graph of order n and f be a meaningful information functional. The (Shannon's) graph entropy of G is defined by*

$$I_f(G) := -\sum_{i=1}^{n} \frac{f(v_i)}{\sum_{j=1}^{n} f(v_j)} \log \frac{f(v_i)}{\sum_{j=1}^{n} f(v_j)}.$$

Definition 4. *Let* $G = (V, E)$ *be a graph of order n. For* $v_i \in V$, *if* $f(v_i) = d_i$, *then*

$$p(v_i) = \frac{d_i}{\sum_{j=1}^{n} d_j}.$$

Therefore, the degree-based graph entropy of G is defined as

$$I_d(G) := -\sum_{i=1}^{n} \frac{d_i}{\sum_{j=1}^{n} d_j} \log \frac{d_i}{\sum_{j=1}^{n} d_j}.$$

3. Generalized Degree-Based Graph Entropy

Many generalizations of entropy measure have been proposed based on the definition of Shannon's entropy [21,22]. For example, Rényi entropy [23], Daròczy's entropy [24] and quadratic entropy [25] are representative generalized entropies. In [13], Dehmer and Mowshowitz introduce a new class of generalized graph entropies that derive from the generalizations of entropy measure mentioned above and present two examples.

Definition 5. *Let* $G = (V, E)$ *be a graph of order n. Then*

$$(1). \quad I_\alpha^1(G) := \frac{1}{1 - \alpha} \log \left[\sum_{i=1}^{n} \left(\frac{f(v_i)}{\sum_{j=1}^{n} f(v_j)} \right)^\alpha \right], \quad \alpha \neq 1;$$

$$(2). \quad I_\alpha^2(G) := \frac{1}{2^{1-\alpha} - 1} \left[\sum_{i=1}^{n} \left(\frac{f(v_i)}{\sum_{j=1}^{n} f(v_j)} \right)^\alpha - 1 \right], \quad \alpha \neq 1;$$

$$(3). \quad I^3(G) := \sum_{i=1}^{n} \frac{f(v_i)}{\sum_{j=1}^{n} f(v_j)} \left[1 - \frac{f(v_i)}{\sum_{j=1}^{n} f(v_j)} \right].$$

Definition 6. *Let* $G = (V, E)$ *be a graph of order n and A its adjacency matrix. Denote by* $\lambda_1, \lambda_2, \cdots, \lambda_n$ *the eigenvalues of G. If* $f(v_i) = |\lambda_i|$, *then the generalized graph entropies are as follows:*

$$(1). \quad {}^\lambda I_\alpha^1(G) := \frac{1}{1 - \alpha} \log \left[\sum_{i=1}^{n} \left(\frac{|\lambda_i|}{\sum_{j=1}^{n} |\lambda_j|} \right)^\alpha \right], \quad \alpha \neq 1;$$

$$(2). \quad {}^\lambda I_\alpha^2(G) := \frac{1}{2^{1-\alpha} - 1} \left[\sum_{i=1}^{n} \left(\frac{|\lambda_i|}{\sum_{j=1}^{n} |\lambda_j|} \right)^\alpha - 1 \right], \quad \alpha \neq 1;$$

$$(3). \quad {}^\lambda I^3(G) := \sum_{i=1}^{n} \frac{|\lambda_i|}{\sum_{j=1}^{n} |\lambda_j|} \left[1 - \frac{|\lambda_i|}{\sum_{j=1}^{n} |\lambda_j|} \right].$$

Definition 7. *Let* $G = (V, E)$ *be a graph of order n. Denote the collection of orbits by* $S = \{V_1, V_2, \cdots, V_k\}$ *and their respective probabilities by* $\frac{|V_1|}{n}, \frac{|V_1|}{n}, \cdots, \frac{|V_k|}{n}$, *where k is the number of orbits. Then another class of generalized graph entropies are derived as*

$$(1). \quad {}^o I_\alpha^1(G) := \frac{1}{1 - \alpha} \log \left[\sum_{i=1}^{k} \left(\frac{|V_i|}{n} \right)^\alpha \right], \quad \alpha \neq 1;$$

$$(2). \quad {}^o I_\alpha^2(G) := \frac{1}{2^{1-\alpha} - 1} \left[\sum_{i=1}^{k} \left(\frac{|V_i|}{n} \right)^\alpha - 1 \right], \quad \alpha \neq 1;$$

$$(3). \quad {}^o I^3(G) := \sum_{i=1}^{k} \frac{|V_i|}{n} \left[1 - \frac{|V_i|}{n} \right].$$

Because it is difficult to obtain the eigenvalues or the collection of orbits of graph G for a large-scale graph, and they may not meet the requirements visually, we focus on the complexity of the graphs or networks determined by the vertices themselves and the relationship between them in this paper. For a given graph G, the vertex degree is a significant graph invariant, which is related to structural properties of the graph. Most other properties of the complex network are based on the degree distribution, such as the clustering coefficient, the community structure and so on. The vertex degree in a graph or network is also intuitional and noticeable. The vertices with varying values of degree chosen as the main construction of the graph or network may decide the complexity of the graph or network. Hence, we study the generalized graph entropies based on the vertex degree and degree powers.

According to the above definitions of generalized graph entropies, let $f(v_i) = d_i$ for $v_i \in V$, then we obtain the generalized degree-based graph entropies as follows:

Definition 8.

(1). $\quad I_{\alpha,d}^1(G) := \dfrac{1}{1-\alpha} \log \left[\sum\limits_{i=1}^n \left(\dfrac{d_i}{\sum_{j=1}^n d_j} \right)^\alpha \right], \quad \alpha \neq 1;$ (1)

(2). $\quad I_{\alpha,d}^2(G) := \dfrac{1}{2^{1-\alpha}-1} \left[\sum\limits_{i=1}^n \left(\dfrac{d_i}{\sum_{j=1}^n d_j} \right)^\alpha - 1 \right], \quad \alpha \neq 1;$ (2)

(3). $\quad I_d^3(G) := \sum\limits_{i=1}^n \dfrac{d_i}{\sum_{j=1}^n d_j} \left[1 - \dfrac{d_i}{\sum_{j=1}^n d_j} \right].$ (3)

4. Properties of the Generalized Degree-Based Graph Entropies

In this section, we will show the relationships among the stated generalized degree-based graph entropies and the degree-based graph entropy. First we will present five simple propositions which can be inferred from the Rényi entropy and [13].

Proposition 1.

$$I_{\alpha,d}^1(G) = \frac{1}{1-\alpha} \log \left[(2^{1-\alpha} - 1) I_{\alpha,d}^2(G) + 1 \right];$$ (4)

$$I_{\alpha,d}^2(G) = \frac{1}{2^{1-\alpha}-1} \left[2^{(1-\alpha) I_{\alpha,d}^1(G)} - 1 \right];$$ (5)

$$I_d^3(G) = 1 - 2^{-I_{2,d}^1(G)} = \frac{1}{2} I_{2,d}^2(G).$$ (6)

Proof. Noticing that

$$\sum_{i=1}^n \left(\frac{d_i}{\sum_{j=1}^n d_j} \right)^\alpha = 2^{(1-\alpha) I_{\alpha,d}^1(G)} = (2^{1-\alpha} - 1) I_{\alpha,d}^2(G) + 1,$$

we can obtain the Equations (4) and (5). Let $\alpha = 2$, we have $\sum\limits_{i=1}^n \left(\frac{d_i}{\sum_{j=1}^n d_j} \right)^2 = 1 - I_d^3(G)$, then the Equation (6) follows. $\square$

Remark 1. *Proposition 1 can be seen as a special case of (12) and (16) in [13] when the value of the information functional is the degree of every vertex.*

Proposition 2.
$$\lim_{\alpha \to 1} I_{\alpha,d}^1(G) = \lim_{\alpha \to 1} I_{\alpha,d}^2(G) = I_d(G)$$ (7)

Proof. Using l'Hôspital's rule, we can obtain the Equation (7). $\square$

Proposition 3. *For $\alpha \in (-\infty, 1) \cup (1, +\infty)$, $I_{\alpha,d}^1(G)$ is monotonically decreasing with respect to α.*

Proof. The derivative of the function $I_{\alpha,d}^1(G)$ is

$$\frac{d I_{\alpha,d}^1(G)}{d\alpha} = -\frac{1}{(1-\alpha)^2} \sum_{i=1}^n q_i \log \frac{q_i}{p_i},$$

where $q_i = \frac{d_i^\alpha}{\sum_{j=1}^n d_j^\alpha}$, $p_i = \frac{d_i}{\sum_{j=1}^n d_j}$. Then $Q = (q_1, q_2, \cdots, q_n)$ and $P = (p_1, p_2, \cdots, p_n)$ are also probability distributions. From the nonnegativity of Kullback-Leibler divergence, we obtain $\frac{dI_{\alpha,d}^1(G)}{d\alpha} \leq 0$. The inequality implies that $I_{\alpha,d}^1(G)$ is monotonically decreasing with respect to α. $\square$

Proposition 4. *For $\alpha < 1$,*

$$I_{\alpha,d}^1(G) \geq I_d(G); \tag{8}$$

and for $\alpha > 1$,

$$I_{\alpha,d}^1(G) \leq I_d(G); \tag{9}$$

Proof. Using Proposition 2 and Proposition 3, we can obtain the equalities above easily. $\square$

Remark 2. *Proposition 2, Proposition 3 and Proposition 4 can be seen as the special cases of the Rényi entropy's properties.*

Proposition 5.

$$I_d^3(G) < \ln 2 \cdot I_d(G). \tag{10}$$

Proof. Using the standard inequality $\ln x < x - 1$ when $x \neq 1$, we have $\ln \frac{d_i}{\sum_{j=1}^n d_j} < \frac{d_i}{\sum_{j=1}^n d_j} - 1$. Therefore,

$$I_d^3(G) = \sum_{i=1}^n \frac{d_i}{\sum_{j=1}^n d_j} \left[1 - \frac{d_i}{\sum_{j=1}^n d_j} \right] = -\sum_{i=1}^n \frac{d_i}{\sum_{j=1}^n d_j} \left[\frac{d_i}{\sum_{j=1}^n d_j} - 1 \right]$$

$$< -\sum_{i=1}^n \frac{d_i}{\sum_{j=1}^n d_j} \ln \frac{d_i}{\sum_{j=1}^n d_j} = -\ln 2 \sum_{i=1}^n \frac{d_i}{\sum_{j=1}^n d_j} \log \frac{d_i}{\sum_{j=1}^n d_j} = \ln 2 \cdot I_d(G).$$

Then the inequality (10) follows. $\square$

Next we define the sum of the α-th degree powers as $D_\alpha := \sum_{i=1}^n d_i^\alpha$, where α is an arbitrary real number.

Theorem 1. *Let $G(n, m)$ be an (n, m)-graph. Then for $\alpha \neq 1$, we have*

$$(1). \quad I_{\alpha,d}^1(G) = \frac{1}{1-\alpha} \log \frac{D_\alpha}{(2m)^\alpha}; \tag{11}$$

$$(2). \quad I_{\alpha,d}^2(G) = \frac{1}{2^{1-\alpha} - 1} \left[\frac{D_\alpha}{(2m)^\alpha} - 1 \right]; \tag{12}$$

$$(3). \quad I_d^3(G) = 1 - \frac{D_2}{(2m)^2}. \tag{13}$$

Proof. By substituting $\sum_{i=1}^n d_i = 2m$ into the Equations (1)–(3), we have

$$I_{\alpha,d}^1(G) = \frac{1}{1-\alpha} \log \left[\sum_{i=1}^n \left(\frac{d_i}{2m} \right)^\alpha \right] = \frac{1}{1-\alpha} \log \left[\sum_{i=1}^n \frac{d_i^\alpha}{(2m)^\alpha} \right]$$

$$= \frac{1}{1-\alpha} \log \left[\frac{1}{(2m)^\alpha} \sum_{i=1}^n d_i^\alpha \right] = \frac{1}{1-\alpha} \log \left[\frac{D_\alpha}{(2m)^\alpha} \right];$$

$$I^2_{\alpha,d}(G) = \frac{1}{2^{1-\alpha}-1}\left[\sum_{i=1}^{n}\left(\frac{d_i}{2m}\right)^\alpha - 1\right] = \frac{1}{2^{1-\alpha}-1}\left[\left(\sum_{i=1}^{n}\frac{d_i^{\;\alpha}}{(2m)^\alpha}\right) - 1\right]$$

$$= \frac{1}{2^{1-\alpha}-1}\left[\frac{1}{(2m)^\alpha}\left(\sum_{i=1}^{n}d_i^{\;\alpha}\right) - 1\right] = \frac{1}{2^{1-\alpha}-1}\left[\frac{D_\alpha}{(2m)^\alpha} - 1\right];$$

$$I^3_d(G) = \sum_{i=1}^{n}\frac{d_i}{2m}\left[1 - \frac{d_i}{2m}\right] = \frac{1}{(2m)^2}\sum_{i=1}^{n}d_i(2m - d_i)$$

$$= \frac{1}{(2m)^2}\left[\left(2m\sum_{i=1}^{n}d_i\right) - \sum_{i=1}^{n}d_i^{\;2}\right] = \frac{1}{(2m)^2}\left[(2m)^2 - D_2\right]$$

$$= 1 - \frac{D_2}{(2m)^2}.$$

So the Equations (11)–(13) hold. □

From the above theorem, we know that the generalized degree-based graph entropies are closely related to the sum of the degree powers D_α. Obviously when $\alpha = 1$, $D_1 = \sum_{i=1}^{n}d_i = 2m$ presents the sum of degrees. The sum of the degree powers as an invariant is called *zeroth order general Randić index* [26–29]. For $\alpha = 2$, D_2 is also called *first Zagreb index* [30–33]. In [34], Chen et al. have reviewed D_α for different values of α and discussed the relationships with some indices such as Zagreb index, graph energies, HOMO-LUMO index, Estrada index [35–43].

Corollary 1. *Let $G(n, m)$ be an (n, m)-graph. Then we have*

$$1 - \frac{2m + (n-1)(n-2)}{4m(n-1)} \leq I^3_d(G) \leq 1 - \frac{1}{n}.$$

Proof. Using Cauchy-Buniakowsky-Schwarz inequality, we obtain

$$D_2 = \sum_{i=1}^{n}d_i^{\;2} \geq \frac{1}{n}\left(\sum_{i=1}^{n}d_i\right)^2 = \frac{(2m)^2}{n}.$$

In [44] de Caen obtains the following inequality

$$D_2 = \sum_{i=1}^{n}d_i^{\;2} \leq m\left(\frac{2m}{n-1} + n - 2\right).$$

So from Equation (13), we have

$$1 - \frac{m}{(2m)^2}\left(\frac{2m}{n-1} + n - 2\right) \leq 1 - \frac{D_2}{(2m)^2} \leq 1 - \frac{(2m)^2}{n}{(2m)^2},$$

or equivalently,

$$1 - \frac{2m + (n-1)(n-2)}{4m(n-1)} \leq I^3_d(G) \leq 1 - \frac{1}{n}.$$

We can also find some conditions for the equalities: If G is a regular graph, then the equality $I^3_d(G) = 1 - \frac{1}{n}$ holds; If G is a tree of order n, then the equality $1 - \frac{2m+(n-1)(n-2)}{4m(n-1)}$ holds. □

Corollary 2. *Let T be a tree of order n. Then we have*

$$I^3_d(S_n) \leq I^3_d(T) \leq I^3_d(P_n),$$

where S_n and P_n denote the star graph and path graph of order n, respectively.

Proof. In [45], Li and Zhao present that among all trees of order n, for $\alpha > 1$ or $\alpha < 0$, the path graph and the star graph attain the minimum and maximum value of D_α respectively; while for $0 < \alpha < 1$, the star graph and the path graph attain the minimum and maximum value of D_α respectively. Then using the Equation (13), the result of the corollary is obtained. □

Theorem 2. *When $\alpha < 1$, we have $I^1_{\alpha,d}(G) < \frac{2^{1-\alpha}-1}{(1-\alpha)\ln 2} I^2_{\alpha,d}(G)$; and when $\alpha > 1$, we have $I^1_{\alpha,d}(G) > \frac{2^{1-\alpha}-1}{(1-\alpha)\ln 2} I^2_{\alpha,d}(G)$. Especially, when $\alpha = 0$, we have $I^1_{\alpha,d}(G) = \frac{\log n}{n-1} I^2_{\alpha,d}(G)$.*

Proof. First we define a new function on α on the set of real numbers $\mathbb{R}$ as follows

$$f(\alpha) = \frac{D_\alpha}{(2m)^\alpha} = \frac{\sum\limits_{i=1}^{n} d_i^\alpha}{(2m)^\alpha}.$$

Because straightforward derivative shows

$$\frac{d}{d\alpha} f(\alpha) = \frac{1}{(2m)^\alpha} \left[\left(\sum_{i=1}^{n} d_i^\alpha \ln d_i \right) - \ln(2m) \left(\sum_{i=1}^{n} d_i^\alpha \right) \right]$$

$$\leq \frac{1}{(2m)^\alpha} \left[\ln(\Delta(G)) \left(\sum_{i=1}^{n} d_i^\alpha \right) - \ln(2m) \left(\sum_{i=1}^{n} d_i^\alpha \right) \right]$$

$$= \frac{\ln(\Delta(G)) - \ln(2m)}{(2m)^\alpha} \sum_{i=1}^{n} d_i^\alpha < 0 \text{ (by } \Delta(G) < 2m),$$

we can claim that $f(\alpha)$ is a strictly decreasing function on α.

For $f(1) = 1$, we have

$$\frac{D_\alpha}{(2m)^\alpha} = \frac{\sum\limits_{i=1}^{n} d_i^\alpha}{(2m)^\alpha} = \begin{cases} > 1, & \alpha < 1; \\ < 1, & \alpha > 1. \end{cases}$$

Using the standard inequality $\ln x < x - 1$ when $x \neq 1$, we find $\ln 2 \cdot \log \frac{D_\alpha}{(2m)^\alpha} < \frac{D_\alpha}{(2m)^\alpha} - 1$ when $\alpha \neq 1$.

Therefore, for $\alpha < 1$, we have

$$\frac{I^1_{\alpha,d}}{I^2_{\alpha,d}} = \frac{\frac{1}{1-\alpha} \log \frac{D_\alpha}{(2m)^\alpha}}{\frac{1}{2^{1-\alpha}-1} \left[\frac{D_\alpha}{(2m)^\alpha} - 1 \right]} < \frac{\frac{1}{(1-\alpha)\ln 2} \left[\frac{D_\alpha}{(2m)^\alpha} - 1 \right]}{\frac{1}{2^{1-\alpha}-1} \left[\frac{D_\alpha}{(2m)^\alpha} - 1 \right]} = \frac{2^{1-\alpha}-1}{(1-\alpha)\ln 2}.$$

For $\alpha > 1$, we have

$$\frac{I^1_{\alpha,d}}{I^2_{\alpha,d}} = \frac{\frac{1}{1-\alpha} \log \frac{D_\alpha}{(2m)^\alpha}}{\frac{1}{2^{1-\alpha}-1} \left[\frac{D_\alpha}{(2m)^\alpha} - 1 \right]} > \frac{\frac{1}{(1-\alpha)\ln 2} \left[\frac{D_\alpha}{(2m)^\alpha} - 1 \right]}{\frac{1}{2^{1-\alpha}-1} \left[\frac{D_\alpha}{(2m)^\alpha} - 1 \right]} = \frac{2^{1-\alpha}-1}{(1-\alpha)\ln 2}.$$

Especially, when $\alpha = 0$, $I^1_{\alpha,d} = \log n$ and $I^2_{\alpha,d} = n - 1$. So $I^1_{\alpha,d}(G) = \frac{\log n}{n-1} I^2_{\alpha,d}(G)$ holds in this case. □

Corollary 3. *When $0 \leq \alpha < 1$, we have $I^1_{\alpha,d}(G) < \frac{1}{\ln 2} I^2_{\alpha,d}(G)$.*

Proof. First we define a new function on α on the set of real numbers $\mathbb{R}$ as follows

$$g(\alpha) = 2^{1-\alpha} - 1 - (1 - \alpha).$$

Because the second order derivative shows

$$\frac{d^2}{d\alpha^2}g(\alpha) = 2^{1-\alpha}(\ln 2)^2 > 0,$$

we can claim that $g(\alpha)$ is a convex function on α. Since $g(0) = g(1) = 0$, we find $2^{1-\alpha} - 1 \leq 1 - \alpha$ for $0 \leq \alpha < 1$, or equivalently $0 < \frac{2^{1-\alpha}-1}{1-\alpha} \leq 1$ for $0 \leq \alpha < 1$. Using Theorem 2, the inequality $I^1_{\alpha,d}(G) < \frac{1}{\ln 2} I^2_{\alpha,d}(G)$ holds. $\square$

Theorem 3. *When $\alpha < 1$ and $1 < \alpha < 2$, we have $I^3_d(G) > (1 - 2^{1-\alpha})I^2_{\alpha,d}(G)$; when $\alpha > 2$, we have $I^3_d(G) < (1 - 2^{1-\alpha})I^2_{\alpha,d}(G)$; and when $\alpha = 2$, we have $I^3_d(G) = (1 - 2^{1-\alpha})I^2_{\alpha,d}(G) = \frac{1}{2}I^2_{\alpha,d}(G)$.*

Proof. First we have

$$I^2_{\alpha,d}(G) - I^3_d(G) = \frac{1}{2^{1-\alpha} - 1}\left[\frac{D_\alpha}{(2m)^\alpha} - 1\right] - \left[1 - \frac{D_2}{(2m)^2}\right]$$

$$= \frac{1}{1 - 2^{1-\alpha}}\left[1 - \frac{D_\alpha}{(2m)^\alpha}\right] - \left[1 - \frac{D_2}{(2m)^2}\right],$$

and $f(\alpha) = \frac{D_\alpha}{(2m)^\alpha}$ is a strictly decreasing function on α.

Therefore, for $\alpha < 1$, $f(\alpha) > f(2)$ and $\frac{1}{1-2^{1-\alpha}} < 0$ are obtained. Then we have

$$I^2_{\alpha,d}(G) - I^3_d(G) > \frac{1}{1 - 2^{1-\alpha}}\left[1 - \frac{D_2}{(2m)^2}\right] - \left[1 - \frac{D_2}{(2m)^2}\right]$$

$$= \left(\frac{1}{1 - 2^{1-\alpha}} - 1\right)\left[1 - \frac{D_2}{(2m)^2}\right] = \left(\frac{1}{1 - 2^{1-\alpha}} - 1\right)I^3_d(G)$$

This implies $I^3_d(G) > (1 - 2^{1-\alpha})I^2_{\alpha,d}(G)$.

For $1 < \alpha < 2$, $f(\alpha) > f(2)$ and $\frac{1}{1-2^{1-\alpha}} > 0$ are obtained. Then we have

$$I^2_{\alpha,d}(G) - I^3_d(G) < \frac{1}{1 - 2^{1-\alpha}}\left[1 - \frac{D_2}{(2m)^2}\right] - \left[1 - \frac{D_2}{(2m)^2}\right]$$

$$= \left(\frac{1}{1 - 2^{1-\alpha}} - 1\right)\left[1 - \frac{D_2}{(2m)^2}\right] = \left(\frac{1}{1 - 2^{1-\alpha}} - 1\right)I^3_d(G)$$

This implies $I^3_d(G) > (1 - 2^{1-\alpha})I^2_{\alpha,d}(G)$.

For $\alpha = 2$, using (6) we have $I^3_d(G) = (1 - 2^{1-\alpha})I^2_{\alpha,d}(G) = \frac{1}{2}I^2_{\alpha,d}(G)$.

For $\alpha > 2$, $f(\alpha) < f(2)$ and $\frac{1}{1-2^{1-\alpha}} > 0$ are obtained. Then we have

$$I^2_{\alpha,d}(G) - I^3_d(G) > \frac{1}{1 - 2^{1-\alpha}}\left[1 - \frac{D_2}{(2m)^2}\right] - \left[1 - \frac{D_2}{(2m)^2}\right]$$

$$= \left(\frac{1}{1 - 2^{1-\alpha}} - 1\right)\left[1 - \frac{D_2}{(2m)^2}\right] = \left(\frac{1}{1 - 2^{1-\alpha}} - 1\right)I^3_d(G)$$

This implies $I^3_d(G) < (1 - 2^{1-\alpha})I^2_{\alpha,d}(G)$. Thus we complete the proof. $\square$

Corollary 4. *When $\alpha \geq 2$, we have $I^3_\alpha(G) < I^2_{\alpha,d}(G)$.*

Proof. For $1 - 2^{1-\alpha} < 1$, we have the result by using Theorem 3 . $\square$

Theorem 4. *When $\alpha < 1$, we have $I^1_{\alpha,d}(G) > \frac{1}{\ln 2} \cdot I^3_d(G)$; when $1 < \alpha < 2$, we have $I^1_{\alpha,d}(G) < \frac{1}{(\alpha-1)\ln 2}$. $\frac{I^3_d(G)}{1 - I^3_d(G)}$; and when $\alpha \geq 2$, we have $I^1_{\alpha,d}(G) > \frac{1}{(\alpha-1)\cdot\ln 2}I^3_d(G)$.*

Proof. For $\alpha < 1$, using (8) and (10) we have $I^1_{\alpha,d}(G) > \frac{1}{\ln 2} \cdot I^3_d(G)$.

For $1 < \alpha < 2$, $f(\alpha) > f(2)$ and $\frac{1}{\alpha-1} > 0$ are obtained. Then we have

$$\frac{I^1_{\alpha,d}(G)}{I^3_d(G)} = \frac{\frac{1}{1-\alpha}\log\frac{D_\alpha}{(2m)^\alpha}}{1 - \frac{D_2}{(2m)^2}} < \frac{\frac{1}{1-\alpha}\log\frac{D_2}{(2m)^2}}{1 - \frac{D_2}{(2m)^2}} = \frac{1}{(\alpha-1)\ln 2}\cdot\frac{\ln\frac{(2m)^2}{D_2}}{1 - \frac{D_2}{(2m)^2}}$$

Using the standard inequality $\ln x < x - 1$ when $x \neq 1$, we find $\ln\frac{(2m)^2}{D_2} < \frac{(2m)^2}{D_2} - 1$. So

$$\frac{I^1_{\alpha,d}(G)}{I^3_d(G)} < \frac{1}{(\alpha-1)\ln 2}\cdot\frac{\frac{(2m)^2}{D_2}-1}{1 - \frac{D_2}{(2m)^2}} = \frac{1}{(\alpha-1)\ln 2}\cdot\frac{1}{\frac{D_2}{(2m)^2}} = \frac{1}{(\alpha-1)\ln 2}\cdot\frac{1}{1 - I^3_d(G)}.$$

This implies $I^1_{\alpha,d}(G) < \frac{1}{(\alpha-1)\ln 2}\cdot\frac{I^3_d(G)}{1-I^3_d(G)}$.

For $\alpha \geq 2$, using Theorem 2 and Theorem 3 we have $I^1_{\alpha,d}(G) > \frac{2^{1-\alpha}-1}{(1-\alpha)\ln 2}I^2_{\alpha,d}(G)$ and $I^3_d(G) \leq (1 - 2^{1-\alpha})I^2_{\alpha,d}(G)$. This implies $I^1_{\alpha,d}(G) > \frac{1}{(\alpha-1)\cdot\ln 2}I^3_d(G)$. Thus we complete the proof. □

5. Numerical Results

In order to illuminate the principle of generalized degree-based graph entropies, we show a network in Figure 1 as an example.

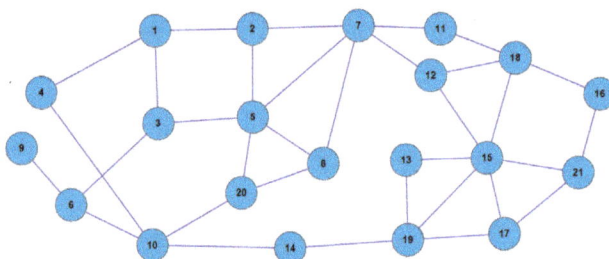

Figure 1. A simple network for example.

The degree of each node of the example network is shown in Table 1.

Table 1. The degree of each node of the example network.

node number	1	2	3	4	5	6	7	8	9	10	11
degree	3	3	3	2	5	3	5	3	1	4	2
node number	12	13	14	15	16	17	18	19	20	21	
degree	3	2	2	6	2	3	4	4	3	3	

We can easily calculate $I^3_d(G) = 0.946$. The details of $I^1_{\alpha,d}(G)$ and $I^2_{\alpha,d}(G)$ with different α are shown in Table 2.

Table 2. The generalized degree-based graph entropies $I^1_{\alpha,d}(G)$ and $I^2_{\alpha,d}(G)$ of the example network.

The Value of α	−1.0	−0.5	0.0	0.5	1.0	1.5	2.0	2.5	3.0	3.5	4.0
$I^1_{\alpha,d}(G)$	4.505	4.447	4.392	4.342	4.294	4.249	4.206	4.165	4.127	4.090	4.056
$I^2_{\alpha,d}(G)$	171.633	55.146	20.000	8.457	4.294	2.631	1.892	1.527	1.329	1.214	1.143

In Table 2 that the value of α is equal to 1.0 means $\alpha \to 1$. Then $I_{\alpha,d}^1(G)$ and $I_{\alpha,d}^2(G)$ are degenerated to the degree-based graph entropy $I_d(G) = 4.294$.

It is clear that following the increase of the value of α, the values of the generalized degree-based graph entropies $I_{\alpha,d}^1(G)$ and $I_{\alpha,d}^2(G)$ of the complex network are decrease. Based on the concept of the entropy, the bigger the value of the entropy is, the more complex of the network is. From the definitions of $I_{\alpha,d}^1(G)$ and $I_{\alpha,d}^2(G)$, the value of the entropic index α can be used to change the construction of the entropies. In other words, the value of α represents the relationship among the nodes in the complex network. Combined with the complex network, the influence of each node's degree on the entropies is changed by the value of α. The relationship between the value of α and the entropies of the complex network is shown as follows:

(1) When $\alpha < 1$, the nodes with small value of degree play an important part in the construction of $I_{\alpha,d}^1(G)$ and $I_{\alpha,d}^2(G)$, or they are chosen as the main construction of the complex networks. Especially when the value of $\alpha = 0$, each node has the same influence on the whole network from the entropic point of view.

(2) When $\alpha \to 1$, the influence of each node on the network is based on the value of degree for each node. The generalized degree-based graph entropies $I_{\alpha,d}^1(G)$ and $I_{\alpha,d}^2(G)$ are degenerated to the degree-based graph entropy $I_d(G)$. So the structure property determined by the node's degree decides the complexity of the complex network.

(3) When $\alpha > 1$, the nodes with big value of degree play an important part in the construction of $I_{\alpha,d}^1(G)$ and $I_{\alpha,d}^2(G)$, or they are chosen as the main construction of the complex networks. The values of the entropies tend to stabilization. The complex network is tended to orderly.

To sum up, according to the definition of the generalized degree-based graph entropies of the complex network, the value of the entropic index α is used to describe the different relationship among the nodes. When the value of α is smaller than 1, the nodes with small value of degree are more important than the nodes with big value of degree. The edges among those nodes with small value of degree become the main part of the complex network. As these nodes with small value of degree are the majority in the complex network, the whole network has greater complexity. When the value of α is equal to 0, the nodes in the network are equal to each other in terms of influence. When the value of α tends to 1, $I_{\alpha,d}^1(G)$ and $I_{\alpha,d}^2(G)$ are degenerated to $I_d(G)$, the level of complexity for the complex network is decided by the structure property. In other words, the complexity of the complex network is decided by the degree sequence and degree distribution. When the value of α is trended to ∞, the construction of the complex network is decided by the node which has a biggest value of degree, the values of $I_{\alpha,d}^1(G)$ and $I_{\alpha,d}^2(G)$ decrease to stable values, and the complex network is more orderly. The complexity of the complex network is not only decided by the structure of the complex network, but also influenced by the kind of the relationship between each node.

From Figure 2, we can see the plotted values of the generalized degree-based graph entropies $I_{\alpha,d}^1(G)$, $I_{\alpha,d}^2(G)$ and $I_d^3(G)$ relative to α ($I_{\alpha,d}^1(G)$, $I_{\alpha,d}^2(G)$ with a pole at $\alpha = 1$). The numerical results can be interpreted as follows: First we observe that the value of $I_{\alpha,d}^1(G)$ is less than that of $I_{\alpha,d}^2(G)$ for $\alpha < 1$, while the value of $I_{\alpha,d}^1(G)$ is larger than that of $I_{\alpha,d}^2(G)$ for $\alpha > 1$. Next, for $I_{\alpha,d}^1(G)$, $I_{\alpha,d}^2(G)$ and $I_d^3(G)$, we can have the values of $I_{\alpha,d}^1(G)$ and $I_{\alpha,d}^2(G)$ is always larger than that of $I_d^3(G)$. Actually, using l'Hôspital's rule we have that the value of $I_{\alpha,d}^1(G)$ tends to 3.459 and the value of $I_{\alpha,d}^2(G)$ tends to 1 when α tends to $+\infty$. At last, all the curves verify the inequalities in the Section 4.

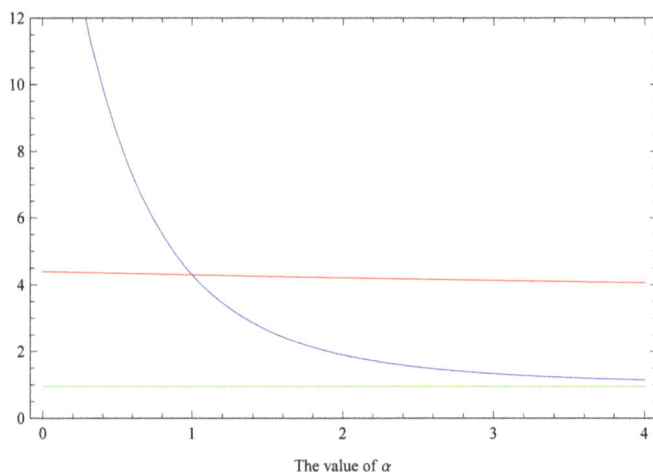

Figure 2. $I_{\alpha,d}^1(G)$(red), $I_{\alpha,d}^2(G)$(blue) and $I_d^3(G)$(green) versus α. ($I_{\alpha,d}^1(G)$, $I_{\alpha,d}^2(G)$ with a pole at $\alpha = 1$).

In addition, generalized graph entropy measures with parameters have been presented to be useful in studying the complexity associated with machine learning. For example, Dehmer et al. have described that the generalized graph entropies can be applied to the graph classification and clustering cases in machine learning. The applications involve optimizing particular parameters associated with graphs or networks in given sets [4,46]. So by applying supervised machine learning methods, the generalized degree-based entropies can be used for classifying the chemical structures, developing methods for characterizing predictive models according to optimal values of relevant parameters in bioinformatics, systems biology, and drug design.

6. Summary and Conclusions

In this paper , we studied the generalized degree-based graph entropies, which are inspired by Dehmer and Mowshowitz in [13] and derived from the Rényi entropy [23], Daròczy's entropy [24] and quadratic entropy [25]. We studied the relationships between the sum of the degree powers and the new entropies. Then we examined the extremal values of the above stated entropies in terms of the sum of the degree powers. We also proved some inequalities between these generalized degree-based graph entropies. Finally, we obtained numerical values for an exemplary complex network for each of the entropies, and concluded that their parameters can influence which kind of nodes contribute to the main part of the network in terms of graph entropy theory. The generalized degree-based graph entropies expand the description methods of the structural complexity of the complex networks. They would play bigger roles in describing structural symmetry and asymmetry in real networks in the future.

Acknowledgments: The author would like to thank the editor and referees for their helpful suggestions and comments on the manuscript.

Conflicts of Interest: The author declares no conflict of interest.

References

1. Solé, R.V.; Montoya, J.M. Complexity and fragility in ecological networks. *Proc. R. Soc. Lond. B* **2001**, *268*, 2039–2045.
2. Mehler, A.; Lücking, A.; Weiß, P. A network model of interpersonal alignment. *Entropy* **2010**, *12*, 1440–1483.
3. Ulanowicz, R.E. Quantitative methods for ecological network analysis. *Comput. Biol. Chem.* **2004**, *28*, 321–339.

4. Dehmer, M.; Barbarini, N.; Varmuza, K.; Graber, A. Novel topological descriptors for analyzing biological networks. *BMC Struct. Biol.* **2010**, *10*, doi:10.1186/1472-6807-10-18.

5. Dehmer, M.; Graber, A. The discrimination power of molecular identification numbers revisited. *MATCH Commun. Math. Comput. Chem.* **2013**, *69*, 785–794.

6. Garrido, A. Symmetry in Complex Networks. *Symmetry* **2011**, *3*, 1–15.

7. Dehmer, M. Information Theory of Networks. *Symmetry* **2011**, *3*, 767–779.

8. Mowshowitz, A. Entropy and the complexity of the graphs: I. An index of the relative complexity of a graph. *Bull. Math. Biophys.* **1968**, *30*, 175–204.

9. Trucco, E. A note on the information content of graphs. *Bull. Math. Biophys.* **1956**, *18*, 129–135.

10. Dehmer, M. Information processing in complex networks: Graph entropy and information functionals. *Appl. Math. Comput.* **2008**, *201*, 82–94.

11. Dehmer, M.; Varmuza, K.; Borgert, S.; Emmert-Streib, F. On entropy-based molecular descriptors: Statistical analysis of real and synthetic chemical structures. *J. Chem. Inf. Model.* **2009**, *49*, 1655–1663.

12. Dehmer, M.; Sivakumar, L.; Varmuza, K. Uniquely discriminating molecular structures using novel eigenvalue-based descriptors. *MATCH Commun. Math. Comput. Chem.* **2012**, *67*, 147–172.

13. Dehmer, M.; Mowshowitz, A. Generalized graph entropies. *Complexity* **2011**, *17*, 45–50.

14. Dehmer, M.; Li, X.; Shi, Y. Connections between generalized graph entropies and graph energy. *Complexity* **2015**, *21*, 35–41.

15. Cao, S.; Dehmer, M.; Shi, Y. Extremality of degree-based graph entropies. *Inform. Sci.* **2014**, *278*, 22–33.

16. Cao, S.; Dehmer, M. Degree-based entropies of networks revisited. *Appl. Math. Comput.* **2015**, *261*, 141–147.

17. Lu, G.; Li, B.; Wang, L. Some new properties for degree-based graph entropies. *Entropy* **2015**, *17*, 8217–8227.

18. Lu, G.; Li, B.; Wang, L. New Upper Bound and Lower Bound for Degree-Based Network Entropy. *Symmetry* **2016**, *8*, 8.

19. Shannon, C.E.; Weaver, W. *The Mathematical Theory of Communication*; University of Illinois Press: Urbana, IL, USA, 1949.

20. Cover, T.M.; Thomas, J.A. *Elements of Information Theory*, 2nd ed.; Wiley & Sons: New York, NY, USA, 2006.

21. Aczél, J.; Daróczy, Z. *On Measures of Information and Their Characterizations*; Academic Press: New York, NY, USA, 1975.

22. Arndt, C. *Information Measures*; Springer: Berlin, Germany, 2004.

23. Rényi, P. On measures of information and entropy. In *Proceedings of the 4th Berkeley Symposium on Mathematics, Statistics and Probability, Volume 1*; University of California Press: Berkeley, CA, USA, 1961; pp. 547–561.

24. Daràczy, Z.; Jarai, A. On the measurable solutions of functional equation arising in information theory. *Acta Math. Acad. Sci. Hungar.* **1979**, *34*, 105–116.

25. Watanabe, S. A new entropic method of dimensionality reduction specially designed for multiclass discrimination (DIRECLADIS). *Pattern Recogn. Lett.* **1983**, *2*, 1–4.

26. Randić, M. Characterization of molecular branching. *J. Am. Chem. Soc.* **1975**, *97*, 6609–6615.

27. Hu, Y.; Li, X.; Shi, Y.; Xu, T.; Gutman, I. On molecular graphs with smallest and greatest zeroth-order general Randić index. *MATCH Commun. Math. Comput. Chem.* **2005**, *54*, 425–434.

28. Hu, Y.; Li, X.; Shi, Y.; Xu, T. Connected (n, m)-graphs with minimum and maximum zeroth-order general Randić index. *Discrete Appl. Math.* **2007**, *155*, 1044–1054.

29. Li, X.; Shi, Y. A survey on the Randić index. *MATCH Commun. Math. Comput. Chem.* **2008**, *59*, 127–156.

30. Arezoomand, M.; Taeri, B. Zagreb indices of the generalized hierarchical product of graphs. *MATCH Commun. Math. Comput. Chem.* **2013**, *69*, 131–140.

31. Gutman, I. An exceptional property of first Zagreb index. *MATCH Commun. Math. Comput. Chem.* **2014**, *72*, 733–740.

32. Vasilyev, A.; Darda, R.; Stevanovic, D. Trees of given order and independence number with minimal first Zagreb index. *MATCH Commun. Math. Comput. Chem.* **2014**, *72*, 775–782.

33. Lin, H. Vertices of degree two and the first Zagreb index of trees. *MATCH Commun. Math. Comput. Chem.* **2014**, *72*, 825–834.

34. Chen, Z.; Dehmer, M.; Shi, Y. Bounds for degree-based network entropies. *Appl. Math. Comput.* **2015**, *265*, 983–993.

35. Estrada, E. Characterization of 3D molecular structure. *Chem. Phys. Lett.* **2000**, *319*, 713–718.

36. Estrada, E.; Rodriguez-Velazquez J.A. Subgraph centrality in complex networks. *Phys. Rev. E.* **2005**, *71*, 056103.
37. De la Peña, J.A.; Gutman, I.; Rada, J. Estimating the Estrada index. *Linear. Algebra. Appl.* **2007**, *427*, 70–76.
38. Das, K.; Xu, K.; Gutman, I. On Zagreb and Harary indices. *MATCH Commun. Math. Comput. Chem.* **2013**, *70*, 301–314.
39. Das, K.C.; Gutman, I.; Cevik, A.; Zhou, B. On Laplacian energy. *MATCH Commun. Math. Comput. Chem.* **2013**, *70*, 689–696.
40. Li, X.; Li, Y.; Shi, Y.; Gutman, I. Note on the HOMO-LUMO index of graphs. *MATCH Commun. Math. Comput. Chem.* **2013**, *70*, 85–96.
41. Khosravanirad, A. A lower bound for Laplacian Estrada index of a graph. *MATCH Commun. Math. Comput. Chem.* **2013**, *70*, 175–180.
42. Abdo, H.; Dimitrov, D.; Reti, T.; Stevanovic, D. Estimating the spectral radius of a graph by the second Zagreb index. *MATCH Commun. Math. Comput. Chem.* **2014**, *72*, 741–751.
43. Bozkurt, S.B.; Bozkurt, D. On incidence energy. *MATCH Commun. Math. Comput. Chem.* **2014**, *72*, 215–225.
44. De Caen, D. An upper bound on the sum of squares of degrees in a graph. *Discrete Math.* **1998**, *185*, 245–248.
45. Li, X.; Zhao, H. Trees with the first three smallest and largest generalized topological indices. *MATCH Commun. Math. Comput. Chem.* **2004**, *50*, 57–62.
46. Dehmer, M.; Mueller, L.A.J.; Emmert-Streib, F. Quantitative network measures as biomarkers for classifying prostate cancer disease states: A systems approach to diagnostic biomarkers. *PLoS ONE* **2013**, *8*, e77602.

symmetry

MDPI

Article

Multi-Class Disease Classification in Brain MRIs Using a Computer-Aided Diagnostic System

Muhammad Faisal Siddiqui [1,2], Ghulam Mujtaba [3,4], Ahmed Wasif Reza [5,*] and Liyana Shuib [3,*]

1　Faculty of Engineering, Department of Electrical Engineering, University of Malaya, Kuala Lumpur 50603, Malaysia; mfaisal_bce@yahoo.com
2　Department of Electrical Engineering, Faculty of Engineering, COMSATS Institute of Information Technology, Islamabad 45550, Pakistan
3　Department of Information Systems, Faculty of Computer Science and Information Technology, University of Malaya, Kuala Lumpur 50603, Malaysia; mujtaba@iba-suk.edu.pk
4　Department of Computer Science, Sukkur Institute of Business Administration, Sukkur 65200, Pakistan
5　Department of Computer Science and Engineering, Faculty of Science and Engineering, East West University, Dhaka 1212, Bangladesh
*　Correspondence: awreza98@yahoo.com (A.W.R.); liyanashuib@um.edu.my (L.S.); Tel.: +880-2-09666775577-395 (A.W.R.)

Academic Editor: Angel Garrido
Received: 29 November 2016; Accepted: 1 March 2017; Published: 8 March 2017

Abstract: Background: An accurate and automatic computer-aided multi-class decision support system to classify the magnetic resonance imaging (MRI) scans of the human brain as normal, Alzheimer, AIDS, cerebral calcinosis, glioma, or metastatic, which helps the radiologists to diagnose the disease in brain MRIs is created. Methods: The performance of the proposed system is validated by using benchmark MRI datasets (OASIS and Harvard) of 310 patients. Master features of the images are extracted using a fast discrete wavelet transform (DWT), then these discriminative features are further analysed by principal component analysis (PCA). Different subset sizes of principal feature vectors are provided to five different decision models. The classification models include the J48 decision tree, k-nearest neighbour (kNN), random forest (RF), and least-squares support vector machine (LS-SVM) with polynomial and radial basis kernels. Results: The RF-based classifier outperformed among all compared decision models and achieved an average accuracy of 96% with 4% standard deviation, and an area under the receiver operating characteristic (ROC) curve of 99%. LS-SVM (RBF) also shows promising results (i.e., 89% accuracy) when the least number of principal features was used. Furthermore, the performance of each classifier on different subset sizes of principal features was (80%–96%) for most performance metrics. Conclusion: The presented medical decision support system demonstrates the potential proof for accurate multi-class classification of brain abnormalities; therefore, it has a potential to use as a diagnostic tool for the medical practitioners.

Keywords: computer aided diagnostic system; neuroimaging; brain magnetic resonance imaging (MRI); multi-classification; medical imaging

1. Introduction

In this modern era, different advanced imaging modalities (e.g., X-rays, computerized tomography (CT) scans, positron emission tomography (PET), single-photon emission computerized tomography (SPECT), and magnetic resonance imaging (MRI)) are used in neurology and basic neuroscience fields. In X-rays and CT scans, the patients are exposed in ionizing radiation waves which may increase the risk of developing cancers, whereas PET and SPECT use radioactive tracers, with minimal exposure to harmful radiation. However, MRI is a non-invasive, dominant, and flexible modality to investigate the pathological conditions of the brain and other body parts. The common practice to identify the brain abnormalities is done by MRI. MRI scans provide high contrast and high spatial resolution images, which enables to differentiate the characteristics of the soft tissues. Magnetic resonance (MR) image texture is used to distinguish between the healthy and diseased anatomy. Brain MR images having any diseased (such as Alzheimer, AIDS dementia, cerebral calcinosis, glioma, or metastatis) are categorized by large cells and high contrast, which can be identified by abrupt changes in the images [1,2]. In recent years, machine learning techniques have been widely employed in the medical domain to support decision-making [3–10]. Moreover, medical decision support systems are in high demand to automatically detect these abrupt changes properly and classify the brain MRI as normal or any class of disease [3,5]. The very large amount of MR imaging data generates complexity to interpret the full pattern of atrophy by the existing visual inspection method. Therefore, it generates the requirement of computer-aided diagnosis (CAD) system to identify the specific condition of the brain MRI and enhance the diagnostic capabilities of the medical personnel. The radiologists can use these automated systems as an instrument for diagnosis, pre-surgical, and post-surgical procedures [4,11–22].

Generally, supervised classification methods are used instead of unsupervised methods, because of better accuracy, for brain MRI classification. There are, commonly, three phases involved in the implementation of a classifier for medical images: (1) features extraction; (2) feature reduction; and (3) training/testing of classification models. In the first step, discriminative features are extracted from brain MR images. Then, in the second step, these features are processed by some feature reduction technique to reduce the dimensionality of the features, such as PCA, linear discriminant analysis (LDA), etc. Finally, the reduced principal features are used to train the classifier model and classify the query images on the bases of these features. Widely-used decision models include naïve Bayes (NB), J48, random forest (RF), k-nearest neighbor (kNN), and support vector machine (SVM).

Recently, various feature selection schemes and machine learning decision models for brain MRI classification have been proposed. In [6,14–19], the authors have used 2D-DWT (two-dimensional discrete wavelet transform) and principal component analysis (PCA) for a features extraction and selection, respectively. Zhang et al. [14–17] proposed different advanced decision models based solutions with DWT and PCA techniques in their research and achieved promising results with some limitations. They used forward neural networks (FNN) with scaled chaotic artificial bee colony (SCABC) [14], back propagation with conjugate gradient method [15], kernel support vector machines (KSVM) with Gaussian radial basis function (GRB) [16] and particle swarm optimization (PSO) [17] as decision models to improve the efficiency of the brain MRI classifier. The schemes proposed in [18,19], have used feed-forward back propagation artificial neural networks (FP-ANN), kNN, feedback pulse-coupled neural network (FBPNN) and achieved an average accuracy of 99% for binary classification of brain MR images. Recently, [20–22] have proposed numerous complex feature engineering techniques with SVM- and NB-based decision models to enhance the classifier performance. In [20,21], the authors proposed Ripplet transform and discrete wavelet packet transform, respectively, instead of DWT for feature extraction. Whereas, the authors proposed wavelet entropy method in [22] for feature reduction. Wang et al. [23] proposed a different classification scheme by using dual-tree complex wavelet transform and twin support vector machine for the classification of pathological brain detection. In [23], the authors have achieved average accuracy of 99.57%. However, usage of a small number of cases and limited number of disease categories in their datasets are the main limitation of these works and also their performance is significantly reduced when large datasets are used. Regardless, the advanced complex methods are used in [20–23], which increases the complexity of the classifier and consumed relatively higher computational time, are not able to perform well on large datasets. Conversely, the technique proposed in [6] achieve accurate results for larger datasets; when using DWT, PCA, and LS-SVM (RBF) for feature extraction, feature reduction and classification, respectively. In addition, all these schemes [6,14–23] have proposed for binary classification and only capable to predict normal and abnormal anatomy of a brain MRI. Furthermore, a multi-class brain MRI classifier has proposed by Zacharaki et al. [24]. The authors extended the scope of the features and included age, tumour shape, and ROI (region of interest) as a part of feature sets. This technique is semiautomatic because ROIs needs to trace manually. In [24], the authors have tested three different disease classes (i.e., meningioma, glioma, and metastasis) and achieved a mean accuracy about 90%. The comparison between linear discriminant analysis (LDA), kNN, and non-linear SVM-based decision models were shown. The main limitation of this technique is that it needs human intervention for classification. Therefore, the research gap in the development of fully automatic multi-class classifier with significant accuracy is generated.

The main motivation behind this study is to develop an accurate multi-class brain MRI classifier, which is capable to diagnose the diseases class in brain MRIs. The proposed multi-class brain MRI classifier has a potential to classify five different brain diseases. These brain diseases include Alzheimer, AIDS dementia, cerebral calcinosis, glioma, and metastatsis. The proposed system technique is composed of three sub-models; master feature extraction, principal feature analysis and decision models. Fast discrete wavelet transform (DWT) used for extracting the master features from the brain MR images. The principal feature analysis was done by PCA and different subsets were used to calculate the efficiency of the multi-class classifier system. In addition, PCA analysis reduced the dimension of the master features, which also decreases the classification time and complexity of the classifier. For a comprehensive comparison of decision models' performance on the multi-classification of brain MRIs, the proposed research compared five different decision models (J48, kNN, RF, and LS-SVM with polynomial (Poly) and radial basis functions (RBF)). For comparative analysis with the proposed system, some of the other published methods from recent literature [6,18,20,21] were also tested using the same large datasets.

2. Materials and Methods

The proposed classifier for multi-classification is composed of master feature extraction, principal feature analysis, and classification model blocks, as shown in Figure 1, which illustrates the methodology of the proposed system. The classifier is constructed and evaluated using two phases: (1) a training phase, and (2) a testing phase. In the training phase, the classifier is trained by randomly selected images from the datasets. Once the classifier is trained, then it is capable to classify the query images. In the testing phase, the query image(s) is/are fed to the trained classifier to classify the image(s) as normal, Alzheimer, AIDS, cerebral calcinosis, metastatic, or glioma. Furthermore, a five-fold cross-validation is used in this work to minimize the generalization error.

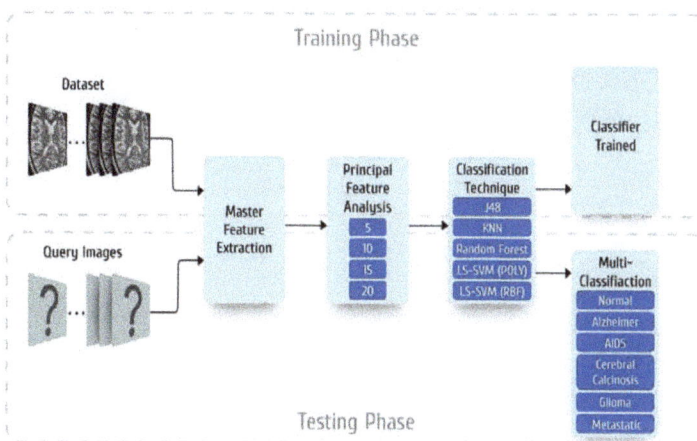

Figure 1. Methodology of the proposed classifier.

2.1. Dataset Collection

The benchmark MRI dataset used in this research was collected from 'Open Access Series of Imaging Studies (OASIS)' and 'Harvard Medical School' MRI databases to validate the proposed classification system. This database consists of human brain MRI images in the axial plane. These datasets were acquired using the following scan parameters: Voxel res: $1.0 \times 1.0 \times 1.25$ (mm^3), Rect. FOV: 256/256, TR: 9.7 (ms), TE: 4.0 (ms), TI: 20.0 (ms), and flip angle: $10°$. The dimensions of the image are 256×256 in a plane-resolution. Three hundred and ten patients' (men and women) brain MRI scans were involved to formulate this database.

The brain MR image dataset is composed of healthy and abnormal images. The abnormal image database has five types of different brain diseases. The abnormal MRI scan images having the following diseases: Alzheimer's disease, AIDS dementia, cerebral calcinosis, glioma and metastatic dementia. A sample image of each class of the images included in the benchmark dataset is shown in Figure 2.

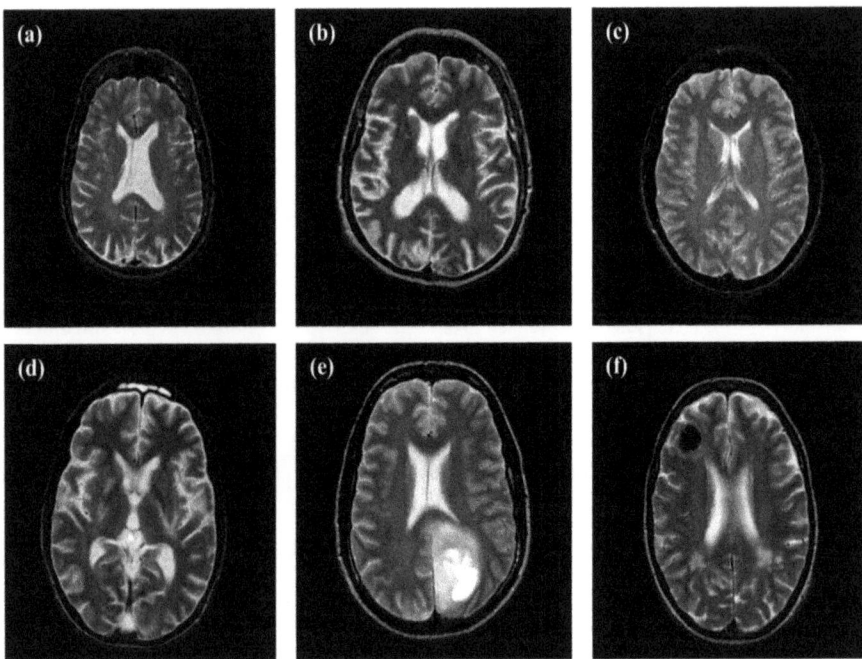

Figure 2. The sample images of healthy and abnormal magnetic resonance imaging (MRI) (**a**) normal/healthy; (**b**) Alzheimer's disease; (**c**) AIDS dementia; (**d**) cerebral calcinosis; (**e**) glioma; and (**f**) metastatic dementia.

The dataset is comprised of 310 brain MR images having 70 healthy (normal), 70 Alzheimer, 50 AIDS, and 40 each for cerebral calcinosis, glioma, and metastasis. The distribution of training and testing images is shown in Table 1. The ratio of training and testing images, i.e., 70% of the dataset is used for training and the remaining 30% of the dataset is used for testing purposes. Training images were used to construct the classifier, whereas testing images were used to evaluate the performance of the multi-class classifier. In addition, the testing images were unknown to the constructed classifier for the sake of unbiased evaluation.

Symmetry **2017**, *9*, 37

Table 1. The distribution of training and testing images.

Class	Total No. of Images	Total No. of Training Images	Total No. of Testing Images	Distribution (%)
Normal	70	49	21	22.58
Alzheimer	70	49	21	22.58
Aids	50	35	15	16.13
Cerebral Calcinosis	40	28	12	12.90
Glioma	40	28	12	12.90
Metastasis	40	28	12	12.90

2.2. Master Feature Extraction

A MATLAB (R2013a, The Mathworks, Inc. Natick, MA, United States) script was written, using discrete wavelet transform, to extract the main features of the brain MR images. To improve the efficiency of a classifier, the master features in the MRI image is needed to be identified properly. In recent literature [6,14–22,24–27], there are many different algorithms (such as DWT and Ripplet transform) used to extract the main features of the images. DWT has some advantages over RT, being less computationally complex and also due to the characteristics of brain MRIs. The sparse nature of MRIs provides an opportunity to identify the major contributed features of the MR image by representing it in some sophisticated domains (such as wavelet domains) [28]. Thus, DWT provides master features, having rich knowledge of the input MR image pattern with less complex implementation. The master features extracted from the MRI database using DWT has a potential to increase the capability of the decision making power and complexity of the classifier. A three-level "Haar" DWT was used to extract the master features of the images in this paper.

2.3. Preparation of the Principal Feature Vector

The main characteristics of any robust and accurate classifier are a selection of the discriminative features from the dataset and reduce the dimensions of the dataset. Large databases increase the feature dimensions, which eventually increase the complexity of the classifier and demands excessive time to classify. Therefore, different feature reduction schemes are used by the researchers to remove the curse of dimensionality problems in the classifier system [19,29–31].

In this article, the PCA technique was applied on the discriminative features of MR image to further reduce the dimension of the master features extracted by DWT. PCA preserved the variance by extracting the linear lower-dimensional representation of the MR image features [32,33]. Therefore, it extracts the major components of the image (data) and forms the principal feature vector. This leads to an increase the efficiency of the classifier system.

2.3.1. Feature Subset Sizes

Principal feature vectors are used for decision modelling. However, subsets of the principal feature vectors were introduced to check the performance trend of the classifier. Subset sizes of 5, 10, 15, and 20 principal components were used to compute the results of the proposed multi-class classifier.

2.4. Classifier Models

Weka toolkit (Version 3.8, University of Waikato, Hamilton, New Zealand) was used to develop decision models. Principal feature vectors were exported from MATLAB in comma separated values (CSV) format. Afterwards the CSV data were loaded into Weka for further analysis. Five different decision models were constructed for performance measures. These five classifiers were J48, k-nearest neighbour (kNN), random forest (RF), least-square support vector machine with polynomial kernel (LS-SVM (Poly)), and least-squares support vector machine with radial basis function kernel (LS-SVM (RBF)).

2.4.1. J48 Classifier (J48)

J48 is a kind of decision tree algorithm [34]. J48 utilizes the entropy to compute the homogeneity of a sample. If entropy is zero, then it means that the sample is completely homogeneous and if the sample is unequally divided, then it has entropy of one. The relative entropy of a given dataset X having positive and negative class instances is mathematically defined as:

$$E(X) = -p(P)\log_2 p(P) - p(N)\log_2 p(N) \tag{1}$$

where $p(P)$ and $p(N)$ are related probabilities of positive and negative class, respectively.

2.4.2. K-Nearest Neighbor (kNN)

kNN is also known as a lazy learning non-parametric algorithm. kNN is the simplest classification algorithm that stores all training instances and uses a Euclidean distance function to classify new instances (shown in Equation (2)) [35,36]:

$$\sqrt{\sum_{i=1}^{k}(x_i - y_i)^2} \tag{2}$$

where x and y are two vectors (trained instance vector and a query vector for classification), and k represents the number of attributes.

2.4.3. Random Forest (RF)

RF was proposed by UC Berkeley visionary Leo Breiman in 1999 [37]. This algorithm works as a large collection of decorrelated decision trees using a bagging technique. The RF creates various sub-training sets from a super training set. A decision tree classifier is constructed from each sub-training set. At the time of testing, each input vector of the test set is classified by all of the decision trees in a forest and, finally, the forest is responsible for choosing the classification results; using either majority votes or averaging the predictions using the equation given below [38]:

$$f = \frac{1}{B}\sum_{b=1}^{B} f_b(x) \tag{3}$$

where B represents the samples/trees, f_b is a predictor, and x corresponds to the test point.

2.4.4. Least Squares-Support Vector Machine (LS-SVM)

The SVM classifier is highly influenced by advances in statistical learning theory [39–41]. SVM plays a vital role in the application of object detection [42], face detection [43], handwriting recognition [44], medical imaging classification [6], and bioinformatics [45]. SVM learns from training examples. An improved version of SVM, i.e., LS-SVM, was used in this article because of its robustness and efficiency. Each training instance consists of n number of attributes $(x_1, x_2, \cdots, x_n)$ with a corresponding class label. The nonlinear function estimation can be mathematically presented as:

$$y = sign\left[W\prime\varphi(x) + b\right] \tag{4}$$

where the high dimensional feature space is represented by $\varphi(x)$, the weight vector is defined by W, and the bias term is denoted by b. Then, the LS-SVM solution of such an optimization problem can be obtained as follows (for a deeper introduction of this method, readers can refer to [46–48]):

$$y(x) = sign\left[\sum_{i=1}^{N} \alpha_i y_i K(x, x_i) + b\right] \tag{5}$$

Table 2 provides some of the choices of kernel functions $K(x_k, x_l)$.

Table 2. Least squares-support vector machine (LS-SVM) kernel functions.

Kernel	Expression
Linear	$K(x,y) = x^T y$
Polynomial	$K(x,y) = \left(1 + \frac{x^T y}{\sigma^2}\right)^d$
RBF	$K(x,y) = \exp\left\{-\frac{\|x - y\|^2}{\sigma^2}\right\}$

2.5. Performance Measures

Recall (sensitivity), precision, F-measure, accuracy, and area under the receiver operating characteristic (ROC) curve are widely used metrics to determine the performance of the classifiers [49]. The possible outcomes of the proposed classifier can be described as:

TP (True Positive): Number of images correctly diagnosed under any specific class;
TN (True Negative): Number of images correctly rejected by the classifier;
FP (False Positive): Number of images incorrectly identified by the classifier;
FN (False Negative): Number of images incorrectly discarded by the classifier.

For multi-class classification, macro-averaged recall, macro-averaged precision and macro-averaged F-measure are used to validate the performance of the classifier [49].

Recall$_M$ is the average of the each class recall (i.e., the probability of the test finding the positive cases among all the positive cases of the respective class):

$$Recall_M = \frac{\sum_{i=1}^{C} \frac{TP_i}{TP_i + FN_i}}{C} \tag{6}$$

Precision$_M$ is the average of the each class precision (i.e., the probability of the test correctly diagnosed as positive cases given that the number of cases labelled by the system as positive):

$$Precision_M = \frac{\sum_{i=1}^{C} \frac{TP_i}{TP_i + FP_i}}{C} \tag{7}$$

F-Measure$_M$ (macro-averaged F-measure) is a weighted combination of the *Recall*$_M$ and *Precision*$_M$. Mathematically, it is defined as:

$$F\text{-}Measure_M = \frac{(\beta^2 + 1)\, Recall_M \times Precision_M}{\beta^2\, Precision_M + Recall_M} \tag{8}$$

Average Accuracy is the fraction of test results predicted as correct among all the classes:

$$Accuracy_{Avg} = \frac{\sum_{i=1}^{C} \frac{TP_i + TN_i}{TP_i + FN_i + TN_i + FP_i}}{C} \tag{9}$$

Area under the ROC curve (*AUC*) is the area occupied by the receiver operating characteristic curve of each class. It is used to analyse how good any classification model predicts the specific class versus all other classes:

$$AUC = \frac{1}{2}\left(\frac{TP}{TP+FN} + \frac{TN}{TN+FP}\right) \qquad (10)$$

where C represents the total number of classes. i.e., $C = 6$. M index represents to macro-averaging. $\beta = 1$ was used in this research.

2.6. Experimental Setup

Separate experiments were conducted on training and testing datasets. PCA with DWT was applied to extract the discriminative principal feature vectors with four different subset sizes (5, 10, 15, and 20). Four plus four (total of eight) principal feature vector sets were extracted from training datasets and testing datasets, respectively. To evaluate the performance of the proposed multi-class classifier using performance metrics; a total of 40 (8 × 5) analyses were performed by applying each of these 8 feature sets to five different classifier models (J48, kNN, RF, LS-SVM (Poly), and LS-SVM (RBF)).

3. Results and Discussion

3.1. Feature Reduction

In order to extract discriminative and reduced features, fast DWT with PCA was used in this research. The fast DWT only computes the approximation component of the wavelet features, which includes the major information of the MR image pattern. By only computing the approximation component of DWT decomposition, it decreases the size of the MRI images, which eventually reduces the computation time and complexity of the classifier. Initially, MRI images were 256 × 256 in size. After applying DWT with a three-level Haar wavelet decomposition (approximation component only) it changes to 32 × 32. Then PCA was applied on these reduced master feature sets, which allows a further decrease in the size of the feature sets by extracting the high variance components. In this article, four different feature subset sizes (5, 10, 15, and 20) were used. For classification purposes, the classifier used only 0.076%, 0.015%, 0.023%, and 0.031% of the original MR image in preparation of principal feature of size 5, 10, 15, and 20, respectively. Therefore, the proposed classification system achieved approximately 99.969% feature reduction while retains the accuracy of the classifier.

3.2. Performance Evaluation

The performance of the proposed multi-class classifier was evaluated in terms of macro-averaged recall, macro-averaged precision, macro-averaged F-measure, overall accuracy, and AUC of each class. Figure 3 shows the comparison of different decision models' performance against the number of principal features used.

Figure 3a illustrates the macro-averaged recall for each classifier model with respect to the feature subset sizes. A majority of the classifier models achieved recall$_M$ values greater than 81% for any number of principal components were used. To observe the effect of the feature subset sizes, the results indicate that the LS-SVM (RBF) classifier model produced fixed 86% recall$_M$ without dependence on feature subset sizes. However, RF and J48 models increase recall$_M$ values as the number of principal feature subset sizes increase and the attained macro-averaged recall values increase to 96% and 87%, respectively, whereas the remaining classifier models (kNN and LS-SVM (Poly)) were not able to increase their performance in terms of recall$_M$.

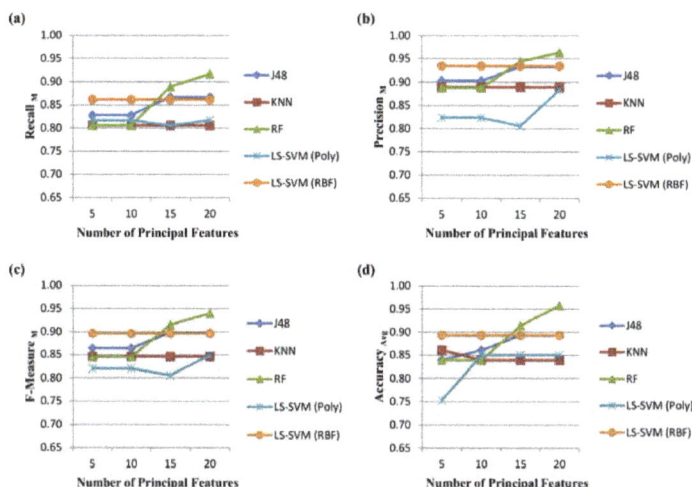

Figure 3. Performance measures of the proposed multi-class classifiers: (**a**) macro-averaged recall; (**b**) macro-averaged precision; (**c**) macro-averaged f-measure; and (**d**) average accuracy.

Figure 3b shows the performance of each classifier model with respect to the number of principal feature components in terms of macro-averaged precision. From the results, it is observed that a feature subset size of 10 or more produced precision$_M$ greater than 90% for all five classifier models. However, the RF model outperformed and achieved precision$_M$ values up to 96% using a feature subset size of 20. The lowest precision$_M$ was observed in LS-SVM (Poly) for any number of given principal features.

The performance evaluation in terms of macro-averaged F-measure, with respect to the number of principal components used by the classifiers, is shown in Figure 3c. The results revealed that the F-measure$_M$ generally exceeded 90% for RF and J48 when feature subset size used 10 or more. However, LS-SVM (RBF) achieved 90% F-measure$_M$ values for any combination of feature subsets. Furthermore, kNN and LS-SVM (Poly) could not able to improve the efficiency significantly in terms of F-measure$_M$ even the number of features was increased.

The overall accuracy of each classifier model was compared (Figure 3d). The average accuracy of each classifier model exceeded 84% for maximum number of principal components was used. However, RF improved the average accuracy with increasing the number of features and achieved the highest accuracy rate (i.e., 96%, standard deviation = 4%) when a plateau of 20 features was reached. The results again show that LS-SVM (RBF) overall average accuracy was stable and not associated with increased feature subset sizes. LS-SVM (RBF) provided the best results when the least number of principal features was used for classification. Furthermore, we found that no significant accuracy improvement was achieved, for any feature subset sizes, in the case of kNN and LS-SVM (Poly) decision models.

In Figure 4, the area under the ROC curve for each class was estimated. The results reveal that all five classifier models achieved AUC 100% with 0% standard deviation for the "Normal" class as shown in Figure 4a. RF, LS-SVM (RBF), and kNN produced significant results (i.e., 100% AUC with 0% standard deviation) for the "Alzheimer" class, as depicted in Figure 4b. However, J48 and LS-SVM (Poly) has a fluctuating trend of AUC for the "Alzheimer" class. The comparison of AUC for the "AIDS" class is shown in Figure 4c. Only the RF decision model achieved AUC 100% for the "AIDS" class when using equal to, and more than, a 15 feature subset. Moreover, the remaining four classifier models exceeded AUC 78% for different sizes of principal feature sets.

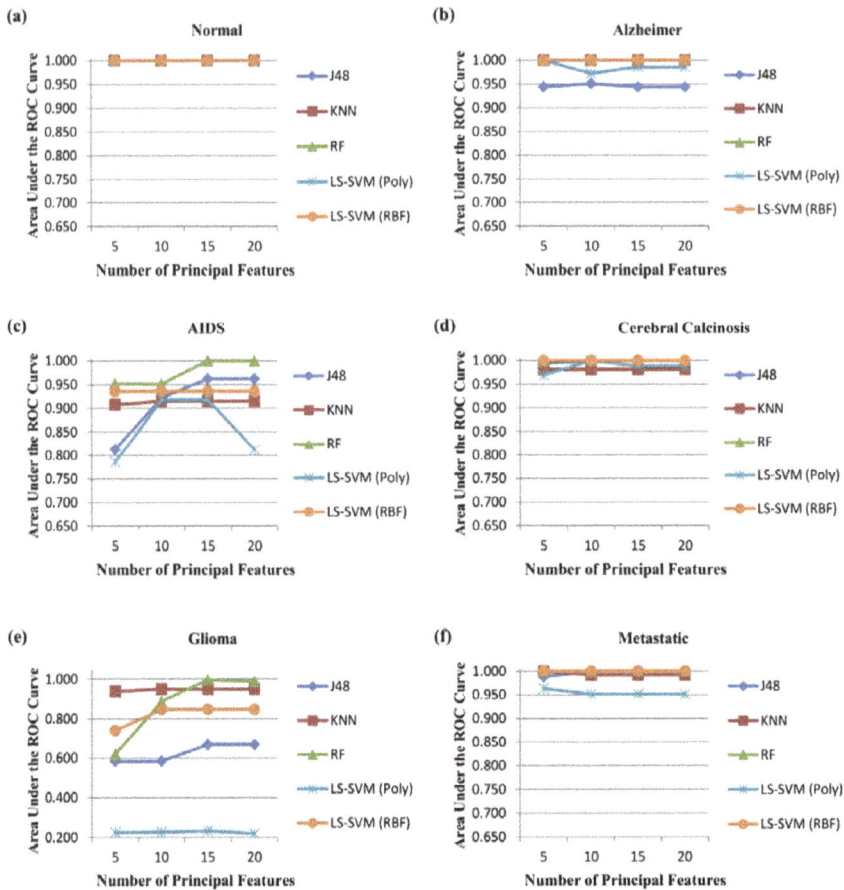

Figure 4. Area under the receiver operating characteristic (ROC) curve for each class: (a) normal; (b) Alzheimer; (c) AIDS; (d) cerebral calcinosis; (e) glioma; and (f) metastasis.

In Figure 4d,e, a majority of decision models attained $AUC > 96\%$ with $< 4\%$ standard deviation for classes "cerebral calcinosis" and "metastasis", respectively. However, RF and LS-SVM (RBF) showed AUC 100% with 0% standard deviation for any feature subset order. Figure 4c shows the AUC performance of the "glioma" class for all analyses. With the exception of RF and kNN models, all AUC measures were below 85% for any number of principal features and decision models. However, the AUC values measured for the RF and kNN models were greater than 95% when keeping the feature subset size more than equal to 15. The maximum AUC value achieved for "glioma" class by the RF model was 99% with 1% standard deviation.

The overall comparison results of decision models with different number of principal features subset showed that RF performed significantly better than other four decision models (J48, kNN, LS-SVM (Poly), and LS-SVM (RBF)). However, RF has not performed well when the lesser number of features is used, but its performance gradually increases as the size of the feature sets increased. It is also notable that increasing the number of principal features may not always be worthy because the complexity of the machine learning classifier may be increased by using larger feature subset sizes. Therefore, from the results it is observed that RF required at least 15 features to achieve better performance in terms of accuracy, precision$_M$, recall$_M$, F-measure$_M$, and AUC. On the other hand, the overall comparison also reveals that LS-SVM (RBF) achieved significant performance regardless of the feature set size. LS-SVM (RBF) achieved a constant performance trend for any number of principal features used, which leads to a decrease in the computational time and complexity of the multi-class classifier.

3.3. Comparison with Existing State-of-the-Art Classification Schemes

Different classification schemes were compared to evaluate the proposed multi-class classifier performance, which were examined for the same MRI dataset and on the same platform for multi-class brain MRI classification. Initially, these methods [6,18,20,21] were proposed for binary classification of brain MR images. The average accuracy results (when using 20 principal features) were gathered for each scheme, as presented in Table 3.

Table 3. Performance comparison with different classification schemes.

Scheme	Proposed in	Average Accuracy (%)
DWT + PCA + kNN	[18]	83.87
RT + PCA + LS-SVM (RBF)	[20]	86.02
DWPT + GEPSVM	[21]	88.92
DWT + PCA + LS-SVM (RBF)	[6]	89.25
DWT + PCA + RF	this paper	95.70

The results reveal that the highest accuracy (i.e., 95.7%) for multi-class brain MRI classification is achieved by the proposed scheme with RF as a decision model. LS-SVM (RBF) classification scheme also provided promising results with 89.25% accuracy rate. The scheme proposed by Das et al. [20] for binary classification attained an average accuracy of 86.02% when applied to multi-class brain MRI classification, regardless of the complex algorithm used (i.e., Ripplet transform) for feature extraction. Furthermore, the complex method used in [21] is not able to increase the accuracy rate up to 90% or more and managed to achieve a correctness rate of 88.92%. The average accuracy of the kNN-based scheme (i.e., 83.87%) is the worst performance among all of the compared state-of-the-art schemes examined for multi-class brain MRI classification. In addition, the method proposed in [24] includes age, tumour shape and ROI as feature sets for multi-class classification of brain MRI diseases. This scheme needs to trace ROIs manually, which makes this scheme semiautomatic. However, our proposed multi-class classifier is automatic and has no need of human intervention for decision-making purposes. In [24], the authors have achieved mean accuracy as 81.1%, 89.8%, and 91.2% for LDA, kNN, and non-linear SVM based decision models, respectively. Regardless of human intervention involved in feature extraction scheme, which is an additional cost, the average accuracy claimed in [24] is almost 4% less than our proposed work.

It is observed from the results and comparisons that the proposed classifier performance is quite remarkable as compared to the existing state-of-the-art techniques. Furthermore, a comprehensive study of different decision models' performances on MRI brain multi-class classification is shown. The comparison of decision models suggests that RF, LS-SVM (RBF), and J48 are more accurate than kNN and LS-SVM (POLY). RF achieved the highest accuracy rate when 20 features were used. Moreover, LS-SVM (RBF) maintained its performance for any number of features, which leads to being more advantageous when the least number of principal features are available. The feature engineering scheme used in this study proves that it reduced the number of discriminating features, which eventually reduces the classifier complexity and enhances its accuracy. Furthermore, the proposed classifier has the potential to classify various disease classes accurately using brain MRIs. The limitation of this study is that the experiments only involved brain MR images. However, the proposed approach has the potential to produce accurate results for different body parts' MR images as well.

4. Conclusions

In this article, a multi-class classifier has been developed to classify brain MR slices as normal, Alzheimer, AIDS, cerebral calcinosis, glioma, or metastatsis. It is composed of fast DWT, PCA, and five different decision models. The proposed medical decision support system yielded better performance in terms of macro-averaged recall, macro-averaged precision, macro-averaged F-measure, overall accuracy and *AUC* for each class, when compared to the other state-of-the-art schemes. This study provides a comprehensive comparison of different decision models performance, which concludes that RF work more accurately than other classification models (J48, kNN, LS-SVM (POLY), and LS-SVM (RBF)). It is evident from the results that the proposed classifier has the potential to classify the brain MR images accurately. Furthermore, the promising results indicate that the general practitioners can use this automated multi-class classifier as a second opinion, which assist them to reach the final decision more quickly. In future, the proposed method can be extended for automated classification of different pathological conditions and disease types, which are manually identified by the MRI scans. Moreover, this work can be employed on other imaging modalities (such as CT-scan, PET, and SPECT) datasets as well.

Acknowledgments: Muhammad Faisal Siddiqui likes to thank the University of Malaya Bright Sparks program. This research work was partially supported by Faculty of Computer Science and Information Technology, University of Malaya under a special allocation of Post Graduate Fund. This research was also partially funded by the University Malaya Research Grant (No: RP028F-14AET).

Author Contributions: Muhammad Faisal Siddiqui and Ghulam Mujtaba conceived and designed the experiments; Muhammad Faisal Siddiqui and Ghulam Mujtaba performed the experiments; Muhammad Faisal Siddiqui, Ghulam Mujtaba, Ahmed Wasif Reza, and Liyana Shuib analysed the data; Muhammad Faisal Siddiqui, Ghulam Mujtaba, Ahmed Wasif Reza, and Liyana Shuib contributed reagents/materials/analysis tools; Muhammad Faisal Siddiqui wrote the paper.

Conflicts of Interest: The authors declare no conflict of interest.

References

1. McKhann, G.; Drachman, D.; Folstein, M.; Katzman, R.; Price, D.; Stadlan, E.M. Clinical diagnosis of Alzheimer's disease report of the nincds-adrda work group* under the auspices of department of health and human services task force on Alzheimer's disease. *Neurology* **1984**, *34*, 939. [CrossRef] [PubMed]
2. Sahu, O.; Anand, V.; Kanhangad, V.; Pachori, R.B. Classification of magnetic resonance brain images using bi-dimensional empirical mode decomposition and autoregressive model. *Biomed. Eng. Lett.* **2015**, *5*, 311–320. [CrossRef]
3. Prasad, P.V. *Magnetic Resonance Imaging: Methods and Biologic Applications*; Springer Science & Business Media: New York, NY, USA, 2006; Volume 124.
4. Maji, P.; Chanda, B.; Kundu, M.K.; Dasgupta, S. Deformation correction in brain MRI using mutual information and genetic algorithm. In Proceedings of the International Conference on Computing: Theory and Applications, Kolkata, India, 5–7 March 2007; pp. 372–376.
5. Scapaticci, R.; Di Donato, L.; Catapano, I.; Crocco, L. A feasibility study on microwave imaging for brain stroke monitoring. *Prog. Electromagn. Res. B Pier B* **2012**, *40*, 305–324. [CrossRef]
6. Siddiqui, M.F.; Reza, A.W.; Kanesan, J. An automated and intelligent medical decision support system for brain MRI scans classification. *PLoS ONE* **2015**, *10*, e0135875.
7. Mujtaba, G.; Shuib, L.; Raj, R.G.; Rajandram, R.; Shaikh, K. Automatic text classification of ICD-10 related CoD from complex and free text forensic autopsy reports. In Proceedings of the 5th IEEE International Conference on Machine Learning and Applications (ICMLA), Anaheim, CA, USA, 18–20 December 2016; pp. 1055–1058.
8. Mujtaba, G.; Shuib, L.; Raj, R.G.; Rajandram, R.; Shaikh, K.; Al-Garadi, M.A. Automatic ICD-10 multi-class classification of cause of death from plaintext autopsy reports through expert-driven feature selection. *PLoS ONE* **2017**, *12*, e0170242. [CrossRef] [PubMed]

9. Jin, Y.; Wee, C.Y.; Shi, F.; Thung, K.H.; Ni, D.; Yap, P.T.; Shen, D. Identification of infants at high-risk for autism spectrum disorder using multiparameter multiscale white matter connectivity networks. *Hum. Brain Mapp.* **2015**, *36*, 4880–4896. [CrossRef]

10. Huang, L.; Jin, Y.; Gao, Y.; Thung, K.-H.; Shen, D.; Initiative, A.S.D.N. Longitudinal clinical score prediction in Alzheimer's disease with soft-split sparse regression based random forest. *Neurobiol. Aging* **2016**, *46*, 180–191. [CrossRef] [PubMed]

11. Mwangi, B.; Ebmeier, K.P.; Matthews, K.; Steele, J.D. Multi-centre diagnostic classification of individual structural neuroimaging scans from patients with major depressive disorder. *Brain* **2012**, *135*, 1508–1521. [CrossRef] [PubMed]

12. Klöppel, S.; Stonnington, C.M.; Barnes, J.; Chen, F.; Chu, C.; Good, C.D.; Mader, I.; Mitchell, L.A.; Patel, A.C.; Roberts, C.C. Accuracy of dementia diagnosis—A direct comparison between radiologists and a computerized method. *Brain* **2008**, *131*, 2969–2974. [CrossRef] [PubMed]

13. Faisal, A.; Parveen, S.; Badsha, S.; Sarwar, H.; Reza, A.W. Computer assisted diagnostic system in tumor radiography. *J. Med. Syst.* **2013**, *37*. [CrossRef] [PubMed]

14. Zhang, Y.; Wu, L.; Wang, S. Magnetic resonance brain image classification by an improved artificial bee colony algorithm. *Prog. Electromagn. Res. Pier* **2011**, *116*, 65–79. [CrossRef]

15. Zhang, Y.; Dong, Z.; Wu, L.; Wang, S. A hybrid method for MRI brain image classification. *Expert Syst. Appl.* **2011**, *38*, 10049–10053. [CrossRef]

16. Zhang, Y.; Wu, L. An MR brain images classifier via principal component analysis and kernel support vector machine. *Prog. Electromagn. Res. Pier* **2012**, *130*, 369–388. [CrossRef]

17. Zhang, Y.; Wang, S.; Ji, G.; Dong, Z. An MR brain images classifier system via particle swarm optimization and kernel support vector machine. *Sci. World J.* **2013**, *2013*. [CrossRef] [PubMed]

18. El-Dahshan, E.-S.A.; Hosny, T.; Salem, A.-B.M. Hybrid intelligent techniques for MRI brain images classification. *Digit. Signal Process.* **2010**, *20*, 433–441. [CrossRef]

19. El-Dahshan, E.-S.A.; Mohsen, H.M.; Revett, K.; Salem, A.-B.M. Computer-aided diagnosis of human brain tumor through MRI: A survey and a new algorithm. *Expert Syst. Appl.* **2014**, *41*, 5526–5545. [CrossRef]

20. Das, S.; Chowdhury, M.; Kundu, M.K. Brain MR image classification using multiscale geometric analysis of Ripplet. *Prog. Electromagn. Res. Pier* **2013**, *137*, 1–17. [CrossRef]

21. Zhang, Y.; Dong, Z.; Wang, S.; Ji, G.; Yang, J. Preclinical diagnosis of magnetic resonance (MR) brain images via discrete wavelet packet transform with Tsallis entropy and generalized eigenvalue proximal support vector machine (GEPSVM). *Entropy* **2015**, *17*, 1795–1813. [CrossRef]

22. Zhou, X.; Wang, S.; Xu, W.; Ji, G.; Phillips, P.; Sun, P.; Zhang, Y. Detection of pathological brain in MRI scanning based on wavelet-entropy and naive Bayes classifier. In Proceedings of the Bioinformatics and Biomedical Engineering, Granada, Spain, 15–17 April 2015; pp. 201–209.

23. Wang, S.; Lu, S.; Dong, Z.; Yang, J.; Yang, M.; Zhang, Y. Dual-tree complex wavelet transform and twin support vector machine for pathological brain detection. *Appl. Sci.* **2016**, *6*, 169. [CrossRef]

24. Zacharaki, E.I.; Wang, S.; Chawla, S.; Soo Yoo, D.; Wolf, R.; Melhem, E.R.; Davatzikos, C. Classification of brain tumor type and grade using MRI texture and shape in a machine learning scheme. *Magn. Reson. Med.* **2009**, *62*, 1609–1618. [CrossRef] [PubMed]

25. Chaplot, S.; Patnaik, L.; Jagannathan, N. Classification of magnetic resonance brain images using wavelets as input to support vector machine and neural network. *Biomed. Signal Process. Control* **2006**, *1*, 86–92. [CrossRef]

26. Zhang, Y.; Wang, S.; Wu, L. A novel method for magnetic resonance brain image classification based on adaptive chaotic pso. *Prog. Electromagn. Res. Pier* **2010**, *109*, 325–343. [CrossRef]

27. Maitra, M.; Chatterjee, A. A slantlet transform based intelligent system for magnetic resonance brain image classification. *Biomed. Signal Process. Control* **2006**, *1*, 299–306. [CrossRef]

28. Lustig, M.; Donoho, D.L.; Santos, J.M.; Pauly, J.M. Compressed sensing MRI. *IEEE Signal Process. Mag.* **2008**, *25*, 72–82. [CrossRef]

29. Blum, A.L.; Langley, P. Selection of relevant features and examples in machine learning. *Artif. Intell.* **1997**, *97*, 245–271. [CrossRef]

30. Kohavi, R.; John, G.H. Wrappers for feature subset selection. *Artif. Intell.* **1997**, *97*, 273–324. [CrossRef]

31. Wettschereck, D.; Aha, D.W.; Mohri, T. A review and empirical evaluation of feature weighting methods for a class of lazy learning algorithms. *Artif. Intell. Rev.* **1997**, *11*, 273–314. [CrossRef]

32. Sengur, A. An expert system based on principal component analysis, artificial immune system and fuzzy k-NN for diagnosis of valvular heart diseases. *Comput. Biol. Med.* **2008**, *38*, 329–338. [CrossRef] [PubMed]
33. Duda, R.O.; Hart, P.E.; Stork, D.G. *Pattern Classification*; John Wiley & Sons: New York, NY, USA, 2012.
34. Zhao, Y.; Zhang, Y. Comparison of decision tree methods for finding active objects. *Adv. Space Res.* **2008**, *41*, 1955–1959. [CrossRef]
35. Bao, Y.; Ishii, N.; Du, X. Combining multiple k-nearest neighbor classifiers using different distance functions. In Proceedings of the International Conference on Intelligent Data Engineering and Automated Learning, Exeter, UK, 25–27 August 2004; pp. 634–641.
36. Fukunaga, K. *Introduction to Statistical Pattern Recognition*; Academic Press: Salk Lake City, UT, USA, 2013.
37. Breiman, L. Random forests. *Mach. Learn.* **2001**, *45*, 5–32. [CrossRef]
38. Liaw, A.; Wiener, M. Classification and regression by random forest. *R News* **2002**, *2*, 18–22.
39. Vapnik, V. *The Nature of Statistical Learning Theory*; Springer Science & Business Media: New York, NY, USA, 2013.
40. Patil, N.; Shelokar, P.; Jayaraman, V.; Kulkarni, B. Regression models using pattern search assisted least square support vector machines. *Chem. Eng. Res. Des.* **2005**, *83*, 1030–1037. [CrossRef]
41. Wang, F.-F.; Zhang, Y.-R. The support vector machine for dielectric target detection through a wall. *Prog. Electromagn. Res. Pier Lett.* **2011**, *23*, 119–128. [CrossRef]
42. Chen, G.-C.; Juang, C.-F. Object detection using color entropies and a fuzzy classifier. *IEEE Comput. Intell. Mag.* **2013**, *8*, 33–45. [CrossRef]
43. Magalhães, F.; Sousa, R.; Araújo, F.M.; Correia, M.V. Compressive sensing based face detection without explicit image reconstruction using support vector machines. In Proceedings of the 10th International Conference on Image Analysis and Recognition, Berlin, Germany, 26–28 June 2013; pp. 758–765.
44. Dasgupta, J.; Bhattacharya, K.; Chanda, B. A holistic approach for off-line handwritten cursive word recognition using directional feature based on arnold transform. *Pattern Recogn. Lett.* **2016**, *79*, 73–79. [CrossRef]
45. Komiyama, Y.; Banno, M.; Ueki, K.; Saad, G.; Shimizu, K. Automatic generation of bioinformatics tools for predicting protein–ligand binding sites. *Bioinformatics* **2015**, *32*, 901–907. [CrossRef] [PubMed]
46. Cristianini, N.; Shawe-Taylor, J. *An introduction to SVM*; Cambridge University Press: Cambridge, UK, 1999.
47. Suykens, J.A.; Vandewalle, J. Least squares support vector machine classifiers. *Neural Process. Lett.* **1999**, *9*, 293–300. [CrossRef]
48. Van Gestel, T.; Suykens, J.A.; Baesens, B.; Viaene, S.; Vanthienen, J.; Dedene, G.; De Moor, B.; Vandewalle, J. Benchmarking least squares support vector machine classifiers. *Mach. Learn.* **2004**, *54*, 5–32. [CrossRef]
49. Sokolova, M.; Lapalme, G. A systematic analysis of performance measures for classification tasks. *Inf. Process. Manag.* **2009**, *45*, 427–437. [CrossRef]

Article

Transmission Power Adaption for Full-Duplex Relay-Aided Device-to-Device Communication

Hui Dun, Fang Ye * and Yibing Li

College of Information and Communication Engineering, Harbin Engineering University, Harbin 150001, Heilongjiang Province, China; m18846198960@163.com (H.D.); yibing0920@126.com (Y.L.)
* Correspondence: yefang0923@126.com; Tel.: +86-133-0460-5678

Academic Editor: Angel Garrido
Received: 1 December 2016; Accepted: 3 March 2017; Published: 9 March 2017

Abstract: Device-to-device (D2D) communications bring significant improvements of spectral efficiency by underlaying cellular networks. However, they also lead to a more deteriorative interference environment for cellular users, especially the users in severely deep fading or shadowing. In this paper, we investigate a relay-based communication scheme in cellular systems, where the D2D communications are exploited to aid the cellular downlink transmissions by acting as relay nodes with underlaying cellular networks. We modeled two-antenna infrastructure relays employed for D2D relay. The D2D transmitter is able to transmit and receive signals simultaneously over the same frequency band. Then we proposed an efficient power allocation algorithm for the base station (BS) and D2D relay to reduce the loopback interference which is inherent due to the two-antenna infrastructure in full-duplex (FD) mode. We derived the optimal power allocation problem in closed form under the independent power constraint. Simulation results show that the algorithm reduces the power consumption of D2D relay to the greatest extent and also guarantees cellular users' minimum transmit rate. Moreover, it also outperforms the existing half-duplex (HD) relay mode in terms of achievable rate of D2D.

Keywords: device-to-device (D2D); relay aided; full-duplex (FD); power allocation; independent power constraint

1. Introduction

Device-to-device (D2D) communication enables users in proximity to exchange information directly, without traversing to the base station (BS) or core network in two hops and has attracted increasing attention from both industrial and academic communities [1–7]. Due to the potential proximity gain, reuse gain, and hop gain, D2D communications can significantly increase system spectral efficiency and energy efficiency, which has been considered a promising candidate technique for next generation cellular networks [8].

Accordingly, many researches have focused on a D2D communications scheme that underlayings the cellular network infrastructure to increase the cellular capacity, improving the users' throughput, and extending the battery lifetime of user equipment (UE) [9–11]. However, the existing D2D communication scheme, such as the underlaying cellular spectrum, bring more complex interference environment to cellular users which have higher priority than the D2D users, and lead to deteriorative performance on quality of service for cellular users [12–14]. In particular, for cellular users in the cell edge who are susceptible to suffered deep fading or shadowing, it would be unadvisable to share an allocated spectrum with D2D peers [15–17].

To avoid the problem mentioned above, we outline a relay-aided D2D communication scheme to fulfill the transmission gap that affects cellular users which are in the cell edge or strongly-shadowed urban environment. Unlike existing works, a D2D transmitter in our scheme is ordered to aid a cellular downlink transmission by acting as a relay node between the BS and the cellular UE. In return, the D2D transmitter is allowed to directly communicate with the intended D2D receiver by underlaying the cellular downlink spectral resource [18–20].

Few works have been carried out so far to explore relaying-based D2D communication schemes. By scanning existing research papers, we can classify this relay-aided D2D communication schemes into two categories in terms of relay mode, namely, half-duplex (HD) and full-duplex (FD). In fact, few works have been carried out so far to explore HD relaying based scheme to transmit forward signals from BS to cellular users in two transmission phases. In this scheme, respective reception and transmission at the relay are allocated to one frame which is divided into double orthogonal time slots which are equivalent to a time division access way for the D2D relay node [21–26]. Then the superimposed signal, which is a linear weighted combination of cellular user's signal and the D2D signal is generated by the relay. The weight factor represents the power allocation for the cellular user which is a fraction of the total transmitting power of D2D relay. Therefore, the residual power is used for broadcasting the D2D signal in the second time slot.

The drawback of HD is lower efficiency in spectrum and energy. Therefore, as a replacer for HD relay, the FD mode is more applicable in practice to transmit and receive signals simultaneously on the same frequency resource. For the sake of communication bi-directional for relay, a two-antenna infrastructure is typically deployed to facilitate spectral efficiency in practice. However, FD relay is more susceptible to loopback interference which is generated from transmitting antenna to receiving antenna of relay nodes. Although few works have been done to eliminate the inter-relay interference, it cannot be ultimately mitigated because even the residual loopback interference is significant when compared to the white Gaussian noise [27–29]. For this reason, the power split factor should be adapted cautiously in FD mode. Moreover, it must be made clear that due to the FD mode being enable relay to forward two components signals simultaneously on the same spectral band, the cellular user's signal and D2D signal will treat each other as interference.

Many works have been carried out to explore relaying-based D2D communication schemes. In [21], the authors outline an HD relaying-based D2D scheme by operating on the same frequency band as cellular user in frequency domain by splitting a frame into double transmitting phases in the time domain. In this scheme, because the BS transmit signal to the cellular user takes place in one half of a frame, the achievable transmission rate is dimidiated in this mode. Then to compensate for the cellular users' capacity loss, the D2D relay has to allocate almost all of its transmitting power to forward cellular signals when the direct link between the BS and cellular users is good. Then this scheme will degenerate to a conventional relaying mode. In contrast, when the direct link is in deep fading or shadowing, it also cannot give a significantly increasing transmission rate. Furthermore, the authors describe explicitly that the BS and the D2D relay should use maximum transmitting power as they can maintain the transmit capacity of the cellular user, which leads to low energy efficiency for the system.

Alternatively, the FD mode as explicitly described in [27–29] relays cellular signals and the D2D component to the respective receiver simultaneously. Unfortunately, the self-interference cannot be eliminated perfectly and, consequently, the potential existence of residual interference is not negligible. The work that has been done in [27] advocates a new FD mode, in which the aggregate power of the system is defined as a constant. That makes the D2D relay much more able to use transmitting power than one when the power is normalized. By this means, they can get a significant capacity increase and a closed form of expression of the optimal power at BS and D2D relay by adding a power equation. However, it is more realistic to consider the BS and D2D relay with their own transmission power constraint.

In contrast to the aforementioned works, the contributions of this paper are summarized as follows:

(1) The main contribution of the article is to draft a novel D2D relaying-aided scheme to improve the transmitting rate of a cellular user who is in deep fading or shadowing. Unlike the works that

have been done before, our scheme guarantees cellular user' quality of service in terms of the minimum rate requirement. This means that the performance of the cellular user is improved rather than deteriorated or just maintained, as done in [13,21].

(2) We consider an independent power constraint for the BS and D2D relay respectively rather than under the aggregate power constraint as in [27]. Although we use two inequalities to replace one equation, we also derived a closed form expression of the optimal allocated power and the power split factor.

(3) By making comparisons with traditional direct communication of cellular mode and HD relaying mode, the FD mode we proposed could use the minimum transmitting power to make sure the cellular' rate requirement with also achieve the D2D rate as far as possible. It also has been demonstrated that the approach we proposed is more prominent in terms of spectral efficiency and power efficiency, in the situation that the residual loopback interference generating the D2D relay is controlled at a tolerable level.

The rest of the paper is organized as follows: we describe the system model and scheme of the proposed FD relaying assisted D2D communication by underlaying the downlink frequency resource of a cellular network in Section 2. The optimal power adaption strategies for guaranteeing the rate requirement of cellular user versus benchmark HD mode are evaluated in Section 3. Simulations have been obtained are presented in Section 4. Finally, the conclusions are drawn in Section 5.

2. System Model

In Figure 1 we consider a two-antenna deployment scenario, where a macro BS is intended to communicate with a cellular user who is in the edge of traditional cell and suffering severe deep fading or shadowing. This means the direct link between macro BS and cellular is weak and unable to guarantee the rate requirement of cellular user. Moreover, there is a pair of D2D users located nearby the cellular user and there is a good channel quality from BS and cellular respectively. We assume that the transmitter of the D2D pair is equipped with isolated receiving and transmitting antennas and caches the date that intends to send to the D2D receiver. We can automatically see that the system performance will be beneficial from deploying D2D transmitter as a relay for BS and the cellular user. We assume that the D2D communication is allowed to share the downlink spectral resource allocated to the cellular user while remaining under the control of the BS. We set the D2D relay to operate in FD mode, suffering from the backhaul interference is inevitable due to the simultaneously receiving and transmitting signals which is an inherent nature of FD mode. It also should be made clear that this scenario corresponds to the situation that radio resource management has taken place and spectrum resources have been allocated for a particular BS to serve a particular cellular user.

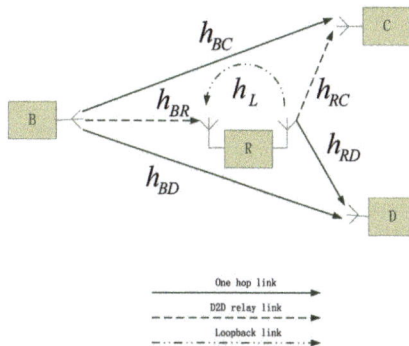

Figure 1. Relay-aided device-to-device (D2D) communication model in full-duplex (FD) mode.

We assume that the system operates in the downlink in frequency division duplex (FDD) mode with a bandwidth of W Hz. We denote the channel impulse response for the link from BS to cellular user, BS to the relay, namely, the D2D transmitter, from BS to the D2D receiver, from the relay to cellular, from the relay to the D2D receiver and the loopback link from the transmitting antenna to the receiving antenna as h_{BC}, h_{BR}, h_{BD}, h_{RC}, h_{RD} and h_L respectively. These are all complex variables. The system is parametrized with channel gain-to-noise ratios, which are defined as:

$$\gamma_{BR} = |h_{BR}|^2/\sigma^2, \; \gamma_{BC} = |h_{BC}|^2/\sigma^2, \; \gamma_{BD} = |h_{BD}|^2/\sigma^2$$
$$\gamma_{RC} = |h_{RC}|^2/\sigma^2, \; \gamma_{RD} = |h_{RD}|^2/\sigma^2, \; \gamma_L = |h_L|^2/\sigma^2 \tag{1}$$

Where σ^2 denotes the variance of the zero-mean additive white Gaussian noise (AWGN). As we can see, each of these variables actually corresponds to the signal-to-noise ratio (SNR) of one-hop transmission over the particular channel with maximum transmission power. In other words, we set $p_B = 1$ and $p_R = 1$. All of the channels we used are modeled to be narrow band quasi-static state and frequency flat fading. Note that we assume that the nodes in our scenario are generally stationary or moving at moderate speed to allow channel state information acquisition.

The transmitting power of BS and relay node are denoted by p_B and p_R respectively. These have been normalized and are subject to independent constraints by $0 \leq p_B \leq 1$ and $0 \leq p_R \leq 1$ instead of aggregate transmitting power constraint $p_B + p_R = 2$ as does in [27]. We assume that the D2D relay adapts amplify-and-forward protocol to aid the downlink transmission of a cellular user and the relay uses a fraction of its transmitting power to forward the cellular user's signals.

FD Relay-Aided D2D Communication Scheme

Before making an explicit description of the proposed relaying based D2D scheme, we review the traditional cellular direct transmission (DT) mode. In a traditional cellular network, the cellular users have no choice but to communicate with the BS even the DT link suffers from severely deep fading or shadowing. Unfortunately, this situation does not happen in a small probability. When the cellular users are located in the cell edge or urban environment, it will be a frequent occurrence. Therefore, DT links from the BS to cellular users in a conventional cellular mode will lead to a poor performance of the system in this situation, and might even cause high outage probability or call drop for cellular users. Assume frequency-division multiple-access technology is used for the cellular downlink transmissions in DT mode. At any time instant, the BS transmits symbols to the cellular user over the allocated frequency band. Therefore, the achievable rate of direct link is given by:

$$R_{BC}^{DT} = W\log_2(1 + p_B\gamma_{BC}) \tag{2}$$

where W corresponds to the width of the subchannel. Let R^{re} represent the rate requirement of the cellular user. When the channel impulse response h_{BC} suffers from deep fading or shadowing, it will lead to the situation that the achievable rate of the DT link will fall below the rate requirement even the BS use maximum transmitting power with $p_B = 1$, namely:

$$R^{re} > W\log_2(1 + \gamma_{BC}) \tag{3}$$

Hence, for the sake of fulfilling the requirement rate of cellular user, we point out that the desired channel to noise ratio γ should satisfy:

$$\gamma \geq \frac{\left(2^{R^{qe}/W} - 1\right)}{p_B} \tag{4}$$

From the analysis above, we conclude that the conventional DT mode can do nothing to satisfy the cellular user's quality of service requirement in terms of transmitting rate when the DT link is weak. Therefore, we aim to find a novelty D2D relaying-aided scheme to guarantee the cellular user's rate requirement.

In order to improve spectral efficiency, throughput and energy efficiency, BS allows UE who are in proximity to communicate directly without the help of BS. Moreover, for the sake of improving the poor performance of conventional cellular mode in which the cellular users suffer severely from deep fading or shadowing, we outline an FD relaying based D2D communication scheme to allow the downlink transmission from BS to cellular user via D2D relay and D2D direct communication from D2D relay to D2D receiver take place at the same instant. In our system model, it is assumed that the D2D relay is equipped with two isolate antennas.

We now describe the relaying based D2D protocol. Let us denote $x(i)$ and $d(i)$ as the signals transmitted by BS and D2D relay, namely, D2D transmitter at time instant i. Then we have:

$$E\left\{|x(i)|^2\right\} = p_B \tag{5}$$

$$E\left\{|d(i)|^2\right\} = (1-\alpha)p_R \tag{6}$$

in which $E\{\cdot\}$ denotes the expectation of variable and α represents the power split factor at relay. We assume that the D2D relay is willing to share $\alpha(0 \leq \alpha \leq 1)$ fraction of its power to assist the cellular user's downlink transmission. Then the residual $(1-\alpha)$ fraction is used for its own signal transmission. In addition, the system will degrade to a conventional relay forward mode or a D2D underlaying mode when the power split factor α is equal to zero or one respectively. Furthermore, considering the loopback interference from the relay transmitting antenna to the receiving antenna, the signal $r(i)$ received at the D2D relay and the transmitting signal $t(i)$ by D2D relay can be formulated as:

$$r(i) = h_{BR}x(i) + h_L t(i) + n(i) \tag{7}$$

$$t(i) = gr(i) + \sqrt{(1-\alpha)p_R}d(i) \tag{8}$$

where $n(i)$ is the AWGN at relay with mean value $E\left\{|n(i)|^2\right\} = \sigma^2$. From this, we know that the power of receiving signal at D2D relay is:

$$E\left\{|r(i)|^2\right\} = p_B|h_{BR}|^2 + p_R|h_L|^2 + \sigma^2 \tag{9}$$

g represents the normalized amplify factor by the D2D relay with adopting amplify forward protocol which means that the relay will amplify the input signal by factor g. It can be given by:

$$g = \sqrt{\frac{\alpha p_R}{p_B|h_{BR}|^2 + p_R|h_L|^2 + \sigma^2}} \tag{10}$$

in which αp_R is denoted as the power used for forwarding cellular user' signals. Naturally, we also have the aggregate power consumption at D2D relay as $E\left\{|t(i)|^2\right\} = p_R$. Thus, the signal-to-interference and noise ratios (SINR) γ_R of the first hop at the relay can be formulated as:

$$\gamma_R = \frac{p_B|h_{BR}|^2}{p_R|h_L|^2 + \sigma^2} \tag{11}$$

Then we let the numerator and denominator be divided by σ^2, the SINR can be rewritten as:

$$\gamma_R = \frac{p_B\gamma_{BR}}{p_R\gamma_L + 1} \tag{12}$$

From the Equation (8), we can see that the transmitting signals of D2D relay are composed by two portions. The first portion contains the signal $x(i)$ transmitting for cellular user and the second portion is the signal component sending to the D2D receiver using the remaining $(1 - \alpha)$ fraction power. When decoding respective signals at the receivers, the cellular user and D2D receiver will treat the signal component designated for the other as interference. We get the receiver signal at cellular user and D2D receiver as follows:

$$y_C(i) = h_{BC}x(i) + h_{RC}t(i) + n(i) \tag{13}$$

$$y_D(i) = h_{BD}x(i) + h_{RD}t(i) + n(i) \tag{14}$$

In FD mode, the cellular user decodes the relayed signal $t(i)$ from $y_C(i)$ and the weak DT link signal $x(i)$ is treated as interference. Therefore, as above, the SINR of the second hop at the cellular user $\gamma_C(i)$ can be formulated as:

$$\gamma_C = \frac{\alpha p_R \gamma_{RC}}{p_B \gamma_{BC} + (1 - a)p_R \gamma_{RC} + 1} \tag{15}$$

The instantaneous end-to-end SINR of one symbol transmitting from BS to cellular user in FD mode can be expressed as [28]:

$$\gamma_{FD} = \frac{\gamma_R \gamma_D}{\gamma_R + \gamma_D + 1} \tag{16}$$

In our model, these expressions we deduce are explicitly including the effects of the self-interference and the overheard direct link. Then to guarantee the rate requirement of the cellular user with the proposed spectrum sharing protocol, the SINR of the system should be greater or equal to the requirement of SINR in the conventional DT mode as follows:

$$\gamma_{FD} \geq \gamma \tag{17}$$

As before, the SINR of D2D receiver can also be formulated as:

$$\gamma_D = \frac{(1 - \alpha)p_R \gamma_{RD}}{p_B \gamma_{BD} + \alpha p_R \gamma_{RD} + 1} \tag{18}$$

Finally, we can get the achievable rates for the cellular and D2D links respectively, as:

$$R_C^{FD} = W \log_2(1 + \gamma_{FD}) \tag{19}$$

$$R_D^{FD} = W \log_2(1 + \gamma_D) \tag{20}$$

3. The Optimal Power Adaption Strategies

Let us then define the objective function as:

$$(p_B^*, p_R^*, \alpha^*) = \arg \max_{(p_B, p_R, \alpha)} R_D^{FD} \tag{21}$$

$$s. t. \quad R_C^{FD} = R^{re} \tag{21a}$$

$$0 \leq p_B \leq 1 \tag{21b}$$

$$0 \leq p_R \leq 1 \tag{21c}$$

$$0 \leq \alpha \leq 1 \tag{21d}$$

Where (p_B^*, p_R^*, α^*) represents the optimal power allocation for BS and the D2D relay and the power split factor for the FD relaying D2D.

Next we propose transmission power adaptation as a technique to mitigate the effect of self-interference on D2D relay. Without optimal transmit power adaptation at the D2D relay, it implies that the D2D transmitter simply uses the maximum allowed power, i.e., $p_R = 1$.

Remark 1: *For extension of cellular coverage, the BS should always use the maximum transmission power, i.e., $p_B = 1$. In this way, it will enlarge the SINR for the first relaying hop, namely $\gamma_R = p_B\gamma_{BR}/(p_R\gamma_L + 1)$. Moreover, because typically we have $\gamma_L \gg \gamma_{BC}$, the worsening effect on $\gamma_C = \alpha p_R\gamma_{RC}/(p_B\gamma_{BC} + (1 - \alpha)p_R\gamma_{RC} + 1)$ will be insignificant with the maximum transmission power usage for BS. Thus, the best performance under constraint $0 \leq p_B \leq 1$ is achieved when $p_B = 1$ and we may concentrate on power adaptation for D2D relay individually. Transmission power adaptation in FD relays is motivated by the observation that the end-to-end performance is limited by the weakest hop. Thus, if the limiting factor is the first hop due to excessive loop interference, the end-to-end performance can be, in fact, improved by decreasing the relay transmit power.*

To solve the optimal problem, we first obtain $\gamma_{FD} = \gamma$ from $R_C^{FD} = R^{re}$.

Remark 2: *from (16), we know that γ_{FD} is decided by γ_R and γ_C jointly. By getting the first class partial differential equation of γ_{FD} in term of γ_R, namely, $\partial\gamma_{FD}/\partial\gamma_R$, we can get the expression as:*

$$\frac{\partial\gamma_{FD}}{\partial\gamma_R} = \frac{\gamma_C^2 + 1}{(\gamma_C + \gamma_R + 1)^2} \tag{22}$$

The equation above is obviously constant positive. Therefore, γ_{FD} is monotonous increasing in terms of γ_R. We can derive that this is the case for γ_C also. It is easy to understand from this result that the end-to-end SINR of two hops FD relay will be improved by either of any one link with better channel quality. So, for the sake of fulfilling the rate requirement of cellular, we should enhance one or both of γ_R and γ_C.

Let us then analyze the corresponding received SINR in the D2D relay and in the destination of the cellular user which are expressed as shown in (12) and (15) respectively. In the first hop between BS and D2D relay, the SINR will be improved by decreasing the transmitting power of the relay due to the excessive loopback interference. In contrast, the SINR at the destination of the cellular user is a monotonous increasing function in terms of the transmitting power of relay. Then we know that there will be an optimal transmitting power of relay. In terms of power splitting factor of α, the SINR at the destination of the cellular user will also be improved following α increasing. On the contrary, the SINR for D2D receiver will decrease as the spilt factor increases.

Now we provide a new insight for the objective function which considers fulfilling the rate requirement of the cellular user in the primary and improves the D2D transmitting rate in the secondary. We have observed that an optimal power at the D2D relay exists such that a higher or lower value will result in the deterioration of the end-to-end SINR of cellular user. Therefore, by getting the optimal transmitting power of relay, the energy consumption for D2D users who usually handle equipment such as a mobile phone will decrease. Moreover, our proposed scheme also makes sense of the demand of power split factor α at the D2D relay. Then the D2D relay can spare more power to transmit its own signals.

It should be made clear that the optimal transmitting power at the relay is not the global optimization as the achievable rate of D2D link is monotonously increased as follow p_R. We can demonstrate the first class partial differential of γ_D in term of p_R:

$$\begin{aligned}\frac{\partial\gamma_D}{\partial p_R} &= \frac{(1-\alpha)\gamma_{RD}(\gamma_{BD}+\alpha p_R\gamma_{RD}+1)-\alpha\gamma_{RD}(1-\alpha)p_R\gamma_{RD}}{(\gamma_{BD}+\alpha p_R\gamma_{RD}+1)^2} \\ &= \frac{(1-\alpha)\gamma_{RD}(\gamma_{BD}+1)}{(\gamma_{BD}+\alpha p_R\gamma_{RD}+1)^2}\end{aligned} \tag{23}$$

In the situation that $0 \leq \alpha \leq 1$, γ_{RD} and γ_{BD} is a constant greater than zero, so $\partial \gamma_D / \partial P_R \geq 0$ is permanently satisfied. It is obvious that the SINR will be greater as the transmitter power is larger when the channel to noise ratio is a constant.

Therefore, in order to improve the energy efficiency of D2D relay, we adapt the optimal power for guaranteeing cellular rate requirement which is equivalent to the minimum transmission power usage for D2D relay. The optimal FD relay transmission power is:

$$p_R^* = \min\left\{1, \sqrt{\frac{(\gamma_{BR}+1)(\gamma_{BC}+1)}{\gamma_L \gamma_{RC}}}\right\} \tag{24}$$

Proof: By substituting Equation (12), (15) into (16), the end-to-end SINR for cellular can be given by (we can convert into the following form):

$$\gamma_{FD} = \frac{\gamma_{BR}\gamma_{RD}p_R}{\gamma_{RD}\gamma_L p_R^2 + \gamma_{RD}p_R + \gamma_{BD}\gamma_L p_R + \gamma_L p_R + \gamma_{BR}\gamma_{BD} + \gamma_{BR} + \gamma_{BD} + 1} \tag{25}$$

by solving the one class partial differential equation of γ_{FD} in terms of p_R, we can get the numerator expression of $\partial \gamma_{FD} / \partial p_R$ as:

$$\gamma_{BR}^2 \gamma_{RC}\gamma_{BC} + \gamma_{BR}^2 \gamma_{RC} + \gamma_{BR}\gamma_{RC}\gamma_{BC} + \gamma_{BR}\gamma_{RC} - \gamma_{BR}\gamma_L \gamma_{RC}^2 p_R^2 = 0 \tag{26}$$

which is equivalent to the one class partial differential equation $\partial \gamma_{FD} / \partial p_R$. The optimal transmission power can be obtained by solving the equation above.

Consequently, Equation (24) reveals that the maximum transmission power will be chosen, namely $p_R^* = 1$ when:

$$\gamma_L \leq \frac{(\gamma_{BR}+1)(\gamma_{BC}+1)}{\gamma_{RC}} \tag{27}$$

the inequalities above show that the maximum transmit power is optimal only with weak loopback interference. Otherwise, the performance can be improved by backing off from the maximum power as reasoned before. In practice, we apply constrained power adaptation for which the SINRs coincide with those of the unconstrained case given the above whenever the conditions in (27) are not satisfied.

By further substituting p_R^* into $\gamma_{FD} = \gamma$ where the equation guarantees the rate requirement of cellular we can obtain the optimal power split factor (also the minimum power the cellular share from the D2D relay) as:

$$\alpha^* = \min\left\{1, \frac{\gamma(\gamma_{BC}+1+\gamma_{RC}p_R^*)(\gamma_{BR}+1+\gamma_L p_R^*)}{2\gamma_{BR}\gamma_{RC}p_R^*}\right\} \tag{28}$$

Up to now, we have solved the optimal transmit powers allocated at the BS and at the D2D transmitter as well as the optimal power splitting factor at the D2D transmitter. The optimal power allocation could be found at the BS in a centralized manner, but the global channel state information is required. Via a dedicated feedback channel, e.g., the cognitive pilot channel proposed by the E2R2/E3 consortium in [12], the channel state information and the optimal power allocation can be conveyed between the BS and the distributed UE reliably.

4. Simulation Results

In this section, we show that the simulation results demonstrate high achievement transmit rate and power efficiency of the proposed FD relay-based D2D communication scheme and guarantee the rate requirement for the cellular user in the primary. We choose the HD relaying mode as a benchmark. It should be made clear that the spectrum band consumed in a traditional cellular system is double that of our underlaying D2D communication due to one orthogonal subchannel that is needed for the cellular user and D2D users respectively. Therefore, the D2D communication we proposed achieves an improvement of spectral efficiency for cellular system. In this section, we study relaying and transmit power adaptation in the case of static channels and, thus, exploiting instantaneous channel state information. Alternatively, the following analysis explains the performance during one instantaneous snapshot within channel coherence time in a slow-fading environment. The parameter of system as rate requirement for cellular is set as $R^{re} = 20$ kb/s, which represents a typical voice signal characteristic and the bandwidth $W = 0.1$ MHz which corresponds to the width of the subchannel allocated to one cellular user.

In Figure 2, we show the impact of loopback interference on the optimal transmission power of D2D relay by setting $\gamma_{BR} = \gamma_{RC} = 10$ dB and $\gamma_{BC} = \gamma_{BD} = -10$ dB which means that channels from BS to cellular user and D2D receiver suffer from deep fading or shadowing. In the simulation, we change the channel gain of the loopback link at the D2D relay γ_L from 5 dB to 30 dB. It can be seen that the optimal transmit power is decreasing as the channel gain of loopback link is stronger. We can also observe that for a particular loopback gain, enlargement of the transmission power of relay is even harmful for the achievable rate for the cellular user which demonstrates our analysis we mentioned before. Therefore, in view of this problem, it is desirable to determine the optimal allocation power for the relay.

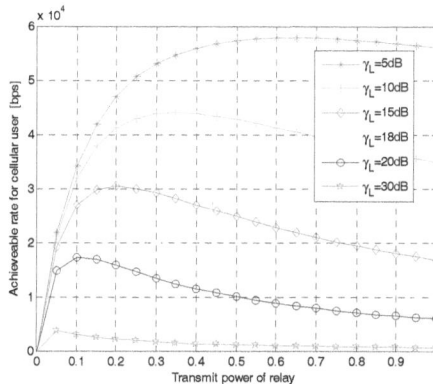

Figure 2. Achievable rate for cellular user with different transmission power usage.

In Figure 3, we observe the impact of power split factor on the system performance when we set $\gamma_L = 10$ dB. As we can observe from Figure 3, in the range of $\alpha \geq 0.24$, the scheme we proposed could satisfy the rate requirement for the cellular while at the same time the achievable rate for D2D communication is appropriate. Furthermore, this scheme provides significant performance improvement in terms of the achievable sum rate of the system. It also can be seen that the choice of power split factor has a great significance on system performance. The transmit rate of the cellular user will increase with a large power split factor. However, enlarging power split factor will also lead to the deterioration of D2D communication performance. Therefore, the selection of power split factor for the relay node is important.

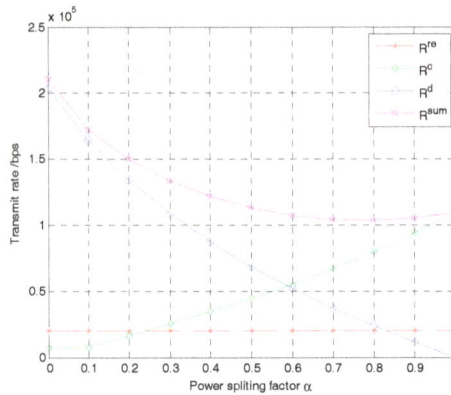

Figure 3. System performance with different power splitting factor.

In Figures 4 and 5, we make a comparison to the HD D2D relay mode in terms of optimal transmit power p_R^* and power splitting factor α. Due to transmitting signals in two orthogonal time slots, D2D relay in HD mode uses the maximum power as the optimal transmit power. In contrast, the scheme we proposed uses less power and guarantees the rate requirement for the cellular user which makes it much better in extending the battery life. It also can be observed that the optimal power is decreasing as the channel gain between D2D relay and cellular is stronger in our proposed scheme. Figure 5 shows that the power splitting factor gets smaller as the self-interference becomes weaker in FD mode, which means the D2D relay could share more of its power to support its own communication demand.

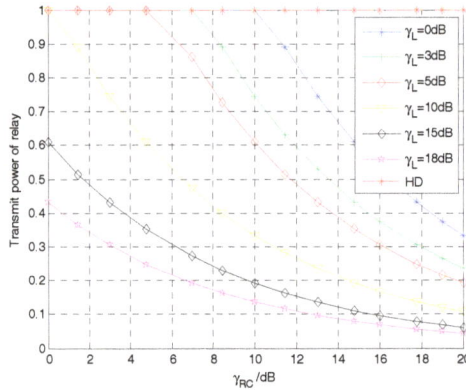

Figure 4. Optimal transmit power of relay.

In Figure 6, with the optimal transmit power and power split factor as we illustrated above, it is obvious that the D2D link can achieve significant rate gain in our proposed FD relaying-based D2D communication scheme. The simulation result also shows that the performance of a D2D achievable rate is significantly improved when $\gamma_L < 15$ dB and it also has a faster increase tendency than the benchmark of HD mode, even in the presence of $\gamma_L = 15$ dB. The reason for this is that the FD relaying-based D2D scheme could transmit and receive signal simultaneously rather than occupy one of two orthogonal time slots for transmitting and receiving signal respectively. For this reason, the FD relaying based D2D communication scheme inherently has a much higher spectrum efficiency in comparison to HD mode.

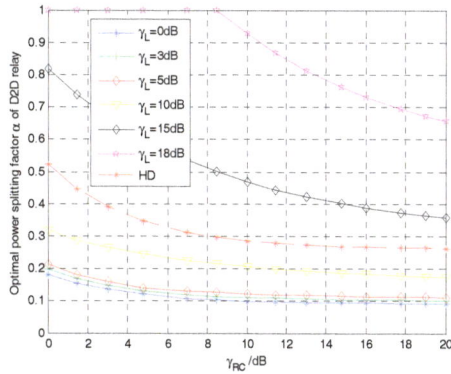

Figure 5. Optimal power splitting factor of relay.

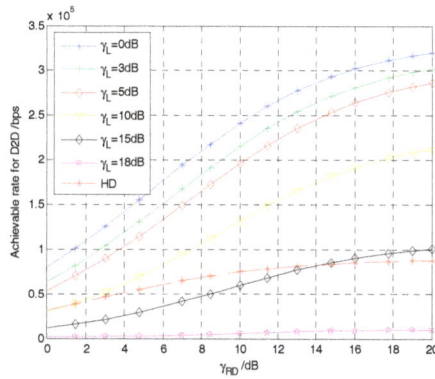

Figure 6. Achievable rate for D2D communication.

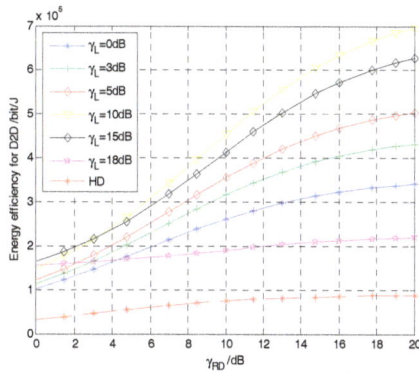

Figure 7. Energy efficient of D2D communication.

In contrast to Figure 6, Figure 7 reveals the energy efficiency of our proposed scheme when compared to the traditional HD mode. We can observe that the proposed scheme improves the energy efficiency of HD mode greatly. The reason for this is that the relay node of the HD mode forwards the cellular signal with maximum power usage for D2D relay, namely, $p_R^* = 1$. However, the optimal power for the FD mode is always less than the maximum power and decreases as the channel gain for the loopback link increases. It also can be seen that the performance of energy efficiency does not improve with the mitigation of the self-interference of the relay. The reason for this result is that, upon efficient mitigation of self-interference of relay, the optimal power at D2D relay is increasing, and as we can see from Figure 4, this ultimately decreases the energy efficiency.

5. Conclusions

In this paper, we proposed a FD relaying based D2D communication scheme to assist the cellular communication by underlaying its downlink spectral resource. To fulfill the rate requirement of the cellular user by using the least power and to maximize the achievable rates for D2D link at the same time, we derive the optimal transmit power and power split factor for the D2D relay in a closed form. The simulation results show that the scheme we proposed brings significant achievable rate gain for D2D communication. Moreover, the scheme decreases transmit power usage by 80% for the D2D relay node in comparison to the HD mode, even in the presence of $\gamma_L = 15$ dB in Figure 4, which extends the lifetime of the battery and improves the energy efficiency of system, as shown in Figure 7.

Acknowledgments: This paper is funded by the International Exchange Program of Harbin Engineering University for Innovation-oriented Talents Cultivation. And this work is partially supported by National Natural Science Foundation of China (Grant No. 51509049), the Natural Science Foundation of Heilongjiang Province China (Grant No. F201345), the Fundamental Research Funds for the Central Universities of China (No. GK2080260140) and the National key research and development program (Grant No. 2016YFF0102806).

Author Contributions: Hui Dun Conceived, designed the experiments and performed the experiments; Fang Ye and Yibing Li analyzed the data; Fang Ye contributed reagents/materials/analysis tools; Hui Dun wrote the paper.

Conflicts of Interest: The authors declare no conflict of interest.

References

1. Lin, X.; Andrwes, J.; Ghosh, A.; Ratasuk, R. An overview of 3GPP device-to-device proximity services. *IEEE Commun. Mag.* **2014**, *52*, 40–48. [CrossRef]
2. Asadi, A.; Wang, Q.; Mancuso, V. A survey on device-to-device communication in cellular networks. *IEEE Commun. Surv. Tutor.* **2014**, *16*, 1801–1819. [CrossRef]
3. Doppler, K.; Rinne, M.; Wijting, C.; Ribeiro, C.; Hugl, K. Device-to device communication as an underlay to LTE-Advanced networks. *IEEE Commun. Mag.* **2009**, *47*, 42–49. [CrossRef]
4. Corson, M.; Laroia, R.; Li, J.; Park, V.; Richardson, T.; Tsirtsis, G. Toward proximity-aware internetworking. *IEEE Wirel. Commun. Mag.* **2010**, *17*, 26–33. [CrossRef]
5. Yang, M.J.; Lim, S.Y.; Park, H.J.; Park, N.H. Solving the data overload: Device-to-device bearer control architecture for cellular data offloading. *IEEE Veh. Technol. Mag.* **2013**, *8*, 31–39. [CrossRef]
6. Phunchongharn, P.; Hossain, E.; Kim, D. Resource allocation for device-to-device communications underlaying LTE-Advanced networks. *IEEE Wirel. Commun. Mag.* **2013**, *20*, 91–100. [CrossRef]
7. Feng, D.Q.; Lu, L.; Yuan-Wu, Y.; Li, G.; Li, S.; Feng, G. Device-to-device communications in cellular networks. *IEEE Commun. Mag.* **2014**, *52*, 49–55. [CrossRef]
8. Yu, C.H.; Doppler, K.; Ribeiro, C.; Tirkkonen, O. Resource sharing optimization for device-to-device communication underlaying cellular networks. *IEEE Trans. Wirel. Commun.* **2011**, *10*, 2752–2763.
9. Yu, G.D.; Xu, L.; Feng, D.; Yin, R.; Li, G.Y.; Jiang, Y. Joint mode selection and resource allocation for device-to-device communications. *IEEE Trans. Commun.* **2014**, *62*, 3814–3824. [CrossRef]
10. Ye, Q.; Al-Shalash, M.; Caramanis, C.; Andrews, J.G. Device-to-device modeling and analysis with a modified Matern hardcore BS location model. In Proceedings of the IEEE GLOBECOM, Atlanta, GA, USA, 9–13 December 2013; pp. 1825–1830.
11. Lin, X.; Andrews, J.G. Optimal spectrum partition and mode selection in device-to-device overlaid cellular networks. In Proceedings of the IEEE GLOBECOM, Atlanta, GA, USA, 9–13 December 2013; pp. 1837–1842.
12. Min, H.; Seo, W.; Lee, J.; Park, S.; Hong, D. Reliability improvement using receive mode selection in the device-to-device uplink period underlaying cellular networks. *IEEE Trans. Wirel. Commun.* **2011**, *10*, 413–418. [CrossRef]

13. Jiang, Y.; Liu, Q.; Zheng, F.; Gao, X.; You, X. Energy-Efficient Joint Resource Allocation and Power Control for D2D Communications. *IEEE Trans. Veh. Technol.* **2016**, *65*, 6119–6127. [CrossRef]

14. Hakola, S.; Chen, T.; Lehtom, J.; Koskela, T. Device-to-device (D2D) communication in cellular network–performance analysis of optimum and practical communication mode selection. In Proceedings of the IEEE Wireless Communications and Networking Conference, Sydney, Australia, 18–21 April 2010.

15. Janis, P.; Koivunen, V.; Ribeiro, C.; Korhonen, J.; Doppler, K.; Hugl, K. Interference-aware resource allocation for device-to-device radio underlaying cellular networks. In Proceedings of the IEEE 69th Vehicular Technology Conference Spring, Barcelona, Spain, 26–29 April 2009.

16. Zhao, W.; Wang, S. Resource Allocation for Device-to-Device Communication Underlaying Cellular Networks: An Alternating Optimization Method. *IEEE Commun. Lett.* **2015**, *19*, 1398–1401. [CrossRef]

17. Xu, S.; Wang, H.; Chen, T.; Huang, Q.; Peng, T. Effective interference cancellation scheme for device-to-device communication underlaying cellular networks. In Proceedings of the IEEE Vehicular Technology Conference Fall, Ottawa, ON, Canada, 6–9 September 2010.

18. Chen, X.; Proulx, B.; Gong, X.; Zhang, J. Exploiting social ties for cooperative D2D communications: A mobile social networking case. *IEEE/ACM Trans. Netw.* **2015**, *23*, 1471–1484. [CrossRef]

19. Laneman, J.N.; Tse, D.N.C.; Wornell, G.W. Cooperative Diversity in Wireless Networks: Efficient Protocols and Outage Behavior. *IEEE Trans. Inf. Theory* **2004**, *50*, 3062–3080. [CrossRef]

20. Zheng, D.; He, C.; Jiang, L.; Ding, J.; Zhang, Q. Power Optimization for D2D Communication Based on Rate Requirement in Relay-Assisted Networks. In Proceedings of the 2015 IEEE International Conference on Communication Workshop (ICCW), London, UK, 8–12 June 2015; pp. 686–691.

21. Han, Y.; Pandharipande, A.; Ho Ting, S. Cooperative Spectrum Sharing via Controlled Amplify-and-Forward Relaying. In Proceedings of the IEEE Wireless Communications and Networking Conference (WCNC), Singapore, 11–13 December 2013; pp. 3346–3351.

22. Pei, Y.; Liang, Y.C. Resource allocation for device-to-device communications overlaying two-way cellular networks. *IEEE Trans. Wirel. Commun.* **2013**, *12*, 3611–3621. [CrossRef]

23. Zhou, B.; Hu, H.; Huang, S.Q.; Chen, H.H. Intracluster device-to-device relay algorithm with optimal resource utilization. *IEEE Trans. Veh. Technol.* **2013**, *62*, 2315–2326. [CrossRef]

24. Zhao, M.; Gu, X.; Wu, D.; Ren, L. A Two-Stages Relay Selection and Resource Allocation Joint Method for D2D Communication System. In Proceedings of the IEEE Wireless Communications and Networking Conference (WCNC), Doha, Qatar, 3–6 April 2016.

25. Han, Y.; Pandharipande, A.; Ho Ting, S. Cooperative Decode-and-Forward Relaying for Secondary Spectrum Access. *IEEE Trans Wirel. Commun.* **2009**, *8*, 4945–4950.

26. Zhang, G.; Yang, K.; Wu, S.; Mei, X.; Zhao, Z. Efficient power control for half-duplex relay based D2D networks under sum power constraints. *Wirel. Netw.* **2015**, *21*, 2345–2355. [CrossRef]

27. Zhang, G.; Yang, K.; Liu, P.; Wei, J. Power Allocation for Full-Duplex Relaying-Based D2D Communication Underlaying Cellular Networks. *IEEE Trans. Veh. Technol.* **2015**, *64*, 4911–4916. [CrossRef]

28. Riihonen, T.; Werner, S.; Wichman, R. Hybrid full-duplex/half-duplex relaying with transmit power adaptation. *IEEE Trans. Wirel. Commun.* **2011**, *10*, 3074–3085. [CrossRef]

29. Kim, T.; Dong, M. An Iterative Hungarian Method to Joint Relay Selection and Resource Allocation for D2D Communications. *IEEE Wirel. Commun. Lett.* **2014**, *3*, 625–628.

Article

Neural Networks for Radar Waveform Recognition

Ming Zhang, Ming Diao, Lipeng Gao * and Lutao Liu

College of Information and Telecommunication, Harbin Engineering University, Harbin 150001, China; zhangming@hrbeu.edu.cn (M.Z.); diaoming@hrbeu.edu.cn (M.D.); liulutao@hrbeu.edu.cn (L.L.)
* Correspondence: gaolipeng2000@163.com or gaolipeng@hrbeu.edu.cn; Tel.: +86-0451-8251-9804

Academic Editor: Angel Garrido
Received: 6 March 2017; Accepted: 12 May 2017; Published: 17 May 2017

Abstract: For passive radar detection system, radar waveform recognition is an important research area. In this paper, we explore an automatic radar waveform recognition system to detect, track and locate the low probability of intercept (LPI) radars. The system can classify (but not identify) 12 kinds of signals, including binary phase shift keying (BPSK) (barker codes modulated), linear frequency modulation (LFM), Costas codes, Frank code, P1-P4 codesand T1-T4 codeswith a low signal-to-noise ratio (SNR). It is one of the most extensive classification systems in the open articles. A hybrid classifier is proposed, which includes two relatively independent subsidiary networks, convolutional neural network (CNN) and Elman neural network (ENN). We determine the parameters of the architecture to make networks more effectively. Specifically, we focus on how the networks are designed, what the best set of features for classification is and what the best classified strategy is. Especially, we propose several key features for the classifier based on Choi–Williams time-frequency distribution (CWD). Finally, the recognition system is simulated by experimental data. The experiments show the overall successful recognition ratio of 94.5% at an SNR of −2 dB.

Keywords: radar countermeasure; waveform recognition; T-F distribution; convolutional neural network

1. Introduction

Modern radars usually have low instantaneous power, called low probability of intercept (LPI) radars, which are used in electronic warfare (EW). For a radar electronic intelligence (ELINT) system (anti-radar system), analyzing and classifying the waveforms of LPI radars is one of the most effective methods to detect, track and locate the LPI radars [1,2]. Therefore, the second order statistics and power spectral density are utilized in the waveforms' recognition earlier to classify phase shift keying (PSK), frequency shift keying (FSK) and amplitude shift keying (ASK) [3]. Dudczyk presents the parameters (such as pulse repetition interval (PRI), pulse width (PW), etc.) to identify different radar signals [4–7]. Nandi introduces the decision theoretic approach to classify different types of modulated signals [8]. Additionally, the ratio of successful recognition (RSR) is over 94% at a signal-to-noise ratio (SNR) $\geq$15 dB. The artificial neural network is also utilized in the recognition system. The multi-layer perceptron (MLP) recognizer reaches more than 99% recognized performance at SNR $\geq$0 dB [9]. Atomic decomposition (AD) is also addressed in the detection and classification of complex radar signals. Additionally, the receiver realizes the interception of four signals (including linear frequency modulation (LFM), PSK, FSK and continuous wave (CW)) [10]. Time-frequency techniques can increase signal processing gain for the low power signals [11]. In [12], López analyzes the differences among LFM, PSK and FSK based on the short-time Fourier transform (STFT). Additionally, the RSR $\geq$90% at SNR $\geq$0 dB. Lundén [13] introduces a wide classification system to classify the intercepted pulse compression waveforms. The system achieves overall RSR $\geq$98% at SNR $\geq$6 dB. Ming improves the system of Lundén and shows the results in [14]. The sparse classification (SC) based on random projections is proposed in [15]. The approach improves efficiency, noise robustness and information completeness. LFM, FSK and PSK are recognized with RSR $\geq$90% at SNR $\geq$0 dB.

We investigate the convolutional neural network (CNN) for radar waveform recognition. CNN has been proposed in image recognition fields [16,17]. Recently, it has been applied for speech recognition [18–21], computer vision [22,23] and handwritten recognition [24–26], etc. Abdel-Hamid introduces the approaches to reduce the further error rate by utilizing CNNs in [27]. Experimental results show that CNNs reduce the error rate by 6%–10% compared with deep neural networks (DNNs) on the speech recognition tasks. In [26], a hybrid model of two superior classifiers CNN and support vector machine (SVM) is discussed. The RSR of the model achieves more than 99.81%, in which SVM performs as a classifier and CNN works as a feature extractor.

In this paper, we explore a wide radar waveform recognition system to classify, but not identify In this paper, the meaning of "classify" is that we distinguish the different types of waveforms. Additionally, "identify" is distinguishing the different individuals of the same type. Twelve types of waveforms (LFM, BPSK, Costas codes, polyphase codes and polytime codes) by using CNN and Elman neural network (ENN) are discussed. We propose time-frequency and statistical characteristic approaches to process detected signals, which transmit in the highly noisy environment. The detected signals are processed into time-frequency images with the Choi–Williams distribution (CWD). CWD has few cross terms for signals, which is a member of the Cohen classes [28]. Time-frequency images show the three main pieces of information of signals: time location, frequency location and amplitude. In the images, time and frequency information is more robust than amplitude. To make the images more suitable for the classifier, a thresholding method is investigated. The method handles the time-frequency images as binary images. After that, binary images are addressed by noise-removing approaches. The final images are used for classification and feature extraction. However, polyphase codes (including Frank code and P1-P4 codes) and LFM are similar to each other. It is difficult to classify them through shapes individually. Therefore, we extract some effective features for further classification of them. Features extraction is from binary images through digital image processing (such as skeleton extraction, Zernike moments [29], principal component analysis (PCA), etc.). The set of features is the input of ENN. Additionally, the output of ENN is the classification result. The entire structure of the classifier consists of two networks, CNN and ENN. CNN is the primary cell of the classifier, and ENN is auxiliary. Binary images are resized for CNN to separate polytime codes (include T1–T4) from the other eight kinds of waveforms. Additionally, we extract features for ENN, which can indicate the eight remaining codes obviously. Only if "others" are selected by the CNN, ENN starts to work (see Figure 2). In the experiments, the recognition system has overall RSR $\geq$94% at SNR $\geq$−2 dB.

In this paper, the major contributions can be summarized as follows: (1) build the framework of signals processing; additionally, establish the label data for testing the system; (2) the proposed recognition system can classify as many as 12 kinds of waveforms, which are described in the context; previous articles can seldom reach such a wide range of classification of radar signals; especially, four kinds of polytime codes are classified together for the first time in the published literature; (3) almost all interested parameters and all features will be estimated by received data without a priori knowledge; (4) propose a hybrid classifier that has two different networks (CNN and ENN).

The paper is organized as follows. The structure of the recognition system is exhibited in Section 2. Section 3 proposes the signal model and preprocessing. Section 4 explores the feature extraction, including signal features and image features. Additionally, it lists all features that we need. After that, Section 5 searches the structure of the classifier and describes it in detail. Section 6 shows the experiments. Section 7 draws the conclusions.

2. System View

The entire classification system mainly consists of three components: preprocessing, feature estimation and recognition; see Figure 1. It is an automatic process from the preprocessing part to the recognition part. In the preprocessing part, the received data are transformed into time-frequency images by utilizing CWD transformation. Then, the time-frequency images are transformed into the binary image through image binarization, image opening operation and noise-removing algorithms. In the feature extraction part, we extract effective features to train and test the classifier. Different kinds of waveforms have different shapes in the images. After image processing, the differences of shapes are more significant. CNN has a powerful ability of classification, which distinguishes polytime codes from others. To classify these similar waveforms (such as polyphase codes), we extract features from detected signals and binary images. In the recognition part, all of the waveforms are classified via the proposed classifier based on the extracted features.

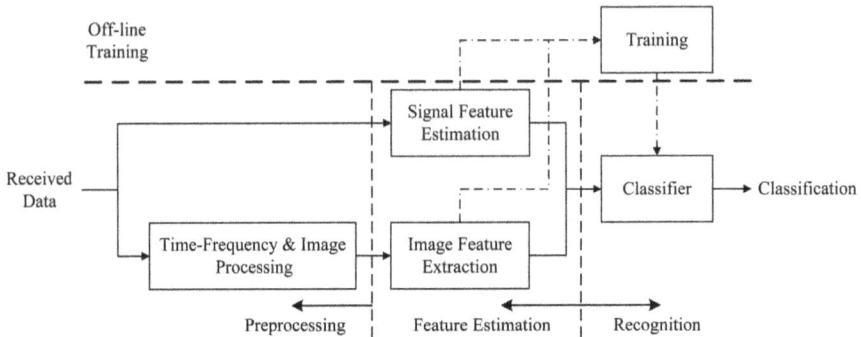

Figure 1. The figure shows the systematic components. Received data is processed in the preprocessing part and feature estimation part to extract features. And the data is classified in the classifier part.

The hybrid classifier consists of two networks, *network*1 and *network*2 ; see Figure 2. The entire classifier can classify 12 different kinds of radar waveforms, which has been mentioned in the writing. *Network*1 is the main network composed of CNN. Its input is a binary image after preprocessing. Additionally, the outputs are five different kinds of classification results. They are four kinds of polytime codes (T1-T4) and others (do not belong to the polytime class). *Network*2 is ENN, which is an auxiliary network. *Network*2 assists the main network (*network*1) in classifying the eight remaining waveforms that do not belong to polytime codes. When the waveform is considered as "others" by *network*1, *network*2 will begin to classify the waveform into one of the eight kinds of waveforms. The proposed structure of the classifier can improve the classified power.

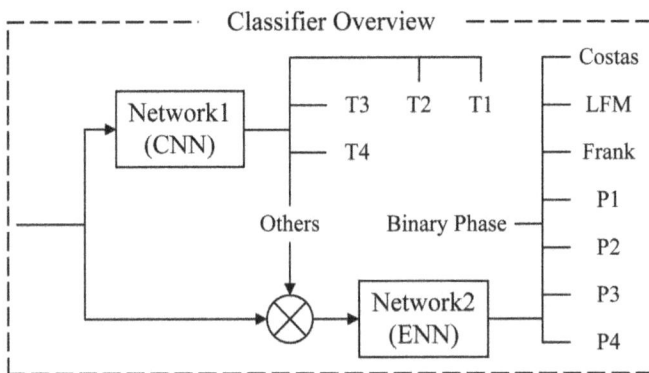

Figure 2. This figure shows the details of the classifier. *Network*1 is the main network composed of CNN and *Network*2 is ENN. *Network*2 assists the main network (*network*1) to complete the classification.

3. Preprocessing

In this section, detected signals are processed into binary images with the Choi–Williams time-frequency distribution.

3.1. Signal Model

We assume the signal is contaminated by additive white Gaussian noise (AWGN). Additionally, the amplitude is constant for time. In summary, the signal model is formulated as follows

$$y(nT) = s(nT) + m(nT) = Ae^{j\phi(nT)} + m(nT) \tag{1}$$

where $s(nT)$ is the n-th sample of complex signals. Additionally, $m(nT)$ is the n-th sample of complex white Gaussian noise (WGN). The variance of WGN equals σ_ε^2. A is the amplitude. However, for the sake of simplicity, we suppose $A = 1$. ϕ is the instantaneous phase of complex signals. To process detected signals from real to complex, Hilbert transform is applied [30].

3.2. Choi–Williams Distribution

The Choi–Williams distribution is a kind of time-frequency distribution, which expresses the details of detected signals. It can reduce the cross terms from the signals obviously.

$$C(t,\omega) = \iiint_\infty e^{j2\pi\xi(s-t)} f(\xi,\tau)$$
$$\cdot x(s+\tau/2)x^*(s-\tau/2)e^{-j\omega\tau} d\xi ds d\tau \tag{2}$$

where ω and t are the axes of frequency and time, respectively. $f(\xi,\tau)$ is a two-dimensional low-pass filter to balance cross terms and resolution. The kernel function is formulated as follows:

$$f(\xi,\tau) = \exp\left[\frac{(\pi\xi\tau)^2}{2\sigma}\right]. \tag{3}$$

σ is the controllable factor. The cross terms will be more obvious with the increase of σ. In this paper, $\sigma = 1$ is applied. In Figure 3, 12 kinds of signals are transformed into time-frequency images through CWD transformation. The work in [31] proposes a new fast calculation of CWD based on standard Fourier transformation (FFT). We could recommend that the number of the sampling is the power of two, such as 256, 512, etc. In this paper, 1024 sampling points are investigated. However, the length of signals is $N < 1024$ for most of the time. Therefore, zero padding is utilized in the process.

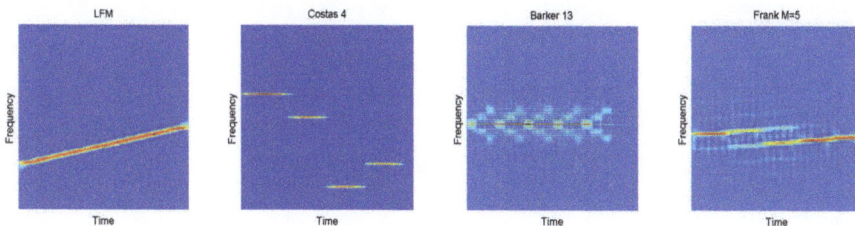

Figure 3. *Cont.*

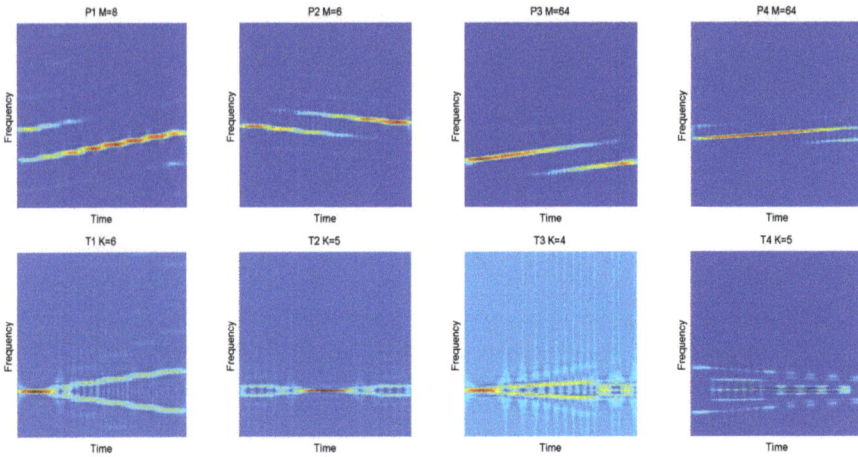

Figure 3. In this figure, different waveform classes are shown, including Linear Frequency Modulation (LFM), Binray Phase Shift Keying (BPSK), Frank, Costas codes, P1-P4 codes and T1-T4 codes. There are significant differences among the Choi-Williams Time-Frequency Distribution (CWD) images.

3.3. Binary Image

In this part, the detected signals are processed into binary images with the global thresholding algorithm [32]. Before it is done, we need to resize the time-frequency images to reduce the computational load, in which the resized size is $N \times N$. The algorithm is organized as follows.

a.   Normalize the resized images $G(x,y) \in [0,1]$, i.e.,

$$G(x,y) = \frac{CWD_{N \times N}(x,y) - \min CWD_{N \times N}(x,y)}{\max(CWD_{N \times N}(x,y) - \min CWD_{N \times N}(x,y))};$$

b.   Estimate the threshold T of $G(x,y)$, i.e.,

$$T = \frac{\max G(x,y) + \min G(x,y)}{2};$$

c.   Separate the image into two pixel groups G_1 and G_2; G_1 includes all pixels that values $> T$, and G_2 includes others;

d.   Calculate the average value μ_1 and μ_2 of two pixel groups G_1 and G_2, respectively;

e.   Update the threshold, i.e.,

$$T = \frac{\mu_1 + \mu_2}{2};$$

f.   Repeat (b)–(e), until the δT is smaller than 0.001, i.e.,

$$\delta T = T_{now} - T_{before};$$

g.   Compute $B(x,y)$ as follows:

$$B(x,y) = \begin{cases} 1 & G(x,y) > T \\ 0 & others \end{cases};$$

h.   Output $B(x,y)$.

3.4. Noise Removed

After the operation of binarization, meanwhile, there are some isolated noises and processed noises in $B(x,y)$. Isolated noises come from the noisy environment. In the binary image, isolated noises are groups of pixels that have no fixed shape. Processed noises are generated from the CWD kernel especially. In the binary image, they are straight lines, long but thin. The length is more than 50 pixels, but the width of lines is less than three pixels. Image opening operation (erosion followed by dilation) algorithms are proposed to remove the processed noises. Additionally, the size of the operational kernel is 3 × 3. It is effective to remove the shape whose width is less than three pixels. For isolated noises, we count the number of pixels of signals or noises. In this paper, the groups are removed, in which sizes are smaller than 10% of the largest one. The content of removing noise is introduced in Figure 4. The finished binary images are used in CNN and feature extraction.

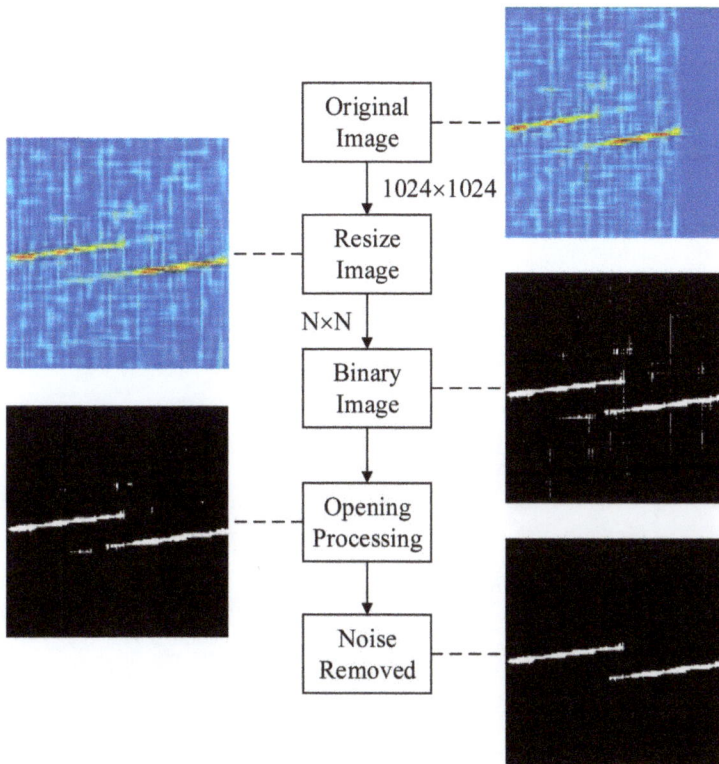

Figure 4. In this figure, we exhibit the processing with P3 code at an signal-to-noise ratio (SNR) of −4 dB.

4. Feature Extraction

In this section, we extract some useful features and build a feature vector for ENN in order to assist the CNN to complete recognition. The section consists of two parts, including signal features and image features. The features, which we can estimate or calculate from detected signals directly, belong to signal features. Similarly, image features include the features that are extract from binary images. Table 1 lists the signal features and image features that are used in *network2*.

Table 1. List of the features for *network2*.

Index	Description	Symbol
1	Moment (1-order)	$\hat{M}_{10}$
2	Moment (2-order)	$\hat{M}_{20}$
3	Cumulant (2-order)	$\hat{C}_{20}$
4	PSD maximum (1-order)	γ_1
5	PSD maximum (2-order)	γ_2
6	Std. of phase	$\hat{\sigma}_\phi$
7	Std. of frequency	$\hat{\sigma}_f$
8	No. of objects (20%)	N_{obj1}
9	No. of objects (50%)	N_{obj2}
10	CWD time peak location	t_{max}
11	Std. of object width	$\hat{\sigma}_{obj}$
12	Maximum of PCA degree	$\hat{\theta}_{max}$
13	Std. of f_n	$\hat{\sigma}_{Wf}$
14	Autocorrelation of f_n	r
15	FFT of correlation f_n	$\hat{a}_{max}$
16	Pseudo-Zernike moment (2-order)	$\hat{Z}_{20}$
17	Pseudo-Zernike moment (3-order)	$\hat{Z}_{30}$
18	Pseudo-Zernike moment (4-order)	$\hat{Z}_{22}$
19	Pseudo-Zernike moment (4-order)	$\hat{Z}_{31}$
20	Pseudo-Zernike moment (5-order)	$\hat{Z}_{32}$
21	Pseudo-Zernike moment (6-order)	$\hat{Z}_{33}$
22	Pseudo-Zernike moment (7-order)	$\hat{Z}_{43}$

PSD: Power Spectral Density. PCA: Principal Component Analysis. FFT: Fast Fourier Transformation. Std.: Standard Deviation.

4.1. Signal Features

In this part, the features are extracted from signals based on signal processing approaches.

4.1.1. Based on the Statistics

We estimate the *n*-order moment of complex signals as follows:

$$\hat{M}_{nm} = \left| \frac{1}{N} \sum_{k=0}^{N-1} y^{n-m}(k)(y^*(k))^m \right| \tag{4}$$

where $(*)$ is the conjugated symbol and N is the sample number. We utilize absolute values to ensure that the estimated values are invariant constants when the signal phase rotates. $\hat{M}_{10}$ and $\hat{M}_{20}$ are calculated by Equation (4).

The *n*-order cumulant is given by [33,34]:

$$\hat{C}_{nm} = \left| \frac{1}{N} \sum_{k=0}^{N-1} (y(k) - \hat{M}_{10})^{n-m} (y^*(k) - \hat{M}_{10})^m \right| \tag{5}$$

where, the same as context, $\hat{M}_{10}$ is from Equation (4).

4.1.2. Based on the Power Spectral Density

Before estimation of Power Spectral Density (PSD), the detected signals should be normalized as follows:

$$\tilde{y}(k) = \frac{y(k)}{\sqrt{\hat{M}_{21} - \sigma_{\varepsilon}^2}} \tag{6}$$

where $\hat{M}_{21}$ is obtained from Equation (4) and $y(k)$ is the k-th sample. The variance of additive noise σ_{ε}^2 can be obtained in [35].

The PSD are calculated as follows:

$$\gamma_m = \frac{1}{N}\max_n \left\{ \frac{1}{N} \left| \sum_{k=0}^{N-1} \tilde{y}^m(k)e^{-j2\pi nk/N} \right|^2 \right\} \tag{7}$$

where $\tilde{y}(k)$ is from Equation (6).

4.1.3. Based on the Instantaneous Properties

Instantaneous properties are the essential characteristics of detected signals. They can distinguish frequency modulated signals from phase modulated signals effectively. In this paper, we estimate the instantaneous frequency and instantaneous phase from samples. The standard deviation of instantaneous phase is addressed in [9]. For brevity, $\phi(k) = \tan^{-1}[\text{Im}(y(k))/\text{Re}(y(k))]$ is applied; where, Re and Im are the real and imaginary parts of complex signals, respectively. The standard deviation of instantaneous phase is given by:

$$\hat{\sigma}_{\phi} = \sqrt{\frac{1}{N}\left(\sum_{k=0}^{N-1}\phi^2(k)\right) - \left(\frac{1}{N}\sum_{k=0}^{N-1}|\phi(k)|\right)^2} \tag{8}$$

where N is the sample number. ϕ is the instantaneous phase with the range of $[-\pi, \pi]$.

Instantaneous frequency estimation is more complex than instantaneous phase. We describe the method in several steps to make it clear.

a. Calculate $\phi(k)$;
b. Calculate $\phi_u(k)^{\star}$ from $\phi(k)$;
c. Calculate $f(k)^{\star\star}$, i.e.,

$$f(k) = \phi_u(k) - \phi_u(k-1);$$

d. Calculate μ_f, i.e.,

$$\mu_f := \frac{1}{N}\sum_{k=0}^{N-1}f(k);$$

e. Normalize the instantaneous frequency $\tilde{f}(k)$,

$$\tilde{f}(k) = (f(k) - \mu_f)/(\max|f(k) - \mu_f|)$$

f. Output the standard deviation of instantaneous frequency $\hat{\sigma}_f$,

$$\hat{\sigma}_f = \sqrt{\frac{1}{N}\left(\sum_{k=0}^{N-1}\tilde{f}^2(k)\right) - \left(\frac{1}{N}\sum_{k=0}^{N-1}|\tilde{f}(k)|\right)^2}. \tag{9}$$

★ $\phi_u(k)$ is the unwrapped phase of $\phi(k)$. When the absolute jumps from $\phi(k)$, we can add $\pm 2\pi$ to recover the consecutive phase.
★★ In the sequence of $f(k)$, some spikes are created by processing. We use the median filter algorithm with window size of five to smooth the spikes.

4.2. Image Features

In this part, we extract the features based on binary images. The number of objects (N_{obj}) is a key feature. For instance, Costas codes have more than three objects, but Frank code and P2 have two. Additionally, P1, P4 and LFM only have one. We estimate two features N_{obj1} and N_{obj2}. N_{obj1} is the number of objects, the sizes of the pixels of which are more than 20% of the size of the largest object. Likewise, $N_{obj2} \geq 50\%$.

The maximum energy location in time domain is also a feature, i.e.,

$$t_{max} = \frac{1}{N-1} \arg\max_t \{C_{N \times N}(t, \omega)\} \tag{10}$$

where $C_{N \times N}(t, \omega)$ is the resized time-frequency image and N is the sample number.

The standard deviation of the width of signal objects ($\hat{\sigma}_{obj}$) can describe the concentration of signal energy. The feature is estimated as follows.

- Repeat for every object, do $k = 1, 2, ..., N_{obj1}$;

  1. Retain the k-th object and remove others, called $B_k(x, y)$;
  2. Estimate the principal components of $B_k(x, y)$;
  3. Rotate* the $B_k(x, y)$ until the principal components are vertical; record as $B_{k'}(x, y)$;
  4. Sum the vertical axis, i.e.,

  $$v(x) = \sum_{y=0}^{N-1} B_{k'}(x, y),$$

  $x = 0, 1, 2, ..., N - 1$;
  5. Normalize $v(x)$ as follows

  $$\hat{v}(x) = \frac{v(x)}{\max\{v(x)\}};$$

  6. Estimate the standard deviation of $\hat{v}(x)$, i.e.,

  $$\hat{\sigma}_{k,obj1} = \sqrt{1/N \sum_x \hat{v}^2(x) - (1/N \sum_x \hat{v}(x))^2},$$

  where N is the sample number;
- Output the rotation degree $\hat{\theta}_{max}$, which performs Step (c) at the maximum object.
- Output the average of the $\hat{\sigma}_{k,obj1}$, i.e.,

$$\hat{\sigma}_{obj} = (1/N_{obj1}) \sum_{k=1}^{N_{obj1}} \hat{\sigma}_{k,obj1}. \tag{11}$$

★ Nearest neighbor interpolation is applied in rotation processing.

P2 has a negative slope in five types of polyphase codes. Therefore, the feature $\hat{\theta}_{max}$ can classify P2 from others easily. The feature shows the angle between the maximum object and the vertical direction. It can be obtained from the calculation of $\hat{\sigma}_{obj}$ easily.

Next, we retain the maximum object in the binary image, but others are removed. The skeleton of the object is extracted by utilizing the image morphology method. Additionally, the linear trend of the object is also estimated based on minimizing the square errors method at the same time. Subtract the linear trend from the skeleton to achieve the difference vector f_n. The standard deviation of f_n is estimated as:

$$\hat{\sigma}_{Wf} = \sqrt{\frac{1}{M-1} \sum_{k=1}^{M} f_n^2(k) - \left(\frac{1}{M-1} \sum_{k=1}^{M} f_n(k)\right)^2} \tag{12}$$

where M is the sample number of f_n.

Some features are extracted by using autocorrelation of f_n, i.e., $c(m) = \sum_k f_n(k)f_n(k-m)$, $m = 0, 1, ..., N-1$. The autocorrelation method makes differences more significant among stepped waveforms (P1, Frank code) and linear waveforms (P3, P4, LFM). See Figure 3 for more details.

The ratio of the maximum value and sidelobe maximum value of $c(m)$ is formulated as:

$$r = \frac{N \max_{m \in [m_0, N-1]} c(m)}{(N-m_1) \max c(m)} \tag{13}$$

where m_0 is the value corresponding to the minimum of $c(m)$ and m_1 is the value corresponding to the maximum of $c(m)$ in the location of $[m_0, N-1]$.

We estimate the maximum of the absolute of FFT operation $\hat{a}_{max}$ as follows:

$$\hat{a}_{max} = \max\{\text{abs}[\text{FFT}(c_{normal}(m))]\} \tag{14}$$

where $c_{normal}(m)$ is normalized from $c(m)$ and $c_{normal}(m) \in [-1, 1]$.

Pseudo-Zernike moments are invariant for topological transformation [36], such as rotation, translation, mirroring and scaling. They are widely applied in pattern recognition [37–39]. The n-order image geometric moments are calculated as:

$$m_{pq} = \sum_x \sum_y B(x, y) x^p y^q \tag{15}$$

where $B(x, y)$ is from Section 3.3. The central geometric moments for scale and translation invariant are given by:

$$G_{pq} = \frac{1}{m_{00}^{(p+q+2)/2}} \sum_x \sum_y B(x, y)(x - \bar{x})^p (y - \bar{y})^q \tag{16}$$

where $\bar{x} = m_{10}/m_{00}$ and $\bar{y} = m_{01}/m_{00}$.

The scale and translation invariant radial geometric moments are shown as:

$$R_{pq} = \frac{1}{m_{00}^{(p+q+3)/2}} \sum_x \sum_y B(x, y)(\tilde{x}^2 + \tilde{y}^2)^{1/2} \tilde{x}^p \tilde{y}^q \tag{17}$$

where $\tilde{x} = x - \bar{x}$ and $\tilde{y} = y - \bar{y}$.

Then, the pseudo-Zernike moments can be estimated as follows:

$$
\begin{aligned}
Z_{nm} = &\frac{n+1}{\pi} \sum_{\substack{s=0 \\ n-s-m=even}}^{n-|m|} \sum_{a=0}^{k} \sum_{b=0}^{m} (-j)^b \\
&\binom{k}{a}\binom{m}{b} D_{nms} G_{2k-2a+m-b, 2a+b} \\
+ &\frac{n+1}{\pi} \sum_{\substack{s=0 \\ n-s-m=odd}}^{n-|m|} \sum_{a=0}^{d} \sum_{b=0}^{m} (-j)^b \\
&\binom{d}{a}\binom{m}{b} D_{nms} R_{2d-2a+m-b, 2a+b}
\end{aligned} \tag{18}
$$

where $k = (n-s-m)/2$, $d = (n-s-m-1)/2$ and:

$$D_{nms} = (-1)^s \frac{(2n+1-s)!}{s!(n-|m|-s)!(n+|m|+1-s)!}. \tag{19}$$

At last, $\hat{Z}_{nm}$ is estimated, i.e., $\hat{Z}_{nm} = \ln|Z_{nm}|$. The members of pseudo-Zernike moments include $\hat{Z}_{20}, \hat{Z}_{22}, \hat{Z}_{30}, \hat{Z}_{31}, \hat{Z}_{32}, \hat{Z}_{33}$ and $\hat{Z}_{43}$.

5. Classifier

In Section 4, we complete the resized binary image labels for CNN and the feature vector extraction with 22 elements for ENN. In this section, we describe the structure of two networks in detail.

5.1. CNN

CNN is a new neural network, which has a special structure for image feature extraction. Different from traditional network, the input of CNN is a two-dimensional feature (image). The convolution layers can extract information, and pooling layers reduce computer load effectively. CNN is not a full connected network, which is similar to the cerebral cortex. The architecture of the CNN model is shown in Figure 5. CNN has the hierarchical architecture [40]. Hence, we describe the neural architecture as follows.

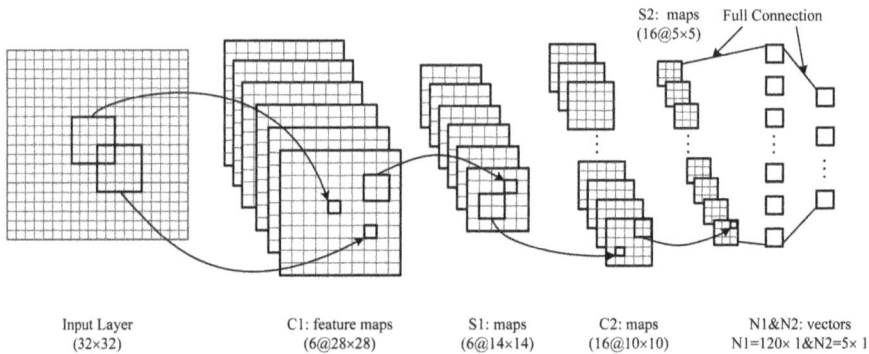

Figure 5. The figure shows the structure of Convolutional Neural Network (CNN). The input image are processed in the hidden layers and classified in the out layer.

a. The input layer is a binary image, which is from Section 3.4. To reduce the computer load, we resize the image to 32×32 with the nearest neighbor interpolation algorithm.

b. The first hidden layer C_1 is a convolutional layer, which has six feature maps. Different feature maps require a different convolutional kernel. C_1 has six convolutional kernels with a size of 5×5. We utilize $C_1(m, n, k)$ to represent the value of the k-th feature map at position (m, n) in the C_1 layer.

c. The second hidden layer S_1 is a down-sampling layer with six feature maps. In S_1, every feature value is the average of four adjacent elements in C_1. We denote $S_1(m, n, k)$ as the context. Further, we have:

$$S_1(m, n, k) = \text{mean}(C_1(2m - 1, 2n - 1, k), C_1(2m - 1, 2n, k), C_1(2m, 2n - 1, k), C_1(2m, 2n, k)). \tag{20}$$

The size of feature maps in S_1 reduces to $1/4$, compared with feature maps of C_1.

d. C_2 is a convolutional layer with 16 different kernels. It is not fully connected with the S_1 layer [41]. The connection details are described in Table 2. $C_2(m, n, k)$ is also utilized to describe the neurons in this layer. For the α-th column in Table 2, we mark row indices by $\beta_{\alpha,0}, \beta_{\alpha,1}, \cdots, \beta_{\alpha,p-1}$. For instance, if $\alpha = 7$, then we will get parameters as follows: $p = 4$, $\beta_{7,0} = 1$, $\beta_{7,1} = 2$,

$\beta_{7,2} = 3$, $\beta_{7,3} = 4$. Further, the size of the convolutional kernel is $p \times 5 \times 5$. K_α is the α-th kernel. Additionally, we have:

$$C_2(m,n,\alpha) = \sum_{r=0}^{p-1} \sum_{m_0=0}^{4} \sum_{n_0=0}^{4} [\, S_1(m + m_0, n + n_0, \beta_{\alpha,r}) \tag{21}$$
$$\times K_\alpha(5 - m_0, 5 - n_0, p - 1 - r) \,].$$

For example, for the zeroth column, $p = 3$, $\alpha_{00} = 0$, $\alpha_{01} = 1$, $\alpha_{02} = 2$, and we also have:

$$C_2(m,n,0) = \sum_{r=0}^{p-1} \sum_{m_0=0}^{4} \sum_{n_0=0}^{4} [\, S_1(m + m_0, n + n_0, \beta_{0,r}) \tag{22}$$
$$\times K_0(5 - m_0, 5 - n_0, p - 1 - r) \,].$$

e. Similar to S_1, this layer is a down-sampling layer, called S_2. S_2 has 16 feature maps. To follow the context in Equation (20), we donate:

$$S_2(m,n,k) = \mathrm{mean}(C_2(2m - 1, 2n - 1, k), C_2(2m - 1, 2n, \tag{23}$$
$$k), C_2(2m, 2n - 1, k), C_2(2m, 2n, k)).$$

f. The connection between S_2 and N_1 is a full connection. Each kernel in N_1 will be connected with all of the feature maps in S_2. There are 120 kernels in this layer. Additionally, the size of the kernel is 5×5, which means the output is a column vector with the size of 120×1. We describe $N_1(\lambda)$ as the λ-th feature map of N_1 and K_λ as the λ-th kernel. Then, we have:

$$N_1(\lambda) = \sum_{r=0}^{15} \sum_{m_0=0}^{4} \sum_{n_0=0}^{4} [\, S_2(m_0, n_0, r) \tag{24}$$
$$\times K_\lambda(5 - m_0, 5 - n_0, 15 - r) \,].$$

g. Finally, the connected style between N_1 and output layer is fully connected. There are five neurons (defined by the classes we want to classify) in the output layer with the *sigmoid* function.

Table 2. Connection detail about S_1 and C_2.

	0	1	2	3	4	5	6	7	8	9	10	11	12	13	14	15
0	✓				✓	✓	✓			✓	✓	✓	✓		✓	✓
1	✓	✓				✓	✓	✓			✓	✓	✓	✓		✓
2	✓	✓	✓				✓	✓	✓			✓		✓	✓	✓
3		✓	✓	✓			✓	✓	✓	✓			✓		✓	✓
4			✓	✓	✓			✓	✓	✓	✓		✓	✓		✓
5				✓	✓	✓			✓	✓	✓	✓		✓	✓	✓

5.2. ENN

The three-layer ENN is utilized in the paper for signal classification. The connections, which connect different hidden layers or output layers, have different weights [42]. At every time step, the input is propagated in a feed-forward fashion and the feedback of the output. Additionally, the error back propagation (BP) learning algorithm is also utilized [43]. The connection results in that the context units always maintain a copy of the previous values of hidden units. Thus, the network can keep the past state, which is useful for applications such as sequence prediction [44–46]. In Figure 6, there are 46 neurons in the hidden layer. For the input and output layer, the number of neurons is determined by the dimension of input and output vectors. Sigmoid function $f(x) = 1/(1 + e^{-x})$ is proposed in every layer. In [47], Sheela discusses the different methods to fix the neurons number H_{num} of hidden layers. In this paper, a simple formula is given by:

$$H_{num} = \frac{C \times X + 0.5 \times C \times (X^2 + X) - 1}{C + X} \tag{25}$$

where X is the dimension of the feature vector. C is the number of categories. The proposed formula cannot determine the optimal number of hidden layer completely. We may fine-tune the number in some situations. Forty-six neurons of hidden layers are applied in this paper.

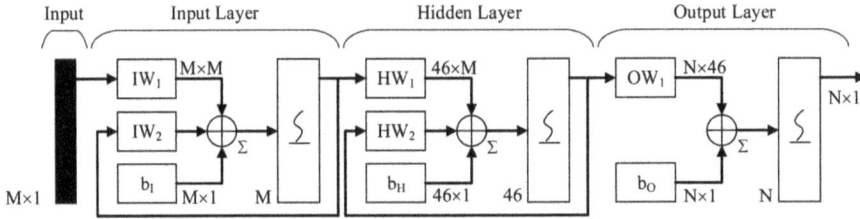

Figure 6. This figure shows the structure of Elman neural network. It is a 3 layers network that has feedback loops. So, the network can keep the past state, which is useful for waveform classification.

6. Simulation Results and Discussion

In this section, the performance of the proposed recognition system is analyzed by utilizing simulated data. The section consists of three parts, including creating the simulated data, discussing the relationship between SNR and RSR, depicting the accurate rate of robustness and summarizing the experiments.

6.1. Production of Simulated Signals

In this part, simulated signals are created. In addition, the SNR is proposed as SNR $= 10\log_{10}(\sigma_s^2)/(\sigma_\varepsilon^2)$; where σ_s^2 and σ_ε^2 are the variances of the signal and noise, respectively. Every signal has different parameters that need to be set. We denote a uniform variable U($\cdot$) based on the sample rate. For example, we assume that the original frequency (f_0) is 1000 Hz and the sample rate (f_s) is 8000 Hz. Then, the uniform result is $f_0 = $ U(f_0/f_s) $=$ U($1/8$). Meanwhile, U($1/8, 1/4$) expresses the random variable that belongs to [U($1/8$), U($1/4$)]. And in this paragraph, [1,3] also represents a set that includes {1,2,3}. Table 3 lists the parameters of the waveforms. For LFM, the sample points change from 500–1000 randomly. Additionally, the range of bandwidth (Δf) is U($1/16, 1/8$), so is the initial frequency (f_0). For BPSK, the cycle number per phase code (cpp) and the code periods' number (N_p) are [1, 5] and [100, 300], respectively. The length of the Barker codes is selected from {7, 11, 13} randomly. The carrier frequency is U($1/8, 1/4$). For the Costas codes, the fundamental frequency (f_{min}) is U($1/24$). Additionally, the frequency changed number is [3, 6]. For the Frank code, frequency steps (M) are in the range of [4, 8]. Polyphase codes have the same types of parameters as the Frank code. For polytime codes, the range of segments number (k) and overall code duration (T) are [4, 6] and [0.07, 0.1], respectively.

Table 3. List of the parameters of simulated signals.

Signal Waveforms	Parameters	Uniform Ranges
-	Sampling rate (f_s)	U(1)
LFM	Samples number (N)	[500, 1000]
	Bandwidth (Δf)	U(1/16, 1/8)
	Initial frequency (f_0)	U(1/16, 1/8)
BPSK	Cycles per phase code (cpp)	[1, 5]
	Number of code periods (N_p)	[100, 300]
	Barker codes	{7, 11, 13}
	Carrier frequency (f_c)	U(1/8, 1/4)
Costas codes	N	[512, 1024]
	Fundamental frequency (f_{min})	U(1/24)
	Number changed	[3, 6]
Frank and P1 codes	f_c	U(1/8, 1/4)
	cpp	[1, 5]
	Frequency steps (M)	[4, 8]
P2 code	f_c	U(1/8, 1/4)
	cpp	[1, 5]
	M	$2 \times [2, 4]$
P3 and P4 codes	f_c	U(1/8, 1/4)
	cpp	[1, 5]
	M	$2 \times [16, 35]$
T1-T4 codes	Number of segments (k)	[4, 6]
	Overall code duration (T)	[0.07, 0.1]

6.2. Experiment with SNR

In this part, we depict the relation between SNR and RSR in Figure 7. There are 1000 labels in each waveform class. Twenty percent of the labels are utilized for testing and 80% for training. The result is compared with Lundén's system [13] and our previous work [14], both of which are wide systems in waveform classification.

Figure 7 plots the experimental results of RSR with different SNR. Twelve kinds of waveforms and the "overall" are provided. The solid line shows the proposed system, and the dotted lines represent others. For LFM and P4, the proposed approach provides better performance than Lundén's, especially at low SNR, but poorer than the previous work, although the difference is not too much. For BPSK and Costas codes, the three RSRs almost have similar results, and all of them are at a high level. For Frank and P2, the results of the proposed method and previous work are alike and higher than Lundén's. In the simulation of P1, the proposed method is the best when the SNR is more than −2 dB. The results of P3 are similar to P1; proposed method performs well at high SNR. For polytime codes, the proposed approach also has excellent RSRs. It benefits from the outstanding design of pre-processing and the high RSR of the classifier. Finally, the overall RSR has been raised by 20% in the proposed approaches, compared to Lundén's and previous work. At SNR of −2 dB, the overall probabilities are still more than 90%. Table 4 exhibits the confusion table of 12 kinds of waveforms at the SNR of −2 dB. As Table 4 shows, the waveforms of P3 and P4 are not "always" classified correctly. For P3, most of the errors are classified into the Frank code. Meanwhile, most of the errors of P4 are classified into LFM. However, the two pairs are very similar; see Figure 3.

Table 4. Confusion matrix for the system at an SNR of −2 dB. The overall Ratio of Successful Recognition (RSR) is 94.5%.

	T1	T2	T3	T4	LFM	Costas	BPSK	Frank	P1	P2	P3	P4
T1	99.5	0	0	0	0	0	0	0	0	0	0	0
T2	0	98.5	0	0	0	0	0	0	0	0	0	0
T3	0.5	0	96	0	0	0	0	0	0	0	0	0
T4	0	0	0	98.5	0	1.5	2	0	0	0	0	0
LFM	0	0	0	0	89.5	0	0	0	1	0	2	15.5
Costas	0	0	0	0	0	97.5	0	0	0	1.5	1	0
BPSK	0	0	0.5	0	0	0	98	1	0	0	0	0.5
Frank	0	0	0	0	1	1	0	90	4	7	11	1.5
P1	0	0	0	0	0.5	0	0	0	87.5	0	0	7.5
P2	0	0	0	0	1	0	0	6.5	5.5	90	3	5.5
P3	0	1.5	3.5	1.5	0	0	0	2.5	1	1.5	78.5	0.5
P4	0	0	0	0	8	0	0	0	1	0	4.5	69

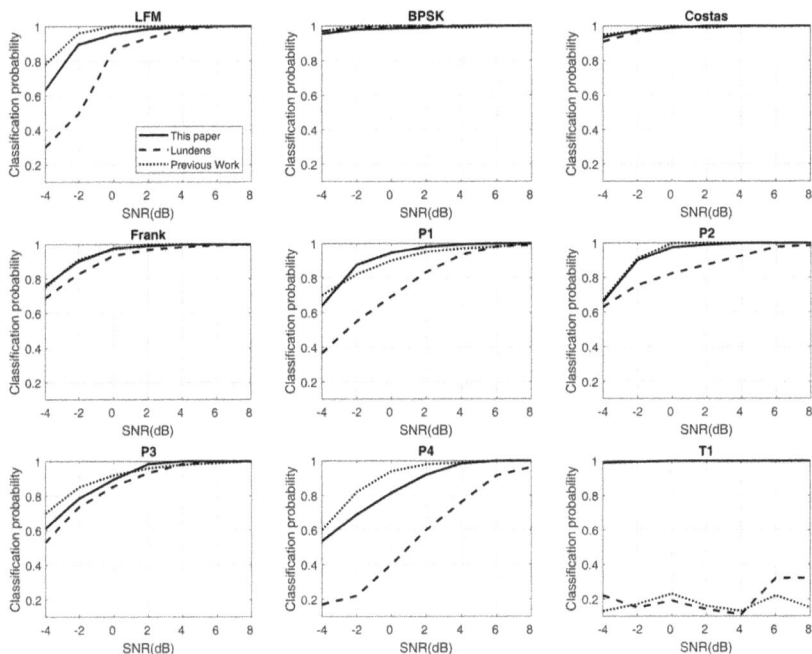

Figure 7. *Cont.*

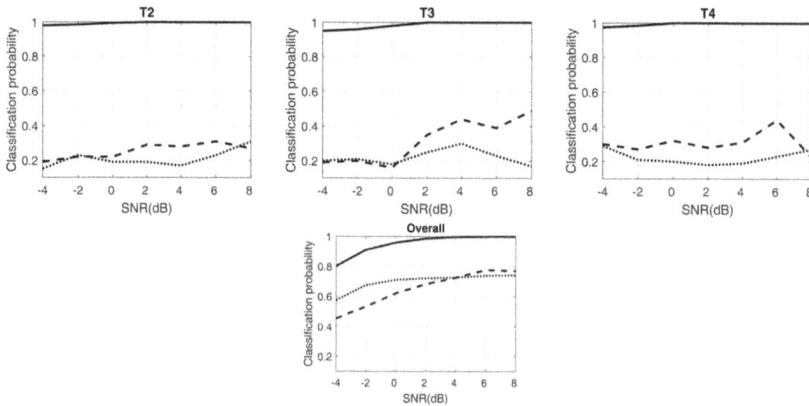

Figure 7. This figure depicts the different probabilities of 12 types of radar waveforms with testing data. SNR: Signal-to-noise ratio.

6.3. Experiment with Robustness

The robustness of proposed approaches is explored in different training samples. There are 900 label samples in each waveform for training and 100 labels for testing. Afterwards, the training samples will be increased from 100–900 with a step of 200. Meanwhile, the experiment will be repeated for three times in the condition of SNR = −4 dB, 0 dB and 6 dB.

Figure 8 plots the impact of training samples on successful recognition with three conditions of SNR. In general, it is positively correlated between training samples and successful recognition. When the samples are less than 500, the successful recognition increases obviously. However, when the samples are more than 500, the successful recognition is substantially retained. It means that the proposed approaches are able to work well in a small number of samples.

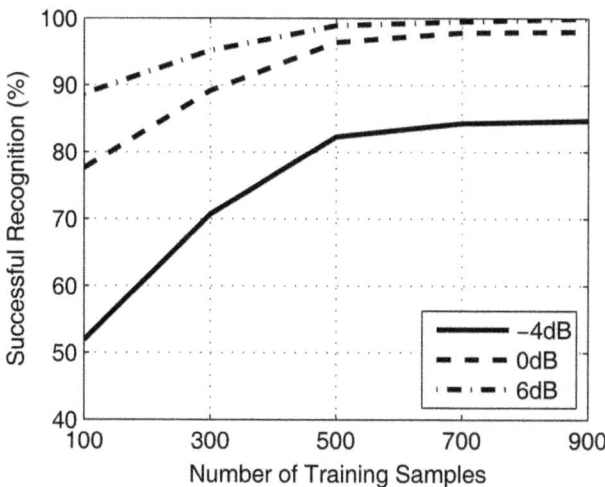

Figure 8. The figure shows the successful recognition ratio of different numbers of samples.

6.4. Experiment with Computational Burden

Computational burden is also an important issue for the classification system. We measure the time of the proposed method and compare it with [13,14] in the same conditions. Three different SNRs, −4 dB, −0 dB and 6 dB are tested, and each test repeats 50 times to calculate the average value. Table 5 shows the testing environment, and Table 6 demonstrates the testing results, respectively.

Table 5. The testing environment.

Item	Model/Version
CPU	E5-1620v2 (Intel)
Memory	16GB (DDR3 @ 1600 MHz)
GPU	NVS315 (Quadro)
MATLAB	R2012a

CPU: Central Processing Unit. GPU: Graphics Processing Unit. MATLAB is a software produced by the MathWorks, Inc. that located in Natick, Massachusetts, United States.

Table 6. Computational burden test (this paper/previous work/Lundén; unit: s).

	LFM	BPSK	Costas	Frank
−4 dB	54.798/55.302/85.331	51.117/51.132/82.553	54.463/54.798/84.735	56.115/56.221/86.132
0 dB	54.336/55.096/85.152	50.860/50.903/82.094	54.108/54.255/84.113	55.704/55.909/85.754
6 dB	53.983/54.887/84.755	50.378/50.875/81.598	53.766/53.842/83.795	55.368/55.806/85.389
	P1	**P2**	**P3**	**P4**
−4 dB	58.887/58.889/88.112	55.559/55.759/86.739	58.386/58.522/87.117	54.105/54.732/85.079
0 dB	58.398/58.519/87.847	55.308/55.431/86.180	58.106/58.310/86.869	53.858/54.338/84.787
6 dB	57.792/58.034/87.106	54.668/55.307/85.848	57.707/57.802/86.478	53.501/54.196/84.503
	T1	**T2**	**T3**	**T4**
−4 dB	53.781/54.086/83.308	52.896/53.117/85.401	55.269/55.887/86.249	56.703/56.861/85.322
0 dB	53.266/53.799/83.011	52.715/52.980/85.166	54.523/55.300/86.093	56.359/56.622/85.054
6 dB	52.823/53.201/82.799	52.107/52.741/84.455	54.396/54.916/85.702	55.993/56.279/84.811

In Table 6, the proposed method and previous work spend less than 60 s, while Lundén's more than 80 s; because Lundén's method has more calculations, we do not need to compute, such as the Wigner–Ville distribution, peak search and data driven, etc. We also improve the effectiveness of the system and reduce the consumption of time compared with previous work. In the same type of waveform, the highest SNR has the least time. In the different types of waveforms, BPSK is easiest to calculate, but P3 code is the opposite. However, overall, the change of cost is not obvious. The proposed method is stable, and different waveforms or SNR also have little effect on the computational burden of the classification system.

7. Conclusions

In this paper, an automatic system to realize the recognition of radar signal waveforms is proposed. We build the processing flow for detected signals by utilizing signal and image processing algorithms. Using these methods, the signal waveforms are fully represented via sets of feature vectors and binary images. The vectors and images are classified into 12 types in the classifier. The simulation results show that the overall RSR is more than 94% at SNR ≥ -2 dB. Additionally, the processes of feature extraction and noise removed make the system robust. When the sample labels are more than 500, the successful recognition is substantially retained. At last, the computational burden is tested. The proposed method is stable in different waveforms or SNR and spends less time than Lundén's method and our previous work.

Acknowledgments: This work was supported by the National Natural Science Foundation of China (NSFC) under Grants 61201410, 61571149 and the Fundamental Research Funds for the Central Universities HEUCF170802. The authors would like to thank the associate editor and reviewers for useful feedback.

Author Contributions: M.Z. and M.D. conceived of and designed the experiments. M.Z. performed the experiments. M.Z. and LT.L. analyzed the data. LP.G. contributed analysis tools. M.Z. wrote the paper.

Conflicts of Interest: All co-authors have seen and agree with the contents of the manuscript and declare no conflict of interest. We certify that the submission is original work and is not under review at any other publication. The founding sponsors have no role in the design of the study; in the collection, analyses or interpretation of data; in the writing of the manuscript; nor in the decision to publish the results.

Abbreviations

The following abbreviations are used in this manuscript:

LPI	Low probability of intercept
EW	Electronic warfare
ELINT	Electronic intelligence
PSK	Phase shift keying
FSK	Frequency shift keying
ASK	Smplitude shift keying
PRI	Pulse repetition interval
PW	Pulse width
RSR	Ratio of successful recognition
SNR	Signal to noise ratio
MLP	Multi-layer perceptron
AD	Atomic decomposition
LFM	Linear frequency modulation
CW	Continuous wave
STFT	Short time Fourier transform
SC	Sparse classification
CNN	Convolutional neural network
DNN	Deep neural network
SVM	Support vector machine
BPSK	Binary phase shift keying
ENN	Elman neural network
CWD	Choi–Williams time-frequency distribution
PCA	Principal component analysis
AWGN	Additive white Gaussian noise
WGN	White gaussian noise
FFT	Fast Fourier transformation
PSD	Power spectral density
CPU	Central Processing Unit
GPU	Graphics Processing Unit

References

1. Adamy, D.L. *EW 102: A Second Course in Electronic Warfare*; Artech House: Norwood, MA, USA, 2004.
2. Adamy, D.L. *EW 101: A First Course in Electronic Warfare*; Artech House: Norwood, MA, USA, 2000.
3. Nandi, A.; Azzouz, E. Automatic analogue modulation recognition. *Signal Process.* **1995**, *46*, 211–222.
4. Dudczyk, J.; Kawalec, A. Specific emitter identification based on graphical representation of the distribution of radar signal parameters. *Bull. Pol. Acad. Sci. Tech. Sci.* **2015**, *63*, 391–396.

5.   Dudczyk, J.; Kawalec, A. Fast-decision identification algorithm of emission source pattern in database. *Bull. Pol. Acad. Sci. Tech. Sci.* **2015**, *63*, 385–389.

6.   Dudczyk, J. A method of feature selection in the aspect of specific identification of radar signals. *Bull. Pol. Acad. Sci. Tech. Sci.* **2017**, *65*, 113–119.

7.   Dudczyk, J. Radar Emission Sources Identification Based on Hierarchical Agglomerative Clustering for Large Data Sets. *J. Sens.* **2016**, *2016*, 1879327.

8.   Nandi, A.K.; Azzouz, E.E. Algorithms for automatic modulation recognition of communication signals. *IEEE Trans. Commun.* **1998**, *46*, 431–436.

9.   Wong, M.D.; Nandi, A.K. Automatic digital modulation recognition using artificial neural network and genetic algorithm. *Signal Process.* **2004**, *84*, 351–365.

10.   López-Risueño, G.; Grajal, J.; Sanz-Osorio, Á. Digital channelized receiver based on time-frequency analysis for signal interception. *IEEE Trans. Aerosp. Electron. Syst.* **2005**, *41*, 879–898.

11.   Cohen, L. Time-frequency distributions-a review. *Proc. IEEE* **1989**, *77*, 941–981.

12.   Lopez-Risueno, G.; Grajal, J.; Yeste-Ojeda, O. Atomic decomposition-based radar complex signal interception. *IEE Proc. Radar Sonar Navig.* **2003**, *150*, 323–331.

13.   Lundén, J.; Koivunen, V. Automatic radar waveform recognition. *IEEE J. Sel. Top. Signal Process.* **2007**, *1*, 124–136.

14.   Ming, Z.; Lutao, L.; Ming, D. LPI Radar Waveform Recognition Based on Time-Frequency Distribution. *Sensors* **2016**, *16*, 1682.

15.   Ma, J.; Huang, G.; Zuo, W.; Wu, X.; Gao, J. Robust radar waveform recognition algorithm based on random projections and sparse classification. *IET Radar Sonar Navig.* **2013**, *8*, 290–296.

16.   Lawrence, S.; Giles, C.L.; Tsoi, A.C.; Back, A.D. Face recognition: A convolutional neural-network approach. *IEEE Trans. Neural Netw.* **1997**, *8*, 98–113.

17.   LeCun, Y.; Huang, F.J.; Bottou, L. Learning methods for generic object recognition with invariance to pose and lighting. In Proceedings of the 2004 IEEE Computer Society Conference on Computer Vision and Pattern Recognition, Washington, DC, USA, 27 June–2 July 2004.

18.   Sainath, T.N.; Kingsbury, B.; Saon, G.; Soltau, H.; Mohamed, A.R.; Dahl, G.; Ramabhadran, B. Deep convolutional neural networks for large-scale speech tasks. *Neural Netw.* **2015**, *64*, 39–48.

19.   Yoshioka, T.; Karita, S.; Nakatani, T. Far-field speech recognition using CNN-DNN-HMM with convolution in time. In Proceedings of the IEEE International Conference on Acoustics, Speech and Signal Processing (ICASSP), Brisbane, QLD, Australia, 19–24 April 2015; pp. 4360–4364.

20.   Swietojanski, P.; Arnab, G.; Steve, R. Convolutional neural networks for distant speech recognition. *IEEE Signal Process. Lett.* **2014**, *21*, 1120–1124.

21.   Abdel-Hamid, O.; Mohamed, A.R.; Jiang, H.; Penn, G. Applying convolutional neural networks concepts to hybrid NN-HMM model for speech recognition. In Proceedings of the Acoustics, Speech and Signal Processing (ICASSP), 2012 IEEE International Conference on Acoustics, Speech and Signal Processing (ICASSP), Kyoto, Japan, 25–30 March 2012; pp. 4277–4280.

22.   Lin, T.Y.; RoyChowdhury, A.; Maji, S. Bilinear CNN models for fine-grained visual recognition. In Proceedings of the IEEE International Conference on Computer Vision, Cambridge, MA, USA, 20–23 June 2015; pp. 1449–1457.

23.   Vaddi, R.S.; Boggavarapu, L.N.P.; Vankayalapati, H.D.; Anne, K.R. Comparative analysis of contrast enhancement techniques between histogram equalization and CNN. In Proceedings of the 2011 Third International Conference on Advanced Computing, Chennai, India, 14–16 December 2011; pp. 106–110.

24.   Zhou, M.K.; Zhang, X.Y.; Yin, F.; Liu, C.L. Discriminative quadratic feature learning for handwritten Chinese character recognition. *Pattern Recognit.* **2016**, *49*, 7–18.

25.   Akhand, M.A.H.; Rahman, M.M.; Shill, P.C.; Islam, S.; Rahman, M.H. Bangla handwritten numeral recognition using convolutional neural network. In Proceedings of the 2015 International Conference on Electrical Engineering and Information Communication Technology (ICEEICT), Dhaka, Bangladesh, 21–23 May 2015; pp. 1–5.

26.   Niu, X.X.; Suen, C.Y. A novel hybrid CNN–SVM classifier for recognizing handwritten digits. *Pattern Recognit.* **2012**, *45*, 1318–1325.

27.   Abdel-Hamid, O.; Mohamed, A.R.; Jiang, H.; Deng, L.; Penn, G.; Yu, D. Convolutional neural networks for speech recognition. *IEEE/ACM Trans. Audio Speech Lang. Process.* **2014**, *22*, 1533–1545.

28. Feng, Z.; Liang, M.; Chu, F. Recent advances in time–frequency analysis methods for machinery fault diagnosis: A review with application examples. *Mech. Syst. Signal Process.* **2013**, *38*, 165–205.

29. Chen, B.; Shu, H.; Zhang, H.; Chen, G.; Toumoulin, C.; Dillenseger, J.L.; Luo, L.M. Quaternion Zernike moments and their invariants for color image analysis and object recognition. *Signal Process.* **2012**, *92*, 308–318.

30. Xu, B.; Sun, L.; Xu, L.; Xu, G. Improvement of the Hilbert method via ESPRIT for detecting rotor fault in induction motors at low slip. *IEEE Trans. Energy Convers.* **2013**, *28*, 225–233.

31. Pace, P.E. *Detecting and Classifying Low Probability of Intercept Radar*, 2nd ed.; Artech House: Norwood, MA, USA, 2009.

32. Gonzalez, R.C. *Digital Image Processing*; Pearson Education: Delhi, India, 2009.

33. Zhu, Z.; Aslam, M.W.; Nandi, A.K. Genetic algorithm optimized distribution sampling test for M-QAM modulation classification. *Signal Process.* **2014**, *94*, 264–277.

34. Ozen, A.; Ozturk, C. A novel modulation recognition technique based on artificial bee colony algorithm in the presence of multipath fading channels. In Proceedings of the 2013 36th International Conference on Telecommunications and Signal Processing (TSP), Rome, Italy, 2–4 July 2013; pp. 239–243.

35. Stoica, P.; Moses, R.L. *Spectral Analysis of Signals*; Pearson/Prentice Hall Upper: Saddle River, NJ, USA, 2005.

36. Bailey, R.R.; Srinath, M. Orthogonal moment features for use with parametric and non-parametric classifiers. *IEEE Trans. Pattern Anal. Mach. Intell.* **1996**, *18*, 389–399.

37. Chen, Z.; Sun, S.K. A Zernike moment phase-based descriptor for local image representation and matching. *IEEE Trans. Image Process.* **2010**, *19*, 205–219.

38. Tan, C.W.; Kumar, A. Accurate iris recognition at a distance using stabilized iris encoding and Zernike moments phase features. *IEEE Trans. Image Process.* **2014**, *23*, 3962–3974.

39. Honarvar, B.; Paramesran, R.; Lim, C.L. Image reconstruction from a complete set of geometric and complex moments. *Signal Process.* **2014**, *98*, 224–232.

40. Lauer, F.; Suen, C.Y.; Bloch, G. A trainable feature extractor for handwritten digit recognition. *Pattern Recognit.* **2007**, *40*, 1816–1824.

41. LeCun, Y.; Bottou, L.; Bengio, Y.; Haffner, P. Gradient-based learning applied to document recognition. *Proc. IEEE* **1998**, *86*, 2278–2324.

42. Johnson, A.E.; Ghassemi, M.M.; Nemati, S.; Niehaus, K.E.; Clifton, D.; Clifford, G.D. Machine learning and decision support in critical care. *Proc. IEEE* **2016**, *104*, 444–466.

43. Bengio, Y.; Courville, A.; Vincent, P. Representation learning: A review and new perspectives. *IEEE Trans. Pattern Anal. Mach. Intell.* **2013**, *35*, 1798–1828.

44. Lin, W.M.; Hong, C.M.; Chen, C.H. Neural-network-based MPPT control of a stand-alone hybrid power generation system. *IEEE Trans. Power Electron.* **2011**, *26*, 3571–3581.

45. Lin, C.M.; Boldbaatar, E.-A. Autolanding control using recurrent wavelet Elman neural network. *IEEE Trans. Syst. Man Cybern.-Syst.* **2015**, *45*, 1281–1291.

46. Yin, S.; Yang, H.; Gao, H.; Qiu, J.; Kaynak, O. An adaptive NN-based approach for fault-tolerant control of nonlinear time-varying delay systems with unmodeled dynamics. *IEEE Trans. Neural Netw. Learn. Syst.* **2016**, *PP*, 1–12.

47. Sheela, K.G.; Deepa, S. Review on methods to fix number of hidden neurons in neural networks. *Math. Probl. Eng.* **2013**, *2013*, 425740.

Communication

Path Embeddings with Prescribed Edge in the Balanced Hypercube Network

Dan Chen, Zhongzhou Lu, Zebang Shen, Gaofeng Zhang, Chong Chen and Qingguo Zhou *

School of Information Science and Engineering, Lanzhou University, Lanzhou, Gansu 730000, China; chend13@lzu.edu.cn (D.C.); luzhzh15@lzu.edu.cn (Z.L.); shenzb12@lzu.edu.cn (Z.S.); zhanggaof@lzu.edu.cn (G.Z.); chench2013@lzu.edu.cn (C.C.)
* Correspondence: zhouqg@lzu.edu.cn

Academic Editor: Angel Garrido
Received: 21 February 2017; Accepted: 15 May 2017; Published: 26 May 2017

Abstract: The balanced hypercube network, which is a novel interconnection network for parallel computation and data processing, is a newly-invented variant of the hypercube. The particular feature of the balanced hypercube is that each processor has its own backup processor and they are connected to the same neighbors. A Hamiltonian bipartite graph with bipartition $V_0 \cup V_1$ is Hamiltonian laceable if there exists a path between any two vertices $x \in V_0$ and $y \in V_1$. It is known that each edge is on a Hamiltonian cycle of the balanced hypercube. In this paper, we prove that, for an arbitrary edge e in the balanced hypercube, there exists a Hamiltonian path between any two vertices x and y in different partite sets passing through e with $e \neq xy$. This result improves some known results.

Keywords: interconnection network; balanced hypercube; Hamiltonian path; passing prescribed edge; data processing

---

1. Introduction

Interconnection networks play an essential role in the performance of parallel and distributed systems. In the event of practice, large multi-processor systems can also be adopted as tools to address complex management and big data problems. It is well-known that an interconnection network is generally modeled by an undirected graph, in which processors are represented by vertices and communication links between them are represented by edges. The hypercube network is recognized as one of the most popular interconnection networks, and it has gained great attention and recognition from researchers both in graph theory and computer science. Nevertheless, the hypercube also has some shortcomings. For example, its diameter is large. Therefore, many variants of the hypercube have been put forward [1–10] to improve performance of the hypercube in some aspects. Among these variants, the balanced hypercube has the following special properties: each vertex of the balanced has a backup (matching) vertex and they have the same neighborhood. Therefore, the backup vertex can undertake tasks that originally run on a faulty vertex. It has been proved that the diameter of an odd-dimensional balanced hypercube BH_n is $2n - 1$ [10], which is smaller than that of the hypercube Q_{2n}.

With regard to the special properties discussed above, the balanced hypercube has been investigated by many researchers. Huang and Wu [11] studied the problem of resource placement of the balanced hypercube. Xu et al. [12] showed that the balanced hypercube is edge-pancyclic and Hamiltonian laceable. It is found that the balanced hypercube is bipanconnected for all $n \geq 1$ by Yang [13]. Huang et al. [14] discussed area efficient layout problems of the balanced hypercube. Yang [15] determined super (edge) connectivity of the balanced hypercube. Lü et al. studied (conditional) matching preclusion, hyper-Hamiltonian laceability, matching extendability and extra connectivity of the balanced hypercube in [16–19], respectively. Some symmetric properties of the balanced hypercube are presented in [20,21]. As stated above, the balanced hypercube possesses some desirable properties that the hypercube does not have, so it is interesting to explore other favorable properties that the balanced hypercube may have.

Since parallel applications such as image and signal processing are originally designed on array and ring architectures, it is important to have path and cycle embeddings in a network. Especially, Hamiltonian path and cycle embeddings and other properties of famous networks are extensively studied by many authors [12,13,22–26]. Xu et al. [12] proved that each edge of the balanced hypercube is on a cycle of even length from 4 to 4^n, that is, the balanced hypercube is *edge-bipancyclic*. They also showed that the balanced hypercube is Hamiltonian laceable for all integers $n \geq 1$. Recently, Lü et al. [17] further obtained that the balanced hypercube is hyper-Hamiltonian laceable for all integers $n \geq 1$.

The rest of this paper is organized as follows. Some necessary definitions are presented as preliminaries in Section 2. The main result of this paper is shown in Section 3. Finally, conclusions are given in Section 4.

2. Preliminaries

Let $G = (V, E)$ be a simple undirected graph, where V is a vertex-set of G and E is an edge-set of G. A *path* P from v_0 to v_n is a sequence of vertices $v_0 v_1 \cdots v_n$ from v_0 to v_n such that every pair of consecutive vertices are adjacent and all vertices are distinct except for v_0 and v_n. We also denote the path $P = v_0 v_1 \cdots v_n$ by $\langle v_0, P, v_n \rangle$. The *length* of a path P is the number of edges in P, denoted by $l(P)$. A *cycle* is a path with at least three vertices such that the first vertex is the same as the last one. A graph is *bipartite* if its vertex-set can be partitioned into two subsets V_0 and V_1 such that each edge has its ends in different subsets. A graph is *Hamiltonian* if it possesses a spanning cycle. A graph is *Hamiltonian connected* if there exists a Hamiltonian path joining any two vertices of it. Obviously, any bipartite graph is not Hamiltonian connected. Simmons [27] proposed Hamiltonian laceability of bipatite graphs: a bipartite graph $G = (V_0 \cup V_1, E)$ is *Hamiltonian laceable* if there exists a Hamiltonian path between any two vertices x and y in different partite sets of G. A graph G is *hyper-Hamiltonian laceable* if it is Hamiltonian laceable and, for any vertex $v \in V_i (i \in \{0, 1\})$, there exists a Hamiltonian path in $G - v$ between any pair of vertices in V_{1-i}. For the graph definitions and notations not mentioned here, we refer the readers to [28,29].

Wu and Huang [10] gave the following definition of BH_n as follows.

Definition 1. *An n-dimensional balanced hypercube, denoted by BH_n, consists of 4^n vertices labelled by $(a_0, a_1, \ldots, a_{n-1})$, where $a_i \in \{0, 1, 2, 3\}$ for each $0 \leq i \leq n - 1$. Any vertex $(a_0, \ldots, a_{i-1}, a_i, a_{i+1}, \ldots, a_{n-1})$ with $1 \leq i \leq n - 1$ of BH_n has the following 2n neighbors:*

1. $((a_0 + 1) \mod 4, a_1, \ldots, a_{i-1}, a_i, a_{i+1}, \ldots, a_{n-1}),$
    $((a_0 - 1) \mod 4, a_1, \ldots, a_{i-1}, a_i, a_{i+1}, \ldots, a_{n-1}),$ *and*
2. $((a_0 + 1) \mod 4, a_1, \ldots, a_{i-1}, (a_i + (-1)^{a_0}) \mod 4, a_{i+1}, \ldots, a_{n-1}),$
    $((a_0 - 1) \mod 4, a_1, \ldots, a_{i-1}, (a_i + (-1)^{a_0}) \mod 4, a_{i+1}, \ldots, a_{n-1}).$

In BH_n, the first coordinate a_0 of vertex $(a_0, \ldots, a_i, \ldots, a_{n-1})$ is called the *inner index* and the other coordinates are known as the a_i $(1 \leq i \leq n - 1)$ *i-dimensional index*. Clearly, each vertex in BH_n has two *inner* neighbors, and $2n - 2$ other neighbors. Note that all of the arithmetic operations on indices of vertices in BH_n are four-modulated.

BH_1 and BH_2 are illustrated in Figures 1 and 2, respectively.

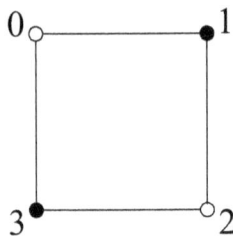

Figure 1. BH_1.

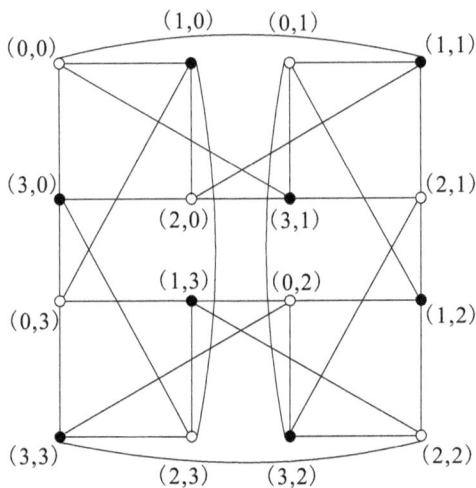

Figure 2. BH_2.

In the following, we give some basic properties of BH_n.

Proposition 1. *[10] The balanced hypercube is bipartite.*

Proposition 2. *[10,20] The balanced hypercube is vertex-transitive and edge-transitive.*

Proposition 3. *[10] The vertices* $(a_0, a_1, \ldots, a_{n-1})$ *and* $((a_0 + 2) \bmod 4, a_1, \ldots, a_{n-1})$ *of* BH_n *have the same neighborhood.*

3. Main Results

Firstly, we characterize edges of the BH_n. Let u and v be two adjacent vertices in BH_n. If u and v differ in only the inner index, then uv is said to be a 0-*dimensional* edge, and u is a 0-dimensional neighbor of v. If u and v differ in not only the inner index, but also some i-dimensional index ($i \neq 0$) of the vertices, then uv is called an i-*dimensional* edge, and u is an i-dimensional neighbor of v. For convenience, we denote the set of all i-dimensional edges by ∂D_i ($0 \leq i \leq n - 1$). Let $BH_{n-1}^{(i)}$ ($0 \leq i \leq 3$) be the subgraph of BH_n induced by the vertices of BH_n with the $(n - 1)$-dimensional index i. That is, the $BH_{n-1}^{(i)}$'s can be obtained from BH_n by deleting all $(n - 1)$-dimensional edges. Therefore, $BH_{n-1}^{(i)} \cong BH_{n-1}$ for each $0 \leq i \leq 3$.

By Proposition 1, we know that BH_n is bipartite. We can use V_0 and V_1 to denote the two partite sets of BH_n such that V_0 and V_1 consist of vertices of BH_n with an even inner index and an odd inner index, respectively. For convenience, the vertices of V_0 and V_1 are colored white and black, respectively. Throughout this paper, we use w_i and u_i (resp. b_i and v_i) to denote white (resp. black) vertices in $BH_{n-1}^{(i)}$ ($i \in \{0, 1, 2, 3\}$).

Lemma 1. *[16] In BH_n, $\partial D_i(0 \leq i \leq n-1)$ can be divided into 4^{n-1} edge-disjoint 4-cycles for $n \geq 1$.*

Lemma 2. *[12] The balanced hypercube BH_n is Hamiltonian laceable and edge-bipancyclic for $n \geq 1$.*

Lemma 3. *[17] The balanced hypercube BH_n is hyper-Hamiltonian laceable for $n \geq 1$.*

Lemma 4. *[30] Assume u and x are two different vertices in V_0, and v and y are two different vertices in V_1. Then, there exist two vertex-disjoint paths P and Q such that P joins x to y, Q joins u to v and $V(P) \cup V(Q) = V(BH_n)$, where $n \geq 1$.*

Lemma 5. *Let $n \geq 2$ be an integer. Suppose that u, v, x and y are four distinct vertices differ only the inner index in BH_n. In addition, $u, x \in V_0$ and $v, y \in V_1$. Then, there exists a Hamiltonian path from u to v in $BH_n - x - y$.*

Proof. We proceed with the proof by the induction on n. First, we consider $n = 2$. Clearly, u, v, x and y are in the same 4-cycle of ∂D_0. A Hamiltonian path of $BH_2 - x - y$ from u to v is shown in Figure 3. Thus, we suppose that the lemma holds for all integers $n - 1$ with $n \geq 3$. Next, we consider BH_n. We split BH_n into four BH_{n-1}s by deleting $(n-1)$-dimensional edges. For convenience, we denote the four BH_{n-1} by B_0, B_1, B_2 and B_3 according to the last position of vertices in BH_n, respectively. Without loss of generality, we may assume that u, v, x and y are in B_0. By an induction hypothesis, there exists a Hamiltonian path P_0 from u to v in $B_0 - x - y$. Let $u_0 v_0 \in E(P_1)$, where u_0 (resp. v_0) are neither end-vertex of P_0. We denote the segment of P_0 from u to v_0 by P_{00}, and the segment of P_0 from u_0 to v by P_{10}. By Definition 1, u_0 (resp. v_0) has an $(n-1)$-dimensional neighbor v_1 (resp. u_3) in B_1 (resp. B_3). Moreover, there exist an edge $v_3 u_2$ from B_3 to B_2, and an edge $v_2 u_1$ from B_2 to B_1. Therefore, there exist a Hamiltonian path P_3 from u_3 to v_3 in B_3, a Hamiltonian path P_2 from u_2 to v_2 in B_2, and a Hamiltonian path P_1 from u_1 to v_1 of B_1. Hence, $\langle u, P_{00}, v_0, u_3, P_3, v_3, u_2, P_2, v_2, u_1, P_1, v_1, u_0, P_{10}, v \rangle$ is a Hamiltonian path of $BH_n - x - y$ (see Figure 4). $\square$

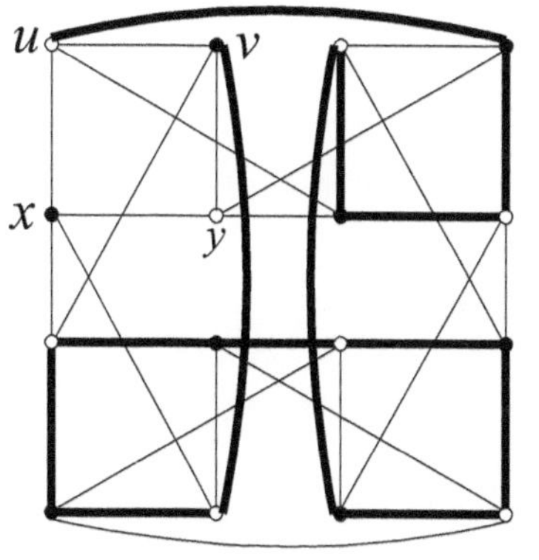

Figure 3. A Hamiltonian path of $BH_2 - x - y$.

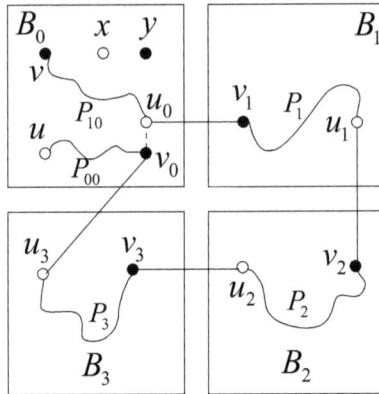

Figure 4. A Hamiltonian path of $BH_n - x - y$.

Next, we present the following lemma as a basis of our main theorem.

Lemma 6. *Let e be an arbitrary edge in* BH_2. *In addition, let* $x \in V_1$ *and* $y \in V_0$ *be any two vertices in* BH_2 *with* $e \neq xy$. *Then, there exists a Hamiltonian path between x and y passing through e.*

Proof. By Proposition 2, BH_2 is vertex-transitive and edge-transitive, and we may suppose that $e = (0,0)(1,0)$. Obviously, if $e = xy$, then there exists no Hamiltonian path of BH_2 from x to y passing e. Thus, at most, one of x and y is the end-vertex of e. We consider the following two cases:

Case 1: Neither x nor y is incident to e. By the relative positions of x and y, and Proposition 3, we consider the following: (1) $x \in V(B_0)$, $y \in V(B_0)$; (2) $x \in V(B_0)$, $y \in V(B_1)$; (3) $x \in V(B_0)$, $y \in V(B_2)$; (4) $x \in V(B_0)$, $y \in V(B_3)$; (5) $x \in V(B_1)$, $y \in V(B_1)$; (6) $x \in V(B_1)$, $y \in V(B_2)$; (7) $x \in V(B_1)$, $y \in V(B_3)$; (8) $x \in V(B_2)$, $y \in V(B_2)$; (9) $x \in V(B_2)$, $y \in V(B_3)$; (10) $x \in V(B_3)$, $y \in V(B_3)$. For simplicity, we list all Hamiltonian paths of the conditions above in Table 1.

Case 2: Either x or y is incident to e. Without loss of generality, suppose that x is incident to e, that is, $x = (1,0)$. By Proposition 3, we need only to consider four conditions of y: (1) $y \in V(B_0)$; (2) $y \in V(B_1)$; (3) $y \in V(B_2)$; and (4) $y \in V(B_3)$. Again, we list Hamiltonian paths of the conditions of x and y in this case in Table 2. $\square$

Table 1. Hamiltonian paths passing through e with neither x nor y being incident to e.

	x	y	Hamiltonian Paths Passing through e with Neither x nor y Being Incident to e
(1)	(3,0)	(2,0)	(3,0)(0,3)(3,3)(2,3)(1,3)(0,2)(3,2)(2,2)(1,2)(2,1)(3,1)(0,1)(1,1)(0,0)(1,0)(2,0)
(2)	(3,0)	(0,1)	(3,0)(0,0)(1,0)(2,3)(3,3)(0,3)(1,3)(0,2)(3,2)(2,2)(1,2)(2,1)(3,1)(2,0)(1,1)(0,1)
(3)	(3,0)	(2,2)	(3,0)(0,3)(3,3)(2,3)(1,0)(0,0)(3,1)(2,0)(1,1)(0,1)(1,2)(2,1)(3,2)(0,2)(1,3)(2,2)
(4)	(3,0)	(0,3)	(3,0)(0,0)(1,0)(2,0)(3,1)(0,1)(1,1)(2,1)(1,2)(2,2)(3,2)(0,2)(1,3)(2,3)(3,3)(0,3)
(5)	(1,1)	(2,1)	(1,1)(0,1)(3,1)(2,0)(1,0)(0,0)(3,0)(0,3)(3,3)(2,3)(1,3)(0,2)(3,2)(2,2)(1,2)(2,1)
(6)	(1,1)	(2,2)	(1,1)(0,1)(3,1)(2,0)(1,0)(0,0)(3,0)(0,3)(3,3)(2,3)(1,3)(0,2)(3,2)(2,1)(1,2)(2,2)
(7)	(1,1)	(2,3)	(1,1)(0,0)(3,1)(0,1)(1,2)(2,1)(3,2)(2,2)(1,3)(0,2)(3,3)(0,3)(1,0)(2,0)(3,0)(2,3)
(8)	(1,2)	(2,2)	(1,2)(2,1)(1,1)(0,1)(3,1)(2,0)(1,0)(0,0)(3,0)(0,3)(3,3)(2,3)(1,3)(0,2)(3,2)(2,2)
(9)	(1,2)	(2,3)	(1,2)(2,1)(1,1)(0,1)(3,1)(2,0)(1,0)(0,0)(3,0)(0,3)(1,3)(2,2)(3,2)(0,2)(3,3)(2,3)
(10)	(1,3)	(2,3)	(1,3)(0,3)(3,0)(0,0)(1,0)(2,0)(1,1)(2,1)(3,1)(0,1)(3,2)(2,2)(1,2)(0,2)(3,3)(2,3)

Table 2. Hamiltonian paths passing through e with x or y being incident to e.

	x	y	Hamiltonian Paths Passing through e with x or y Being Incident to e
(1)	(1,0)	(2,0)	(1,0)(0,0)(3,0)(0,3)(3,3)(2,3)(1,3)(0,2)(3,2)(2,2)(1,2)(2,1)(1,1)(0,1)(3,1)(2,0)
(2)	(1,0)	(0,1)	(1,0)(0,0)(3,0)(0,3)(3,3)(2,3)(1,3)(0,2)(3,2)(2,2)(1,2)(2,1)(1,1)(2,0)(3,1)(0,1)
(3)	(1,0)	(0,2)	(1,0)(0,0)(3,0)(0,3)(1,3)(2,3)(3,3)(2,2)(3,2)(2,1)(1,1)(2,0)(3,1)(0,1)(1,2)(0,2)
(4)	(1,0)	(0,3)	(1,0)(0,0)(3,0)(2,0)(3,1)(0,1)(1,1)(2,1)(1,2)(2,2)(3,2)(0,2)(1,3)(2,3)(3,3)(0,3)

Now, we are ready to state the main theorem of this paper.

Theorem 1. *Let $n \geq 2$ be an integer and e be an arbitrary edge in BH_n. In addition, let $x \in V_1$ and $y \in V_0$ be any two vertices in BH_n with $e \neq xy$. Then, there exists a Hamiltonian path of BH_n between x and y passing through e.*

Proof. We prove this theorem by induction on n. By Lemma 6, we know that the theorem is true for $n = 2$. Therefore, we suppose that the theorem holds for $n - 1$ with $n \geq 3$. Next, we consider BH_n. Firstly, we divide BH_n into $BH_{n-1}^{(i)}$ ($0 \leq i \leq 3$) by deleting all $(n - 1)$-dimensional edges. For convenience, we denote $BH_{n-1}^{(i)}$ by B_i according to the last position of the vertices in BH_n for each $i \in \{0, 1, 2, 3\}$. Similarly, suppose that $e \in E(B_0)$. Let $x \in V_1$ and $y \in V_0$ be two distinct vertices in BH_n. By relative positions of x and y, we consider the following cases:

Case 1: $x \in V(B_0)$, $y \in V(B_0)$. By an induction hypothesis, there exists a Hamiltonian path P_0 from x to y of B_0 passing through e. Thus, there is an edge $u_0 v_0$ on P_0 such that $u_0 v_0$ is not adjacent to e and $u_0 v_0$ divides P_0 into two sections P_{00} and P_{10}, where P_{00} connects x to u_0 and P_{10} connects v_0 to y. Let v_1 (resp. u_3) be an $(n - 1)$-dimensional neighbor of u_0 (resp. v_0). By Definition 1, there exist an edge $u_1 v_2$ from B_1 to B_2, and an edge $u_2 v_3$ from B_2 to B_3. Thus, by Lemma 2, there exist a Hamiltonian path P_1 from v_1 to u_1 in B_1, a Hamiltonian path P_2 from v_2 to u_2 in B_2, and a Hamiltonian path P_3 from v_3 to u_3 in B_3. Hence, $\langle x, P_{00}, u_0, v_1, P_1, u_1, v_2, P_2, u_2, v_3, P_3, u_3, v_0, P_{10}, y \rangle$ is a Hamiltonian path of BH_n from x to y passing through e (see Figure 5).

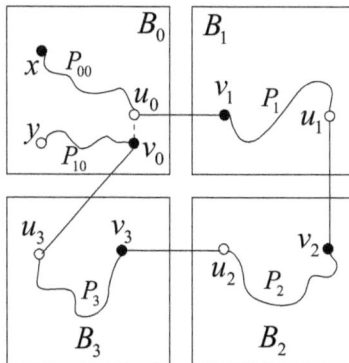

Figure 5. Illustration for Case 1.

Case 2: $x \in V(B_0)$, $y \in V(B_1)$. Let $u_0 \in V(B_0)$ be a white vertex such that u_0 is not incident to e. By an induction hypothesis, there exists a Hamiltonian path P_0 of B_0 from x to u_0 passing through e. Supposing that v_0 is a black vertex adjacent to u_0 on P_0, we denote the segment of the path P_0 from x to v_0 by P_{00}. Let the two $(n - 1)$-dimensional neighbors of u_0 be b_1 and v_1. By Lemma 2, there exists a Hamiltonian path P_1 of B_1 from b_1 to y. Let u_1 be the neighbor of v_1 in the section of P_1 from b_1 to v_1. Then $P_1 - u_1 v_1$ consists of two subpaths P_{01} and P_{11}, which connect u_1 to b_1 and v_1 to y, respectively.

Let u_3 (resp. v_2) be an $(n-1)$-dimensional neighbor of v_0 (resp. u_1). Furthermore, there exists an edge v_3u_2 from B_3 to B_2. Then, there exist a Hamiltonian path P_2 from u_2 to v_2 in B_2, and a Hamiltonian path P_3 from u_3 to v_3 in B_3. Hence, $\langle x, P_{00}, v_0, u_3, P_3, v_3, u_2, P_2, v_2, u_1, P_{01}, b_1, u_0, v_1, P_{11}, y \rangle$ is a Hamiltonian path of BH_n from x to y passing through e (see Figure 6).

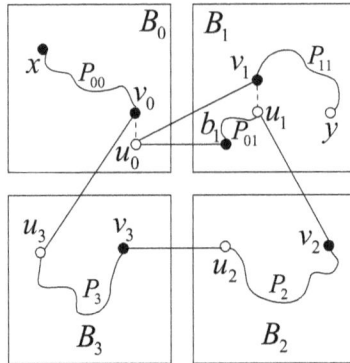

Figure 6. Illustration for Case 2.

Case 3: $x \in V(B_0)$, $y \in V(B_2)$. Let u_0 be a white vertex in B_0 not incident to e, and b_1 and v_1 be two $(n-1)$-dimensional neighbors of u_0. In addition, assume that w_1 is an arbitrary white vertex in B_1. There exists a Hamiltonian path of B_1 from b_1 to w_1. Thus, there exists an edge $u_1v_1 \in E(P_1)$ whose removal will lead to two disjoint subpaths P_{01} and P_{11}, where P_{01} connects u_1 to b_1 and P_{11} connects v_1 to w_1. Let v_2 (resp. b_2) be an $(n-1)$-dimensional neighbor of u_1 (resp. w_1). There also exists a Hamiltonian path P_2 of B_2 from y to b_2 via the edge v_2u_2. Deleting v_2u_2 results in two disjoint paths P_{02} and P_{12}, where P_{02} connects u_2 to b_2 and P_{12} connects v_2 to y. By an induction hypothesis, there exists a Hamiltonian path P_0 of B_0 from x to u_0 via the edge v_0u_0. For convenience, denote $P_0 - u_0$ by P_{00}, that is, P_{00} connects x to v_0. Let u_3 (resp. v_3) be an $(n-1)$-dimensional neighbor of v_0 (resp. u_2). Again, there exists a Hamiltonian path P_3 of B_3 from u_3 to v_3. Hence, $\langle x, P_{00}, v_0, u_3, P_3, v_3, u_2, P_{02}, b_2, w_1, P_{11}, v_1, u_0, b_1, P_{01}, u_1, v_2, P_{12}, y \rangle$ is a Hamiltonian path of BH_n from x to y passing through e (see Figure 7).

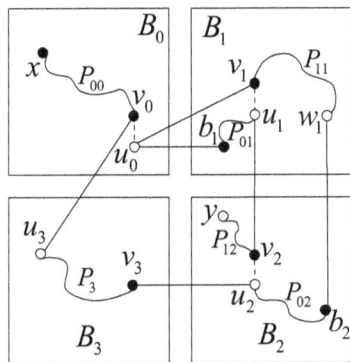

Figure 7. Illustration for Case 3.

Case 4: $x \in V(B_0)$, $y \in V(B_3)$. Let u_0 (resp. v_3) be a white (resp. black) vertex in B_0 (resp. B_3). There exist an edge u_0v_1 from B_0 to B_1, an edge u_1v_2 from B_1 to B_2, and an edge u_2v_3 from B_2 to B_3.

By Lemma 2, there exist a Hamiltonian path P_1 of B_1 from v_1 to u_1, a Hamiltonian path P_2 of B_2 from v_2 to u_2, and a Hamiltonian path P_3 of B_3 from v_3 to u_3. By an induction hypothesis, there exists a Hamiltonian path P_0 of B_0 from x to u_0 passing through e. Hence, $\langle x, P_0, u_0, v_1, P_1, u_1, v_2, P_2, u_2, v_3, P_3, y \rangle$ is a Hamiltonian path of BH_n from x to y passing through e (see Figure 8).

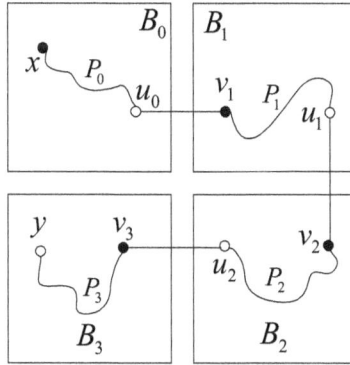

Figure 8. Illustration for Case 4.

Case 5: $x \in V(B_1), y \in V(B_1)$. Let $v_1 \neq x$ be a black vertex in B_1. By Lemma 3, there exists a Hamiltonian path P_1 of $B_1 - y$ from x to v_1. Furthermore, there exist an edge $v_1 u_0$ from B_1 to B_0, an edge $v_0 u_3$ from B_0 to B_3, an edge $v_3 u_2$ from B_3 to B_2, and an edge $v_2 y$ from B_2 to B_1. Moreover, there exist a Hamiltonian path P_0 of B_0 from u_0 to v_0 passing through e, a Hamiltonian path P_3 of B_3 from u_3 to v_3, and a Hamiltonian path P_2 of B_2 from u_2 to v_2. Hence, $\langle x, P_1, v_1, u_0, P_0, v_0, u_3, P_3, v_3, u_2, P_2, v_2, y \rangle$ is a Hamiltonian path of BH_n from x to y passing through e (see Figure 9).

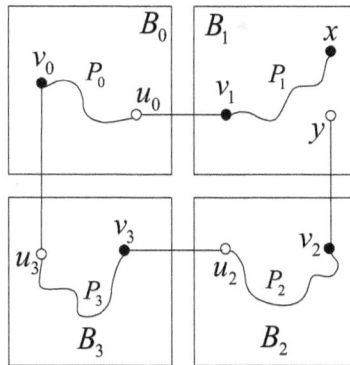

Figure 9. Illustration for Case 5.

Case 6: $x \in V(B_1), y \in V(B_2)$. Let $v_1 \neq x$ (resp. u_1) be a black (resp. white) vertex in B_1. By Lemma 3, there exists a Hamiltonian path P_1 of $B_1 - u_1$ from x to v_1. In addition, suppose that v_2 and b_2 are two $(n-1)$-dimensional neighbors of u_1. By Lemma 2, there exists a Hamiltonian P_2 of B_2 from v_2 to y via the edge $u_2 b_2$. Thus, P_2 can be divided into three sections: P_{02}, $u_2 v_2$ and P_{12}, where P_{02} connects u_2 to v_2 and P_{12} connects b_2 to y. Furthermore, there exist an edge $v_1 u_0$ from B_1 to B_0, an edge $v_0 u_3$ from B_0 to B_3, and an edge $v_3 u_2$ from B_3 to B_2. Therefore, there exist a Hamiltonian path P_0 of B_0 from u_0 to v_0 passing through e, and a Hamiltonian path P_3 of B_3 from u_3 to v_3. Hence, $\langle x, P_1, v_1, u_0, P_0, v_0, u_3, P_3, v_3, u_2, P_{02}, v_2, u_1, b_2, P_{12}, y \rangle$ is a Hamiltonian path of BH_n from x to y passing through e (see Figure 10).

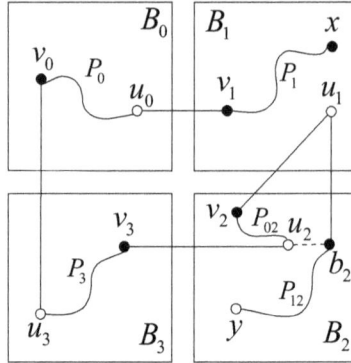

Figure 10. Illustration for Case 6.

Case 7: $x \in V(B_1)$, $y \in V(B_3)$. Let v_3 and b_3 be two black vertices in B_3. Suppose that u_2 and w_2 are $(n-1)$-dimensional neighbors of v_2 and b_2, respectively. By Lemma 3, there exists a Hamiltonian path P_3 of $B_3 - y$ from b_3 to v_3. By Definition 1, there exist two edges $v_2 u_1$ and $b_2 w_1$ from B_2 to B_1, an edge $v_1 u_0$ from B_1 to B_0, and an edge $v_0 y$ from B_0 to B_3, where $x \neq v_1$. By Lemma 4, there exist two vertex-disjoint paths P_{01} and P_{11} such that P_{01} joins v_1 and u_1, P_{11} joins x and w_1, and $V(P_{01}) \cup V(P_{11}) = V(B_1)$. Similarly, there exist two vertex-disjoint paths P_{02} and P_{12} such that P_{02} joins v_2 and u_2, P_{12} joins b_2 and w_2, and $V(P_{02}) \cup V(P_{12}) = V(B_2)$. By an induction hypothesis, there exists a Hamiltonian path P_0 of B_0 from u_0 to v_0 passing through e. Hence, $\langle x, P_{11}, w_1, b_2, P_{12}, w_2, b_3, P_3, v_3, u_2, P_{02}, v_2, u_1, P_{01}, v_1, u_0, P_0, v_0, y \rangle$ is a Hamiltonian path of BH_n from x to y passing through e (see Figure 11).

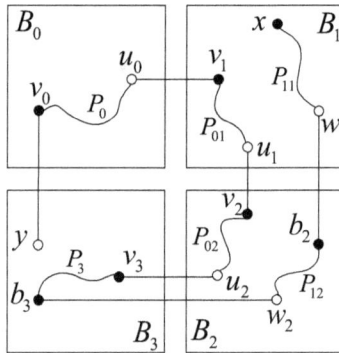

Figure 11. Illustration for Case 7.

Case 8: $x \in V(B_2)$, $y \in V(B_2)$. Let $u_2 \in V(B_2)$ be an arbitrary white vertex. By Lemma 3, there exists a Hamiltonian path P_2 of $B_2 - x$ from u_2 to y. By Definition 1, there exist an edge $x u_1$ from B_2 to B_1, an edge $v_1 u_0$ from B_1 to B_0, an edge $v_0 u_3$ from B_0 to B_3, and an edge $v_3 u_2$ from B_3 to B_2. Following Lemma 2, we can obtain a Hamiltonian path P_1 of B_1 from u_1 to v_1, and a Hamiltonian path P_3 of B_3 from u_3 to v_3. By an induction hypothesis, there exists a Hamiltonian path P_0 of B_0 from u_0 to v_0 passing through e. Therefore, $\langle x, u_1, P_1, v_1, u_0, P_0, v_0, u_3, P_3, v_3, u_2, P_2, y \rangle$ is a Hamiltonian path of BH_n from x to y passing through e (see Figure 12).

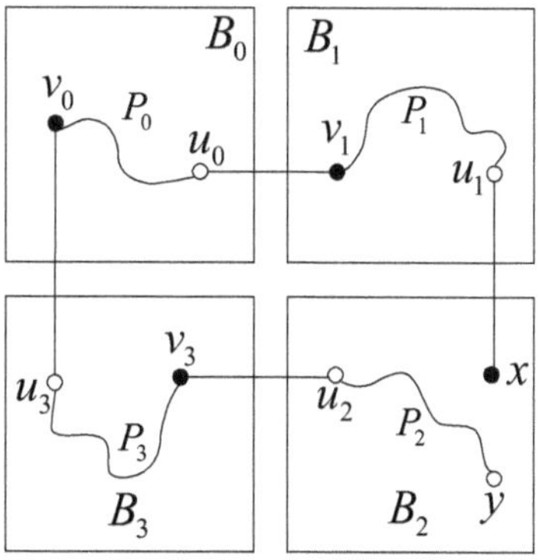

Figure 12. Illustration for Case 8.

Case 9: $x \in V(B_2)$, $y \in V(B_3)$. Let u_2 and w_2 be two distinct white vertices in B_2, and v_3 and b_3 be $(n-1)$-dimensional neighbors of u_2 and w_2, respectively. By Lemma 3, there exists a Hamiltonian path P_2 of $B_2 - x$ from u_2 to w_2. By Lemma 2, there exists a Hamiltonian path P_3 of B_3 from v_3 to y via the edge u_3b_3. By deleting u_3b_3, we can obtain two disjoint subpaths: P_{03} and P_{13}, where P_{03} connects u_3 to v_3 and P_{13} connects b_3 to y. Furthermore, there exist an edge xu_1 from B_2 to B_1, an edge v_1u_0 from B_1 to B_0, and an edge v_0u_3 from B_0 to B_3. By Lemma 2, there exists a Hamiltonian path P_1 of B_1 from u_1 to v_1. By an induction hypothesis, there exists a Hamiltonian path P_0 of B_0 from u_0 to v_0 passing through e. Hence, $\langle x, u_1, P_1, v_1, u_0, P_0, v_0, u_3, P_{03}, v_3, u_2, P_2, w_2, b_3, P_{13}, y \rangle$ is a Hamiltonian path of BH_n from x to y passing through e (see Figure 13).

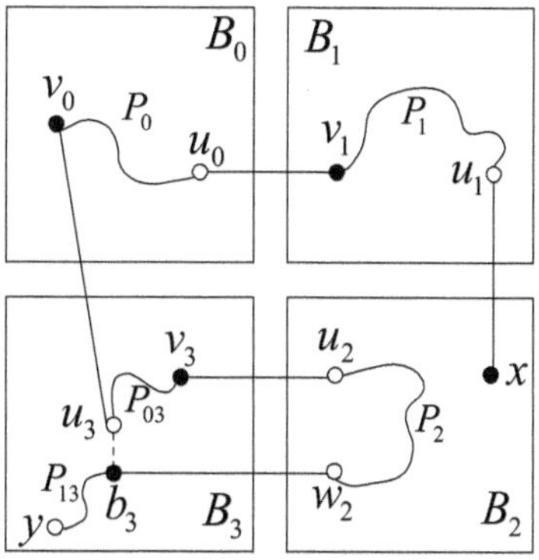

Figure 13. Illustration for Case 9.

Case 10: $x \in V(B_3)$, $y \in V(B_3)$. The proof is analogous to that of Case 5, and we omit it. □

4. Conclusions

In this paper, we study a type of path embedding of the balanced hypercube, and show that, for an arbitrary edge $e \neq xy$, there exists a Hamiltonian path between any two vertices x and y in different partite sets passing through e. This result also implies that each edge is on a Hamiltonian cycle of the balanced hypercube, which is part of the results of edge bipancyclicity of the balanced hypercube.

Acknowledgments: This work was supported by National Natural Science Foundation of China under Grant Nos. 61402210 and 60973137, Program for New Century Excellent Talents in University under Grant No. NCET-12-0250,

Major Project of High Resolution Earth Observation System with Grant No. 30-Y20A34-9010-15/17, "Strategic Priority Research Program" of the Chinese Academy of Sciences with Grant No. XDA03030100, Gansu Sci.&Tech. Program under Grant Nos. 1104GKCA049, 1204GKCA061 and 1304GKCA018, The Fundamental Research Funds for the Central Universities under Grant No. lzujbky-2016-140, Gansu Telecom Cuiying Research Fund under Grant No. lzudxcy-2013-4, Google Research Awards and Google Faculty Award, China.

Author Contributions: Dan Chen and Chong Chen initiated the research idea and developed the models with contributions from Gaofeng Zhang. Zhongzhou Lu and Zebang Shen offered help while constructing the model. The manuscript was written by Dan Chen and Chong Chen with Rui Zhou and Qingguo Zhou providing review and comments. All the authors were engaged in the final manuscript preparation and agreed to the publication of this paper.

Conflicts of Interest: The authors declare no conflict of interest.

References

1. Abraham, S.; Padmanabhan, K. The twisted cube topology for multiprocessors: A study in network asymmetry. *J. Parall. Distrib. Comput.* **1991**, *13*, 104–110, doi:10.1016/0743-7315(91)90113-N.
2. Choudum, S.A.; Sunitha, V. Augmented cubes. *Networks* **2002**, *40*, 71–84.
3. Cull, P.; Larson, S.M. The Möbius cubes. *IEEE Trans. Comput.* **1995**, *44*, 647–659.
4. Dally, W.J. Performance analysis of *k*-ary *n*-cube interconnection networks. *IEEE Trans. Comput.* **1990**, *39*, 775–785.
5. Efe, K. The crossed cube architecture for parallel computation. *IEEE Trans. Parall. Distr. Syst.* **1992**, *3*, 513–524.
6. El-amawy, A.; Latifi, S. Properties and performance of folded hypercubes. *IEEE Trans. Parall. Distrib. Syst.* **1991**, *2*, 31–42.
7. Li, T.K.; Tan, J.J.M.; Hsu, L.H.; Sung, T.Y. The shuffle-cubes and their generalization. *Inform. Process. Lett.* **2001**, *77*, 35–41.
8. Preparata, F.P.; Vuillemin, J. The cube-connected cycles: A versatile network for parallel computation. *Comput. Arch. Syst.* **1981**, *24*, 300–309.
9. Xiang, Y.; Stewart, I.A. Augmented *k*-ary *n*-cubes. *Inform. Sci.* **2011**, *181*, 239–256.
10. Wu, J.; Huang, K. The balanced hypercube: A cube-based system for fault-tolerant applications. *IEEE Trans. Comput.* **1997**, *46*, 484–490.
11. Huang, K.; Wu, J. Fault-tolerant resource placement in balanced hypercubes. *Inform. Sci.* **1997**, *99*, 159–172.
12. Xu, M.; Hu, H.; Xu, J. Edge-pancyclicity and Hamiltonian laceability of the balanced hypercubes. *Appl. Math. Comput.* **2007**, *189*, 1393–1401.
13. Yang, M. Bipanconnectivity of balanced hypercubes. *Comput. Math. Appl.* **2010**, *60*, 1859–1867.
14. Huang, K.; Wu, J. Area efficient layout of balanced hypercubes. *Int. J. High Speed Electr. Syst.* **1995**, *6*, 631–645.
15. Yang, M. Super connectivity of balanced hypercubes. *Appl. Math. Comput.* **2012**, *219*, 970–975.
16. Lü, H.; Li, X.; Zhang, H. Matching preclusion for balanced hypercubes. *Theor. Comput. Sci.* **2012**, *465*, 10–20, doi:10.1016/j.tcs.2012.09.020.
17. Lü, H.; Zhang, H. Hyper-Hamiltonian laceability of balanced hypercubes. *J. Supercomput.* **2014**, *68*, 302–314, doi:10.1007/s11227-013-1040-6.
18. Lü, H.; Gao, X.; Yang, X. Matching extendability of balanced hypercubes. *Ars Combinatoria* **2016**, *129*, 261–274.
19. Lü, H. On extra connectivity and extra edge-connectivity of balanced hypercubes. *Int. J. Comput. Math.* **2017**, *94*, 813–820.
20. Zhou, J.-X.; Wu, Z.-L.; Yang, S.-C.; Yuan, K.-W. Symmetric property and reliability of balanced hypercube. *IEEE Trans. Comput.* **2015**, *64*, 876–881.
21. Zhou, J.-X.; Kwak, J.; Feng, Y.-Q.; Wu, Z.-L. Automorphism group of the balanced hypercube. *Ars Math. Contemp.* **2017**, *12*, 145–154.
22. Jha, P.K.; Prasad, R. Hamiltonian decomposition of the rectangular twisted torus. *IEEE Trans. Comput.* **2012**, *23*, 1504–1507
23. Chang, N.-W.; Tsai, C.-Y.; Hsieh, S.-Y. On 3-extra connectivity and 3-extra edge connectivity of folded hypercubes. *IEEE Trans. Comput.* **2014**, *63*, 1594–1600.
24. Hsieh, S.-Y.; Yu, P.-Y. Fault-free mutually independent Hamiltonian cycles in hypercubes with faulty edges. *J. Combin. Optim.* **2007**, *13*, 153–162.

25. Park, C.; Chwa, K. Hamiltonian properties on the class of hypercube-like networks. *Inform. Process. Lett.* **2004**, *91*, 11–17.

26. Wang, S.; Li, J.; Wang, R. Hamiltonian paths and cycles with prescribed edges in the 3-ary n-cube. *Inform. Sci.* **2011**, *181*, 3054–3065

27. Simmons, G. Almost all n-dimensional rectangular lattices are Hamilton–Laceable. *Congr. Numer.* **1978**, *21*, 103–108.

28. West, D.B. *Introduction to Graph Theory*, 2nd ed.; Prentice Hall: Englewood Cliffs, NJ, USA, 2001.

29. Xu, J.M. *Topological Structure and Analysis of Interconnection Networks*; Kluwer Academic Publishers: Dordrecht, The Netherlands, 2001.

30. Cheng, D.; Hao, R.-X.; Feng, Y.-Q. Two node-disjoint paths in balanced hypercubes. *Appl. Math. Comput.* **2014**, *242*, 127–142.

symmetry

MDPI

Article

Fuel Consumption Estimation System and Method with Lower Cost

Chi-Lun Lo [1,2], Chi-Hua Chen [1,3,4,*] ⓘ, Ta-Sheng Kuan [1], Kuen-Rong Lo [1] and Hsun-Jung Cho [2]

[1] Telecommunication Laboratories, Chunghwa Telecom Co., Ltd., Taoyuan 326, Taiwan;
    cllo@cht.com.tw (C.-L.L.); ditto@cht.com.tw (T.-S.K.); lo@cht.com.tw (K.-R.L.)
[2] Department of Transportation and Logistics Management, National Chiao Tung University, Hsinchu 300,
    Taiwan; hjcho001@gmail.com
[3] Department of Industrial Engineering and Engineering Management, National Tsing Hua University,
    Hsinchu 300, Taiwan
[4] Department of Electrical and Computer Engineering, National Chiao Tung University, Hsinchu 300, Taiwan
* Correspondence: chihua0826@gmail.com; Tel.: +886-3-424-4091

Academic Editor: Angel Garrido
Received: 16 May 2017; Accepted: 27 June 2017; Published: 3 July 2017

Abstract: This study proposes a fuel consumption estimation system and method with lower cost. On-board units can report vehicle speed, and user devices can send fuel information to a data analysis server. Then the data analysis server can use the proposed fuel consumption estimation method to estimate the fuel consumption based on driver behaviours without fuel sensors for cost savings. The proposed fuel consumption estimation method is designed based on a genetic algorithm which can generate gene sequences and use crossover and mutation for retrieving an adaptable gene sequence. The adaptable gene sequence can be applied as the set of fuel consumption in accordance with the pattern of driver behaviour. The practical experimental results indicated that the accuracy of the proposed fuel consumption estimation method was about 95.87%.

Keywords: fuel consumption estimation; driver behavior; genetic algorithm

1. Introduction

With the development of the economic environment and the evolution of mobile communications, the intelligent transportation system has been more and more popular for obtaining fleet management services for the logistics industries and bus carriers [1]. A logistics company may have hundreds or thousands of trucks to provide freight services. However, the fuel cost of these industries is the most important challenge of fleet management services. For instance, Taiwan Institute of Economic Research (TIER) reported that the fuel costs of logistics industries and bus carriers were about 35.8 billion dollars [2] and 3.4 billion dollars [3], respectively, in Taiwan in 2015. Therefore, monitoring and saving fuel consumption efficiently can improve the profits for the logistics industries and bus carriers and reduce the air pollution from carbon dioxide (i.e., CO_2) for city governance [4,5].

For the measurement of fuel consumption, some studies used fuel sensors to detect the remaining quantity of fuel and calculated the differences among the remaining quantities. Furthermore, the data from the on-board diagnostics (OBD) could also be retrieved to obtain the fuel system status, vehicle speed, and engine revolutions per minute (RPM) [6–9]. Although these methods can measure the fuel consumption, fuel sensors and OBD devices should be equipped to report the fuel quantity data periodically, albeit with higher cost. For instance, the cost of a fuel level sensor is about 150,000 dollars, and the cost of data communications is about 13,000 dollars per month in Taiwan for a fleet of 1000 vehicles. Moreover, these methods cannot support estimating fuel consumption in accordance with driver behaviours.

For the estimation of fuel consumption based on driver behaviours, some studies have proposed using gravity sensors, accelerometers, and OBD devices to collect and analyse the data of azimuth, acceleration, movement records, and fuel quantities [10–13]. Some studies used a genetic algorithm (GA) and a neural network to analyse fuel consumption and classify driver behaviours [14–16]. Although the relation between fuel consumption and driver behaviour can be estimated, sensors and OBD devices are required in these methods. Furthermore, the measurement errors of sensors and OBD devices have not been discussed, and signal interference may lead to large estimation errors.

Therefore, a lower cost solution for a system is proposed and implemented to estimate the fuel consumption for the logistics industries. In this system, On-board units (OBUs) can send the information of movement to a data analysis server, and users can input and send the information of fuel quantity through user devices. Then the data analysis server can analyse the movement information and the fuel quantity information to estimate the fuel consumption based on driver behaviours without fuel sensors for saving costs. The proposed fuel consumption estimation method is designed based on a GA [17–19] which can generate gene sequences and use crossover and mutation to retrieve an adaptable gene sequence. The adaptable gene sequence can be applied as the set of fuel consumption in accordance with the pattern of driver behaviour.

In the next section, the architecture of the proposed consumption estimation system is described in detail. Section 3 proposes a consumption estimation method based on GA. In Section 4, practical experiments are designed to evaluate the proposed methods, and the results of these experiments are also analyzed and discussed in this section. The conclusions and future work of this paper are presented in Section 5.

2. Fuel Consumption Estimation System

The proposed fuel consumption estimation system includes OBUs, user devices, a data analysis server, and a database server (shown in Figure 1). The OBU can send the movement information which includes the timestamp, location (i.e., longitude and latitude), and vehicle speed to the data analysis server. Furthermore, a user can use his user device to input the fuel quantity after refuelling. The data analysis server can store these data in the database server and perform a fuel consumption estimation method to estimate fuel consumption in accordance with the pattern of driver behaviour.

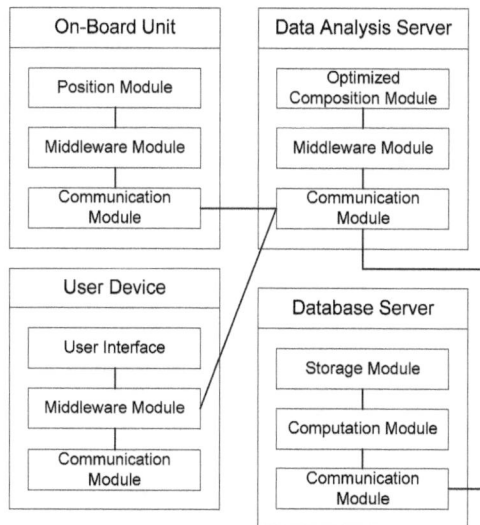

Figure 1. The architecture of the fuel consumption estimation system.

2.1. On-Board Unit

The OBU includes a position module, a middleware module, and a communication module as follows.

1. The position module can support a global positioning system (GPS) to receive and analyze satellite signals for estimating the location (i.e., longitude and latitude) and speed of the vehicle.
2. The communication module can support the techniques of long term evolution (LTE), and the OBU can connect with the data analysis server through the communication module and cellular networks.
3. The middleware module can support hypertext transfer protocol (HTTP) and representational state transfer (REST), and the OBU can periodically call application program interfaces (APIs) and send the movement information (e.g., OBU ID, car type, driver ID, timestamp, longitude, latitude, and vehicle speed) to the data analysis server.

In this study, an OBU stores an OBU ID, a car type, and a driver ID. There are C_N OBUs, T_N car types, and D_N drivers. The OBU can send the movement information to the data analysis server every 30 s. For instance, Driver 1 drove the car which was equipped with OBU 1 on 1 January 2015, and the type of this car was Type 1. The position module of OBU 1 was used to estimate the location (i.e., 120.5423383° E and 24.09490167° N) and speed (i.e., 44 km/h) at 06:00:00, and the middleware module was used to call REST APIs of the data analysis server for the transmission of the movement information (shown in Table 1).

Table 1. Movement information.

OBU ID	Car Type	Driver ID	Time	Longitude	Latitude	Speed (km/h)
OBU 1	Type 1	Driver 1	1 January 2015 06:00:00	120.5423383	24.09490167	44
OBU 1	Type 1	Driver 1	1 January 2015 06:00:30	120.5361317	24.09120167	39
OBU 1	Type 1	Driver 1	1 January 2015 06:01:00	120.5360417	24.09114667	2
OBU 1	Type 1	Driver 1	1 January 2015 06:01:30	120.5360383	24.09115	0
OBU 1	Type 1	Driver 1	1 January 2015 06:02:00	120.536035	24.09113833	0
OBU 1	Type 1	Driver 1	1 January 2015 06:02:30	120.5356167	24.09070333	7
OBU 1	Type 1	Driver 1	1 January 2015 06:03:00	120.53052	24.09449167	48
OBU 1	Type 1	Driver 1	1 January 2015 06:03:30	120.52868	24.09591167	30
...						
OBU C_N	Type T_N	Driver D_N	31 December 2015 22:00:00	121.0601083	24.75685833	102

OBU: On-board unit.

2.2. User Device

The user device includes a user interface, a middleware module, and a communication module as follows.

1. The user interface can be used to input the OBU ID, timestamp, and fuel quantity after refuelling.
2. The communication module can support the techniques of LTE, and the connection between a user device and the data analysis server can be built through the communication module.
3. The middleware module can support the techniques of HTTP and REST, and the user device can send the information such as OBD ID, timestamp, and fuel quantity to the data analysis server through the middleware module.

In this study, a user can input the information of fuel quantity (e.g., OBU ID, timestamp, and fuel quantity) through the user interface of the user device after refuelling, and the fuel quantity information can be sent to the data analysis server through the middleware module. For instance, the car which was equipped with OBU 1 was refueled with 43.04 L of gas at 18:51:00 on 5 January 2015. Then a user inputted the OBU ID (i.e., OBU 1), timestamp (i.e., 5 January 2015 18:51:00), and fuel quantity (i.e., 43.03 L) through the user interface of the user device, and the middleware module was used to send the inputted data (i.e., the fuel quantity information) to the data analysis server (shown in Table 2).

Table 2. Fuel quantity information.

OBU ID	Time	Fuel Quantity (L)
OBU 1	5 January 2015 18:51:00	43.04
OBU 1	6 January 2015 21:11:00	47.11
OBU 1	8 January 2015 17:49:00	31.81
OBU 1	10 January 2015 20:35:00	21.50
OBU 1	12 January 2015 19:59:00	41.16
OBU 1	14 January 2015 11:36:00	34.43
OBU 1	15 January 2015 19:18:00	27.75
OBU 1	16 January 2015 19:15:00	38.26
	. . .	
OBU C_N	31 December 2015 23:00:00	51.79

2.3. Data Analysis Server

The data analysis server includes a middleware module, communication module, and an optimized composition module as follows.

1. The middleware module can obtain several REST APIs to receive the movement information (e.g., OBU ID, car type, driver ID, timestamp, longitude, latitude, and vehicle speed) and the fuel quantity information (e.g., OBU ID, timestamp, and fuel quantity) from the OBUs and user devices through HTTP. These data can be stored in a database server.
2. The communication module can support Ethernet and build the connections among the data analysis server and other devices (e.g., OBUs, user devices, and a database server).
3. The optimized composition module can use the proposed fuel consumption estimation method to collect and analyse the movement information and fuel quantity information for generating the estimated results of fuel consumption.

In this study, the data analysis server can request Google Maps [20] or Chunghwa Telecom GeoWeb [21] to find the corresponding road type of the location through the middleware module. For instance, the road type of the location which is positioned at 120.5423383° E and 24.09490167° N is an urban road. The data analysis server can add the column of road type into the movement information (shown in Table 3), and the modified movement information can be stored in a database server.

Table 3. The modified movement information.

OBU ID	Car Type	Driver ID	Time	Longitude	Latitude	Road Type	Speed (km/h)
OBU 1	Type 1	Driver 1	1 January 2015 06:00:00	120.5423383	24.09490167	Urban	44
OBU 1	Type 1	Driver 1	1 January 2015 06:00:30	120.5361317	24.09120167	Urban	39
OBU 1	Type 1	Driver 1	1 January 2015 06:01:00	120.5360417	24.09114667	Urban	2
OBU 1	Type 1	Driver 1	1 January 2015 06:01:30	120.5360383	24.09115	Urban	0
OBU 1	Type 1	Driver 1	1 January 2015 06:02:00	120.536035	24.09113833	Urban	0
OBU 1	Type 1	Driver 1	1 January 2015 06:02:30	120.5356167	24.09070333	Urban	7
OBU 1	Type 1	Driver 1	1 January 2015 06:03:00	120.53052	24.09449167	Urban	48
OBU 1	Type 1	Driver 1	1 January 2015 06:03:30	120.52868	24.09591167	Urban	30
			. . .				
OBU C_N	Type T_N	Driver D_N	31 December 2015 22:00:00	121.0601083	24.75685833	Highway	102

2.4. Database Server

The database server includes a storage module, a computation module, and a communication module as follows.

1. The communication module can support Ethernet, and the connection between the database server and the data analysis server can be built by this module.
2. The computation module can receive the requests from the data analysis server through the communication module and access the storage module in accordance with the requests.
3. The storage module can perform the operations of creation, update, deletion, and query.

3. Fuel Consumption Estimation Method

This study proposes a fuel consumption estimation method which includes a movement information collection method, a fuel information collection method, and an optimized composition method. The details of each method are illustrated in the following subsections.

3.1. Movement Information Collection Method

The process of the movement information collection method includes: (1) receiving the movement information from the OBUs; (2) analyzing and storing the movement information; and (3) calculating the amount of each vehicle speed interval (i.e., driver behaviour in this study) for each OBU and each driver during a time interval.

In this study, the movement information collection method can be performed by the optimized composition module of the data analysis server to retrieve the modified movement information (shown in Table 3). The vehicle speed (v) of each movement record can be converted into a vehicle speed interval, and the amount of each vehicle speed interval for each OBU and each driver during a time interval can be calculated. This study chooses 10 km/h as a vehicle speed interval and one year as a time interval for the estimation of driver behaviour. For instance, Driver 1 drove a car which was equipped with OBU 1 during 2015; $c_{1,1,1}^{C,D}$ records which include idle speed (i.e., the value of v is zero) are reported by OBU 1 during 2015; $c_{1,1,2}^{C,D}$ records which include the speed between 0 km/h and 10 km/h are reported by OBU 1 during 2015; consequently, $c_{1,1,14}^{C,D}$ records which include the speed higher than 120 km/h are reported by OBU 1 during 2015. Furthermore, Driver D_N drove a car which was equipped OBU C_N during 2015; $c_{C_N,D_N,1}^{C,D}$ records which include idle speed are reported by OBU C_N during 2015; $c_{C_N,D_N,2}^{C,D}$ records which include the speed between 0 km/h and 10 km/h are reported by OBU C_N during 2015; consequently, $c_{C_N,D_N,14}^{C,D}$ records which include the speed higher than 120 km/h are reported by OBU C_N during 2015 (shown in Table 4).

Table 4. The movement information of each OBU and each driver during 2015 (The unit of v is km/h).

Movement / OBU ID and Driver ID	$v = 0$	$0 < v \leq 10$	$10 < v \leq 20$	$\cdots$	$110 < v \leq 120$	$120 < v$
Driver 1 drove OBU 1	$c_{1,1,1}^{C,D}$	$c_{1,1,2}^{C,D}$	$c_{1,1,3}^{C,D}$	$\cdots$	$c_{1,1,13}^{C,D}$	$c_{1,1,14}^{C,D}$
Driver 2 drove OBU 1	$c_{1,2,1}^{C,D}$	$c_{1,2,2}^{C,D}$	$c_{1,2,3}^{C,D}$	$\cdots$	$c_{1,2,13}^{C,D}$	$c_{1,2,14}^{C,D}$
$\cdots$	$\cdots$	$\cdots$	$\cdots$	$\cdots$	$\cdots$	$\cdots$
Driver 1 drove OBU 2	$c_{2,1,1}^{C,D}$	$c_{2,1,2}^{C,D}$	$c_{2,1,3}^{C,D}$	$\cdots$	$c_{2,1,13}^{C,D}$	$c_{2,1,14}^{C,D}$
$\cdots$	$\cdots$	$\cdots$	$\cdots$	$\cdots$	$\cdots$	$\cdots$
Driver D_N drove OBU C_N	$c_{C_N,D_N,1}^{C,D}$	$c_{C_N,D_N,2}^{C,D}$	$c_{C_N,D_N,3}^{C,D}$	$\cdots$	$c_{C_N,D_N,13}^{C,D}$	$c_{C_N,D_N,14}^{C,D}$

For precise estimation of driver behaviour, this study also chooses one month as a time interval. For instance, Driver 1 drove a car which was equipped with OBU 1; $c1_{1,1,1}^{C,D}$ records which include idle speed are reported by OBU 1 during January 2015; consequently, $cM_{1,1,1}^{C,D}$ records which include idle speed are reported by OBU 1 during the M-th month of 2015. Therefore, the summary of each monthly record of 2015 is equal to the yearly record (i.e., $\sum_{M=1}^{12} cM_{1,1,1}^{C,D} = c_{1,1,1}^{C,D}$).

3.2. Fuel Information Collection Method

The process of the fuel information collection method includes: (1) receiving the fuel quantity information from the user devices; (2) analyzing and storing the fuel quantity information; and (3) calculating the amount of fuel quantities for each OBU and each driver during a time interval.

In this study, the fuel information collection method can retrieve the fuel quantity information (shown in Table 2) and analyse the amount of fuel quantities for each OBU and each driver during a time interval. This study chooses one month or one year as a time interval. For instance, Driver 1 drove a car which was equipped with OBU 1 during 2015; the summary of fuel quantities of OBU 1 during January 2015 is $Q1_{1,1}^{C,D}$ L; the summary of fuel quantities of OBU 1 during the M-th of 2015 is $QM_{1,1}^{C,D}$ L; consequently, the summary of fuel quantities of OBU 1 during 2015 is $Q_{1,1}^{C,D}$ L (i.e., $\sum_{M=1}^{12} QM_{1,1}^{C,D} = Q_{1,1}^{C,D}$). Furthermore, Driver D_N drove a car which was equipped with OBU C_N during 2015; the summary of fuel quantities of OBU C_N during January 2015 is $Q1_{C_N,D_N}^{C,D}$ L; the summary of fuel quantities of OBU C_N during the M-th of 2015 is $QM_{C_N,D_N}^{C,D}$ L; consequently, the summary of fuel quantities of OBU C_N during 2015 is $Q_{C_N,D_N}^{C,D}$ L (i.e., $\sum_{M=1}^{12} QM_{C_N,D_N}^{C,D} = Q_{C_N,D_N}^{C,D}$).

3.3. Optimized Composition Method

In this subsection, the design of the optimized composition method is presented in Section 3.3.1, and a case study of this method is given in Section 3.3.2.

3.3.1. The Process of the Method

The process of the optimized composition method includes: (1) receiving the patterns of driver behaviours from the movement information collection method; (2) receiving the patterns of fuel consumption from the fuel information collection method; and (3) performing a GA (shown in Figure 2) to analyse the set of fuel consumption in accordance with the pattern of driver behaviour. The steps of the optimized composition method are illustrated as follows.

1. The values of the parameters including the amount of initial maternal DNA (deoxyribonucleic acid) sequences (*count$_g$*), the number of evolution times (*count$_c$*), the maximum number of iterations (*count$_i$*), crossover rate (α), and mutation rate (β) are initially given in this step.
2. The model of fitness function is designed for finding the cost of each DNA sequence which includes several chromosomes.
3. The process of the initial population can generate *count$_g$* maternal DNA sequences.
4. Each DNA sequence can be adopted into the model of fitness function for the cost calculation.
5. The process of the convergence check can be performed to check the values of the number of evolution times (*count$_c$*) and the maximum number of iterations (*count$_i$*). If the number of evolution times (*count$_c$*) is equal to the maximum number of iterations (*count$_i$*), the adaptable DNA sequence with the lowest cost is outputted as the estimated results of the fuel consumption based on the driver behaviour; otherwise the number of evolution times (*count$_c$*) is increased by one.
6. The process of gene selection can select two of the maternal DNA sequences for crossover and mutation.
7. The process of gene crossover can generate a child's DNA sequences in the first generation in accordance with the crossover rate (α) and the maternal DNA sequences in the first generation.
8. The process of gene mutation can generate a child's DNA sequences in the second generation in accordance with the mutation rate (β) and the maternal DNA sequences in the second generation.
9. The process of gene reproduction can support two new generated child's DNA sequences being substituted for two original maternal DNA sequences which have the highest cost.

10. The costs of the two reproduced DNA sequences can be measured by using the model of fitness function, and the GA is performed repeatedly.

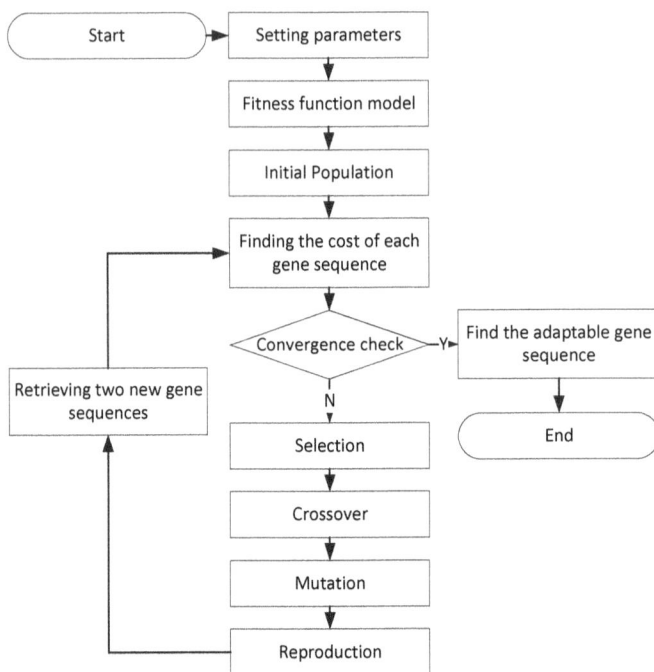

Figure 2. The process of the fuel consumption estimation method.

3.3.2. A Case Study

The parameters of GA in this study are adopted as follows: the amount of initial maternal DNA sequences ($count_g$) is 14; the initial number of evolution times ($count_c$) is 0; the maximum number of iterations ($count_i$) is 1000; the crossover rate (α) is 100%; the mutation rate (β) is 7%. A DNA sequence $q_{j,i}^{C,D} = \left\{ q_{j,i,1}^{C,D}, q_{j,i,2}^{C,D}, ..., q_{j,i,14}^{C,D} \right\}$ includes 14 chromosomes (i.e., $\left| q_{j,i}^{C,D} \right| = 14$) for the j-th OBU driven by the i-th driver. Each chromosome is encoded as a float. For instance, the parameter $q_{j,i,1}^{C,D}$ can be used to estimate the quantity of fuel consumption at idle speed during each 30 s period for the j-th OBU driven by the i-th driver. Furthermore, the model of fitness function $s = \left| \left[\sum_{k=1}^{14} \left(c_{j,i,k}^{C,D} \times q_{j,i,k}^{C,D} \right) \right] - Q_{j,i}^{C,D} \right|$ (shown in Figure 3) is adopted to estimate the cost of each DNA sequence. The unit of cost is a liter in this model. The best DNA sequence has the lowest cost (i.e., $s = \left| \left[\sum_{k=1}^{14} \left(c_{j,i,k}^{C,D} \times q_{j,i,k}^{C,D} \right) \right] - Q_{j,i}^{C,D} \right| = 0$) in this study.

For the initial population, 14 maternal DNA sequences are randomly generated, and each DNA sequence includes 14 chromosomes. For instance, the first maternal DNA sequence is $q1_{j,i}^{C,D} = \left\{ q1_{j,i,1}^{C,D}, q1_{j,i,2}^{C,D}, ..., q1_{j,i,14}^{C,D} \right\}$ for the j-th OBU driven by the i-th driver; the second maternal DNA sequence is $q2_{j,i}^{C,D} = \left\{ q2_{j,i,1}^{C,D}, q2_{j,i,2}^{C,D}, ..., q2_{j,i,14}^{C,D} \right\}$ for the j-th OBU driven by the i-th driver; consequently, the fourteenth maternal DNA sequence is $q14_{j,i}^{C,D} = \left\{ q14_{j,i,1}^{C,D}, q14_{j,i,2}^{C,D}, ..., q14_{j,i,14}^{C,D} \right\}$ for the j-th OBU driven by the i-th driver (shown in Table 5).

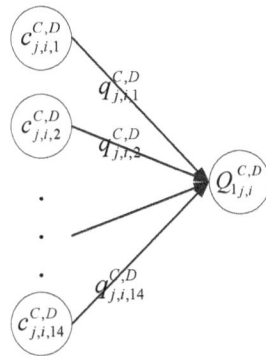

Figure 3. The model of the fitness function.

Table 5. Maternal DNA sequences.

DNA Sequences	Chromosome 1	Chromosome 2	$\cdots$	Chromosome 14
DNA Sequence 1 ($q1_{j,i}^{C,D}$)	$q1_{j,i,1}^{C,D}$	$q1_{j,i,2}^{C,D}$	$\cdots$	$q1_{j,i,14}^{C,D}$
DNA Sequence 2 ($q2_{j,i}^{C,D}$)	$q2_{j,i,1}^{C,D}$	$q2_{j,i,2}^{C,D}$	$\cdots$	$q2_{j,i,14}^{C,D}$
DNA Sequence 14 ($q14_{j,i}^{C,D}$)	$q14_{j,i,1}^{C,D}$	$q14_{j,i,2}^{C,D}$	$\cdots$	$q14_{j,i,14}^{C,D}$

This study gives a case study of OBU 1 driven by Driver 1 to explain the process of the optimized composition method. The set of movement information of OBU 1 driven by Driver 1 during 2015 is recorded as $C_{1,1}^{C,D} = \left\{ c_{1,1,1}^{C,D}, c_{1,1,2}^{C,D}, ..., c_{1,1,14}^{C,D} \right\} = \{103100, 66752, ..., 4\}$, and the fuel quantity of OBU 1 driven by Driver 1 during 2015 is recorded as 10921.364 L (i.e., $Q_{1,1}^{C,D} = 10921.364$). Furthermore, the 14 maternal DNA sequences are randomly generated as shown in Table 6. For example, the first maternal DNA sequence is $q1_{1,1}^{C,D} = \{0.013249146, 0.018487159, ..., 0.551971137\}$; the second maternal DNA sequence is $q2_{1,1}^{C,D} = \{0.016574516, 0.02331678, ..., 0.553625064\}$; consequently, the fourteenth maternal DNA sequence is $q14_{1,1}^{C,D} = \{0.01539256, 0.021892833, ..., 0.555117159\}$.

For finding the cost of each DNA sequence, the set of chromosomes in a maternal DNA sequence is adopted into the fitness function. In this study, the cost of the *h*-th maternal DNA sequence is defined as $s_h = \left| \left[\sum\limits_{k=1}^{14} \left(c_{j,i,k}^{C,D} \times qh_{j,i,k}^{C,D} \right) \right] - Q_{j,i}^{C,D} \right|$ for the *j*-th OBU driven by the *i*-th driver. For instance, the cost of the first maternal DNA sequence is calculated by Equation (1) for OBU 1 driven by Driver 1; the cost of the second maternal DNA sequence is calculated by Equation (2) for OBU 1 driven by Driver 1; consequently, the cost of the fourteenth maternal DNA sequence is calculated by Equation (3) for OBU 1 driven by Driver 1.

$$\begin{aligned} s_1 &= |(103100 \times 0.013249146 + 66752 \times 0.018487159 + ... + 4 \times 0.551971137) - 10921.364| \\ &= 260.2534752 \end{aligned} \tag{1}$$

$$\begin{aligned} s_2 &= |(103100 \times 0.016574516 + 66752 \times 0.02331678 + ... + 4 \times 0.553625064) - 10921.364| \\ &= 1062.546744 \end{aligned} \tag{2}$$

$$\begin{aligned} s_{14} &= |(103100 \times 0.01539256 + 66752 \times 0.021892833 + ... + 4 \times 0.555117159) - 10921.364| \\ &= 1009.53678 \end{aligned} \tag{3}$$

Table 6. The maternal DNA sequence for OBU 1 driven by Driver 1.

Chromosome / DNA Sequences	Chromosome 1	Chromosome 2	$\cdots$	Chromosome 14
DNA Sequence 1 ($q_{1\,j,i}^{C,D}$)	0.013249146	0.018487159	$\cdots$	0.551971137
DNA Sequence 2 ($q_{2\,j,i}^{C,D}$)	0.016574516	0.02331678	$\cdots$	0.553625064
DNA Sequence 14 ($q_{14\,j,i}^{C,D}$)	0.01539256	0.021892833	$\cdots$	0.555117159

For the convergence check, the adaptable DNA sequence with the lowest cost is output as the estimated results of the fuel consumption based on driver behaviour if the number of evolution times ($count_c$) is equal to the maximum number of iterations ($count_i$); otherwise the number of evolution times ($count_c$) is increased by one.

For gene selection, this study uses the roulette wheel selection method to select two of the maternal DNA sequences. For instance, DNA Sequence 1 ($q_{11,1}^{C,D} = \{0.013249146, 0.018487159, ..., 0.551971137\}$) and DNA Sequence 2 ($q_{21,1}^{C,D} = \{0.016574516, 0.02331678, ..., 0.553625064\}$) (shown in Table 7) are selected as the maternal DNA sequences in the first generation for OBU 1 driven by Driver 1.

Table 7. The maternal DNA sequences in the first generation.

Chromosome / DNA Sequences	Chromosome 1	Chromosome 2	$\cdots$	Chromosome 14
DNA Sequence 1 ($q_{11,1}^{C,D}$)	0.013249146	0.018487159	$\cdots$	0.551971137
DNA Sequence 2 ($q_{21,1}^{C,D}$)	0.016574516	0.02331678	$\cdots$	0.553625064

For gene crossover, the 1-point crossover method is performed in accordance with the crossover rate (α). For instance, the value of the crossover point is randomly determined as 2. Then the two child's DNA sequences (shown in Table 8) are generated as Equations (4) and (5) according to the two maternal DNA sequences (shown in Table 7). The line in Equations (4) and (5) is the crossover point.

$$\text{DNA Sequence 1} \quad \left(q_{11,1}^{C,D}\right)' = \left\{\left(q_{11,1,1}^{C,D}\right)', \left(q_{11,1,2}^{C,D}\right)', ..., \left(q_{11,1,14}^{C,D}\right)'\right\} \tag{4}$$
$$= \{0.013249146, |0.02331678, ..., 0.553625064\}$$

$$\text{DNA Sequence 2} \quad \left(q_{21,1}^{C,D}\right)' = \left\{\left(q_{21,1,1}^{C,D}\right)', \left(q_{21,1,2}^{C,D}\right)', ..., \left(q_{21,1,14}^{C,D}\right)'\right\} \tag{5}$$
$$= \{0.016574516, |0.018487159, ..., 0.551971137\}$$

Table 8. The child's DNA sequences in the first generation.

Chromosome / DNA Sequences	Chromosome 1	Chromosome 2	$\cdots$	Chromosome 14
DNA Sequence 1 $\left(q_{11,1}^{C,D}\right)'$	0.013249146	0.02331678	$\cdots$	0.553625064
DNA Sequence 2 $\left(q_{21,1}^{C,D}\right)'$	0.016574516	0.018487159	$\cdots$	0.551971137

For gene mutation, the set of binary vectors ($\eta = \{\eta_1, \eta_2, ..., \eta_{14}\}$) is randomly generated in accordance with the mutation rate (β). If the value of η_n is equal to one, the value of the n-th chromosome will be changed after the process of gene mutation. For instance, the child's DNA sequences in the first generation (shown in Table 8) are selected as the maternal DNA sequences in the second generation (shown in Table 9). Furthermore, the set of binary vectors is randomly generated as $\eta = \{1, 0, 0, 0, 0, 0, 0, 0, 0, 0, 0, 0, 0\}$. Then the two child's DNA sequences (shown in Table 10) are determined according to the two maternal DNA sequences (shown in Table 9).

Table 9. The maternal DNA sequences in the second generation.

DNA Sequences \ Chromosome	Chromosome 1	Chromosome 2	$\cdots$	Chromosome 14
DNA Sequence 1 $\left(q1_{1,1}^{C,D}\right)$	0.013249146	0.02331678	$\ldots$	0.553625064
DNA Sequence 2 $\left(q2_{1,1}^{C,D}\right)$	0.016574516	0.018487159	$\ldots$	0.551971137

Table 10. The child's DNA sequences in the second generation.

DNA Sequences \ Chromosome	Chromosome 1	Chromosome 2	$\cdots$	Chromosome 14
DNA Sequence 1 $\left(q1_{1,1}^{C,D}\right)'$	0.011241019	0.02331678	$\ldots$	0.553625064
DNA Sequence 2 $\left(q2_{1,1}^{C,D}\right)'$	0.012500034	0.018487159	$\ldots$	0.551971137

For gene reproduction, the two new generated child's DNA sequences are substituted for two of the original maternal DNA sequences which have the highest cost. For instance, the costs of DNA Sequence 2 ($q2_{j,i}^{C,D}$) and DNA Sequence 14 ($q14_{j,i}^{C,D}$) are higher than the costs of the other maternal DNA sequences in Table 6 for OBU 1 driven by Driver 1. Therefore, the two new generated child's DNA sequences in Table 10 are substituted for DNA Sequence 2 ($q2_{j,i}^{C,D}$) and DNA Sequence 14 ($q14_{j,i}^{C,D}$) (shown in Table 11). Furthermore, the costs of these two reproduced DNA sequences are determined by Equations (6) and (7), respectively.

$$
\begin{aligned}
s_2 &= |(103100 \times 0.011241019 + 66752 \times 0.02331678 + \ldots + 4 \times 0.553625064) - 10921.364| \\
&= 512.663178
\end{aligned}
\tag{6}
$$

$$
\begin{aligned}
s_{14} &= |(103100 \times 0.01539256 + 66752 \times 0.021892833 + \ldots + 4 \times 0.555117159) - 10921.364| \\
&= 183.020039
\end{aligned}
\tag{7}
$$

Table 11. The maternal DNA sequences after the first iteration.

DNA Sequences \ Chromosome	Chromosome 1	Chromosome 2	$\cdots$	Chromosome 14
DNA Sequence 1 $(q1_{1,1}^{C,D})$	0.013249146	0.018487159	$\ldots$	0.551971137
DNA Sequence 2 $(q2_{1,1}^{C,D} = \left(q1_{1,1}^{C,D}\right)')$	0.011241019	0.02331678	$\ldots$	0.553625064
$\cdots$				
DNA Sequence 14 $(q14_{1,1}^{C,D} = \left(q2_{1,1}^{C,D}\right)')$	0.012500034	0.018487159	$\ldots$	0.551971137

After gene reproduction, the convergence check is performed repeatedly. The adaptable DNA sequence with the lowest cost is output as the estimated results of the fuel consumption based on the driver behaviour when the number of evolution times ($count_c$) is equal to the maximum number of iterations ($count_i$). In the case of OBU 1 driven by Driver 1 in this study, the adaptable DNA sequence with the lowest cost, DNA Sequence 14 ($q14_{j,i}^{C,D}$), is output as the estimated results of the fuel consumption based on the driver behaviour (shown in Equation (8)). Therefore, the parameter $q1_{1,1}^{C,D}$ (i.e., 0.0125003 L) is used to estimate the quantity of fuel consumption at idle speed during each 30 s period; consequently, the quantity of fuel consumption at higher speed (>120 km/h) during each 30 s period is estimated as $q_{1,1,14}^{C,D}$ (i.e., 0.551971137 L).

$$
\begin{aligned}
q_{1,1}^{C,D} &= q14_{1,1}^{C,D} \\
&= \left\{ q_{1,1,1}^{C,D}, q_{1,1,2}^{C,D}, \dots, q_{1,1,14}^{C,D} \right\} \\
&= \left\{ \begin{array}{l}
0.012500034, 0.018487159, 0.035458125, 0.064768478, 0.088150596, \\
0.036826864, 0.070565879, 0.095323361, 0.154325441, 0.149250046, \\
0.063605949, 0.045212032, 0.132129733, 0.551971137
\end{array} \right\}
\end{aligned}
\tag{8}
$$

4. Practical Experimental Results and Discussions

This section describes the practical experimental environments and data, and some experimental designs are given to evaluate the proposed method in the following subsections.

4.1. Experimental Environments

The practical experimental data including the movement information and fuel quantity of fifteen trucks during November and December 2016 were collected from an eFMS (e-Fleet Management Service) system which was built by Chunghwa Telecom [22] for the evaluation of the proposed method. This study used the Package 'GA' [23] to implement the GA algorithm for fuel consumption estimation. The movement information and fuel quantity information during November 2016 were used as training data, and the movement information and fuel quantity information during December 2016 were used as testing data to evaluate the performance of the proposed fuel consumption estimation system and method. In the training data, 420,673 movement records were retrieved, and 105,569 km were driven by the trucks; in the testing data, 414,798 movement records were retrieved, and 107,651 km were driven by the trucks (shown in Table 12). Table 13 shows that the fuel quantity was about 31,687.786 L from 280 refuelling times in the training data, and the fuel quantity was about 33,164.136 L from 286 refuelling times in the testing data.

Table 12. The movement information in practical experimental environments.

OBU ID	November 2016 (i.e., Training Data)		December 2016 (i.e., Testing Data)	
	The Number of Movement Records	Driving Mileage (km)	The Number of Movement Records	Driving Mileage (km)
1	29,115	10,636	33,546	13,548
2	39,957	11,598	39,573	12,500
3	29,289	7897	28,097	6825
4	18,980	2844	15,313	2211
5	23,514	5928	22,015	5418
6	26,422	4880	28,820	5964
7	32,345	9404	34,440	10,258
8	30,859	6707	29,066	6195
9	36,165	10,316	34,445	10,229
10	30,046	6419	28,199	6367
11	26,074	5884	26,301	5917
12	19,258	5340	19,191	5337
13	26,106	5720	26,353	6067
14	24,620	5475	24,252	5066
15	27,923	6520	25,187	5749
Summary	420,673	105,569	414,798	107,651

Table 13. The fuel quantity information in practical experimental environments.

OBU ID	November 2016 (i.e., Training Data)		December 2016 (i.e., Testing Data)	
	The Number of Refuelling Times	Fuel Quantity (L)	The Number of Refuelling Times	Fuel Quantity (L)
1	20	3619.020	25	4582.252
2	21	3820.000	24	4180.871
3	16	2704.016	15	2425.483
4	15	581.154	11	479.641
5	23	1031.828	25	975.490
6	26	838.163	30	896.126
7	16	3222.007	18	3578.000
8	14	2440.178	14	2331.264
9	23	3662.796	22	3691.831
10	17	2334.016	18	2350.214
11	11	2100.003	12	2241.000
12	30	897.420	25	874.300
13	15	2166.003	16	2296.017
14	20	1690.028	20	1782.006
15	13	581.154	11	479.641
Summary	280	31,687.786	286	33,164.136

4.2. Experimental Designs

Three cases were designed to evaluate the performances of the fuel consumption estimation with or without road types. Furthermore, this study compared the performances of the proposed method and the fuel economy guide [24]. Each case is illustrated as follows.

- Case 1: The fuel consumption estimation method considered the different road types (e.g., urban road and highway). In this case, each DNA sequence included 28 chromosomes.
- Case 2: The fuel consumption estimation method did not consider the different road types. In this case, each DNA sequence included 14 chromosomes.
- Case 3: The significant differences between the results of Case 1 and Case 2 were evaluated.

In this study, the mean absolute percentage error (MAPE) was used to measure the accuracy of the fuel consumption estimation method by Equation (9). Furthermore, the *t*-test and F-test were adopted to evaluate the significant differences between the results of Case 1 and Case 2.

$$accuracy = 100\% - MAPE = 100\% - \frac{|real\_value - estimated\_value|}{real\_value} \qquad (9)$$

4.3. Experimental Results and Discussions

This subsection used the records during November 2016 (i.e., training data) with the proposed method to retrieve the adaptable DNA sequence as the estimated fuel consumption based on the driver behaviour. Then the records during December 2016 (i.e., testing data) were used to test the performance of the proposed method. In Case 1 (i.e., 28 chromosomes), the testing results indicated that the accuracies of the proposed method and fuel economy guide were 95.54% and 24.06% (shown in Table 14), respectively. Furthermore, Table 15 shows that the accuracy of the proposed method was 95.87% in Case 2 (i.e., 14 chromosomes). The Bureau of Energy did not consider driver behaviour, traffic condition, and weather when it performed its experiments, so larger errors of fuel consumption estimation were generated in accordance with the fuel economy guide [24]. Therefore, the proposed method can provide precise fuel consumption estimation in practical environments for the logistics industries.

Table 14. The accuracy of fuel consumption estimation with road types (i.e., Case 1).

OBU ID	The Proposed Method		Fuel Economy Guide [24]	
	Accuracy of Training Data	Accuracy of Testing Data	Accuracy of Training Data	Accuracy of Testing Data
1	99.73%	91.55%	16.89%	16.99%
2	99.70%	92.66%	17.45%	17.18%
3	99.28%	93.60%	16.78%	16.17%
4	99.49%	94.90%	28.12%	26.49%
5	99.95%	97.12%	33.02%	31.92%
6	98.76%	95.48%	33.46%	38.25%
7	99.34%	97.29%	16.77%	16.48%
8	99.80%	99.37%	15.80%	15.27%
9	99.62%	97.89%	16.19%	15.92%
10	99.00%	95.30%	15.81%	15.57%
11	99.68%	96.09%	16.10%	15.17%
12	99.47%	97.22%	34.20%	35.08%
13	99.65%	95.41%	15.18%	15.19%
14	94.36%	96.56%	18.62%	16.34%
15	99.39%	92.71%	64.48%	68.89%
Mean	**99.15%**	**95.54%**	**23.92%**	**24.06%**

Table 15. The accuracy of fuel consumption estimation without road types (i.e., Case 2).

OBU ID	The proposed Method		Fuel Economy Guide [24]	
	Accuracy of Training Data	Accuracy of Testing Data	Accuracy of Training Data	Accuracy of Testing Data
1	99.54%	92.19%	16.89%	16.99%
2	98.99%	91.59%	17.45%	17.18%
3	98.86%	96.81%	16.78%	16.17%
4	97.26%	97.70%	28.12%	26.49%
5	99.79%	98.71%	33.02%	31.92%
6	98.24%	95.61%	33.46%	38.25%
7	99.77%	95.74%	16.77%	16.48%
8	99.71%	98.44%	15.80%	15.27%
9	98.90%	96.59%	16.19%	15.92%
10	99.02%	97.81%	15.81%	15.57%
11	99.04%	93.26%	16.10%	15.17%
12	99.88%	99.05%	34.20%	35.08%
13	96.85%	89.22%	15.18%	15.19%
14	97.23%	99.28%	18.62%	16.34%
15	98.77%	96.12%	64.48%	68.89%
Mean	**98.79%**	**95.87%**	**23.92%**	**24.06%**

The practical data were collected from 15 trucks to evaluate the accuracy of the proposed method in Cases 1 and 2. In Case 3, a two-tailed t-test was performed to determine the differences between the mean accuracy of the proposed method in Case 1 and Case 2. The p-value of this two-tailed t-test was measured as 0.6396 which is higher than the alpha level of 0.05, so the H0 (i.e., null hypothesis) was accepted under 14 degrees of freedom (i.e., $14 = 15 - 1$). Furthermore, this study performed a two-tailed F-test to determine the differences between the variance of the accuracy of the proposed method in Case 1 and Case 2. The p-value of this two-tailed F-test was measured as 0.1160 which is higher than the alpha level of 0.05, so the H0 (i.e., null hypothesis) was accepted under 14 degrees of freedom (i.e., $14 = 15 - 1$). There were no significant differences between the results of Case 1 and Case 2. Therefore, the fuel consumption estimation without road types can be adopted for reducing the computation costs in the future.

5. Conclusions and Future Work

As monitoring and reducing fuel costs is the greatest challenge for the logistics industries and bus carriers, a fuel consumption estimation system and method based on a GA is proposed to collect the movement information and fuel quantity information for analyzing the relation of fuel consumption and driver behaviour. This study models the pattern of driver behaviour as a gene sequence, and the GA including crossover and mutation processes is used to retrieve an adaptable gene sequence. The adaptable gene sequence can be applied as the set of fuel consumption in accordance with the pattern of driver behaviour. In practical experimental environments, 835,471 movement records from 15 trucks were collected for the evaluation of the proposed method. The practical results indicated that the accuracy of the proposed fuel consumption estimation method was 95.87%, which was higher than the fuel economy guide from the Bureau of Energy in Taiwan.

In the future, the evaluation model of the driver performance can be designed based on the proposed fuel consumption estimation method. Furthermore, some suggestions and guidelines can be made based on the driver behaviours for the driver's reference.

Author Contributions: Chi-Lun Lo and Chi-Hua Chen conceived and designed the experiments; Chi-Lun Lo and Chi-Hua Chen performed the experiments; Chi-Lun Lo, Chi-Hua Chen, Ta-Sheng Kuan, Kuen-Rong Lo, and Hsun-Jung Cho analyzed the data; Chi-Hua Chen contributed reagents/materials/analysis tools; Chi-Lun Lo, Chi-Hua Chen, Ta-Sheng Kuan, Kuen-Rong Lo, and Hsun-Jung Cho wrote the paper.

Conflicts of Interest: The authors declare no conflict of interest.

References

1. Iodice, P.; Senatore, A. Road transport emission inventory in a regional area by using experimental two-wheelers emission factors. In Proceedings of the World Congress on Engineering, London, UK, 3–5 July 2013.
2. Chan, S.Y. The Basic Information of Truck Freight Transportation. Taiwan Institute of Economic Research, 2016. Available online: https://goo.gl/7zZ9ZY (accessed on 2 February 2017).
3. Chan, S.Y. The Basic Information of Bus Transportation. Taiwan Institute of Economic Research, 2016. Available online: https://goo.gl/hJnzXe (accessed on 2 February 2017).
4. Iodice, P.; Senatore, A. Exhaust emissions of new high-performance motorcycles in hot and cold conditions. *Int. J. Environ. Sci. Technol.* **2015**, *12*, 3133–3144. [CrossRef]
5. Iodice, P.; Senatore, A. *Analysis of a Scooter Emission Behavior in Cold and Hot Conditions: Modelling and Experimental Investigations*; SAE Technical Paper 2012-01-0881; Society of Automotive Engineers: Warrendale, PN, USA, 2012. [CrossRef]
6. Zhang, Z.; Yu, X.; Han, W. Investigation of real-time fuel consumption measuring system. *Int. J. Hybrid Inf. Technol.* **2015**, *8*, 235–242. [CrossRef]
7. Zhang, J.; Feng, Y.; Shi, F.; Wang, G.; Ma, B.; Li, R.; Jia, X. Vehicle routing in urban areas based on the oil consumption weight-Dijkstra algorithm. *IET Intell. Transp. Syst.* **2016**, *10*, 495–502. [CrossRef]
8. Zhang, S.; Wu, Y.; Liu, H.; Huang, R.; Un, P.; Zhou, Y.; Fu, L.; Hao, J. Real-world fuel consumption and CO$_2$ (carbon dioxide) emissions by driving conditions for light-duty passenger vehicles in China. *Energy* **2014**, *69*, 247–257. [CrossRef]
9. Chang, C.W.; Kuo, Y.C.; Tsai, Y.Y.; Kuo, Y.L. Vehicle Fuel Consumption Detection System and Detection Method. Patent CN105,628,125, 1 June 2016.
10. Toledo, G.; Shiftan, Y. Can feedback from in-vehicle data recorders improve driver behavior and reduce fuel consumption. *Transp. Res. Part A Policy Pract.* **2016**, *94*, 194–204. [CrossRef]
11. Silva, J.A.; Moura, F.; Garcia, B.; Vargas, R. Influential vectors in fuel consumption by an urban bus operator: Bus route, driver behavior or vehicle type. *Transp. Res. Part D Transp. Environ.* **2015**, *38*, 94–104. [CrossRef]
12. Castignani, G.; Derrmann, T.; Frank, R.; Engel, T. Driver behavior profiling using smartphones: A low-cost platform for driver monitoring. *IEEE Intell. Transp. Syst. Mag.* **2015**, *7*, 91–102. [CrossRef]
13. Yao, H.H.; Kuo, Y.C.; Shang, T.J. Method for Calculating Fuel Consumption during Driving and Driving Fuel Consumption Calculation System. U.S. Patent 20,130,253,813, 26 September 2013.

14. Magaña, V.C.; Organero, M.M. GATSF: Genetic algorithm to save fuel. In Proceedings of the 2012 IEEE International Conference on Consumer Electronics-Berlin, Berlin, Germany, 3–5 September 2012.

15. Bao, Y.; Chen, W. A personalized route search method based on joint driving and vehicular behavior recognition. In Proceedings of the 2016 IEEE MTT-S International Wireless Symposium, Shanghai, China, 14–16 March 2016.

16. Rocha, T.V.; Jeanneret, B.; Trigui, R.; Leclercq, L. How simplifying urban driving cycles influence fuel consumption estimation? *Procedia-Soc. Behav. Sci.* **2012**, *48*, 1000–1009. [CrossRef]

17. Whitley, D. A genetic algorithm tutorial. *Stat. Comput.* **1994**, *4*, 65–85. [CrossRef]

18. Rajasekar, N.; Basil, J.; Balasubramanian, K.; Priya, K.; Sangeetha, K.; Babu, T.S. Comparative study of PEM fuel cell parameter extraction using genetic algorithm. *Ain Shams Eng. J.* **2015**, *6*, 1187–1194. [CrossRef]

19. Lee, W.C.; Tsang, K.F.; Chi, H.R.; Hung, F.H.; Wu, C.K.; Chui, K.T.; Lau, W.H.; Leung, Y.W. A high fuel consumption efficiency management scheme for PHEVs using an adaptive genetic algorithm. *Sensors* **2015**, *15*, 1245–1251. [CrossRef] [PubMed]

20. Google. *Google Maps*; Google Inc.: Mountain View, CA, USA, 2005. Available online: https://maps.google.com (accessed on 30 June 2017).

21. The Telecommunication Laboratories. *Chunghwa GeoWeb Service*; The Telecommunication Laboratories, Chunghwa Telecom Co., Ltd.: Taoyuan, Taiwan, 2007. Available online: http://demo.map.hinet.net/ (accessed on 30 June 2017).

22. The Telecommunication Laboratories. *eFMS (e-Fleet Management Service)*; The Telecommunication Laboratories, Chunghwa Telecom Co., Ltd.: Taoyuan, Taiwan, 2014. Available online: http://efms.hinet.net/ (accessed on 9 March 2017).

23. Scrucca, L. Package 'GA', the Comprehensive R Archive Network. 2016. Available online: https://cran.r-project.org/web/packages/GA/ (accessed on 26 June 2017).

24. The Bureau of Energy. *Fuel Economy Guide*; The Bureau of Energy Ministry of Economic Affairs: Taipei, Taiwan, 2016. Available online: https://goo.gl/YpgQ9Q (accessed on 9 March 2017).

symmetry

MDPI

Article

Forecasting Purpose Data Analysis and Methodology Comparison of Neural Model Perspective

Sungju Lee [1] and Taikyeong Jeong [2,*]

[1] Department of Computer Convergence Software, Korea University, Korea; peacfeel@korea.ac.kr
[2] Department of Computer Science and Engineering, Seoul Women's University, Korea
* Correspondence: ttjeong@swu.ac.kr; Tel.: +82-2-970-5759

Academic Editor: Angel Garrido
Received: 16 May 2017; Accepted: 29 June 2017; Published: 5 July 2017

Abstract: The goal of this paper is to compare and analyze the forecasting performance of two artificial neural network models (i.e., MLP (multi-layer perceptron) and DNN (deep neural network)), and to conduct an experimental investigation by data flow, not economic flow. In this paper, we investigate beyond the scope of simple predictions, and conduct research based on the merits and data of each model, so that we can predict and forecast the most efficient outcomes based on analytical methodology with fewer errors. In particular, we focus on identifying two models of neural networks (NN), a multi-layer perceptron (i.e., MLP) model and an excellent model between the neural network (i.e., DNN) model. At this time, predictability and accuracy were found to be superior in the DNN model, and in the MLP model, it was found to be highly correlated and accessible. The major purpose of this study is to analyze the performance of MLP and DNN through a practical approach based on an artificial neural network stock forecasting method. Although we do not limit S&P (i.e., Standard&Poor's 500 index) to escape other regional exits in order to see the proper flow of capital, we first measured S&P data for 100 months (i.e., 407 weeks) and found out the following facts: First, the traditional artificial neural network (ANN) model, according to the specificity of each model and depending on the depth of the layer, shows the model of the prediction well and is sensitive to the index data; Second, comparing the two models, the DNN model showed better accuracy in terms of data accessibility and prediction accuracy than MLP, and the error rate was also shown in the weekly and monthly data; Third, the difference in the prediction accuracy of each model is not statistically significant. However, these results are correlated with each other, and are considered robust because there are few error rates, thanks to the accessibility to various other prediction accuracy measurement methodologies.

Keywords: data analytics; forecasting; neural network; stock exchange; time-series; econometrics

1. Introduction

The stock market index, which is shaking many capital markets around the world, can be used as an inflection point in the economy for a measure of important changes [1–4]. There are many ways of predicting objects as data. However, as a part of traditional supervised learning, we used an artificial neural network (i.e., ANN) model that uses labels as data. There are many examples [5–7], based on forecasting purpose data research, which are sometimes correct. However, this approach is worth investigating even the results are not correct. Numerous model design and performance analysis studies for stock forecasting have been carried out with various economic and statistical approaches [8–12]. From the viewpoint of prediction and forecasting, it is most important to establish each model by identifying the neural networks (NN) model and data learning method.

Therefore, due to the uniqueness and volatility of the stock market, many researchers and scientists seek prediction using models similar to neural network (NN) models. Since neural networks models have a long history of development regarding the number of layers (in this case, *x*), that is, deep-learning can occur, we aim to determine how accurately predictions and forecasting can be made possible through simple model comparisons [13–15].

Although the neural network-based method for stock forecasting has already been studied [13,16,17], it is necessary to conduct a scientific analysis of the stock forecasting method by comparing neural network models of single-layer and multi-layer networks. A comparison of the two above models, multi-layer perceptron (MLP) and deep neural network (DNN), reveals that it is meaningful to compare the data at the stage of establishing the two models. In addition, the stock market of a country can be seen as a flow of data rather than as a relation of capital flow. The greatest difference between MLP and DNN is to avoid overfitting, which includes random error or noise. For example, when MLP is applied, overfitting occurs and it is difficult to compare performance with DNN. In order to avoid this statistical error phenomenon, [18–20] studies use dropout, early stopping, and weights regulation for MLP. In this study, one statistical solution, dropout, is applied as a method to avoid overfitting. In addition, the MLP assumes that there is one hidden layer (i.e., 1-HL), and that the DNN has two or more hidden layers (i.e., 3-HL).

To compare the two artificial neural network models for stock forecasting, S&P 500 (i.e., Standard&Poor's 500 index) was selected because it could see data flow for the week, month, and year according to accessibility and change. We also want to analyze in more detail the success of forecasting and forecasting in the areas of statistics and econometrics to create basic data for actual data usage analysis.

The following is a list of the composition of this paper. First, Section 2 will explain the underlying data and Section 3 will explain the forecasting methodology. Next, Section 4 will show the empirical measurement results, and in Section 5, the conclusions will be discussed.

2. Data Acquisition

It is difficult to prove the excellence of objective data in the absence of local specificity. However, in order to collect the most efficient and objective data, S&P 500 data are accessed in this text. However, these statistical results are based on real measurements and show in the public domain a good prediction for an individual as well as for a large economic impact on society. In fact, the dataset spans from 31 May 2009 to 12 March 2017, totaling 100 months (i.e., 407 weeks) of observations. Statistical analysis for these eight years is a condition that should not be excluded or influenced by other data.

Actual data can be subdivided for viewing and forecasting, and thus divided into several ranges. As a result of efforts to determine the potential impact and efficiency of a given index data for the ultimate goal of prediction accuracy, the following results were obtained: 20% of the total number of observations were in the long-term range, 13–8% were in the intermediate range, and 6% were in the short-term range. The actual data tends to be grouped by scope, and if they are outside of this range, they cannot prove to be effective. Of the first long-range predictions, 330 weeks of data were used for model identification and estimation, and the remaining 77 weeks of data (about 20% of 407 weeks) were used to evaluate the performance of the MLP and DNN models. The intermediate range and short range were also defined in a similar way. After obtaining the middle range (31–50 weeks ahead) and the short range (23 weeks ahead) forecast horizon, it becomes necessary to check once again the soundness and efficiency of the data.

Once again, we present three predicted horizons (20%: long-term range, 13%: intermediate range, 8%: short-term range) and acknowledge that all the data used are obtained for the purpose of prediction and forecasting. At the same time, all results can be viewed in real time in order to see the flow of data.

3. Forecasting Methodology

3.1. MLP and DNN Model Methodology

In this section, we discuss the accuracy prediction methodology along with the basics of the neural network theory that can be found by increasing the number of layers (x) by elements of the traditional NN model. We also present the prediction efficiency of each model when the NN models are developed as back-propagation neural networks. The main reason for adopting MLP is because it is a partial differential of the network error function. Thus, it is an efficient way to develop a network model that minimizes discrepancies between the actual data and the output from the network model. In this case, the number of outputs per layer is increased and the output has an important scientific meaning in terms of efficiency, and it is a factor used to measure the quality of learning. In fact, MLP and DNN can have more than two hidden layers. However, in this study, we designed a stock forecasting model based on DNN with one hidden layer, as well as a model with two or more hidden layers. In the case of MLP, a method of avoiding overfitting is applied.

Figure 1 shows an illustration of the artificial neural network model methodology, in particular a Back-Propagation Neural Network (MLP and DNN). We designed an artificial neural network based on MLP and DNN models that predicts the future by using past S&P 500 data as input data. In addition, input and output data are normalized based on the highest index of S&P 500.

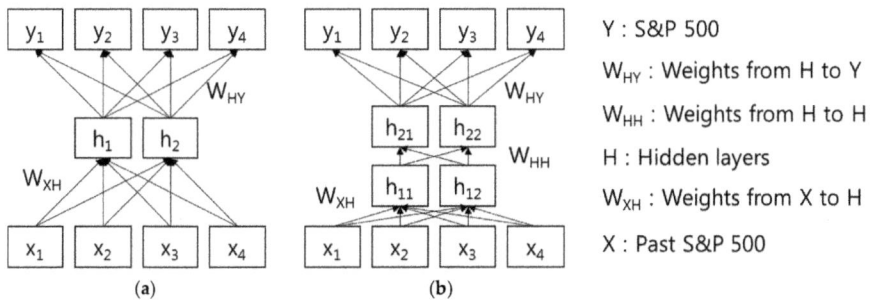

Figure 1. Illustration of the MLP and DNN model methodology. (**a**) MLP with 1-HL; (**b**) DNN with 2-HL.

In particular, MLP and DNN can be trained using the historical data of a time series in order to capture the non-linear characteristics of the specific time series. Each model parameter will be adjusted iteratively by a process of minimizing the forecasting errors. For time series forecasting, the relationship between the output (Y) and the inputs (X) can be described by the following mathematical formula:

$$y_k = a_0 + \sum_{j=1}^{n} \sum_{l=1}^{d} a_j h\left(W_{ol} + \sum_{i=1}^{m} W_{il} y_{k-1}\right) + E_K, \tag{1}$$

where a_j ($j = 0, 1, 2, \ldots, n$) is a bias on the jth unit; w_{ij} ($i = 0, 1, 2, \ldots, m; j = 0, 1, 2, \ldots, n$) is the connection weights between layers of the model; $h(\cdot)$ is the transfer function of the hidden layer; m is the number of input nodes; and n is the number of hidden nodes. The MLP model performs a nonlinear functional mapping from past observations ($y_{k-1}, y_{k-2}, \ldots, y_{k-p}$), to future values ($y_k$), i.e.,

$$y_k = \varnothing\left(y^{k-1}, y^{k-2}, \ldots, y^{k-m}\right) + E_k, \tag{2}$$

where w is a vector of all parameters and $\varnothing$ is a function determined by the network structure and connection weights.

In addition, it is necessary to explain the expanded method of neural network theory in order to increase the number of layers (x) of existing ANN models, and increase the forecast and prediction by the time series methodology. Therefore, we investigated the DNN model in the same context but with a different approach. In order to obtain the corresponding estimates of S&P 500, we used the multi-layer neural network [20] model to find the best fit of a time series model by using past values, and this method generated the optimal weights in hidden layers.

As these forecasting methodologies are important, there is a need to check for inadequate outlooks and susceptibility to errors. Therefore, the forecast error (*er* (k)) of each model is another important factor. It is mainly expressed as the method of subtracting the forecasted value from the actual value (price or profit) of our S&P 500 data and subtracting the difference from the absolute value of the actual data as follows:

$$er\ (t) = \frac{\text{ABS}(k) - \text{FORECASTING}(k)}{|\text{ABS}(k)|}, \tag{3}$$

where ABS(k) = actual value of S&P 500 in period (k) function, and FORECASTING (k) = forecasted value of S&P 500 in period (k) function.

3.2. Implementation and Avioding the Overfitting Problem

In this paper, DNN is implemented as shown in Algorithm 1 using Python. Algorithm 1 shows the implementation method when there are three hidden layers. In Step 1, the values of the input node and output node are stored in x, y. In addition, layer-1 to layer-*n* are initialized for *n* hidden layers. To calculate the hidden layer, we define two variables: layer and synapse. The layer stores the objective value of the hidden layer, and the synapse stores the value of the weights. Note that, since layer-0 is defined as the input layer, the value of x is substituted. It also generates and initializes variables from synapse_0 to synapse_*n*. In Step 2, the non-linear sigmoid is used to update the result between layers. In Step 3, we apply the dropout operation to avoid the overfitting problem. Step 4 calculates the error rate of the input and output between layers. In Step 5, we update each synapse with an error using the correction value. The same operation is repeated through 500 iterations from Step 2 to Step 5 to optimize the synapse values.

In the ANN proposed in this work, the dropout, early stopping, and weight regularization techniques are applied to avoid the overfitting phenomenon and to ensure good performance and the best generalization of the implemented model. Since it is already verified that dropout can solve the problem of overfitting that frequently occurs in the neural network model, we apply the dropout method between the hidden layer and the final classification layer of MLP and DNN models to apply the effect of regularization. Also, for every learning process, the epoch for each data is fixed as 500.

Note that we developed the MLP and DNN by using Python programming [21–23], which is used to train data for model developments and test data for the forecasting accuracy of the models developed. Here, the stop criteria for the supervised training of the networks are specified as follows. The maximum epochs specify how many iterations (over the training set) will be done if no other criterion kicks in. The training terminates when one of the following four conditions is met: (i) when the mean square error of the validation set begins to rise, indicating that overfitting might be happening; (ii) when the threshold is less than 0.01, i.e., we are on effectively flat ground; (iii) when the training time has reached 500 epochs; or (iv) when the goal (the difference between output and target is less than 0.01) has been met. After the training was completed, its epochs and the simulation procedure were completed successfully, which indicates that the network was trained and was predicting the output as desired. This is similar to the predictions we are currently pursuing in the Convolution Neural Network (CNN) of Artificial Intelligence (AI).

In particular, it is necessary to apply the principle of Ockham's razor [24], which is one of the many tools used in the development of philosophical and theoretical models of scientists. It is very important to identify the fundamental principles at this time. Aristotle's scientific curiosity was conveyed by Stephen Hawking [25] in a modernized sense as a measure of the path and process theories determined by the amount of neural weight delivered, which states that neural network theory is intelligent. The simplest network topology is used to produce the correct result of the prediction. As you increase the number of layers, it should be noted that you have to create the layers of the network in the order of input 2, layer 1, and output 1. In other words, this means that two input layers, one hidden layer, and one output layer are related to the accuracy of the prediction regardless of the order of propagation. At the same time, the hidden layer value in the middle, x, can be changed and can be set to various variables such as 2, 3, 4, and 5. Of course, 1 means simply that the value of the hidden layer is 1, but sometimes there is a best result when there is a hidden layer. The topology of the propagated network becomes an important factor.

Algorithm 1. MLP and DNN by using Python Language

Step 1:	x <- Input nodes y <- Output nodes INIT(layer1, layer2, layer3) # <- random values with normalized 0 to 1 dropout_percent, do_dropout = (0.2, True) layer_0 <- x INIT(synapse_0, synapse_1, synapse_2, synapse_3) # <- random values with normalized 0 to 1 alpha <- 0.1 for j in range(500):

Step 2:	for i in range (len(x)):
	        layer_0 = x[i]
	        layer_1 = sigmoid(np.dot(layer_0,synapse_0))
	        layer_2 = sigmoid(np.dot(layer_1,synapse_1))
	        layer_3 = sigmoid(np.dot(layer_2,synapse_2))
	        if(do_dropout):

Step 3:	layer_1*=np.random.binomial([np.ones((len(X),hidden_dim))],
	                1-dropout_percent)[0] * (1.0/(1-dropout_percent)
	            layer_2*=np.random.binomial([np.ones((len(H),hidden_dim))],
	                1-dropout_percent)[0] * (1.0/(1-dropout_percent))

Step 4:	layer_3_error = layer_3 – y[i]
	#layer_3_error = y [i]- layer_3
	layer_3_delta = layer_3_error * sigmoid_output_to_derivative(layer_3)
	layer_2_error = layer_3_delta.dot(synapse_2.T)
	layer_2_delta = layer_2_error * sigmoid_output_to_derivative(layer_2)
	layer_1_error = layer_2_delta.dot(synapse_1.T)
	layer_1_delta = layer_1_error * sigmoid_output_to_derivative(layer_1)

Step 5:	#layer_3_error = layer_3 – y synapse_2 –= alpha * np.reshape(layer_2,(–1,1))*layer_3_delta synapse_1 –= alpha * np.reshape(layer_1,(–1,1))*layer_2_delta synapse_0 –= alpha * np.reshape(layer_0,(–1,1))*layer_1_delta #layer_3_error = y - layer_3 synapse_2 += alpha * np.reshape(layer_2,(−1,1))*layer_3_delta synapse_1 += alpha * np.reshape(layer_1,(−1,1))*layer_2_delta synapse_0 += alpha * np.reshape(layer_0,(−1,1))*layer_1_delta

---

4. Analysis and Results

4.1. Optimal Paramters

We first experimentally determined the number of hidden nodes to compare the performance of MLP and DNN by using weekly and monthly data. The number of hidden nodes in MLP is selected as 150%, 100%, 50%, and 25% of the number of input nodes. In addition, the number of hidden nodes in DNN is three times the number of hidden nodes in MLP. Finally, overfitting was avoided using dropout, and the epoch was fixed at 500.

Table 1 shows the results using MLP-based weekly and monthly data. The number of hidden nodes in the input nodes 77, 50, 31, 23, 17, 11, and 7 is 35, 25, 15, 23, 17, and 11, respectively. In addition, the results using DNN-based weekly and monthly data are shown in Table 2. In the input nodes 77, 50, 31, 23, 17, 11, and 7, the number of hidden nodes is 105, 36, 21, 33, 24, and 15, respectively.

Table 1. MLP (1-HL) by using weekly and monthly data.

# of Input	# of HL	Mean Square Error (MSE)	# of HL	MSE
		Weekly		
77	115	0.005	35	**0.004**
	77	0.005	19	0.005
50	75	0.007	25	**0.005**
	50	0.006	12	0.006
31	46	0.004	15	**0.001**
	31	0.002	7	0.003
23	34	0.003	11	0.002
	23	**0.001**	5	0.003
		Monthly		
17	25	0.021	8	0.022
	17	**0.017**	4	0.029
11	16	0.005	5	0.004
	11	**0.003**	2	0.006
7	10	**0.003**	3	0.018
	7	0.006	1	0.222

Table 2. DNN (3-HL) by using weekly and monthly data.

# of Input	# of HL	Mean Square Error (MSE)	# of HL	MSE
		Weekly		
77	345	0.006	105	**0.004**
	231	0.005	57	0.005
50	225	0.007	75	0.005
	150	0.006	**36**	**0.005**
31	138	0.004	45	0.001
	93	0.003	**21**	**0.001**
23	102	0.003	33	**0.001**
	69	0.002	15	0.002
		Monthly		
17	75	0.021	**24**	**0.018**
	51	0.022	12	0.023
11	48	0.007	**15**	**0.006**
	33	0.007	6	0.008
7	30	0.002	9	0.018
	21	**0.001**	3	0.022

4.2. Analysis and Results

In this section, descriptive statistics of forecast errors (*er*) from the MLP model and DNN model using S&P data are presented in Table 1. Those statistics of forecast errors (FE) for four different forecasting horizons such as 77 weeks ahead (long range), 50 weeks ahead (upper mid range), 31 weeks ahead (lower mid range), and 23 weeks ahead (short range) are presented in Table 1. As shown, statistical results such as average Mean, MAE (mean absolute error), MSE were applied with the measuring horizons of 77, 50, 31 and 23, respectively. What is unusual is that the DNN model provides a smaller MAE and MSE than the MLP model does for all forecast horizons, with the exception of the short forecast horizon (23 weeks ahead). These results were statistically significant at all predicted horizon measurement points by nonparametric Wilcoxon test [15]. The point that can be deduced, as mentioned by Wang et al. [16], is that the DNN model provides a more accurate prediction than the MLP model.

In particular, in Tables 3 and 4, the hidden nodes of MLP and DNN are set to about half the number of inputs and outputs. In the case of DNN, the hidden layers are all set to six. Therefore, the number of hidden nodes is six times larger than that of MLP.

Table 3. Statistical analysis and forecasting accuracy of each model by actual weekly data.

Forecasting Horizon (Weeks)	Model	Mean	MAE	MSE
77	MLP(1-HL)	0.194	0.136	0.021
	DNN(3-HL)	0.07	0.053	0.004
50	MLP(1-HL)	0.097	0.529	0.005
	DNN(3-HL)	0.046	0.034	0.001
31	MLP(1-HL)	0.055	0.034	0.002
	DNN(3-HL)	0.034	0.027	0.001
23	MLP(1-HL)	0.033	0.027	0.001
	DNN(3-HL)	0.031	0.023	0.001

Mean = $1/n \Sigma$ FE. Forecast errors (FE) are defined as: $(A-F)/|A|$; MAE = $1/n \Sigma |FE|$; MSE = $1/n \Sigma (FE)^2$.

Table 4. Statistical analysis and forecasting accuracy of each model by actual monthly data.

Forecasting Horizon (Months)	Model	Mean	MAE	MSE
17	MLP(1-HL)	0.185	0.130	0.017
	DNN(3-HL)	0.064	0.046	0.003
11	MLP(1-HL)	0.096	0.070	0.005
	DNN(3-HL)	0.010	0.033	0.001
7	MLP(1-HL)	0.072	0.054	0.003
	DNN(3-HL)	0.033	0.025	0.001

Mean = $1/n \Sigma$ FE. Forecast errors (FE) are defined as: $(A-F)/|A|$; MAE = $1/n \Sigma |FE|$; MSE = $1/n \Sigma (FE)^2$.

In terms of the prediction horizon, the NN model, which is a time series rather than the MLP model, has a small MAE and MSE. However, the difference was statistically significant for 11 months compared to the values initially predicted. Because $\alpha < 0.01$, the statistical observations were reasonably judged. In conclusion, the results of the weekly data and monthly data are consistent.

Tables 5 and 6 show the results of MLP and DNN with the number of hidden nodes increased. Although increasing the number of hidden nodes increases the learning time, it can confirm the relative increase in the accuracy. The execution time required for the learning time can be reduced through parallel processing or cloud computing [24–26]. The results of Tables 5 and 6 show that the hidden nodes are twice as many as the input and output numbers, compared to Tables 3 and 4, in which the hidden nodes are about half of the input and output counts. As a result, a slight improvement in the accuracy performance was confirmed as the result of increased hidden nodes in the weekly and monthly data-based MLP and DNN.

As described above, annual data can be inferred from the data of the week and month taken from the time series data. As the forecasting horizon changes, it is meaningless in the Neural Network Perspective. Below, we can see the recent yearly data from 2015 to 2017 for comparison. Figure 2 shows that a comparison of the recent yearly data by layer $x = 0$, which includes real data, 20%, 13%, 8%, and 6 %, respectively. Each colored line shows five layer-by-layer predictions, and the legend on the left shows the stock price index of the S&P 500 [27,28] in Figure 3. It can be seen that the main inflection point and the amount of network applied are very similar to the real data, that is, the actual measured data.

Table 5. Statistical analysis and forecasting accuracy of each model by actual weekly data.

Forecasting Horizon (Weeks)	Model	Mean	MAE	MSE
77	MLP(1HL)	0.132	0.132	0.018
	DNN(3HL)	0.139	0.139	0.020
50	MLP(1HL)	0.070	0.070	0.005
	DNN(3HL)	0.079	0.079	0.006
31	MLP(1HL)	0.044	0.044	0.002
	DNN(3HL)	0.040	0.040	0.001
23	MLP(1HL)	0.026	0.026	0.001
	DNN(3HL)	0.011	0.012	0.001

Mean = $1/n \Sigma$ FE. Forecast errors (FE) are defined as: $(A–F)/ |A|$; MAE = $1/n \Sigma |FE|$; MSE = $1/n \Sigma (FE)^2$.

Table 6. Statistical analysis and forecasting accuracy of each model by actual monthly data.

Forecasting Horizon (Months)	Model	Mean	MAE	MSE
17	MLP(1HL)	0.128	0.128	0.017
	DNN(3HL)	0.135	0.135	0.018
11	MLP(1HL)	0.071	0.071	0.005
	DNN(3HL)	0.081	0.081	0.006
7	MLP(1HL)	0.028	0.028	0.001
	DNN(3HL)	0.011	0.011	0.001

Mean = $1/n \Sigma$ FE. Forecast errors (FE) are defined as: $(A–F)/ |A|$; MAE = $1/n \Sigma |FE|$; MSE = $1/n \Sigma (FE)^2$.

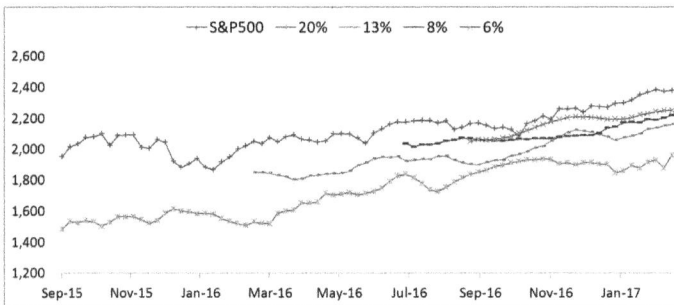

(a)

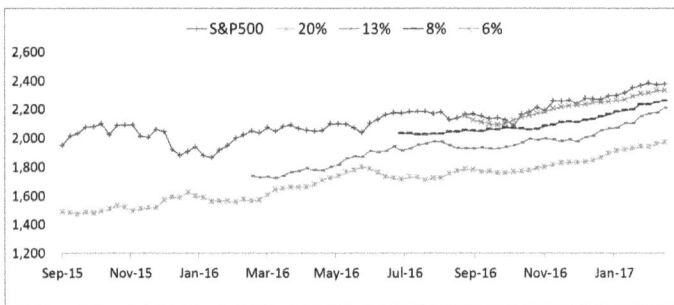

(b)

Figure 2. Comparison of recent yearly data by x = 0 values from 20%, 10%, 8%, 6%, respectively. (a) MLP(1-HL); (b) DNN (3-HL).

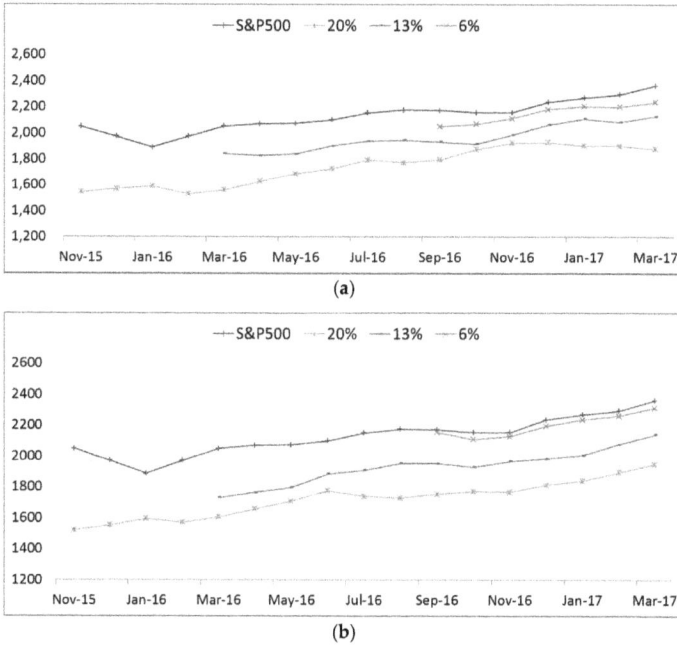

Figure 3. Comparison of recent yearly data by x = 0 values from 20%, 13%, 6%, respectively. (**a**) MLP (1-HL); (**b**) DNN (3-HL).

Furthermore, in the case of the real data, that is, the actually measured data, the layer is 6%, and the reason for this is as follows. First, the error rate tends to be lower than other data. On the other hand, the higher the percentage of the layer, the more likely it is that the initial value will be the same as the real data.

5. Conclusions

The ultimate goal of this paper was to compare and analyze the forecasting performance of two artificial neural network (ANN) models, and to conduct scientific analysis by data flow, not economic flow. In particular, we investigated which is the better model between the multi-layer perceptron (MLP) model and the deep neural network (DNN) model by forecasting the Stock Price Index (S&P 500) during a certain time frame. Particularly, forecasting performance was measured by the forecast accuracy metrics, such as the absolute forecasting errors and square forecasting errors of each model.

As a result, DNN was found to perform better than MLP. In addition, although the method of input data is limited to the past stock index data in this study, monthly data learning results provided better prediction performance than weekly data learning results. It should be noted that the use of an artificial neural network, which is a scientific approach, can grasp the flow of stock trends, though it is difficult to predict detailed stock changes.

There is a first task of comparing and analyzing the models by limiting the object of S&P 500, as well as showing the flow of data and the viewpoint of analysis. S&P 500 data and its return data over a period of 100 months (i.e., 407 weeks), extending from 2009 to 2017, were analyzed. We found the followings: first, the DNN model generally provided more accurate forecasts for the S&P 500 than the MLP model. Although the ANN models have evolved continuously, it may be good for a deep running perspective, but since the approach between the models is different and the utility is different, it was not statistically reliable to predict on S&P 500 returns. It was observed that these results are applicable to both weekly, monthly, and annual data shown in DNN and are useful for accurate prediction without error rate. Since the accuracy of forecasting values is dependent on the developing process of forecasting models, the results of this study may also be sensitive to the developing process of the MLP model and DNN model.

Acknowledgments: This work has been supported by the Basic Science Research through NRF (Korea) grant 2015R1C1A1A02037688, 2017R1A1A2059115. This work was supported by Institute for IITP (Korea) grant funded by the Korea government, MSIP, 2017-0-00403.

Author Contributions: Sungju Lee developed stimuli, and interpreted the results; Taikyeong Jeong supervised the project, conducted model's data analysis and wrote the paper. Authorship must be limited to those who have contributed substantially to the research reported.

Conflicts of Interest: The authors declare no conflict of interest.

References

1. Coch, D.; Ansari, D. Thinking about mechanisms is crucial to connecting neuroscience and education. *Cortex* **2009**, *45*, 546–547. [CrossRef] [PubMed]
2. Hamid, S.A.; Iqbal, Z. Using neural networks for forecasting volatility of S&P 500 Index futures prices. *J. Bus. Res.* **2004**, *57*, 1116–1125.
3. Huang, G.B.; Saratchandran, P.; Sundararajan, N. A generalized growing and pruning RBF (GGAP-RBF) neural network for function approximation. *IEEE Trans. Neural Netw.* **2005**, *16*, 57–67. [CrossRef] [PubMed]
4. Jain, A.; Kumar, A.M. Hybrid neural network models for hydrologic time series forecasting. *Appl. Soft Comput.* **2007**, *7*, 585–592. [CrossRef]
5. Singh, K.P.; Basant, A.; Malik, A.; Jain, G. Artificial neural network modeling of the river water quality—A case study. *Ecol. Model.* **2009**, *220*, 888–895. [CrossRef]
6. Eakins, S.G.; Stansell, S.R. Can value-based stock selection criteria yield superior risk-adjusted returns: An application of neural networks. *Int. Rev. Financ. Anal.* **2003**, *12*, 83–97. [CrossRef]
7. Trinkle, B.S.; Baldwin, A.A. Interpretable credit model development via artificial neural networks. *Intell. Syst. Account. Financ. Manag.* **2007**, *15*, 123–147. [CrossRef]
8. Leung, M.T.; Daouk, H.; Chen, A.-S. Forecasting stock indices: A comparison of classification and level estimation models. *Int. J. Forecast.* **2000**, *16*, 173–190. [CrossRef]
9. Enke, D.; Thawornwong, S. The use of data mining and neural networks for forecasting stock market returns. *Expert Syst. Appl.* **2005**, *29*, 927–940. [CrossRef]
10. Atsalakis, G.S.; Valavanis, K.P. Surveying stock market forecasting techniques—Part II: Soft computing methods. *Expert Syst. Appl.* **2009**, *36*, 5932–5941. [CrossRef]
11. Avellaneda, M.; Lee, J.-H. Statistical arbitrage in the US equities market. *Quant. Financ.* **2010**, *10*, 761–782. [CrossRef]
12. Fama, E.F.; French, K.R. A five-factor asset pricing model. *J. Financ. Econ.* **2015**, *116*, 1–22. [CrossRef]
13. Dixon, M.; Klabjan, D.; Bang, J.H. Implementing deep neural networks for financial market prediction on the Intel Xeon Phi. In Proceedings of the 8th Workshop on High Performance Computational Finance, New York, NY, USA, 15 November 2015; pp. 1–6.
14. Hagan, M.T.; Menhaj, M.B. Training feedforward networks with the Marquardt algorithm. *IEEE Trans. Neural Netw.* **1994**, *5*, 989–993. [CrossRef] [PubMed]
15. Lozano, A.M.; Lang, A.E.; Galvez-Jimenez, N.; Miyasaki, J.; Duff, J.; Hutchinson, W.D.; Dostrovsky, J.O. Effect of GPi pallidotomy on motor function in Parkinson's disease. *Lancet* **1995**, *346*, 1383–1387. [CrossRef]
16. Donaldson, R.G.; Kamstra, M. Forecast combining with neural networks. *J. Forecast.* **1996**, *15*, 49–61. [CrossRef]
17. Zhang, Y.; Wu, L. Stock market prediction of S&P 500 via combination of improved BCO approach and BP neural network. *Expert Syst. Appl.* **2009**, *36*, 8849–8854.
18. Mario, C.; Giovanni, P. An innovative approach for forecasting of energy requirements to improve a smart home management system based on BLE. *IEEE Trans. Green Commun. Netw.* **2017**, *1*, 112–120.
19. Manabe, Y.; Chakraborty, B. Estimating embedding parameters using structural learning of neural network. In Proceedings of the IEEE 2005 International Workshop on Nonlinear Signal and Image Processing (NSIP 2005), Sapporo, Japan, 18–20 May 2005.

20. Wu, S.L.; Li, K.L.; Huang, T.Z. Exponential stability of static neural networks with time delay and impulses. *J. IET Control Theory Appl.* **2011**, *5*, 943–951. [CrossRef]

21. Hornik, K.; Stinchcombe, M.; White, H. Multi-layered feedforward networks are universal approximators. *Neural Netw.* **1989**, *2*, 359–366. [CrossRef]

22. Huang, W.; Nakamori, Y.; Wang, S. Forecasting stock market movement direction with support vector machine. *Comput. Oper. Res.* **2005**, *32*, 2513. [CrossRef]

23. Nils, O.; Alessandra, N.; Pierre, C. MLP Tools: A PyMOL plugin for using the molecular lipophilicity potential in computer-aided drug design. *J. Comput. Aided Mol. Des.* **2014**, *28*, 587–596.

24. Jefferys, W.H.; Berger, J.O. Ockham's razor and Bayesian analysis. *Am. Sci.* **1992**, *80*, 64–72.

25. Gardner, H. *Frames of Mind: The Theory of Multiple Intelligences*; Basic books: New York, NY, USA, 2011.

26. Lee, S.; Kim, H.; Park, D.; Chung, Y.; Jeong, T. CPU-GPU hybrid computing for feature extraction from video stream. *IEICE Electron. Express* **2014**, *11*. [CrossRef]

27. Lee, S.; Kim, H.; Sa, J.; Park, B.; Chung, Y. Real-time processing for intelligent-surveillance applications. *IEICE Electron. Express* **2017**, *14*. [CrossRef]

28. Lee, S.; Jeong, T. Cloud-based parameter-driven statistical services and resource allocation in a heterogeneous platform on enterprise environment. *Symmetry* **2016**, *8*, 103. [CrossRef]

MDPI AG

St. Alban-Anlage 66

4052 Basel, Switzerland

Tel. +41 61 683 77 34

Fax +41 61 302 89 18

http://www.mdpi.com

Symmetry Editorial Office

E-mail: symmetry@mdpi.com

http://www.mdpi.com/journal/symmetry